Dictionary of
Science

Dictionary of Science

Collins

HarperCollins*Publishers*
Westerhill Road, Glasgow G64 2QT

www.collins.co.uk

First published 2003

© 2003 Research Machines Plc
Helicon Publishing is a division of Research Machines

Reprint 10 9 8 7 6 5 4 3 2 1

ISBN 0 00 712670 0

A catalogue record for this book is available from the British Library.

Typeset by Davidson Pre-Press Graphics Ltd, Glasgow

Printed and bound in Great Britain by Clays Ltd, St Ives plc.

INTRODUCTION

The Collins Dictionary of Science has been created to provide a guide to the vocabulary of modern science, and, by giving clear explanations of terms and concepts, it aims to enhance the reader's understanding of science.

Choosing entries for such a dictionary is a challenging enterprise as science covers such a large area of human endeavour. The dictionary provides entries that give a thorough background to the central sciences of physics, chemistry and biology, and also spreads the net more widely with a selection of significant entries from mathematics, computing and the internet, earth sciences, environmental sciences, astronomy and medicine.

A feature of this dictionary is the inclusion of long explanatory entries on a range of key topics – such as **DNA**, **molecule** or **relativity**. These review entries provide the reader with fuller understanding of the significance of these important areas of science.

The book is organized alphabetically, with each headword appearing in bold. The order is decided as if there are no spaces between words, for example:

> **acid**
> **acid amide**
> **acid–base balance**
> **acidic oxide**
> **acidosis**
> **acid rain**

Cross references are indicated by an asterisk * in front of the entry being cross-referenced. Cross referencing is selective; a cross reference is shown when another entry contains material directly relevant to the subject matter of an entry, and to where the reader may not otherwise think of looking.

The text and illustrations for this book have been prepared by Helicon Publishing, a division of Research Machines Plc. The entries have been taken from their major educational encyclopedic database. Collins would like to thank Alice Goldie and Catherine Gaunt for their assistance in the selection of entries.

A A

A symbol for *mass number, the number of neutrons and protons in an atomic nucleus.

A in physics, symbol for *ampere, a unit of electrical current.

aberration in biology, an abnormal structure or deviation from the type.

aberration, optical any of a number of defects that impair the image in an optical instrument. Aberration occurs because of minute variations in lenses and mirrors, and because different parts of the light *spectrum are reflected or refracted by varying amounts.

abiotic factor non-living variable within the ecosystem, affecting the life of organisms. Examples include temperature, light, and water. Abiotic factors can be harmful to the environment, as when sulphur dioxide emissions from power stations produce acid rain.

ablation the loss of snow and ice from a *glacier by melting and evaporation. It is the opposite of *accumulation.

abscissa in *coordinate geometry, the *x*-coordinate of a point – that is, the horizontal distance of that point from the vertical or *y*-axis. For example, a point with the coordinates (4, 3) has an abscissa of 4. The *y*-coordinate of a point is known as the *ordinate.

abscissin or **abscissic acid**, plant hormone found in all higher plants. It is involved in the process of *abscission and also inhibits stem elongation, germination of seeds, and the sprouting of buds.

abscission in botany, the controlled separation of part of a plant from the main plant body – most commonly, the falling of leaves or the dropping of fruit controlled by *abscissin. In *deciduous plants the leaves are shed before the winter or dry season, whereas *evergreen plants drop their leaves continually throughout the year. Fruitdrop, the abscission of fruit while still immature, is a naturally occurring process.

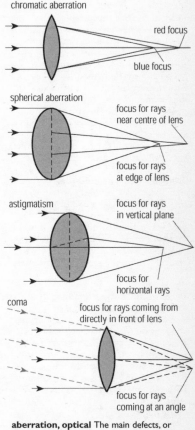

aberration, optical The main defects, or aberrations, of optical systems. Chromatic aberration, or coloured fringes around images, arises because light of different colours is focused at different points by a lens, causing a blurred image. Spherical aberration arises because light that passes through the centre of the lens is focused at a different point from light passing through the edge of the lens. Astigmatism arises if a lens has different curvatures in the vertical and horizontal directions. Coma arises because light passing directly through a lens is focused at a different point to light entering the lens from an angle.

absolute (of a value) in computing, real and unchanging. For example, an **absolute address** is a location in memory and an **absolute cell reference** is a single fixed cell in a spreadsheet display. The opposite of absolute is *relative.

absolute temperature temperature given by an absolute scale which is independent of the properties of thermometric substances. Absolute thermodynamic temperature, proposed by Irish physicist William Kelvin (1824–1907), is defined according to the principles of thermodynamics alone, but for experimental work is closely approximated by the International Temperature Scale of 1990 (ITS-90). It is usually expressed in units of degrees Celsius (or centigrade) from *absolute zero, and freezing point is then 273.15 K.

absolute value or **modulus**, in mathematics, the value, or magnitude, of a number irrespective of its sign. The absolute value of a number n is written $|n|$ (or sometimes as mod n), and is defined as the positive square root of n^2. For example, the numbers –5 and 5 have the same absolute value:

$$|5| = |-5| = 5$$

For a *complex number, the absolute value is its distance to the origin when it is plotted on an *Argand diagram, and can be calculated (without plotting) by applying *Pythagoras' theorem. By definition, the absolute value of any complex number $a + ib$ (where a and b are real numbers and i is the square root of –1) is given by the expression:

$$|a + ib| = \sqrt{(a^2 + b^2)}$$

absolute weight weight of a body considered apart from all modifying influences such as the atmosphere. To determine its absolute weight, the body must, therefore, be weighed in a vacuum or allowance must be made for buoyancy.

absolute zero lowest temperature theoretically possible according to kinetic theory, zero kelvin (0 K), equivalent to –273.15°C/–459.67°F, at which molecules are in their lowest energy state. Although the third law of *thermodynamics indicates the impossibility of reaching absolute zero, in practice temperatures of less than a billionth of a degree above absolute zero have been reached. Near absolute zero, the physical properties of some materials change substantially; for example, some metals lose their electrical resistance and become superconducting.

absolutism in physics, the theories that space and time are absolutes. These theories were held by English physicist and mathematician Isaac Newton (1642–1727) and were defended on his behalf by Samuel Clarke in his controversy with German mathematician Gottfried Leibniz (1646–1716) who was in many ways a precursor of Albert Einstein (1879–1955) and his theory of *relativity.

absorption in physics, taking up of matter or energy of one substance by another, such as a liquid by a solid (ink by blotting paper) or a gas by a liquid (ammonia by water). In physics, absorption is the phenomenon by which a substance retains the energy of radiation of particular wavelengths; for example, a piece of blue glass absorbs all visible light except the wavelengths in the blue part of the spectrum; it also refers to the partial loss of energy resulting from light and other electromagnetic waves passing through a medium. In nuclear physics, absorption is the capture by elements, such as boron, of neutrons produced by fission in a reactor.

absorption of light phenomenon whereby light energy striking a medium may be partially or totally lost by not being transmitted, reflected, or scattered. For example, red wine appears red because it transmits the red part of the spectrum but absorbs the yellow, green, and blue parts.

absorption spectroscopy or **absorptiometry**, in analytical chemistry, a technique for determining the identity or amount present of a chemical substance by measuring the amount of electromagnetic radiation the substance absorbs at specific wavelengths; see *spectroscopy.

abzyme in biotechnology, an artificially created antibody that can be used like an enzyme to accelerate reactions.

AC in physics, abbreviation for *alternating current.

Acanthocephala (Greek *akantha* 'thorn', *cephale* 'head') phylum of about 500 species of parasitic pseudocoelomate worms of which *Echinorhynchus* is the chief genus. The largest species, *E. gigantorhynchus*, a parasite of pigs, may attain a length of more than a metre, but most species are less than 1 cm/0.4 in long.

Accelerated Graphics Port AGP, dedicated port that links the graphics controller on a personal computer directly to the computer's memory, instead of data having to be fetched via the expansion bus. AGP can handle at least twice the throughput of a standard *PCI bus. The AGP was developed by Intel, and AGP support was added to later versions of Microsoft Windows 95.

acceleration rate of change of the velocity of a moving body. For example, an object falling towards the ground covers more distance in each successive time interval. Therefore, its velocity is changing with time and the object is accelerating. It is usually measured in metres per second per second (m s^{-2}) or feet per second per second (ft s^{-2}). Acceleration = change in velocity/time taken. Because velocity is a *vector quantity (possessing both magnitude and direction), a body travelling at constant speed may be said to be accelerating if its direction of motion changes. According to Newton's second law of motion (see *Newton's laws of motion), a body will accelerate only if it is acted upon by an unbalanced, or resultant, *force. Acceleration of free fall is the acceleration of a body falling freely under the influence of the Earth's gravitational field; it varies slightly at different latitudes and altitudes. The value adopted internationally for gravitational acceleration is 9.806 m s^{-2}/32.174 ft s^{-2}.

acceleration, uniform in physics, acceleration in which the velocity of a body changes by equal amounts in successive time intervals. Uniform acceleration is represented by a straight line on a *speed–time graph.

accelerator in physics, a device to bring charged particles (such as protons and electrons) up to high speeds and energies, at which they can be of use in industry, medicine, and pure physics. At low energies, accelerated particles can be used to produce the image on a television screen and (by means of a *cathode-ray tube) generate X-rays, destroy tumour cells, or kill bacteria. When high-energy particles collide with other particles, the fragments formed reveal the nature of the fundamental forces.

accelerometer apparatus, either mechanical or electromechanical, for measuring *acceleration or deceleration – that is, the rate of increase or decrease in the *velocity of a moving object.

accent in mathematics, symbol used to express feet and inches, for example $2'6'' = 2$ ft 6 in, and minutes and seconds as subdivisions of an angular *degree, for example $60' = 60$ minutes, $30'' = 30$ seconds.

access privilege in computer networking, authorized access to files and other shared resources. The ability to authorize or restrict access selectively to files or directories, including separate privileges such as reading, writing, or changing data, is a key element in computer security systems. This kind of system ensures that, for example, a company's employees cannot read its personnel files or alter payroll data unless they work for the appropriate departments, or that freelance or temporary staff can be given access to some areas of the computer system but not others. Access privileges may also exist for shared printers and devices such as tape drives and modems.

access provider in computing, another term for *Internet Service Provider.

accommodation in biology, the ability of the *eye to focus on near or far objects by changing the shape of the lens. (See diagram, page 4.)

account on a computer network, a *user ID issued to a specific individual to enable access to the system and the files and other shared resources that

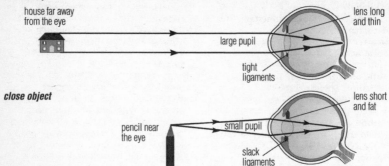

distant object

house far away
from the eye

lens long
and thin

large pupil

tight
ligaments

close object

lens short
and fat

pencil near
the eye

small pupil

slack
ligaments

accommodation The mechanism by which the shape of the lens in the eye is changed so that clear images of objects, whether distant or near, can be focused on the retina.

they require. The existence of accounts allows system administrators to assign *access privileges to specific individuals (which in turn enables those individuals to receive private messages such as e-mail) and also to track the use of the computer system and its resources.

accumulation in earth science, the addition of snow and ice to a *glacier. It is the opposite of *ablation. Snow is added through snowfall and avalanches, and is gradually compressed to form ice. Although accumulation occurs at all parts of a glacier, it is most significant at higher altitudes near the glacier's start where temperatures are lower.

accumulator in computing, a special register, or memory location, in the *arithmetic and logic unit of the computer processor. It is used to hold the result of a calculation temporarily or to store data that is being transferred.

accumulator in electricity, a storage *battery – that is, a group of rechargeable secondary cells. A familiar example is the lead–acid car battery. An ordinary 12-volt car battery consists of six lead–acid cells which are continually recharged when the motor is running by the car's alternator or dynamo. It has electrodes of lead and lead oxide in an electrolyte of sulphuric acid. Another common type of accumulator is the 'nife' or NiFe cell, which has electrodes of nickel and iron

in a potassium hydroxide electrolyte.

accuracy in mathematics, a measure of the precision of a number. The degree of accuracy depends on how many figures or *decimal places are used in *rounding off the number. For example, the result of a calculation or measurement (such as 13.429314) might be rounded off to three decimal places (13.429), to two decimal places (13.43), to one decimal place (13.4), or to the nearest whole number (13). The first answer is more accurate than the second, the second more accurate than the third, and so on.

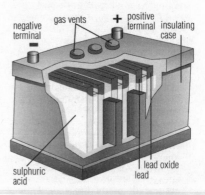

negative
terminal

gas vents

positive
terminal

insulating
case

sulphuric
acid

lead oxide

lead

accumulator The lead–acid car battery is a typical example of an accumulator. The battery has a set of grids immersed in a sulphuric acid electrolyte. One set of grids is made of lead (Pb) and acts as the anode and the other set made of lead oxide (PbO_2) acts as the cathode.

acetaldehyde   common name for *ethanal.

acetamide   common name for *ethanamide.

acetate   common name for *ethanoate.

acetic acid   common name for *ethanoic acid.

acetic anhydride   common name for *ethanoic anhydride.

acetic ester   common name for *ethyl acetate (ethyl ethanoate).

acetone   common name for *propanone.

acetonitrile   formerly **methyl cyanide**; CH_3CN, clear liquid, miscible with water, which, unless freshly purified (when it has a faint, pleasant smell), retains the mouselike odour of ethanamide, from which it is prepared by dehydration, using phosphorus pentoxide as the dehydrating reagent. Its reactions are typical of volatile, reactive nitriles (containing the monovalent group –CN).

acetyl   organic group that would result from the elimination of hydroxyl from ethanoic acid; it therefore corresponds to the formula CH_3CO. It is not stable in the free state, but is looked upon as the radical of such compounds as acetyl chloride (CH_3COCl).

acetylacetone $CH_3COCH_2COCH_3$, colourless pungent liquid, boiling point 194°C/381.2°F. This diketone is unusual in that it exists to the extent of about 85% as the enol form, the enol form being considerably stabilized by hydrogen bonding between the enol OH proton and the carbonyl oxygen. Acetylacetone is an important reagent for the preparation of *chelate compounds with a wide range of metals.

acetylcholine   **ACh**, chemical that serves as a *neurotransmitter, communicating nerve impulses between the cells of the nervous system. It is largely associated with the transmission of impulses across the *synapse (junction) between nerve and muscle cells, causing the muscles to contract.

ACh is produced in the synaptic knob (a swelling at the end of a nerve cell) and stored in vesicles until a nerve impulse triggers its discharge across the synapse. When the ACh reaches the membrane of the receiving cell it binds with a specific site and brings about depolarization – a reversal of the electric charge on either side of the membrane – causing a fresh impulse (in nerve cells) or a contraction (in muscle cells). Its action is shortlived because it is quickly destroyed by the enzyme cholinesterase.

*Anticholinergic drugs have a number of uses in medicine to block the action of ACh, thereby disrupting the passage of nerve impulses and relaxing certain muscles, for example in premedication before surgery.

acetyl coenzyme A   or **acetyl CoA**, compound active in processes of metabolism. It is a heat-stable coenzyme with an acetyl group ($-COCH_3$) attached by a sulphur linkage. This linkage is a high-energy bond and the acetyl group can easily be donated to other compounds. Acetyl groups donated in this way play an important part in glucose breakdown as well as in fatty acid and steroid synthesis.

It is involved in the *Krebs cycle, the cyclical pathway involved in the intracellular metabolism of foodstuffs.

acetylene   common name for *ethyne.

acetylsalicylic acid   chemical name for the painkilling drug *aspirin.

achene   dry, one-seeded *fruit that develops from a single *ovary and does not split open to disperse the seed. Achenes commonly occur in groups – for example, the fruiting heads of buttercup *Ranunculus* and clematis. The outer surface may be smooth, spiny, ribbed, or tuberculate, depending on the species.

Achilles tendon   tendon at the back of the ankle attaching the calf muscles to the heel bone. It is one of the largest tendons in the human body, and can resist great tensional strain, but is sometimes ruptured by contraction of the muscles in sudden extension of the foot.

Ancient surgeons regarded wounds in this tendon as fatal, probably because of the Greek legend of Achilles, which relates how the mother of the hero Achilles dipped him when an infant into the River Styx, so that he became invulnerable except for the heel by which she held him.

achromatic lens   combination of lenses made from materials of different refractive indexes, constructed in such a way as to minimize chromatic aberration (which in a single lens causes coloured fringes around images because the lens diffracts the different wavelengths in white light to slightly different extents).

acid in chemistry, compound that releases hydrogen ions (H^+ or protons) in the presence of an ionizing solvent (usually water). Acids react with *bases to form salts, and they act as solvents. Strong acids are corrosive; dilute acids have a sour or sharp taste, although in some organic acids this may be partially masked by other flavour characteristics. The strength of an acid is measured by its hydrogen-ion concentration, indicated by the *pH value. All acids have a pH below 7.0.

Acids can be classified as monobasic, dibasic, tribasic, and so forth, according to their basicity (the number of hydrogen atoms available to react with a base) and degree of ionization (how many of the available hydrogen atoms dissociate in water). Dilute sulphuric acid is classified as a strong (highly ionized), dibasic acid.

Inorganic acids include boric, carbonic, hydrochloric, hydrofluoric, nitric, phosphoric, and sulphuric. Organic acids include ethanoic (acetic), benzoic, citric, methanoic (formic), lactic, oxalic, and salicylic, as well as complex substances such as *nucleic acids and *amino acids.

Sulphuric, nitric, and hydrochloric acids are sometimes referred to as the mineral acids. Most naturally occurring acids are found as organic compounds, such as the fatty acids $R-COOH$ and sulphonic acids $R-SO_3H$, where R is an organic group.

acid amide any organic compound that may be regarded as being derived from ammonia by the substitution of acid or acyl groups for atoms of hydrogen. They are described as primary, secondary, or tertiary, according to the number of atoms of hydrogen displaced. Thus the general formula for a primary amide is $R-CONH_2$. The main acid amides are *ethanamide and methanamide.

acid–base balance essential balance between the amount of carbonic acid

and bicarbonate (an alkali) in the blood. It must be kept constant so that the hydrogen ion concentration in the blood plasma is in turn kept constant. Any deviation in acid–base balance can have profound effects on physiological functioning.

acidic oxide oxide of a *non-metal. Acidic oxides are covalent compounds. Those that dissolve in water, such as sulphur dioxide, give acidic solutions.

$$SO_2 + H_2O \leftrightarrow H_2SO_{3(aq)}$$
$$\leftrightarrow H^+_{(aq)} + HSO_3^-_{(aq)}$$

All acidic oxides react with alkalis to form salts.

$$CO_2 + NaOH \rightarrow NaHCO_3$$

acidosis in medicine, a condition characterized by the body's production of abnormal acids or by a reduction in the alkali reserve of the blood. A mild form of acidosis may occur in children with severe fevers. Symptoms include vomiting, thirst, restlessness, and lassitude. Acidosis may be detected by the presence of *ketones in the urine.

acid rain acidic precipitation thought to be caused mainly by the release into the atmosphere of sulphur dioxide (SO_2) and oxides of nitrogen (NO_x), which dissolve in pure rainwater making it acidic. Sulphur dioxide is formed by the burning of fossil fuels, such as coal, that contain high quantities of sulphur; nitrogen oxides are produced by various industrial activities and are present in car exhaust fumes.

Acidity is measured on the *pH scale, where the value of 0 represents liquids and solids that are completely acidic and 14 represents those that are highly alkaline. Distilled water is neutral and has a pH of 7. Normal rain has a value of 5.6. It is slightly acidic due to the presence of carbonic acid formed by the mixture of CO_2 and rainwater. Acid rain has values of 5.6 or less on the pH scale.

Acid deposition occurs not only as **wet precipitation** (mist, snow, or rain), but also comes out of the atmosphere as dry particles (**dry deposition**) or is absorbed directly by lakes, plants, and masonry as gases. Acidic gases can travel over 500 km/310 mi a day, so acid rain can be considered an example of transboundary (international) pollution.

Acid rain is linked with damage to and the death of forests and lake organisms in Scandinavia, Europe, and eastern North America. It is increasingly common in countries such as China and India that are industrializing rapidly. It also results in damage to buildings and statues. According to the UK Department of the Environment figures, emissions of sulphur dioxide from power stations would have to be decreased by 81% in order to stop the damage.

acid rock igneous rock that contains more than 60% by weight silicon dioxide, SiO_2, such as a granite or rhyolite. Along with the terms **basic rock** and **ultrabasic rock** it is part of an outdated classification system based on the erroneous belief that silicon in rocks is in the form of silicic acid. Geologists today are more likely to use the descriptive term *felsic rock or report the amount of SiO_2 in percentage by weight.

acid salt chemical compound formed by the partial neutralization of a dibasic or tribasic *acid (one that contains two or three replaceable hydrogen atoms). Although a salt, it still contains replaceable hydrogen, so it may undergo the typical reactions of an acid. Examples are sodium hydrogen sulphate ($NaHSO_4$) and acid phosphates.

ack abbreviation for acknowledgement, special type of message sent between computers on a network. When a computer has received data correctly from another it sends back an ack message to prompt the sending machine to transmit its next piece of data. If data isn't received properly, then a negative acknowledgement message, or a **nak**, is sent. The sending computer then knows to try and send the data again.

acoustic coupler device that enables computer data to be transmitted and received through a normal telephone handset; the handset rests on the coupler to make the connection. A small speaker within the device is used to convert the computer's digital output data into sound signals, which are then picked up by the handset and transmitted through the telephone system. At the receiving telephone, a second acoustic coupler or modem converts the sound signals back into digital data for input into a computer.

acoustic ohm c.g.s. unit of acoustic impedance (the ratio of the sound pressure on a surface to the sound flux through the surface). It is analogous to the ohm as the unit of electrical *impedance.

acoustics in general, the experimental and theoretical science of sound and its transmission; in particular, that branch of the science that has to do with the phenomena of sound in a particular space such as a room or theatre. In architecture, the sound-reflecting character of an internal space.

acquired character feature of the body that develops during the lifetime of an individual, usually as a result of repeated use or disuse, such as the enlarged muscles of a weightlifter.

acquired immune deficiency syndrome full name for the disease *AIDS.

acre traditional English land measure equal to 4,840 square yards (4,047 sq m/ 0.405 ha). Originally meaning a field, it was the area that a yoke of oxen could plough in a day.

acre-foot unit sometimes used to measure large volumes of water, such as the capacity of a reservoir (equal to its area in acres multiplied by its average depth in feet). One acre-foot equals 1,233.5 cu m/43,560 cu ft or the amount of water covering one acre to a depth of one foot.

acronym abbreviation that can be pronounced as a word, for example **RISC** (Reduced Instruction Set Computer) and **MUD** (multi-user dungeon). People in the computer industry often incorrectly refer to all abbreviations as acronyms. Both are frequently used as industry jargon and as shorthand to save typing on the Net.

acrophobia *phobia involving fear of heights.

acrylic acid common name for *propenoic acid.

acrylonitrile or **vinyl cyanide**; $CH_2{=}CHCN$, colourless liquid compound used in the production of acrylic fibres and synthetic rubbers. It polymerizes by a free radical mechanism in the presence of an appropriate catalyst. Commercially, the polymer is woven, and the yarn is known as Orlon.

Acrylonitrile is made by reacting propene, ammonia, and air with a catalyst at 450°C/842°F, a process developed in the 1960s from research into ways of utilizing a worldwide surplus of propene derived from ethene manufacture.

ACTH abbreviation of **adrenocorticotrophic hormone**, *hormone secreted by the anterior lobe of the *pituitary gland. It controls the production of corticosteroid hormones by the *adrenal gland and is commonly produced as a response to stress.

actinide any of a series of 15 radioactive metallic chemical elements with atomic numbers 89 (actinium) to 103 (lawrencium). Elements 89 to 95 occur in nature; the rest of the series are synthesized elements only. Actinides are grouped together because of their chemical similarities (for example, they are all bivalent), the properties differing only slightly with atomic number. The series is set out in a band in the *periodic table of the elements, as are the *lanthanides.

actinium chemical symbol Ac, (Greek *aktis* 'ray') white, radioactive, metallic element, the first of the actinide series, atomic number 89, relative atomic mass 227; it is a weak emitter of high-energy alpha particles.

Actinium occurs with uranium and radium in *pitchblende and other ores, and can be synthesized by bombarding radium with neutrons. The longest-lived isotope, Ac-227, has a half-life of 21.8 years (all the other isotopes have very short half-lives). Chemically, it is exclusively trivalent, resembling in its reactions the lanthanides and the other actinides. Actinium was discovered in 1899 by the French chemist André Debierne (1874–1949).

actinium K original name of the radioactive element *francium, given 1939 by its discoverer, the French scientist Marguerite Perey (1909–1975).

action and reaction in physical mechanics, equal and opposite forces which act together. For example, the pressure of expanding gases from the burning of fuel in a rocket engine produces an equal and opposite reaction, which causes the rocket to move. This is Newton's third law of motion (see *Newton's laws of motion).

action potential in biology, a change in the *potential difference (voltage) across the membrane of a nerve cell when an impulse passes along it. A change in potential (from about −60 to +60 millivolts) accompanies the passage of sodium and potassium ions across the membrane.

activation analysis in analytical chemistry, a technique used to reveal the presence and amount of minute impurities in a substance or element. A sample of a material that may contain traces of a certain element is irradiated with *neutrons, as in a reactor. Gamma rays emitted by the material's radioisotopes have unique energies and relative intensities, similar to the spectral lines from a luminous gas. Measurements and interpretation of the gamma-ray spectrum, using data from standard samples for comparison, provide information on the amounts of impurities present.

activation energy in chemistry, the minimum energy required in order to start a chemical reaction. Some elements and compounds will react together merely by bringing them into contact (spontaneous reaction). For others it is necessary to supply energy (heat, radiation, or electrical charge) in order to start the reaction, even if there is ultimately a net output of energy. This initial energy is the activation energy.

The point at which the reaction begins is known as the energy barrier. When the energy barrier is reached, the chemical bonds in the reactants are broken, enabling them to proceed from reactants to products.

In some reactions, such as the *combustion of *fuels, the activation energy required for the chemical reaction to take place is very small, resulting in a rapid reaction. Other chemical reactions, such as the *rusting of iron (a type of *oxidation) have a very large energy barrier and take place slowly.

A chemical equation only describes the *energy of reaction; the activation energy is not shown. The total chemical energy involved can be represented in an **energy level diagram**; this also shows whether a reaction is *exothermic (giving off energy) or *endothermic (absorbing energy).

Active Desktop optional feature of Microsoft Windows that changes the desktop into the equivalent of a Web page. This feature enables programs to be run with a single click of the mouse instead of the usual double-clicking. Active Desktop was introduced with Microsoft's Internet Explorer 4 Web browser for Windows. It is an integral component of Windows XP.

active matrix LCD or **TFT (thin film transistor) display**, type of colour *liquid crystal display (LCD) commonly used in laptop computers. Active matrix displays are made by sandwiching a film containing tiny transistors between two plates of glass. They achieve high contrast and brightness by applying voltage across the horizontal and vertical wires between the two glass plates, balanced by using a small transistor inside each

*pixel to amplify the voltage when so instructed.

Active Server Page ASP, in computing, Web page created upon request by a Web server. When the Web browser requests an Active Server Page, the server responds by sending a page dynamically generated in HTML code. Active Server Pages have the file extension .asp instead of .html.

active transport movement of molecules or ions across a cell membrane using *energy provided by *respiration. Examples of substances that can be actively transported across membranes are sodium ions and glucose.

Energy is needed because the movement occurs against a concentration gradient, with substances being moved from an area of low concentration to an area where there is a higher concentration. Active transport is therefore quite different from *diffusion, which requires no input of energy. In diffusion the movement is in the opposite direction – from an area of high concentration to an area where the concentration is low. An example of diffusion is the movement of oxygen into the blood vessels of the lungs.

● water molecule
■ sugar molecule

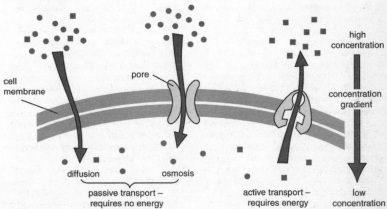

active transport Different types of transport across a cell membrane. Diffusion and osmosis are passive modes of transport, requiring no energy, moving from areas of high concentration to areas of low concentration. Active transport requires energy to transport molecules from low concentration to high concentration.

active window on graphical operating systems, the *window containing the program actually in use at any given time. Usually active windows are easily identified by the use of colour schemes that assign a different colour to the window's title bar (a thin strip along the top of each window bearing the name of the window's specific program or function) from that of the title bars of inactive windows.

ActiveX in computing, Microsoft's umbrella name for a collection of technologies used to create applications that run on the World Wide Web or on *intranets.

activity in physics, number of particles emitted in one second by a radioactive source. The term is used to describe the radioactivity or the potential danger of that source. The unit of activity is the becquerel (Bq), named after the French physicist Henri Becquerel (1852–1908), who discovered radioactivity.

activity series in chemistry, alternative name for *reactivity series.

acute angle angle between 0° and 90°; that is, an amount of turn that is less than a quarter of a circle.

Ada high-level computer-programming language, developed and owned by the US Department of Defense, designed for use in situations in which a computer directly controls a process or machine, such as a military aircraft. The language took more than five years to specify, and became commercially available only in the late 1980s. It is named after English mathematician Ada Augusta Byron (1815–1852), daughter of the poet Lord Byron.

adaptation (Latin *adaptare* 'to fit to') in biology, any feature in the structure or function of an organism that allows it to survive and reproduce more effectively in its environment. Much adaptation is inherited (see *inheritance) and is the result of many thousands of years of *evolution. It is thought to occur as a result of random variation in the genetic make-up of organisms coupled with *natural selection. Species become extinct when they are no longer adapted to their environment.

Adaptations to cope with changing seasons can be quite different from one organism to another. Some plants die back to survive the winter and have specially protected structures such as tight buds on a twig or a bulb underground. These structures may store food so that growth can start rapidly in spring. A carrot, for example, stores sugars for this reason. Some animals, such as swallows, may migrate, while other animals, such as the dormouse, may hibernate, allowing the body temperature and heart rate to drop in the winter.

Adaptations also occur as a result of an animal evolving into an efficient predator. A predator may have forward facing eyes, acute vision and sense of smell, and have claws, talons, or a beak for killing. The prey also adapt as a result of evolution. Prey may have eyes on the side of their heads for a wide field of view, acute hearing and sense of smell, be easily startled, be nocturnal, and be camouflaged.

In evolution, adaptation is the result of *natural selection. Inherited variation usually occurs in a species. If the environment changes in some way, such as the winters becoming more severe, those individuals possessing genes that allow them to survive and reproduce more successfully pass their genes to their offspring more than those that do not. Some genes therefore become more common in the species, leading to an evolutionary change that enables the species to be better adapted to the changed environment.

adaptive radiation in evolution, the formation of several species, with *adaptations to different ways of life, from a single ancestral type. Adaptive radiation is likely to occur whenever members of a species migrate to a new habitat with unoccupied ecological niches. It is thought that the lack of competition in such niches allows sections of the migrant population to develop new adaptations, and eventually to become new species.

The colonization of newly formed volcanic islands has led to the development of many unique species. The 13 species of Darwin's finch (named after English naturalist Charles Darwin (1809–1882) on the Galapagos Islands, for example, are probably descended from a single species from the South American mainland. The parent stock evolved into different species that now occupy a range of diverse niches.

ADC in electronics, abbreviation for *analogue-to-digital converter.

adder electronic circuit in a computer or calculator that carries out the process of adding two binary numbers. A separate adder is needed for each pair of binary *bits to be added. Such circuits are essential components of a computer's *arithmetic and logic unit (ALU).

addition in arithmetic, the operation of combining two numbers to form a sum; thus, $7 + 4 = 11$. It is one of the four basic operations of arithmetic (the others are subtraction, multiplication, and division).

addition polymerization
*polymerization reaction in which a single monomer gives rise to a single polymer, with no other reaction products. Addition polymerization occurs in *alkenes, hydrocarbons containing double bonds. The alkenes, such as *ethene, are the monomers, the small starting molecules. In addition polymerization, ethene undergoes an addition reaction with itself. As one molecule joins to a second, a long molecular chain is built up. The long molecular chain is called a polymer. In the case of ethene, *polythene is formed. It is made up of repeating units of the monomer.

addition reaction chemical reaction in which the atoms of an element or compound react with a double bond or triple bond in an organic compound by opening up one of the bonds and becoming attached to it, for example:

$$CH_2=CH_2 + HCl \rightarrow CH_3CH_2Cl$$

Another example is the addition of hydrogen atoms to *unsaturated compounds in vegetable oils to produce margarine. Addition reactions are used to make polymers from *alkenes.

add-on small program written to extend the features of a larger one. The earliest successful add-on for personal computer users in the UK was a small routine which allowed the original version of the spreadsheet Lotus 1-2-3 to print out a pound sign (£), something the program's US developers had thought unnecessary.

adenine in biochemistry, one of four base molecules that form part of the *base pairs in the DNA molecule, where it is always paired with the purine *thymine.

Adenine is also part of the genetic code of RNA, where it pairs with the pyrimidine uracil.

adenosine triphosphate energy-rich compound present in cells. See *ATP.

adenovirus any of a group of DNA-containing viruses that cause respiratory infections and conjunctivitis.

ADH abbreviation for antidiuretic hormone, in biology, part of the system maintaining a correct salt/water balance in vertebrates.

adhesion in medicine, the abnormal binding of two tissues as a result of inflammation or damage. The moving surfaces of joints or internal organs may merge together if they have been inflamed and tissue fluid has been present between the surfaces.

adhesive substance that sticks two surfaces together. Natural adhesives (glues) include gelatin in its crude industrial form (made from bones, hide fragments, and fish offal) and vegetable gums. Synthetic adhesives include thermoplastic and thermosetting resins, which are often stronger than the substances they join; mixtures of epoxy resin and hardener that set by chemical reaction; and elastomeric (stretching) adhesives for flexible joints. Superglues are fast-setting adhesives used in very small quantities.

adipic acid common name for *hexanedioic acid.

adipose tissue type of *connective tissue of vertebrates that serves as an energy reserve, and also pads some organs. It is commonly called fat tissue, and consists of large spherical cells filled with fat. In mammals, major layers are in the inner layer of skin and around the kidneys and heart.

adjacent angles   pair of angles meeting at a common vertex (corner) and sharing a common arm. Two adjacent angles lying on the same side of a straight line add up to 180° and are said to be supplementary.

adjacent side   in a *right-angled triangle, the side that is next to a given angle but is not the hypotenuse (the side opposite the right angle). The third side is the **opposite side** to the given angle.

ADP   abbreviation for adenosine diphosphate, the chemical product formed in cells when *ATP breaks down to release energy.

adrenal gland   or **suprarenal gland**, triangular endocrine gland situated on top of the *kidney. The adrenals are soft and yellow, and consist of two parts: the cortex and medulla. The **cortex** (outer part) secretes various steroid hormones and other hormones that control salt and water metabolism and regulate the use of carbohydrates, proteins, and fats. The **medulla** (inner part) secretes the hormones adrenaline and noradrenaline which, during times of stress, cause the heart to beat faster and harder, increase blood flow to the heart and muscle cells, and dilate airways in the lungs, thereby delivering more oxygen to cells throughout the body and in general preparing the body for 'fight or flight'.

adrenaline   or **epinephrine**, hormone secreted by the medulla of the *adrenal glands. Adrenaline is synthesized from a closely related substance, noradrenaline, and the two hormones are released into the bloodstream in situations of fear or stress.

adrenocorticotrophic hormone hormone secreted by the anterior lobe of the *pituitary gland; see *ACTH.

ADSL   abbreviation for asymmetric digital subscriber line, loop, or link, standard for transmitting data through existing copper telephone wires. ADSL was developed by US telephone companies as a way of competing with cable television companies in delivering both TV and telephone services. By 1996 it was developing into a possible alternative means for high-speed Internet access. ADSL is one of several

types of digital subscriber loops (DSLs) in use.

adsorption   taking up of a gas or liquid at the surface of another substance, most commonly a solid (for example, activated charcoal adsorbs gases). It involves molecular attraction at the surface, and should be distinguished from *absorption (in which a uniform solution results from a gas or liquid being incorporated into the bulk structure of a liquid or solid).

advanced gas-cooled reactor   AGR, type of *nuclear reactor favoured in Western Europe. The AGR uses a fuel of enriched uranium dioxide in stainless-steel cladding and a moderator of graphite. Carbon dioxide gas is pumped through the reactor core to extract the heat produced by the *fission of the uranium. The heat is transferred to water in a steam generator, and the steam drives a turbogenerator to produce electricity.

Advanced Technology Attachment Packet Interface   in computing, enhancement to integrated drive electronics (IDE), usually abbreviated to *ATAPI.

aerated water   water that has had air (oxygen) blown through it. Such water supports aquatic life and prevents the growth of putrefying bacteria. Polluted waterways may be restored by artificial aeration.

aerial   or **antenna**, in radio and television broadcasting, a conducting device that radiates or receives electromagnetic waves. The design of an aerial depends principally on the wavelength of the signal. Long waves (hundreds of metres in wavelength) may employ long wire aerials; short waves (several centimetres in wavelength) may employ rods and dipoles; microwaves may also use dipoles – often with reflectors arranged like a toast rack – or highly directional parabolic dish aerials. Because microwaves travel in straight lines, requiring line-of-sight communication, microwave aerials are usually located at the tops of tall masts or towers.

aerial oxidation   in chemistry, a reaction in which air is used to oxidize another substance, as in the contact process for the manufacture of sulphuric acid:

$$2SO_2 + O_2 \leftrightarrow 2SO_3$$

and in the souring of wine, where alcohol (ethanol) is oxidized by atmospheric oxygen to produce vinegar (ethanoic, or acetic, acid).

aerobic in biology, describing those organisms that require *oxygen in order to survive. Aerobic organisms include all plants and animals and many micro-organisms. They use oxygen (usually dissolved in water) to release the energy contained in food molecules such as glucose in a process called aerobic respiration. Oxygen is used to break down carbohydrates into *carbon dioxide and *water, releasing *energy, which is used to drive many processes within the cells.

aerobic respiration in plant and animal cells, a complex process of chemical reactions in which oxygen is used to break *glucose into *carbon dioxide and *water. Energy is released in this process. The starting and finishing points of this process would be the same if glucose was burned. During burning the energy that is released is all in the form of heat. In aerobic respiration, however, the energy is released in a controlled way and less is released as heat. Most of the released energy is used to drive various processes in the cell, such as growth or movement.

If insufficient oxygen reaches human muscles in severe exercise, such as a sprint, the muscles can carry on respiring using *anaerobic respiration for a short time.

aerodynamics branch of fluid physics that studies the forces exerted by air or other gases in motion. Examples include the airflow around bodies moving at speed through the atmosphere (such as land vehicles, bullets, rockets, and aircraft), the behaviour of gas in engines and furnaces, air conditioning of buildings, the deposition of snow, the operation of air-cushion vehicles (hovercraft), wind loads on buildings and bridges, bird and insect flight, musical wind instruments, and meteorology. For maximum efficiency, the aim is usually to design the shape of an object to produce a streamlined flow, with a minimum of turbulence in the moving air. The behaviour of aerosols or the pollution of the atmosphere by foreign particles are other aspects of aerodynamics.

aerogel light, transparent, highly porous material composed of more than 90% air. Such materials are formed from silica, metal oxides, and organic chemicals, and are produced by drying gels – networks of linked molecules suspended in a liquid – so that air fills the spaces previously occupied by the liquid. They are excellent heat insulators and have unusual optical, electrical, and acoustic properties.

Aerogels were first produced by US scientist Samuel Kristler in the early 1930s by drying silica gels at high temperatures and pressures.

aestivation in zoology, a state of inactivity and reduced metabolic activity, similar to *hibernation, that occurs during the dry season in species such as lungfish and snails. In botany, the term is used to describe the way in which flower petals and sepals are folded in the buds. It is an important feature in *plant classification.

aether alternative form of *ether, the hypothetical medium once believed to permeate all of space.

aetiology in medicine, the systematic investigation into the causes of disease.

affine geometry geometry that preserves parallelism and the ratios between intervals on any line segment.

affinity in chemistry, the force of attraction (see *bond) between atoms that helps to keep them in combination in a molecule. The term is also applied to attraction between molecules, such as those of biochemical significance (for example, between *enzymes and substrate molecules). This is the basis for affinity *chromatography, by which biologically important compounds are separated.

The atoms of a given element may have a greater affinity for the atoms of one element than for another (for example, hydrogen has a great affinity for chlorine, with which it easily and rapidly combines to form hydrogen chloride, but has little or no affinity for argon).

afforestation planting of trees in areas that have not previously held forests. (**Reafforestation** is the planting of trees in deforested areas.) Trees may be planted (1) to provide timber and wood pulp; (2) to provide firewood in countries where this is an energy source; (3) to bind soil together and prevent soil erosion; and (4) to act as windbreaks.

agar jellylike carbohydrate, obtained from seaweeds. It is used mainly in microbiological experiments as a culture medium for growing bacteria and other micro-organisms. The agar is resistant to breakdown by micro-organisms, remaining a solid jelly throughout the course of the experiment.

agate cryptocrystalline (with crystals too small to be seen with an optical microscope) silica, SiO_2, composed of cloudy and banded chalcedony, sometimes mixed with *opal, that forms in rock cavities.

Agate stones, being hard, are also used to burnish and polish gold applied to glass and ceramics and as clean vessels for grinding rock samples in laboratories.

Agenda 21 non-binding treaty, signed by representatives of 178 countries in 1992, which sets out a framework of recommendations designed to protect the environment and achieve sustainable development. The treaty highlights the importance of international cooperation, but also discusses the role of individuals, communities, and local authorities in achieving the goals it sets out.

agonist in biology, a *muscle that contracts and causes a movement. Contraction of an agonist is complemented by relaxation of its *antagonist. For example, the biceps (in the front of the upper arm) bends the elbow while the triceps (lying behind the biceps) straightens the arm.

agonist in medicine, a drug or other substance that has a similar effect to normal chemical messengers in the body through its actions at receptor sites of cells. Examples are sympatho-mimetic drugs that mimic the actions of adrenaline and are used in the treatment of certain heart disorders.

agoraphobia *phobia involving fear of open spaces and public places. The anxiety produced can be so severe that some sufferers are unable to leave their homes for many years.

Agoraphobia affects 1 person in 20 at some stage in their lives. The most common time of onset is between the ages of 18 and 28.

AGP abbreviation for *Accelerated Graphics Port.

AGR abbreviation for **advanced gas-cooled reactor**, a type of nuclear reactor.

agranulocytosis virtual absence from the blood of the white cells known as neutrophils that destroy bacteria. It is a life-threatening condition that results from toxic damage to the bone marrow by some drugs. It is treated with antibiotics and sometimes transfusion of white blood cells.

agrochemical artificially produced chemical used in modern, intensive agricultural systems. Agrochemicals include nitrate and phosphate fertilizers, pesticides, some animal-feed additives, and pharmaceuticals. Many are responsible for pollution and almost all are avoided by organic farmers.

AI(D) abbreviation for *artificial insemination (by donor). AI(H) is artificial insemination by husband.

AIDS acronym for **acquired immune deficiency syndrome**, most serious of all the *sexually transmitted diseases (STDs). It is caused by the *retrovirus human immunodeficiency virus (*HIV), and is transmitted in body fluids, such as blood, salvia, semen, and vaginal secretions. AIDS is the world's most deadly infectious disease, and the fourth leading global cause of death.

air the mixture of gases making up the Earth's *atmosphere.

airlock airtight chamber that allows people to pass between areas of different pressure; also an air bubble in a pipe that impedes fluid flow. An airlock may connect an environment at ordinary pressure and an environment that has high air pressure (such as a submerged caisson used for tunnelling or building dams or bridge foundations).

air mass large body of air with particular characteristics of temperature and humidity. An air mass forms when air rests over an area long enough to pick up the conditions of that area. When an air mass moves to another area it affects the *weather of that area, but its own characteristics become modified in the process. For example, an air mass formed over the Sahara will be hot and dry, becoming cooler as it moves northwards. Air masses that meet form **fronts**.

There are four types of air mass. **Tropical continental** (Tc) air masses form over warm land and **Tropical maritime** (Tm) masses form over warm seas. Air masses that form over cold land are called **Polar continental** (Pc) and those forming over cold seas are called **Polar** or **Arctic maritime** (Pm or Am).

The weather of the UK is affected by a number of air masses which, having different characteristics, bring different weather conditions. For example, an Arctic air mass brings cold conditions whereas a Tropical air mass brings hot conditions.

air pollution contamination of the atmosphere caused by the discharge, accidental or deliberate, of a wide range of toxic airborne substances. Often the amount of the released substance is relatively high in a certain locality, so the harmful effects become more noticeable. The cost of preventing any discharge of pollutants into the air is prohibitive, so attempts are more usually made to reduce the amount of discharge gradually and to disperse it as quickly as possible by using a very tall chimney, or by intermittent release. One major air pollutant is sulphur dioxide (SO_2), produced from the burning of *fossil fuels. It dissolves in atmospheric moisture to produce sulphurous acid (*acid rain):

$$SO_2 + H_2O \rightarrow H_2SO_3$$

air resistance force tending to oppose the motion of a body as it moves through air. It is a form of *friction. An object falling through the air experiences a force due to air particles opposing its motion, which has the effect of slowing down the object. Air resistance is greater for objects moving faster and with greater areas.

air sac in birds, a thin-walled extension of the lungs. There are nine of these and they extend into the abdomen and bones, effectively increasing lung capacity.

albinism rare hereditary condition in which the body has no tyrosinase, one of the enzymes that form the pigment *melanin, normally found in the skin, hair, and eyes. As a result, the hair is white and the skin and eyes are pink. The skin and eyes are abnormally sensitive to light, and vision is often impaired. The condition occurs among all human and animal groups.

albumin any of a group of sulphur-containing *proteins. The best known is in the form of egg white (albumen); others occur in milk, and as a major component of serum. Many vegetables and fluids also contain albumins. They are soluble in water and dilute salt solutions, and are coagulated by heat.

alchemy (Arabic *al-Kimya*) supposed technique of transmuting base metals, such as lead and mercury, into silver and gold by the philosopher's stone, a hypothetical substance, to which was also attributed the power to give eternal life.

alcohol in chemistry, any member of a group of organic chemical compounds characterized by the presence of one or more aliphatic OH (hydroxyl) groups in the molecule, and which form *esters with acids. The main uses of alcohols are as solvents for gums, resins, lacquers, and varnishes; in the making of dyes; for essential oils in perfumery; and for medical substances in pharmacy. The alcohol produced naturally in the *fermentation process and consumed as part of alcoholic beverages is called *ethanol. When consumed the effects of alcohol include poisoning at high concentrations, and changes in the functioning of human nerve cells.

Alcohols may be liquids or solids, according to the size and complexity of the molecule. A **monohydric alcohol** contains only one *hydroxyl group in each molecule. The five simplest alcohols form a series in which the number of carbon and hydrogen atoms in the molecule increases progressively, each one having an extra CH_2 (methylene) group: methanol or wood spirit (methyl alcohol, CH_3OH); ethanol (ethyl alcohol, C_2H_5OH); propanol (propyl alcohol, C_3H_7OH); butanol (butyl alcohol, C_4H_9OH); and pentanol (amyl alcohol, $C_5H_{11}OH$). The lower alcohols are liquids that mix with water; the higher alcohols, such as pentanol, are oily liquids immiscible with water; and the highest are waxy solids – for example, hexadecanol (cetyl alcohol, $C_{16}H_{33}OH$) and melissyl alcohol ($C_{30}H_{61}OH$), which occur in sperm-whale oil and beeswax, respectively. Alcohols containing the CH_2OH group are primary; those containing CHOH are secondary; while those containing COH are tertiary.

alcoholic solution solution produced when a solute is dissolved in ethanol.

alcoholometry determination of the proportion of alcohol in a liquid, using a specially calibrated hydrometer.

Sikes's hydrometer is used for determining the strength of spirits, and Atkins's has a series of scales for any 'wort', or beer in the making.

aldehyde any of a group of organic chemical compounds prepared by oxidation of primary alcohols, so that the OH (hydroxyl) group loses its hydrogen to give an oxygen atom joined by a double bond to a carbon atom (*carbonyl group). The aldehyde group has the formula CHO. Ethanal, CH_3CHO, is an example of an aldehyde.

algae singular **alga**, highly varied group of plants, ranging from single-celled forms to large and complex seaweeds. They live in both fresh and salt water, and in damp soil. Algae do not have true roots, stems, or leaves.

Marine algae help combat *global warming by removing carbon dioxide from the atmosphere during *photosynthesis.

algebra branch of mathematics in which the general properties of numbers are studied by using symbols, usually letters, to represent variables

and unknown quantities. For example, the algebraic statement:

$$(x + y)^2 = x^2 + 2xy + y^2$$

is true for all values of x and y. For instance, the substitution $x = 7$ and $y = 3$ gives:

$$(7 + 3)^2 = 7^2 + 2(7 \times 3) + 3^2 = 100$$

In ordinary algebra the same operations are carried on as in *arithmetic, but, as the symbols are capable of a more generalized and extended meaning than the figures used in arithmetic, it facilitates calculation where the numerical values are not known, or are inconveniently large or small, or where it is desirable to keep them in an analysed form.

order of calculation The *simplification of an algebraic *equation or expression must be completed in a set order. The procedure follows the rules of *BODMAS.

simultaneous equations If there are two or more algebraic equations that contain two or more unknown quantities that may have a unique solution, they can be solved simultaneously as *simultaneous equations.

An algebraic *expression that has one or more variables (denoted by letters) is a *polynomial equation.

quadratic equation A *quadratic equation is a polynomial equation of second degree (that is, an equation containing as its highest power the square of a variable, such as x^2).

Algebra is used in many areas of mathematics – for example, *arithmetic progressions, or number sequences, and Boolean algebra (the latter is used in working out the logic for computers).

'Algebra' was originally the name given to the study of equations. In the 9th century, the Arab mathematician Muhammad ibn-Musa al-Khwarizmi used the term *al-jabr* for the process of adding equal quantities to both sides of an equation. When his treatise was later translated into Latin, *al-jabr* became 'algebra' and the word was adopted as the name for the whole subject.

The basics of algebra were familiar in ancient Babylonia (*c.* 18th century BC).

Numerous tablets giving sets of problems and their answers, evidently classroom exercises, survive from that period. The subject was also considered by mathematicians in ancient Egypt, China, and India. A comprehensive treatise on the subject, entitled *Arithmetica*, was written in the 3rd century AD by Diophantus of Alexandria. In the 9th century, al-Khwarizmi drew on Diophantus' work and on Hindu sources to produce his influential work *Hisab al-jabr wa'l-muqabalah/Calculation by Restoration and Reduction*.

the development of symbolism From ancient times until the Middle Ages, equation-solving depended on expressing everything in words or in geometric terms. It was not until the 16th century that the modern symbolism began to be developed (notably by François Viète) in response to the growing complexity of mathematical statements that were impossibly cumbersome when expressed in words. Further research in algebra was aided not only because the symbolism was a convenient 'shorthand' but also because it revealed the similarities between different problems and pointed the way to the discovery of generally applicable methods and principles.

quarternions and the idempotent law In the mid-19th century, algebra was raised to a completely new level of abstraction. In 1843 Sir William Rowan Hamilton discovered a three-dimensional extension of the number system, which he called 'quaternions', in which the commutative law of multiplication is not generally true; that is, $ab \neq ba$ for most quaternions a and b. In 1854 George Boole applied the symbolism of algebra to logic and found it fitted perfectly except that he had to introduce a 'special law' that $a^2 = a$ for all a (called the idempotent law).

algebraic structures Discoveries like this led to the realization that there are many possible 'algebraic structures', which can be described as one or more operations acting on specified objects and satisfying certain laws. (Thus the number system has the operations of addition and multiplication acting on numbers and obeying the commutative, associative, and distributive laws.)

In modern terminology, an algebraic structure consists of a *set, A, and one or more binary operations (that is, *functions mapping $A \times A$ into A) which satisfy prescribed 'axioms'. A typical example is a structure that had been studied from the 18th century onwards and is known as a *group. This structure had turned up in the study of the solvability of polynomial equations, but it also appears in numerous other problems (for example, in geometry), and even has applications in modern physics.

modern algebra The objective of modern algebra is to study each possible structure in turn, in order to establish general rules for each structure that can be applied in any situation in which the structure occurs. Numerous structures have been studied, and since 1930 a greater level of generality has been achieved by the study of 'universal algebra', which concentrates on properties that are common to all types of algebraic structure.

algebraic fraction in mathematics, fraction in which letters are used to represent numbers – for example, a/b, xy/z^2, and $1/(x+y)$. Like numerical fractions, algebraic fractions may be simplified or factorized. Two equivalent algebraic fractions can be cross-multiplied; for example, if

$$\frac{a}{b} = \frac{c}{d}$$

then $ad = bc$

(In the same way, the two equivalent numerical fractions $2/3$ and $4/6$ can be cross-multiplied to give cross-products that are both 12.)

factorization Algebraic fractions can be simplified by factorization, that is by taking out those factors that are common to both the numerator and denominator. For example, the algebraic fraction:

$$\frac{(x^2 - 2x - 8)(x^2 + 5x + 4)}{(x^2 - 16)(x + 1)}$$

can be simplified as follows:

$(x^2 - 2x - 8)$ factorizes as $(x - 4)(x + 2)$

$(x^2 + 5x + 4)$ factorizes as $(x + 4)(x + 1)$

$(x^2 - 16)$ factorizes as $(x - 4)(x + 4)$

so the $(x - 4)$, $(x + 4)$, and $(x + 1)$ from the denominator cancel out those in the numerator, leaving only $(x + 2)$.

addition and subtraction As with numerical fractions, to add or subtract algebraic fractions a common denominator must be found. The easiest way to do this is to multiply the denominators together. For example:

$$\frac{3}{x} - \frac{4}{x + 2} = \frac{3(x + 2)}{x(x + 2)} - \frac{4x}{x(x + 2)}$$

which can be simplified to:

$$\frac{3x + 6 - 4x}{x(x + 2)} = \frac{6 - x}{x^2 + 2x}$$

alginate in chemistry, salt of alginic acid, $(C_6H_8O_6)_n$, obtained from brown seaweeds and used in textiles, paper, food products, and pharmaceuticals.

ALGOL contraction of algorithmic language, in computing, an early high-level programming language, developed in the 1950s and 1960s for scientific applications. A general-purpose language, ALGOL is best suited to mathematical work and has an algebraic style. Although no longer in common use, it has greatly influenced more recent languages, such as *Ada and *Pascal.

algorithm procedure or series of steps that can be used to solve a problem. In computer science, it describes the logical sequence of operations to be performed by a program. A flow chart is a visual representation of an algorithm.

alias name representing a particular user or group of users in e-mail systems. This feature, which is not available on all systems, is a matter of convenience as it allows a user to substitute shorter or easier-to-remember real names for e-mail addresses. In 1995 CompuServe announced a system of named aliases for its long, numbered addresses.

alimentary canal tube through which food passes in animals – it extends from the mouth to the anus and forms a large part of the digestive system. In human adults, it is about 9 m/30 ft long, consisting of the mouth cavity, pharynx, oesophagus, stomach, and the small and large intestines. It is also known as the gut. It is a complex organ, specifically adapted for *digestion and the absorption of food. Enzymes from the wall of the canal and from other associated organs, such as the pancreas, speed up the digestive process.

aliphatic compound any organic chemical compound in which the carbon atoms are joined in straight chains, as in hexane (C_6H_{14}), or in branched chains, as in 2-methylpentane ($CH_3CH(CH_3)CH_2CH_2CH_3$).

Aliphatic compounds have bonding electrons localized within the vicinity of the bonded atoms. *Cyclic compounds that do not have delocalized electrons are also aliphatic, as in the alicyclic compound cyclohexane (C_6H_{12}) or the heterocyclic piperidine ($C_5H_{11}N$). Compare *aromatic compound.

alizarin or **1,2-dihydroxy-anthraquinone**; $C_6H_4(CO)_2C_6H_2(OH)_2$, derivative from anthraquinone. It is now prepared synthetically from anthracene. Alizarin crystallizes in dark red prisms and sublimes in orange-coloured needles, melting at 290°C/554°F. It is almost insoluble in water, but dissolves in alcohol.

It yields with metallic oxides magnificently coloured insoluble compounds called 'lakes', to which it owes its great value for dyeing purposes. Ferric oxide with alizarin gives a violet-black compound, chromium oxide a claret, calcium oxide a blue, and aluminium and tin give different shades of red.

alk resin obtained from the turpentine tree *Pistacia terebinthus*, which grows chiefly in the Mediterranean region. A yellow to green liquid, Chian or Chio turpentine, is distilled from it.

alkali in chemistry, a *base that is soluble in water. Alkalis neutralize acids, and solutions of alkalis are soapy to the touch. The strength of an alkali is measured by its hydrogen-ion concentration, indicated by the *pH value. They may be divided into strong and weak alkalis: a strong alkali (for example, potassium hydroxide, KOH) ionizes completely when dissolved in water, whereas a weak alkali (for example, ammonium hydroxide, NH_4OH) exists in a partially ionized state in solution. All alkalis have a pH above 7.0.

The hydroxides of metals are alkalis. Those of sodium and potassium are corrosive; both were historically derived from the ashes of plants.

The four main alkalis are sodium hydroxide (caustic soda, NaOH); potassium hydroxide (caustic potash, KOH); calcium hydroxide (slaked lime or limewater, $Ca(OH)_2$); and aqueous ammonia ($NH_{3(aq)}$). Their solutions all contain the hydroxide ion OH⁻, which gives them a characteristic set of properties.

alkali metal any of a group of six metallic elements with similar chemical properties: *lithium, *sodium, *potassium, *rubidium, *caesium, and *francium. They form a linked group (Group 1) in the *periodic table of the elements. They each have a valency of one and have very low densities (lithium, sodium, and potassium float on water); in general they are reactive, soft, low-melting-point metals. Because of their reactivity they are only found as compounds in nature.

alkaline-earth metal any of a group of six metallic elements with similar bonding properties: beryllium, magnesium, calcium, strontium, barium, and radium. They form a linked group in the *periodic table of the elements. They are strongly basic, bivalent (have a valency of two), and occur in nature only in compounds.

alkaloid any of a number of physiologically active and frequently poisonous substances contained in some plants. They are usually organic bases and contain nitrogen. They form salts with acids and, when soluble, give alkaline solutions.

Substances in this group are included by custom rather than by scientific rules. Examples include morphine, cocaine, quinine, caffeine, strychnine, nicotine, and atropine.

In 1992, epibatidine, a chemical extracted from the skin of an Ecuadorean frog, was identified as a member of an entirely new class of alkaloid. It is an organochlorine compound, which is rarely found in animals, and a powerful painkiller, about 200 times as effective as morphine.

alkane member of a group of *hydrocarbons having the general formula C_nH_{2n+2}, commonly known as **paraffins**. As they contain only single *covalent bonds, alkanes are said to be saturated. Lighter alkanes, such as methane, ethane, propane, and butane, are colourless gases; heavier ones are liquids or solids. In nature

Name	Molecular formula	Structural formula
methane	CH_4	
uses: domestic fuel (natural gas)		
ethane	C_2H_6	
uses: industrial fuel and chemical feedstock		
propane	C_3H_8	
uses: bottled gas (camping gas)		
butane	C_4H_{10}	
uses: bottled gas (lighter fuel, camping gas)		

alkane The lighter alkanes methane, ethane, propane, and butane, showing the aliphatic chains, where a hydrogen atom bonds to a carbon atom at all available sites.

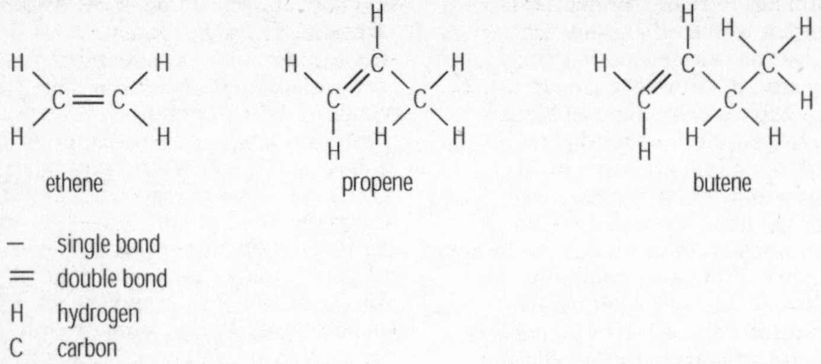

ethene   propene   butene

—   single bond
=   double bond
H   hydrogen
C   carbon

alkene The alkenes ethene (C_2H_4), propene ($CH_3CH=CH_2$), and butene (C_4H_8). Alkenes all have the general formula C_nH_{2n}.

they are found in natural gas and petroleum.

alkene member of the group of *hydrocarbons having the general formula C_nH_{2n}, formerly known as **olefins**. Alkenes are unsaturated compounds, characterized by one or more double bonds between adjacent carbon atoms. Lighter alkenes, such as *ethene and propene, are gases, obtained from the *cracking of oil fractions. Alkenes react by addition, and many useful compounds, such as *polythene and bromoethane, are made from them.

alkyl any organic radical of the formula C_nH_{2n+1}; the chief members are methyl, ethyl, propyl, butyl, and amyl. These radicals are not stable in the free state but are found combined in a large number of types of organic compounds such as alcohols, esters, aldehydes, ketones, and halides.

alkylphenolethoxylate APEO, chemical used mainly in detergents, but also in herbicides, cleaners, packaging, and paints. Nonylphenol, a breakdown product of APEOs is a significant river pollutant; 60% of APEOs end up in the water. Nonylphenol, and other APEO breakdown products, have a feminizing effect on fish: male fish start to produce yolk protein and the growth of testes is slowed.

alkyne member of the group of *hydrocarbons with the general formula C_nH_{2n-2}, formerly known as the **acetylenes**. They are unsaturated compounds, characterized by one or more triple bonds between

adjacent carbon atoms. Lighter alkynes, such as ethyne, are gases; heavier ones are liquids or solids.

allele one of two or more alternative forms of a *gene at a given position (locus) on a *chromosome, caused by a difference in the sequence of *DNA. This is best explained with examples. A gene which controls eye colour in humans may have two alternative forms – an allele that can produce blue eyes, and an allele that produces brown eyes. In a plant that occurs in tall and short forms, there may be an allele that tends to produce tall plants and an alternative allele that produces short plants.

The individual genes that form a pair of alleles are located at exactly the same point along a chromosome. Organisms with two sets of chromosomes (diploids), such as animals and plants, have chromosomes that are found as matching pairs in the *nucleus of each cell. This means that there will always be two genes for a characteristic in a cell. If the same allele is present twice, the organism is said to be *homozygous for this characteristic. If, however, one chromosome contains one allele and the other chromosome a contrasting allele, the organism is said to be *heterozygous.

In a heterozygous organism the appearance of the organism (phenotype) may be determined by one allele and not the other. The allele that determines the phenotype is said to be dominantly expressed; it shows *dominance over other alleles. The expression of the other allele is described as being recessive.

allene any of a class of dienes (hydrocarbons with two double bonds) with adjacent double bonds. The simplest example is $CH_2=C=CH_2$, allene itself. Because of the stereochemistry of the double bonds, the terminal hydrogen atoms lie in planes mutually at right angles. Allenes behave mainly as typical unsaturated compounds, but are less stable than dienes with nonadjacent double bonds.

allergy special sensitivity of the body that makes it react with an exaggerated response of the natural immune defence mechanism to the introduction of an otherwise harmless foreign substance (**allergen**).

allogamy or **cross-fertilization**, (Greek *allos* 'other', *gamos* 'marriage') transfer of the pollen of one flower to the pistil of another in the fertilization of flowering plants.

allometry in biology, a regular relationship between a given feature (for example, the size of an organ) and the size of the body as a whole, when this relationship is not a simple proportion of body size. Thus, an organ may increase in size proportionately faster, or slower, than body size does. For example, a human baby's head is much larger in relation to its body than is an adult's.

alloparental care in animal behaviour, the care of another animal's offspring. 'Fostering' is common in some birds, such as pigeons, and social mammals, such as meerkats. Usually both the adoptive parent and the young benefit.

allopathy (Greek *allos* 'other', *pathos* 'suffering') in *homeopathy, a term used for orthodox medicine, using therapies designed to counteract the manifestations of the disease. In strict usage, allopathy is the opposite of homeopathy.

allosteric effect regulatory effect on an enzyme that takes place at a site distinct from that enzyme's catalytic site. For example, in a chain of enzymes the end product may act on an enzyme in the chain to regulate its own production.

allotropy property whereby an element can exist in two or more forms (allotropes), each possessing different physical properties but the same state of matter (gas, liquid, or solid). The allotropes of carbon are diamond, fullerene, and graphite. Sulphur has several allotropes (flowers of sulphur, plastic, rhombic, and monoclinic). These solids have different crystal structures, as do the white and grey forms of tin and the black, red, and white forms of phosphorus.

Oxygen exists as two gaseous allotropes: one used by organisms for respiration (O_2), and the other a poisonous pollutant, ozone (O_3).

alloxan or **mesoxalylurea**, $C_4H_2O_4N_2$, an important decomposition product of uric acid. It is obtained by oxidizing uric acid with nitric acid. On treatment with alkali, it takes up two molecules of water, producing urea and mesoxalic acid. It is used in medical research to produce experimental diabetes, since it selectively destroys the cells of the pancreas that secrete insulin.

alloy metal blended with some other metallic or non-metallic substance to give it special qualities, such as resistance to corrosion, greater hardness, or tensile strength. The atoms in a *metal are held together by the *metallic bond. In a pure metal the atoms are all the same size and can slip over each other if a force is applied. In an alloy, the presence of different sized atoms prevents such dislocations from weakening the metal. Useful alloys include bronze, brass, cupronickel, duralumin, German silver, gunmetal, pewter, solder, steel, and stainless steel.

Among the oldest alloys is bronze (mainly an alloy of copper and tin), the widespread use of which ushered in the Bronze Age. Complex alloys are now common; for example, in dentistry, where a cheaper alternative to gold is made of chromium, cobalt, molybdenum, and titanium. Among the most recent alloys are superplastics: alloys that can stretch to double their length at specific temperatures, permitting, for example, their injection into moulds as easily as plastic.

alluvial deposit layer of broken rocky matter, or sediment, formed from material that has been carried in suspension by a river or stream and dropped as the velocity of the current decreases. River plains and deltas are made entirely of alluvial deposits, but smaller pockets can be found in the beds of upland torrents.

Alluvial deposits can consist of a whole range of particle sizes, from boulders down through cobbles, pebbles, gravel, sand, silt, and clay. The raw materials are the rocks and soils of upland areas that are loosened by erosion and washed away by mountain streams. Much of the world's richest farmland lies on alluvial deposits. These deposits can also provide an economic source of minerals. River currents produce a sorting action, with particles of heavy material deposited first while lighter materials are washed downstream.

Hence heavy minerals such as gold and tin, present in the original rocks in small amounts, can be concentrated and deposited on stream beds in commercial quantities. Such deposits are called 'placer ores'.

alluvial fan roughly triangular sedimentary formation found at the base of slopes. An alluvial fan results when a sediment-laden stream or river rapidly deposits its load of gravel and silt as its speed is reduced on entering a plain.

The surface of such a fan slopes outward in a wide arc from an apex at the mouth of the steep valley. A small stream carrying a load of coarse particles builds a shorter, steeper fan than a large stream carrying a load of fine particles. Over time, the fan tends to become destroyed piecemeal by the continuing headward and downward erosion levelling the slope.

alluvium sediments laid down by streams and rivers. The most common constituents are clay, silt, and gravel. The loose, unconsolidated material forms features such as river terraces, *flood plains, and *deltas.

Alluvium is deposited along the river channel where the water's velocity is too low to transport the river's load – for example, on the inside bend of a *meander.

allyl in chemistry, the **propenyl group**, an unsaturated organic radical corresponding to the formula $CH_2=CHCH_2-$.

The chief compounds are allyl alcohol (propenol), with the general properties of a primary alcohol; allyl iodide, a colourless liquid with an odour of garlic; allyl bromide, a heavy liquid obtained by treating allyl alcohol with phosphorus tribromide; allyl sulphide, or 'oil of garlic', obtained by macerating garlic; and allyl isothiocyanate, or 'oil of mustard', occurring in black mustard seeds.

almanac book or table containing a calendar of the days, weeks, and months of the year, with the addition of notices of astronomical phenomena, ecclesiastical feasts, and similar useful information.

alpha in computing, the first version of a new software program. Developing modern software requires much testing and many versions before the definitive product is achieved. The first versions of any new product are typically full of *bugs, and are tested by the developers and their assistants. Later versions, known as *beta versions, are given to outside users to test.

alpha decay spontaneous alteration of the nucleus of a radioactive atom, which transmutes the atom from one atomic number to another through the emission of a helium nucleus (known as an *alpha particle). As a result, the atomic number decreases by two and the atomic weight decreases by four. See also *radioactivity.

alphanumeric data data made up of any of the letters of the alphabet and any digit from 0 to 9. The classification of data according to the type or types of character contained enables computer *validation systems to check the accuracy of data: a computer can be programmed to reject entries that contain the wrong type of character. For example, a person's name would be rejected if it contained any numeric data, and a bank-account number would be rejected if it contained any alphabetic data. A car's registration number, by comparison, would be expected to contain alphanumeric data but no punctuation marks.

alpha particle or **alpha ray**, positively charged (2+), high-energy particle emitted from the nucleus of a radioactive atom, discovered by New-Zealand-born British physicist Ernest Rutherford (1871–1937). It is one of the products of the spontaneous disintegration of radioactive elements (see *radioactivity) such as radium and thorium, and is identical to the nucleus of a helium atom (^{4}He) – that is, it consists of two protons and two neutrons. The process of emission, **alpha decay**, transforms one element into another, decreasing the atomic number by two and the atomic mass by four. Plutonium-239 (^{239}Pu) is an example of a material that emits alpha particles.

Because of their large mass, alpha particles have a short range of only a few centimetres in air. They have a low penetrating power and can be stopped by a sheet of paper or aluminium. They have a strongly ionizing effect (see *ionizing radiation) on the molecules that they strike, and are therefore capable of damaging living cells. Alpha particles travelling in a vacuum are deflected slightly by magnetic and electric fields.

Alps, Lunar see *Lunar Alps.

alternate angles pair of angles that lie on opposite sides and at opposite ends of a transversal (a line that cuts two or more lines in the same plane). The alternate angles formed by a transversal of two parallel lines are equal.

alternating current AC, electric current that flows for an interval of time in one direction and then in the opposite direction; that is, a current that flows in alternately reversed directions through or around a circuit. Electric energy is usually generated as alternating current in a power station, and alternating currents may be used for both power and lighting.

If half an amp moves in one direction past a point in half a second and in the next half a second half an amp reverses and moves past the same point in the opposite direction, then in total one coulomb of electrons or one

amp has passed the point in one second. The current flow is 1 amp AC.

The advantage of alternating current over direct current (DC), as from a battery, is that its voltage can be raised or lowered economically by a transformer; high voltage for generation and transmission, and low voltage for safe utilization. Railways, factories, and domestic appliances, for example, use alternating current.

alternation of generations typical life cycle of terrestrial plants and some seaweeds, in which there are two distinct forms occurring alternately: **diploid** (having two sets of chromosomes) and **haploid** (one set of chromosomes). The diploid generation produces haploid spores by *meiosis, and is called the sporophyte, while the haploid generation produces gametes (sex cells), and is called the gametophyte. The gametes fuse to form a diploid *zygote which develops into a new sporophyte; thus the sporophyte and gametophyte alternate.

alternative energy see *energy, alternative.

alternator electricity *generator that produces an alternating current.

alt hierarchy on Usenet, the 'alternative' set of newsgroups, set up so that anyone can start a newsgroup on any topic. Most areas of Usenet, such as the Big Seven hierarchies, allow the creation of newsgroups only after structured discussion and a vote to demonstrate that demand for the newsgroup exists. The alt hierarchy was created to allow users to bypass this process.

altimetry process of measuring altitude, or elevation. Satellite altimetry involves using an instrument – commonly a laser – to measure the distance between the satellite and the ground.

altitude in geometry, the perpendicular distance from a *vertex (corner) of a figure, such as a triangle, to the base (the side opposite the vertex).

altitude measurement of height, usually given in metres (or feet) above sea level.

altitude sickness or **mountain sickness**, condition resulting from sudden exposure to low atmospheric pressure and reduced oxygen supply found at high altitudes (above 3,000m/10,000ft). Symptoms include nausea, breathlessness and exhaustion; with increasing altitude pulmonary and cerebral oedema may develop. It is treated with rest, analgesics, and when necessary, supplementary oxygen, *diuretics and a return to lower altitudes.

altruism in biology, helping another individual of the same species to reproduce more effectively, as a direct result of which the altruist may leave fewer offspring itself. Female honey bees (workers) behave altruistically by rearing sisters in order to help their mother, the queen bee, reproduce, and forgo any possibility of reproducing themselves.

ALU abbreviation for *arithmetic and logic unit.

alum any double sulphate of a monovalent metal or radical (such as sodium, potassium, or ammonium) and a trivalent metal (such as aluminium, chromium, or iron). The commonest alum is the double sulphate of potassium and aluminium, $K_2Al_2(SO_4)_4.24H_2O$, a white crystalline powder that is readily soluble in water. It is used in curing animal skins. Other alums are used in papermaking and to fix dyes in the textile industry.

alumina or **corundum**; Al_2O_3, oxide of aluminium, widely distributed in clays, slates, and shales. It is formed by the decomposition of the feldspars in granite and used as an abrasive. Typically it is a white powder, soluble in most strong acids or caustic alkalis but not in water. Impure alumina is called 'emery'. Rubies, sapphires, and topaz are corundum gemstones. It is the chief component of *bauxite.

aluminium chemical symbol Al, lightweight, silver-white, ductile and malleable, metallic element, atomic number 13, relative atomic mass 26.9815, melting point 658°C/1,216°F. It is the third most abundant element (and the most abundant metal) in the Earth's crust, of which it makes up about 8.1% by mass. It is non-magnetic, an excellent conductor of electricity, and oxidizes easily, the layer

of oxide on its surface making it highly resistant to tarnish.

pure aluminium Aluminium is a reactive element which forms stable compounds, so a great deal of energy is needed in order to separate aluminium from its ores, and the pure metal was not readily obtainable until the middle of the 19th century. Commercially, it is prepared by the electrolysis of alumina (aluminium oxide), which is obtained from the ore *bauxite. In its pure state aluminium is a weak metal, but when combined with elements such as copper, silicon, or magnesium it forms alloys of great strength.

aluminium chloride $AlCl_3$, white solid made by direct combination of aluminium and chlorine.

$$2Al + 3Cl_2 \rightarrow 2AlCl_3$$

The anhydrous form is a typical covalent compound.

aluminium hydroxide or **alumina cream**; $Al(OH)_3$, gelatinous precipitate formed when a small amount of alkali solution is added to a solution of an aluminium salt.

$$Al_{(aq)} + 3OH_{(aq)} \rightarrow Al(OH)_{3(s)}$$

It is an *amphoteric compound as it readily reacts with both acids and alkalis.

aluminium ore raw material from which aluminium is extracted. The main ore is *bauxite, a mixture of minerals, found in economic quantities in Australia, Guinea, the West Indies, and several other countries.

aluminium oxide or **alumina**; Al_2O_3, white solid formed by heating aluminium hydroxide. It is an *amphoteric oxide, since it reacts readily with both acids and alkalis, and it is used as a refractory (furnace lining) and in column *chromatography.

alveolus plural **alveoli**, one of the many thousands of tiny air sacs in the *lungs. By the process of *diffusion *oxygen is transported from air in the alveoli into the blood, while *carbon dioxide diffuses out of the blood and into the air in the alveoli. This process is called gas exchange. To aid this process, the alveoli have very thin walls, which are folded in order to increase their surface area and allow rapid gas exchange, and are surrounded by blood capillaries. The air in the alveoli is replenished as a result of *breathing or ventilation. Smoking and exposure to other pollutants can cause the loss of the folds in the alveoli, which causes difficulties in breathing and the condition called emphysema.

Alzheimer's disease common manifestation of *dementia, thought to afflict 1 in 20 people over 65. After heart disease, cancer, and strokes it is the most common cause of death in the Western world. Attacking the brain's 'grey matter', it is a disease of mental processes rather than physical function, characterized by memory loss and progressive intellectual impairment. It was first described by Alois Alzheimer in 1906. Dementia affects nearly 18 million people worldwide, 66% of whom live in developed countries; this includes some 4 million people in the USA, and over 750,000 in Britain (2001). Numbers are expected to rise with the world's ageing population, reaching an estimated 34 million worldwide by 2025.

AM in physics, abbreviation for *amplitude modulation.

amalgam any alloy of mercury with other metals. Most metals will form amalgams, except iron and platinum. Amalgam is used in dentistry for filling teeth, and usually contains copper, silver, and zinc as the main alloying ingredients. This amalgam is pliable when first mixed and then sets hard, but the mercury leaches out and may cause a type of heavy-metal poisoning.

amalgamation the process of forming an amalgam, is a technique sometimes used to extract gold and silver from their ores. The ores are ground to a fine sand and brought into contact with mercury, which dissolves the gold and silver particles. The amalgam is then heated to distil off the mercury, leaving a residue of silver and gold. The mercury is recovered and reused.

Amalgamation to extract gold from its ore has been in use since Roman times.

American National Standards Institute ANSI, US national standards body. It sets official procedures in (among other areas) computing and electronics. The ANSI character set is the standard set of characters used by Windows-based computers.

americium chemical symbol Am, radioactive metallic element of the *actinide series, atomic number 95, relative atomic mass 243; it was first synthesized in 1944. It occurs in nature in minute quantities in *pitchblende and other uranium ores, where it is produced from the decay of neutron-bombarded plutonium, and is the element with the highest atomic number that occurs in nature. It is synthesized in quantity only in nuclear reactors by the bombardment of plutonium with neutrons. Its longest-lived isotope is Am-243, with a half-life of 7,650 years.

The element was named by US nuclear chemist Glenn Seaborg, one of the team who first synthesized it. Ten isotopes are known.

amethyst variety of *quartz, SiO_2, coloured violet by the presence of small quantities of impurities such as manganese or iron; used as a semi-precious stone. Amethysts are found chiefly in the Ural Mountains, India, the USA, Uruguay, and Brazil.

amide any organic chemical derived from a fatty acid by the replacement of the hydroxyl group (–OH) by an amino group (–NH_2).

One of the simplest amides is ethanamide (acetamide, CH_3CONH_2), which has a strong odour.

amine any of a class of organic chemical compounds in which one or more of the hydrogen atoms of ammonia (NH_3) have been replaced by other groups of atoms.

Methyl amines have unpleasant ammonia odours and occur in decomposing fish. They are all gases at ordinary temperature.

Aromatic amine compounds include aniline, which is used in dyeing.

amino acid water-soluble organic *molecule, mainly composed of carbon, oxygen, hydrogen, and nitrogen, containing both a basic amino group (NH_2) and an acidic carboxyl (COOH)

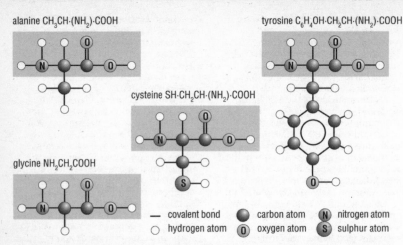

alanine $CH_3CH \cdot (NH_2) \cdot COOH$

tyrosine $C_6H_4OH \cdot CH_2CH \cdot (NH_2) \cdot COOH$

cysteine $SH \cdot CH_2CH \cdot (NH_2) \cdot COOH$

glycine NH_2CH_2COOH

— covalent bond   ● carbon atom   Ⓝ nitrogen atom
○ hydrogen atom   ⓞ oxygen atom   Ⓢ sulphur atom

amino acid Amino acids are natural organic compounds that make up proteins and can thus be considered the basic molecules of life. There are 20 different common amino acids. They consist mainly of carbon, oxygen, hydrogen, and nitrogen. Each amino acid has a common core structure (consisting of two carbon atoms, two oxygen atoms, a nitrogen atom, and four hydrogen atoms, shown in grey above) to which is attached a variable group, known as the R group. In glycine, the R group is a single hydrogen atom; in alanine, the R group consists of a carbon and three hydrogen atoms (methyl group).

group. They are small molecules able to pass through membranes. When two or more amino acids are joined together, they are known as *peptides; *proteins are made up of peptide chains folded or twisted in characteristic shapes.

Many different proteins are found in the cells of living organisms, but they are all made up of the same 20 amino acids, joined together in varying combinations (although other types of amino acid do occur infrequently in nature). Eight of these, the **essential amino acids**, cannot be synthesized by humans and must be obtained from the diet. Children need a further two amino acids that are not essential for adults. Other animals also need some preformed amino acids in their diet, but green plants can manufacture all the amino acids they need from simpler molecules, relying on energy from the Sun and minerals (including nitrates) from the soil.

aminoglycoside any of a group of antibiotics, including gentamicin, neomycin, and streptomycin, that are effective against a wide range of bacteria. As they are not absorbed from the intestine, they are usually given by injection. Drugs in this group are quite toxic, with side effects including damage to the ears and kidneys. For this reason they are only used for infections that do not respond to other antibiotics.

amino group the $-NH_2$ functional group, a defining feature of *amins and *amino acids.

ammeter instrument that measures electric current (flow of charge per unit time), usually in *amperes, through a conductor. It should not to be confused with a *voltmeter, which measures potential difference between two points in a circuit. The ammeter is placed in series (see *series circuit) with the component through which current is to be measured, and is constructed with a low internal resistance in order to prevent the reduction of that current as it flows through the instrument itself. A common type is the *moving-coil meter, which measures direct current (DC), but can, in the presence of a rectifier (a device which converts alternating current to direct current), measure alternating current (AC) also. Hot-wire, moving-iron, and dynamometer ammeters can be used for both DC and AC.

ammonia NH_3, colourless pungent-smelling gas, lighter than air and very soluble in water. It is made on an industrial scale by the *Haber (or Haber–Bosch) process, and used mainly to produce nitrogenous fertilizers, nitric acid, and some explosives.

In aquatic organisms and some insects, nitrogenous waste (from the breakdown of amino acids) is excreted in the form of ammonia, rather than as urea in mammals.

ammoniacal solution in chemistry, a solution produced by dissolving a solute in aqueous ammonia.

ammonium carbonate $(NH_4)_2CO_3$, white, crystalline solid that readily sublimes at room temperature into its constituent gases: ammonia, carbon dioxide, and water. It was formerly known as sal volatile acid and used in *smelling salts.

ammonium chloride or **sal ammoniac**; NH_4Cl, volatile salt that forms white crystals around volcanic craters. It is prepared synthetically for use in 'dry-cell' batteries, fertilizers, and dyes.

ammonium nitrate NH_4NO_3, colourless, crystalline solid, prepared by *neutralization of nitric acid with ammonia; the salt is crystallized from the solution. It sublimes on heating.

amnesia loss or impairment of memory. As a clinical condition it may be caused by disease or injury to the brain, by some drugs, or by shock; in some cases it may be a symptom of an emotional disorder.

amniotic fluid in mammals, fluid consisting mainly of water that is produced by the amnion, a fibrous membrane that lines the cavity of the uterus during pregnancy, in which the fetus floats and is protected from external pressure. It is swallowed by the fetus and excreted by the kidneys back into the amniotic sac. In humans, there is about 0.5–1 1/0.8–1.75 pt of amniotic fluid. The amniotic sac normally ruptures in early labour to release the fluid (the 'waters').

amoebiasis ongoing infection of the intestines, caused by the amoeba *Entamoeba histolytica*, resulting in chronic *dysentery and consequent

weakness and dehydration. Endemic in the developing world, it is now occurring in North America and Europe.

amp in physics, abbreviation for *ampere, a unit of electrical current.

ampere symbol A, SI unit of electrical current, named after French physicist André-Marie Ampère (1775–1836). Electrical current (a flow of negative charge) is measured in a similar way to water current, in terms of an amount per unit time; one ampere (amp) represents a flow of one coulomb per second, which is about 6.28×10^{18} *electrons per second.

Ampère's rule rule developed by French physicist André-Marie Ampère (1775–1836) connecting the direction of an electric current and its associated magnetic currents. It states that around a wire carrying a current towards the observer, the magnetic field curls in the anticlockwise direction. This assumes the conventional direction of current flow (from the positive to the negative terminal of a battery).

amphoteric of a chemical compound, being able to behave either as an *acid or as a *base depending on their environment. For example, the metals aluminium and zinc, and the oxides and hydroxides, act as bases in acidic solutions and as acids in alkaline solutions.

Amino acids and proteins are also amphoteric, as they contain both a basic (amino, $-NH_2$) and an acidic (carboxyl, $-COOH$) group.

amplifier electronic device that increases the strength of a signal, such as a radio signal. The ratio of output signal strength to input signal strength is called the gain of the amplifier. As well as achieving high gain, an amplifier should be free from distortion and able to operate over a range of frequencies. Practical amplifiers are usually complex circuits, although simple amplifiers can be built from single transistors or valves.

amplitude or **argument**, in mathematics, the angle in an *Argand diagram between the line that represents the complex number and the real (positive horizontal) axis. If the complex number is written as $r(\cos \theta + i \sin \theta)$, where r is radius and $i = \sqrt{(-1)}$, the amplitude is the angle θ (theta).

amplitude in physics, maximum displacement of an oscillation from the equilibrium position. For a *transverse wave motion, as in electromagnetic waves, it is the height of a crest (or the depth of a trough). For a *longitudinal wave, such as a sound wave, amplitude is the maximum distance a particle is pushed (due to compression) or pulled (due to rarefaction) from its resting position. A quiet sound has a lower amplitude and a loud sound has a higher amplitude. For a louder sound, more sound energy enters the ear every second. Amplitude is generally denoted by a.

amplitude modulation AM, method of transmitting audio-frequency radio waves that would otherwise not travel very far in space. The amplitude of the low-frequency audio signal is made to modulate (vary slightly) the amplitude of a continuously transmitted radio **carrier wave** of a higher frequency. In this way the **modulating signal** is imprinted on the carrier wave.

amygdala almond-shaped region of the *brain adjacent to the hippocampus, that links the cortex, responsible for conscious thought, with the regions controlling emotions. It is involved in interpreting fear-provoking information and linking it to fear responses. For example, where the amygdala is damaged, patients are unable to recognize fearful expressions.

amygdalin $C_{20}H_{27}O_{11}N$, glucoside found in bitter almonds, cherry kernels, and other vegetable products.

amyl alcohol former name for *pentanol.

amylase *enzyme that breaks down starch into a complex sugar that can be used in the body. It occurs widely in both plants and animals. In humans it is found in saliva and in the pancreatic digestive juices that drain into the *alimentary canal.

anabolic steroid any *hormone of the *steroid group that stimulates muscular tissue growth. Its use in medicine is limited to the treatment of some anaemias and breast cancers; it may help to break up blood clots. Side effects include aggressive behaviour, masculinization in women, and, in children, reduced height.

It is used in sports, such as weightlifting and athletics, to increase muscle bulk for greater strength and stamina, but it is widely condemned because of the side effects. In 1988 the Canadian sprinter Ben Johnson was stripped of an Olympic gold medal for having taken anabolic steroids.

anabolism process of building up body tissue, promoted by the influence of certain hormones. It is the constructive side of *metabolism, as opposed to *catabolism.

anaemia condition caused by a shortage of haemoglobin, the oxygen-carrying component of red blood cells. The main symptoms are fatigue, pallor, breathlessness, palpitations, and poor resistance to infection. Treatment depends on the cause.

anaerobic not requiring oxygen for the release of energy from food molecules such as glucose. An organism is described as anaerobic if it does not require *oxygen in order to survive. Instead, anaerobic organisms use *anaerobic respiration to obtain energy from food. Most anaerobic organisms are micro-organisms such as bacteria, yeasts, and internal parasites that live in places where there is never much oxygen, such as in the mud at the bottom of a lake or pond, or in the alimentary canal. Anaerobic organisms release much less of the available energy from their food than do *aerobic organisms.

anaerobic respiration in plant and animal cells, a process in which energy is released from food molecules such as glucose without requiring oxygen. Some aerobic plants and animals are able to use anaerobic respiration for short periods of time. For example, during a sprint, human muscles can respire anaerobically. Unfortunately, *lactic acid is produced and accumulates until the muscles cannot continue working. Anaerobic respiration in humans is less efficient than *aerobic respiration at releasing energy, but releases energy faster (see *respiration). This explains why humans can run faster in a sprint than over longer distances. When humans stop after a sprint, they have to continue breathing more heavily for a while. This is to take in 'extra' oxygen in order to break down the accumulated lactic acid on top of the 'normal' breakdown of sugar in aerobic respiration. The body is paying back the *oxygen debt built up during the sprint.

In plants, yeasts, and bacteria, anaerobic respiration results in the production of alcohol and carbon dioxide, a process that is exploited by both the brewing and the baking industries (see *fermentation).

Although anaerobic respiration is a primitive and inefficient form of energy release, deriving from the period when oxygen was missing from the atmosphere, it can also be seen as an *adaptation. To survive in some habitats, such as the muddy bottom of a polluted river, an organism must be to a large extent independent of oxygen; such habitats are said to be **anoxic**.

anaesthetic   drug that produces loss of sensation or consciousness; the resulting state is **anaesthesia**, in which the patient is insensitive to stimuli. Anaesthesia may also happen as a result of nerve disorder.

analgesic   agent for relieving pain. *Opiates alter the perception or appreciation of pain and are effective in controlling 'deep' visceral (internal) pain. Non-opiates, such as *aspirin, *paracetamol, and NSAIDs (nonsteroidal anti-inflammatory drugs), relieve musculoskeletal pain and reduce inflammation in soft tissues.

analogous   in biology, term describing a structure that has a similar function to a structure in another organism, but not a similar evolutionary path. For example, the wings of bees and of birds have the same purpose – to give powered flight – but have different origins. Compare *homologous.

analogue computer   computing device that performs calculations through the interaction of continuously varying physical quantities, such as voltages (as distinct from the more common *digital computer, which works with discrete quantities). An analogue computer is said to operate in real time

(corresponding to time in the real world), and can therefore be used to monitor and control other events as they happen.

analogue signal   in electronics, current or voltage that conveys or stores information, and varies continuously in the same way as the information it represents (compare digital signal). Analogue signals are prone to interference and distortion.

analogue-to-digital converter   ADC, electronic circuit that converts an analogue signal into a digital one. Such a circuit is needed to convert the signal from an analogue device into a digital signal for input into a computer. For example, many *sensors designed to measure physical quantities, such as temperature and pressure, produce an analogue signal in the form of voltage and this must be passed through an ADC before a computer can process the information. A *digital-to-analogue converter performs the opposite process.

analysis   in chemistry, the determination of the composition of substances; see *analytical chemistry.

analysis   branch of mathematics concerned with limiting processes on axiomatic number systems; *calculus of variations and infinitesimal calculus is now called analysis.

analytical chemistry   branch of chemistry that deals with the determination of the chemical composition of substances. **Qualitative analysis** determines the identities of the substances in a given sample; **quantitative analysis** determines how much of a particular substance is present.

Simple qualitative techniques exploit the specific, easily observable properties of elements or compounds – for example, the flame test makes use of the different flame colours produced by metal cations when their compounds are held in a hot flame. More sophisticated methods, such as those of *spectroscopy, are required where substances are present in very low concentrations or where several substances have similar properties.

Most quantitative analyses involve initial stages in which the substance to be measured is extracted from the test sample, and purified. The final analytical stages (or 'finishes') may involve measurement of the substance's mass (gravimetry) or volume (volumetry, titrimetry), or a number of techniques initially developed for qualitative analysis, such as fluorescence and absorption spectroscopy, chromatography, electrophoresis, and polarography. Many modern methods enable quantification by means of a detecting device that is integrated into the extraction procedure (as in gas–liquid chromatography).

analytical engine programmable computing device designed by English mathematician Charles Babbage (1792–1871) in 1833. It was based on the *difference engine but was intended to automate the whole process of calculation. It introduced many of the concepts of the digital computer but, because of limitations in manufacturing processes, was never built.

analytical geometry another name for *coordinate geometry.

anamorphic projection technique used in film and in *virtual reality to squeeze wide-frame images so that they fit into the dimensions of a 35-mm frame of film. In film projection, the projector has a complementary lens which reverses the process. In virtual reality, the computer must calculate the amount of deformation and reverse it.

anaphylaxis in medicine, a severe allergic response. Typically, the air passages become constricted, the blood pressure falls rapidly, and the victim collapses. A rare condition, anaphylaxis most often occurs following wasp or bee stings, or treatment with some drugs.

AND gate in electronics, a type of *logic gate.

androgen general name for any male sex hormone, of which *testosterone is the most important. They are all *steroids and are principally involved in the production of male *secondary sexual characteristics (such as beard growth).

Andromeda galaxy galaxy 2.2 million light years away from Earth in the constellation Andromeda, and the most distant object visible to the naked eye. It is the largest member of the Local Group of galaxies. Like the Milky Way, it is a spiral orbited by several companion galaxies but contains about twice as many stars as the Milky Way. It is about 200,000 light years across.

AND rule rule used for finding the combined probability of two or more independent events both occurring. If two events E_1 and E_2 are independent (have no effect on each other) and the probabilities of their taking place are p_1 and p_2, respectively, then the combined probability p that both E_1 and E_2 will happen is given by:

$$p = p_1 \times p_2$$

For example, if a blue die and a red die are thrown together, the probability of a blue six is $1/6$, and the probability of a red six is $1/6$. Therefore, the probability of both a red six and a blue six being thrown is $1/6 \times 1/6 = 1/36$.

By contrast, the **OR rule** is used for finding the probability of either one event or another taking place.

anencephaly absence of the *cerebral hemispheres and the surrounding skull plates. Anencephalic babies do not survive more than a few hours. A *neural tube defect, it can be detected prenatally.

aneurysm weakening in the wall of an artery, causing it to balloon outwards with the risk of rupture and serious, often fatal, blood loss. If detected in time, some accessible aneurysms can be repaired by bypass surgery, but such major surgery carries a high risk for patients in poor health.

angina or **angina pectoris**, severe pain in the chest due to impaired blood supply to the heart muscle because a coronary artery is narrowed. Faintness and difficulty in breathing accompany the pain. Treatment is by drugs or bypass surgery.

angiography technique for X-raying major blood vessels. A radiopaque dye is injected into the bloodstream so that the suspect vessel is clearly silhouetted on the X-ray film.

angiosperm flowering plant in which the seeds are enclosed within an ovary, which ripens into a fruit. Angiosperms are divided into *monocotyledons (single seed leaf in the embryo) and *dicotyledons (two seed leaves in the embryo). They include the majority of flowers, herbs, grasses, and trees except conifers.

There are over 250,000 different species of angiosperm, found in a wide range of habitats. Like *gymnosperms, they are seed plants, but differ in that ovules and seeds are protected within the carpel. Fertilization occurs by male gametes passing into the ovary from a pollen tube. After fertilization the ovule develops into the seed while the ovary wall develops into the fruit.

There is evidence of fossil angiosperms from the Jurassic era, and genera that seem very similar to modern examples have been found from the Cretaceous period.

angle   in mathematics, the amount of turn or rotation; it may be defined by a pair of rays (half-lines) that share a common endpoint (*vertex) but do not lie on the same line. Angles are measured in *degrees (°) or *radians (rads or ^{c}) – a complete turn or circle being 360° or 2π rads.

All angles around a point on a straight line add up to 180°. All angles around a point add up to 360°.

Angles are classified generally by their degree measures: **acute angles** are less than 90°; **right angles** are exactly 90° (a quarter turn) and are created by two *perpendicular lines crossing; **obtuse angles** are greater than 90° but less than 180° (a straight line); **reflex angles** are greater than 180° but less than 360°; **supplementary angles** add up to 180°.

angle brackets   symbols <, >, in documentation, brackets that indicate places where the user should input information of the type described between the brackets. Angle brackets are also used in online services and on the Internet to indicate that the name used is a user-ID rather than a real name, and on CompuServe as part of certain *emoticons.

angle of declination   angle at a particular point on the Earth's surface between the direction of the true or geographic North Pole and the magnetic north pole. The angle of declination has varied over time because of the slow drift in the position of the magnetic north pole.

angle of dip   or **angle of inclination**, angle at a particular point on the Earth's surface between the direction of the Earth's magnetic field and the horizontal; see *magnetic dip.

angle of incidence   angle between a ray of light striking a surface (incident ray) and the normal to that surface. For a mirror it is equal to the *angle of reflection.

angle of reflection   angle between a ray of light reflected from a mirror and the normal (perpendicular) to that mirror. The law of reflection states that the angle of reflection is equal to the *angle of incidence.

angle of refraction   angle between a refracted ray of light and the normal to the surface at which *refraction occurred. When a ray passes from air into a denser medium such as glass, it is bent towards the normal so that the angle of refraction is less than the *angle of incidence.

angstrom   symbol Å, unit of length equal to 10^{-10} metres or one-ten-millionth of a millimetre, used for atomic measurements and the wavelengths of electromagnetic radiation. It is named after the Swedish scientist Anders Ångström.

angular momentum   in physics, a type of *momentum.

anhydride   chemical compound obtained by the removal of water from another compound; usually a dehydrated acid. For example, sulphur(VI) oxide (sulphur trioxide, SO_3) is the anhydride of sulphuric acid (H_2SO_4).

anhydrite   naturally occurring anhydrous calcium sulphate ($CaSO_4$). It is used commercially for the manufacture of plaster of Paris and builders' plaster.

anhydrous   of a chemical compound, containing no water. If the water of crystallization is removed from blue crystals of copper(II) sulphate, a white powder (anhydrous copper sulphate) results. Liquids from which all traces of water have been removed are also described as being anhydrous.

aniline or **phenylamine**; $C_6H_5NH_2$, (Portuguese *anil* 'indigo') one of the simplest aromatic chemicals (a substance related to benzene, with its carbon atoms joined in a ring). When pure, it is a colourless oily liquid; it has a characteristic odour, and turns brown on contact with air. It occurs in coal tar, and is used in the rubber industry and to make drugs and dyes. It is highly poisonous.

Aniline was discovered in 1826, and was originally prepared by the dry distillation of indigo, hence its name.

animal or **metazoan**, (Latin *anima* 'breath', 'life') member of the *kingdom Animalia, one of the major categories of living things, the science of which is **zoology**. Animals are all multicellular *heterotrophs (they obtain their energy from organic substances produced by other organisms); they have eukaryotic cells (the genetic material is contained within a distinct nucleus) which are bounded by a thin cell membrane rather than the thick cell wall of plants. Most animals are capable of moving around for at least part of their life cycle.

animal behaviour scientific study of the behaviour of animals, either by comparative psychologists (with an interest mainly in the psychological processes involved in the control of behaviour) or by ethologists (with an interest in the biological context and relevance of behaviour).

animal cell *cell in an animal, which, like all cells, has a *cell membrane, *cytoplasm, and a *nucleus, but in which the surface consists of the cell surface membrane only – it does not have a cell wall. The cell surface membrane keeps the cell together by being strong, even though it is very thin and flexible. The membrane also controls what enters and leaves the cell. For example, nutrients need to be able to enter and waste materials to leave. The cell contains cytoplasm, in which is located a nucleus. This controls the activities of the cell.

animism in anthropology, the belief that everything, whether animate or inanimate, possesses a soul or spirit. It is a fundamental system of belief in certain religions, particularly those of some pre-industrial societies. Linked with this is the worship of natural objects such as stones and trees, thought to harbour spirits (naturism); fetishism; and ancestor worship. In psychology and physiology, animism is the view of human personality that attributes human life and behaviour to a force distinct from matter. In developmental psychology, an animistic stage in the early thought and speech of the child has been described, notably by Swiss psychologist Jean Piaget (1896–1980).

In philosophy, the view that in all things consciousness or something mindlike exists.

anion *ion carrying a negative charge. An anion is formed from an *atom by the gain of electrons, a process known as ***ionic bonding**. *Non-metallic elements form anions. During *electrolysis, anions in the electrolyte move towards the anode (positive electrode).

An electrolyte, such as the salt zinc chloride ($ZnCl_2$), is dissociated in aqueous solution or in the molten state into doubly charged Zn^{2+} *cations and singly-charged Cl^- anions. During electrolysis, the zinc cations flow to the cathode (to become discharged and liberate zinc metal) and the chloride anions flow to the anode (to become discharged and form chlorine gas).

annealing controlled cooling of a material to increase ductility and strength. The process involves first heating a material (usually glass or metal) for a given time at a given temperature, followed by slow cooling. It is a common form of *heat treatment. Normalizing is a type of annealing where a metal is heated to a point above its *critical temperature before being cooled. This helps to refine the grain.

annelid any segmented worm of the phylum Annelida. Annelids include earthworms, leeches, and marine worms such as lugworms.

annual rings or **growth rings**, concentric rings visible on the wood of a cut tree trunk or other woody stem. Each ring represents a period of growth when new *xylem is laid down to replace tissue being converted into wood (secondary xylem). The wood formed from xylem produced in the spring and early summer has larger and more numerous vessels than the wood formed from xylem produced in autumn when growth is slowing down. The result is a clear boundary between the pale spring wood and the denser, darker autumn wood. Annual rings may be used to estimate the age of the plant (see *dendrochronology), although occasionally more than one growth ring is produced in a given year.

annular eclipse solar *eclipse in which the Moon does not completely obscure the Sun and a thin ring of sunlight remains visible. Annular eclipses occur when the Moon is at its farthest point from the Earth.

anode positive electrode of an electrolytic cell, towards which negative particles (anions), usually in solution, are attracted. See *electrolysis. In a thermionic valve, cathode ray tube, or similar device, electrons are drawn to the anode after being emitted from the *cathode.

anodizing process that increases the resistance to *corrosion of a metal, such as aluminium, by building up a protective oxide layer on the surface. The natural corrosion resistance of aluminium is provided by a thin film of aluminium oxide; anodizing increases the thickness of this film and thus the corrosion protection.

It is so called because the metal becomes the *anode in an electrolytic bath containing a solution of, for example, sulphuric or chromic acid as the *electrolyte. During *electrolysis oxygen is produced at the anode, where it combines with the metal to form an oxide film.

anomalous expansion of water expansion of water as it is cooled from 4°C to 0°C. This behaviour is unusual because most substances contract when they are cooled. It means that water has a greater density at 4°C than at 0°C. Hence ice floats on water, and the water at the bottom of a pool in winter is warmer than at the surface. As a result lakes and ponds freeze slowly from the surface downwards, usually remaining liquid near the

bottom, where aquatic life is more likely to survive.

anonymous FTP abbreviation for anonymous file transfer protocol, method of retrieving a file from a remote computer without having an account on that computer. Many organizations, such as universities and software companies, maintain publicly accessible archives of files that may be retrieved across the Internet via *FTP. An ordinary user who is not affiliated to the organization may retrieve files by entering the FTP address and then typing in either 'anonymous' or 'ftp' when asked for a user-ID or log-in name, followed by the user's e-mail address in place of a password. These users are typically offered *access privileges to only a small part of the company's stored files, and the rest may be cordoned off from access by a *firewall.

anorexia lack of desire to eat, or refusal to eat, especially the pathological condition of **anorexia nervosa**, most often found in adolescent girls and young women. Compulsive eating, or *bulimia, distortions of body image, and depression often accompany anorexia.

anoxia or **hypoxia**, in biology, deprivation of oxygen, a condition that rapidly leads to collapse or death, unless immediately reversed.

Antabuse proprietary name for disulfiram, a synthetic chemical used in the treatment of alcoholism. It produces unpleasant side effects if combined with alcohol, such as nausea, headaches, palpitations, and collapse. The 'Antabuse effect' is produced coincidentally by certain antibiotics.

antagonist in biology, a *muscle that relaxes in response to the contraction of its agonist muscle. See *antagonistic muscles.

antagonist in medicine, a drug or body chemical with the reverse effect of another drug or chemical. The drug naloxone is an antagonist of *morphine.

antagonistic muscles in the body, a pair of *muscles working together to allow coordinated movement of the skeletal joints. Muscles can only exert a force and do work by contracting. An

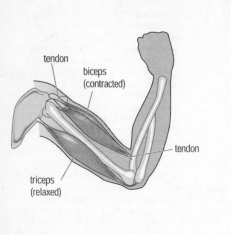

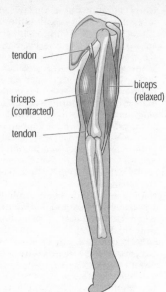

antagonistic muscle Even simple movements such as bending and straightening the arm require muscle pairs to contract and relax synchronously.

example is the antagonistic pair of muscles used in bending and straightening the arm. To bend the arm requires one set of muscles – the *biceps – to contract, while another set – the triceps – relaxes. If the arm is to be straightened, the reverse happens. The individual components of antagonistic pairs can be classified into extensors (muscles that straighten a limb) and flexors (muscles that bend a limb).

anterior in biology, the front of an organism, usually the part that goes forward first when the animal is moving. In higher organisms the anterior end of the nervous system, over the course of evolution, has developed into a brain with associated receptor organs able to detect stimuli including light and chemicals.

anther in a flower, the terminal part of a stamen in which the *pollen grains are produced. It is usually borne on a slender stalk or filament, and has two lobes, each containing two chambers, or pollen sacs, within which the pollen is formed.

anthracene white, glistening, crystalline, tricyclic, aromatic hydrocarbon with a faint blue fluorescence when pure. Its melting point is about 216°C/421°F and its boiling point 351°C/664°F. It occurs in the high-boiling-point fractions of coal tar, where it was discovered in 1832 by the French chemists Auguste Laurent (1808–1853) and Jean Dumas (1800–1884).

anthracite (from Greek *anthrax* 'coal') hard, dense, shiny variety of *coal, containing over 90% carbon and a low percentage of ash and impurities, which causes it to burn without flame, smoke, or smell. Because of its purity, anthracite gives off relatively little sulphur dioxide when burnt.

anthrax disease of livestock, occasionally transmitted to humans, usually via infected hides and fleeces. It may also be used as a weapon in biological warfare. It develops as black skin pustules or severe pneumonia. Treatment is possible with antibiotics, and vaccination is effective.

anthropic principle idea that 'the universe is the way it is because if it were different we would not be here to observe it'. The principle arises from the observation that if the laws of science were even slightly different, it would have been impossible for intelligent life to evolve. For example, if the strengths of the fundamental forces were only slightly different, stars would have been unable to burn hydrogen and produce the chemical elements that make up our bodies.

anthropometry science dealing with the measurement of the human body, particularly stature, body weight, cranial capacity, and length of limbs, in samples of living populations, as well as the remains of buried and fossilized humans.

antibiotic drug that kills or inhibits the growth of bacteria and fungi. The earliest antibiotics, the *penicillins, came into use from 1941 and were quickly joined by chloramphenicol, the *cephalosporins, erythromycins, tetracyclines, and aminoglycosides. A range of broad-spectrum antibiotics, the 4-quinolones, was developed in 1989, of which ciprofloxacin was the first. Each class and individual antibiotic acts in a different way and may be effective against either a broad spectrum or a specific type of disease-causing agent. Use of antibiotics has become more selective as side effects, such as toxicity, allergy, and resistance, have become better understood. Bacteria have the ability to develop resistance following repeated or subclinical (insufficient) doses, so more advanced antibiotics and synthetic antimicrobials are continually required to overcome them.

antibody protein molecule produced in the blood by *lymphocytes in response to the presence of foreign or invading substances (*antigens); such substances include the proteins carried on the surface of infecting micro-organisms. Antibody production is only one aspect of *immunity in vertebrates.

Each antibody acts against only one kind of antigen, and combines with it to form a 'complex'. This action may render antigens harmless, or it may destroy micro-organisms by setting off chemical changes that cause them to self-destruct. In other cases, the formation of a complex will cause antigens to form clumps that can then be detected and engulfed by white blood cells, such as *macrophages and *phagocytes.

Each bacterial or viral infection will bring about the manufacture of a specific antibody, which will then fight the disease. Many diseases can only be contracted once because antibodies remain in the blood after the infection has passed, preventing any further invasion. Vaccination boosts a person's resistance by causing the production of antibodies specific to particular infections. Large quantities of specific antibodies can now be obtained by the monoclonal technique (see *monoclonal antibody).

Antibodies were discovered in 1890 by the German physician Emil von Behring and the Japanese bacteriologist Shibasaburo Kitasato.

anticholinergic any drug that blocks the passage of certain nerve impulses in the *central nervous system by inhibiting the production of acetylcholine, a neurotransmitter.

anticoagulant substance that inhibits the formation of blood clots. Common anticoagulants are heparin, produced by the liver and some white blood cells, and derivatives of coumarin, such as warfarin. Anticoagulants are used medically in the prevention and treatment of thrombosis and heart attacks. Anticoagulant substances are also produced by blood-feeding animals, such as mosquitoes, leeches, and vampire bats, to keep the victim's blood flowing.

anticonvulsant any drug used to prevent epileptic seizures (convulsions or fits) or reduce their severity; see *epilepsy.

antidepressant any drug used to relieve symptoms in depressive illness. The main groups are the *selective serotonin-reuptake inhibitors (SSRIs), the tricyclic antidepressants (TCADs), and the monoamine oxidase inhibitors (MAOIs). They all act by altering chemicals available to the central nervous system. All may produce serious side effects.

antidiuretic hormone pituitary hormone that prevents excessive fluid loss. See *ADH.

antidote drug used to counteract a poison. For instance, acetylcysteine is an antidote to *paracetamol and is administered to prevent liver damage following overdosage.

antiferromagnetic material material with a very low magnetic susceptibility that increases with temperature up to a certain temperature, called the Néel temperature. Above the Néel temperature, the material is only weakly attracted to a strong magnet.

antifreeze substance added to a water-cooling system (for example, that of a car) to prevent it freezing in cold weather.

antigen any substance that causes the production of *antibodies by the body's immune system. Common antigens include the proteins carried on the surface of bacteria, viruses, and pollen grains. The proteins of incompatible blood groups or tissues also act as antigens, which has to be taken into account in medical procedures such as blood transfusions and organ transplants.

antihistamine any substance that counteracts the effects of *histamine. Antihistamines may occur naturally or they may be synthesized.

antiknock substance added to *petrol to reduce knock in car engines, caused by premature combusting. It is a mixture of dibromoethane and tetraethyl lead. Its use in leaded petrol has resulted in atmospheric pollution by lead compounds.

antilogarithm or **antilog**, inverse of *logarithm, or the number whose logarithm to a given base is a given number. If $y = \log_a x$, then $x = \text{antilog}_a y$.

antimatter in physics, form of matter in which most of the attributes (such as electrical charge, magnetic moment, and spin) of *elementary particles are reversed. These *antiparticles can be created in particle accelerators, such as those at *CERN in Geneva, Switzerland, and at *Fermilab in the USA. In 1996 physicists at CERN created the first atoms of antimatter: nine atoms of antihydrogen survived for 40 nanoseconds (40 billionths of a second).

antimony chemical symbol Sb, silver-white, brittle, semimetallic element (a metalloid), atomic number 51, relative atomic mass 121.76. Its chemical symbol comes from Latin *stibium*. It occurs chiefly as the ore stibnite, and is used to make alloys harder; it is also used in photosensitive substances in colour photography, optical electronics, fireproofing, pigments, and medicine. It was employed by the ancient Egyptians in a mixture to protect the eyes from flies.

antinode in physics, the position in a *standing wave pattern at which the amplitude of vibration is greatest (compare *node). The standing wave of a stretched string vibrating in the fundamental mode has one antinode at its midpoint. A vibrating air column in a pipe has an antinode at the pipe's open end and at the place where the vibration is produced.

antiparticle in nuclear physics, a particle corresponding in mass and properties to a given *elementary particle but with the opposite electrical charge, magnetic properties, or coupling to other fundamental forces. For example, an electron carries a negative charge whereas its anti-particle, the positron, carries a positive one. When a particle and its anti-particle collide, they destroy each other, in the process called 'annihilation', their total energy being converted to lighter particles and/or photons. A substance consisting entirely of antiparticles is known as *antimatter.

antipsychotic or **neuroleptic**, any drug used to treat the symptoms of severe mental disorder.

antiseptic any substance that kills or inhibits the growth of micro-organisms. The use of antiseptics was pioneered by English surgeon Joseph Lister (1827–1912). He used carbolic acid (*phenol), which is a weak antiseptic; antiseptics such as TCP are derived from this.

antivirus software computer software that detects *viruses and/or cleans viruses from an infected computer system. Antivirus programs can be designed to catch all viruses entering a system, or to catch only those viruses delivered by a particular method, such as e-mail. In order for the software to remain effective it must be updated regularly with new signature files to enable the software to recognize the latest viruses; this can usually be done over the Internet.

anxiety unpleasant, distressing emotion usually to be distinguished from fear. Fear is aroused by the perception of actual or threatened danger; anxiety arises when the danger is imagined or cannot be identified or clearly perceived. It is a normal response in stressful situations, but is frequently experienced in many mental disorders.

Anxiety is experienced as a feeling of suspense, helplessness, or alternating hope and despair together with excessive alertness and characteristic bodily changes such as tightness in the throat, disturbances in breathing and heartbeat, sweating, and diarrhoea.

In psychiatry, an anxiety state is a type of neurosis in which the anxiety either seems to arise for no reason or else is out of proportion to what may have caused it. 'Phobic anxiety' refers to the irrational fear that characterizes *phobia.

anxiolytic any drug that relieves an anxiety state.

aorta the body's main *artery, arising from the left ventricle of the heart in birds and mammals. Carrying freshly oxygenated blood, it arches over the top of the heart and descends through the trunk, finally splitting in the lower abdomen to form the two iliac arteries. Arteries branching off the arch of the aorta carry blood to the upper body. Loss of elasticity in the aorta provides evidence of *atherosclerosis, which may lead to heart disease.

In fish a ventral aorta carries deoxygenated blood from the heart to the gills, and the dorsal aorta carries oxygenated blood from the gills to other parts of the body.

a.p. in physics, abbreviation for **atmospheric pressure**.

apex the highest point of a triangle, cone, or pyramid – that is, the vertex (corner) opposite a given base.

aphasia general term for the many types of disturbance in language that are due to brain damage, especially in the speech areas of the dominant hemisphere.

aphid any of the family of small insects, Aphididae, in the order Hemiptera,

suborder Homoptera, that live by sucking sap from plants. There are many species, often adapted to particular plants; some are agricultural pests.

apnoea temporary cessation of breathing due to reduction of the carbon dioxide content in the blood. Newborns are prone to apnoeic attacks.

apogamy reproduction in plants without fusion of gametes. See *apomixis.

apomixis reproduction in flowering plants without the fusion of gametes. The term covers all types of asexual reproduction, including vegetative reproduction in which a plant is propagated by organs other than the sex organs in the flower, and agamospermy, in which seeds are produced without fertilization.

apoptosis or **cell suicide**, self-destruction of a cell. All cells contain genes that cause them to self-destruct if damaged, diseased, or as part of the regulation of cell numbers during the organism's normal development. Many cancer cells have mutations in genes controlling apoptosis, so understanding apoptosis may lead to new cancer treatments where malfunctioning cells can be instructed to destroy themselves.

During apoptosis, a cell first produces the enzymes needed for self-destruction before shrinking to a characteristic spherical shape with balloon-like bumps on its outer surface. The enzymes break down its contents into small fragments which are easily digestible by surrounding cells.

Appleton layer or **F-layer**, band containing ionized gases in the Earth's upper atmosphere, at a height of 150–1,000 km/94–625 mi, above the *E-layer (formerly the Kennelly-Heaviside layer). It acts as a dependable reflector of radio signals as it is not affected by atmospheric conditions, although its ionic composition varies with the sunspot cycle.

application in mathematics, a curved line that connects a series of points (or 'nodes') in the smoothest possible way. The shape of the curve is governed by a series of complex mathematical formulae. Applications are used in computer graphics and *CAD.

application service provider ASP, company providing access to computer applications via the Internet. By renting applications through an ASP, companies can avoid the task of procuring and implementing complex systems themselves, and, in some cases, cut down considerably on their in-house *information technology infrastructure.

applications package in computing, set of programs and related documentation (such as instruction manuals) used in a particular application. For example, a typical payroll applications package would consist of separate programs for the entry of data, updating the master files, and printing the pay slips, plus documentation in the form of program details and instructions for use.

Applications Program Interface API, in computing, specific method through which a programmer writing an application can interface with an operating system or another application. An API ensures that all applications are consistent with the operating system and have a similar *user interface.

approximation rough estimate of a given value. For example, for *pi (which has a value of 3.1415926 correct to seven decimal places), 3 is an approximation to the nearest whole number.

aqua fortis former name for *nitric acid.

aquamarine blue variety of the mineral *beryl. A semi-precious gemstone, it is used in jewellery.

aqua regia (Latin 'royal water') mixture of three parts concentrated hydrochloric acid and one part concentrated nitric acid, which dissolves 'noble' metals such as gold and platinum.

aqua vitae (Latin 'water of life') the elixir of life of the alchemists who discovered distillation. They identified it with the distilled quintessence of wine, but in French *eau-de-vie* applies to spirits distilled from other substances than wine. In Gaelic its equivalent *uisge beatha* provides the derivation of whisky.

aqueous solution solution in which the solvent is water.

arc in geometry, a section of a curved line or circle. A circle has three types of arc: a **semicircle**, which is exactly half of the circle; **minor arcs**, which are less than the semicircle; and **major arcs**, which are greater than the semicircle.

Archaean or **Archaeozoic**, widely used term for the earliest era of geological time; the first part of the Precambrian **Eon**, spanning the interval from the formation of Earth to about 2,500 million years ago.

Archimedes' principle in physics, the principle that the weight of the liquid displaced by a floating body is equal to the weight of the body. The principle is often stated in the form: 'an object totally or partially submerged in a fluid displaces a volume of fluid that weighs the same as the apparent loss in weight of the object (which, in turn, equals the upwards force, or upthrust, experienced by that object).' It was discovered by the Greek mathematician Archimedes (c. 287–212 BC).

Archimedes screw one of the earliest kinds of pump, associated with the Greek mathematician Archimedes (c. 287–212 BC). It consists of an enormous spiral screw revolving inside a close-fitting cylinder. It is used, for example, to raise water for irrigation.

archipelago group of islands, or an area of sea containing a group of islands. The islands of an archipelago are usually volcanic in origin, and they sometimes represent the tops of peaks in areas around continental margins flooded by the sea.

architecture in computing, overall design of a computer system, encompassing both hardware and software. The architecture of a particular system includes the specifications of individual components and the ways they interact. Because the operating system defines how these elements interact with each other and with application software, it is also included in the term.

arc lamp or **arc light**, electric light that uses the illumination of an electric arc maintained between two electrodes. The English chemist Humphry Davy demonstrated the electric arc in 1802 and electric arc lighting was first introduced by English electrical engineer W E Staite in 1846. The lamp consists of two carbon electrodes, between which a very high voltage is maintained. Electric current arcs (jumps) between the two electrodes, creating a brilliant light. Its main use in recent years has been in cinema projectors.

arc minute symbol ', unit for measuring small angles, used in geometry, surveying, map-making, and astronomy. An arc minute is one-sixtieth of a degree and is divided into 60 arc seconds (symbol "). Small distances in the sky, as between two close stars or the apparent width of a planet's disc, are expressed in minutes and seconds of arc.

arctic climate Antarctica, all of Greenland, the north of Alaska, Canada, and Russia are areas with continuous *permafrost. Winters are severe and the sea freezes; summers have continuous periods of daylight but the monthly temperatures struggle to rise above freezing point. Nearer each *pole the climate is constant frost. Precipitation is light and falls mainly as snow.

area size of a surface. It is measured in square units, usually square centimetres (cm^2), square metres (m^2), or square kilometres (km^2). **Surface area** is the area of the outer surface of a solid. *Integration may be used to determine the area of shapes enclosed by curves.

arenavirus medium-sized, spherical, RNA-containing enveloped virus that causes various haemorrhagic fevers, including Lassa fever and Argentine haemorrhagic fever. A number of different rodent species are carriers for the virus, for example, cotton rats and vesper mice in South America. It is transmitted to humans during contact with dust containing dried particles of infected rodent excreta.

It derives its name from the Latin *arena* ('sand') because of its grainy, sandy appearance under the electron microscope.

Argand burner burner for an oil lamp, in which the wick is in the form of a hollow cylinder, so that air rises within and without the flame, procuring more complete oxidation and therefore a brighter light. The addition of a cylindrical chimney creates a greater draught, at the same time promoting steadiness of the flame by preventing side draughts. It was devised by Pierre Ami Argand (1750–1803).

Argand diagram in mathematics, a method for representing complex numbers by Cartesian coordinates (x, y). Along the x-axis (horizontal axis) are plotted the real numbers, and along the y-axis (vertical axis) the nonreal, or *imaginary, numbers.

argon chemical symbol Ar, (Greek *argos* 'idle') colourless, odourless, non-metallic, gaseous element, atomic number 18, relative atomic mass 39.948. It is grouped with the *noble gases (rare gases) in Group 0 of the *periodic table of the elements. It was long believed not to react with other substances, but observations now indicate that it can be made to combine with boron fluoride to form compounds. It constitutes almost 1% of the Earth's atmosphere, and was discovered in 1894 by British chemists John Rayleigh (1842–1919) and William Ramsay (1852–1916) after all oxygen and nitrogen had been removed chemically from a sample of air. It is used in electric discharge tubes and argon lasers.

argument in astronomy, angular distance of an orbiting spacecraft or other body from the object being orbited. For example, the **argument of periapsis** is the angle between the periapsis (closest point of the orbiting body to the object it orbits) and the ascending node (point where the orbit intersects the plane of the *ecliptic, moving from south to north of the ecliptic). The **argument of perihelion** is the angle between the ascending node and the perihelion (closest point between the Sun and an orbiting body).

argument in computing, value on which a *function operates. For example, if the argument 16 is operated on by the function 'square root', the answer 4 is produced.

argument in mathematics, a specific value of the independent variable of a *function of x. It is also another name for *amplitude.

arithmetic branch of mathematics concerned with the study of numbers and their properties. The fundamental operations of arithmetic are addition, subtraction, multiplication, and division. Raising to powers (for example, squaring or cubing a number), the extraction of roots (for example, square roots), percentages, fractions, and ratios are developed from these operations.

arithmetic and logic unit ALU, in a computer, the part of the *central processing unit (CPU) that performs the basic arithmetic and logic operations on data.

arithmetic mean average of a set of n numbers, obtained by adding the numbers and dividing by n. For example, the arithmetic mean of the set of 5 numbers 1, 3, 6, 8, and 12 is $(1 + 3 + 6 + 8 + 12) \div 5 = 30 \div 5 = 6$.

The term *average is often used to refer only to the arithmetic mean, even though the mean is in fact only one form of average (the others include *median and *mode). In addition it is useful to know the *range or spread of the data. In grouped data, the mean can only be estimated.

arithmetic progression or **arithmetic sequence**, sequence of numbers or terms that have a common difference between any one term and the next in the sequence. For example, 2, 7, 12, 17, 22, 27, ... is an arithmetic sequence with a common difference of 5.

The nth term in any arithmetic progression can be found using the formula:

nth term $= a + (n - 1)d$

where a is the first term and d is the common difference.

An **arithmetic series** is the sum of the terms in an arithmetic sequence. The sum S of n terms is given by:

$$S = \frac{n}{2}[2a + (n - 1)d]$$

armature in a motor or generator, the wire-wound coil that carries the current and rotates in a magnetic field.

(In alternating-current machines, the armature is sometimes stationary.) The pole piece of a permanent magnet or electromagnet and the moving, iron part of a *solenoid, especially if the latter acts as a switch, may also be referred to as armatures.

Arnica genus of the daisy family Compositae, found in cold and temperate climates. *A. montana*, the common arnica or mountain tobacco, is a perennial herb common in woods in the foothills of the Alps, and contains an acrid resin and a volatile oil. A tincture, formerly much used for contusions, is prepared from all parts of the plant.

aromatic compound organic chemical compound in which some of the bonding electrons are delocalized (shared among several atoms within the molecule and not localized in the vicinity of the atoms involved in bonding). The commonest aromatic compounds have ring structures, the atoms comprising the ring being either all carbon or mostly carbon with one or more different atoms (usually nitrogen, sulphur, or oxygen). Typical examples are benzene (C_6H_6) and pyridine (C_6H_5N).

array in computer programming, a list of values that can all be referred to by a single *variable name. Separate values are distinguished by using a **subscript** with each variable name.

array collection of numbers (or letters representing numbers) arranged in rows and columns. A *matrix is an array shown inside a pair of brackets; it indicates that the array should be treated as a single entity.

arrhythmia disturbance of the normal rhythm of the heart. There are various kinds of arrhythmia, some benign, some indicative of heart disease. In extreme cases, the heart may beat so fast as to be potentially lethal and surgery may be used to correct the condition.

arsenic chemical symbol As, brittle, greyish-white, semimetallic element (a metalloid), atomic number 33, relative atomic mass 74.92. It occurs in many ores and occasionally in its elemental state, and is widely distributed, being present in minute quantities in the soil,

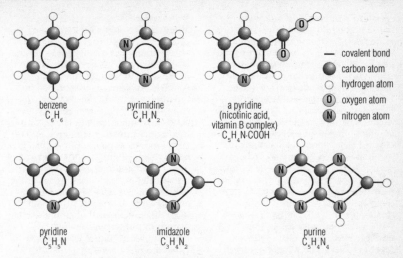

benzene
C_6H_6

pyrimidine
$C_4H_4N_2$

a pyridine
(nicotinic acid,
vitamin B complex)
$C_5H_4N \cdot COOH$

— covalent bond
● carbon atom
○ hydrogen atom
Ⓞ oxygen atom
Ⓝ nitrogen atom

pyridine
C_5H_5N

imidazole
$C_3H_4N_2$

purine
$C_5H_4N_4$

aromatic compound Compounds whose molecules contain the benzene ring, or variations of it, are called aromatic. The term was originally used to distinguish sweet-smelling compounds from others.

the sea, and the human body. In larger quantities, it is poisonous. The chief source of arsenic compounds is as a by-product from metallurgical processes. It is used in making semiconductors, alloys, and solders.

arsenic poisoning As it is a cumulative poison, its presence in food and drugs is very dangerous. The symptoms of arsenic poisoning are vomiting, diarrhoea, tingling and possibly numbness in the limbs, and collapse. It featured in some drugs, including Salvarsan, the first specific treatment for syphilis. Its name derives from the Latin *arsenicum*. The maximum safe level for arsenic in drinking water, as recommended by the World Health Organization, is 10 micrograms per litre.

arsine arsenic hydride, AsH_3, prepared by the action of dilute sulphuric acid on sodium or zinc arsenides. Arsine is a colourless, highly poisonous gas with an unpleasant odour, and is a very powerful reducing agent. It burns in air with a clear blue flame, giving water vapour and a black deposit of arsenic.

arsphenamine organic compound of arsenic discovered by Paul Ehrlich in 1909, used to treat syphilis. It was given the proprietary name Salvarsan. Neoarsphenamine (neosalvarsan)

followed arsphenamine and was used until 1945, when penicillin superseded it.

arteriosclerosis hardening of the arteries, with thickening and loss of elasticity. It is associated with smoking, ageing, and a diet high in saturated fats. The term is used loosely as a synonym for *atherosclerosis.

artery blood vessel that carries *blood from the *heart to any part of the body. It is built to withstand considerable pressure, having thick walls that contain *muscle and elastic fibres. As blood pulses out of the heart, arteries expand to allow for the increase in pressure – this elasticity helps the blood to flow evenly. The *pulse or pressure wave generated can be felt at the wrist. Not all arteries carry oxygenated (oxygen-rich) blood – the pulmonary arteries convey deoxygenated (oxygen-poor) blood from the heart to the lungs.

Arteries are flexible, elastic tubes, consisting of three layers. The middle layer is muscular and its rhythmic contraction aids the pumping of blood around the body. In middle and old age, artery walls become damaged. Some of this damage is caused by the build-up of fatty deposits (see *fats). This reduces elasticity and narrows the space (or bore) through which the blood can flow. This condition is known as hardening of the arteries or *atherosclerosis and can lead to high blood pressure, loss of circulation, and death. When it affects the arteries supplying blood to the muscle of the heart wall it is called heart disease. People with heart disease run an increased risk of having a heart attack. In a heart attack blood vessels supplying the heart muscles are blocked by a blood clot and this can result in sudden death. Research indicates that for some people a typical Western diet, high in fat from animals, increases the chances of heart disease developing.

artesian well well that is supplied with water rising naturally from an underground water-saturated rock layer (aquifer). The water rises from the aquifer under its own pressure. Such a well may be drilled into an aquifer that is confined by impermeable rocks both above and below. If the water table (the top of the region of water saturation) in that aquifer is above the level of the well head, hydrostatic pressure will force the water to the surface.

Artesian wells are often overexploited because their water is fresh and easily available, and they eventually become unreliable. There is also some concern that pollutants such as pesticides or nitrates can seep into the aquifers.

arthritis inflammation of the joints, with pain, swelling, and restricted motion. Many conditions may cause arthritis, including gout, infection, and trauma to the joint. There are three main forms of arthritis: *rheumatoid arthritis; osteoarthritis; and septic arthritis.

arthropod member of the phylum Arthropoda; an invertebrate animal with jointed legs and a segmented body with a horny or chitinous casing (exoskeleton), which is shed periodically and replaced as the animal grows. Included are arachnids such as spiders and mites, as well as crustaceans, millipedes, centipedes, and insects.

artificial insemination AI, introduction by instrument of semen from a sperm bank or donor into the female reproductive tract to bring about fertilization. Originally used by animal breeders to improve stock with sperm from high-quality males, in the 20th century it was developed for use in humans, to help the infertile.

artificial intelligence AI, branch of science concerned with creating computer programs that can perform actions comparable with those of an intelligent human. AI research covers such areas as planning (for robot behaviour), language understanding, pattern recognition, and knowledge representation.

artificial life or **Alife**, in computing, area of scientific research that attempts to simulate biological phenomena via computer programs. The first ALife workshop was held at Los Alamos, USA, in 1987. Research in this area is being conducted all around the world; the most significant centres are the Santa Fe Institute and the MIT Media Lab.

artificial radioactivity natural and spontaneous radioactivity arising from radioactive isotopes or elements that are formed when elements are bombarded with subatomic particles – protons, neutrons, or electrons – or small nuclei.

artificial respiration emergency procedure to restart breathing once it has stopped; in cases of electric shock or apparent drowning, for example, the first choice is the expired-air method, the kiss of life by mouth-to-mouth breathing until natural breathing is restored.

asbestos any of several related minerals of fibrous structure that offer great heat resistance because of their nonflammability and poor conductivity. Commercial asbestos is generally either made from serpentine ('white' asbestos) or from sodium iron silicate ('blue' asbestos). The fibres are woven together or bound by an inert material. Over time the fibres can work loose and, because they are small enough to float freely in the air or be inhaled, asbestos usage is now strictly controlled; exposure to its dust can cause cancer.

asbestosis lung disease prevalent in people chronically exposed to asbestos. It may give rise to *mesothelioma and is associated with an increased risk of lung cancer, particularly in smokers.

ASCII acronym for American Standard Code for Information Interchange, in computing, coding system in which numbers are assigned to letters, digits, and punctuation symbols. Although computers work in code based on the *binary number system, ASCII numbers are usually quoted as decimal or hexadecimal numbers. For example, the decimal number 45 (binary 0101101) represents a hyphen, and 65 (binary 1000001) a capital A. The first 32 codes are used for control functions, such as carriage return and backspace.

ascorbic acid or **vitamin C**; $C_6H_8O_6$, relatively simple organic acid found in citrus fruits and vegetables. It is soluble in water and destroyed by prolonged boiling, so soaking or overcooking of vegetables reduces their vitamin C content. Lack of ascorbic acid results in scurvy.

In the human body, ascorbic acid is necessary for the correct synthesis of *collagen. Lack of vitamin C causes skin sores or ulcers, tooth and gum problems, and burst capillaries (scurvy symptoms) owing to an abnormal type of collagen replacing the normal type in these tissues.

The Australian billygoat plum, *Terminalia ferdiandiana*, is the richest natural source of vitamin C, containing 100 times the concentration found in oranges.

asepsis practice of ensuring that bacteria are excluded from open sites during surgery, wound dressing, blood sampling, and other medical procedures. Aseptic technique is a first line of defence against infection.

asexual reproduction reproduction that does not involve the manufacture and fusion of sex cells (gametes) from two parents. Asexual reproduction has advantages in that there is no need to search for a mate; every asexual organism can reproduce on its own. Asexual reproduction can therefore lead to a rapid population build-up. However, every new organism produced by asexual reproduction is genetically identical to the parent – a *clone. (See diagram, page 44.)

In evolutionary terms, the disadvantage of asexual reproduction arises from the fact that only identical individuals (clones) are produced – there is no variation. In agriculture and horticulture, where standardized production is needed, this is useful. Taking cuttings of a good *variety of fruit tree is an example of artificial asexual reproduction. However, in the wild, an asexual population that cannot adapt to a changing environment or evolve defences against a new disease is at risk of extinction. Many asexually reproducing organisms are therefore capable of reproducing sexually as well.

Asexual reproduction is very common in micro-organisms. But there are also many plants that use it naturally. The blackberry or bramble spreads by allowing its stems to root where they touch the ground. However, the blackberry also reproduces sexually using its flowers.

ASIC abbreviation for application-specific integrated circuit, integrated circuit built for a specific application.

ASP abbreviation for *Active Server Pages or *application service provider.

aspartame noncarbohydrate sweetener used in foods under the tradename Nutrasweet. It is about 200 times as sweet as sugar and, unlike saccharine, has no aftertaste.

The aspartame molecule consists of two amino acids (aspartic acid and phenylalanine) linked by a methylene ($-CH_2-$) group. It breaks down slowly at room temperature and rapidly at higher temperatures. It is not suitable for people who suffer from *phenylketonuria.

asphalt mineral mixture containing semisolid brown or black *bitumen, used in the construction industry. Asphalt is mixed with rock chips to form paving material, and the purer varieties are used for insulating material and for waterproofing masonry. It can be produced artificially by the distillation of *petroleum.

asphyxia suffocation; a lack of oxygen that produces a potentially lethal build-up of carbon dioxide waste in the tissues.

aspiration withdrawal of fluid from the body using a suction instrument.

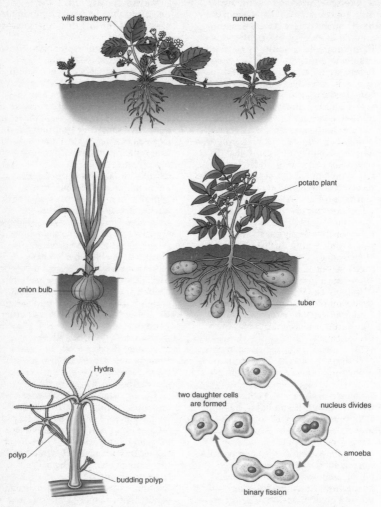

asexual reproduction Examples of asexual reproduction. Asexual reproduction is the simplest form of reproduction, occurring in many plants and simple animals. Strawberry plants can reproduce by sending out runners; onion plants form bulbs; and potato plants form tubers. Amoebas divide into two (binary fission) and hydra form new hydra by budding. The offspring are always genetically identical to the parent.

aspirin acetylsalicylic acid, a popular pain-relieving drug (*analgesic) developed in the late 19th century as a household remedy for aches and pains. It relieves pain and reduces inflammation and fever. It is the world's most widely used drug.

Aspirin was first refined from salicylic acid by German chemist Felix Hoffman, and marketed in 1899. Although salicylic acid occurs naturally in bark of willow *Salix alba* (and has been used for pain relief since 1763) the acetyl derivative is less bitter and less likely to cause vomiting.

assay in chemistry, the determination of the quantity of a given substance present in a sample. Usually it refers to determining the purity of precious metals.

The assay may be carried out by 'wet' methods, when the sample is wholly or partially dissolved in some reagent (often an acid), or by 'dry' or 'fire' methods, in which the compounds present in the sample are combined with other substances.

assembly language low-level computer-programming language closely related to a computer's internal codes. It consists chiefly of a set of short sequences of letters (mnemonics), which are translated, by a program called an assembler, into *machine code for the computer's *central processing unit (CPU) to follow directly. In assembly language, for example, 'JMP' means 'jump' and 'LDA' means 'load accumulator'. Assembly code is used by programmers who need to write very fast or efficient programs.

assimilation in animals, the process by which absorbed food molecules, circulating in the blood, pass into the cells and are used for growth, tissue repair, and other metabolic activities. The actual destiny of each food molecule depends not only on its type, but also on the body requirements at that time.

Association for Computing Machinery **ACM**, US organization made up of computer professionals of all types. Its monthly journal, the *Communications of the Association for Computing Machinery*, is peer-reviewed. Its subsidiary special interest groups, or **SIGs**, focus on areas such as graphics and human–computer interaction. Several of these run major conferences for their areas such as SIGGRAPH (graphics) and SIGCHI (human–computer interaction).

The equivalent UK organization is the British Computer Society (BCS).

associative operation in mathematics, an operation in which the outcome is independent of the grouping of the numbers or symbols concerned. For example, multiplication is associative, as $4 \times (3 \times 2) = (4 \times 3) \times 2 = 24$;

however, division is not, as $12 \div (4 \div 2) = 6$, but $(12 \div 4) \div 2 = 1.5$. Compare *commutative operation and distributive operation.

astatine chemical symbol At, (Greek *astatos* 'unstable') non-metallic, radioactive element, atomic number 85, relative atomic mass 210. It is a member of the *halogen group, and is found at the bottom of Group 7 of the *periodic table of the elements. It is very rare in nature. Astatine is highly unstable, with at least 19 isotopes; the longest lived has a *half-life of about eight hours.

asteroid any of many thousands of small bodies, made of rock and minerals, that orbit the Sun. Most lie in a region called the **asteroid belt** between the orbits of Mars and Jupiter, and are thought to be fragments left over from the formation of the Solar System. About 100,000 asteroids may exist, but their total mass is only a few hundredths of the mass of the Moon. These rocky fragments range in size from 1 km/0.6 mi to 900 km/560 mi in diameter.

asthma chronic condition characterized by difficulty in breathing due to spasm of the bronchi (air passages) in the lungs. Attacks may be provoked by allergy, infection, and stress. The incidence of asthma may be increasing as a result of air pollution and occupational hazard. Treatment is with bronchodilators to relax the bronchial muscles and thereby ease the breathing, and in severe cases by inhaled *steroids that reduce inflammation of the bronchi.

astigmatism aberration occurring in the lens of the eye. It results when the curvature of the lens differs in two perpendicular planes, so that rays in one plane may be in focus while rays in the other are not. With astigmatic eyesight, the vertical and horizontal cannot be in focus at the same time; correction is by the use of a cylindrical lens that reduces the overall focal length of one plane so that both planes are seen in sharp focus.

astronomical unit AU, unit equal to the mean distance of the Earth from the Sun: 149.6 million km/92.96 million mi. It is used to describe planetary distances. Light travels this distance in approximately 8.3 minutes.

astronomy science of the celestial bodies: the Sun, the Moon, and the planets; the stars and galaxies; and all other objects in the universe. It is concerned with their positions, motions, distances, and physical conditions and with their origins and evolution. Astronomy thus divides into fields such as astrophysics, celestial mechanics, and *cosmology. See also *radio astronomy, *ultraviolet astronomy, and *X-ray astronomy.

astrophysics study of the physical nature of stars, galaxies, and the universe. It began with the development of spectroscopy in the 19th century, which allowed astronomers to analyse the composition of stars from their light. Astrophysicists view the universe as a vast natural laboratory in which they can study matter under conditions of temperature, pressure, and density that are unattainable on Earth.

asymmetric digital subscriber loop in computing, standard for transmitting video data; see *ADSL.

asymptote in *coordinate geometry, a straight line that a curve approaches progressively more closely but never reaches. The x and y axes are asymptotes to the graph of $xy =$ constant (a rectangular *hyperbola).

If a point on a curve approaches a straight line such that its distance from the straight line is d, then the line is an asymptote to the curve if limit d tends to zero as the point moves towards infinity. Among *conic sections (curves obtained by the intersection of a plane and a double cone), a hyperbola has two asymptotes, which in the case of a rectangular hyperbola are at right angles to each other.

asynchronous transfer mode ATM, in computing, high-speed computer *networking standard suitable for all types of data, including voice and video, that can be used on both private and public networks. ATM is used mainly on the core 'backbones' of large communications networks and in wide-area networks.

asystole failure of the heart to beat. It is seen in *cardiac arrest.

ATAPI acronym for Advanced Technology Attachment Packet Interface, in computing, enhancement to integrated drive electronics (IDE) that allows easier installation and support of CD-ROM drives and other devices. Part of the Enhanced IDE standard introduced by hard disk manufacturer Western Digital in 1994, ATAPI uses a standard software device driver and does away with the need for older, proprietary interfaces.

atavism (Latin *atavus* 'ancestor') in genetics, the reappearance of a characteristic not apparent in the immediately preceding generations; in psychology, the manifestation of primitive forms of behaviour.

ataxia loss of muscular coordination due to neurological damage or disease.

AT command set abbreviation for attention command set, set of standard commands allowing a *modem to be controlled via software. These commands are used via special communications software to control a modem's actions from the computer console. The most common are ATZ to reset the modem and ATH to hang the modem up at the end of a call. The set was invented by Hayes Computer Products for its earliest modems.

atherosclerosis thickening and hardening of the walls of the arteries, associated with atheroma.

athlete's foot commonest type of *ringworm, affecting the skin between the toes. A fungal infection, it is highly contagious. The medical name is *tinea pedis*.

ATM in computing, abbreviation for *asynchronous transfer mode, **automated teller machine**, or Adobe Type Manager, depending on context.

atmosphere mixture of gases surrounding a planet. Planetary atmospheres are prevented from escaping by the pull of gravity. On Earth, atmospheric pressure decreases with altitude. In its lowest layer, the atmosphere consists of nitrogen (78%) and oxygen (21%), both in molecular form (two atoms bonded together) and argon (1%). Small quantities of other gases are important to the chemistry and physics of the Earth's atmosphere, including water, carbon dioxide, and traces of other noble gases (rare gases), as well as

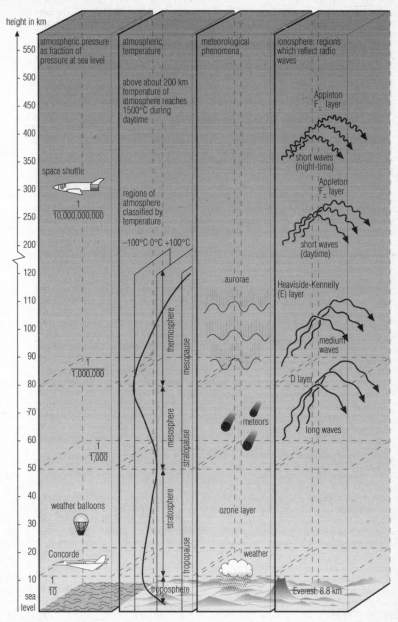

atmosphere All but 1% of the Earth's atmosphere lies in a layer 30 km/19 mi above the ground. At a height of 5,500 m/18,000 ft, air pressure is half that at sea level. The temperature of the atmosphere varies greatly with height; this produces a series of layers, called the troposphere, stratosphere, mesosphere, and thermosphere.

ozone. The atmosphere plays a major part in the various cycles of nature (the *water cycle, the *carbon cycle, and the *nitrogen cycle). It is the principal industrial source of nitrogen, oxygen, and argon, which are obtained by the *fractional distillation of liquid air.

atmosphere symbol atm; or **standard atmosphere**, in physics, a unit of pressure equal to 760 mmHg, 1013.25 millibars, or 1.01325×10^5 pascals, or newtons per square metre. The actual pressure exerted by the atmosphere fluctuates around this value, which is assumed to be standard at sea level and 0°C/32°F, and is used when dealing with very high pressures.

atmospheric pollution contamination of the atmosphere with the harmful by-products of human activity; see *air pollution.

atmospheric pressure pressure at any point on the Earth's surface that is due to the weight of the column of air above it; it therefore decreases as altitude increases, because there is less air above. Particles in the air exert a force (pressure) against surfaces; when large numbers of particles press against a surface, the overall effect is known as air pressure. At sea level the average pressure is 101 kilopascals (1,013 millibars, or 760 mm Hg, or 14.7 lb per sq in, or 1 atmosphere). Changes in atmospheric pressure, measured with a barometer, are used in weather forecasting. Areas of relatively high pressure are called anticyclones; areas of low pressure are called *depressions.

atom (Greek *atomos* 'undivided') smallest unit of matter that can take part in a chemical reaction, and which cannot be broken down chemically into anything simpler. An atom is made up of protons and neutrons in a central nucleus (except for hydrogen, which has a single proton in its nucleus) surrounded by electrons (see *atomic structure). The atoms of the various elements differ in atomic number, relative atomic mass, and chemical behaviour.

Atoms are much too small to be seen by even the most powerful optical microscope (the largest, caesium, has a diameter of 0.0000005 mm/0.00000002

in), and they are in constant motion. However, modern electron microscopes, such as the *scanning tunnelling microscope (STM) and the *atomic force microscope (AFM), can produce images of individual atoms and molecules.

atom, electronic structure of arrangement of electrons around the nucleus of an atom, in distinct energy levels, also called atomic *orbitals, or shells. These shells can be regarded as a series of concentric spheres, each of which can contain a certain maximum number of electrons; the noble gases (rare gases) have an arrangement in which every shell contains this number (see *noble gas structure). The energy levels are usually numbered beginning with the shell nearest to the nucleus. The outermost shell is known as the *valency shell as it contains the valence electrons.

The lowest energy level, or innermost shell, can contain no more than two electrons. Outer shells are considered to be stable when they contain eight electrons but additional electrons can sometimes be accommodated provided that the outermost shell has a stable configuration. Electrons in unfilled shells are available to take part in chemical bonding, giving rise to the concept of valency. In ions, the electron shells contain more or fewer electrons than are required for a neutral atom, resulting in negative or positive charges.

The atomic number of an element indicates the number of electrons in a

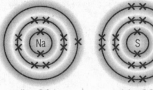

sodium 2.8.1   sulphur 2.8.6

atom, electronic structure The arrangement of electrons in a sodium atom and a sulphur atom. The number of electrons in a neutral atom gives that atom its atomic number: sodium has an atomic number of 11 and sulphur has an atomic number of 16.

neutral atom. From this it is possible to deduce its electronic structure. For example, sodium has atomic number 11 ($Z = 11$) and its electronic arrangement (configuration) is two electrons in the first energy level, eight electrons in the second energy level and one electron in the third energy level – generally written as 2.8.1. Similarly for sulphur ($Z = 16$), the electron arrangement is 2.8.6. The electronic structure dictates whether two elements will combine by ionic or covalent bonding (see *bond) or not at all.

atomic energy   another name for *nuclear energy.

atomic force microscope   AFM, microscope developed in the late 1980s that produces a magnified image using a diamond probe, with a tip so fine that it may consist of a single atom, dragged over the surface of a specimen to 'feel' the contours of the surface. In effect, the tip acts like the stylus of a record player, reading the surface. The tiny up-and-down movements of the probe are converted to an image of the surface by computer and displayed on a screen. The AFM is useful for examination of biological specimens since, unlike the *scanning tunnelling microscope, the specimen does not have to be electrically conducting.

atomicity   number of atoms of an *element that combine together to form a molecule. A molecule of oxygen (O_2) has atomicity 2; sulphur (S_8) has atomicity 8.

atomic mass   see *relative atomic mass.

atomic mass unit   or dalton; symbol u, unit of mass that is used to measure the relative mass of atoms and molecules. It is equal to one-twelfth of the mass of a carbon-12 atom, which is approximately the mass of a proton or 1.66×10^{-27} kg. The *relative atomic mass of an atom has no units; thus oxygen-16 has an atomic mass of 16 daltons but a relative atomic mass of 16.

atomic number   or proton number; symbol Z, number of protons in the nucleus of an atom. It is equal to the positive charge on the nucleus. In a neutral atom, it is also equal to the

number of electrons surrounding the nucleus. The chemical elements are arranged in the *periodic table of the elements according to their atomic number. *Nuclear notation is used to label an atom according to the composition of its nucleus.

atomic physics   former name for *nuclear physics.

atomic radiation   energy given out by disintegrating atoms during *radioactive decay, whether natural or synthesized. The energy may be in the form of fast-moving particles, known as *alpha particles and *beta particles, or in the form of high-energy electromagnetic waves known as *gamma radiation. Overlong exposure to atomic radiation can lead to *radiation sickness.

atomic size   or **atomic radius**, size of an atom expressed as the radius in *angstroms or other units of length. The sodium atom has an atomic radius of 1.57 angstroms (1.57×10^{-8} cm). For metals, the size of the atom is always greater than the size of its ion. For non-metals the reverse is true.

atomic structure   internal structure of an *atom.

   the nucleus The core of the atom is the **nucleus**, a dense body only one ten-thousandth the diameter of the atom itself. The simplest nucleus, that of hydrogen, comprises a single stable positively charged particle, the **proton**. Nuclei of other elements contain more protons and additional particles, called **neutrons**, of about the same mass as the proton but with no electrical charge. Each element has its own characteristic nucleus with a unique number of protons, the atomic number. The number of neutrons may vary. Where atoms of a single element have different numbers of neutrons, they are called *isotopes. Although some isotopes tend to be unstable and exhibit *radioactivity, all those of a single element have identical chemical properties.

   electrons The nucleus is surrounded by a number of moving **electrons**, each of which has a negative charge equal to the positive charge on a proton, but which has a mass of only $1/_{1,836}$ times as much. In a neutral atom, the nucleus is surrounded by the same number of electrons as it contains protons. According to *quantum theory, the position of an electron is uncertain; it may be found at any point. However, it is more likely to be found in some places than others. The region of space in which an electron is most likely to be found is called an atomic *orbital. The chemical properties of an element are determined by the ease with which its atoms can gain or lose electrons. See *atom, electronic structure.

attraction and repulsion According to the theory of fundamental *forces, atoms are held together by the electrical forces of attraction between each negative electron and the positive protons within the nucleus. The latter repel one another with enormous forces; a nucleus holds together only because an even stronger force, called the **strong nuclear force**, attracts the protons and neutrons to one another. The strong force acts over a very short range – the protons and neutrons must be in virtual contact with one another. If, therefore, a fragment of a complex nucleus, containing some protons, becomes only slightly loosened from the main group of neutrons and protons, the natural repulsion between the protons will cause this fragment to fly apart from the rest of the nucleus at high speed. It is by such fragmentation of atomic nuclei (nuclear *fission) that nuclear energy is released.

atomic time time as given by atomic clocks, which are regulated by natural resonance frequencies of particular atoms, and display a continuous count of seconds.

atomic weight another name for *relative atomic mass.

ATP abbreviation for **adenosine triphosphate**, nucleotide molecule found in all cells. It can yield large amounts of energy, and is used to drive the thousands of biological processes needed to sustain life, growth, movement, and reproduction. Green plants use light energy to manufacture ATP as part of the process of *photosynthesis. In animals, ATP is formed by the breakdown of glucose molecules, usually obtained from the carbohydrate component of a diet, in a

series of reactions termed *respiration. It is the driving force behind muscle contraction and the synthesis of complex molecules needed by individual cells.

atrium either of the two upper chambers of the heart. The left atrium receives freshly oxygenated blood from the lungs via the pulmonary vein; the right atrium receives deoxygenated blood from the *vena cava. Atrium walls are thin and stretch easily to allow blood into the heart. On contraction, the atria force blood into the thick-walled ventricles, which then give a second, more powerful beat.

atrophy in medicine, a diminution in size and function, or output, of a body tissue or organ. It is usually due to nutritional impairment, disease, or disuse (muscle).

atropine alkaloid derived from *belladonna, a plant with toxic properties. It acts as an *anticholinergic, inhibiting the passage of certain nerve impulses. It is used in premedication, to reduce bronchial and gastric secretions. It is also administered as a mild antispasmodic drug, and to dilate the pupil of the eye.

attachment way of incorporating a file into an e-mail message for transmission. Within a single system, such as a corporate local area network (LAN) or a commercial online service, *binary files can be sent intact. Over the Internet, attached files must be encoded into *ASCII characters and then decoded by the receiver. See *MIME.

attention-deficit hyperactivity disorder ADHD, psychiatric condition occurring in young children characterized by impaired attention and hyperactivity. The disorder, associated with disruptive behaviour, learning difficulties, and under-achievement, is more common in boys. It is treated with methylphenidate (Ritalin). There was a 50% increase in the use of the drug in the USA 1994–96, with an estimated 5% of school-age boys diagnosed as suffering from ADHD. In 1998 the number of children and adults in the USA taking medication for ADHD (mostly Ritalin) was approximately 4 million. In the

UK the prescription of Ritalin doubles each year.

Audio–Video Interleave in computing, *file format for video clips.

auditory canal tube leading from the outer *ear opening to the eardrum. It is found only in animals whose eardrums are located inside the skull, principally mammals and birds.

audit trail record of computer operations, showing what has been done and, if available, who has done it. The term is taken from accountancy, but audit trails are now widely used to check many aspects of computer security, in addition to use in accounts programs.

augmented reality use of computer systems and data to overlay video or other real-life representations. For example, a video of a car engine with the mechanical drawings overlaid.

aurora coloured light in the night sky near the Earth's magnetic poles, called **aurora borealis** ('northern lights') in the northern hemisphere and **aurora australis** ('southern lights') in the southern hemisphere. Although auroras are usually restricted to the polar skies, fluctuations in the *solar wind occasionally cause them to be visible at lower latitudes. An aurora is usually in the form of a luminous arch with its apex towards the magnetic pole, followed by arcs, bands, rays, curtains, and coronae, usually green but often showing shades of blue and red, and sometimes yellow or white. Auroras are caused at heights of over 100 km/60 mi by a fast stream of charged particles from solar flares and low-density 'holes' in the Sun's corona. These are guided by the Earth's magnetic field towards the north and south magnetic poles, where they enter the upper atmosphere and bombard the gases in the atmosphere, causing them to emit visible light.

australopithecines see *human species, origins of

authentication in computing, system for certifying the origin of an electronic communication. In the non-electronic world, a handwritten signature authenticates a document, for example a contract, as coming from a particular person. In the electronic world,

encryption systems provide the same function via *digital signatures and other techniques.

authoring development of multimedia presentations. Authoring includes pulling together the necessary audio, video, graphics, and text files and formatting them for display.

authoring tool software that allows developers to create multimedia presentations or World Wide Web pages. Typically, these tools automate some of the more difficult parts of generating program source codes so that developers can work on a higher, more abstract level. Popular authoring tools for the World Wide Web include Microsoft FrontPage, Macromedia Dreamweaver, and Adobe PageMill.

authorization permission to access a particular system. Unauthorized access to private computer systems was made illegal in many countries during the late 1980s.

autism rare disorder, generally present from birth, characterized by a withdrawn state and a failure to develop normally in language or social behaviour. Although the autistic person may, rarely, show signs of high intelligence (in music or with numbers, for example), many have impaired intellect. The cause is unknown, but is thought to involve a number of factors, possibly including an inherent abnormality of the brain. Special education may bring about some improvement.

AutoCAD leading computer-aided design (CAD) software package. It is published by the specialist US company AutoDesk (founded 1982). Users include engineers, architects, and designers.

autoclave pressurized vessel that uses superheated steam to sterilize materials and equipment such as surgical instruments. It is similar in principle to a pressure cooker.

autoexec.bat in computing, a file in the MS-DOS operating system that is automatically run when the computer is booted.

autogenics in alternative medicine, system developed in the 1900s by German physician Johannes Schultz, designed to facilitate mental control of biological and physiological functions generally considered to be involuntary. Effective in inducing relaxation, assisting healing processes and relieving psychosomatic disorders, autogenics is regarded as a precursor of biofeedback.

autogenous in medicine, self-generated by the body. For example, autogenous blood transfusion involves the collection of blood from a patient, prior to undergoing surgery, for use during and after the operation. It is a valuable procedure for operations that may require large transfusions or where a person has a rare blood group.

autoimmunity in medicine, condition in which the body's immune responses are mobilized not against 'foreign' matter, such as invading germs, but against the body itself. Diseases considered to be of autoimmune origin include *myasthenia gravis, *rheumatoid arthritis, and *lupus erythematosus.

autolysis in biology, the destruction of a *cell after its death by the action of its own *enzymes, which break down its structural molecules.

automatic fallback in computing, feature allowing *modems to drop to a slower speed if conditions such as line noise make it necessary. Modem speeds are typically rated according to one or another *CCITT standard (known as a **V number**). All modems rated for a specific standard are *backwards compatible.

automatism performance of actions without awareness or conscious intent. It is seen in sleepwalking and in some (relatively rare) psychotic states.

autonomic nervous system in mammals, the part of the nervous system that controls those functions not controlled voluntarily, including the heart rate, activity of the intestines, and the production of sweat.

There are two divisions of the autonomic nervous system. The **sympathetic** system responds to stress, when it speeds the heart rate, increases blood pressure, and generally prepares the body for action. The **parasympathetic** system is more important when the body is at rest, since it slows the heart rate, decreases

blood pressure, and stimulates the digestive system.

At all times, both types of autonomic nerves carry signals that bring about adjustments in visceral organs. The actual rate of heartbeat is the net outcome of opposing signals. Today, it is known that the word 'autonomic' is misleading – the reflexes managed by this system are actually integrated by commands from the brain and spinal cord (the central nervous system).

autopsy or **postmortem**, examination of the internal organs and tissues of a dead body, performed to try to establish the cause of death.

autoresponder on the Internet, a *server that responds automatically to specific messages or input. A common use for autoresponders is to automate the dispatch of sales information via e-mail. A user requesting such information typically sends a message with specified words such as 'send info' in the subject line or the body of the message. The words trigger the autoresponder to send the prepared information file.

autosome any *chromosome in the cell other than a sex chromosome. Autosomes are of the same number and kind in both males and females of a given species.

autotroph any living organism that synthesizes organic substances from inorganic molecules by using light or chemical energy. Autotrophs are the primary producers in all food chains since the materials they synthesize and store are the energy sources of all other organisms. All green plants and many planktonic organisms are autotrophs, using sunlight to convert carbon dioxide and water into sugars by *photosynthesis.

The total *biomass of autotrophs is far greater than that of animals, reflecting the dependence of animals on plants, and the ultimate dependence of all life on energy from the Sun – green plants convert light energy into a form of chemical energy (food) that animals can exploit. Some bacteria use the chemical energy of sulphur compounds to synthesize organic substances. It is estimated that

10% of the energy in autotrophs can pass into the next stage of the *food chain, the rest being lost as heat or indigestible matter. See also *heterotroph.

auxin plant *hormone that regulates stem and root growth in plants. Auxins influence many aspects of plant growth and development, including cell enlargement, inhibition of development of axillary buds, *tropisms, and the initiation of roots. Auxin affects cell division mainly at the tip, because it is here that cell division in a stem or root mainly occurs. Just behind the tip the cells grow in size under the influence of auxins, causing the stem or root to grow longer. Auxin therefore affects the amount of elongation here too.

avatar computer-generated character that represents a human in on-screen interaction. In the mid-1990s, avatars were primarily used in computer games, but because they take up much less memory or bandwidth than full video, companies such as British Telecom were researching the possibility of building multiparty videoconferencing systems using this technology.

average in statistics, a term used inexactly to indicate the typical member of a set of data. It usually refers to the *arithmetic mean. The term is also used to refer to the middle member of the set when it is sorted in ascending or descending order (the *median), and the most commonly occurring item of data (the *mode), as in 'the average family'.

aversion therapy type of behaviour therapy designed to inhibit the occurrence of certain undesirable habits. It involves delivering an unpleasant stimulus, such as a weak electric shock, in association with some manifestation of the deviant behaviour – for example, filling a glass in alcoholism. It is hoped that repetition of this treatment will set up an aversion to the undesired behaviour.

AVI abbreviation for audio-visual interleave, file format capable of storing moving images (such as video) with accompanying sound. AVI files can be replayed by any multimedia PC with Microsoft Windows and a *sound card. AVI files are frequently very large (around 50 Mbyte for a five-minute rock video, for example), so they are usually stored on *CD-ROM.

Avogadro's hypothesis in chemistry, the law stating that equal volumes of all gases, when at the same temperature and pressure, have the same numbers of molecules. One *mole of any gas contains 6.023×10^{23} particles and occupies 24 dm^3 at room temperature and pressure. The type of gas does not make any difference. The law was first put forward by Italian chemist Amedeo Avogadro (1776–1856).

Avogadro's number or **Avogadro's constant**, number of carbon atoms in 12 g of the carbon-12 isotope (6.022045 $\times 10^{23}$). It is named after Italian chemist Amedeo Avogadro (1776–1856). The *relative atomic mass of any element, expressed in grams, contains this number of atoms and is called a *mole. For example, one mole of any substance contains 6.022×10^{23} particles. One mole of carbon has a mass of 12 g.

axil upper angle between a leaf (or bract) and the stem from which it grows. Organs developing in the axil, such as shoots and buds, are termed axillary, or lateral.

axilla anatomical term for the armpit.

axiom in mathematics, a statement that is assumed to be true and upon which theorems are proved by using logical deduction; for example, two straight lines cannot enclose a space. The Greek mathematician Euclid (c. 330– c. 260 BC) used a series of axioms that he considered could not be demonstrated in terms of simpler concepts to prove his geometrical theorems.

axis plural **axes**, in geometry, one of the reference lines by which a point on a graph may be located. The horizontal axis is usually referred to as the x-axis, and the vertical axis as the y-axis. The term is also used to refer to the imaginary line about which an object may be said to be symmetrical (**axis of symmetry**) – for example, the diagonal of a square – or the line about which an object may revolve (**axis of rotation**).

axon long threadlike extension of a *nerve cell that conducts electrochemical impulses away from the cell body towards other nerve cells, or towards an effector organ such as a muscle. Axons terminate in *synapses, junctions with other nerve cells, muscles, or glands.

azo dye synthetic dye containing the azo group of two nitrogen atoms (–N=N–) connecting aromatic ring compounds. Azo dyes are usually red, brown, or yellow, and make up about half the dyes produced. They are manufactured from aromatic *amines.

AZT drug used in the treatment of AIDS; see *zidovudine.

Babbit metal soft, white metal, an *alloy of tin, lead, copper, and antimony, used to reduce friction in bearings, developed by the US inventor Isaac Babbit in 1839.

bacillus genus of rod-shaped *bacteria that occur everywhere in the soil and air. Some are responsible for diseases such as *anthrax, or for causing food spoilage.

backbone in networking, a high-*bandwidth trunk to which smaller networks connect. The original backbone of the Internet was NSFnet, funded by the US National Science Foundation, which linked together the five regional supercomputing centres.

backcross cross between an offspring and one of its parents, or an individual that is genetically identical to one of its parents. It is a breeding technique used to determine the genetic makeup of an individual organism.

back-up system in computing, duplicate computer system that can take over the operation of a main computer system in the event of equipment failure. A large interactive system, such as an airline's ticket-booking system, cannot be out of action for even a few hours without causing considerable disruption. In such cases a complete duplicate computer system may be provided to take over and run the system should the main computer develop a fault or need maintenance.

backwards compatible in computing, term describing a product that is designed to be compatible with its predecessors. In software, a word processor is backwards compatible if it can read and write the files of earlier versions of the same software, and an operating system is backwards compatible if it can run programs designed for earlier versions of the operating system. Similarly, all modems are compatible with all the standards (V numbers) which precede the fastest one they can handle.

bacteria singular **bacterium**, microscopic single-celled organisms lacking a membrane-bound nucleus. Bacteria, like fungi and *viruses, are micro-organisms – organisms that are so small they can only be seen using a microscope. They are organisms that are more simple than the cells of animals, plants, and fungi in that they lack a nucleus. Bacteria are widespread, being present in soil, air, and water, and as parasites on and in other living things. In fact, they occur anywhere life can exist. Some parasitic bacteria cause disease by producing toxins, but others are harmless and can even benefit their hosts. Bacteria usually reproduce by *binary fission (dividing into two equal parts), and, on average, this occurs every 20 minutes. Only around 4,000 species of bacteria are known, although bacteriologists believe that around 3 million species may actually exist. Certain types of bacteria are vital in many food and industrial processes, while others play an essential role in the *nitrogen cycle, which maintains soil fertility. They can be the first organism of a food chain, by acting as decomposers of dead plant and animal remains. This helps to recycle nutrients.

Bacteria can be grown on the surface of agar jelly in dishes in the laboratory and have been studied in detail. This has led to the development of antibiotics, chemicals that kill or inhibit other micro-organisms such as bacteria. Sometimes the chemical is not harmful to humans and so can be used to treat disease. Some antibiotics work against a range of bacteria and in many situations, such as penicillins. Others are quite specific. For example, neomycin sulphate is especially active against bacteria that cause infections of the middle ear. Unfortunately, there are increasing numbers of bacteria that are resistant to antibiotics. New antibiotics need to be discovered so that antibiotic-resistant bacteria can be destroyed, but this is only occurring very slowly.

bacteriology   study of *bacteria.

bacteriophage   virus that attacks *bacteria, commonly called a phage. Such viruses are now useful vectors in genetic engineering for introducing modified DNA.

Bakelite   first synthetic *plastic, a phenol-formaldehyde resin, created by Belgian-born US chemist Leo Baekeland in 1909. Bakelite is hard, tough, and heatproof, and is used as an electrical insulator. It is made by *condensation polymerization of phenol with methanal (formaldehyde), producing a powdery resin that sets solid when heated. Objects are made by subjecting the resin to compression moulding (simultaneous heat and pressure in a mould). It is one of the thermosetting plastics, which do not remelt when heated, and is often used for electrical fittings as it has high electrical resistance.

baking powder   mixture of *bicarbonate of soda, an acidic compound (such as tartaric acid, cream of tartar, sodium or calcium acid phosphates, or glucono-delta-lactone), and a nonreactive filler (usually starch or calcium sulphate),

used in baking as a raising agent. It gives a light open texture to cakes and scones, and is used as a substitute for yeast in making soda bread. The acidic compounds react with the sodium hydrogencarbonate in the presence of water and heat, to release the carbon dioxide that causes the cake mix or dough to rise.

balance   apparatus for weighing or measuring mass. The various types include the **beam balance**, consisting of a centrally pivoted lever with pans hanging from each end, and the **spring balance**, in which the object to be weighed stretches (or compresses) a vertical coil spring fitted with a pointer that indicates the weight on a scale. Kitchen and bathroom scales are balances.

ball-and-socket joint   joint between bones that allows considerable movement in three dimensions, for instance the joint between the pelvis and the femur (hip joint). To facilitate movement, such joints are rimmed with cartilage and lubricated by synovial fluid. The bones are kept in place by ligaments and moved by muscles.

ballistics   study of the motion and impact of projectiles such as bullets, bombs, and missiles. For projectiles from a gun, relevant exterior factors include temperature, barometric pressure, and wind strength; and for nuclear missiles these extend to such factors as the speed at which the Earth turns.

ball valve   valve that works by the action of external pressure raising a ball and thereby opening a hole.

bandwidth   in computing and communications, the rate of data transmission, measured in *bits per second (bps).

bang   in Unix, exclamation mark (!). It appears in some older types of Internet addresses and is used in dictating the commands necessary to run Unix systems.

bang path   list of routing that appears in the header of a message sent across the Internet, showing how it travelled from the sender to its destination. It is named after the *bangs separating the sites in the list.

bar in earth sciences, deposit of sand or silt formed in a river channel, or a long ridge of sand or pebbles running parallel to a coastline (see *coastal erosion). Coastal bars can extend across estuaries to form **bay bars** and are formed in one of two ways. *Longshore drift can transport material across a bay and deposit it, thereby closing off the bay. Alternatively, an offshore bar (formed where waves touch the seabed and disturb the sediments, causing a small ridge to be formed) may be pushed towards the land as the sea level rises. These bars are greatly affected by the beach cycle. The high tides and high waves of winter erode the beach and deposit the sand as offshore bars. These are known as barrier beaches in the USA.

bar in physics, unit of pressure equal to 10^5 pascals or 10^6 dynes/cm^2, approximately 750 mmHg or 0.987 atm. Its diminutive, the **millibar** (one-thousandth of a bar), is commonly used by meteorologists.

barbiturate hypnosedative drug, commonly known as a 'sleeping pill', consisting of any salt or ester of barbituric acid $C_4H_4O_3N_2$. It works by depressing brain activity. Most barbiturates, being highly addictive, are no longer prescribed and are listed as controlled substances.

bar chart in statistics, a way of displaying data, using horizontal or vertical bars. The heights or lengths of the bars are proportional to the quantities they represent.

barium chemical symbol Ba, (Greek *barytes* 'heavy') soft, silver-white, metallic element, atomic number 56, relative atomic mass 137.32. It is one of the alkaline-earth metals, found in nature as barium carbonate and barium sulphate. As the sulphate it is used in medicine: taken as a suspension (a 'barium meal'), its movement along the gut is followed using X-rays. The barium sulphate, which is opaque to X-rays, shows the shape of the gut, revealing any abnormalities of the alimentary canal. Barium is also used in alloys, pigments, and safety matches and, with strontium, forms the emissive surface in cathode-ray tubes. It was first discovered in baryte or heavy spar.

bark protective outer layer on the stems and roots of woody plants, composed mainly of dead cells. To allow for expansion of the stem, the bark is continually added to from within, and the outer surface often becomes cracked or is shed as scales. Trees deposit a variety of chemicals in their bark, including poisons. Many of these chemical substances have economic value because they can be used in the manufacture of drugs. Quinine, derived from the bark of the *Cinchona* tree, is used to fight malarial infections; curare, an anaesthetic used in medicine, comes from the *Strychnus toxifera* tree in the Amazonian rainforest.

barometer instrument that measures atmospheric pressure as an indication of weather. Most often used are the mercury barometer and the aneroid barometer.

barrel unit of liquid capacity, the value of which depends on the liquid being measured. It is used for petroleum, a barrel of which contains 159 litres/35 imperial gallons; a barrel of alcohol contains 189 litres/41.5 imperial gallons.

barrier reef coral reef that lies offshore, separated from the mainland by a shallow lagoon.

baryon A subatomic particle consisting of three quarks. *Protons, *neutrons and *hyperons are all baryons. See *hadron.

basal cell carcinoma in medicine, another term for rodent ulcer. A slow-growing malignant tumour on the face, usually occurring at the edge of the eyelids, lips, or nostrils.

basal ganglia several masses of grey matter located deep in the *cerebrum, including the caudate nucleus, the globus pallidus and putamen. They are involved with the subconscious regulation of movement. See also *brain.

basal metabolic rate BMR, minimum amount of energy needed by the body to maintain life. It is measured when the subject is awake but resting, and includes the energy required to keep the heart beating, sustain breathing, repair tissues, and keep the brain and nerves functioning. The rate varies depending on the height, weight, age, and activity of the person. Measuring the subject's consumption of oxygen gives an accurate value for BMR, because oxygen is needed to release energy from food.

basalt   commonest volcanic *igneous rock in the Solar System. Basalt is an *extrusive rock, created by the outpouring of volcanic magma. The magma cools quickly, allowing only small crystals to form. Much of the surfaces of the terrestrial planets Mercury, Venus, Earth, and Mars, as well as the Moon, are composed of basalt. Earth's ocean floor is virtually entirely made of basalt. Basalt is mafic, that is, it contains relatively little *silica: about 50% by weight. It is usually dark grey but can also be green, brown, or black. Its essential constituent minerals are calcium-rich feldspar, and calcium- and magnesium-rich *pyroxene.

base   in chemistry, a substance that accepts hydrogen ions, or protons. A base reacts with an *acid, neutralizing it to form a *salt: acid + base → salt + water. Metal oxides and metal hydroxides are bases; examples include copper oxide and sodium hydroxide. Bases can contain negative ions such as the hydroxide ion (OH^-), which is the strongest base, or be molecules such as ammonia (NH_3). Ammonia is a weak base, as only some of its molecules accept protons.

$$OH^- + H^+_{(aq)} \rightarrow H_2O_{(l)}$$

$$NH_3 + H_2O \rightarrow NH_4^+ + OH^-$$

Bases that dissolve in water are called *alkalis.

Inorganic bases are usually oxides or hydroxides of metals, which react with dilute acids to form a salt and water. Many carbonates also react with dilute acids, additionally giving off carbon dioxide.

base   in mathematics, the number of different single-digit symbols used in a particular number system. In our usual (decimal) counting system of numbers (with symbols 0, 1, 2, 3, 4, 5, 6, 7, 8, 9) the base is 10. In the *binary number system, which has only the symbols 1

and 0, the base is two. A base is also a number that, when raised to a particular power (that is, when multiplied by itself a particular number of times as in $10^2 = 10 \times 10 = 100$), has a *logarithm equal to the power. For example, the logarithm of 100 to the base ten is 2.

In geometry, the term is used to denote the line or area on which a polygon or solid stands.

baseband   in computing, type of *network that transmits a computer signal without modulation (conversion of digital signals to analogue). To be able to send a computer's signal over the analogue telephone network, a *modem is required to convert – or modulate – the signal. On baseband networks, which include the most popular standards such as *Ethernet, the signal can be sent directly, without such processing.

base pair   in biochemistry, the linkage of two base (purine or pyrimidine) molecules that join the complementary strands of *DNA. Adenine forms a base pair with thymine (or uracil in RNA) and cytosine pairs with guanine in a double-stranded nucleic acid molecule.

One base lies on one strand of the DNA double helix and one on the other, so that the base pairs link the two strands like the rungs of a ladder. In DNA, there are four bases: adenine and guanine (purines) and cytosine and thymine (pyrimidines). Adenine always pairs with thymine and cytosine with guanine.

BASIC   acronym for Beginner's All-purpose Symbolic Instruction Code, high-level computer-programming language, developed in 1964, originally designed to take advantage of *multiuser systems (which can be used by many people at the same time). The language is relatively easy to learn and is popular among microcomputer users.

basicity   number of replaceable hydrogen atoms in an acid. Nitric acid (HNO_3) is monobasic, sulphuric acid (H_2SO_4) is dibasic, and phosphoric acid (H_3PO_4) is tribasic.

basic–oxygen process   most widely used method of steelmaking, involving the blasting of oxygen at high pressure into molten pig iron.

Pig iron from a blast furnace, together with steel scrap, is poured into a converter, and a jet of oxygen is then projected into the mixture. The excess carbon in the mix and other impurities quickly burn out or form a slag, and the converter is emptied by tilting. It takes only about 45 minutes to refine 350 tonnes/400 tons of steel. The basic–oxygen process was developed 1948 at a steelworks near the Austrian towns of Linz and Donawitz. It is a version of the *Bessemer process.

basic rock igneous rock with relatively low silica contents of 45–52% by weight, such as gabbro and basalt. Along with the terms **acid rock** and **ultrabasic rock** it is part of an outdated classification system based on the erroneous belief that silicon in rocks is in the form of silicic acid. Geologists today are more likely to use the descriptive term *mafic rock or report the amount of SiO_2 in percentage by weight.

basidiocarp spore-bearing body, or 'fruiting body', of all basidiomycete fungi (see *fungus), except the rusts and smuts. A well known example is the edible mushroom *Agaricus brunnescens*. Other types include globular basidiocarps (puffballs) or flat ones that project from tree trunks (brackets). They are made up of a mass of tightly packed, intermeshed *hyphae.

Basidiomycetes large class of fungi, with perhaps 12,000 species, which includes mushrooms, toadstools, bracket fungi, puffballs, smuts, and rusts. The *thallus, the body of the plant, is a mass of branching threads. The reproductive spores are borne on a club-shaped organ called a basidium (plural basidia), and are called basidiospores. There are usually four on each basidium. The basidia are grouped into conspicuous fruiting bodies, from which they are dispersed by the wind. Mushrooms and brackets are fruiting bodies.

batch file in computing, file that runs a group (batch) of commands. The most commonly used batch file is the *DOS start-up file *AUTOEXEC.BAT.

batch processing in computing, system for processing data with little or no operator intervention. Batches of data are prepared in advance to be processed during regular 'runs' (for example, each night). This allows efficient use of the computer and is well suited to applications of a repetitive nature, such as a company payroll, or the production of utility bills.

battery any energy-storage device allowing release of electricity on demand. It is made up of one or more electrical *cells. Electricity is produced by a chemical reaction in the cells. There are two types of battery: primary-cell batteries, which are disposable; and secondary-cell batteries, or *accumulators, which are rechargeable. Primary-cell batteries are an extremely uneconomical form of energy, since they produce only 2% of the power used in their manufacture. It is dangerous to try to recharge a primary-cell battery.

primary cell The **dry cell** is the most common type of primary-cell battery, based on the Leclanché cell. Dry cells are used in batteries to power, for example, torches, electronic toys, and compact stereo systems. Dry cells consist of a zinc case used as a negative electrode, and a carbon rod as a positive electrode suspended in the centre of the case and immersed in a

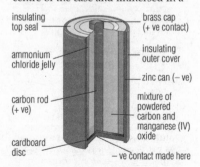

battery The common dry cell relies on chemical changes occurring between the electrodes – the central carbon rod and the outer zinc casing – and the ammonium chloride electrolyte to produce electricity. The mixture of carbon and manganese is used to increase the life of the cell.

paste of manganese dioxide and ammonium chloride acting as an electrolyte.

The cell depends on the difference in electronegativity between the zinc and carbon to produce electricity. Carbon strongly holds onto its electrons; it is more electronegative. Zinc weakly holds onto its electrons; it is less electronegative. Zinc dissolves in the ammonium chloride and loses two electrons for each atom of zinc. The electrons move towards the carbon electrode:

$$Zn \rightarrow Zn^{2+} + 2e^-$$

A small amount of charge is produced and zinc becomes negatively charged and carbon becomes positively charged. When a connection is made externally between the positive carbon terminal and the negative zinc terminal, electrons flow to the carbon from the zinc. The charge is neutralized on both electrodes and more of the zinc metal dissolves in the electrolyte to produce more electrons. A dry cell provides a current until the zinc electrode is completely used up.

secondary cell A storage battery of secondary cells gives large amounts of power for a short time and can be recharged. The lead–acid car battery is a secondary-cell battery. The electrolyte is sulphuric acid (*battery acid), the positive electrode is lead peroxide, and the negative electrode is lead. A typical lead–acid battery consists of six lead–acid cells in a case. Each cell produces 2 volts, so the whole battery produces a total of 12 volts.

Hydrogen cells and sodium–sulphur batteries were developed in 1996 to allow cars to run entirely on battery power for up to 160 km/100 mi.

The introduction of rechargeable nickel–cadmium batteries has revolutionized portable electronic news gathering (sound recording, video) and information processing (computing). These batteries offer a stable, short-term source of power free of noise and other electrical hazards.

battery acid *sulphuric acid of approximately 70% concentration used in lead–acid cells (as found in car batteries).

The chemical reaction within the battery that is responsible for generating electricity also causes a change in the acid's composition. This can be detected as a change in its specific gravity: in a fully charged battery the acid's specific gravity is 1.270–1.290; in a half-charged battery it is 1.190–1.210; in a flat battery it is 1.110–1.130.

Baudot code five-bit code developed in France by engineer Emil Baudot (1845–1903) in the 1870s. It is used for telex.

bauxite principal ore of *aluminium, consisting of a mixture of hydrated aluminium oxides and hydroxides, generally contaminated with compounds of iron, which give it a red colour. It is formed by the *chemical weathering of rocks in tropical climates. Chief producers of bauxite are Australia, Guinea, Jamaica, Russia, Kazakhstan, Suriname, and Brazil.

Bayesian statistics form of statistics that uses the knowledge of prior probability together with the probability of actual data to determine posterior probabilities, using *Bayes' theorem. Named after English mathematician Thomas Bayes (1702–1761)

Bayes' theorem in statistics, a theorem relating the *probability of particular events taking place to the probability that events conditional upon them have occurred. Named after English mathematician Thomas Bayes (1702–1761)

B cell or **B lymphocyte**, type of lymphocyte (white blood cell) that develops in the bone marrow and aids the immune system by producing antibodies. Each B cell produces just one type of *antibody, specific to a single *antigen. Lymphocytes are related to *T cells.

BCG abbreviation for **bacille Calmette-Guérin**, bacillus injected as a vaccine to confer active immunity to *tuberculosis (TB).

beam balance instrument for measuring mass (or weight). A simple form consists of a beam pivoted at its midpoint with a pan hanging at each end. The mass to be measured, in one pan, is compared with a variety of standard masses placed in the other. When the beam is balanced, the masses' turning effects or moments under gravity, and hence the masses themselves, are equal.

bearing device used in a machine to allow free movement between two parts, typically the rotation of a shaft in a housing. **Ball bearings** consist of two rings, one fixed to a housing, one to the rotating shaft. Between them is a set, or race, of steel balls. They are widely used to support shafts, as in the spindle in the hub of a bicycle wheel.

bearing direction of a fixed point, or the path of a moving object, from a point of observation. Bearings are angles measured in degrees (°) from the north line in a clockwise direction. A bearing must always have three figures. For instance, north is 000°, northeast is 045°, south is 180°, and southwest is 225°.

True north differs slightly from magnetic north (the direction in which a compass needle points), hence northeast may be denoted as 045M or 045T, depending on whether the reference line is magnetic (M) or true (T) north. True north also differs slightly from grid north since it is impossible to show a sphere on a flat map.

beat frequency in musical acoustics, fluctuation produced when two notes of nearly equal pitch or *frequency are heard together. Beats result from the *interference between the sound waves of the notes. The frequency of the beats equals the difference in frequency of the notes.

Beaufort scale system of recording wind velocity, devised by British admiral Francis Beaufort (1774–1857) in 1806. It is a numerical scale ranging from 0 to 17, calm being indicated by 0 and a hurricane by 12; 13–17 indicate degrees of hurricane force.

In 1874 the scale received international recognition; it was modified in 1926. Measurements are made at 10 m/33 ft above ground level.

becquerel symbol Bq, SI unit of *radioactivity, equal to one radioactive disintegration (change in the nucleus of an atom when a particle or ray is given off) per second.

The becquerel is much smaller than the previous standard unit, the *curie (3.7×10^{10} Bq). It is named after French physicist Henri Becquerel (1852–1908).

behaviourism school of psychology originating in the USA, of which the leading exponent was John B Watson.

Behaviourists maintain that all human activity can ultimately be explained in terms of conditioned reactions or reflexes and habits formed in consequence. Leading behaviourists include Russian psychologist Ivan Pavlov (1849–1936) and US psychologist B F Skinner (1904–1990).

behaviour therapy in psychology, the application of behavioural principles, derived from learning theories, to the treatment of clinical conditions such as *phobias, *obsessions, and sexual and interpersonal problems.

bel unit of sound measurement equal to ten *decibels. It is named after Scottish-born US scientist and inventor Alexander Graham Bell (1847–1922).

belladonna or **deadly nightshade**, poisonous plant belonging to the nightshade family, found in Europe and Asia. It grows to 1.5 m/5 ft in height, with dull green leaves growing in unequal pairs, up to 20 cm/8 in long, and single purplish flowers that produce deadly black berries. Drugs are made from the leaves. (*Atropa belladonna*, family Solanaceae.)

benchmark in computing, measure of the performance of a piece of equipment or software, usually consisting of a standard program or suite of programs. Benchmarks can indicate whether a computer is powerful enough to perform a particular task, and so enable machines to be compared. However, they provide only a very rough guide to practical performance, and may lead manufacturers to design systems that get high scores with the artificial benchmark programs but do not necessarily perform well with day-to-day programs or data.

bends or **compressed-air sickness** or **caisson disease**, popular name for a syndrome seen in deep-sea divers, arising from too rapid a release of nitrogen from solution in their blood. If a diver surfaces too quickly, nitrogen that had dissolved in the blood under increasing water pressure is suddenly released, forming bubbles in the bloodstream and causing pain (the 'bends') and paralysis. Immediate treatment is gradual decompression in a decompression chamber, whilst breathing pure oxygen.

benign   term used to describe a non-cancerous growth, such as a polyp.

It is also applied to any medical condition which is not actually harmful to the patient.

bentonite   soft porous rock consisting mainly of clay minerals, such as montmorillonite, and resembling fuller's earth, which swells when wet. It is formed by the chemical alteration of glassy volcanic material, such as tuff. Bentonite is used in papermaking, moulding sands, drilling muds for oil wells, and as a decolorant in food processing.

benzaldehyde   (benzene carbaldehyde) C_6H_5CHO, colourless liquid with the characteristic odour of almonds. It is used as a solvent and in the making of perfumes and dyes. It occurs in certain leaves, such as the cherry, laurel, and peach, and in a combined form in certain nuts and kernels. It can be extracted from such natural sources, but is usually made from *toluene.

benzene C_6H_6, clear liquid hydrocarbon of characteristic odour, occurring in coal tar. It is used as a solvent and in the synthesis of many chemicals.

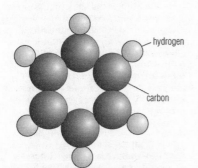

hydrogen

carbon

benzene The molecule of benzene consists of six carbon atoms arranged in a ring, with six hydrogen atoms attached. The benzene ring structure is found in many naturally occurring organic compounds.

The benzene molecule consists of a ring of six carbon atoms, all of which are in a single plane, and it is one of the simplest *cyclic compounds. Benzene is the simplest of a class of compounds collectively known as **aromatic compounds**. Some are considered carcinogenic (cancer-inducing).

benzoic acid   (benzene carboxylic acid) C_6H_5COOH, white crystalline solid, sparingly soluble in water, that is used as a preservative for certain foods and as an antiseptic. It is obtained chemically by the direct oxidation of benzaldehyde and occurs in certain natural resins, some essential oils, and as hippuric acid.

benzpyrene   (benzopyrene) $C_{20}H_{12}$, one of a number of organic compounds associated with a particular polycyclic ring structure. Benzpyrene is present in coal tar at low levels and is considered carcinogenic (cancer-inducing). Traces of benzpyrene is present in wood smoke, and this has given rise to some concern about the safety of naturally smoked foods.

beriberi   nutritional disorder occurring mostly in the tropics and resulting from a deficiency of vitamin B_1 (*thiamine). The disease takes two forms: in one oedema (waterlogging of the tissues) occurs; in the other there is severe emaciation. There is nerve degeneration in both forms and many victims succumb to heart failure.

berkelium   chemical symbol Bk, synthesized, radioactive, metallic element of the actinide series, atomic number 97, relative atomic mass 247. It was first produced in 1949 by US nuclear chemist Glenn Seaborg and his team, at the University of California at Berkeley, California, after which it is named.

Bernoulli's principle   law stating that the pressure of a fluid varies inversely with speed, an increase in speed producing a decrease in pressure (such as a drop in hydraulic pressure as the fluid speeds up flowing through a constriction in a pipe) and vice versa. The principle also explains the pressure differences on each surface of an aerofoil, which gives lift to the wing of an aircraft. The principle is named after Swiss mathematician and physicist Daniel Bernoulli (1700–1782).

beryl in full **beryllium aluminium silicate**, $3BeO.Al_2O_3.6SiO_2$, mineral that forms crystals chiefly in granite. It is the chief ore of beryllium. Two of its gem forms are aquamarine (light-blue crystals) and emerald (dark-green crystals).

beryllium chemical symbol Be, hard, lightweight, silver-white, metallic element, atomic number 4, relative atomic mass 9.012. It is one of the *alkaline-earth metals, with chemical properties similar to those of magnesium. In nature it is found only in combination with other elements and occurs mainly as beryl ($3BeO.Al_2O_3.6SiO_2$). It is used to make sturdy, light alloys and to control the speed of neutrons in nuclear reactors. Beryllium oxide was discovered in 1798 by French chemist Louis-Nicolas Vauquelin (1763–1829), but the element was not isolated until 1828, by Friedrich Wöhler and Antoine-Alexandre-Brutus Bussy independently.

In 1992 large amounts of beryllium were unexpectedly discovered in six old stars in the Milky Way.

Bessemer process first cheap method of making *steel, invented by English engineer and inventer Henry Bessemer (1813–1898) in England in 1856. It has since been superseded by more efficient steel-making processes, such as the *basic–oxygen process. In the Bessemer process compressed air is blown into the bottom of a converter, a furnace shaped like a cement mixer, containing molten pig iron. The excess carbon in the iron burns out, other impurities form a slag, and the furnace is emptied by tilting.

beta-blocker any of a class of drugs that block impulses that stimulate certain nerve endings (beta receptors) serving the heart muscle. This reduces the heart rate and the force of contraction, which in turn reduces the amount of oxygen (and therefore the blood supply) required by the heart. Beta-blockers may be useful in the treatment of angina, arrhythmia (abnormal heart rhythms), and raised blood pressure, and following heart attacks. They must be withdrawn from use gradually.

beta decay disintegration of the nucleus of an atom to produce a beta particle, or high-speed electron, and an electron antineutrino. During beta decay, a neutron in the nucleus changes into a proton, thereby increasing the atomic number by one while the mass number stays the same. The mass lost in the change is converted into kinetic (movement) energy of the beta particle. Beta decay is caused by the weak nuclear force, one of the fundamental *forces of nature operating inside the nucleus.

beta particle or **beta ray**, discovered by New-Zealand-born British physicist Ernest Rutherford (1871–1937). Electron ejected with great velocity from a radioactive atom that is undergoing spontaneous disintegration. Beta particles are created in the nucleus on disintegration, beta decay, when a neutron converts to a proton (the atomic number increases by one while the atomic mass stays the same) by emitting an electron. The mass lost in the change is converted into *kinetic energy of the beta particle. Strontium-90 (^{90}Sr) is an example of a material that emits beta particles.

Beta particles are more penetrating than *alpha particles, but less so than *gamma radiation; they can travel several metres in air, but are stopped by 2–3 mm/0.08–0.1 in of aluminium. They are less strongly ionizing than alpha particles. Owing to their low mass, beta particles, like cathode rays, are easily deflected by magnetic and electric fields. Beta decay is caused by the weak nuclear force.

beta version in computing, pre-release version of *software or an application program, usually distributed to a limited number of expert users (and often reviewers). Distribution of beta versions allows user testing and feedback to the developer, so that any necessary modifications can be made before release.

bhp abbreviation for **brake horsepower**.

bicarbonate common name for *hydrogencarbonate.

bicarbonate indicator pH indicator sensitive enough to show a colour change as the concentration of the gas carbon dioxide increases. The indicator is used in photosynthesis and respiration experiments to find out whether carbon dioxide is being liberated. The initial red colour changes to yellow as the pH becomes more acidic.

bicarbonate of soda or **baking soda**; technical name **sodium hydrogencarbonate**, $NaHCO_3$, white crystalline solid that neutralizes acids and is used in medicine to treat acid indigestion. It is also used in baking powders and effervescent drinks.

biceps (Latin *bis* 'twice', *caput* 'head') anatomical term for two *muscles of the human body, one of the arm and one of the leg, although in popular use it generally denotes the muscle of the arm. The **biceps brachii** is the muscle on the upper arm, which flexes the shoulder, the elbow, and supinates the forearm. To extend the arm its *antagonistic muscle – the triceps – has to contract.

bicuspid in biology, a structure with two cusps. Examples include the premolars (bicuspid teeth) and the mitral (bicuspid) valve of the heart.

bicuspid valve or **mitral valve**, in the left side of the *heart, a flap of tissue that prevents blood flowing back into the atrium when the ventricle contracts.

Big Bang in astronomy, the hypothetical 'explosive' event that marked the origin of the universe as we know it. At the time of the Big Bang, the entire universe was squeezed into a hot, superdense state. The Big Bang explosion threw this compact material outwards, producing the expanding universe seen today (see *red shift). The cause of the Big Bang is unknown; observations of the current rate of expansion of the universe suggest that it took place about 10–20 billion years ago. The Big Bang theory began modern *cosmology.

According to a modified version of the Big Bang, called the **inflationary theory**, the universe underwent a rapid period of expansion shortly after the Big Bang, which accounts for its current large size and uniform nature.

The inflationary theory is supported by the most recent observations of the *cosmic background radiation.

Scientists have calculated that 10^{-36} seconds (equivalent to one million-million-million-million-million-millionth of a second) before the Big Bang, the universe was the size of a pea, and the temperature was 10 billion million million million°C/18 billion million million million°F. One second after the Big Bang, the temperature was about 10 billion°C/18 billion°F.

According to theory, one-tenth of a second after the Big Bang, the temperature and pressure had decreased by many millions of degrees, thus allowing the formation of subatomic particles. After 10 seconds, neutrons had combined with protons to form nuclei of deuterium (an isotope of hydrogen). The nuclei of deuterium then joined together to form helium nuclei. As the universe continued to expand for the next 300,000 years, the temperature cooled to 10,000°C/18,000°F. Under these conditions helium nuclei were able to join with electrons to form helium atoms. Also, hydrogen nuclei joined to form lithium nuclei and thence lithium atoms. After millions of years, at lower temperature and pressure, the force of gravity was able to attract particles together. After millions more years, the universe formed into clumped matter joined together to form galaxies, stars, planets, and moons.

The first detailed images of the universe as it existed 300,000 years after the Big Bang were released by the US National Aeronautics and Space Administration (NASA) in April 2000. The images were created by mapping cosmic background radiation.

bile brownish alkaline fluid produced by the *liver. Bile is stored in the gall bladder and is intermittently released into the small intestine (the duodenum), which is part of the *gut, in order to help *digestion. Bile contains chemicals that *emulsify fats. In other words it acts to disperse fat globules into tiny droplets, which speeds up their digestion.

Bile consists of bile salts, bile pigments, cholesterol, and lecithin. The

bile salts assist in the breakdown and absorption of fats, while **bile pigments** are the breakdown products of old red blood cells, which are passed into the gut to be eliminated with the faeces.

bilharzia or **schistosomiasis**, disease that causes anaemia, inflammation, formation of scar tissue, dysentery, enlargement of the spleen and liver, cancer of the bladder, and cirrhosis of the liver. It is contracted by bathing in water contaminated with human sewage. Some 200 million people are thought to suffer from this disease in the tropics, and 750,000 people a year die.

bimetallic strip strip made from two metals each having a different coefficient of *thermal expansion; it therefore bends when subjected to a change in temperature. Such strips are used widely for temperature measurement and control, for instance in the domestic thermostat.

bimodal in statistics, having two distinct peaks of *frequency distribution.

binary file any file that is not plain text. Program (.EXE or .COM), sound, video, and graphics files are all types of binary files. Such files require special treatment for inclusion in e-mail sent across the Internet, which can transmit only *ASCII text and imposes a size limit of 64Kb per message. Several programs have been developed to code binary files into ASCII for transmission, splitting them into smaller parts as necessary. The most commonly used such program is *UUencode, but there are others including base64 and BinHex. See also *MIME.

binary fission in biology, a form of *asexual reproduction, whereby a single-celled organism, such as a bacterium or amoeba, divides into two smaller 'daughter' cells.

binary large object BLOB, in computing, any large single block of data stored in a database, such as a picture or sound file. A BLOB does not include record fields, and so cannot be directly searched by the database's search engine.

binary number system system of numbers to *base two, using

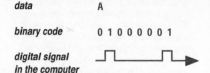

data	A
binary code	0 1 0 0 0 0 0 1
digital signal in the computer	⌐_⌐⌐_⌐→

binary number system The capital letter A represented in binary form.

combinations of the digits 1 and 0. Codes based on binary numbers are used to represent instructions and data in all modern digital computers, the values of the binary digits (contracted to 'bits') being stored or transmitted as, for example, open/closed switches, magnetized/unmagnetized disks and tapes, and high/low voltages in circuits.

binary search in computing, rapid technique used to find any particular record in a list of records held in sequential order. The computer is programmed to compare the record sought with the record in the middle of the ordered list. This being done, the computer discards the half of the list in which the record does not appear, thereby reducing the number of records left to search by half. This process of selecting the middle record and discarding the unwanted half of the list is repeated until the required record is found.

binding energy in physics, the amount of energy needed to break the nucleus of an atom into the neutrons and protons of which it is made.

binoculars optical instrument for viewing an object in magnification with both eyes; for example, field glasses and opera glasses. Binoculars consist of two telescopes containing lenses and prisms, which produce a stereoscopic effect as well as magnifying the image.

Use of prisms has the effect of 'folding' the light path, allowing for a compact design.

The first binocular telescope was constructed by the Dutch inventor Hans Lippershey in 1608. Later development was largely due to the German Ernst Abbe (1840–1905) of Jena, who at the end of the 19th century designed prism binoculars that foreshadowed the instruments of today, in which not only magnification but also stereoscopic effect is obtained.

binomial in mathematics, an expression consisting of two terms, such as $a + b$ or $a - b$.

binomial system of nomenclature in biology, the system in which all organisms are identified by a two-part Latinized name. Devised by the Swedish biologist Carl Linnaeus (1707–1778), it is also known as the Linnaean system. The first name is capitalized and identifies the *genus; the second identifies the *species within that genus, for example the bear genus *Ursus* includes *Ursus arctos*, the grizzly bear, and *Ursus maritimus*, the polar bear.

binomial theorem formula whereby any power of a binomial quantity may be found without performing the progressive multiplications.

It was discovered by English physicist and mathematician Isaac Newton (1642–1727) and first published in 1676.

biochemical oxygen demand BOD, amount of dissolved oxygen taken up by micro-organisms in a sample of water. Since these micro-organisms live by decomposing organic matter, and the amount of oxygen used is proportional to their number and metabolic rate, BOD can be used as a measure of the extent to which the water is polluted with organic compounds.

biochemistry science concerned with the chemistry of living organisms: the structure and reactions of proteins (such as enzymes), nucleic acids, carbohydrates, and lipids.

Its study has led to an increased understanding of life processes, such as those by which organisms synthesize essential chemicals from food materials, store and generate energy, and pass on their characteristics through their genetic material. A great deal of medical research is concerned with the ways in which these processes are disrupted. Biochemistry also has applications in agriculture and in the food industry (for instance, in the use of enzymes).

biodiversity contraction of biological diversity, measure of the variety of the Earth's animal, plant, and microbial species, of genetic differences within species, and of the *ecosystems that support those species. High biodiversity means there are lots of different species in an area. The maintenance of biodiversity is important for ecological stability and as a resource for research into, for example, new drugs and crops.

Estimates of the number of species vary widely because many species-rich ecosystems, such as tropical forests, contain unexplored and unstudied habitats. Among small organisms in particular many are unknown. For example, it is thought that less than 1% of the world's bacterial species have been identified.

The most significant threat to biodiversity comes from the destruction of rainforests and other habitats. It is estimated that 7% of the Earth's surface hosts 50–75% of the world's biological diversity. Costa Rica, for example, has an area less than 10% of the size of France but possesses three times as many vertebrate species.

bioenergetics in alternative medicine, extension of Reichian therapy principles developed in the 1960s by US physician Alexander Lowen, and designed to promote, by breathing, physical exercise, and the elimination of muscular blockages, the free flow of energy in the body and thus restore optimum health and vitality.

bioengineering application of engineering to biology and medicine. Common applications include the design and use of artificial limbs, joints, and organs, including hip joints and heart valves.

biofeedback in biology, modification or control of a biological system by its results or effects. For example, a change in the position or *trophic level of one species affects all levels above it. Many biological systems are controlled by negative feedback. When enough of the hormone thyroxine has been released into the blood, the hormone adjusts its own level by 'switching off' the gland that produces it. In ecology, as the numbers in a species rise, the food supply available to each individual is reduced. This acts to reduce the population to a sustainable level.

biofeedback in medicine, the use of electrophysiological monitoring devices to 'feed back' information about internal processes and thus facilitate conscious control. Developed in the USA in the 1960s, independently by neurophysiologist Barbara Brown and neuropsychiatrist Joseph Kamiya, the technique is effective in alleviating hypertension and preventing associated organic and physiological dysfunctions.

biofuel any solid, liquid, or gaseous fuel produced from organic (once living) matter, either directly from plants or indirectly from industrial, commercial, domestic, or agricultural wastes. There are three main methods for the development of biofuels: the burning of dry organic wastes (such as household refuse, industrial and agricultural wastes, straw, wood, and peat); the fermentation of wet wastes (such as animal dung) in the absence of oxygen to produce biogas (containing up to 60% methane), or the fermentation of sugar cane or maize to produce alcohol and esters; and energy forestry (producing fast-growing wood for fuel).

Fermentation produces two main types of biofuels: alcohols and esters. These could theoretically be used in place of fossil fuels but, because major alterations to engines would be required, biofuels are usually mixed with fossil fuels. The EU allows 5% ethanol, derived from wheat, beet, potatoes, or maize, to be added to fossil fuels. A quarter of Brazil's transportation fuel in 1994 was ethanol.

biogenesis biological term coined in 1870 by English scientist Thomas Henry Huxley to express the hypothesis that living matter always arises out of other similar forms of living matter. It superseded the opposite idea of *spontaneous generation or abiogenesis (that is, that living things may arise out of nonliving matter).

biogeochemistry emerging branch of *geochemistry involving the study of

how chemical elements and their isotopes move between living organisms and geological materials. For example, the analysis of carbon in bone gives biogeochemists information on how the animal lived, its diet, and its environment.

biological clock regular internal rhythm of activity, produced by unknown mechanisms, and not dependent on external time signals. Such clocks are known to exist in almost all animals, and also in many plants, fungi, and unicellular organisms; the first biological clock gene in plants was isolated in 1995 by a US team of researchers. In higher organisms, there appears to be a series of clocks of graded importance. For example, although body temperature and activity cycles in human beings are normally 'set' to 24 hours, the two cycles may vary independently, showing that two clock mechanisms are involved.

biological control control of pests such as insects and fungi through biological means, rather than the use of chemicals. This can include breeding resistant crop strains; inducing sterility in the pest; infecting the pest species with disease organisms; or introducing the pest's natural predator. Biological control tends to be naturally self-regulating, but as ecosystems are so complex, it is difficult to predict all the consequences of introducing a biological controlling agent.

biological shield shield around a nuclear reactor that is intended to protect personnel from the effects of *radiation. It usually consists of a thick wall of steel and concrete.

biological weathering or **organic weathering**, form of *weathering caused by the activities of living organisms – for example, the growth of roots or the burrowing of animals. Tree roots are probably the most significant agents of biological weathering as they are capable of prising apart rocks by growing into cracks and joints. Plants also give off organic acids that help to break down rocks chemically (see *chemical weathering).

biology (Greek *bios* 'life', *logos* 'discourse') science of life. Biology includes all the life sciences – for example, anatomy and physiology (the study of the structure of living things), cytology (the study of cells), zoology (the study of animals), botany (the study of plants), ecology (the study of habitats and the interaction of living species), animal behaviour, embryology, and taxonomy (classification), and plant breeding. Increasingly biologists have concentrated on molecular structures: biochemistry, biophysics, and genetics (the study of inheritance and variation).

bioluminescence   production of light by living organisms. It is a feature of many deep-sea fishes, crustaceans, and other marine animals. On land, bioluminescence is seen in some nocturnal insects such as glow-worms and fireflies, and in certain bacteria and fungi. Light is usually produced by the oxidation of luciferin, a reaction catalysed by the *enzyme luciferase. This reaction is unique, being the only known biological oxidation that does not produce heat. Animal luminescence is involved in communication, camouflage, or the luring of prey, but its function in some organisms is unclear.

biomass   total mass of living organisms present in a given area. It may be used to describe the mass of a particular species (such as earthworm biomass), for a general category (such as herbivore biomass – animals that eat plants), or for everything in a *habitat. Estimates also exist for the entire global plant biomass. Biomass can be the mass of the organisms as they are – wet biomass – or the mass of the organisms after they have been dried to remove all the water – dry biomass. Measurements of biomass can be used to study interactions between organisms, the stability of those interactions, and variations in population numbers. *Growth results in an increase in biomass, so biomass is a good measure of the extent to which organisms thrive in particular habitats. For a plant, biomass increase occurs as a result of the process of *photosynthesis. For a herbivore, biomass increase depends on the availability of plant food. Studying

biomass in a habitat is a useful way to see how food is passed from organism to organism along *food chains and through food webs.

biome   broad natural assemblage of plants and animals shaped by common patterns of vegetation and climate. Examples include the *tundra biome, the *rainforest biome, and the *desert biome.

biomechanics   application of mechanical engineering principles and techniques in the field of medicine and surgery, studying natural structures to improve those produced by humans. For example, mother-of-pearl is structurally superior to glass fibre, and deer antlers have outstanding durability because they are composed of microscopic fibres. Such natural structures may form the basis of 'high-tech' composites. Biomechanics has been responsible for many advances in *orthopaedics, anaesthesia, and intensive care. Biomechanical assessment of the requirements for replacement of joints, including evaluation of the stresses and strains between parts, and their reliability, has allowed development of implants with very low friction and long life.

biometrics   in computing, term applied loosely to the measurement of biological (human) data, usually for security purposes, rather than the statistical analysis of biological data. For example, when someone wants to enter a building or cash a cheque, their finger or eyeball may be scanned and compared with a fingerprint or eyeball scan stored earlier. Biometrics saves people from having to remember PINs (personal identification numbers) and passwords.

bionics   (from 'biological electronics') design and development of electronic or mechanical artificial systems that imitate those of living things. The bionic arm, for example, is an artificial limb (prosthesis) that uses electronics to amplify minute electrical signals generated in body muscles to work electric motors, which operate the joints of the fingers and wrist.
The first artificial eye that connects directly into the optic nerve was due to be implanted in mid-2000 for an initial

trial. The eye is composed of a video camera, radio antenna, and microchip. It stimulates different parts of the optic nerve enabling visual sensations in the brain. Bionic ears work by replacing with electrodes the hairs in the ear that naturally convert sounds into electrical impulses. The first person to receive two bionic ears was Peter Stewart, an Australian journalist, who received his right ear in 1984 and his left in 1989.

biopsy removal of a living tissue sample from the body for diagnostic examination.

bioreactor sealed vessel in which microbial reactions can take place. The simplest bioreactors involve the slow decay of vegetable or animal waste, with the emission of methane that can be used as fuel. Laboratory bioreactors control pH, acidity, and oxygen content and are used in advanced biotechnological operations, such as the production of antibiotics by genetically engineered bacteria.

BIOS acronym for Basic Input/Output System, part of a computer's operating system that handles the basic input and output operation for standard computer *hardware. For example, the BIOS reads the keystrokes from the keyboard, puts information on the display, and sends information to the printer. The small computer programs within the BIOS that carry out these tasks are called device drivers.

biosensor device based on microelectronic circuits that can directly measure medically significant variables for the purpose of diagnosis or monitoring treatment. One such device measures the blood sugar level of diabetics using a single drop of blood, and shows the result on a liquid crystal display within a few seconds.

bistable circuit or **flip-flop**, a basic electronic circuit which remains in one of two possible states until it receives a signal. It then switches over to the other state and waits for the next signal before switching back. In one state it can be considered to represent 0 and in the other state a 1, and is used as the basis of a computer memory.

biosynthesis synthesis of organic chemicals from simple inorganic ones by living cells – for example, the conversion of carbon dioxide and water to glucose by plants during *photosynthesis. Other biosynthetic reactions produce cell constituents including proteins and fats.

biotechnology industrial use of living organisms. Examples of its uses include fermentation, *genetic engineering (gene technology), and the manipulation of reproduction. The brewing and baking industries have long relied on the yeast micro-organism for *fermentation purposes, while the dairy industry employs a range of bacteria and fungi to convert milk into cheeses and yoghurts. *Enzymes, whether extracted from cells or produced artificially, are central to most biotechnological applications. Recent advances include genetic engineering, in which single-celled organisms with modified *DNA are used to produce insulin and other drugs.

In 1993, 70% of biotechnology companies were concentrating on human health developments, while only 10% were concerned with applications for food and agriculture. There are many medical and industrial applications of the use of micro-organisms, such as drug production. One important area is the production of antibiotics such as penicillin.

It is thought that biotechnology may be helpful in reducing world food shortages. Micro-organisms grow very quickly in suitable conditions and they often take substances that humans cannot eat and use them to produce foods that we can eat.

biotic factor organic variable affecting an ecosystem – for example, the changing population of elephants and its effect on the African savannah.

biotic potential the total theoretical reproductive capacity of an individual organism or species under ideal environmental conditions. The biotic potential of many small organisms such as bacteria, annual plants, and small mammals is very high but rarely reached, as other elements of the ecosystem such as predators and nutrient availability keep the population growth in check.

biotin or **vitamin H**, vitamin of the B complex, found in many different kinds of food; egg yolk, liver, legumes, and yeast contain large amounts. Biotin is essential to the metabolism of fats. Its absence from the diet may lead to dermatitis.

birth act of producing live young from within the body of female animals. Both viviparous and ovoviviparous animals give birth to young. In viviparous animals, embryos obtain nourishment from the mother via a *placenta or other means.

In ovoviviparous animals, fertilized eggs develop and hatch in the oviduct of the mother and gain little or no nourishment from maternal tissues. See also *pregnancy.

birth control the use of *contraceptives prevent pregnancy. It is part of the general practice of family planning.

bisect to divide a line or angle into two equal parts.

bisector a line that bisects an angle or another line (known as a **perpendicular bisector** when it bisects at right angles).

bismuth chemical symbol Bi, hard, brittle, pinkish-white, metallic element, atomic number 83, relative atomic mass 208.98. It has the highest atomic number of all the stable elements (the elements from atomic number 84 up are radioactive). Bismuth occurs in ores and occasionally as a free metal (*native metal). It is a poor conductor of heat and electricity, and is used in alloys of low melting point and in medical compounds to soothe gastric ulcers. The name comes from the Latin *besemutum*, from the earlier German *Wismut*.

bisulphate another term for *hydrogen sulphate.

bit contraction of binary digit, in computing, a single binary digit, either 0 or 1. A bit is the smallest unit of data stored in a computer; all other data must be coded into a pattern of individual bits. A *byte represents sufficient computer memory to store a single *character of data, and usually contains eight bits. For example, in the *ASCII code system used by most microcomputers the capital letter A would be stored in a single byte of memory as the bit pattern 01000001.

bit map in computing, pattern of *bits used to describe the organization of data. Bit maps are used to store typefaces or graphic images (bit-mapped or *raster graphics), with 1 representing black (or a colour) and 0 white.

bit-mapped font font held in computer memory as sets of bit maps.

Bitnet acronym for Because It's Time NETwork, news *network developed in 1983 at the City University of New York, USA. Bitnet operates as a collection of mailing lists using Listserv, which was picked up by the rest of the Internet and is widely used, although Bitnet itself is falling into disuse.

bit pad computer input device; see *graphics tablet.

bits per second bps, commonly used measure of the speed of transmission of a *modem. In 2003 the fastest modems were rated at 56,600 bps. Modem speeds should conform to standards, known as *V numbers, laid down by the Comité Consultatif International Téléphonique et Télégraphique (CCITT) so that modems from different manufacturers can connect to each other. Many modems transfer data much faster than their nominal speeds via techniques such as *data compression.

bitumen involatile, tarry material, containing a mixture of *hydrocarbons (mainly *alkanes), that is the residue from the *fractional distillation of crude oil (unrefined *petroleum). Sometimes the term is restricted to a soft kind of pitch resembling asphalt.

Naturally occurring solid bitumen may have arisen as a residue from the evaporation of petroleum. If evaporation took place from a pool or lake of petroleum, the residue might form a pitch or asphalt lake, such as Pitch Lake in Trinidad. Bitumen was used in ancient times as a mortar, and by the Egyptians for embalming.

bivalent in biology, a name given to the pair of homologous chromosomes during reduction division (*meiosis). In chemistry, the term is sometimes used to describe an element or group with a *valency of two, although the term 'divalent' is more common.

black body in physics, a hypothetical object that completely absorbs all electromagnetic radiation striking it. It is also a perfect emitter of thermal radiation.

Black Death great epidemic of *plague, mainly the bubonic variant, that ravaged Europe in the mid-14th century. Contemporary estimates that it killed between one-third and half of the population (about 75 million people) are probably accurate. The cause of the plague was the bacterium *Yersinia pestis*, transmitted by fleas that infested migrating Asian black rats. Originating in China, the disease followed the trade routes through India into Europe. The name Black Death was first used in England in the early 19th century.

black hole object in space whose gravity is so great that nothing can escape from it, not even light. It is thought to form when a massive star shrinks at the end of its life. A black hole sucks in more matter, including other stars, from the space around it. Matter that falls into a black hole is squeezed to infinite density at the centre of the hole. Black holes can be detected because gas falling towards them becomes so hot that it emits X-rays.

Black holes containing the mass of millions of stars are thought to lie at the centres of *quasars. Satellites have detected X-rays from a number of objects that may be black holes, but only a small number of likely black holes have been identified in our Galaxy.

blast freezing industrial method of freezing substances such as foods by blowing very cold air over them. See *deep freezing.

blast furnace smelting furnace used to extract metals from their ores, chiefly pig iron from iron ore. The temperature is raised by the injection of an air blast.

In the extraction of iron the ingredients of the furnace are iron ore (iron(III) oxide), coke (carbon), and limestone (calcium carbonate). The coke is the fuel and provides the agent (carbon monoxide) for the *reduction of the iron ore.

$$C + O_2 \rightarrow CO_2$$

$$CO_2 + C \rightarrow 2CO$$

At the high temperature of the furnace the iron oxide is reduced to the metal.

$$Fe_2O_3 + 3CO \rightarrow 2Fe + 3CO_2$$

The limestone is decomposed to quicklime (calcium oxide), which combines with the acidic impurities of the ore to form a molten mass known as slag (calcium silicate).

$$CaCO_3 \rightarrow CaO + CO_2$$

$$CaO + SiO_2 \rightarrow CaSiO_3$$

The principle of reducing ferrous oxides by carbon has been known for thousands of years, but the present blast furnace was introduced around 1400. The fuel was originally charcoal and the resulting iron was either used for casting or converted to wrought iron or steel. Production increased with the use of coke as fuel in the 18th century.

blastocyst in mammals, the hollow ball of cells which is an early stage in the development of the *embryo, roughly equivalent to the *blastula of other animal groups.

blastomere in biology, a cell formed in the first stages of embryonic development, after the splitting of the fertilized ovum, but before the formation of the *blastula or blastocyst.

blastula early stage in the development of a fertilized egg, when the egg changes from a solid mass of cells (the morula) to a hollow ball of cells (the blastula), containing a fluid-filled cavity (the blastocoel). See also *embryology.

bleaching decolorization of coloured materials. The two main types of bleaching agent are **oxidizing bleaches**, which bring about the *oxidation of pigments and include the ultraviolet rays in sunshine, hydrogen peroxide, and chlorine in household bleaches; and **reducing bleaches**, which bring about *reduction and include sulphur dioxide.

Bleach is used in industry to lighten or whiten fabrics, yarns, or fibres. Bleaching processes have been known from antiquity, mainly those acting through sunlight. Both natural and synthetic pigments usually possess highly complex molecules, the colour property often being due to only a part of the molecule. Bleaches usually attack only that small part, yielding another substance similar in chemical structure but colourless.

bleeding loss of blood from the circulation; see *haemorrhage.

blight any of a number of plant diseases caused mainly by parasitic species of *fungus, which produce a whitish appearance on leaf and stem surfaces; for example, **potato blight** *Phytophthora infestans*. General damage caused by aphids or pollution is sometimes known as blight.

blindness complete absence or impairment of sight. It may be caused by heredity, accident, disease, or deterioration with age.

blind spot area where the optic nerve and blood vessels pass through the retina of the *eye. No visual image can be formed as there are no light-sensitive cells in this part of the retina. Thus the organism is blind to objects that fall in this part of the visual field.

bloatware disparaging term for personal computer software that takes up lots of hard drive space and/or memory. In the early days of computing, both were very expensive, and expert programmers attached most kudos to the ability to fit programs into very small spaces – sometimes with unfortunate consequences, as the century date change or year 2000 problem shows. Now that PC memory chips and hard drives are extremely cheap, programmers have different concerns, but some people brought up in more frugal times find it hard to adjust.

BLOB in computing, acronym for *binary large object.

block in computing, a group of records treated as a complete unit for transfer to or from backing storage. For example, many disk drives transfer data in 512-byte blocks.

block and tackle type of *pulley.

blocking software any of various software programs that work on the World Wide Web to block access to categories of information considered offensive or dangerous. Typically used by parents or teachers to ensure that children do not see pornographic or other adult material, some blocking products additionally allow the blocking of personal information such as home addresses and telephone numbers; some people regard this as censorship.

blood fluid pumped by the *heart, that circulates in the *arteries, *veins, and *capillaries of vertebrate animals forming the *bloodstream. The term also refers to the corresponding fluid in those invertebrates that possess a closed *circulatory system. Blood carries nutrients and oxygen to each body cell and removes waste products such as carbon dioxide. It is also important in the immune response and,

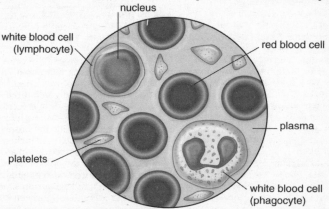

blood Composition of blood. Human blood contains red blood cells, white blood cells (phagocytes and lymphocytes), and platelets, suspended in plasma.

in many animals, in the distribution of heat throughout the body.

The adult human body contains about 5.5 l/10 pt of blood (about 5% of the body weight). It is composed of a fluid called *plasma, in which are suspended microscopic cells of three main varieties:

*Red blood cells (erythrocytes) form nearly half the volume of the blood, with about 6 million red blood cells in every millilitre of an adult's blood. They transport *oxygen around the body. Oxygen is absorbed into the millions of blood capillaries surrounding the tiny air sacs of the *lungs and is carried in the blood by *haemoglobin, a red protein within the red blood cells.

*White blood cells (leucocytes) are of various kinds. Some (phagocytes) ingest invading bacteria and so protect the body from disease; these also help to repair injured tissues. Others (lymphocytes) produce antibodies, which help provide immunity by binding to disease-causing *bacteria and destroying them. Yet others can kill cells infected by *viruses.

Blood *platelets (thrombocytes) assist in the clotting of blood.

Blood cells constantly wear out and die and are replaced from the bone marrow. Red blood cells die at the rate of 200 billion per day but the body produces new cells at an average rate of 9,000 million per hour.

Many different substances are transported by the blood around the body. Following *digestion small food molecules are absorbed into the blood from the *gut. The blood transports these nutrients (for example *glucose) to the cells of the body. Waste products produced by the cells, such as carbon dioxide, are carried by the blood to the lungs to be exhaled. *Hormones (such as oestrogen) are transported by the blood from glands to body cells in order to regulate various processes.

blood–brain barrier   theoretical term for the defence mechanism that prevents many substances circulating in the bloodstream (including some germs) from invading the brain.

The blood–brain barrier is not a single entity, but a defensive complex comprising various physical features and chemical reactions to do with the permeability of cells. It ensures that 'foreign' proteins, carried in the blood vessels supplying the brain, do not breach the vessel walls and enter the brain tissue. Many drugs are unable to cross the blood–brain barrier.

blood clotting   complex series of events (known as the blood clotting cascade) resulting from a series of enzymatic reactions in the blood that prevents excessive bleeding after injury. The result is the formation of a meshwork of protein fibres (fibrin) and trapped blood cells over the cut blood vessels.

When platelets (cell fragments) in the bloodstream come into contact with a damaged blood vessel, they and the vessel wall itself release the enzyme **thrombokinase**, which brings about the conversion of the inactive enzyme **prothrombin** into the active **thrombin**. Thrombin in turn catalyses the conversion of the soluble protein **fibrinogen**, present in blood plasma, to the insoluble **fibrin**. This fibrous protein forms a net over the wound that traps red blood cells and seals the wound; the resulting jellylike clot hardens on exposure to air to form a scab. Calcium, vitamin K, and a variety of enzymes called factors are also necessary for efficient blood clotting. *Haemophilia is one of several diseases in which the clotting mechanism is impaired.

blood group   any of the types into which blood is classified according to the presence or otherwise of certain *antigens on the surface of its red cells. Red blood cells of one individual may carry molecules on their surface that act as antigens in another individual whose red blood cells lack these molecules. The two main antigens are designated A and B. These give rise to four blood groups: having A only (A), having B only (B), having both (AB), and having neither (O). Each of these groups may or may not contain the *rhesus factor. Correct typing of blood groups is vital in transfusion, since incompatible types of donor and recipient blood will result in coagulation, with possible death of the recipient.

The ABO system was first described by Austrian scientist Karl Landsteiner in 1902. Subsequent research revealed at least 14 main types of blood group systems, 11 of which are involved with induced *antibody production. Blood typing is also of importance in forensic medicine, cases of disputed paternity, and in anthropological studies.

blood poisoning   presence in the bloodstream of quantities of bacteria or bacterial toxins sufficient to cause serious illness.

blood pressure   pressure, or tension, of the blood against the inner walls of blood vessels, especially the arteries, due to the muscular pumping activity of the heart. Abnormally high blood pressure (*hypertension) may be associated with various conditions or arise with no obvious cause; abnormally low blood pressure (hypotension) occurs in shock and after excessive fluid or blood loss from any cause.

In mammals, the left ventricle of the *heart pumps blood into the arterial system. This pumping is assisted by waves of muscular contraction by the arteries themselves, but resisted by he elasticity of the inner and outer walls of the same arteries. Pressure is greatest when the heart ventricle contracts (**systole**) and lowest when the ventricle relaxes (**diastole**), and pressure is solely maintained by the elasticity of the arteries. Blood pressure is measured in millimetres of mercury (the height of a column on the measuring instrument, a

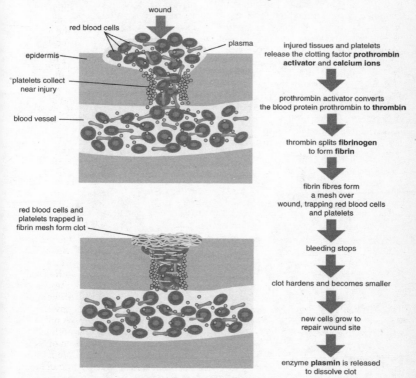

blood clotting The blood clotting cascade. When there is a wound in the skin, the clotting factor prothrombin activator is released by the injured tissues and platelets. This converts the blood protein prothrombin to thrombin. Thrombin splits fibrinogen to form fibrin, which forms a mesh over the wound trapping red blood cells and forming a clot.

sphygmomanometer). Normal human blood pressure varies with age, but in a young healthy adult it is around 120/80mm Hg; the first number represents the systolic pressure and the second the diastolic. Large deviations from this reading usually indicate ill health.

bloodstream   in animals, the means of carrying substances around the body. Oxygen needed by the cells of the body for *respiration is absorbed into the blood at the lungs by *diffusion and is transported to every part of the body by the blood stream. The oxygen is carried by specialized cells – the red blood cells. Inside each cell there are lots of haemoglobin molecules, which can combine with oxygen and release it when it is needed. Cells that respire produce carbon dioxide. The blood stream carries carbon dioxide from cells to the lungs where it is lost from the body by diffusion. The bloodstream also has an important role in carrying food around the body to cells that need it. This food is absorbed into the blood following *digestion in the *gut.

blood test   laboratory evaluation of a blood sample. There are numerous blood tests, from simple typing to establish the *blood group to sophisticated biochemical assays of substances, such as hormones, present in the blood only in minute quantities. The majority of tests fall into one of three categories: **haematology** (testing the state of the blood itself), **microbiology** (identifying infection), and **blood chemistry** (reflecting chemical events elsewhere in the body). Before operations, a common test is haemoglobin estimation to determine how well a patient might tolerate blood loss during surgery.

blood transfusion   see *transfusion.

blood vessel   tube that conducts blood either away from or towards the heart in multicellular animals. The principal types are *arteries, which conduct blood away from the heart, *veins, which conduct blood towards the heart, and *capillaries, which conduct blood from arteries to veins. Arteries always carry oxygenated blood and veins deoxygenated blood, with the exception of the **pulmonary artery** which carries deoxygenated blood from the heart to the lungs, and the **pulmonary vein**, which carries oxygenated blood from the lungs to the heart.

blow-out   hollow or depression of bare sand in an area of dunes on which vegetation grows. Blow-outs are common in coastal dune complexes and are formed by wind erosion, which can be triggered by the

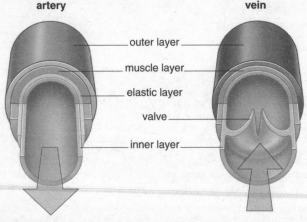

blood vessel Cross sections of an artery and a vein. Arteries have thicker walls than veins as they have to withstand a higher blood pressure than do veins. Veins have valves to prevent blood from flowing backwards.

destruction of small areas of vegetation by people or animals; lack of sand supply from a beach; and localized dryness.

blue-green algae or **cyanobacteria**, single-celled, primitive organisms that resemble bacteria in their internal cell organization, sometimes joined together in colonies or filaments. Blue-green algae are among the oldest known living organisms and, with bacteria, belong to the kingdom Monera; remains have been found in rocks up to 3.5 billion years old. They are widely distributed in aquatic habitats, on the damp surfaces of rocks and trees, and in the soil.

Blue-green algae and bacteria are prokaryotic organisms. Some can fix nitrogen and thus are necessary to the nitrogen cycle, while others follow a symbiotic existence – for example, living in association with fungi to form lichens. Fresh water can become polluted by nitrates and phosphates from fertilizers and detergents. This eutrophication, or overenrichment, of the water causes multiplication of the algae in the form of algae blooms. The algae multiply and cover the water's surface, remaining harmless until they give off toxins as they decay. These toxins kill fish and other wildlife and can be harmful to domestic animals, cattle, and people.

BMP in Windows, a file extension indicating a graphics file in *bit-map format. Bit-mapped files are commonly used for icons and wallpaper.

BODMAS mnemonic for **brackets, of, division, multiplication, addition, subtraction**, order in which an arithmetical expression should be calculated.

Any elements in **b**rackets should always be calculated first, followed by power **of** (or index), **d**ivision, **m**ultiplication, **a**ddition, and **s**ubtraction.

For example, to solve the equation:

$$3(2x - x - 1) = 2(x + 3 + 4)$$

collect the like terms and work out the brackets:

$$3(x - 1) = 2(x + 7)$$

multiply out the brackets:

$$3x - 3 = 2x + 14$$

collect the xs on the left-hand side of the equation:

$$3x - 3 - 2x = 14$$

then solve for x:

$$x - 3 = 14 \quad x = 14 + 3 \quad x = 17$$

Inequations or *inequalities may be solved using similar rules. When multiplying or dividing by a negative value, however, the direction of the inequality must be reversed, for example: $-x > 5$ is equivalent to $x < -5$.

body mass index BMI, in medicine, calculation of weight to height ratio that is a useful method of assessing whether an individual is underweight or obese. It is calculated using the formula:

$$BMI = \frac{weight\ (kg)}{height^2\ (m^2)}$$

Normal values for the BMI in adults are between 20 and 25. Values below 20 indicate that an individual is underweight and values above 30 indicate that an individual is obese. These values can be compared with extensive data obtained from life insurance companies to determine the risk of morbidity or mortality of a particular individual.

body temperature temperature deep in the tissues (core temperature). It is often measured on a thermometer. In a healthy adult it is normally 36.9°C/98.4°F. Any significant departure from this norm is potentially serious (although *fever is a necessary response to many types of infection). Hyperthermia (temperature above 41°C/106°F) and *hypothermia (below 35°C/95°F) are life-threatening. Excessive exposure to heat can cause heatstroke, where the body temperature rises.

bog type of wetland where decomposition is slowed down and dead plant matter accumulates as *peat. Bogs develop under conditions of low temperature, high acidity, low nutrient supply, stagnant water, and oxygen deficiency. Typical bog plants are sphagnum moss, rushes, and cotton grass; insectivorous plants such as sundews and bladderworts are common in bogs (insect prey make up for the lack of nutrients).

bohrium chemical symbol Bh, synthesized, radioactive element of the *transactinide series, atomic number 107, relative atomic mass 264. It was first synthesized by the Joint Institute for Nuclear Research in Dubna, Russia, in 1976; in 1981 the Laboratory for Heavy-Ion Research in Darmstadt, Germany, confirmed its existence. It was named in 1997 after Danish physicist Niels Bohr (1885–1962).

The first chemical study of bohrium was published in 2000, after experiments by Swiss researchers produced six atoms of bohrium-267 (half-life 17 seconds). It behaved like a typical group VII element, with chemical similarities to technetium and rhenium.

Bohr model model of the atom conceived by Danish physicist Niels Bohr (1885–1962) in 1913. It assumes that the following rules govern the behaviour of electrons: (1) electrons revolve in orbits of specific radius around the nucleus without emitting radiation; (2) within each orbit, each electron has a fixed amount of energy; electrons in orbits farther away from the nucleus have greater energies; (3) an electron may 'jump' from one orbit of high energy to another of lower energy causing the energy difference to be emitted as a *photon of electro-magnetic radiation such as light; (4) an electron may absorb a photon of radiation and jump from a lower-energy orbit to a higher-energy one. The Bohr model has been superseded by wave mechanics (see *quantum theory).

boiling process of changing a liquid into its vapour, by heating it at the maximum possible temperature for that liquid (see *boiling point) at atmospheric pressure.

boiling point for any given liquid, the temperature at which the application of heat raises the temperature of the liquid no further, but converts it into vapour.

The boiling point of water under normal pressure is 100°C/212°F. The lower the pressure, the lower the boiling point and vice versa.

bolson basin without an outlet, found in desert regions. Bolsons often contain temporary lakes, called playa lakes, and become filled with sediment from inflowing intermittent streams.

Boltzmann constant symbol k, in physics, the constant that relates the kinetic energy (energy of motion) of a gas atom or molecule to temperature, named after Austrian physicist Ludwig Boltzmann (1844–1906). Its value is 1.38066×10^{-23} joules per kelvin. It is equal to the gas constant R, divided by *Avogadro's number.

bond in chemistry, the result of the forces of attraction that hold together atoms in an element or compound. The principal types of bonding are *ionic, *covalent, *metallic, and *intermolecular (such as hydrogen bonding).

The type of bond formed depends on the elements concerned and their electronic structure. In an **ionic** or **electrovalent bond**, common in inorganic compounds, the combining atoms gain or lose electrons to become ions; for example, sodium (Na) loses an electron to form a sodium ion (Na^+) while chlorine (Cl) gains an electron to form a chloride ion (Cl^-) in the ionic bond of sodium chloride (NaCl).

In a **covalent bond**, the atomic orbitals of two atoms overlap to form a molecular orbital containing two electrons, which are thus effectively shared between the two atoms. Covalent bonds are common in organic compounds, such as the four carbon–hydrogen bonds in methane (CH_4). In a dative covalent or coordinate bond, one of the combining atoms supplies both of the valence electrons in the bond.

A **metallic bond** joins metals in a crystal lattice, the atoms occupy lattice positions as positive ions, and valence electrons are shared between all the ions in an 'electron gas'.

In a **hydrogen bond**, a hydrogen atom joined to an electronegative atom, such as nitrogen or oxygen, becomes partially positively charged, and is weakly attracted to another electronegative atom on a neighbouring molecule.

bond energy in chemistry, amount of energy that is needed to break a chemical bond between atoms, usually measured in joules. The energy required to break a specific chemical bond is identical to that generated by the formation of the same bond.

bone hard connective tissue comprising the *skeleton of most vertebrate animals. Bone is composed of a network of collagen fibres impregnated with mineral salts (largely calcium phosphate and calcium carbonate), a combination that gives it great density and strength, comparable in some cases with that of reinforced concrete. Enclosed within this solid matrix are bone cells, blood vessels, and nerves. The interior of the long bones of the limbs consists of a spongy matrix filled with a soft marrow that produces blood cells.

bone marrow substance found inside the cavity of bones. In early life it produces red blood cells but later on lipids (fat) accumulate and its colour changes from red to yellow.

bookmark facility for marking a specific place in electronic documentation to enable easy return to it. It is used in several types of software, including electronic help files and tutorials. Bookmarks are especially important on the World Wide Web, where it can be difficult to remember a uniform resource locator (*URL) in order to return to it. Most Web browsers therefore have built-in bookmark facilities, whereby the browser stores the URL with the page name attached. To return directly to the site, the user picks the page name from the list of saved bookmarks.

Boolean algebra set of algebraic rules, named after English mathematician George Boole, in which TRUE and FALSE are equated to 0 and 1. Boolean algebra includes a series of operators (AND, OR, NOT, NAND (NOT AND), NOR, and XOR (exclusive OR)), which can be used to manipulate TRUE and FALSE values. It is the basis of computer logic because the truth values can be directly associated with *bits.

boot disk or **emergency disk**, floppy disk containing the necessary files to boot a computer without needing to access its hard disk. Boot disks are vital in recovering from virus attacks, when it is not known which files on a computer's hard disk may be infected; in recovering from a system crash which has corrupted existing files; or

section through a long bone (the femur)

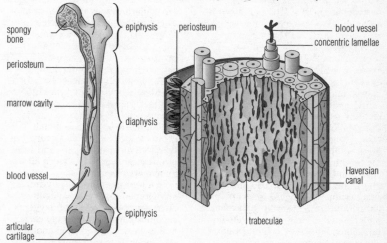

spongy bone

periosteum

marrow cavity

blood vessel

epiphysis

articular cartilage

epiphysis

diaphysis

periosteum

blood vessel

concentric lamellae

Haversian canal

trabeculae

bone The upper end of the thighbone or femur is made up of spongy bone, which has a fine lacework structure designed to transmit the weight of the body. The shaft of the femur consists of hard compact bone designed to resist bending. Fine channels carrying blood vessels, nerves, and lymphatics in correcting mistakes introduced into files necessary for starting up the computer by newly installed software programs.

borax hydrous sodium borate, $Na_2B_4O_7.10H_2O$, found as soft, whitish crystals or encrustations on the shores of hot springs and in the dry beds of salt lakes in arid regions, where it occurs with other borates, halite, and *gypsum. It is used in bleaches and washing powders.

A large industrial source is Borax Lake, California. Borax is also used in glazing pottery, in soldering, as a mild antiseptic, and as a metallurgical flux.

Bordeaux mixture solution made up of equal quantities of copper(II) sulphate and lime in water, used in horticulture and in the wine industry as a *fungicide.

bore surge of tidal water up an estuary or a river, caused by the funnelling of the rising tide by a narrowing river mouth. A very high tide, possibly fanned by wind, may build up when it is held back by a river current in the river mouth. The result is a broken wave, a metre or a few feet high, that rushes upstream.

boric acid or **boracic acid**; $B(OH)_3$, acid formed by the combination of hydrogen and oxygen with non-metallic boron. It is a weak antiseptic and is used in the manufacture of glass and enamels. It is also an efficient insecticide against ants and cockroaches.

boron chemical symbol B, non-metallic element, atomic number 5, relative atomic mass 10.811. In nature it is found only in compounds, as with sodium and oxygen in borax. It exists in two allotropic forms (see *allotropy): brown amorphous powder and very hard, brilliant crystals. Its compounds are used in the preparation of boric acid, water softeners, soaps, enamels, glass, and pottery glazes. In alloys it is used to harden steel. Because it absorbs slow neutrons, it is used to make boron carbide control rods for nuclear reactors. It is a necessary trace element in the human diet. The element was named by Humphry Davy, who isolated it in 1808, from *bor*ax + -on, as in carb*on*.

boson in physics, an elementary particle whose spin can only take values that are whole numbers or zero. Bosons may be classified as *gauge bosons (carriers of the four fundamental forces) or *mesons. All elementary particles are either bosons or *fermions.

'bot (or agent) abbreviation for robot, on the Internet, automated piece of software that performs specific tasks. 'Bots are also known as agents or intelligent agents. On the World Wide Web, 'bots automate maintenance tasks such as indexing Web pages and tracing broken links. Search engines send out 'bots (also called 'spiders') to crawl through the Internet, compiling information on the Web addresses they visit. The information compiled by the spiders is used to rank Web sites in order of relevance when a user carries out a search using a search engine.

botany (Greek *botane* 'herb') study of living and fossil *plants, including form, function, interaction with the environment, and classification.

botulism rare, often fatal type of *food poisoning. Symptoms include vomiting, diarrhoea, muscular paralysis, breathing difficulties and disturbed vision.

It is caused by a toxin produced by the bacterium *Clostridium botulinum*, found in soil and sometimes in improperly canned foods. Thorough cooking destroys the toxin, which otherwise suppresses the cardiac and respiratory centres of the brain.

The toxin acts to block the release of chemicals which cause muscles to contract. It does have medical uses; in neurology, botulinum toxin is sometimes used to treat rare movement disorders and in cosmetic surgery it is used for the short term removal of wrinkles (botox treatment).

bounce in computing, system by which an e-mail message that cannot be delivered to its addressee is returned ('bounced back') to the sender, with a note advising of its failure to reach its destination. Failed delivery is usually due to an incorrect e-mail address or a network problem.

boundary line around the edge of an area; a perimeter. The boundary of a circle is known as the **circumference**. The boundary which marks the limit of land may be indicated by a post, ditch, hedge, march of stones, road, or river, or it may be indicated by reference to a plan, or to possession of tenants, or by actual measurement.

Bourdon gauge   instrument for measuring pressure, patented by French watchmaker Eugène Bourdon in 1849. The gauge contains a C-shaped tube, closed at one end. When the pressure inside the tube increases, the tube uncurls slightly causing a small movement at its closed end. A system of levers and gears magnifies this movement and turns a pointer, which indicates the pressure on a circular scale. Bourdon gauges are often fitted to cylinders of compressed gas used in industry and hospitals.

bovine spongiform encephalopathy BSE; or **mad cow disease**, disease of cattle, related to scrapie in sheep, that attacks the nervous system, causing aggression, lack of coordination, and collapse. First identified in 1985, by 1996 it had claimed the lives of 158,000 British cattle. After safety measures were put in place, British beef was declared safe (by the British government) in 1999. Following outbreaks of BSE in French, German, and Spanish cattle in late 2000, European Union (EU) agriculture ministers agreed to ban, as of 1 January 2001, the use of meat-and-bone meal from animal feed and to ban all cattle over 30 months old from the food chain unless tested for BSE.

   BSE is one of a group of diseases known as the transmissible spongiform encephalopathies, since they are characterized by the appearance of spongy changes in brain tissue. Some scientists believe that all these conditions, including *Creutzfeldt–Jakob disease (CJD) in humans, are in effect the same disease.

   The cause of these universally fatal diseases is not fully understood, but they may be the result of a rogue protein called a *prion. A prion may be inborn or it may be transmitted in contaminated tissue.

Bowman's capsule   in the vertebrate kidney, a cup-shaped structure enclosing the glomerulus, which is the initial site of filtration of the blood leading to urine formation. Named after English physician William Bowman (1816–1892).

box plot   or **box and whisker diagram**, in statistics, figure used to represent *quartile information. The box represents the central 50% of the data and the whiskers extend from the box to the largest and smallest values. This method gives an overall spread of the data.

Boyle's law   law stating that the volume of a given mass of gas at a constant temperature is inversely proportional to its pressure. For example, if the pressure on a gas doubles, its volume will be reduced by a half, and vice versa. The law was discovered in 1662 by Irish physicist and chemist Robert Boyle (1627–1691). See also *gas laws.

   If a gas is compressed in a cylinder the volume of the gas decreases. The number of particles of gas in the cylinder remains the same. The particles get closer together, collide with each other more frequently, and the pressure of the gas increases due to the force of the particles colliding. If the pressure is P and the volume is V, then $P = 1/V$. Therefore, as the volume decreases the pressure increases.

   Boyle's law can be investigated by using an apparatus consisting of a foot pump attached to a pressure gauge. This is attached to a glass tube (with a scale) containing an oil and trapped air. As more air is pumped into the apparatus from the foot pump and into the oil reservoir, the oil in the glass tube is forced in and the pressure on the trapped air gets higher. A series of pressure and volume readings are taken. It is found that multiplying pressure by volume for each of the readings produces the same result. Therefore, $P \times V =$ constant. The results of P and $1/V$ plotted as a graph give a straight line. Readings from the experiment show that:

$$P_1 V_1 = P_2 V_2$$

where P_1 and V_1 are the initial pressure

and volume of a gas, and P_2 and V_2 are its final pressure and volume.

bozo filter facility to eliminate messages from irritating users. It is also known as a *killfile.

bps abbreviation for *bits per second*, measure used in specifying data transmission rates.

Bq in physics, symbol for *becquerel, the SI unit of radioactivity (equal to the average number of disintegrations per second in a given time).

brackets pairs of signs that show which part of a calculation should be worked out first. For example, $4(7 + 3)$ indicates that 4 is to be multiplied by the result obtained from adding 7 and 3. The mnemonic *BODMAS can help one to remember the order in which an arithmetic expression should be calculated. Brackets may be nested, for example, $4(20 - (7 + 3))$, in which case the expression $7 + 3$ within the innermost pair of brackets is evaluated first, the result subtracted from 20 and that result multiplied by 4.

bract leaflike structure in whose *axil a flower or inflorescence develops. Bracts are generally green and smaller than the true leaves. However, in some plants they may be brightly coloured and conspicuous, taking over the role of attracting pollinating insects to the flowers, whose own petals are small; examples include poinsettia *Euphorbia pulcherrima* and bougainvillea.

brain in higher animals, a mass of interconnected *nerve cells forming the anterior part of the *central nervous system, whose activities it coordinates and controls. In *vertebrates, the brain is contained by the skull. It is composed of three main regions. At the base of the *brainstem, the *medulla oblongata* contains centres for the control of respiration, heartbeat rate and strength, blood pressure, and thermoregulatory control (the control of body temperature). Overlying this is the *cerebellum*, which is concerned with coordinating complex muscular processes such as maintaining posture and moving limbs, and the control of balance. The *cerebrum* (cerebral hemispheres) are paired outgrowths of the front end of the forebrain, in early vertebrates mainly concerned with the

senses, but in higher vertebrates greatly developed and involved in the integration of all sensory input and motor output, and in thought, emotions, memory, and behaviour. Sensory information arrives in the cerebrum in the form of nerve impulses that come from receptors – these may be found in sense organs, such as cones (light sensitive cells) in the retina of the eye, which send impulses to the brain along the optic nerve. The cerebrum processes the information received and can cause impulses to be sent out to the body to produce a response, such as moving towards an object that has been seen. Because a decision is made whether to make this kind of movement it is said to be voluntary.

In vertebrates, many of the nerve fibres from the two sides of the body cross over as they enter the brain, so that the left cerebral hemisphere is associated with the right side of the body and vice versa. In humans, a certain asymmetry develops in the two halves of the cerebrum. In right-handed people, the left hemisphere seems to play a greater role in controlling verbal and some mathematical skills, whereas the right hemisphere is more involved in spatial perception. In general, however, skills and abilities are not closely localized. In the brain, nerve impulses are passed across *synapses by neurotransmitters, in the same way as in other parts of the nervous system.

In mammals the cerebrum is the largest part of the brain, carrying the **cerebral cortex**. This consists of a thick surface layer of cell bodies (grey matter), below which fibre tracts (white matter) connect various parts of the cortex to each other and to other points in the central nervous system. As cerebral complexity grows, the surface of the brain becomes convoluted into deep folds. In higher mammals, there are large unassigned areas of the brain that seem to be connected with intelligence, personality, and higher mental faculties. Language is controlled in two special regions usually in the left side of the brain: **Broca's area** governs the ability to talk, and **Wernicke's area** is responsible for the comprehension of spoken and written words. In 1990 scientists at Johns Hopkins University, Baltimore, succeeded in culturing human brain cells.

brainstem   region where the top of the spinal cord merges with the undersurface of the brain, consisting largely of the medulla oblongata and midbrain.

The oldest part of the brain in evolutionary terms, the brainstem is the body's life-support centre, containing regulatory mechanisms for vital functions such as breathing, heart rate, and blood pressure. It is also involved in controlling the level of consciousness by acting as a relay station for nerve connections to and from the higher centres of the brain.

In many countries, death of the brainstem is now formally recognized as death of the person as a whole. Such cases are the principal donors of organs for transplantation. So-called 'beating-heart donors' can be maintained for a limited period by life-support equipment.

brainstem death   in medicine, criterion for determining that 'death' has occurred when brain damage results in the irreversible loss of brain function, including that of the brainstem (which is important in the control of many processes which are essential to life,

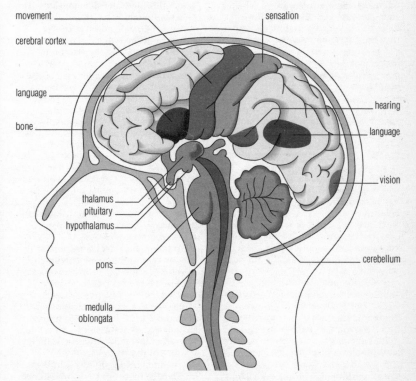

brain The structure of the human brain. At the back of the skull lies the cerebellum, which coordinates reflex actions that control muscular activity. The medulla controls respiration, heartbeat, and blood pressure. The hypothalamus is concerned with instinctive drives and emotions. The thalamus relays signals to and from various parts of the brain. The pituitary gland controls the body's hormones. Distinct areas of the large convoluted cerebral hemispheres that fill most of the skull are linked to sensations, such as hearing and sight, and voluntary activities, such as movement.

such as respiration and heart rate), so that the individual is unable to live without the aid of a ventilator. Many countries allow artificial ventilation to be stopped once brainstem death has been determined.

Artificial ventilation and heart beat are maintained after brainstem death has been established if organs are to be removed for use in transplantation procedures. This ensures that the organs are in the best condition for transplantation.

brass metal *alloy of copper and zinc, with not more than 5% or 6% of other metals. The zinc content ranges from 20% to 45%, and the colour of brass varies accordingly from coppery to whitish yellow. Brasses are characterized by the ease with which they may be shaped and machined; they are strong and ductile, resist many forms of corrosion, and are used for electrical fittings, ammunition cases, screws, household fittings, and ornaments.

brazing method of joining two metals by melting an *alloy or metal into the joint. It is similar to soldering but takes place at a much higher temperature. Copper and silver alloys are widely used for brazing, at temperatures up to about 900°C/1,650°F.

breadth thickness, another name for width. The area of a rectangle is given by the formula: area = length times breadth.

breast one of a pair of organs on the chest of the adult human female, also known as a mammary gland. Each of the two breasts contains milk-producing cells and a network of tubes or ducts that lead to openings in the nipple.

breast cancer in medicine, *cancer of the *breast. It is usually diagnosed following the detection of a painless lump in the breast (either through self-examination or mammography). Other, less common symptoms, include changes in the shape or texture of the breast and discharge from the nipple. It is the commonest cancer amongst women: there are 28,000 new cases of breast cancer in Britain each year and 185,700 in the USA.

breast-feeding in medicine, the process by which a baby gains its nourishment from milk secreted from the breast, by sucking on the nipple. Breast milk provides complete nutrition and additional protection from infection for the infant, and helps to establish 'bonding' between the mother and the baby. It also helps the mother to return to her previous body weight and encourages the uterus to shrink to normal size following childbirth. The early establishment of breast feeding maintains and improves the milk supply.

breathing or **ventilation**, movement of air into and out of the air passages of an animal, brought about by muscle contraction. It is a form of *gas exchange. Breathing is sometimes referred to as external respiration, for true respiration is a cellular (internal) process. In a mammal, breathing involves the action of the muscles of the diaphragm and the intercostal muscles (between the ribs). When a mammal breathes in, the diaphragm muscles contract, which lowers the diaphragm. The external intercostal muscles contract, which raises the ribs. Lowering the diaphragm and raising the ribs increases the volume of the *thorax. This lowers the pressure of the air inside the lungs in the thorax. The pressure is now lower than atmospheric pressure so air flows into the air passages and the lungs inflate.

breed recognizable group of domestic animals, within a species, with distinctive characteristics that have been produced by artificial selection.

breeder reactor or **fast breeder**, alternative name for *fast reactor, a type of nuclear reactor.

breeding in nuclear physics, a process in a reactor in which more fissionable material is produced than is consumed in running the reactor.

brewing making of beer, ale, or other alcoholic beverage, from malt and barley by steeping (mashing), boiling, and fermenting.

Mashing the barley releases its sugars. Yeast is then added, which contains the enzymes needed to convert the sugars into ethanol (alcohol) and carbon dioxide. Hops are added to give a bitter taste.

brewster symbol B, unit for measuring the reaction of optical materials to stress, defined in terms of the slowing down of light passing through the material when it is stretched or compressed.

bridge in computing, device that connects two similar local area networks (LANs). Bridges transfer data in packets between the two networks, without making any changes or interpreting the data in any way. See also *router.

brine common name for a solution of *sodium chloride (NaCl) in water. Brines are used extensively in the food-manufacturing industry for canning vegetables, pickling vegetables (sauerkraut manufacture), and curing meat. Industrially, brine is the source from which *chlorine, caustic soda (*sodium hydroxide), and sodium carbonate are made.

The chlor-alkali industry is based on the *electrolysis of brine, which produces chlorine, hydrogen, and sodium hydroxide. In the Solvay process, sodium carbonate is produced from sodium chloride.

Brinell hardness test test of the hardness of a substance according to the area of indentation made by a 10 mm/0.4 in hardened steel or sintered tungsten carbide ball under standard loading conditions in a test machine. The resulting Brinell number is equal to the load (kg) divided by the surface area (mm²) and is named after its inventor, Swedish metallurgist Johann Brinell.

British Standards Institution or BSI, UK national standards body. Although government funded, the institution is independent. The BSI interprets international technical standards for the UK, and also sets its own.

British thermal unit symbol Btu, imperial unit of heat, now replaced in the SI system by the *joule (one British thermal unit is approximately 1,055 joules). Burning one cubic foot of natural gas releases about 1,000 Btu of heat.

brittle material material that breaks suddenly under stress at a point just beyond its elastic limit (see *elasticity). Brittle materials may also break suddenly when given a sharp knock. Pottery, glass, and cast iron are examples of brittle materials. Compare *ductile material.

broadband in computing, a type of data transmission in which a single circuit can carry several channels at once, used for example in cable television. Broadband networking is one way of supplying much greater Internet *bandwidth over the existing telephone system. See also *ADSL.

bromide salt of the halide series containing the Br⁻ ion, which is formed when a bromine atom gains an electron. The term 'bromide' is sometimes used to describe an organic compound containing a bromine atom, even though it is not ionic. Modern naming uses the term 'bromo-' in such cases. For example, the compound C_2H_5Br is now called bromoethane; its traditional name, still used sometimes, is ethyl bromide.

bromine chemical symbol Br, (Greek *bromos* 'stench') dark, reddish-brown, non-metallic element, a volatile liquid at room temperature, atomic number 35, relative atomic mass 79.904. It is a member of the *halogen group, has an unpleasant odour, and is very irritating to mucous membranes. Its salts are known as bromides.

Bromine was formerly extracted from salt beds but is now mostly obtained from sea water, where it occurs in small quantities. Its compounds are used in photography and in the chemical and pharmaceutical industries.

bronchiectasis dilation of the airways (bronchi) due to infection or obstruction. The patient coughs up large quantities of sputum, sometimes containing blood, and suffers repeated infections. It is treated with antibiotics and postural drainage.

bronchiole small-bore air tube found in the vertebrate lung responsible for delivering air to the main respiratory surfaces. Bronchioles lead off from the larger bronchus and branch extensively before terminating in the many thousand alveoli that form the bulk of lung tissue.

bronchitis inflammation of the bronchi (air passages) of the lungs, usually caused initially by a viral infection, such as a cold or flu. It is aggravated by environmental pollutants, especially smoking, and results in a persistent cough, irritated mucus-secreting glands, and large amounts of sputum.

bronze alloy of copper and tin, yellow or brown in colour. It is harder than pure copper, more suitable for casting, and also resists *corrosion. Bronze may contain as much as 25% tin, together with small amounts of other metals, mainly lead.

brouter device for connecting computer networks that incorporates the facilities of both a *bridge and a *router. Brouters usually offer routing over a limited number of *protocols, operating by routing where possible and bridging the remaining protocols.

brown dwarf in astronomy, object less massive than a star but denser than a planet. Brown dwarfs do not have enough mass to ignite nuclear reactions at their centres, but shine by heat released during their contraction from a gas cloud. Groups of brown dwarfs have been discovered, and some astronomers believe that vast numbers of them exist throughout the Galaxy.

Brownian motion continuous random motion of particles in a fluid medium (gas or liquid) as they are subjected to impact from the molecules of the medium. The phenomenon was explained by German-born US physicist Albert Einstein (1879–1955) in 1905 but was observed as long ago as 1827 by the Scottish botanist Robert Brown. Brown was looking at pollen grains in water under a microscope when he noticed the pollen grains were in constant, haphazard motion. The motion of these particles was due to the impact of moving water molecules. It provides evidence for the *kinetic theory of matter.

In order for the irregular motion to be observed, the particles in the medium must be sufficiently small relative to the bombarding molecules for the impact of the bombarding molecules to have an effect. A tennis ball in air, for instance, would not show Brownian motion because the impacts of the moving air molecules

on one side of the tennis ball would be balanced by impacts of the molecules on the other side. In other words, the *resultant force of the impacts would be too small.

Einstein's explanation of Brownian motion and its subsequent experimental confirmation was one of the most important pieces of evidence for the hypothesis that matter is composed of atoms.

brown ring test in analytical chemistry, a test for the presence of *nitrates. To an aqueous solution containing the test substance is added iron(II) sulphate. Concentrated sulphuric acid is then carefully poured down the inside wall of the test tube so that it forms a distinct layer at the bottom. The formation of a brown colour at the boundary between the two layers indicates the presence of nitrate.

browse to explore a computer system or network for particular files or information. To browse in Windows is to search for a particular file to open or run. On the World Wide Web, browsing is the activity of moving from site to site to view information. This is sometimes also called 'surfing'.

browser in computing, any program that allows the user to search for and view data. Browsers are usually limited to a particular type of data, so, for example, a graphics browser will display graphics files stored in many different file formats. Browsers usually do not permit the user to edit data, but are sometimes able to convert data from one file format to another. Web browsers allow access to the *World Wide Web. Netscape Navigator and Microsoft's Internet Explorer were the leading Web browsers in 2001. They act as a *graphical user interface to information available on the Internet – reading *HTML documents and displaying them as graphical documents which may include images, video, sound, and *hypertext links to other documents.

bryophyte member of the Bryophyta, a division of the plant kingdom containing three classes: the Hepaticae (liverwort), Musci (moss), and Anthocerotae (hornwort). Bryophytes are generally small, low-growing, terrestrial plants with no vascular (water-conducting) system as in higher plants. Their life cycle shows a marked *alternation of generations. Bryophytes chiefly occur in damp habitats and require water for the dispersal of the male gametes (antherozoids).

BSE abbreviation for *bovine spongiform encephalopathy.

BSI abbreviation for *British Standards Institution.

Btu symbol for *British thermal unit: originally it was defined as the heat required to raise the temperature of one pound of water by one degree Fahrenheit. Now defined as equal to 1,055.6 joules.

bubble chamber in physics, a device for observing the nature and movement of atomic particles, and their interaction with radiation. It is a vessel filled with a superheated liquid through which ionizing particles move and collide. The paths of these particles are shown by strings of bubbles, which can be photographed and studied. By using a pressurized liquid medium instead of a gas, it overcomes drawbacks inherent in the earlier *cloud chamber. It was invented by US physicist Donald Glaser in 1952. See *particle detector.

bubble memory in computing, a memory device based on the creation of small 'bubbles' on a magnetic surface. Bubble memories typically store up to 4 megabits (4 million *bits) of information. They are not sensitive to shock and vibration, unlike other memory devices such as disk drives, yet, like magnetic disks, they are nonvolatile and do not lose their information when the computer is switched off.

bubble sort in computing, technique for *sorting data. Adjacent items are continually exchanged until the data are in sequence.

bubonic plague epidemic disease of the Middle Ages; see *plague and *Black Death.

buckminsterfullerene an allotrope (see *allotropy) of carbon, made up of molecules (buckyballs) consisting of 60 carbon atoms arranged in 12 pentagons and 20 hexagons to form a perfect sphere. It was named after the US architect and engineer Richard Buckminster Fuller because of its structural similarity to the geodesic dome that he designed.

buckyballs popular name for molecules of *buckminsterfullerene.

bud undeveloped shoot usually enclosed by protective scales; inside is a very short stem and numerous undeveloped leaves, or flower parts, or both. Terminal buds are found at the tips of shoots, while axillary buds develop in the *axils of the leaves, often remaining dormant unless the terminal bud is removed or damaged. Adventitious buds may be produced anywhere on the plant, their formation sometimes stimulated by an injury, such as that caused by pruning.

budding type of *asexual reproduction in which an outgrowth develops from a cell to form a new individual. Most yeasts reproduce in this way.

buffer in chemistry, mixture of compounds chosen to maintain a steady *pH. The commonest buffers consist of a mixture of a weak organic acid and one of its salts or a mixture of acid salts of phosphoric acid. The addition of either an acid or a base causes a shift in the *chemical equilibrium, thus keeping the pH constant.

buffer in computing, a part of the *memory used to store data temporarily while it is waiting to be used. For example, a program might store data in a printer buffer until the printer is ready to print it.

bug in computing, an error in a program. It can be an error in the logical structure of a program or a syntax error, such as a spelling mistake. Some bugs cause a program to fail immediately; others remain dormant, causing problems only when a particular combination of events occurs. The process of finding and removing errors from a program is called **debugging**.

bulb underground bud with fleshy leaves containing a reserve food supply and with roots growing from its base. Bulbs function in vegetative reproduction and are characteristic of many monocotyledonous plants such as the daffodil, snowdrop, and onion. Bulbs are grown on a commercial scale in temperate countries, such as England and the Netherlands.

bulimia (Greek 'ox hunger') eating disorder in which large amounts of food are consumed in a short time ('binge'), usually followed by depression and self-criticism. The term is often used for **bulimia nervosa**, an emotional disorder in which eating is followed by deliberate vomiting and purging. This may be a chronic stage in *anorexia nervosa.

bulletin board in computing, a centre for the electronic storage of messages, usually accessed over the telephone network via a *modem but also sometimes accessed via *Telnet across the Internet. Bulletin board systems (often abbreviated to BBSs) are usually dedicated to specific interest groups, and may carry public and private messages, notices, and programs.

bundling computer industry practice of selling different, often unrelated, products in a single package. Bundles may consist of hardware or software or both; for example, a modem or a selection of software may be bundled with a personal computer to make the purchase of the computer seem more attractive.

Bunsen burner gas burner used in laboratories, consisting of a vertical metal tube through which a fine jet of fuel gas is directed. Air is drawn in through airholes near the base of the tube and the mixture is ignited and burns at the tube's upper opening.

The invention of the burner is attributed to German chemist Robert von Bunsen in 1855, but English chemist and physicist Michael Faraday (1791–1867) is known to have produced a similar device at an earlier date. A later refinement was the metal collar that can be turned to close or partially close the airholes, thereby regulating the amount of air sucked in and hence the heat of the burner's flame.

buoyancy lifting effect of a fluid on a body wholly or partly immersed in it. This was studied by the Greek mathematician and philosopher

Archimedes (*c.* 287–212). in the 3rd century BC.

bur or **burr**, in botany, a type of 'false fruit' or *pseudocarp, surrounded by numerous hooks; for instance, that of burdock *Arctium*, where the hooks are formed from bracts surrounding the flowerhead. Burs catch in the feathers or fur of passing animals, and thus may be dispersed over considerable distances.

burette in chemistry, a piece of apparatus, used in *titration, for the controlled delivery of measured variable quantities of a liquid. It consists of a long, narrow, calibrated glass tube, with a tap at the bottom, leading to a narrow-bore exit.

burning common name for *combustion.

bus in computing, the electrical pathway through which a computer processor communicates with some of its parts and/or peripherals. Physically, a bus is a set of parallel tracks that can carry digital signals; it may take the form of copper tracks laid down on the computer's *printed circuit boards (PCBs), or of an external cable or connection.

butadiene or **buta-1,3-diene**; $CH_2=CHCH=CH_2$, inflammable gas derived from petroleum, used in making synthetic rubber and resins.

butane C_4H_{10}, *alkane (saturated hydrocarbon) derived from *natural gas and as a product of the *fractional distillation of crude oil (unrefined *petroleum). Liquefied under pressure, it is used as a *fuel for industrial and domestic purposes (for example, in portable cookers).

butene C_4H_8, fourth member of the *alkene series of hydrocarbons. It is an unsaturated compound, containing one double bond.

byte sufficient computer memory to store a single *character of data, such as a letter of the alphabet. The character is stored in the byte of memory as a pattern of *bits (binary digits), using a code such as *ASCII. A byte usually contains eight bits – for example, the capital letter F can be stored as the bit pattern 01000110.

°C symbol for degrees *Celsius, sometimes called centigrade.

C in physics, symbol for *coulomb, the SI unit of electrical charge.

C in computing, a high-level, general-purpose programming language popular on minicomputers and microcomputers. Developed in the early 1970s from an earlier language called BCPL, C was first used as the language of the operating system *Unix, though it has since become widespread beyond Unix. It is useful for writing fast and efficient systems programs, such as operating systems (which control the operations of the computer).

C++ in computing, a high-level programming language used in *object-oriented applications. It is derived from the language C.

cable modem box supplied by cable companies to provide television and telephone services, including Internet. The advantages of cable modems over traditional *modems, which operate over standard telephone lines, are greatly increased speed of communications as well as the ability to transmit video and two-way audio, and lower costs.

cache memory in computing, a reserved area of the *immediate access memory used to increase the running speed of a computer program.

CAD acronym for **computer-aided design**, use of computers in creating and editing design drawings. CAD also allows such things as automatic testing of designs and multiple or animated three-dimensional views of designs. CAD systems are widely used in architecture, electronics, and engineering, for example in the motor-vehicle industry, where cars designed with the assistance of computers are now commonplace. With a CAD system, picture components are accurately positioned using grid lines. Pictures can be resized, rotated, or mirrored without loss of quality or proportion.

cadmium chemical symbol Cd, soft, silver-white, ductile, and malleable metallic element, atomic number 48, relative atomic mass 112.41. Cadmium occurs in nature as a sulphide or carbonate in zinc ores. It is a toxic metal that, because of industrial dumping, has become an environmental pollutant. It is used in batteries, electroplating, and as a constituent of alloys used for bearings with low coefficients of friction; it is also a constituent of an alloy with a very low melting point.

Cadmium is also used in the control rods of nuclear reactors, because of its high absorption of neutrons. It was named in 1817 by the German chemist Friedrich Strohmeyer (1776–1835) after the Greek mythological character Cadmus.

caecum in the *digestive system of animals, a blind-ending tube branching off from the first part of the large intestine, terminating in the appendix. It has no function in humans but is used for the digestion of cellulose by some grass-eating mammals.

Caesarean section surgical operation to deliver a baby by way of an incision in the mother's abdominal and uterine walls. It may be recommended for almost any obstetric complication implying a threat to mother or baby.

caesium chemical symbol Cs, (Latin *caesius* 'bluish-grey') soft, silvery-white, ductile metallic element, atomic number 55, relative atomic mass 132.905. It is one of the *alkali metals that form Group 1 of the periodic table of the elements. The alkali metals increase in reactivity down the group, and caesium, with only the short-lived radioactive francium below it, is the most reactive of them all. In air it ignites spontaneously, and it reacts violently with water. It is the most electropositive of all the elements. It is used in the manufacture of photocells.

The rate of vibration of caesium atoms has been used as the standard of measuring time. Its radioactive isotope Cs-137 (half-life 30.17 years) is a product of fission in nuclear explosions and in nuclear reactors; it is one of the most dangerous waste products of the nuclear industry, being a highly radioactive biological analogue of potassium.

cal symbol for *calorie.

CAL acronym for **computer-assisted learning**, use of computers in education and training: the computer displays instructional material to a student and asks questions about the information given; the student's answers determine the sequence of the lessons.

calamine $ZnCO_3$, zinc carbonate, an ore of zinc. The term also refers to a pink powder made of a mixture of zinc oxide and iron(II) oxide used in lotions and ointments as an astringent for treating, for example, sunburn, eczema, measles rash, and insect bites and stings.

In the USA the term refers to zinc silicate $Zn_4Si_2O_7(OH)_2.H_2O$ (hemimorphite).

calcification in medicine, the deposition of calcium salts in bone cells as part of the growth process. Bone is comprised of fibrous tissue and a matrix containing calcium phosphate and calcium carbonate. Bones grow in thickness from the fibrous tissue and the calcium salts deposited in their cells by calcification.

calcination *oxidation of substances by roasting in air.

calcite colourless, white, or light-coloured common rock-forming mineral, calcium carbonate, $CaCO_3$. It is the main constituent of *limestone and marble and forms many types of invertebrate shell.

calcium chemical symbol Ca, (Latin *calcis* 'lime') soft, silvery-white metallic element, atomic number 20, relative atomic mass 40.08. It is one of the *alkaline-earth metals. It is the fifth most abundant element (the third most abundant metal) in the Earth's crust. It is found mainly as its carbonate $CaCO_3$, which occurs in a fairly pure condition as chalk and limestone (see *calcite). Calcium is an essential component of bones, teeth, shells, milk, and leaves, and it forms 1.5% of the human body by mass.

Calcium ions in animal cells are involved in regulating muscle contraction, blood clotting, hormone secretion, digestion, and glycogen metabolism in the liver. It is acquired mainly from milk and cheese, and its uptake is facilitated by vitamin D. Calcium deficiency leads to chronic muscle spasms (tetany); an excess of calcium may lead to the formation of stones (calculi) in the kidney or gall bladder.

The element was discovered and named by the English chemist Humphry Davy in 1808. Its compounds include slaked lime (calcium hydroxide, $Ca(OH)_2$); plaster of Paris (calcium sulphate, $CaSO_4.^1/_2H_2O$); calcium phosphate ($Ca_3(PO_4)_2$), the main constituent of animal bones; calcium hypochlorite ($CaOCl_2$), a bleaching agent; calcium nitrate ($Ca(NO_3)_2.4H_2O$), a nitrogenous fertilizer; calcium carbide (CaC_2), which reacts with water to give ethyne (acetylene); calcium cyanamide ($CaCN_2$), the basis of many pharmaceuticals, fertilizers, and plastics, including melamine; calcium cyanide ($Ca(CN)_2$), used in the extraction of gold and silver and in electroplating; and others used in baking powders and fillers for paints.

calcium carbonate $CaCO_3$, white solid, found in nature as limestone, marble, and chalk. It is a valuable resource, used in the making of iron, steel, cement, glass, slaked lime, bleaching powder, sodium carbonate and bicarbonate, and many other industrially useful substances.

calcium-channel blocker any of a group of drugs that interfere with the entry of calcium into smooth muscle cells. They influence the cells of the heart muscle and conducting tissue, reducing the strength of contraction and slowing the beat, and also the cells within the walls of the arteries, causing vasodilation. They are used to treat *angina, *hypertension and cardiac *arrhythmias. One drug in the group, nimodipine, is used to prevent spasm of the cerebral arteries following subarachnoid haemorrhage.

calcium hydrogencarbonate
$Ca(HCO_3)_2$, substance found in *hard water, formed when rainwater passes over limestone rock:

$$CaCO_{3\,(s)} + CO_{2\,(g)} + H_2O_{\,(l)} \rightarrow Ca(HCO_3)_{2\,(aq)}$$

When this water is boiled it reforms calcium carbonate, removing the hardness; this type of hardness is therefore known as temporary hardness.

calcium hydrogenphosphate
$Ca(H_2PO_4)_2$, substance made by heating calcium phosphate with 70% sulphuric acid. It is more soluble in water than calcium phosphate, and is used as a fertilizer.

calcium hydroxide or **slaked lime**; $Ca(OH)_2$, white solid, slightly soluble in water. A solution of calcium hydroxide is called *limewater and is used in the laboratory to test for the presence of carbon dioxide.

It is manufactured industrially by adding water to calcium oxide (quicklime) in a strongly *exothermic reaction:

$$CaO + H_2O \rightarrow Ca(OH)_2$$

It is used to reduce soil acidity and as a cheap alkali in many industrial processes.

calcium nitrate $Ca(NO_3)_2$, white crystalline compound that is very soluble in water. The solid decomposes on heating to form oxygen, brown nitrogen(IV) oxide gas, and the white solid calcium oxide:

$$2Ca(NO_3)_2 \rightarrow 2CaO + 4NO_2 + O_2$$

calcium oxide or **quicklime**; CaO, white solid compound, formed by heating *calcium carbonate:

$$CaCO_3 \rightarrow CaO + CO_2$$

When water is added it forms calcium hydroxide (slaked lime) in an *exothermic reaction:

$$CaO + H_2O \rightarrow Ca(OH)_2$$

It is a typical basic oxide, turning litmus blue.

calcium phosphate or **calcium orthophosphate**; $Ca_3(PO_4)_2$, white solid, the main constituent of animal bones. It occurs naturally as the mineral apatite and in rock phosphate, and is used in the preparation of phosphate fertilizers.

calcium sulphate $CaSO_4$, white solid compound, found in nature as gypsum and anhydrite. It dissolves slightly in water to form *hard water; this hardness is not removed by boiling, and hence is sometimes called permanent hardness.

calculus (Latin 'pebble') branch of mathematics which uses the concept of a derivative (see *differentiation) to analyse the way in which the values of a *function vary. Calculus is probably the most widely used part of mathematics. Many real-life problems are analysed by expressing one quantity as a function of another – position of a moving object as a function of time, temperature of an object as a function of distance from a heat source, force on an object as a function of distance from the source of the force, and so on – and calculus is concerned with such functions.

There are several branches of calculus. Differential and integral calculus, both dealing with small quantities which during manipulation are made smaller and smaller, compose the **infinitesimal calculus**. **Differential equations** relate to the derivatives of a set of variables and may include the variables. Many give the mathematical models for physical phenomena such as *simple harmonic motion. Differential equations are solved generally by *integration, depending on their degree. If no analytical processes are available, integration can be performed numerically. Other branches of calculus include calculus of variations and calculus of errors.

Calculus methods have been developed slowly since the ancient Greek mathematicians. In the 17th century English physicist and mathematician Isaac Newton (1642–1727) and German mathematician Gottfried Leibniz (1646–1716) were the first to give (independently) general rules for calculus but it was very difficult to put the subject on a secure logical basis, mainly because of the difficult concepts of *limit and *continuity involved. Instead of using the idea of limit, 18th- and 19th-century mathematicians sought to base calculus on the ideas of 'infinitesimals' (roughly, very small quantities) and 'differentials' and the subject has in the past been known as 'infinitesimal calculus' or 'differential calculus'. The first complete presentation of calculus using limits was given by Augustin Cauchy in 1821, but his ideas were not generally adopted (particularly in Britain) for many years.

calibration   the preparation of a usable scale on a measuring instrument. A mercury *thermometer, for example, can be calibrated with a Celsius scale by noting the heights of the mercury column at two standard temperatures – the freezing point ($0°C$) and boiling point ($100°C$) of water – and dividing the distance between them into 100 equal parts and continuing these divisions above and below.

californium   chemical symbol Cf, synthesized, radioactive, metallic element of the actinide series, atomic number 98, relative atomic mass 251. It is produced in very small quantities and used in nuclear reactors as a neutron source. The longest-lived isotope, Cf-251, has a half-life of 800 years.

   It is named after the state of California, where it was first synthesized in 1950 by US nuclear chemist Glenn Seaborg and his team at the University of California at Berkeley.

callipers   measuring instrument used, for example, to measure the internal and external diameters of pipes. Some callipers are made like a pair of compasses, having two legs, often curved, pivoting about a screw at one end. The ends of the legs are placed in contact with the object to be measured, and the gap between the ends is then measured against a rule. The slide calliper looks like an adjustable spanner, and carries a scale for direct measuring, usually with a vernier scale for accuracy.

callus   growth of healing tissue, also containing blood and bone-forming cells, that forms around the ends of a bone following a fracture. Callus formation is an important factor in the union of the fracture.

callus   in botany, a tissue that forms at a damaged plant surface. Composed of large, thin-walled *parenchyma cells, it grows over and around the wound, eventually covering the exposed area.

   In animals, a callus is a thickened pad of skin, formed where there is repeated rubbing against a hard surface. In humans, calluses often develop on the hands and feet of those involved in heavy manual work.

calorie   c.g.s. unit of heat, now replaced by the *joule (one calorie is approximately 4.2 joules). It is the heat required to raise the temperature of one gram of water by $1°C$. In dietetics, the Calorie or kilocalorie is equal to 1,000 calories.

calorific value   amount of heat generated by a given mass of fuel when it is completely burned. It is measured in joules per kilogram. Calorific values are measured experimentally with a bomb calorimeter.

calorimeter   instrument used in physics to measure various thermal properties, such as heat capacity or the heat produced by fuel. A simple calorimeter consists of a heavy copper vessel that is polished (to reduce heat losses by radiation) and covered with insulating material (to reduce losses by convection and conduction).

calyx   collective term for the *sepals of a flower, forming the outermost whorl of the perianth. It surrounds the other flower parts and protects them while in bud. In some flowers, for example, the campions *Silene*, the sepals are fused along their sides, forming a tubular calyx.

CAM   acronym for **computer-aided manufacturing**, use of computers to control production processes; in particular, the control of machine tools and robots in factories. In some factories, the whole design and production system is automated by linking *CAD (computer-aided design) to CAM.

cam part of a machine that converts circular motion to linear motion or vice versa. The **edge cam** in a car engine is in the form of a rounded projection on a shaft, the camshaft. When the camshaft turns, the cams press against linkages (plungers or followers) that open the valves in the cylinders.

cambium in botany, a layer of actively dividing cells (lateral *meristem), found within stems and roots, that gives rise to *secondary growth in perennial plants, causing an increase in girth. There are two main types of cambium: **vascular cambium**, which gives rise to secondary *xylem and *phloem tissues, and **cork cambium** (or phellogen), which gives rise to secondary cortex and cork tissues (see *bark).

Cambrian period period of geological time roughly 570–510 million years ago; the first period of the Palaeozoic Era. All invertebrate animal life appeared, and marine algae were widespread. The **Cambrian Explosion** 530–520 million years ago saw the first appearance in the fossil record of modern animal phyla; the earliest fossils with hard shells, such as trilobites, date from this period.

The name comes from Cambria, the medieval Latin name for Wales, where Cambrian rocks are typically exposed and were first described.

Campylobacter genus of bacteria that cause serious outbreaks of gastroenteritis. The bacteria grow best at 43°C, and so are well suited to the digestive tract of birds. Poultry is therefore the most likely source of a *Campylobacter* outbreak, although the bacteria can also be transmitted via beef or milk. *Campylobacter* can survive in water for up to 15 days, so may be present in drinking water if supplies are contaminated by sewage or reservoirs are polluted by seagulls.

cancel in mathematics, to simplify a fraction or ratio by dividing both numerator and denominator by the same number (which must be a *common factor of both of them). For example, $^{5x}/_{25}$ cancels to $^{x}/_{5}$ when divided top and bottom by five.

cancelbot automated software program (see *bot) that cancels messages on Usenet. The arrival of *spamming (advertising) on the Net prompted the development of technology to use features built into Usenet to cancel messages. While single messages are easily cancelled manually, an automated routine is needed to handle mass postings, which may go out to more than 14,000 newsgroups. Cancelbot is activated by the CancelMoose.

cancer group of diseases characterized by abnormal proliferation of cells. Cancer (malignant) cells are usually degenerate, capable only of reproducing themselves (tumour formation). Malignant cells tend to spread from their site of origin by travelling through the bloodstream or lymphatic system (see *metastasis). Cancer kills about 6 million people a year worldwide.

candela symbol cd, SI unit of luminous intensity, which replaced the former units of candle and standard candle. It measures the brightness of a light itself rather than the amount of light falling on an object, which is called **illuminance** and measured in *lux.

Candida albicans yeastlike fungus present in the human digestive tract and in the vagina, which causes no harm in most healthy people. However, it can cause problems if it multiplies excessively, as in vaginal candidiasis or *thrush, the main symptom of which is intense itching.

The most common form of thrush is oral, which often occurs in those taking steroids or prolonged courses of antibiotics.

candidiasis in medicine, infections due to the fungus *Candida albicans*.

canine in mammalian carnivores, any of the long, often pointed teeth found at the front of the mouth between the incisors and premolars. Canine teeth are used for catching prey, for killing, and for tearing flesh. They are absent in herbivores such as rabbits and sheep, and are much reduced in humans.

canyon (Spanish *cañon* 'tube') deep, narrow valley or gorge running through mountains. Canyons are formed by stream down-cutting, usually in arid areas, where the rate of down-cutting is greater than the rate of weathering, and where the stream or river receives water from outside the area.

There are many canyons in the western USA and in Mexico, for example the Grand Canyon of the Colorado River in Arizona, the canyon in Yellowstone National Park, and the Black Canyon in Colorado.

capacitance, electrical property of a capacitor that determines how much charge can be stored in it for a given potential difference between its terminals. It is equal to the ratio of the electrical charge stored to the potential difference. The SI unit of capacitance is the *farad, but most capacitors have much smaller capacitances, and the microfarad (a millionth of a farad) is the commonly used practical unit.

capacitor or **condenser**, device for storing electric charge, used in electronic circuits; it consists of two or more metal plates separated by an insulating layer called a dielectric (see *capacitance).

capacity alternative term for *volume, generally used to refer to the amount of liquid or gas that may be held in a container. Units of capacity include litre and millilitre (metric); pint and gallon (imperial).

capillarity spontaneous movement of liquids up or down narrow tubes, or capillaries. The movement is due to unbalanced molecular attraction at the boundary between the liquid and the tube. If liquid molecules near the boundary are more strongly attracted to molecules in the material of the tube than to other nearby liquid molecules, the liquid will rise in the tube. If liquid molecules are less attracted to the material of the tube than to other liquid molecules, the liquid will fall.

capillary in biology, narrowest blood vessel in vertebrates measuring 0.008–0.02 mm in diameter, barely wider than a *red blood cell. Capillaries are distributed as **beds**, complex networks connecting *arteries and *veins. The function of capillaries is to exchange materials with their surroundings. For this reason, capillary walls are extremely thin, consisting of a single layer of cells through which nutrients, dissolved gases, and waste products can easily pass. This makes the capillaries the main area of exchange between the fluid (*lymph) bathing body tissues and the blood. They provide a large surface area in order to maximize *diffusion.

All body cells lie close to capillaries so they can receive the food and oxygen they require from the capillaries. Networks of capillaries are especially extensive in certain parts of the body. For example, in the *lungs they surround the air sacs (alveoli), taking up *oxygen and releasing *carbon dioxide. They are also found in the lining of the *gut – in little finger-like projections called villi – where they absorb the products of the *digestion of food.

capillary in physics, a very narrow, thick-walled tube, usually made of glass, such as in a thermometer. Properties of fluids, such as surface tension and viscosity, can be studied using capillary tubes.

capsule in botany, a dry, usually many-seeded fruit formed from an ovary

cross section of a capillary

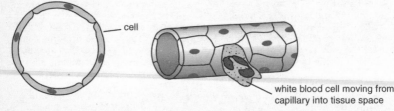

cell

white blood cell moving from capillary into tissue space

capillary The structure of a blood capillary. Capillaries are only one cell thick and white blood cells, which can move independently of the other blood components, can squeeze through them into tissue spaces.

composed of two or more fused *carpels, which splits open to release the seeds. The same term is used for the spore-containing structure of mosses and liverworts; this is borne at the top of a long stalk or seta.

captured rotation or **synchronous rotation**, in astronomy, circumstance in which one body in orbit around another, such as the moon of a planet, rotates on its axis in the same time as it takes to complete one orbit. As a result, the orbiting body keeps one face permanently turned towards the body about which it is orbiting. An example is the rotation of our own *Moon, which arises because of the tidal effects of the Earth over a long period of time.

carat (Arabic *quirrat* 'seed') unit for measuring the mass of precious stones; it is equal to 0.2 g/0.00705 oz, and is part of the troy system of weights. It is also the unit of purity in gold (US 'karat'). Pure gold is 24-carat; 22-carat (the purest used in jewellery) is 22 parts gold and two parts alloy (to give greater strength); 18-carat is 75% gold.

carbide compound of carbon and one other chemical element, usually a metal, silicon, or boron.

Calcium carbide (CaC_2) can be used as the starting material for many basic organic chemical syntheses, by the addition of water and generation of ethyne (acetylene). Some metallic carbides are used in engineering because of their extreme hardness and strength. Tungsten carbide is an essential ingredient of carbide tools and high-speed tools. The 'carbide process' was used during World War II to make organic chemicals from coal rather than from oil.

carbohydrate chemical compounds composed of carbon, hydrogen, and oxygen, with the basic formula $C_m(H_2O)_n$, and related compounds

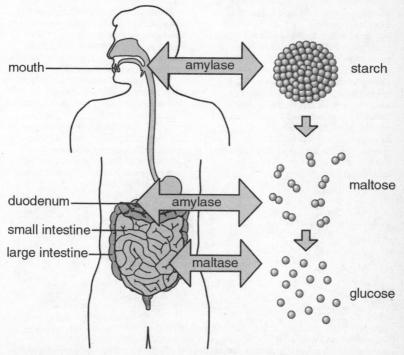

mouth

duodenum

small intestine

large intestine

amylase

starch

maltose

amylase

maltase

glucose

carbohydrate The digestion of carbohydrate. The complex polysaccharide starch is broken down into glucose by the enzymes amylase and maltase (secreted by the small intestine).

with the same basic structure but modified *functional groups. They are important to living organisms and, as sugar and starch, are an important part of a balanced human diet, providing energy for life processes including growth and movement. Excess carbohydrate intake can be converted into fat and stored in the body.

The simplest carbohydrates are sugars (**monosaccharides**, such as glucose and fructose, and **disaccharides**, such as sucrose), which are soluble compounds, some with a sweet taste. When these basic sugar units are joined together in long chains or branching structures they form **polysaccharides**, such as starch and glycogen.

The simple sugar called *glucose is used in living organisms in *respiration to release energy that can be used for life processes. The sugar we use to sweeten foods is a complex sugar called sucrose. Starch is used by plants as a way of storing energy-rich food, and animals that eat plants take advantage of this. It is a common part of the human diet because it is major constituent of wheat, rice, potatoes, and maize. Humans use a carbohydrate called glycogen as an energy store. This is also made from glucose. It is stored in the *liver and *muscles and can be broken down to supply body cells with glucose. The hormone *insulin regulates this process, causing glucose to be taken in the bloodstream to the liver to be converted to glycogen.

carbolic acid common name for the aromatic compound *phenol.

carbon chemical symbol C, (Latin *carbo, carbonaris* 'coal') non-metallic element, atomic number 6, relative atomic mass 12.011. It occurs on its own as diamond, graphite, and as fullerenes (the allotropes), as compounds in carbonaceous rocks such as chalk and limestone, as carbon dioxide in the atmosphere, as hydrocarbons in petroleum, coal, and natural gas, and as a constituent of all organic substances.

In its amorphous form, it is familiar as coal, charcoal, and soot. The atoms of carbon can link with one another in rings or chains, giving rise to innumerable complex compounds. Of the inorganic carbon compounds, the chief ones are carbon dioxide (CO_2), a colourless gas formed when carbon is burned in an adequate supply of air; and carbon monoxide (CO), formed when carbon is oxidized in a limited supply of air. Carbon disulphide (CS_2) is a dense liquid with a sweetish odour when pure. Another group of compounds is the carbon halides, including carbon tetrachloride (*tetrachloromethane, CCl_4).

When added to steel, carbon forms a wide range of alloys with useful properties. In pure form, it is used as a moderator in nuclear reactors; as colloidal graphite it is a good lubricant and, when deposited on a surface in a vacuum, reduces photoelectric and secondary emission of electrons. Carbon is used as a fuel in the form of

graphite   diamond   buckminsterfullerene

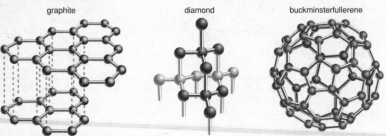

carbon Carbon has three allotropes: diamond, graphite, and the fullerenes. Diamond is strong because each carbon atom is linked to four other carbon atoms. Graphite is made up of layers that slide across one another (giving graphite its qualities as a lubricator); each layer is a giant molecule. In the fullerenes, the carbon atoms form spherical cages. Buckminsterfullerene (shown here) has 60 atoms. Other fullerenes, with 28, 32, 50, 70, and 76 carbon atoms, have also been identified.

coal or coke. The radioactive isotope carbon-14 (half-life 5,730 years) is used as a tracer in biological research and in radiocarbon dating. Analysis of interstellar dust has led to the discovery of discrete carbon molecules, each containing 60 carbon atoms. The C_{60} molecules have been named *buckminsterfullerenes because of their structural similarity to the geodesic domes designed by US architect and engineer Richard Buckminster Fuller.

carbonate CO_3^{2-}, ion formed when carbon dioxide dissolves in water; any salt formed by this ion and another chemical element, usually a metal.

Carbon dioxide (CO_2) dissolves sparingly in water (for example, when rain falls through the air) to form carbonic acid (H_2CO_3), which unites with various basic substances to form carbonates. Calcium carbonate ($CaCO_3$) (chalk, limestone, and marble) is one of the most abundant carbonates known, being a constituent of mollusc shells and the hard outer skeletons of crustaceans.

Carbonates give off carbon dioxide when heated or treated with dilute acids.

$$CaCO_{3(s)} \rightarrow CaO_{(s)} + CO_{2(g)}$$

$$CaCo_{3(s)} + 2HCl_{(aq)} \rightarrow CaCl_{2(aq)} + H_2O_{(l)} + CO_{2(g)}$$

The latter reaction is used as the laboratory test for the presence of the ion, as it gives an immediate effervescence, with the gas turning limewater (a solution of calcium hydroxide, $Ca(OH)_2$) milky. See *calcium carbonate.

carbonation   in earth science, a form of *chemical weathering caused by rainwater that has absorbed carbon dioxide from the atmosphere and formed a weak carbonic acid. The slightly acidic rainwater is then capable of dissolving certain minerals in rocks. Water can also pick up acids when it passes through soil. This water – enriched with organic acid – is also capable of dissolving rock. *Limestone is particularly vulnerable to this form of weathering.

carbon cycle   sequence by which *carbon circulates and is recycled through the natural world. Carbon is usually found in a carbon compound of one sort or another and so the carbon cycle is really about the cycling of carbon compounds. Carbon dioxide is released into the atmosphere by most living organisms as a result of *respiration. The CO_2 is taken up and converted into high-energy chemicals – *glucose and other *carbohydrates – during *photosynthesis by plants; the oxygen component is released back into the atmosphere. Some glucose is used by the plant and some is converted into other carbon compounds, making new tissues. However, some of these compounds can be transferred to other organisms. An animal may eat the plant and that animal may be eaten and so on down the food chain. Carbon is also released through the *decomposition of dead plant and animal matter, and the burning of *fossil fuels such as *coal and *oil, which produce carbon dioxide that is released into the atmosphere. The oceans absorb 25–40% of all carbon dioxide released into the atmosphere.

The carbon cycle is in danger of being disrupted by the increased burning of fossil fuels, and the destruction of large areas of tropical forests. The rising levels of carbon dioxide in the atmosphere are probably increasing the temperature on Earth (enhanced *greenhouse effect). It is thought that by limiting the production of carbon dioxide through human activities we can slow the rate at which temperatures on Earth will rise. (See diagram, page 96.)

carbon dioxide CO_2, colourless, odourless gas, slightly soluble in *water, and denser than air. Discovered by Scottish scientist Joseph Black (1727–1799) in 1754. It is formed by the complete oxidation of carbon. Carbon dioxide is produced by living things during the processes of *respiration and the *decomposition of organic matter, and it is used up during *photosynthesis. It therefore plays a vital role in the *carbon cycle. Solid carbon dioxide is called dry ice, as it changes directly from a solid to a gas (sublimes) on warming. It is used as a coolant in its solid form and in the chemical industry.

Its increasing quantity in the atmosphere contributes to the *greenhouse effect and *global warming. Britain has 1% of the world's population, yet it produces 3% of CO_2 emissions; the USA has 5% of the world's population and produces 25% of CO_2 emissions.

carbon fibre fine, black, silky, continuous filament of pure carbon produced by heating organic fibres, such as cellulose, in an inert atmosphere, and used for reinforcing plastics, epoxy, and polyester resins. The resulting composite is very stiff and, weight for weight, has four times the strength of high-tensile steel. It is used in the aerospace industry, cars, and electrical and sports equipment.

carbonic acid H_2CO_3, weak, dibasic acid formed by dissolving carbon

dioxide in water:

$$H_2O + CO_2 \leftrightarrow H_2CO_3$$

It forms two series of salts: *carbonates and *hydrogencarbonates. Fizzy drinks are made by dissolving carbon dioxide in water under pressure; soda water is a solution of carbonic acid.

Carboniferous period period of geological time roughly 362.5 to 290 million years ago, the fifth period of the Palaeozoic Era. In the USA it is divided into two periods: the Mississippian (lower) and the Pennsylvanian (upper).

Typical of the lower-Carboniferous rocks are shallow-water *limestones, while upper-Carboniferous rocks have *delta deposits with *coal (hence the name). Amphibians were abundant, and reptiles evolved during this period.

carbon monoxide CO, colourless, odourless gas formed when carbon is

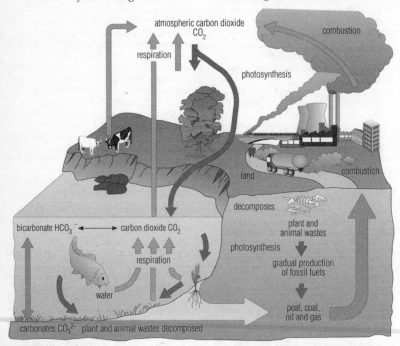

carbon cycle The carbon cycle is necessary for the continuation of life. Since there is only a limited amount of carbon in the Earth and its atmosphere, carbon must be continuously recycled if life is to continue. Other chemicals necessary for life – nitrogen, sulphur, and phosphorus, for example – also circulate in natural cycles.

oxidized in a limited supply of air. It is a poisonous constituent of car exhaust fumes, forming a stable compound with haemoglobin in the blood, thus preventing the haemoglobin from transporting oxygen to the body tissues.

In industry, carbon monoxide is used as a reducing agent in metallurgical processes – for example, in the extraction of iron in *blast furnaces – and is a constituent of cheap fuels such as water gas. It burns in air with a luminous blue flame to form carbon dioxide.

carbon sequestration disposal of carbon dioxide waste in solid or liquid form. From 1993 energy conglomerates such as Shell, Exxon, and British Coal have been researching ways to reduce their carbon dioxide emissions by developing efficient technologies to trap the gas and store it securely – for example, by burying it or dumping it in the oceans. See also *greenhouse effect.

carbon tetrachloride former name for the chlorinated organic compound *tetrachloromethane.

carbonyl group in organic chemistry, the functional group $>C=O$, which is found in *aldehydes and *ketones.

Carborundum trademark for a very hard, black abrasive, consisting of silicon carbide (SiC), an artificial compound of carbon and silicon. It is harder than corundum but not as hard as *diamond.

It was first produced in 1891 by US chemist Edward Acheson (1856–1931).

carboxyl group in organic chemistry, the acidic functional group $-C=OOH$ that determines the properties of fatty acids (carboxylic acids) and amino acids. One of the oxygen atoms has a double bond with carbon and the other has a single bond.

carboxylic acid organic acid containing the carboxyl group $(-COOH)$ attached to another group (R), which can be hydrogen (giving methanoic acid, $HCOOH$) or a larger group (up to 24 carbon atoms). When R is a straight-chain alkyl group (such as CH_3 or CH_3CH_2), the acid is known as a *fatty acid. Examples of carboxylic acids include ethanoic (acetic) acid, found in vinegar, malic acid, found in

rhubarb, and citric acid, contained in oranges and lemons.

carburation any process involving chemical combination with carbon, especially the mixing or charging of a gas, such as air, with volatile compounds of carbon (petrol, kerosene, or fuel oil) in order to increase potential heat energy during combustion. Carburation applies to combustion in the cylinders of reciprocating petrol engines of the types used in aircraft, road vehicles, or marine vessels. The device by which the liquid fuel is atomized and mixed with air is called a **carburettor**.

carcinogen any agent that increases the chance of a cell becoming cancerous (see *cancer), including various chemical compounds, some viruses, X-rays, and other forms of ionizing radiation. The term is often used more narrowly to mean chemical carcinogens only.

carcinogenesis in medicine, the means by which the changes responsible for the development of *cancer are brought about.

carcinoma malignant *tumour arising from the skin, the glandular tissues, or the mucous membranes that line the gut and lungs.

cardiac arrest sudden failure of the pumping action of the heart. The victim loses consciousness and stops breathing. Without resuscitation, death follows within a few minutes. Most hospitals have cardiac arrest teams, specially trained in cardiopulmonary resuscitation and capable of an immediate response. If breathing and heartbeat are not restarted within three minutes, the brain is irreversibly damaged.

cardinal number in mathematics, one of the series of numbers 0, 1, 2, 3, 4, ... Cardinal numbers relate to quantity, whereas ordinal numbers (first, second, third, fourth, ...) relate to order.

cardioid heart-shaped curve traced out by a point on the circumference of a circle, resulting from the circle rolling around the edge of another circle of the same diameter.

cardiology medical speciality concerned with diagnosis and treatment of heart disease.

cardiomyopathy chronic disorder of the heart muscle. It may be inherited or it may arise from a range of conditions, including viral infection, *muscular dystrophy and alcoholism. It may cause enlargement of the heart as well as *heart failure and cardiac *arrhythmia, both of which can be treated. Some patients may be suitable for a heart transplant.

caries decay and disintegration, usually of the substance of teeth (cavity) or bone, caused by acids produced when the bacteria that live in the mouth break down sugars in the food. Fluoride, a low sugar intake, and regular brushing are all protective. Caries form mainly in the 45 minutes following consumption of sugary food.

carnivore organism that eats other animals. In zoology, a mammal of the order Carnivora. Carnivores have the greatest range of body size of any mammalian order, from the 100 g/3.5 oz weasel to the 800 kg/1,764 lb polar bear. The characteristics of the Carnivora are sharp teeth, small incisors, a well-developed brain, a simple stomach, reduced or absent caecum, and incomplete or absent clavicles (collarbones); there are never less than four toes on each foot; the scaphoid and lunar bones are fused in the hand; and the claws are generally sharp and powerful.

Carnot cycle series of changes in the physical condition of a gas in a reversible heat engine, necessarily in the following order: (1) isothermal expansion (without change of temperature), (2) adiabatic expansion (without change of heat content), (3) isothermal compression, and (4) adiabatic compression. Named after French scientist and military engineer (Nicholas Léonard) Sadi Carnot (1796–1832).

carotene naturally occurring pigment of the carotenoid group. Carotenes produce the orange, yellow, and red colours of carrots, tomatoes, oranges, and crustaceans.

carotid artery one of a pair of major blood vessels, one on each side of the neck, supplying blood to the head.

carpal tunnel syndrome compression of the median nerve at the wrist. It causes pain and numbness in the index and middle fingers and weakness in the thumb. It may require surgery.

carpel female reproductive unit in flowering plants (*angiosperms). It usually comprises an *ovary containing one or more ovules, the stalk or style, and a *stigma at its top which receives the pollen. A flower may have one or more carpels, and they may be separate or fused together. Collectively the carpels of a flower are known as the *gynoecium.

carpus (Greek *karpos* 'wrist') the eight bones of the wrist, articulating between the bones of the lower arm (radius and ulna) and the metacarpals of the hand. In humans there are eight small bones in two irregular rows of four. The upper row articulates with the radius, the lower with the metacarpal bones of the hand. Rudiments of carpal bones are found in all mammals.

carrier in medicine, anyone who harbours an infectious organism without ill effects but can pass the infection to others. The term is also applied to those who carry a recessive gene for a disease or defect without manifesting the condition.

Cartesian coordinates in *coordinate geometry, components used to define the position of a point by its perpendicular distance from a set of two or more axes, or reference lines. For a two-dimensional area defined by two axes at right angles (a horizontal x-axis and a vertical y-axis), the coordinates of a point are given by its perpendicular distances from the y-axis and x-axis, written in the form (x, y). For example, a point P that lies three units from the y-axis and four units from the x-axis has Cartesian coordinates (3, 4) (see *abscissa and *ordinate).

The Cartesian coordinate system can be extended to any finite number of dimensions (axes), and is used thus in theoretical mathematics. Coordinates can be negative numbers, or a positive and a negative; for example (-4, -7), where the point would be to the left of and below zero on the axes. In three-dimensional coordinate geometry, points are located with reference to a third, z-axis, mutually at right angles to the x and y axes.

Cartesian coordinates are named after the French mathematician, René Descartes (1596–1650). The system is useful in creating technical drawings of machines or buildings, and in computer-aided design (*CAD).

cartilage　flexible bluish-white *connective tissue made up of the protein collagen. In cartilaginous fish it forms the skeleton; in other vertebrates it forms the greater part of the embryonic skeleton, and is replaced by *bone in the course of development, except in areas of wear such as bone endings, and the discs between the backbones. It also forms structural tissue in the larynx, nose, and external ear of mammals.

cartography　art and practice of drawing maps, originally with pens and drawing boards, but now mostly with computer-aided drafting programs.

casein　main protein of milk, from which it can be separated by the action of acid, the enzyme rennin, or bacteria (souring); it is also the main protein in cheese. Casein is used as a protein supplement in the treatment of malnutrition. It is used commercially in cosmetics, glues, and as a sizing for coating paper.

case-sensitive　term describing a system that distinguishes between capitals and lower-case letters. Domain names and Internet addresses are typically not case-sensitive; however, a particular system may be case-sensitive for user IDs.

cast iron　cheap but invaluable constructional material, most commonly used for car engine blocks. Cast iron is partly refined pig (crude) *iron, which is very fluid when molten and highly suitable for shaping by casting; it contains too many impurities (for example, carbon) to be readily shaped in any other way. Solid cast iron is heavy and can absorb great shock but is very brittle.

castration　removal of the sex glands (either ovaries or testes). Male domestic animals may be castrated to prevent reproduction, to make them larger or more docile, or to eradicate disease.

catabolism　in biology, the destructive part of *metabolism where living tissue is changed into energy and waste products. It is the opposite of *anabolism. It occurs continuously in the body, but is accelerated during many disease processes, such as fever, and in starvation.

cataclastic rock　metamorphic rock, such as a breccia, containing angular fragments of preexisting rock produced by the grinding and crushing action (cataclasis) of *faults.

catalyst　substance that alters the speed of, or makes possible, a chemical or biochemical reaction but remains unchanged at the end of the reaction. *Enzymes are natural biochemical catalysts. In practice most catalysts are used to speed up reactions. The catalysts used in the chemical industry are often *transition metals or their compounds. According to the *collision theory, particles must collide before they can react, and the colliding particles must contain enough energy to cause bonds to break. Catalysts increase the *rate of reaction by lowering the amount of energy needed for a successful collision. Therefore, more collisions will be successful.

catalytic converter　device fitted to the exhaust system of a motor vehicle in order to reduce toxic emissions from the engine. It converts the harmful exhaust products that cause *air pollution to relatively harmless ones. It does this by passing them over a mixture of catalysts coated on a metal or ceramic honeycomb (a structure that increases the surface area and therefore the amount of active catalyst with which the exhaust gases will come into contact). **Oxidation catalysts** (small amounts of precious palladium and platinum metals) convert hydrocarbons (unburnt fuel) and carbon monoxide into carbon dioxide and water, but do not affect nitrogen oxide emissions. **Three-way catalysts** (platinum and rhodium metals) also convert nitrogen oxide gases into nitrogen and oxygen.

cataract　eye disease in which the crystalline lens or its capsule becomes cloudy, causing blindness. Fluid accumulates between the fibres of the lens and gives place to deposits of *albumin. These coalesce into rounded bodies, the lens fibres break down, and areas of the lens or the lens capsule become filled with opaque products of degeneration. The condition is estimated to have blinded more than 25 million people worldwide, and 150,000 in the UK.

catastrophe theory mathematical theory developed by René Thom in 1972, in which he showed that the growth of an organism proceeds by a series of gradual changes that are triggered by, and in turn trigger, large-scale changes or 'catastrophic' jumps. It also has applications in engineering – for example, the gradual strain on the structure of a bridge that can eventually result in a sudden collapse – and has been extended to economic and psychological events.

catastrophism theory that the geological features of the Earth were formed by a series of sudden, violent 'catastrophes' beyond the ordinary workings of nature. The theory was largely the work of Georges Cuvier. It was later replaced by the concepts of uniformitarianism and *evolution.

catatonia condition characterized by a range of abnormal movements or postures associated with mental illness. The condition may be described as **excited**, with constant restlessness and repetition of purposeless actions, or **inhibited**, with stupor, fixed posture and catalepsy. It is sometimes a feature of schizophrenia but is also seen in other disorders.

catecholamine chemical that functions as a *neurotransmitter or a *hormone. Dopamine, adrenaline (epinephrine), and noradrenaline (norepinephrine) are catecholamines.

catechu extract of the leaves and shoots of *Acacia catechu*, an East Indian plant. It is rich in tannic acid, which is released slowly, a property that makes it a useful intestinal astringent in diarrhoea.

cathode in chemistry, the negative electrode of an electrolytic cell, towards which positive particles (cations), usually in solution, are attracted. See *electrolysis. A cathode is given its negative charge by connecting it to the negative side of an external electrical supply.

cathode in electronics, the part of an electronic device in which electrons are generated. In a thermionic valve, electrons are produced by the heating effect of an applied current; in a photocell, they are produced by the interaction of light and a semiconducting material. The cathode is kept at a negative potential relative to the device's other electrodes (anodes) in order to ensure that the liberated electrons stream away from the cathode and towards the anodes.

cathode ray stream of fast-moving electrons that travel from a cathode (negative electrode) towards an anode (positive electrode) in a vacuum tube. They carry a negative charge and can be deflected by electric and magnetic fields. Cathode rays focused into fine beams of fast electrons are used in cathode-ray tubes, the electrons' *kinetic energy being converted into light energy as they collide with the tube's fluorescent screen.

cathode-ray oscilloscope CRO, instrument used to measure electrical potentials or voltages that vary over time and to display the waveforms of electrical oscillations or signals. Readings are displayed graphically on the screen of a *cathode-ray tube.

The CRO is used as a *voltmeter with any voltage change shown on screen by an up (positive) or down (negative) movement of a bright dot. This dot is produced by a beam of electrons hitting the phospor layer on the inside of the screen. Over time the dot traces a graph across the screen showing voltage change against time. If voltage remains constant the graph consists of a horizontal line.

cathode-ray tube CRT, vacuum tube in which a beam of electrons is produced and focused onto a fluorescent screen. The electrons' kinetic energy is converted into light energy as they collide with the screen. It is an essential component of television receivers, computer visual display units, and *oscilloscopes.

The screen of the CRT is coated on the inside with phosphor, which emits light whe   ruck by an electron beam. The tube itself is glass and coated inside with a black graphite conducting paint, which is connected to one of three anodes. A heated filament heats a metal-oxide coated cathode that emits electrons which pass through a positively charged anode that is held at several thousand volts and accelerates the electrons to a high speed beam. The electrons accumulate on the phosphor of the screen and repel each other back to the conducting graphite paint completing a circuit.

cation *ion carrying a positive charge. During *electrolysis, cations in the electrolyte move to the cathode (negative electrode). Cations are formed from *atoms by loss of electrons during *ionic bonding. *Metals form cations.

CAT scan or **CT scan**; acronym for **computerized axial tomography scan**, sophisticated method of X-ray imaging. Quick and noninvasive, CAT scanning is used in medicine as an aid to diagnosis, helping to pinpoint problem areas without the need for exploratory surgery. It is also used in archaeology to investigate mummies.

caustic soda former name for *sodium hydroxide (NaOH).

Cavendish experiment measurement of the gravitational attraction between large and small lead spheres, which enabled English physicist and chemist Henry Cavendish (1731–1810) to calculate in 1798 a mean value for the mass and density of Earth, using Isaac Newton's law of universal gravitation.

cavitation in hydraulics, *erosion of rocks caused by the forcing of air into cracks. Cavitation results from the pounding of waves on the coast and the swirling of turbulent river currents, and exerts great pressure, eventually causing rocks to break apart. The process is particularly common at waterfalls, where the turbulent falling water contains many air bubbles, which burst and send shock waves into the rocks of the river bed and banks. In addition, as water is forced into cracks in the rock, air within the crack is compressed and literally

explodes, helping to break down the rock.

cavitation in earth science, formation of cavities containing a partial vacuum in fluids at high velocities, produced by propellers or other machine parts in hydraulic engines, in accordance with *Bernoulli's principle. When these cavities collapse, pitting, vibration, and noise can occur in the metal parts in contact with the fluids.

cavity in dentistry, decay (*caries) of tooth enamel by the acids produced by mouth bacteria. Continuing decay undermines the inner tooth and attacks the nerve, causing toothache. Measures can be taken to save teeth by cleaning out the decay (drilling) and filling the tooth with a plastic substance such as silver amalgam or covering the cavity with an inlay or crown.

cc symbol for *cubic centimetre; abbreviation for carbon copy.

CCITT abbreviation for Comité Consultatif International Téléphonique et Télégraphique.

cd in physics, symbol for *candela, the SI unit of luminous intensity.

CD-I or **CD-i**; abbreviation for compact disc-interactive, compact disc developed by Philips for storing a combination of video, audio, text, and pictures. It was intended principally for the consumer market to be used in systems using a combination of computer and television. It flopped as a consumer system but is still used in education and training.

CD-quality sound digitized sound at 44.1 KHz and 16 bits, the standard defined in ISO 10149, known as the Red Book. CD-quality sound was designed to be the minimum standard required to reproduce every sound the human ear can hear. Most audio CDs are recorded to this level.

CD-R abbreviation for compact disc-recordable, compact disc on which data can be recorded (compare with *CD-ROM).

CD-ROM acronym for Compact-Disc Read-Only Memory, computer storage device developed from the technology of the audio compact disc. It consists of a plastic-coated metal disk, on which binary digital information is etched in the form of microscopic pits. This can then be read optically by passing a laser beam over the disk. CD-ROMs typically hold over 600 *megabytes of data, and are used in distributing large amounts of text, graphics, audio, and video, such as encyclopedias, catalogues, technical manuals, and games.

CD-ROM XA abbreviation for CD-ROM extended architecture, in computing, a set of standards for storing multimedia information on CD-ROM. Developed by Philips, Sony, and Microsoft, it is a partial development of the *CD-I standard. It interleaves data (as in CD-I) so that blocks of audio data are sandwiched between blocks of text, graphics, or video. This allows parallel streams of data to be handled, so that information can be seen and heard simultaneously.

CD-RW abbreviation for compact disc-rewritable, compact disc on which data can be recorded (compare with *CD-ROM, compact disc read-only memory); unlike *CD-R disks, stored data can be erased and CD-RW disks reused.

celestial sphere imaginary sphere surrounding the Earth, on which the celestial bodies seem to lie. The positions of bodies such as stars, planets, and galaxies are specified by their coordinates on the celestial sphere. The equivalents of latitude and longitude on the celestial sphere are

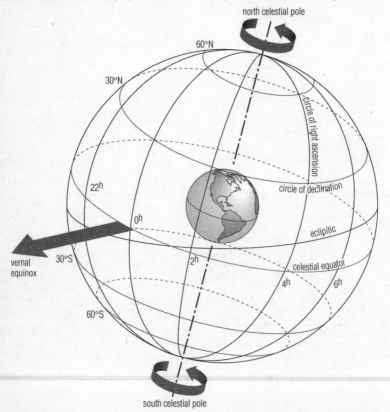

celestial sphere The main features of the celestial sphere. Declination runs from 0° at the celestial equator to 90° at the celestial poles. Right ascension is measured in hours eastwards from the vernal equinox, one hour corresponding to 15° of longitude.

called declination and right ascension (which is measured in hours from 0 to 24). The **celestial poles** lie directly above the Earth's poles, and the **celestial equator** lies over the Earth's Equator. The celestial sphere appears to rotate once around the Earth each day, actually a result of the rotation of the Earth on its axis.

cell in biology, the basic unit of a living organism. It is the smallest unit capable of independent existence. In organisms, other than the smallest ones, the body of the organism is made up of several cells or many cells. A single cell, therefore, is the smallest unit that shows characteristic features of life, such as *reproduction, *growth, *respiration, response to environmental stimuli, and the ability to take in *mineral salts. *Viruses are particles that are not cells. A virus can only reproduce by 'taking over' a cell from another organism. This cell often dies as a result of making many new virus particles.

Some organisms are composed of only one cell, including many bacteria and some fungi, such as yeast. Single-cell organisms are termed **unicellular**, while plants and animals which contain many cells are termed **multicellular** organisms. Organisms such as human beings consist of billions of cells. In organisms made of many cells, groups of cells are specialized to carry out specific functions and are organized into tissues and organs.

Cells always have a *cell membrane around them and *cytoplasm inside, and normally a *nucleus. There are differences between plant cells and animal cells; for example, plant cells have a *cell wall made of cellulose outside their cell membrane. This helps to explain why plants look so different from animals.

*Cell division in plants and animals takes place by *mitosis (growth or asexual reproduction) or by *meiosis (sexual reproduction). New cells produced by mitosis are needed to replace cells that die and some cells live only a short time, such as white blood cells which live for only a few days. Within the human body, about 3 billion cells die every minute. Many cells may be lost in normal activities – for example, human skin cells are

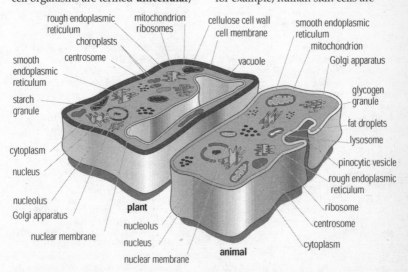

cell structure Typical plant and animal cell. Plant and animal cells share many structures, such as ribosomes, mitochondria, and chromosomes, but they also have notable differences: plant cells have chloroplasts, a large vacuole, and a cellulose cell wall. Animal cells do not have a rigid cell wall but have an outside cell membrane only.

constantly being worn off and have to be replaced by new cells produced by mitosis. The new cells needed during growth are also produced by mitosis.

In respiration a cell uses chemicals that it has taken in, as nutrients, and breaks them down to release *energy that the cell can use for its life processes.

cell in computing, a single reference area on a *spreadsheet program grid. It contains information, a label, or a formula. Each cell reference is unique and is determined according to its position in the spreadsheet grid, using letters to find the column and numbers to find the row.

cell differentiation in developing embryos, the process by which cells acquire their specialization, such as heart cells, muscle cells, skin cells, and brain cells. The seven-day-old human pre-embryo consists of thousands of individual cells, each of which is destined to assist in the formation of individual organs in the body.

Research has shown that the eventual function of a cell, in for example, a chicken embryo, is determined by the cell's position. The embryo can be mapped into areas corresponding with the spinal cord, the wings, the legs, and many other tissues. If the embryo is relatively young, a cell transplanted from one area to another will develop according to its new position. As the embryo develops the cells lose their flexibility and become unable to change their destiny.

cell division process by which a *cell divides. Cells are the basic units of life and they carry out basic functions that are characteristic of living organisms, such as growth and reproduction. Both growth and reproduction usually involve cell division. In plants and animals reproduction can be either *sexual reproduction or *asexual reproduction. Both involve cell division. In sexual reproduction the cell division that is used is *meiosis. In asexual reproduction it is *mitosis that is used. In both forms of cell division, the chemical carrying inherited information, *DNA, has to be copied before division. Rarely, mistakes occur

causing *mutations, but it is normally done accurately. Copying the DNA results in the duplication of structures called *chromosomes in the nucleus. In cell division, the duplicated chromosomes are separated from each other into daughter cells.

In sexual reproduction male and female gametes combine. Gametes are produced by meiosis cell division. Usually the male and female gametes are produced by two different individuals. If so, sexual reproduction combines inherited information from the two parents. Most animals and plants reproduce sexually, though many plants also reproduce asexually.

The male and female sex organs of a plant are usually found in the flower. Many flowers contain both male and female organs. The male gametes of a plant are inside the pollen grains, and the female gametes are inside the ovules.

In mammals, the male gametes are called *sperm and are made in the testes. The female gametes are eggs (see *ovum) and are made in the ovaries.

cell, electrical or **voltaic cell** or **galvanic cell**, device in which chemical energy is converted into electrical energy; the popular name is *'battery', but this strictly refers to a collection of cells in one unit. The reactive chemicals of a **primary cell** cannot be replenished, whereas **secondary cells** – such as storage batteries – are rechargeable: their chemical reactions can be reversed and the original condition restored by applying an electric current. It is dangerous to attempt to recharge a primary cell.

cell, electrolytic device to which electrical energy is applied in order to bring about a chemical reaction; see *electrolysis.

cell membrane or **plasma membrane**, thin layer of protein and fat surrounding cells that keeps the cells together and controls substances passing between the cytoplasm and the intercellular space. The cell membrane is semipermeable, allowing some substances to pass through and some not. Generally, small molecules such as water, glucose, and amino acids can

basic principles

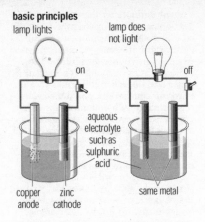

lamp lights

lamp does not light

on

off

aqueous electrolyte such as sulphuric acid

copper anode   zinc cathode

same metal

a simple cell

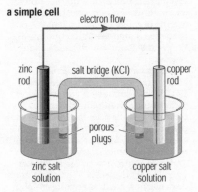

electron flow

zinc rod   salt bridge (KCl)   copper rod

porous plugs

zinc salt solution

copper salt solution

cell, electrical When electrical energy is produced from chemical energy using two metals acting as electrodes in an aqueous solution, it is sometimes known as a galvanic cell or voltaic cell. Here the two metals copper (+) and zinc (−) are immersed in dilute sulphuric acid, which acts as an electrolyte. If a light bulb is connected between the two, an electric current will flow with bubbles of gas being deposited on the electrodes in a process known as polarization.

penetrate the membrane, while large molecules, such as starch, cannot. Substances often cross the membrane by *diffusion, a spontaneous passage of molecules. Water movement across the membrane is a special case of diffusion known as *osmosis.

Membranes also play a part in *active transport, hormonal response (see *hormones), and cell metabolism.

cellular modem type of *modem that connects to a mobile phone for the wireless transmission of data.

celluloid transparent or translucent, highly flammable, plastic material (a *thermoplastic) made from cellulose nitrate and camphor. It was once used for toilet articles, novelties, and photographic film, but has now been replaced by the nonflammable substance *cellulose acetate.

cellulose complex *carbohydrate composed of long chains of glucose units, joined by chemical bonds called glycosidic links. It is the principal constituent of the cell wall of higher plants, and a vital ingredient in the diet of many *herbivores. Molecules of cellulose are organized into long, unbranched microfibrils that give support to the cell wall. No mammal produces the enzyme cellulase, necessary for digesting cellulose; mammals such as rabbits and cows are only able to digest grass because the bacteria present in their gut can manufacture it.

Cellulose is the most abundant substance found in the plant kingdom. It has numerous uses in industry: in rope-making; as a source of textiles (linen, cotton, viscose, and acetate) and plastics (cellophane and celluloid); in the manufacture of nondrip paint; and in such foods as whipped dessert toppings.

Japanese chemists produced the first synthetic cellulose in 1996 and the gene for the plant enzyme that makes cellulose was identified by Australian biologists in 1998.

cellulose acetate or **cellulose ethanoate**, chemical (an *ester) made by the action of acetic acid (ethanoic acid) on cellulose. It is used in making transparent film, especially photographic film; unlike its predecessor, celluloid, it is not flammable.

cellulose nitrate or **nitrocellulose**, series of esters of cellulose with up to three nitrate (NO_3) groups per monosaccharide unit. It is made by the action of concentrated nitric acid on cellulose (for example, cotton waste) in the presence of concentrated sulphuric acid. Fully nitrated cellulose (gun cotton) is explosive, but esters with fewer nitrate groups were once used in making lacquers, rayon, and plastics, such as coloured and photographic film, until replaced by the nonflammable cellulose acetate. *Celluloid is based on cellulose nitrate.

cell wall tough outer surface of the cell in plants. It is constructed from a mesh of *cellulose and is very strong and only very slightly elastic so that it protects the cell and holds it in shape. Most living cells are turgid (swollen with water; see *turgor). Water is absorbed by osmosis causing the cell to expand and develop an internal hydrostatic pressure (wall pressure) that acts against the cellulose wall. The result of this turgor pressure is to give the cell, and therefore the plant, rigidity. Plants, or sections of plants, that are not woody are particularly reliant on this form of support.

The cellulose in cell walls plays a vital role in global nutrition. No vertebrate is able to produce cellulase, the enzyme necessary for the breakdown of cellulose into sugar. However, most mammalian herbivores rely on cellulose, using secretions from micro-organisms living in the gut to break it down. Humans cannot digest the cellulose of the cell walls; they possess neither the correct gut micro-organisms nor the necessary grinding teeth. However, cellulose still forms a necessary part of the human diet as fibre (roughage). Because it is not broken down it acts as a signal that the gut needs to contract to move the contents on.

The cell walls of *bacteria and fungi are not made of cellulose, but are just as strong. Some antibiotics kill bacteria by weakening the cell wall. Penicillins work in this way.

Celsius scale of temperature, previously called centigrade, in which the range from freezing to boiling of water is divided into 100 degrees, freezing point being 0 degrees and boiling point 100 degrees.

The degree centigrade (°C) was officially renamed Celsius in 1948 to avoid confusion with the angular measure known as the centigrade (one hundredth of a grade). The Celsius scale is named after the Swedish astronomer Anders Celsius, who devised it in 1742 but in reverse (freezing point was 100°; boiling point 0°).

cement any bonding agent used to unite particles in a single mass or to cause one surface to adhere to another. **Portland cement** is a powder which when mixed with water and sand or gravel turns into mortar or concrete.

In geology, cement refers to a chemically precipitated material such as carbonate that occupies the interstices of clastic rocks.

The term 'cement' covers a variety of materials, such as fluxes and pastes, and also bituminous products obtained from tar. In 1824 English bricklayer Joseph Aspdin (1779–1855) created and patented the first Portland cement, so named because its colour in the hardened state resembled that of Portland stone, a limestone used in building.

Cement is made by heating limestone (calcium carbonate) with clay (which contains a variety of silicates along with aluminium). This produces a grey powdery mixture of calcium and aluminium silicates. On addition of water, a complex series of reactions occurs and calcium hydroxide is produced. Cement sets by losing water.

Cenozoic Era or **Caenozoic**, era of geological time that began 65 million years ago and continues to the present day. It is divided into the Tertiary and Quaternary periods. The Cenozoic marks the emergence of mammals as a dominant group, and the rearrangement of continental masses towards their present positions.

centigrade former name for the *Celsius temperature scale.

centimetre symbol cm, unit of length equal to one-hundredth of a metre.

central nervous system CNS, brain and spinal cord, as distinct from other components of the *nervous system. The CNS integrates all nervous function.

In invertebrates it consists of a paired ventral nerve cord with concentrations of nerve-cell bodies, known as *ganglia in each segment, and a small brain in the head. Some simple invertebrates, such as sponges and jellyfishes, have no CNS but a simple network of nerve cells called a **nerve net**.

central processing unit CPU, main component of a computer, the part that executes individual program instructions and controls the operation of other parts. It is sometimes called the central processor or, when contained on a single integrated circuit, a microprocessor.

centre of gravity the point in an object about which its weight is evenly balanced. In a uniform gravitational field, this is the same as the *centre of mass.

centre of mass point in or near an object at which the whole mass of the object may be considered to be concentrated. A symmetrical homogeneous object such as a sphere or cube has its centre of mass at its geometrical centre; a hollow object

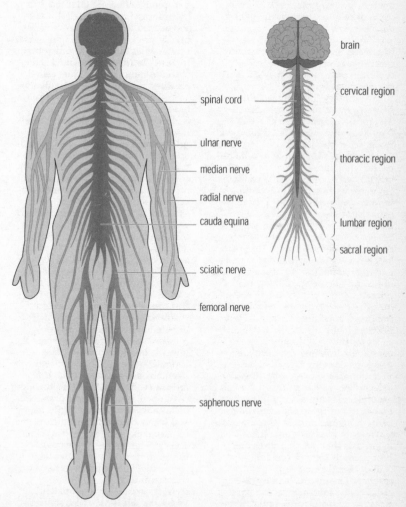

central nervous system The central nervous system (CNS) with its associated nerves.

(such as a cup) may have its centre of mass in space inside the hollow.

For an object to be in stable equilibrium, a vertical line down through its centre of mass must run within the boundaries of its base; if tilted until this line falls outside the base, the object becomes unstable and topples over.

centrifugal force in physics, apparent force arising for an observer moving with a rotating system. For an object of mass m moving with a velocity v in a circle of radius r, the centrifugal force F equals mv^2/r (outwards).

centrifuge apparatus that rotates containers at high speeds, creating centrifugal forces. One use is for separating mixtures of substances of different densities.

A laboratory centrifuge is used to separate small amounts of suspension. Test tubes of suspension are spun around very fast so that the solid gets flung to the bottom. The mixtures are usually spun horizontally in balanced containers ('buckets'), and the rotation sets up centrifugal forces, causing their components to separate according to their densities. A common example is the separation of the lighter plasma from the heavier blood corpuscles in certain blood tests. The **ultracentrifuge** is a very high-speed centrifuge, used in biochemistry for separating *colloids and organic substances; it may operate at several million revolutions per minute. The centrifuges used in the industrial separation of cream from milk, and yeast from fermented wort (infused malt), operate by having mixtures pumped through a continually rotating chamber, the components being tapped off at different points. Large centrifuges are used for physiological research – for example, in astronaut training where bodily responses to gravitational forces many times the normal level are tested.

centriole structure found in the *cells of animals that plays a role in the processes of *meiosis and *mitosis (cell division).

centripetal force force that acts radially inwards on an object moving in a curved path. For example, with a weight whirled in a circle at the end of a length of string, the centripetal force is the tension in the string. For an object of mass m moving with a velocity v in a circle of radius r, the centripetal force F equals mv^2/r (inwards). The reaction to this force is the *centrifugal force.

centromere part of the *chromosome where there are no *genes. Under the microscope, it usually appears as a constriction in the strand of the chromosome, and is the point at which the spindle fibres are attached during *meiosis and *mitosis (cell division).

centrosome cell body that contains the *centrioles. During cell division the centrosomes organize the microtubules to form the spindle that divides the chromosomes into daughter cells. Centrosomes were first described in 1887, independently by German biologist Theodor Boveri (1862–1915) and Belgian biologist Edouard van Beneden.

cephalosporin any of a class of broad-spectrum antibiotics derived from a fungus (genus *Cephalosporium*). They are similar to penicillins and are used on penicillin-resistant infections.

The first cephalosporin was extracted from sewage-contaminated water, and other naturally occurring ones have been isolated from moulds taken from soil samples. Side effects include allergic reactions and digestive upsets. Synthetic cephalosporins can be designed to be effective against a particular *pathogen.

cerebellum part of the brain of *vertebrate animals which controls muscle tone, movement, balance, and coordination. It is relatively small in lower animals such as newts and lizards, but large in birds since flight demands precise coordination. The human cerebellum is also well developed, because of the need for balance when walking or running, and for finely coordinated hand movements.

cerebral pertaining to the brain, especially the part known as the *cerebrum, concerned with higher brain functions.

cerebral cortex or **grey matter**, fissured outer layer of the *cerebrum in the brain. Some 3 mm thick in the adult, it is the most sophisticated part of the brain, responsible for all higher functions and for initiating voluntary movement. Anatomists divide it into four lobes, named after the skull plates beneath which they lie: frontal, parietal, temporal, and occipital.

cerebral haemorrhage or **apoplectic fit**, in medicine, a form of *stroke in which there is bleeding from a cerebral blood vessel into the surrounding brain tissue. It is generally caused by degenerative disease of the arteries and high blood pressure. Depending on the site and extent of bleeding, the symptoms vary from transient weakness and numbness to deep coma and death. Damage to the brain is permanent, though some recovery can be made. Strokes are likely to recur.

cerebral hemisphere one of the two halves of the *cerebrum.

cerebral palsy any nonprogressive abnormality of the brain occurring during or shortly after birth. It is caused by oxygen deprivation, injury during birth, haemorrhage, meningitis, viral infection, or faulty development. Premature babies are at greater risk of being born with cerebral palsy, and in 1996 US researchers linked this to low levels of the thyroid hormone thyroxine. The condition is characterized by muscle spasm, weakness, lack of coordination, and impaired movement; or there may be spastic paralysis, with fixed deformities of the limbs. Intelligence is not always affected.

cerebrospinal fluid CSF, clear fluid that buffers the brain and spinal cord. It is present in the subarachnoid space, the four ventricles of the brain and the central canal of the spinal cord. It is secreted in clusters of blood vessels in the ventricles known as choroid plexuses, and is eventually absorbed into the bloodstream. Its normal contents include glucose, salts, proteins and some white cells. Small amounts of CSF may be withdrawn by lumbar puncture for diagnostic tests.

cerebrum part of the vertebrate *brain, formed from the two paired cerebral hemispheres, separated by a central fissure. In birds and mammals it is the largest and most developed part of the

brain. It is covered with an infolded layer of grey matter, the cerebral cortex, which integrates brain functions. The cerebrum coordinates all voluntary activity.

cerium chemical symbol Ce, malleable and ductile, grey, metallic element, atomic number 58, relative atomic mass 140.12. It is the most abundant member of the lanthanide series, and is used in alloys, electronic components, nuclear fuels, and lighter flints. It was discovered in 1804 by the Swedish chemists Jöns Berzelius and Wilhelm Hisinger (1766–1852), and, independently, by German chemist Martin Klaproth. The element was named after the then recently discovered asteroid Ceres.

cermet contraction of ceramics and metal, bonded material containing ceramics and metal, widely used in jet engines and nuclear reactors. Cermets behave much like metals but have the great heat resistance of ceramics. Tungsten carbide, titanium, zirconium bromide, and aluminium oxide are among the ceramics used; iron, cobalt, nickel, and chromium are among the metals.

CERN particle physics research organization founded in 1954 as a cooperative enterprise among European governments. It has laboratories at Meyrin, near Geneva, Switzerland. It houses the world's largest particle *accelerator, the *Large Electron Positron Collider (LEP), operational 1989–2000, with which notable advances were made in *particle physics.

CERT abbreviation for Computer Emergency Response Team.

cerumen or **ear wax**, a protective substance secreted by sebaceous glands in the external ear canal.

cervical cancer *cancer of the cervix (neck of the womb). It can be detected at an early stage through screening by cervical smear.

cervical smear in medicine, removal of a small sample of tissue from the cervix (neck of the womb) to screen for changes implying a likelihood of cancer. The procedure is also known as the **Pap test** after its originator, George Papanicolaou.

cervix (Latin 'neck') abbreviation for **cervix uteri**, the neck of the womb; see *uterus.

CET abbreviation for **Central European Time**.

CFC abbreviation for *chlorofluorocarbon.

CGA abbreviation for colour graphics adapter, in computing, the first colour display system for IBM PCs and compatible machines. It has been superseded by EGA, *VGA, *SVGA, *XGA, and *SXGA.

CGI abbreviation for *common gateway interface.

c.g.s. system system of units based on the centimetre, gram, and second, as units of length, mass, and time, respectively. It has been replaced for scientific work by the *SI units to avoid inconsistencies in definition of the thermal calorie and electrical quantities.

chain reaction in chemistry, a succession of reactions, usually involving *free radicals, where the products of one stage are the reactants of the next. A chain reaction is characterized by the continual generation of reactive substances.

A chain reaction comprises three separate stages: **initiation** – the initial generation of reactive species; **propagation** – reactions that involve reactive species and generate similar or different reactive species; and **termination** – reactions that involve the reactive species but produce only stable, nonreactive substances. Chain reactions may occur slowly (for example, the oxidation of edible oils) or accelerate as the number of reactive species increases, ultimately resulting in explosion.

chain reaction in nuclear physics, a fission reaction that is maintained because neutrons released by the splitting of some atomic nuclei themselves go on to split others, releasing even more neutrons. Such a reaction can be controlled (as in a nuclear reactor) by using moderators to absorb excess neutrons. Uncontrolled, a chain reaction produces a nuclear explosion (as in an atom bomb).

chain rule in mathematics, rule for differentiating the composition of two functions. If $y = f(u)$ and $u = g(x)$ then:

$$\frac{dy}{dx} = \frac{dy}{du} \times \frac{du}{dx}$$

chalk soft, fine-grained, whitish sedimentary rock composed of calcium carbonate, $CaCO_3$, extensively quarried for use in cement, lime, and mortar, and in the manufacture of cosmetics and toothpaste. **Blackboard chalk** in fact consists of *gypsum (calcium sulphate, $CaSO_4.2H_2O$).

Chalk was once thought to derive from the remains of microscopic animals or foraminifera. In 1953, however, it was seen under the electron microscope to be composed chiefly of coccolithophores, unicellular lime-secreting algae, and hence primarily of plant origin. It is formed from deposits of deep-sea sediments called oozes.

Chalk was laid down in the later *Cretaceous period and covers a wide area in Europe. In England it stretches in a belt from Wiltshire and Dorset continuously across Buckinghamshire and Cambridgeshire to Lincolnshire and Yorkshire, and also forms the North and South Downs, and the cliffs of S and SE England.

chance likelihood, or *probability, of an event taking place, expressed as a fraction or percentage. For example, the chance that a tossed coin will land heads up is 50%.

change of state in science, change in the physical state (solid, liquid, or gas) of a material. For instance, melting, boiling, and evaporation and their opposites, solidification and condensation, are changes of state. The former set of changes are brought about by heating or decreased pressure; the latter by cooling or increased pressure.

These changes involve the absorption or release of heat energy, called *latent heat, even though the temperature of the material does not change during the transition between states. See also *states of matter. Changes of state can be explained by the *kinetic theory of matter. In the unusual change of state called **sublimation**, a solid changes directly to a gas without passing through the

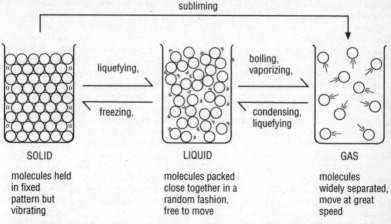

subliming

liquefying,

freezing,

boiling,
vaporizing,

condensing,
liquefying

SOLID

molecules held
in fixed
pattern but
vibrating

LIQUID

molecules packed
close together in a
random fashion,
free to move

GAS

molecules
widely separated,
move at great
speed

change of state The state (solid, liquid, or gas) of a substance is not fixed but varies with changes in temperature and pressure.

liquid state. For example, solid carbon dioxide (dry ice) sublimes to carbon dioxide gas.

chaos theory   or **chaology** or **complexity theory**, branch of mathematics that attempts to describe irregular, unpredictable systems – that is, systems whose behaviour is difficult to predict because there are so many variable or unknown factors. Weather is an example of a chaotic system.

character   one of the symbols that can be represented in a computer. Characters include letters, numbers, spaces, punctuation marks, and special symbols.

characteristic   in mathematics, the integral (whole-number) part of a *logarithm. The fractional part is the mantissa.

character type check   in computing, a *validation check to ensure that an input data item does not contain invalid characters. For example, an input name may be checked to ensure that it contains only letters of the alphabet or an input six-figure date may be checked to ensure it contains only numbers.

charge   see *electric charge.

charge-coupled device   CCD, device for forming images electronically, using a layer of silicon that releases electrons when struck by incoming light. The electrons are stored in

*pixels and read off into a computer at the end of the exposure. CCDs are used in digital cameras, and have now almost entirely replaced photographic film for applications such as astrophotography, where extreme sensitivity to light is paramount.

charged particle beam   high-energy beam of electrons or protons. Such beams are being developed as weapons.

Charles's law   law stating that the volume of a given mass of gas at constant pressure is directly proportional to its absolute temperature (temperature in kelvin). It was discovered by French physicist Jacques Charles in 1787, and independently by French chemist Joseph Gay-Lussac in 1802.

The gas increases by 1/273 (0.003663) of its volume at 0°C for each °C rise of temperature. This means that the coefficient of expansion of all gases is the same. The law is only approximately true.

chelate   chemical compound whose molecules consist of one or more metal atoms or charged ions joined to chains of organic residues by coordinate (or dative covalent) chemical *bonds.

The parent organic compound is known as a **chelating agent** – for example, EDTA (ethylene-diaminetetra-acetic acid), used in chemical analysis.

Chelates are used in analytical chemistry, in agriculture and horticulture as carriers of essential trace metals, in water softening, and in the treatment of thalassaemia by removing excess iron, which may build up to toxic levels in the body. Metalloproteins (natural chelates) may influence the performance of enzymes or provide a mechanism for the storage of iron in the spleen and plasma of the human body.

chemical change change that occurs when two or more substances (reactants) interact with each other, resulting in the production of different substances (products) with different chemical compositions. A simple example of chemical change is the burning of carbon in oxygen to produce carbon dioxide (*combustion). Other types of chemical change include *decomposition, *oxidation, and *reduction.

chemical element alternative name for *element.

chemical equation method of indicating the reactants and products of a chemical reaction by using chemical symbols and formulae. A chemical equation gives two basic pieces of information: (1) the reactants (on the left-hand side) and products (right-hand side); and (2) the reacting proportions (stoichiometry) – that is, how many units of each reactant and product are involved. The equation must balance; that is, the total number of atoms of a particular element on the left-hand side must be the same as the number of atoms of that element on the right-hand side.

$$Na_2CO_3 + 2HCl \rightarrow 2NaCl + CO_2 + H_2O$$

reactants → products

This equation states that one molecule of sodium carbonate combines with two molecules of hydrochloric acid to form two molecules of sodium chloride, one of carbon dioxide, and one of water. Double arrows indicate that the reaction is reversible – in the formation of ammonia from hydrogen and nitrogen, the direction of the reaction depends on the temperature and pressure of the reactants.

$$3H_2 + N_2 \leftrightarrow 2NH_3$$

chemical equilibrium condition in which the products of a chemical reaction are formed at the same rate at which they decompose back into the reactants, so that the concentration of each reactant and product remains constant. It is a *reversible reaction; the reaction can happen in both directions.

For example, in the *Haber process, nitrogen and hydrogen combine to form ammonia, but as the ammonia is formed, it breaks down again into nitrogen and hydrogen.

$$N_2 + 3H_2 \leftrightarrow 2NH_3$$

At equilibrium, the forward and back reactions occur at the same rate. As fast as ammonia is being formed it is broken down into hydrogen and nitrogen, therefore, the amounts of nitrogen, hydrogen, and ammonia remain constant at equilibrium.

The amounts of reactant and product present at equilibrium are defined by the **equilibrium constant** for that reaction and specific temperature.

chemical family collection of elements that have very similar chemical and physical properties. In the *periodic table of the elements such collections are to be found in the vertical columns (groups). The groups that contain the most markedly similar elements are group I, the *alkali metals; group II, the *alkaline-earth metals; group VII, the *halogens; and group 0, the noble or *rare gases.

chemical oxygen demand COD, measure of water and effluent quality, expressed as the amount of oxygen (in parts per million) required to oxidize the reducing substances present.

Under controlled conditions of time and temperature, a chemical oxidizing agent (potassium permanganate or dichromate) is added to the sample of water or effluent under consideration, and the amount needed to oxidize the reducing materials present is measured. From this the chemically equivalent amount of oxygen can be calculated. Since the reducing substances typically include remains of living organisms, COD may be regarded as reflecting the extent to which the sample is polluted. Compare *biological oxygen demand.

chemical symbol see *symbol, chemical.

chemical weathering form of *weathering brought about by chemical attack on rocks, usually in the presence of water. Chemical weathering involves the breakdown of the original minerals within a rock to produce new minerals (such as *clay minerals, *bauxite, and *calcite). The breakdown of rocks occurs because of chemical reactions between the minerals in the rocks and substances in the environment, such as water, oxygen, and weakly acidic rainwater. Some chemicals are dissolved and carried away from the weathering source, while others are brought in.

chemiluminescence the emission of light from a substance as a result of a chemical reaction (rather than raising its temperature). See *luminescence.

chemisorption attachment, by chemical means, of a single layer of molecules, atoms, or ions of gas to the surface of a solid or, less frequently, a liquid. It is the basis of catalysis (see *catalyst) and is of great industrial importance.

chemistry branch of science concerned with the study of the structure and composition of the different kinds of matter, the changes that matter may undergo, and the phenomena which occur in the course of these changes.

*Inorganic chemistry deals with the description, properties, reactions, and preparation of elements and their compounds. *Organic chemistry is the branch of chemistry that deals with the more complex covalent compounds of carbon.*Physical chemistry is concerned with the quantitative explanation of chemical phenomena and reactions, and the measurement of data required for such explanations. This branch studies in particular the movement of molecules and the effects of temperature and pressure, often with regard to gases and liquids.

chemosynthesis method of making *protoplasm (contents of a cell) using the energy from chemical reactions, in contrast to the use of light energy employed for the same purpose in *photosynthesis. The process is used

by certain bacteria, which can synthesize organic compounds from carbon dioxide and water using the energy from special methods of *respiration.

Nitrifying bacteria are a group of chemosynthetic organisms that change free nitrogen into a form that can be taken up by plants; nitrobacteria, for example, oxidize nitrites to nitrates. This is a vital part of the *nitrogen cycle. As chemosynthetic bacteria can survive without light energy, they can live in dark and inhospitable regions, including the hydrothermal vents of the Pacific Ocean. Around these vents, where temperatures reach up to $350°C/662°F$, the chemosynthetic bacteria are the basis of a food web supporting fishes and other marine life.

chemotaxis in biology, the property that certain cells have of attracting or repelling other cells. For example, white blood cells are attracted to the site of infection by the release of substances during certain types of immune response.

chemotherapy any medical treatment with chemicals. It usually refers to treatment of cancer with cytotoxic and other drugs. The term was coined by the German bacteriologist Paul Ehrlich for the use of synthetic chemicals against infectious diseases.

chemotropism movement by part of a plant in response to a chemical stimulus. The response by the plant is termed 'positive' if the growth is towards the stimulus or 'negative' if the growth is away from the stimulus.

chickenpox or **varicella**, common, usually mild disease, caused by a virus of the *herpes group and transmitted by airborne droplets. Chickenpox chiefly attacks children under the age of ten. The incubation period is two to three weeks. One attack normally gives immunity for life.

chimera or **chimaera**, in biology, an organism composed of tissues that are genetically different. Chimeras can develop naturally if a *mutation occurs in a cell of a developing embryo, but are more commonly produced artificially by implanting cells from one organism into the embryo of another.

china clay commercial name for *kaolin.

Chinese medicine a system based on the theory of energy currents in the body associated with the physical organs; there are a number of pathways, called **meridians**, along which energy is considered to flow. Each organ, or group of organs, has its own meridian; and each has a number of associations; for example, one of the five Chinese elements, as well as a time of day, so that if a person is restless or depleted at a particular time daily, identification of the location of the organic imbalance is facilitated. Vitality is identified as *chi* (pronounced 'kee'); and promotion of the regular and vigorous flow of *chi* in the body is the object of medical practice.

chip or **silicon chip**, another name for an *integrated circuit, a complete electronic circuit on a slice of silicon (or other semiconductor) crystal only a few millimetres square.

chitin complex long-chain compound, or *polymer; a nitrogenous derivative of glucose. Chitin is widely found in invertebrates. It forms the *exoskeleton of insects and other arthropods. It combines with protein to form a covering that can be hard and tough, as in beetles, or soft and flexible, as in caterpillars and other insect larvae. It is insoluble in water and resistant to acids, alkalis, and many organic solvents. In crustaceans such as crabs, it is impregnated with calcium carbonate for extra strength.

chlamydia viruslike bacteria which live parasitically in animal cells, and cause disease in humans and birds. Chlamydiae are thought to be descendants of bacteria that have lost certain metabolic processes. In humans, a strain of chlamydia causes trachoma, a disease found mainly in the tropics (a leading cause of blindness); venereally transmitted chlamydiae cause genital and urinary infections.

chlorate any salt derived from an acid containing both chlorine and oxygen and possessing the negative ion ClO^-, ClO_2^-, ClO_3^-, or ClO_4^-. Common chlorates are those of sodium, potassium, and barium. Certain chlorates are used in weedkillers.

chloride Cl^-, negative ion formed when hydrogen chloride dissolves in water, and any salt containing this ion, commonly formed by the action of hydrochloric acid (HCl) on various metals or by direct combination of a metal and chlorine. Sodium chloride (NaCl) is common table salt.

chlorination treatment of water with chlorine in order to disinfect it; also, any chemical reaction in which a chlorine atom is introduced into a chemical compound.

chlorine chemical symbol Cl, (Greek *chloros* 'green') greenish-yellow, gaseous, non-metallic element with a pungent odour, atomic number 17, relative atomic mass 35.453. It is a member of the *halogen group and is widely distributed, in combination with the *alkali metals, as chlorides.

Chlorine was discovered in 1774 by the German chemist Karl Scheele, but English chemist Humphry Davy first proved it to be an element in 1810 and named it after its colour. In nature it is always found in the combined form, as in hydrochloric acid, produced in the mammalian stomach for digestion. Chlorine is obtained commercially by the electrolysis of concentrated brine and is an important bleaching agent and germicide, used for sterilizing both drinking water and swimming pools. As an oxidizing agent it finds many applications in organic chemistry. The pure gas (Cl_2) is a poison and was used in gas warfare in World War I, where its release seared the membranes of the nose, throat, and lungs, causing pneumonia. Chlorine is a component of chlorofluorocarbons (CFCs) and is partially responsible for the depletion of the *ozone layer; it is released from the CFC molecule by the action of ultraviolet radiation in the upper atmosphere, making it available to react with and destroy the ozone. The concentration of chlorine in the atmosphere in 1997 reached just over 3 parts per billion. It was expected to reach its peak in 1999 and then start falling rapidly due to international action to curb ozone-destroying chemicals.

chlorofluorocarbon CFC, a class of **synthetic chemicals** that are odourless, nontoxic, nonflammable, and chemically inert. The first CFC was synthesized in 1892, but no use was found for it until the 1920s. Their stability and apparently harmless properties made CFCs popular as propellants in aerosol cans, as refrigerants in refrigerators and air conditioners, as degreasing agents, and in the manufacture of foam packaging. They are now known to be partly responsible for the destruction of the *ozone layer. In June 1990 representatives of 93 nations, including the UK and the USA, agreed to phase out production of CFCs and various other ozone-depleting chemicals by the end of the 20th century.

When CFCs are released into the atmosphere, they drift up slowly into the stratosphere, where, under the influence of ultraviolet radiation from the Sun, they react with ozone (O_3) to form free chlorine (Cl) atoms and molecular oxygen (O_2), thereby destroying the ozone layer that protects Earth's surface from the Sun's harmful ultraviolet rays. The chlorine liberated during ozone breakdown can react with still more ozone, making the CFCs particularly dangerous to the environment. CFCs can remain in the atmosphere for more than 100 years. Replacements for CFCs are being developed, and research into safe methods for destroying existing CFCs is being carried out.

chloroform or **trichloromethane**; $CHCl_3$, clear, colourless, toxic, carcinogenic liquid with a characteristic pungent, sickly sweet smell and taste, formerly used as an anaesthetic (now superseded by less harmful substances). It is used as a solvent and in the synthesis of organic chemical compounds.

chlorophyll group of pigments including chlorophyll a and chlorophyll b, the green pigments present in *chloroplasts in most plants; it is responsible for the absorption of *light energy during *photosynthesis. The pigment absorbs the red and blue-violet parts of sunlight but reflects the green, thus giving plants their characteristic colour. Other chlorophylls include chlorophyll c (in brown algae) and chlorophyll d (found in red algae).

Chlorophyll is found within chloroplasts, present in large numbers in leaves. Cyanobacteria (blue-green algae) and other photosynthetic bacteria also have chlorophyll, though of a slightly different type. Chlorophyll is similar in structure to *haemoglobin, but with magnesium instead of iron as the reactive part of the molecule. Because magnesium is contained in chlorophyll it is considered an essential plant *mineral salt. Magnesium can often be a part of a mixture of *minerals used as a fertilizer.

chloroplast structure (*organelle) within a plant cell containing the green pigment *chlorophyll. Chloroplasts occur in most cells of green plants that are exposed to light, often in large numbers. Typically, they are shaped like a flattened disc, with a double membrane enclosing the stroma, a gel-like matrix. Within the stroma are stacks of fluid-containing cavities, or vesicles, where *photosynthesis occurs, creating *glucose from carbon dioxide and water to be used in the plant's life processes. Sunlight is absorbed by chlorophyll, providing energy which is transferred to the glucose. The glucose may be converted to starch and stored. Starch can then be converted back to glucose to provide energy for the plant at a later stage.

It is thought that the chloroplasts were originally free-living cyanobacteria (blue-green algae) which invaded larger, non-photosynthetic cells and developed a symbiotic relationship with them. Like *mitochondria, they contain a small amount of DNA and divide by fission. Chloroplasts are a type of *plastid.

choke coil in physics, a coil employed to limit or suppress alternating current without stopping direct current, particularly the type used as a 'starter' in the circuit of fluorescent lighting.

cholecalciferol or **vitamin D**, fat-soluble chemical important in the uptake of calcium and phosphorous for bones. It is found in liver, fish oils, and margarine. It can be produced in the skin, provided that the skin is adequately exposed to sunlight. Lack of vitamin D leads to rickets and other bone diseases.

cholecystectomy surgical removal of the *gall bladder. It is carried out when gallstones or infection lead to inflammation of the gall bladder, which may then be removed either by conventional surgery or by a 'keyhole' procedure.

cholera disease caused by infection with various strains of the bacillus *Vibrio cholerae*, transmitted in contaminated water and characterized by violent diarrhoea and vomiting. It is prevalent in many tropical areas.

cholesterol white, crystalline *sterol found throughout the body, especially in fats, blood, nerve tissue, and bile; it is also provided in the diet by foods such as eggs, meat, and butter. A high level of cholesterol in the blood is thought to contribute to atherosclerosis (hardening of the arteries).

Cholesterol is an integral part of all cell membranes and the starting point for steroid hormones, including the sex hormones. It is broken down by the liver into bile salts, which are involved in fat absorption in the digestive system, and it is an essential component of **lipoproteins**, which transport fats and fatty acids in the blood. **Low-density lipoprotein cholesterol** (LDL-cholesterol), when present in excess, can enter the tissues and become deposited on the surface of the arteries, causing *atherosclerosis. **High-density lipoprotein cholesterol** (HDL-cholesterol) acts as a scavenger, transporting fat and cholesterol from the tissues to the liver to be broken down. The composition of HDL-cholesterol can vary and some forms may not be as effective as others. Blood cholesterol levels can be altered by reducing the amount of alcohol and fat in the diet and by substituting some of the saturated fat for polyunsaturated fat, which gives a reduction in LDL-cholesterol. A 1999 US study of children with high levels of cholesterol found no evidence that controlling cholesterol levels through diet is harmful. Another 1999 US study suggested that cholesterol-lowering

drugs are as beneficial for older men and women as they are for the middle-aged. HDL-cholesterol can be increased by exercise.

choline in medicine, one of the members of the vitamin B complex. The daily requirement in adults, approximately 500 mg, is present in egg yolk, liver, and meat and sufficient choline can be obtained from a normal diet. Choline can be synthesized by the body and deficiency results in a fatty liver.

cholinergic in biology, activity of nerve fibres to release the neurotransmitter *acetylcholine that mediates the transmission of nerve impulses between nerves or between nerves and muscles. Anticholinergic agents, such as pilocarpine, prolong the action of acetylcholine and have a role in the treatment of glaucoma.

chord in geometry, a straight line joining any two points on a curve. The chord that passes through the centre of a circle (its longest chord) is the diameter. The longest and shortest chords of an ellipse (a regular oval) are called the major and minor axes, respectively.

chordate animal belonging to the phylum Chordata, which includes vertebrates, sea squirts, amphioxi, and others. All these animals, at some stage of their lives, have a supporting rod of tissue (notochord or backbone) running down their bodies.

chorea condition featuring involuntary movements of the face muscles and limbs. It is seen in a number of neurological diseases, including *Huntington's chorea.

chromatography (Greek *chromos* 'colour') technique for separating or analysing a mixture of gases, liquids, or dissolved substances. This is brought about by means of two immiscible substances, one of which (**the mobile phase**) transports the sample mixture through the other (**the stationary phase**). The mobile phase may be a gas or a liquid; the stationary phase may be a liquid or a solid, and may be in a column, on paper, or in a thin layer on a glass or plastic support. The components of the mixture are adsorbed or impeded by the stationary phase to different extents and therefore become separated. The technique is used for both qualitative and quantitive analyses in biology and chemistry.

In **paper chromatography**, the mixture separates because the components have differing solubilities in the solvent flowing through the paper and in the chemically bound water of the paper.

In **thin-layer chromatography**, a wafer-thin layer of adsorbent medium on a glass plate replaces the filter paper. The mixture separates because of the differing solubilities of the components in the solvent flowing up the solid layer, and their differing tendencies to stick to the solid (adsorption). The same principles apply in **column chromatography**.

In **gas–liquid chromatography**, a gaseous mixture is passed into a long, coiled tube (enclosed in an oven) filled with an inert powder coated in a liquid. A carrier gas flows through the tube. As the mixture proceeds along the tube it separates as the components dissolve in the liquid to differing extents or stay as a gas. A detector locates the different components as they emerge from the tube. The technique is very powerful, allowing tiny quantities of substances (fractions of parts per million) to be separated and analysed.

Preparative chromatography is carried out on a large scale for the purification and collection of one or more of a mixture's constituents; for example, in the recovery of protein from abattoir wastes.

Analytical chromatography is carried out on far smaller quantities, often as little as one microgram (one-millionth of a gram), in order to identify and quantify the component parts of a mixture. It is used to determine the identities and amounts of amino acids in a protein, and the alcohol content of blood and urine samples. The technique was first used in the separation of coloured mixtures into their component pigments.

chromite $FeCr_2O_4$, iron chromium oxide, the main chromium ore. It is one of the *spinel group of minerals, and crystallizes in dark-coloured octahedra of the cubic system. Chromite is usually found in association with ultrabasic and basic rocks; in Cyprus, for example, it occurs with serpentine, and in South Africa it forms continuous layers in a layered intrusion.

chromium chemical symbol Cr, (Greek *chromos* 'colour') hard, brittle, grey-white, metallic element, atomic number 24, relative atomic mass 51.996. It takes a high polish, has a high melting point, and is very resistant to corrosion. It is used in chromium electroplating, in the manufacture of stainless steel and other alloys, and as a catalyst. Its compounds are used for tanning leather and for *alums. In human nutrition it is a vital trace element. In nature, it occurs chiefly as chrome iron ore or chromite ($FeCr_2O_4$). Kazakhstan, Zimbabwe, and Brazil are sources.

The element was named in 1797 by the French chemist Louis Vauquelin after its brightly coloured compounds.

chromium ore essentially the mineral *chromite, $FeCr_2O_4$, from which chromium is extracted. South Africa and Zimbabwe are major producers.

chromosome structures in a cell *nucleus that carry the many thousands of *genes, in sequence, that determine the characteristics of an organism. There are 46 chromosomes in a normal human cell. Each chromosome normally consists of one very long strand (or molecule) of DNA, coiled and folded to produce a compact structure. The exception is just before cell division when each chromosome contains two strands of DNA, a result of the copying of each molecule of DNA. The point on a chromosome where a particular gene occurs is known as its locus. Most higher organisms have two copies of each chromosome, together known as a **homologous pair** (they are *diploid) but some have only one (they are *haploid). See also *mitosis and *meiosis.

In a working cell chromosomes exist in a less dense form called chromatin and cannot be seen using a laboratory light microscope. However, during cell division they get shorter and fatter and so become visible. When pictures are seen of chromosomes, they have usually been taken in this state during mitosis. In such pictures of human chromosomes it can be seen that most chromosomes in a cell have a matching chromosome of exactly the same size. These are known as homologous pairs. However, one pair is not matched. These are the sex chromosomes – a short, male determining one called the *Y chromosome and the *X chromosome. Males have an X and a Y chromosome and females have two Xs.

In a sex cell (gamete) that has been produced by meiosis, the number of chromosomes is halved. Only one chromosome from each pair of homologous chromosomes is found in a gamete.

artificial chromosomes The first artificial human chromosome was built by US geneticists in 1997. They constructed telomeres, centromeres, and DNA containing genetic information, which they removed from white blood cells, and inserted into human cancer cells. The cells assembled the material into chromosomes. The artificial chromosome was successfully passed onto all daughter cells.

DNA sequencing The first human gene to have its DNA sequenced was chromosome 22. Its 545 genes and 33.5 million bases were sequenced by UK researchers by the end of 1999, leaving only a few gaps. It is one of the smallest human chromosomes, and has been linked to a number of disorders,

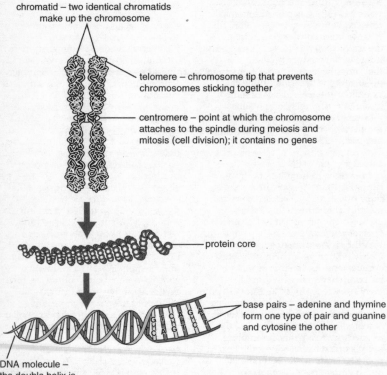

chromatid – two identical chromatids make up the chromosome

telomere – chromosome tip that prevents chromosomes sticking together

centromere – point at which the chromosome attaches to the spindle during meiosis and mitosis (cell division); it contains no genes

protein core

base pairs – adenine and thymine form one type of pair and guanine and cytosine the other

DNA molecule – the double helix is connected by base pairs

chromosome The structure of a chromosome. The chromosome is made up of two identical chromatids joined by a centromere. Each chromatid is made up of coiled DNA.

including schizophrenia and trisomy 22 (a common cause of miscarriage).

chronic in medicine, term used to describe a condition that is of slow onset and then runs a prolonged course, such as rheumatoid arthritis or chronic bronchitis. In contrast, an **acute** condition develops quickly and may be of relatively short duration.

chronic fatigue syndrome CFS; or **myalgic encephalomyelitis (ME)** or **postviral fatigue syndrome**, common debilitating condition characterized by a diffuse range of symptoms present for at least six months including extreme fatigue, muscular pain, weakness, depression, poor balance and coordination, joint pains, and gastric upset. It is usually diagnosed after exclusion of other diseases and frequently follows a flulike illness.

chyme general term for the stomach contents. Chyme resembles a thick creamy fluid and is made up of partly digested food, hydrochloric acid, and a range of enzymes.

cilia singular **cilium**, small hairlike organs on the surface of some cells, particularly the cells lining the upper respiratory tract. Their wavelike movements waft particles of dust and debris towards the exterior. Some single-celled organisms move by means of cilia. In multicellular animals, they keep lubricated surfaces clear of debris. They also move food in the digestive tracts of some invertebrates.

ciliary body in biology, the part of the eye that connects the iris and the choroid. The *ciliary muscles contract and relax to change the shape of the lens during accommodation. The ciliary body is lined with cells that secrete aqueous humour into the anterior chamber of the eye.

ciliary muscle ring of muscle surrounding and controlling the lens inside the vertebrate eye, used in *accommodation (focusing). Suspensory ligaments, resembling spokes of a wheel, connect the lens to the ciliary muscle and pull the lens into a flatter shape when the muscle relaxes. When the muscle is relaxed the lens has its longest *focal length and focuses rays from distant objects. On contraction, the lens returns to its normal spherical state and therefore has a shorter focal length and focuses images of near objects.

circadian rhythm metabolic rhythm found in most organisms, which generally coincides with the 24-hour day. Its most obvious manifestation is the regular cycle of sleeping and waking, but body temperature and the concentration of *hormones that influence mood and behaviour also vary over the day. In humans, alteration of habits (such as rapid air travel round the world) may result in the circadian rhythm being out of phase with actual activity patterns, causing malaise until it has had time to adjust.

In mammals the circadian rhythm is controlled by the suprachiasmatic nucleus in the *hypothalamus. US researchers discovered a second circadian control mechanism in 1996; they found that cells within the retina also produced the hormone *melatonin. In 1997 US geneticists identified a gene, *clock*, in chromosome 5 in mice, that regulated the circadian rhythm.

circle perfectly round shape, the path of a point that moves so as to keep a constant distance from a fixed point (the **centre**). A circle has a *radius (the distance from any point on the circle to the centre), a *circumference (the boundary of the circle, part of which is called an *arc), *diameters (straight lines crossing the circle through the centre), *chords (lines joining two points on the circumference), *tangents (lines that touch the circumference at one point only), *sectors (regions inside the circle between two radii), and *segments (regions between a chord and the circumference).

circuit in physics or electrical engineering, an arrangement of electrical components connected by a conducting material through which a current can flow. There are two basic circuits, series and parallel. In a *series circuit, the components are connected end to end so that the current flows through all components one after the other. In a *parallel circuit, components are connected side by side so that part of the current passes through each

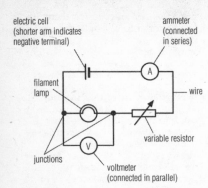

electric cell
(shorter arm indicates
negative terminal)

ammeter
(connected
in series)

filament
lamp

wire

variable resistor

junctions

voltmeter
(connected in parallel)

circuit Electrical symbols commonly used in circuit diagrams.

component. A circuit diagram shows in graphical form how components are connected together, using standard symbols for the components. If the circuit is unbroken, it is a closed circuit and current flows. If the circuit is broken, it becomes an open circuit and no current flows.

circuit breaker switching device designed to protect an electric circuit from overloads such as excessive current flows and voltage failures. It has the same action as a *fuse, and many houses now have a circuit breaker between the incoming mains supply and the domestic circuits. Circuit breakers usually work by means of magnetic-type relays or *solenoids. Those at electricity-generating stations have to be specially designed to prevent dangerous arcing (the release of luminous discharge) when the high-voltage supply is switched off. They may use an air blast or oil immersion to quench the arc.

circulatory system system of vessels in an animal's body that transports essential substances (*blood or other circulatory fluid) to and from the different parts of the body. It was first discovered and described by English physician William Harvey (1578–1657). All animals except for the most simple – such as sponges, jellyfish, sea anemones, and corals – have some type of circulatory system. Some invertebrates (animals without a backbone), such as insects, spiders, and most shellfish, have an 'open'

circulatory system which consists of a simple network of tubes and hollow spaces. Other invertebrates have pump-like structures that send blood through a system of blood vessels. All vertebrates (animals with a backbone), including humans, have a 'closed' circulatory system which principally consists of a pumping organ – the *heart–and a network of blood vessels.

Fish have a single circulatory system in which blood passes once around the body before returning to a two-chambered heart. In birds and mammals, there is a *double circulatory system – the lung or pulmonary circuit and the body or systemic circuit. Blood is first pumped from the heart to the *lungs and back to the heart, before being pumped to the remainder of the body and back. The heart is therefore a double pump and is divided into two halves. In all vertebrates, blood flows in one direction. Valves in the heart, large arteries, and veins prevent backflow, and the muscular walls of the arteries assist in pushing the blood around the body. A network of tiny *capillaries carries the blood from arteries to veins. It is through the walls of capillaries that materials are transported to and from the blood.

Although most animals have a heart or hearts to pump the blood, in some small invertebrates normal body movements circulate the fluid. In the **open system**, found in snails and other molluscs, the blood (more correctly called *haemolymph) passes from the arteries into a body cavity (haemocoel), and from here is gradually returned by other blood vessels to the heart, via the gills. Insects and other arthropods have an open system with a heart. In the **closed system** of earthworms, blood flows directly from the main artery to the main vein, via smaller lateral vessels in each body segment.

The human circulatory system performs a number of functions: it supplies the cells of the body with the food and oxygen they need to survive; it carries carbon dioxide and other waste products away from the cells; it helps to regulate the temperature of the body; and protects the body from

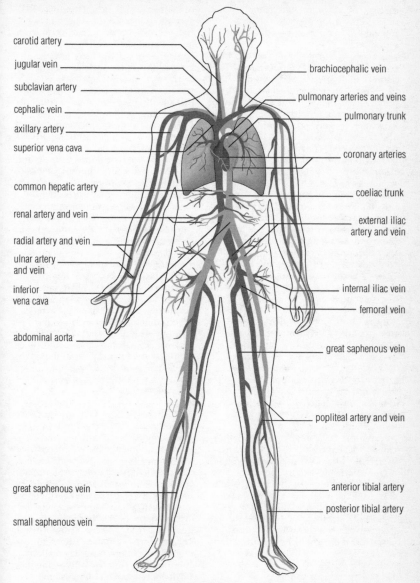

carotid artery

jugular vein

subclavian artery

cephalic vein

axillary artery

superior vena cava

common hepatic artery

renal artery and vein

radial artery and vein

ulnar artery and vein

inferior vena cava

abdominal aorta

great saphenous vein

small saphenous vein

brachiocephalic vein

pulmonary arteries and veins

pulmonary trunk

coronary arteries

coeliac trunk

external iliac artery and vein

internal iliac vein

femoral vein

great saphenous vein

popliteal artery and vein

anterior tibial artery

posterior tibial artery

circulatory system Blood flows through 96,500 km/60,000 mi of arteries and veins, supplying oxygen and nutrients to organs and limbs. Oxygen-poor blood (dark grey) circulates from the heart to the lungs where oxygen is absorbed. Oxygen-rich blood (light grey) flows back to the heart and is then pumped round the body through the aorta, the largest artery, to smaller arteries and capillaries. Here oxygen and nutrients are exchanged with carbon dioxide and waste products and the blood returns to the heart via the veins. Waste products are filtered by the liver, spleen, and kidneys, and nutrients are absorbed from the stomach and small intestine.

disease. In addition, the system transports hormones, which help to regulate the activities of various parts of the body.

circumcision surgical removal of all or part of the foreskin (prepuce) of the penis, usually performed on the newborn; it is practised among Jews (b'rit milah) and Muslims as a sign of God's covenant with the prophet Abraham. In some societies in Africa and the Middle East, female circumcision or clitoridectomy (removal of the labia minora and/or clitoris) is practised on adolescents as well as babies; it is illegal in the West.

circumference in geometry, the curved line that encloses a curved plane figure, for example a *circle or an ellipse. Its length varies according to the nature of the curve, and may be ascertained by the appropriate formula. The circumference of a circle is πd or $2\pi r$, where d is the diameter of the circle, r is its radius, and π is the constant pi, approximately equal to 3.1416.

circumscribe in geometry, to surround a figure with a circle which passes through all the vertices of the figure. Any triangle may be circumscribed and so may any regular polygon. Only certain quadrilaterals may be circumscribed (their opposite angles must add up to 180°).

cirrhosis any degenerative disease in an organ of the body, especially the liver, characterized by excessive development of connective tissue, causing scarring and painful swelling. Cirrhosis of the liver may be caused by an infection such as viral hepatitis, chronic obstruction of the common bile duct, chronic alcoholism or drug use, blood disorder, heart failure, or malnutrition. However, often no cause is apparent. If cirrhosis is diagnosed early, it can be arrested by treating the cause; otherwise it will progress to coma and death.

CISC acronym for Complex Instruction-Set Computer, in computing, a term referring to the design of the CPU and its instruction set. CISC computers are characterized by having large numbers of instructions, varying a lot in size and complexity. Usually, the various stages of instruction execution are followed, and only when all the stages have been finished can a new instruction start to be executed. RISC CPUs, by comparison, usually have few instructions in the instruction set, and the instructions are smaller and simpler. Complex problems are solved by executing several of the smaller instructions.

CITES acronym for Convention on International Trade in Endangered Species, international agreement under the auspices of the World Conservation Union with the aim of regulating trade in *endangered species of animals and plants. The agreement came into force in 1975 and has now been signed by 160 states. It prohibits any trade in a category of 8,000 highly endangered species and controls trade in a further 30,000 species.

Animals and plants listed in Appendix 1 of CITES are classified endangered, and all trade in that species is banned; those listed in Appendix 2 are classified vulnerable, and trade in the species is controlled without a complete ban; those listed in Appendix 3 are subject to domestic controls while national governments request help in controlling international trade. For full details see www.cites.org.

citric acid HOOCCH₂C(OH)(COOH)CH₂COOH, organic acid widely distributed in the plant kingdom; it is found in high concentrations in citrus fruits and has a sharp, sour taste. At one time it was commercially prepared from concentrated lemon juice, but now the main source is the fermentation of sugar with certain moulds.

citric acid cycle in biochemistry, another term for *Krebs cycle.

citrulline an *amino acid not commonly found in proteins. It was first isolated from the water melon *Citrullus vulgaris* and may also be isolated after digestion of the *enzyme caseinogen by *trypsin. Citrulline is involved in the process of *urea excretion.

CJD abbreviation for the fatal brain disorder *Creutzfeldt–Jakob disease.

cladistics method of biological

classification that uses a formal step-by-step procedure for objectively assessing the extent to which organisms share particular characteristics, and for assigning them to taxonomic groups called **clades**. Clades comprise all the species descended from a known or inferred common ancestor plus the ancestor itself, and may be large – consisting of a hierarchy of other clades.

class in biological classification, a subdivision of *phylum and forms a group of related *orders. For example, all mammals belong to the class Mammalia and all birds to the class Aves. Among plants, all class names end in 'idae' (such as Asteridae) and

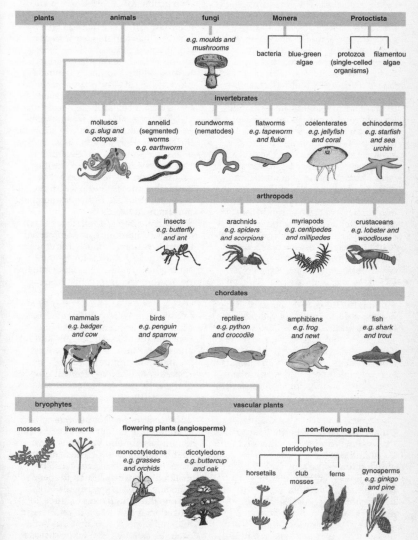

classification The classification of the plant and animal kingdoms.

among fungi in 'mycetes'; there are no equivalent conventions among animals. Related classes are grouped together in a phylum.

class in mathematics another name for a *set.

classification in biology, the arrangement of organisms into a hierarchy of groups on the basis of their similarities. The basic grouping is a *species, several of which may constitute a *genus, which in turn are grouped into families, and so on up through orders, classes, phyla (in plants, sometimes called divisions), and finally to kingdoms. The system that is used is one that reflects the evolutionary origin of the organisms. In other words, organisms belonging one group are thought to have evolved from a common ancestor at some time in the past. (See diagram, page 123.)

class interval in statistics, the range of each class of data, used when arranging large amounts of raw data into grouped data. To obtain an idea of the distribution, the data are broken down into convenient classes (commonly 6–16), which are mutually exclusive and are usually equal in width to enable *histograms to be drawn. The class boundaries should clearly define the range of each class. When dealing with *discrete data, suitable intervals would be, for example, 0–2, 3–5, 6–8, and so on. When dealing with *continuous data, suitable intervals might be $170 \leq X < 180$, $180 \leq X < 190$, $190 \leq X < 200$, and so on.

clavicle (Latin *clavis* 'key') the collar bone of many vertebrates. In humans it is vulnerable to fracture, since falls involving a sudden force on the arm may result in very high stresses passing into the chest region by way of the clavicle and other bones. It is connected at one end with the sternum (breastbone), and at the other end with the shoulder-blade, together with which it forms the arm socket. The wishbone of a chicken is composed of its two fused clavicles.

clay very fine-grained *sedimentary deposit that has undergone a greater or lesser degree of consolidation. When moistened it is plastic, and it hardens on heating, which renders it impermeable. It may be white, grey, red, yellow, blue, or black, depending on its composition. Clay minerals consist largely of hydrous silicates of aluminium and magnesium together with iron, potassium, sodium, and organic substances. The crystals of clay minerals have a layered structure, capable of holding water, and are responsible for its plastic properties. According to international classification, in mechanical analysis of soil, clay has a grain size of less than 0.002 mm/0.00008 in.

clay mineral one of a group of hydrous silicate minerals that form most of the fine-grained particles in clays. Clay minerals are normally formed by weathering or alteration of other silicate minerals. Virtually all have sheet silicate structures similar to the *micas. They exhibit the following useful properties: loss of water on heating; swelling and shrinking in different conditions; cation exchange with other media; and plasticity when wet. Examples are kaolinite, illite, and montmorillonite.

Kaolinite $Al_2Si_2O_5(OH)_4$ is a common white clay mineral derived from alteration of aluminium silicates, especially feldspars. Illite contains the same constituents as kaolinite, plus potassium, and is the main mineral of clay sediments, mudstones, and shales; it is a weathering product of feldspars and other silicates. Montmorillonite contains the constituents of kaolinite plus sodium and magnesium; along with related magnesium- and iron-bearing clay minerals, it is derived from alteration and weathering of mafic igneous rocks. Kaolinite (the mineral name for kaolin or china clay) is economically important in the ceramic and paper industries. Illite, along with other clay minerals, may also be used in ceramics. Montmorillonite is the chief constituent of fuller's earth, and is also used in drilling muds (muds used to cool and lubricate drilling equipment). Vermiculite (similar to montmorillonite) will expand on heating to produce a material used in insulation.

Clean Air Act legislation designed to improve the quality of air by enforcing pollution controls on industry and households. The first Clean Air Act in the UK was passed in 1956 after the London Smog killed 4,000 people. The USA enacted a Clean Air Act in 1970, further amended in 1990.

In Europe, national legislation is supplemented by EC and UN agreements and conventions such as EC regulations on levels of ozone, nitrogen oxide, and sulphur dioxide. In addition the UK has signed protocols drawn up by the UN Economic Commission for Europe to reduce transboundary pollution.

cleartext or **plaintext**, in encryption, the original, unencrypted message.

ClearType font technology developed by Microsoft in 1998 for use in electronic book readers and *personal digital assistants, and first used on Pocket PC devices in 2000. By addressing an area smaller than a *pixel, ClearType makes screen type easier to read, especially on the *liquid-crystal displays found on *laptop computers.

cleavage in geology and mineralogy, the tendency of a rock or mineral to split along defined, parallel planes related to its internal structure; the clean splitting of slate is an example. It is a useful distinguishing feature in rock and mineral identification. Cleavage occurs as a result of realignment of component minerals during deformation or metamorphism. It takes place where bonding between atoms is weakest, and cleavages may be perfect, good, or poor, depending on the bond strengths; a given rock or mineral may possess one, two, three, or more orientations along which it will cleave.

Some minerals have no cleavage, for example, quartz will fracture to give curved surfaces similar to those of broken glass. Some other minerals, such as apatite, have very poor cleavage that is sometimes known as a parting. Micas have one perfect cleavage and therefore split easily into very thin flakes. Pyroxenes have two good cleavages and break (less perfectly) into long prisms. Galena has three perfect cleavages parallel to the cube edges, and readily breaks into

smaller and smaller cubes. Baryte has one perfect cleavage plus good cleavages in other orientations.

cleft palate fissure of the roof of the mouth, often accompanied by a harelip, the result of the two halves of the palate failing to join properly during embryonic development. It can be remedied by plastic surgery. Approximately 1 child in 2,000 is born with a cleft palate.

client in *client–server architecture, software that enables a user to access a store of data or programs on a *server. On the Internet, client software is the software that users need to run on home computers in order to be able to use services such as the World Wide Web. Examples of client programs on a personal computer are Web browser clients, such as Netscape Navigator, Microsoft Internet Explorer, or Opera. The Web browser client communicates with a Web server to send and retrieve data.

client–server architecture in computing, a system in which the mechanics of looking after data are separated from the programs that use the data. For example, the 'server' might be a central database, typically located on a large computer that is reserved for this purpose. The 'client' would be an ordinary program that requests data from the server as needed.

climacteric period during the lifespan when an important physiological change occurs, usually referring to menopause.

climate combination of weather conditions at a particular place over a period of time – usually a minimum of 30 years. A *climate classification encompasses the averages, extremes, and frequencies of all meteorological elements such as temperature, atmospheric pressure, precipitation, wind, humidity, and sunshine, together with the factors that influence them.

The primary factors that influence the climate of an area are: latitude (as a result of the Earth's rotation and orbit); ocean currents; large-scale movements of wind belts and air masses over the Earth's surface; temperature differences between land and sea surfaces; topography; continent positions; and vegetation. In the long term, changes in the Earth's orbit and the angle of its axis inclination also affect climate. Climatologists are especially concerned with the influences of human activity on climate change, among the most important of which, at both local and global levels, are those currently linked with *ozone depleters, the *greenhouse effect, and *global warming.

climate change change in the *climate of an area or of the whole world over an appreciable period of time. That is, a single winter that is colder than average does not indicate climate change. It is the change in average weather conditions from one period of time (30–50 years) to the next. Climate fluctuations are natural phenomena, but there is increasing evidence to suggest that human industrial activity affects the global climate, particularly in terms of *global warming.

climate classification description of different types of *climate, taking into account the averages, extremes, and frequencies of all meteorological elements such as temperature, atmospheric pressure, precipitation, wind, humidity, and sunshine, together with the factors that influence them.

The different types of climate can be described as *tropical (hot), *warm temperate (or subtropical), *cool temperate, *cold (or polar), and *arctic.

climate model computer simulation, based on physical and mathematical data, of a climate system, usually the global (rather than local) climate. It is used by researchers to study such topics as the possible long-term disruptive effects of the greenhouse gases, or of variations in the amount of radiation given off by the Sun.

climatic region geographical area which experiences certain *weather conditions. No two localities on Earth may be said to have exactly the same *climate, but widely separated areas of the world can possess similar climates. These climatic regions have some comparable physical and environmental features as well as

having similar weather patterns. Climatic regions are differentiated by weather conditions (including temperature, humidity, precipitation type and amount, wind speed and direction, atmospheric pressure, sunshine, cloud types, and cloud coverage), and by weather phenomena (such as thunderstorms, *fog, and *frost) that have prevailed there over a long period of time, usually 30 years.

climatology study of climate, its global variations and causes.

climax community group of plants and animals that is best able to exploit the environment in which it exists. It is brought about by *succession (a change in the species present) and represents the point at which succession ceases to occur.

climax vegetation the plants in a *climax community.

climograph diagram that shows both the average monthly temperature and precipitation of a place.

clinical ecology in medicine, ascertaining environmental factors involved in illnesses, particularly those manifesting nonspecific symptoms such as fatigue, depression, allergic reactions, and immune-system malfunctions, and prescribing means of avoiding or minimizing these effects.

clinical trial in medicine, the evaluation of the effectiveness of medical treatment in a systematic fashion. Clinical trials compare new treatments with established treatments or placebos under standardized conditions. Such treatments may be drugs or surgical procedures. Ethical standards are maintained by ethics committees and they also ensure that the clinical trial procedure is explained to patients. The system was established with the development of new medicines in the 1940s and 1950s.

Clipper chip controversial encryption hardware system that contains built-in facilities to allow authorized third parties access to the encrypted data. Adopted as a US government standard in 1994, the Clipper chip was a chip that used *public-key cryptography and a proprietary *algorithm called Skipjack, and could be built into any communications device, such as a telephone or modem. It was developed by the US National Security Agency as part of its Capstone project.

clitoris (Greek *kleitoris* 'little hill') in anatomy, part of the female reproductive system. The glans of the clitoris is visible externally. It connects to a pyramid-shaped pad of erectile tissue. Attached to this are two 'arms' that extend backwards into the body towards the anus and are approximately 9 cm/3.5 in in length.

Between these arms are the clitoral bulbs, lying one on each side of the vaginal cavity.

clone exact replica – in genetics, any one of a group of genetically identical cells or organisms. An identical *twin is a clone; so too are bacteria living in the same colony. 'Clone' also describes genetically engineered replicas of DNA sequences.

British scientists confirmed in February 1997 that they had cloned an

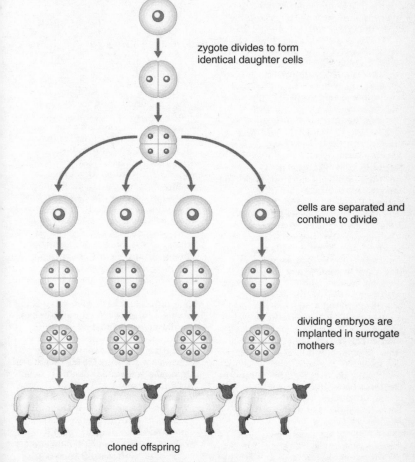

zygote divides to form identical daughter cells

cells are separated and continue to divide

dividing embryos are implanted in surrogate mothers

cloned offspring

clone The production of a clone (an exact replica) happens naturally when a zygote undergoes division. This is the process that brings about multiple births such as identical twins and triplets. Zygote division can be induced in vitro with the resulting embryos then implanted into surrogate mothers. The offspring are all clones of each other but not of their parents.

adult sheep from a single cell to produce a lamb with the same genes as its mother. A cell was taken from the udder of the mother sheep, and its DNA combined with an unfertilized egg that had had its DNA removed. The fused cells were grown in a laboratory and then implanted into the uterus of a surrogate mother sheep. The resulting lamb, Dolly, came from an animal that was six years old. However, in 1999, Dolly was revealed not to be an exact clone – research showed her mitochondria to have come mainly from the egg cell rather than the udder cell. She died in 2003.

This was the first time cloning had been achieved using cells other then reproductive cells. The cloning breakthrough has ethical implications, as the same principle could be used with human cells and eggs. The news was met with international calls to prevent the cloning of humans. The UK, Spain, Germany, Canada, and Denmark already have laws against cloning humans, as do some individual states in the USA (legislators introduced bills to ban human cloning and associated research nationally in March 1997). France and Portugal also have very restrictive laws on cloning.

clone copy of hardware or software that may not be identical to the original design but provides the same functions. All personal computers (PCs) are to some extent clones of the original IBM PC and PC AT launched by IBM in 1981 and 1984, respectively – including IBM's current machines. Clones typically compete by being cheaper and are sometimes less well made than the branded product but this is not always the case. Compaq, for example, competed with IBM by producing the first portable PC and by building better desktop machines, while Dell competed by building PCs to individual orders and supplying customers direct.

closed in mathematics, descriptive of a set of data for which an operation (such as addition or multiplication) done on any members of the set gives a result that is also a member of the set.

Clostridium genus of spore-forming bacteria that are anaerobic (unable to survive in the presence of oxygen). They are widely distributed in soil and in the intestines of mammals.

Clostridium species cause some serious diseases, including *botulism, *tetanus and gas *gangrene.

cloud water vapour condensed into minute water particles that float in masses in the atmosphere. Clouds, like fogs or mists, that occur at lower levels, are formed by the cooling of air containing water vapour, which generally condenses around tiny dust particles.

Clouds are classified according to the height at which they occur, and by their shape. **Cirrus** and **cirrostratus** clouds occur at around 10 km/33,000 ft. The former, sometimes called mares'-tails, consist of minute specks of ice and appear as feathery white wisps, while cirrostratus clouds stretch across the sky as a thin white sheet. Three types of cloud are found at 3–7 km/ 10,000–23,000 ft: cirrocumulus, altocumulus, and altostratus. **Cirrocumulus** clouds occur in small or large rounded tufts, sometimes arranged in the pattern called mackerel sky. **Altocumulus** clouds are similar, but larger, white clouds, also arranged in lines. **Altostratus** clouds are like heavy cirrostratus clouds and may stretch across the sky as a grey sheet. **Stratocumulus** clouds are generally lower, occurring at 2–6 km/ 6,500–20,000 ft. They are dull grey clouds that give rise to a leaden sky that may not yield rain. Two types of clouds, **cumulus** and **cumulonimbus**, are placed in a special category because they are produced by daily ascending air currents, which take moisture into the cooler regions of the atmosphere. Cumulus clouds have a flat base generally at 1.4 km/4,500 ft where condensation begins, while the upper part is dome-shaped and extends to about 1.8 km/6,000 ft. Cumulonimbus clouds have their base at much the same level, but extend much higher, often up to over 6 km/ 20,000 ft. Short heavy showers and sometimes thunder may accompany them. **Stratus** clouds, occurring below 1–2.5 km/3,000–8,000 ft, have the appearance of sheets parallel to the horizon and are like high-level fogs.

In addition to their essential role in the water cycle, clouds are important in the regulation of radiation in the Earth's atmosphere. They reflect short-wave radiation from the Sun, and absorb and re-emit long-wave radiation from the Earth's surface.

cloud chamber apparatus, now obsolete, for tracking ionized particles. It consists of a vessel fitted with a piston and filled with air or other gas, saturated with water vapour. When the volume of the vessel is suddenly expanded by moving the piston outwards, the vapour cools and a cloud of tiny droplets forms on any nuclei, dust, or ions present. As fast-moving ionizing particles collide with the air or gas molecules, they show as visible tracks.

Much information about interactions between such particles and radiations has been obtained from photographs of these tracks. The system was improved upon by the use of liquid hydrogen or helium instead of air or gas (see *particle detector) and by the *spark chamber. The cloud chamber was devised in 1897 by Charles Thomson Rees Wilson (1869–1959) at Cambridge University.

Cluster European Space Agency two-year project to study the interaction of the *solar wind, a stream of atomic particles from the Sun's corona, with the Earth's *magnetosphere, from an array of four identical satellites that can provide a three-dimensional view. The first pair of the Cluster II space weather satellites was launched in July 2000, and the second pair the following month.

cluster several computers joined together by a high speed network or *backbone, and presented to users as one unit. Building clusters of computers improves the system's overall tolerance to faults and availability to users. If one computer in the cluster fails, its programs and users are transferred to another computer in the cluster with minimal interruption.

cm symbol for *centimetre.

CMOS abbreviation for complementary metal-oxide semiconductor, family of integrated circuits (chips) widely used in building electronic systems.

CNS abbreviation for *central nervous system.

coagulation in biology, another term for *blood clotting, the process by which bleeding is stopped in the body.

coal black or blackish mineral substance formed from the compaction of ancient plant matter in tropical swamp conditions. It is used as a fuel and in the chemical industry. Coal is classified according to the proportion of carbon it contains. The main types are *anthracite (shiny, with about 90% carbon), **bituminous coal** (shiny and dull patches, about 75% carbon), and **lignite** (woody, grading into peat, about 50% carbon). Coal can be burned to produce heat energy, for example in power stations to produce electricity. Coal burning is one of the main causes of *acid rain, which damages buildings and can be detrimental to aquatic and plant life.

coal gas gas produced when coal is destructively distilled or heated out of contact with the air. Its main constituents are methane, hydrogen, and carbon monoxide. Coal gas has been superseded by *natural gas for domestic purposes.

coal mining extraction of coal from the Earth's crust. Coal mines may be opencast, adit, or deepcast. The least expensive is opencast but this may result in scars on the landscape.

coastal deposition the laying down of sediment (*deposition) in a low-energy environment with constructive *waves. Coastal deposition occurs where there is a large supply of material from cliffs, rivers, or beaches, *longshore drift, and an irregular coastline. Geographical features include the spit, *bar, beach, foreland, and tombolo, such as at Chesil Beach, England.

coastal erosion the erosion of the land by the constant battering of the sea, primarily by the processes of hydraulic action, corrasion, attrition, and corrosion. *Hydraulic action occurs when the force of the waves compresses air pockets in coastal rocks and cliffs. The air expands explosively, breaking the rocks apart. It is also the force of the water on the cliff. During severe gales this can be as high as 6 tonnes/cm³ – the force of a bulldozer. Rocks and pebbles flung by waves against the cliff face wear it away by the process of corrasion, or abrasion as it is also known. Chalk and limestone coasts are often broken down by *solution (also called *corrosion). Attrition is the process by which the eroded rock particles themselves are worn down, becoming smaller and more rounded.

coaxial cable electric cable that consists of a solid or stranded central conductor insulated from and surrounded by a solid or braided conducting tube or sheath. It can transmit the high-frequency signals used in television, telephone, and other telecommunications transmissions.

cobalt chemical symbol Co, (German *Kobalt* 'evil spirit') hard, lustrous, grey, metallic element, atomic number 27, relative atomic mass 58.933. It is found in various ores and occasionally as a free metal, sometimes in metallic meteorite fragments. It is used in the preparation of magnetic, wear-resistant, and high-strength alloys; its compounds are used in inks, paints, and varnishes.

The isotope Co-60 is radioactive (half-life 5.3 years) and is produced in large amounts for use as a source of gamma rays in industrial radiography, research, and cancer therapy. Cobalt was named in 1730 by Swedish chemist Georg Brandt (1694–1768); the name derives from the fact that miners considered its ore malevolent because it interfered with copper production.

cobalt-60 radioactive (half-life 5.3 years) isotope produced by neutron radiation of cobalt in heavy-water reactors, used in large amounts for gamma rays in cancer therapy, industrial radiography, and research, substituting for the much more costly radium.

cobalt ore cobalt is extracted from a number of minerals, the main ones being **smaltite**, $(CoNi)As_3$; **linnaeite**, Co_3S_4; **cobaltite**, CoAsS; and **glaucodot**, $(CoFe)AsS$.

COBOL acronym for **common business-oriented language**, high-level computer-programming language, designed in the late 1950s for commercial data-processing problems; it has become the major language in this field. COBOL features powerful facilities for file handling and business arithmetic. Program instructions written in this language make extensive use of words and look very much like English sentences. This makes COBOL one of the easiest languages to learn and understand.

COD abbreviation for *chemical oxygen demand, a measure of water and effluent quality.

code expression of an *algorithm in a *programming language. The term is also used as a verb, to describe the act of programming.

codec contraction of coder/decoder or compression/decompression, device or technology that codes and decodes an analogue stream to or from digital data. It is used in applications such as remote broadcast-quality voiceovers recorded in a remote studio and transmitted via codecs and *Integrated Services Digital Network (ISDN) lines to a central studio for final mixing. Codecs are also used to compress and decompress digital audio and video for multimedia presentations, especially when delivered over the Internet using *streaming technology. Formats such as Real Audio and Video, Windows Media Player, and Apple's Quicktime use codecs to enable the efficient transfer of audio and video information. *MPEG is another example of a popular codec.

codeine opium derivative that provides *analgesia in mild to moderate pain. It also suppresses the cough centre of the brain. It is an alkaloid, derived from morphine but less toxic and addictive.

codominance in genetics, the failure of a pair of alleles, controlling a particular characteristic, to show the classic recessive-dominant relationship. Instead, aspects of both alleles may show in the phenotype. For example, the human blood group AB shows the phenotype effect of both A and B codominant genes.

codon or **coding triplet**, in genetics, a triplet of bases (see *base pair) in a molecule of DNA or RNA that directs the placement of a particular amino acid during the process of protein (polypeptide) synthesis. There are 64 codons in the *genetic code.

coefficient number part in front of an algebraic term, signifying multiplication. For example, in the expression $4x^2 + 2xy - x$, the coefficient of x^2 is 4 (because $4x^2$ means $4 \times x^2$), that of xy is 2, and that of x is -1 (because $-1 \times x = -x$).

In general algebraic expressions, coefficients are represented by letters that may stand for numbers; for example, in the equation $ax^2 + bx + c = 0$, a, b, and c are coefficients, which can take any number.

coefficient of relationship probability that any two individuals share a given gene by virtue of being descended from a common ancestor. In sexual reproduction of *diploid species, an individual shares half its genes with each parent, with its offspring, and (on average) with each sibling; but only a quarter (on average) with its grandchildren or its siblings' offspring; an eighth with its great-grandchildren, and so on.

coelom in all but the simplest animals, the fluid-filled cavity that separates the body wall from the gut and associated organs, and allows the gut muscles to contract independently of the rest of the body.

coenzyme small organic nonprotein compound that attaches to an *enzyme and is necessary for its correct functioning. Tightly bound coenzymes are known as prosthetic groups; more loosely bound ones are called cofactors. The coenzyme itself is not usually changed during a reaction. If it is, it is usually converted rapidly back to its original form. Well-known coenzymes include NAD, ATP, and coenzyme A.

coevolution evolution of those structures and behaviours within a species that can best be understood in relation to another species. For example, some insects and flowering plants have evolved together: insects have produced mouthparts suitable for collecting pollen or drinking nectar, and plants have developed chemicals and flowers that will attract insects to

them. Parasites often evolve and speciate with their hosts.

Coevolution occurs because both groups of organisms, over millions of years, benefit from a continuing association, and will evolve structures and behaviours that maintain this association.

cognition in psychology, a general term covering the functions involved in synthesizing information – for example, perception (seeing, hearing, and so on), attention, memory, and reasoning.

cognitive dissonance state of psychological tension occurring when a choice has to be made between two equally attractive or equally unpleasant alternatives. The dissonance is greater the closer the alternatives are in attractiveness or unpleasantness.

cognitive psychology study of information-processing functions in humans and animals, covering their role in learning, memory, reasoning, and language development. Cognitive psychologists use a number of experimental techniques, including laboratory-based research with normal and brain-damaged subjects, as well as computer and mathematical models to test and validate theories.

The study of cognition was largely neglected by psychologists for the early part of the 20th century after the demise of *introspection as a method of investigation and the rise of *behaviourism. However, several influential theorists, such as US psychologist Edward Chase Tolman (1886–1959), continued to argue that in order to comprehend fully the determinants of behaviour, cognitive processes must be studied and understood, and in 1957 Noam Chomsky's examination of behaviourist approaches to language acquisition appeared. With the rise of telecommunications technology and digital computing, such theorists as English psychologist Donald Broadbent (1926–) developed information-processing models of the brain, later elaborated, for example, by German-born US psychologist Ulrich Neisser (1928–). More recently, the limitations of these approaches, for example, in elaborating the role of emotion and motivation in cognitive processes, have become the focus of attention.

coherence in physics, property of two or more waves of a beam of light or other electromagnetic radiation having the same frequency and the same *phase, or a constant phase difference.

coil in medicine, another name for an intrauterine device.

cold, common minor disease of the upper respiratory tract, caused by a variety of viruses. Symptoms are headache, chill, nasal discharge, sore throat, and occasionally cough. Research indicates that the virulence of a cold depends on psychological factors and either a reduction or an increase of social or work activity, as a result of stress, in the previous six months. There is little immediate hope of an effective cure since the viruses transform themselves so rapidly.

cold-blooded of animals, dependent on the surrounding temperature; see *poikilothermy.

cold dark matter in cosmology, unseen material consisting of slow-moving particles believed by some cosmologists to make up the bulk of the matter in the universe. See *dark matter.

cold fusion in nuclear physics, the fusion of atomic nuclei at room temperature. If cold fusion were possible it would provide a limitless, cheap, and pollution-free source of energy, and it has therefore been the subject of research around the world.

cold or polar climate The subarctic regions 60° north of North America, Europe, and Asia are said to have a cold or polar climate. This type of climate also occurs at higher altitudes in more temperate latitudes and in southern Chile. For over six months of the year the mean temperature remains below 6°C/43°F. There is a period each year in the area north of the Arctic Circle during which time the Sun never rises. Winters are long and cold; the minimum mean monthly temperatures may be as low as −30°C/−22°F. The *wind-chill factor is high with strong winds. Summers are short but the long hours of daylight and clear skies mean they

are relatively warm. Precipitation is light as the cold air can only hold a limited amount of moisture, and the small amount of winter snowfall is frequently blown about in blizzards.

cold-working method of shaping metal at or near atmospheric temperature.

collagen protein that is the main constituent of *connective tissue. Collagen is present in skin, cartilage, tendons, and ligaments. Bones are made up of collagen, with the mineral calcium phosphate providing increased rigidity.

Collagen is made by fibroblast cells that are found in connective tissue. The collagen molecule consists of three protein chains, each about 1,000 amino acids in length, that are entwined in a triple helix. There are 15 known types of collagen that can be divided into fibrillar and nonfibrillar. **Fibrillar collagens** are found in cartilage, tendons, skin, and bones and the fibrillar structure gives these great strength. **Nonfibrillar collagens** are found separating the basement membranes, the epithelial layers that line tube and in the cornea. They form networks and sheets.

collagen therapy Collagen is used cosmetically to enhance lips and smooth out wrinkles. The collagen, which is extracted mainly from cow hides, is injected into the skin. For wrinkle treatment it is injected into the dermis over six sessions, which have to be repeated after several months as the implants degrade.

collapsed backbone in computing, a method by which network *backbone cables are replaced with powerful network routers or network switches that effectively collapse the backbone network cable into one 'black box'. All of the devices that used to plug onto the backbone cable then plug into the router or switch. Significant improvements in performance can be achieved using this technique, and it provides a more manageable solution. The disadvantage of using this technique is that it introduces a single point of failure into the network, so if the collapsed backbone device fails, the whole network fails.

collective unconscious in psychology, a shared pool of memories, ideas, modes of thought, and so on, which, according to the Swiss psychiatrist Carl Jung (1875–1961), comes from the life experience of one's ancestors, indeed from the entire human race. It coexists with the personal *unconscious, which contains the material of individual experience, and may be regarded as an immense depository of ancient wisdom.

collenchyma plant tissue composed of relatively elongated cells with thickened cell walls, in particular at the corners where adjacent cells meet.

It is a supporting and strengthening tissue found in nonwoody plants, mainly in the stems and leaves.

colligative property property that depends only on the concentration of particles in a solution and not on the nature of the particles. Such properties include osmotic pressure (see *osmosis), elevation of *boiling point, depression of *freezing point, and lowering of *vapour pressure.

collimator optical device for producing a nondivergent beam of light, or any device for limiting the size and angle of spread of a beam of radiation or particles. The term is also used for a small optical telescope attached to a larger optical instrument to fix its line of sight.

collinear in mathematics, lying on the same straight line.

collision detection in *virtual reality, the ability of software to detect when two on-screen objects make contact.

collision theory theory that explains how chemical reactions take place and why rates of reaction alter. For a reaction to occur the reactant particles must collide. Only a certain fraction of the total collisions cause chemical change; these are called **successful collisions**. The successful collisions have sufficient energy (activation energy) at the moment of impact to break the existing bonds and form new bonds, resulting in the products of the reaction. Increasing the concentration of the reactants and raising the temperature bring about more collisions and therefore more successful collisions, increasing the rate of reaction.

When a *catalyst undergoes collision with the reactant molecules, less energy is required for the chemical change to take place, and hence more collisions have sufficient energy for reaction to occur. The reaction rate therefore increases.

colloid substance composed of extremely small particles of one material (the dispersed phase) evenly and stably distributed in another material (the continuous phase). The size of the dispersed particles (1–1,000 nanometres across) is less than that of particles in suspension but greater than that of molecules in true solution.

a fruitful collision

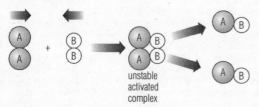

unstable
activated
complex

an unfruitful collision

collision theory Collision theory explains how chemical reactions occur and why rates of reaction differ. For a reaction to occur, particles must collide. If the collision causes a chemical change it is referred to as a fruitful collision.

Colloids involving gases include **aerosols** (dispersions of liquid or solid particles in a gas, as in fog or smoke) and **foams** (dispersions of gases in liquids).

Those involving liquids include **emulsions** (in which both the dispersed and the continuous phases are liquids) and **sols** (solid particles dispersed in a liquid). Sols in which both phases contribute to a molecular three-dimensional network have a jellylike form and are known as **gels**; gelatin, starch 'solution', and silica gel are common examples.

Steel is a solid colloid.

Milk is a natural emulsion of liquid fat in a watery liquid; synthetic emulsions such as some paints and cosmetic lotions have chemical emulsifying agents to stabilize the colloid and stop the two phases from separating out. Colloids were first studied thoroughly by the British chemist Thomas Graham, who defined them as substances that will not diffuse through a semipermeable membrane (as opposed to what he termed crystalloids, solutions of inorganic salts, which will diffuse through).

colon in anatomy, the main part of the large intestine, between the caecum and rectum. Water and mineral salts are absorbed from undigested food in the colon, and the residue passes as faeces towards the rectum.

colonization in ecology, the spread of species into a new habitat, such as a freshly cleared field, a new motorway verge, or a recently flooded valley. The first species to move in are called **pioneers**, and may establish conditions that allow other animals and plants to move in (for example, by improving the condition of the soil or by providing shade). Over time a range of species arrives and the habitat matures; early colonizers will probably be replaced, so that the variety of animal and plant life present changes. This is known as *succession.

colour in physics, quality or wavelength of light emitted or reflected from an object. Visible white light consists of electromagnetic radiation of various wavelengths, and

if a beam is refracted through a prism, it can be spread out into the visible spectrum (that can be detected by the human eye), in which the various colours correspond to different wavelengths. From long to short wavelengths (from about 780 to 380 nanometres) the colours are red (780–650), orange(650–600), yellow(600–550), green(550–500), blue(500–450), and violet(450–380).

The colour of grass is green because grass absorbs all the colours from the spectrum and only transmits or reflects the wavelength corresponding to green. A sheet of white paper reflects all the colours of the spectrum from its surface; black objects absorb all the colours of the spectrum.

All colours can be obtained from mixing proportions of red, green, and blue light. These are known as primary colours. Different colour filters can also produce light of different colours. For example, a red filter only transmits red light, the remaining colours of the spectrum being absorbed by the filter.

Mixing red, green, and blue light in the correct proportions produces white light. When these colours are mixed in different proportions, secondary colours are formed, such as cyan, magenta, and yellow. For example, blue + red = magenta, red + green = yellow, and green + blue = cyan. Yellow light is reflected from the surfaces of some flowers as the petals absorb blue light. Red and green light are reflected back, and these mix to give the sensation of yellow.

colour in elementary particle physics, a property of *quarks analogous to electric charge but having three states, denoted by red, green, and blue. The colour of a quark is changed when it emits or absorbs a *gluon. The term has nothing to do with colour in its usual sense. See *quantum chromodynamics.

colour blindness hereditary defect of vision that reduces the ability to discriminate certain colours, usually red and green. The condition is sex-linked, affecting men more than women.

In the most common types of colour blindness there is confusion among the red–yellow–green range of colours; for example, many colour-blind observers are unable to distinguish red from yellow or yellow from green. The physiological cause of congenital colour blindness is not known, although it probably arises from some defect in the retinal receptors. Lead poisoning and toxic conditions caused by excessive smoking can lead to colour blindness. Between 2% and 6% of men and less than 1% of women are colour-blind.

colour vision   the ability of the eye to recognize different frequencies in the visible spectrum as colours. In most vertebrates, including humans, colour vision is due to the presence on the *retina of three types of light-sensitive cone cell, each of which responds to a different primary colour (red, green, or blue).

Colour vision is one of the ways in which the brain can acquire knowledge of the unchanging characteristics of objects. Perceived colours are functions of the state of the brain, as well as of physical features of objects. They remain more or less stable, and objects remain recognizable, in spite of the continuously changing illumination in which they are seen, a phenomenon known as **colour constancy**.

columbium   symbol Cb, former name for the chemical element *niobium. The name is still used occasionally in metallurgy.

column   vertical list of numbers or terms, especially in matrices.

COM   acronym for *Component Object Model; computer output on microfilm/microfiche.

coma   in medicine, a state of deep unconsciousness from which the subject cannot be roused. Possible causes include head injury, brain disease, liver failure, cerebral haemorrhage, and drug overdose.

coma   in optics, one of the geometrical aberrations of a lens, whereby skew rays from an object make a comet-shaped spot on the image plane instead of meeting at a point.

combination   in mathematics, a selection of a number of objects from some larger number of objects when no account is taken of order within any one arrangement. For example, 123,

213, and 312 are regarded as the same combination of three digits from 1234. Combinatorial analysis is used in the study of *probability.

The number of ways of selecting r objects from a group of n is given by the formula: $n!/[r!(n-r)!]$ (see *factorial). This is usually denoted by ^nC_r.

combine   in *probability theory, to work out the chances of two or more events occurring at the same time.

combustion   burning, defined in chemical terms as the rapid combination of a substance with oxygen, accompanied by the evolution of heat and usually light. A slow-burning candle flame and the explosion of a mixture of petrol vapour and air are extreme examples of combustion. Combustion is an *exothermic reaction as heat energy is given out.

comet   small, icy body orbiting the Sun, usually on a highly elliptical path that takes it beyond the planet Pluto. A comet consists of a central nucleus a few kilometres across and is made mostly of ice mixed with gas and dust. As a comet approaches the Sun its nucleus heats up, releasing gas and dust, which form a coma (comet head) up to 100,000 km/60,000 mi wide. Gas and dust stream away from the coma to form one or more tails, which may extend for millions of kilometres. Some comets, such as Halley's comet, stay within Pluto's orbit for most of the time. Comets are normally visible during sunset or sunrise.

Comets are of many different types, characterized by their orbits, their composition (the ratio of ice to dust, and the amount of frozen volatiles other than water ice, such as methane and carbon monoxide), and their size. Most comets approach the Sun on a hyperbolic orbit and are seen only once; others (the periodic comets) return regularly in elliptical orbits. Famous examples of the periodic comets are *Halley's comet, which has a period of 76 years and is one of the largest comets known, with a nucleus about $15 \times 7 \times 7$ km/$9 \times 4 \times 4$ mi across, and comet Encke, which has one of the shortest periods at only 3.3 years. Current thinking is that the nonperiodic comets, and those with very long periods, mostly originate in the Oort cloud, which lies far beyond the orbit of Pluto and which may contain billions of protocomets, only a few of which are gravitationally perturbed into the inner Solar System each decade. The orbits of the periodic comets suggest a different source, and this is the Kuiper belt, a zone just beyond the orbit of Neptune. This, too, contains a huge number of bodies, most of them too small to be detected by current techniques, although the count is rising steadily. A dozen or more comets are discovered every year, some by amateur astronomers.

command language in computing, a set of commands and the rules governing their use, by which users control a program. For example, an *operating system may have commands such as SAVE and DELETE, or a payroll program may have commands for adding and amending staff records.

command line interface CLI, in computing, a character-based interface in which a prompt is displayed on the screen at which the user types a command, followed by carriage return, at which point the command, if valid, is executed. Additional options may be used after the command in a line; these are known as **command line parameters**.

common denominator denominator that is a common multiple of, and hence exactly divisible by, all the denominators of a set of fractions, and which therefore enables their sums or differences to be found.

For example, $2/3$ and $3/4$ can both be converted to equivalent fractions of denominator 12, $2/3$ being equal to $8/12$ and $3/4$ to $9/12$. Hence their sum is $17/12$ and their difference is $1/12$. The **lowest common denominator** (lcd) is the smallest common multiple of the denominators of a given set of fractions.

common difference the difference between any number and the next in an *arithmetic progression. For example, in the set 1, 4, 7, 10, ... , the common difference is 3.

common factor number that will divide two or more others without

leaving a remainder. For example, the *factors of 8 are 1, 2, 4, and 8; the factors of 12 are 1, 2, 3, 4, 6 and 12; and the factors of 16 are 1, 2, 4, 8 and 16. The numbers 2 and 4 are factors of all three numbers (8, 12, and 16) and are known as the common factors; 4 is known as the **highest common factor**. A *prime number can only have factors of one and itself.

common gateway interface CGI, on the World Wide Web, a facility for adding scripts to handle user input. It allows a Web *server to communicate with other programs running on the same server in order to process data input by visitors to the Web site. CGI scripts 'parse' the input data, identifying each element and feeding it to the correct program for action, normally a *search engine or e-mail program. The results are then fed back to the user in the form of search results or sent by e-mail. The CGI also describes how CGI applications should present their output, so that it can be displayed properly by the Web browser. CGI scripts are written in a variety of langauges, one of the most common being *Perl.

common logarithm another name for a *logarithm to the base ten.

communicable disease any disease that can be passed from one person to another, either by direct or indirect contact, including droplet infection (as in sneezing, for example).

communication in biology, the signalling of information by one organism to another, usually with the intention of altering the recipient's behaviour. Signals used in communication may be **visual** (such as the human smile or the display of colourful plumage in birds), **auditory** (for example, the whines or barks of a dog), **olfactory** (such as the odours released by the scent glands of a deer), **electrical** (as in the pulses emitted by electric fish), or **tactile** (for example, the nuzzling of male and female elephants).

communications program or **comms program**, in the early days of electronic communication, general-purpose program for accessing older online systems and *bulletin board systems which used a *command-line interface;

also known as a terminal emulator. Programs such as Internet Explorer, Netscape, Outlook Express, and Eudora have made the complexities of setting up such programs a thing of the past.

communications satellite relay station in space for sending telephone, television, telex, and other messages around the world. Messages are sent to and from the satellites via ground stations. Most communications satellites are in geostationary orbit, appearing to hang fixed over one point on the Earth's surface.

community in ecology, an assemblage (group) of plants, animals, and other organisms living within a defined area. Communities are usually named by reference to a dominant feature, such as characteristic plant species (for example, a beech-wood community), or a prominent physical feature (for example, a freshwater-pond community).

commutative operation in mathematics, an operation that is independent of the order of the numbers or symbols concerned. For example, addition is commutative: the result of adding $4 + 2$ is the same as that of adding $2 + 4$; subtraction is not as $4 - 2 = 2$, but $2 - 4 = -2$. Compare *associative operation and distributive operation.

commutator device in a DC (direct-current) electric motor that reverses the current flowing in the armature coils as the armature rotates.

A DC generator, or *dynamo, uses a commutator to convert the AC (alternating current) generated in the armature coils into DC. A commutator consists of opposite pairs of conductors insulated from one another, and contact to an external circuit is provided by carbon or metal brushes.

comparative psychology branch of psychology concerned with differences in the behaviour of various animal species; also the study of animal psychology in general. The most important area of research has been that of learning, covering topics such as *conditioning, *behaviourism, and the effects of reward and punishment on performance.

The effects of various drugs on psychological processes and behaviour has also been an important area of study, as has maternal behaviour and interactions between mothers and offspring, particularly in mammals, together with the insights gained in our understanding of infant development. A number of experimental techniques are used, including research under laboratory conditions and field studies where the behaviour of animals is observed under natural conditions.

compensation point in biology, the point at which there is just enough light for a plant to survive. At this point all the food produced by *photosynthesis is used up by *respiration. For aquatic plants, the compensation point is the depth of water at which there is just enough light to sustain life (deeper water = less light = less photosynthesis).

competition in ecology, the interaction between two or more organisms, or groups of organisms, that use a common resource in short supply. There can be competition between members of the same species and competition between members of different species. Competition invariably results in a reduction in the numbers of one or both competitors, and in *evolution contributes both to the decline of certain species and to the evolution of *adaptations.

The resources in short supply for which organisms compete may be obvious things, such as *mineral salts for animals and plants, or light for plants. However, there are less obvious resources. For example, competition for suitable nesting sites is important in some species of birds. Competition results in a reduction in breeding success for one or other organism(s). Because of this it is one of the most important aspects of *natural selection, which may result in evolutionary change if the *environment is changing. Competition also results in the distribution of organisms we see in *habitats. It is believed that organisms tend to occur where the pressures of competition are not as great as in other areas. In agriculture cultivation methods are designed to reduce competition. For example, a crop of wheat is sown at a density that minimizes competition within the same species. The plants are grown far enough apart to reduce competition between the roots of neighbouring wheat plants for soil *mineral nutrients. The spraying of the ground to kill weeds reduces competition between the wheat and weed plants. Some weeds would grow taller than the wheat and deprive it of light.

compiler computer program that translates programs written in a *high-level language into machine code (the form in which they can be run by the computer). The compiler translates each high-level instruction into several machine-code instructions – in a process called **compilation** – and produces a complete independent program that can be run by the computer as often as required, without the original source program being present.

complement in mathematics, the set of the elements within the universal set that are not contained in the designated set. For example, if the universal set is the set of all positive whole numbers and the designated set S is the set of all even numbers, then the complement of S (denoted S') is the set of all odd numbers.

complementarity The theory that a fundamental particle is neither a wave nor a particle, because these are complementary modes of description. Theory proposed by Danish physicist Niels Bohr (1885–1962).

complementary angles two angles that add up to 90°.

complementary medicine in medicine, systems of care based on methods of treatment or theories of disease that differ from those taught in most western medical schools. See *medicine, alternative.

complementary metal-oxide semiconductor CMOS, in electronics, a particular way of manufacturing integrated circuits (chips). The main advantage of CMOS chips is their low power requirement and heat dissipation, which enables them to be used in electronic watches and portable microcomputers. However, CMOS circuits are expensive to manufacture and have lower operating speeds than have circuits of the *transistor–transistor logic (TTL) family.

complementary number in number theory, the number obtained by subtracting a number from its base. For example, the complement of 7 in numbers to base 10 is 3. Complementary numbers are necessary in computing, as the only mathematical operation of which digital computers (including pocket calculators) are directly capable is addition. Two numbers can be subtracted by adding one number to the complement of the other; two numbers can be divided by using successive subtraction (which, using complements, becomes successive addition); and multiplication can be performed by using successive addition.

complementation in genetics, the interaction that can occur between two different mutant alleles of a gene in a *diploid organism, to make up for each other's deficiencies and allow the organism to function normally.

completing the square solving a quadratic equation by replacing one quadratic expression $x^2 + bx + c$ by $(x + b/2)^2 + c - (b/2)^2$ and so obtaining a purely quadratic equation with no linear term.

complex number in mathematics, a number written in the form $a + ib$, where a and b are *real numbers and i is the square root of -1 (that is, $i^2 = -1$); i used to be known as the

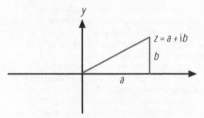

complex number A complex number can be represented graphically as a line whose end-point coordinates equal the real and imaginary parts of the complex number. This type of diagram is called an Argand diagram after the French mathematician Jean Argand (1768–1822) who devised it.

'imaginary' part of the complex number. Some equations in algebra, such as those of the form

$$x^2 + 5 = 0$$

cannot be solved without recourse to complex numbers, because the real numbers do not include square roots of negative numbers.

The sum of two or more complex numbers is obtained by adding separately their real and imaginary parts, for example:

$$(a + bi) + (c + di) = (a + c) + (b + d)i$$

Complex numbers can be represented graphically on an *Argand diagram, which uses rectangular *Cartesian coordinates in which the x-axis represents the real part of the number and the y-axis the imaginary part. Thus the number $z = a + ib$ is plotted as the point (a, b). Complex numbers have applications in various areas of science, such as the theory of alternating currents in electricity.

component in mathematics, one of the vectors produced when a single vector is resolved into two or more parts. The vector sum of the components gives the original vector.

component in physics, one of two or more vectors, normally at right angles to each other, that add together to produce the same effect as a single resultant vector. Any *vector quantity, such as force, velocity, or electric field, can be resolved into components chosen for ease of calculation. For example, the weight of a body resting on a slope can be resolved into two force components (see *resolution of forces); one normal to the slope and the other parallel to the slope.

Component Object Model COM, Microsoft's framework for creating object-oriented software, and a rival to the industry standard Corba. COM is the underlying foundation for Microsoft's *ActiveX.

COM port abbreviation for communication port, on a personal computer (PC), one of the serial *ports through which data communications take place. PCs may have up to four COM ports. However, these cannot all be used simultaneously as COM1 and COM3 share an interrupt, as do COM2 and COM4. A modem added to a machine with a mouse on COM1 must be attached to COM2 or COM4.

composite in industry, any purpose-designed engineering material created by combining materials with complementary properties into a composite form. Most composites have a structure in which one component consists of discrete elements, such as fibres, dispersed in a continuous matrix. For example, lengths of asbestos, glass, or carbon steel, or 'whiskers' (specially grown crystals a few millimetres long) of substances such as silicon carbide may be dispersed in plastics, concrete, or steel.

composite function in mathematics, a function made up of two or more other functions carried out in sequence, usually denoted by *, as in the relation $(f * g) x = f [g(x)]$.

Usually, composition is not commutative: $(f * g)$ is not necessarily the same as $(g * f)$.

composite volcano steep-sided conical *volcano formed above a *subduction zone at a destructive *plate margin. It is made up of alternate layers of ash and lava. The *magma (molten rock) associated with composite volcanoes is very thick and often clogs up the vent. This can cause a tremendous build-up of pressure, which, once released, causes a very violent eruption. Examples of composite volcanoes are Mount St Helens in the USA, and Stromboli and Vesuvius in Italy.

compound chemical substance made up of two or more *elements bonded together, so that they cannot be separated by physical means. Compounds are held together by ionic or covalent bonds.

The name of a compound may give a clue to its composition. If the name ends in -ide (with the notable exceptions of hydroxides and ammonium chloride) it usually contain two elements. For example calcium oxide is a compound of calcium and oxygen.

If the name ends in -ate or -ite the compound contains oxygen; compounds ending in -ate have a greater proportion of oxygen than those ending in -ite. For example, sodium sulphate (Na_2SO_4) contains more oxygen than does sodium sulphite (Na_2SO_3).

If the name starts with per- the compound contains extra oxygen. For example, hydrogen peroxide H_2O_2 contains one more oxygen than hydrogen oxide (water) H_2O.

The prefix thio- indicates that the compound contains an atom of sulphur in place of oxygen. For example, sodium thiosulphate ($Na_2S_2O_3$) contains one more sulphur and one less oxygen than the more common sodium sulphate (Na_2SO_4).

The proportions of the different elements in a compound are shown by the chemical formula of that compound. For example, a molecule of sodium sulphate, represented by the formula Na_2SO_4 contains two atoms of sodium, one of sulphur, and four of oxygen.

Compressed Serial Line Internet Protocol in computing, protocol usually abbreviated to CSLIP.

compression in computing, see *data compression.

compression region of a *sound wave where the particles of the medium through which it is travelling have been pushed close together, initially by the vibrating object that is the source of the sound. Sound waves consist of alternate regions of compressions and *rarefactions travelling away from the source.

Compton effect The increase in wavelength of X-rays scattered by such light elements as carbon, observed by US physicist Arthur Holly Compton, (1892–1962) in 1923. He concluded from this unexpected result that the X-rays were displaying both wavelike and particlelike properties.

computerized axial tomography medical technique, usually known as *CAT scan, for noninvasive investigation of disease or injury.

computer program coded instructions for a computer; see *program.

computing device any device built to perform or help perform computations, such as the abacus, slide rule, or computer.

concave of a surface, curving inwards, or away from the eye. For example, a

bowl appears concave when viewed from above. In geometry, a concave polygon is one that has an interior angle greater than 180°. Concave is the opposite of *convex.

concave lens lens that possesses at least one surface that curves inwards. It is a diverging lens, spreading out those light rays that have been refracted through it. A concave lens is thinner at its centre than at its edges, and is used to correct short-sightedness (myopia).

concave mirror curved mirror that reflects light from its inner surface, the curve being inward. It may be either circular or parabolic in section. A concave mirror converges parallel light rays inward to the point of principal focus. The image formed by a concave mirror is real (reduced and inverted) if the object is not too close to the mirror. A real image is formed at the point of convergence. If the object is close to the mirror then the image formed will be virtual, enlarged, and upright, as the rays of light cannot converge to a point.

concentration in chemistry, the amount of a substance (*solute) present in a specified amount of a solution. Either amount may be specified as a mass or a volume (liquids only). Common units used are *moles per cubic decimetre, grams per cubic decimetre, grams per 100 cubic centimetres, and grams per 100 grams. The term also refers to the process of increasing the concentration of a solution by removing some of the substance (*solvent) in which the solute is dissolved. In a **concentrated solution**, the solute is present in large quantities. Concentrated brine is around 30% sodium chloride in water; concentrated caustic soda (caustic liquor) is around 40% sodium hydroxide; and concentrated sulphuric acid is 98% acid.

concentration gradient change in the concentration of a substance from one area to another. Particles, such as sugar molecules, in a fluid move over time so that they become evenly distributed throughout the fluid. In particular, they move from an area of high concentration to an area of low concentration; that is, they diffuse along the concentration gradient (see *diffusion).

This explains why oxygen in the lungs will diffuse into the blood supply. The oxygen molecules are more concentrated in the lungs than they are in the capillaries surrounding the *alveoli (air sacs). As it diffuses along the concentration gradient, oxygen will tend to pass into the blood. Gas exchange therefore depends on the maintenance of a concentration gradient, so that oxygen will continue to diffuse across the respiratory surface.

concentric circles two or more circles that share the same centre.

concrete building material composed of cement, stone, sand, and water. It has been used since Roman times. Since the late 19th century, it has been increasingly employed as an economical alternative to materials such as brick and wood, and has been combined with steel to increase its tension capacity.

concurrent lines two or more lines passing through a single point; for example, the diameters of a circle are all concurrent at the centre of the circle.

concussion temporary unconsciousness resulting from a blow to the head. It is often followed by amnesia for events immediately preceding the blow.

condensation in organic chemistry, a reaction in which two organic compounds combine to form a larger molecule, accompanied by the removal of a smaller molecule (usually water). This is also known as an addition–elimination reaction. Polyamides (such as nylon) and polyesters (such as Terylene) are made by *condensation polymerization.

condensation conversion of a vapour to a liquid. This is frequently achieved by letting the vapour come into contact with a cold surface. It is the process by which water vapour turns into fine water droplets to form a *cloud.

Condensation in the atmosphere occurs when the air becomes completely saturated and is unable to hold any more water vapour. As air rises it cools and contracts – the cooler it becomes the less water it can hold. Rain is frequently associated with warm weather fronts because the air rises and cools, allowing the water vapour to condense as rain. The temperature at which the air becomes saturated is known as the ***dew point**. Water vapour will not condense in air if there are not enough condensation nuclei (particles of dust, smoke, or salt) for the droplets to form on. It is then said to be supersaturated.

Condensation is an important part of the *water cycle.

condensation polymerization *polymerization reaction in which one or more monomers, with more than one reactive functional group, combine to form a polymer with the elimination of water or another small molecule.

condenser laboratory apparatus used to condense vapours back to liquid so that the liquid can be recovered. It is used in *distillation and in reactions where the liquid mixture can be kept boiling without the loss of solvent.

condenser in electronic circuits, a former name for a *capacitor.

condenser in optics, a *lens or combination of lenses with a short focal length used for concentrating a light source onto a small area, as used in a slide projector or microscope substage lighting unit. A condenser can also be made using a concave mirror.

conditioning in psychology, two major principles of behaviour modification.

In **classical conditioning**, described by Russian psychologist Ivan Pavlov (1849–1936), a new stimulus can evoke an automatic response by being repeatedly associated with a stimulus that naturally provokes that response. For example, the sound of a bell repeatedly associated with food will eventually trigger salivation, even if sounded without food being presented. In **operant conditioning**, described by US psychologists Edward Lee Thorndike (1874–1949) and B F Skinner (1904–1990), the frequency of a voluntary response can be increased by following it with a reinforcer or reward.

conductance ability of a material to carry an electrical current, usually given the symbol G. For a direct

current, it is the reciprocal of
*resistance: a conductor of resistance R
has a conductance of $1/R$. For an
alternating current, conductance is the
resistance R divided by the
*impedance Z: $G = R/Z$. Conductance
was formerly expressed in reciprocal
ohms (or mhos); the SI unit is the
*siemens (S).

conduction, electrical flow of charged
particles through a material giving rise
to electric current. Conduction in
metals involves the flow of negatively
charged free *electrons. Conduction in
gases and some liquids involves the
flow of *ions that carry positive
charges in one direction and negative
charges in the other. Conduction in a
*semiconductor such as silicon
involves the flow of electrons and
positive holes.

conduction, heat flow of heat energy
through a material without the
movement of any part of the material
itself (compare *conduction, electrical).
Heat energy is present in all materials
in the form of the *kinetic energy of
their constituent vibrating particles, and
may be conducted from one particle to
the next in the form of this vibration.

Different materials conduct heat at
different rates. This rate is called the
thermal conductivity. A good
conductor of heat, such as steel, will
have a high thermal conductivity and
poor conductor of heat, such as air,
will have a low thermal conductivity.
In general, non-metals, such as wood
or glass, are poor conductors of heat.

In the construction industry, the
thermal conductivities of the materials
to be installed need to be known. A
low thermal conductivity indicates that
a material is a poor conductor of heat
and therefore a good insulator. For
example, foam is used to insulate lofts,
the air trapped in the foam making it a
good insulator. Bricks used in outer
walls have tiny air spaces, again
allowing air to be trapped and making
the bricks good insulators. Insulation
can also be achieved with double-
glazed windows, with the space
between the two glass panes
containing air or a vacuum.

conductivity, thermal (W m⁻¹ K⁻¹)
measure of how well a material
conducts heat. A good conductor, such
as a metal, has has a high conductivity;
a poor conductor, called an insulator,
has a low conductivity. The
measurement of a material's heat-
conducting properties is given as its
U-value.

conductor any material that conducts
heat or electricity (as opposed to an
insulator, or nonconductor). A good
conductor has a high electrical or heat
conductivity, and is generally a
substance rich in loosely-held free
electrons, such as a metal. Copper and
aluminium are good conductors. A
poor conductor (such as the non-
metals glass, porcelain, and rubber)
has few free electrons and resists the
flow of electricity or heat. *Carbon is
exceptional in being non-metallic and
yet (in some of its forms) a relatively
good conductor of heat and electricity.
Substances such as *silicon and
*germanium, with intermediate
conductivities that are improved by
heat, light, or impurities, are known as
*semiconductors.

Liquids (including water) can also
be electrical conductors. Electricity
(current) can flow by the movement of
charged ions through a solution or a
molten salt (electrolyte). This process is
called electrolysis.

cone in zoology, type of light-sensitive
cell found in the retina of the *eye.

cone in geometry, a pyramid with a
circular base. If the point (vertex) is
directly above the centre of the circle,
it is known as a **right circular cone**.
The *volume (V) of this cone is given
by the formula

$$V = \frac{1}{3}\pi p r^2 h$$

where h is the perpendicular height
and r is the base radius.

config.sys in computing, the
*configuration file used by the MS-
DOS and OS/2 *operating systems.
It is read when the system is booted.
It is used to load device drivers and set
certain parameters used by the
operating system.

configuration in chemistry, the
arrangement of atoms in a molecule or
of electrons in atomic orbitals.

configuration in computing, the way in which a system, whether it be *hardware and/or *software, is set up. A minimum configuration is often referred to for a particular application, and this will usually include a specification of processor, disk and memory size, and peripherals required.

confluence point at which two rivers join, for example the River Thames and River Cherwell at Oxford, England, or the White Nile and Blue Nile at Omdurman, Sudan.

congenital disease in medicine, a disease that is present at birth. It is not necessarily genetic in origin; for example, congenital herpes may be acquired by the baby as it passes through the mother's birth canal.

conglomerate in geology, a type of *sedimentary rock composed of rounded fragments ranging in size from pebbles to boulders, cemented together in a finer sand or clay material.

congruent in geometry, having the same shape and size (and area), as applied to two-dimensional or solid figures. With plane congruent figures, one figure will fit on top of the other exactly, though this may first require *rotation, *translation, or *reflection of one of the figures.

conical having the shape of a *cone.

conic section curve obtained when a conical surface is intersected by a plane. If the intersecting plane cuts both extensions of the cone, it yields a *hyperbola; if it is parallel to the side of the cone, it produces a *parabola. Other intersecting planes produce *circles or *ellipses.

conjugate in mathematics, a term indicating that two elements are connected in some way; for example, $(a + ib)$ and $(a - ib)$ are conjugate complex numbers.

conjugate angles two angles that add up to 360°.

conjugation in biology, temporary union of two single cells (or hyphae in fungi) with at least one of them receiving genetic material from the other: the bacterial equivalent of sexual reproduction. A fragment of the *DNA from one bacterium is passed along a thin tube, the pilus, into another bacterium.

conjugation in organic chemistry, the alternation of double (or triple) and single carbon–carbon bonds in a molecule – for example, in penta-1,3-diene, $H_2C=CH–CH=CH–CH_3$. Conjugation imparts additional stability as the double bonds are less reactive than isolated double bonds.

conjunctiva membrane covering the front of the vertebrate *eye. It is continuous with the epidermis of the eyelids, and lies on the surface of the cornea.

conjunctivitis inflammation of the conjunctiva, the delicate membrane that lines the inside of the eyelids and covers the front of the eye. Symptoms include redness, swelling, and a watery or pus-filled discharge. It may be caused by infection, allergy, or other irritant.

connective tissue in animals, tissue made up of a noncellular substance, the *extracellular matrix, in which some cells are embedded. Skin, bones, tendons, cartilage, and adipose tissue (fat) are the main connective tissues. There are also small amounts of connective tissue in organs such as the brain and liver, where they maintain shape and structure.

connectivity in mathematics, measure of how well connected places are in a transport network. A simple measure is the beta index.

conservation in the life sciences, action taken to protect and preserve the natural world, usually from pollution, overexploitation, and other harmful features of human activity. The late 1980s saw a great increase in public concern for the environment, with membership of conservation groups, such as Friends of the Earth, Greenpeace, and the US Sierra Club, rising sharply and making the *green movement an increasingly-powerful political force. Globally the most important issues include the depletion of atmospheric ozone by the action of *chlorofluorocarbons (CFCs), the build-up of carbon dioxide in the atmosphere (thought to contribute to the *greenhouse effect), and *deforestation.

conservation of energy principle that states that in a chemical reaction, the total amount of energy in the system remains unchanged. Energy can be transferred from one form into another but cannot be created or destroyed.

In a chemical reaction, for each component there may be changes in energy due to change of physical state, changes in the nature of chemical bonds, and either an input or output of energy. However, there is no net gain or loss of energy. This is true outside of chemical reactions; for example the chemical energy of a battery is transferred into the electrical energy in a circuit, which can then be transferred into the heat and light energy of a bulb, but the total amount of energy is constant.

conservation of mass in chemistry, the principle that states that in a chemical reaction the sum of all the masses of the substances involved in the reaction (reactants) is equal to the sum of all of the masses of the substances produced by the reaction (products) – that is, no matter is gained or lost.

conservation of momentum in mechanics, a law that states that total *momentum is conserved (remains constant) in all collisions, providing no external resultant force acts on the colliding bodies. The principle may be expressed as an equation used to solve numerical problems: total momentum before collision = total momentum after collision.

constant in mathematics, a fixed quantity or one that does not change its value in relation to *variables. For example, in the algebraic expression $y^2 = 5x - 3$, the numbers 3 and 5 are constants. In physics, certain quantities are regarded as universal constants, such as the speed of light in a vacuum.

constant composition, law of in chemistry, the law that states that the proportions of the amounts of the elements in a pure compound are always the same and are independent of the method by which the compound was produced.

constellation one of the 88 areas into which the sky is divided for the purposes of identifying and naming celestial objects. The first constellations were simple, arbitrary patterns of stars in which early civilizations visualized gods, sacred beasts, and mythical heroes.

construction in geometry, an addition to a figure, which is drawn to help solve a problem or produce a proof. The term is also applied to the accurate drawing of shapes.

constructive margin in *plate tectonics, a boundary between two lithospheric plates, along which new crust is being created. The term usually refers to *divergent margins, where two oceanic plates are moving away from each other. As they diverge, magma (molten rock) wells up to fill the open space, and solidifies, forming new oceanic crust. Similar processes occur where a continent plate is beginning to split apart.

consumption (Latin *consumptio* 'wasting') in medicine, former popular name for the disease *tuberculosis.

contact force force or push produced when two objects are pressed together and their surface atoms try to keep them apart. Contact forces always come in pairs – for example, the downward force exerted on a floor by the sole of a person's foot is matched by an equal upward force exerted by the floor on that sole.

contact process main industrial method of manufacturing the chemical *sulphuric acid. Sulphur dioxide is produced by burning sulphur: $S + O_2 \rightarrow SO_2$. Sulphur dioxide and air are passed over a hot (450°C) *catalyst of vanadium(V) oxide: $2SO_2 + O_2 \leftrightarrow 2SO_3$. This reaction is *reversible and so reaches *chemical equilibrium; the conditions are chosen to give the best yield in a short time. The sulphur trioxide produced is absorbed in concentrated sulphuric acid to make fuming sulphuric acid, or oleum ($H_2S_2O_7$). The oleum is then diluted with water to give concentrated sulphuric acid (98%): $H_2S_2O_7 + H_2O \rightarrow 2H_2SO_4$. Unreacted gases are recycled.

contaminated land land that is considered to pose a health risk to humans because of pollution; usually land that has been the site of industrial activity.

continent any one of the seven large land masses of the Earth, as distinct from the oceans. They are Asia, Africa, North America, South America, Europe, Australia, and Antarctica. Continents are constantly moving and evolving (see *plate tectonics). A continent does not end at the coastline; its boundary is the edge of the shallow continental shelf, which may extend several hundred kilometres out to sea.

Continental crust, as opposed to the crust that underlies the deep oceans, is composed of a wide variety of igneous, sedimentary, and metamorphic rocks. The rocks vary in age from recent (currently forming) to almost 4,000 million years old. Unlike the ocean crust, the continents are not only high standing, but extend to depths as great at 70 km/45 mi under high mountain ranges. Continents, as high, dry masses of rock, are present on Earth because of the density contrast between them and the rock that underlies the oceans. Continental crust is both thick and light, whereas ocean crust is thin and dense. If the crust were the same thickness and density everywhere, the entire Earth would be covered in water.

At the centre of each continental mass lies a shield or craton, a deformed mass of old *metamorphic rocks dating from Precambrian times. The shield is thick, compact, and solid (the Canadian Shield is an example), and is usually worn flat. Around the shield is a concentric pattern of fold mountains, with older ranges, such as the Rockies, closest to the shield, and younger ranges, such as the coastal ranges of North America, farther away. This general concentric pattern is modified when two continental masses have drifted together and they become welded with a great mountain range along the join, the way Europe and northern Asia are joined along the Urals. If a continent is torn apart, the new continental edges have no mountains; for instance, South America has mountains (the Andes) along its western flank, but none along the east where it tore away from Africa 200 million years ago.

continental climate   type of climate typical of a large, mid-latitude land mass. This type of climate is to be found in the centre of continents between approximately 40° and 60° north of the Equator.

The two main areas are the North American prairies and the Russian steppes (both grasslands). Areas with a continental climate are a long way from the oceans, which affects their climate in two ways: they have very low annual rainfall because little atmospheric moisture is available; and their temperature range is very large over a year, because the temperature-moderating effect of the sea has been lost. Maximum mean monthly summer temperatures for continental climates are around 20°C/68°F. During the winter there are several months when the temperature remains below freezing point. Precipitation decreases rapidly towards the east in Russia as distance from the sea increases, whereas in North America totals are lowest to the west in the rain shadow (area sheltered from rain by hills) of the Rockies. In both areas, annual rainfall averages 500 mm/20 in, and there is a threat of drought. The ground is snow-covered for several months between October and April.

continental drift   in geology, the theory that, about 250–200 million years ago, the Earth consisted of a single large continent (*Pangaea), which subsequently broke apart to form the continents known today. The theory was first proposed in 1912 by German meteorologist Alfred Wegener, but such vast continental movements could not be satisfactorily explained or even accepted by geologists until the 1960s. (See diagram, page 145.)

The theory of continental drift gave way to the theory of *plate tectonics. Whereas Wegener proposed that continents pushed their way through underlying mantle and ocean floor, plate tectonics states that continents are just part of larger lithospheric plates (which include ocean crust as well) that move laterally over the Earth's surface.

continental rise   portion of the ocean floor rising gently from the abyssal plain toward the steeper continental slope. The continental rise is a depositional feature formed from sediments transported down the slope mainly by turbidity currents. Much of the continental rise consists of coalescing submarine alluvial fans bordering the continental slope.

continental shelf submerged edge of a continent, a gently sloping plain that extends into the ocean. It typically has a gradient of less than 1°. When the angle of the sea bed increases to 1°–5° (usually several hundred kilometres away from land), it becomes known as the *continental slope.

continental slope sloping, submarine portion of a continent. It extends downward from the edge of the continental shelf. In some places, such as south of the Aleutian Islands of Alaska, continental slopes extend directly to the ocean deeps or abyssal plain. In others, such as the east coast

Upper Carboniferous period

Eocene

Lower Quaternary

continental drift The continents are slowly shifting their positions, driven by fluid motion beneath the Earth's crust. Over 200 million years ago, there was a single large continent called Pangaea. By 200 million years ago, the continents had started to move apart. By 50 million years ago, the continents were approaching their present positions.

of North America, they grade into the gentler continental rises that in turn grade into the abyssal plains.

continuous data in mathematics, data that can take any of an infinite number of values between whole numbers and so may not be measured completely accurately. This type of data contrasts with *discrete data, in which the variable can only take one of a finite set of values. For example, the sizes of apples on a tree form continuous data, whereas the numbers of apples form discrete data.

continuous variation or **quantitative variation**, slight difference of an individual characteristic, such as height, across a sample of the population. Although there are very tall and very short humans, there are also many people with an intermediate height. The same applies to weight. Continuous variation can be due to the influence of many, rather than a single, genes, or from the influence of individuals' environments.

continuum in mathematics, a *set that is infinite and everywhere continuous, such as the set of points on a line.

contraceptive any drug, device, or technique that prevents pregnancy. The contraceptive pill (the Pill) contains female hormones that interfere with egg production or the first stage of pregnancy. The 'morning-after' pill can be taken up to 72 hours after unprotected intercourse. Barrier contraceptives include condoms (sheaths), femidoms (a female condom), and diaphragms, also called caps or Dutch caps; they prevent the sperm entering the cervix (neck of the womb).

Intrauterine devices, also known as IUDs or coils, cause a slight inflammation of the lining of the womb; this prevents the fertilized egg from becoming implanted.

contractile vacuole tiny organelle found in many single-celled freshwater organisms. It slowly fills with water, and then contracts, expelling the water from the cell.

control in biology, the process by which a tissue, an organism, a population, or an ecosystem maintains itself in a balanced, stable state. Blood sugar must be kept at a stable level if the brain is to function properly, and this steady state is maintained by an interaction between the liver, the hormone insulin, and a detector system in the pancreas.

control bus in computing, the electrical pathway, or *bus, used to communicate control signals.

control character in computing, any character produced by depressing the control key (Ctrl) on a keyboard at the same time as another (usually alphabetical) key. The control characters form the first 32 *ASCII characters and most have specific meanings according to the operating system used. They are also used in combination to provide formatting control in many word processors, although the user may not enter them explicitly.

control experiment essential part of a scientifically valid experiment, designed to show that the factor being tested is actually responsible for the effect observed. In the control experiment all factors, apart from the one under test, are exactly the same as in the test experiments, and all the same measurements are carried out. In drug trials, a placebo (a harmless substance) is given alongside the substance being tested in order to compare effects.

control total in computing, a *validation check in which an arithmetic total of a specific field from a group of records is calculated. This total is input together with the data to which it refers. The program recalculates the control total and compares it with the one entered to ensure that no entry errors have been made.

convection transfer of heat energy that involves the movement of a fluid (gas or liquid). Fluid in contact with the source of heat expands and tends to rise within the bulk of the fluid. Cooler fluid sinks to take its place, setting up a convection current. This is the principle of natural convection in many domestic hot-water systems and space heaters.

Hot-air balloons use convection currents in order to rise into the air. The air inside the balloon is heated. As the hot air rises, so does the balloon. For the balloon to descend, the air in the balloon is cooled or allowed to escape.

Gravity and air currents allow gliders to fly. The glider falls by the force of gravity acting on it. It gains height in the air from rising, warm currents of air known as thermals. These currents are formed by air being heated by the ground; the heated air becomes less dense and rises.

convectional rainfall rainfall associated with hot climates, resulting from the uprising of convection currents of warm air. Air that has been warmed by the extreme heating of the ground surface rises to great heights and is abruptly cooled. The water vapour

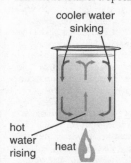
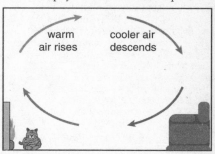

convection Convection is the transfer of heat energy in a liquid or a gas. The fluid near the heat source is heated, becomes less dense, and rises upwards. Once away from the heat source, the fluid starts to cool, becomes denser, and starts to sink. A kettle uses the principle of convection currents. As the water at the bottom is heated by the element, it rises and brings cool water closer to the element.

carried by the air condenses and rain falls heavily. Convectional rainfall is often associated with a *thunderstorm.

convection current current caused by the expansion of a liquid, solid, or gas as its temperature rises. The expanded material, being less dense, rises, while colder, denser material sinks. Material of neutral buoyancy moves laterally. Convection currents arise in the atmosphere above warm land masses or seas, giving rise to *sea breezes and land breezes, respectively. In some heating systems, convection currents are used to carry hot water upwards in pipes.

conventional current direction in which an electric current is considered to flow in a circuit. By convention, the direction is that in which positive-charge carriers would flow – from the positive terminal of a cell to its negative terminal. It is opposite in direction to the flow of electrons. In circuit diagrams, the arrows shown on symbols for components such as diodes and transistors point in the direction of conventional current flow.

convergence in mathematics, the property of a series of numbers in which the difference between consecutive terms gradually decreases. The sum of a converging series approaches a limit as the number of terms tends to *infinity.

convergent evolution or **convergence**, in biology, the independent evolution of similar structures in species (or other taxonomic groups) that are not closely related, as a result of living in a similar way. Thus, birds and bees have wings, not because they are descended from a common winged ancestor, but because their respective ancestors independently evolved flight.

converging lens lens that converges or brings to a focus those light rays that have been refracted by it. It is a *convex lens, with at least one surface that curves outwards, and is thicker towards the centre than at the edge. Converging lenses are used to form real images in many optical instruments, such as cameras and projectors. A converging lens that forms a virtual, magnified image may be used as a magnifying glass or to correct long-sightedness (*hypermetropia).

converse in mathematics, the reversed order of a conditional statement; the converse of the statement 'if a, then b' is 'if b, then a'. The converse does not always hold true; for example, the converse of 'if $x = 3$, then $x^2 = 9$' is 'if $x^2 = 9$, then $x = 3$', which is not true, as x could also be -3.

convex of a surface, curving outwards, or towards the eye. For example, the outer surface of a ball appears convex. In geometry, the term is used to describe any polygon possessing no interior angle greater than 180°. Convex is the opposite of *concave.

convex lens lens that possesses at least one surface that curves outwards. It causes light to deviate inward, bringing the rays of light to a focus, and is thus called a *converging lens. A convex lens is thicker at its centre than at its edges, and is used to correct long-sightedness (hypermetropism).

convex mirror curved mirror that reflects light from its outer surface, the curve being outward. Rays of light are caused to diverge outward on reflection. It forms a reduced, upright, virtual image. Convex mirrors give a wide field of view and are therefore particularly suitable for car wing mirrors and surveillance purposes in shops.

cookie on the World Wide Web, a short piece of text that a Web site stores in a Cookies folder or a cookie.txt file on the user's computer, either for tracking or configuration purposes, for example, to gather information about individuals and improve the targeting of banner advertisements. Cookies can also store user preferences and passwords. Special instructions are issued to the browser by the server when data held in a cookie is required. The transaction is a hidden one, with the recipient often unaware that the cookie has been delivered or accessed. Later versions of Netscape and Microsoft browsers allow the user to block acceptance of cookies. However, some Web sites will not admit surfers who will not accept a cookie file.

cookie recipe urban legend that circulates around the Net. The story is about a protagonist who ate some delicious cookies for dessert after a

meal in a fancy department store or
restaurant. When asked for the recipe,
the waiter refuses, but finally relents
saying it will cost 'two fifty'. When the
bill comes, the protagonist discovers
the restaurant has charged $250.
Feeling stung, he/she posts the recipe
to the Net to ensure the maximum
distribution (and therefore revenge)
possible.

cool temperate climate   There are two
main cool temperate areas, which can
be classified as *continental climate
and western margins. Periods of one to
five months are below 6°C/43°F in
these climate areas.

   Western margin climates are often
referred to as 'northwest European',
and are found on west coasts between
approximately 40° and 60° north and
south of the equator. Other areas with
similar climatic characteristics are
northwest USA and British Columbia,
southern Chile, New Zealand's South
Island, and Tasmania. Summers are
cool with the warmest month at a
temperature of 15–20°C/59–68°F. This
is due to the low angle of the sun in
the sky combined with frequent cloud
cover and the cooling influence of the
sea. Winters are mild in comparison.
Mean monthly temperatures remain a
few degrees above freezing due to the
warming effect of the sea. Autumn is
usually warmer than spring; seasonal
temperature variations depend on
prevailing air masses. Precipitation
often exceeds 2,000 mm/6.5 ft annually
and falls throughout the year, the
highest amount falling during winter
when depressions are more frequent
and intense. Snow is common in the
mountains.

coordinate   in geometry, a number that
defines the position of a point relative
to a point or axis (reference line).
*Cartesian coordinates define a point
by its perpendicular distances from two
or more axes drawn through a fixed
point mutually at right angles to each
other. *Polar coordinates define a point
in a plane by its distance from a fixed
point and direction from a fixed line.

coordinate geometry   or **analytical
geometry**, system of geometry in
which points, lines, shapes, and
surfaces are represented by algebraic

Cartesian coordinates

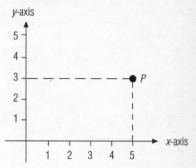

the Cartesian coordinates of *P* are (5,3)

Polar coordinates

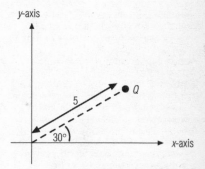

the Polar coordinates of *Q* are (5,30°)

coordinate Coordinates are numbers that
define the position of points in a plane or in
space. In the Cartesian coordinate system, a
point in a plane is charted based upon its
location along intersecting horizontal and
vertical axes. In the polar coordinate system, a
point in a plane is defined by its distance from a
fixed point and direction from a fixed line.

expressions. In plane (two-dimensional)
coordinate geometry, the plane is
usually defined by two axes at right
angles to each other, the horizontal
x-axis and the vertical y-axis, meeting
at O, the origin. A point on the plane
can be represented by a pair of
*Cartesian coordinates, which define
its position in terms of its distance
along the x-axis and along the y-axis

from O. These distances are, respectively, the x and y coordinates of the point.

Lines are represented as equations; for example, $y = 2x + 1$ gives a straight line, and $y = 3x^2 + 2x$ gives a *parabola (a curve). The graphs of varying equations can be drawn by plotting the coordinates of points that satisfy their equations, and joining up the points. One of the advantages of coordinate geometry is that geometrical solutions can be obtained without drawing but by manipulating algebraic expressions. For example, the coordinates of the point of intersection of two straight lines can be determined by finding the unique values of x and y that satisfy both of the equations for the lines, that is, by solving them as a pair of *simultaneous equations. The curves studied in simple coordinate geometry are the *conic sections (circle, ellipse, parabola, and hyperbola), each of which has a characteristic equation.

Coordinate geometry was founded by French philosopher and mathematician René Descartes (1596–1650).

coplanar in geometry, describing lines or points that all lie in the same plane.

copper chemical symbol Cu, red-brown, very malleable and ductile, metallic element, atomic number 29, relative atomic mass 63.546. Its symbol comes from the Latin *cuprum*. It is one of the *transition metals in the *periodic table. Copper is used for its durability, pliability, high thermal and electrical conductivity, and resistance to corrosion. It is used in electrical wires and cables, and water pipes and tanks.

Copper was the first metal used systematically for tools by humans; when mined and worked into utensils it formed the technological basis for the Copper Age in prehistory. When alloyed with tin it forms bronze, which is stronger than pure copper and may hold a sharp edge; the systematic production and use of this *alloy was the basis for the prehistoric Bronze Age. Brass, another hard copper alloy, includes zinc. The element's name comes from the Latin *cyprium* ('Cyprus metal'), due to Cyprus being a major source of copper.

copper(II) carbonate $CuCO_3$, green solid that readily decomposes to form black copper(II) oxide on heating:

$$CuCO_3 \rightarrow CuO + CO_2$$

It dissolves in dilute acids to give blue solutions with effervescence caused by the giving off of carbon dioxide.

$$CuCO_3 + H_2SO_4 \rightarrow CuSO_4 + CO_2 + H_2O$$

copper ore any mineral from which copper is extracted, including native copper, Cu; chalcocite, Cu_2S; chalcopyrite, $CuFeS_2$; bornite, Cu_5FeS_4; azurite, $Cu_3(CO_3)_2(OH)_2$; malachite, $Cu_2CO_3(OH)_2$; and chrysocolla, $CuSiO_3.2H_2O$.

copper(II) oxide CuO, black solid that is readily reduced to copper by carbon, carbon monoxide, or hydrogen if heated with any of these:

$$CuO + C \rightarrow Cu + CO$$

$$CuO + CO \rightarrow Cu + CO_2$$

$$CuO + H_2 \rightarrow Cu + H_2O$$

It is usually made in the laboratory by heating copper(II) carbonate, nitrate, or hydroxide:

$$2Cu(NO_3)_2 \rightarrow 2CuO + 4NO_2 + O_2$$

Copper(II) oxide is a typical basic oxide, dissolving readily in most dilute acids.

copper(II) sulphate $CuSO_4$, substance usually found as a blue, crystalline, hydrated salt $CuSO_4.5H_2O$ (also called blue vitriol). It is made from the action of dilute sulphuric acid on copper(II) oxide, hydroxide, or carbonate.

$$CuO + H_2SO_4 + 4H_2O \rightarrow CuSO_4.5H_2O$$

When the hydrated salt is heated gently it loses its water of crystallization and the blue crystals turn to a white powder. The reverse reaction is used as a chemical test for water.

$$CuSO_4.5H_2O \leftrightarrow CuSO_4 + 5H_2O$$

copulation act of mating in animals with internal *fertilization. Male mammals have a *penis or other organ that is used to introduce spermatozoa into the reproductive tract of the female. Most birds transfer sperm by pressing their cloacas (the openings of their reproductive tracts) together.

cordierite silicate mineral, $(Mg,Fe)_2Al_4Si_5O_{18}$, blue to purplish in colour. It is characteristic of metamorphic rocks formed from clay sediments under conditions of low pressure but moderate temperature; it is the mineral that forms the spots in spotted slate and spotted hornfels.

core in earth science, the innermost part of the Earth. It is divided into an outer core, which begins at a depth of 2,900 km/1,800 mi, and an inner core, which begins at a depth of 4,980 km/ 3,100 mi. Both parts are thought to consist of iron-nickel alloy. The outer core is liquid and the inner core is solid.

Coriolis effect effect of the Earth's rotation on the atmosphere, oceans, and theoretically all objects moving over the Earth's surface. In the northern hemisphere it causes moving objects and currents to be deflected to the right; in the southern hemisphere it causes deflection to the left. The effect is named after its discoverer, French mathematician Gaspard de Coriolis (1792–1843).

cornea transparent front section of the vertebrate *eye. The cornea is curved and behaves as a fixed lens, so that light entering the eye is partly focused before it reaches the lens.

corolla collective name for the petals of a flower. In some plants the petal margins are partly or completely fused to form a **corolla tube**, for example in bindweed *Convolvulus arvensis*.

corona faint halo of hot (about 2,000,000°C/3,600,000°F) and tenuous gas around the Sun, which boils from the surface. It is visible at solar *eclipses or through a **coronagraph**, an instrument that blocks light from the Sun's brilliant disc. Gas flows away from the corona to form the *solar wind. NASA's High-Energy Solar Spectroscopic Imager mission, launched in 2001, was to study the evolution of energy in the corona.

coronary in biology, describing any of several structures in the body that encircle an organ in the manner of a crown. For example, the coronary arteries arise from the aorta and deliver blood to the heart muscle.

coronary artery disease (Latin *corona* 'crown', from the arteries encircling the heart) condition in which the fatty deposits of *atherosclerosis form in the coronary arteries that supply the heart muscle, narrowing them and restricting the blood flow.

These arteries may already be hardened (arteriosclerosis). If the heart's oxygen requirements are increased, as during exercise, the blood supply through the narrowed arteries may be inadequate, and the pain of *angina results. A *heart attack occurs if the blood supply to an area of the heart is cut off, for example because a blood clot (thrombus) has blocked one of the coronary arteries. The subsequent lack of oxygen damages the heart muscle (infarct), and if a large area of the heart is affected, the attack may be fatal. Coronary artery disease tends to run in families and is linked to smoking, lack of exercise, and a diet high in saturated (mostly animal) fats, which tends to increase the level of blood *cholesterol. It is a common cause of death in many industrialized countries; older men are the most vulnerable group. The condition is treated with drugs or bypass surgery.

coronary thrombosis in medicine, acute manifestation of coronary heart disease causing a *heart attack.

corpuscle in biology, a small body. The cellular components of blood are sometimes referred to as corpuscles; see *red blood cell and *white blood cell.

corpuscular theory hypothesis about the nature of light championed by English physicist and mathematician Isaac Newton (1642–1727), who postulated that it consists of a stream of particles or corpuscles. The theory was superseded at the beginning of the 19th century by English physicist Thomas Young's wave theory. *Quantum theory and wave mechanics embody both concepts.

corpus luteum glandular tissue formed in the mammalian *ovary after ovulation from the Graafian follicle, a group of cells associated with bringing the egg to maturity. It secretes the hormone progesterone in anticipation of pregnancy.

correlation degree of relationship between two sets of information. If one set of data increases at the same time

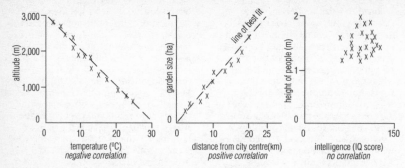

correlation Scattergraphs showing different kinds of correlation. In this way, a causal relationship between two variables may be proved or disproved, provided there are no hidden factors.

as the other, the relationship is said to be **positive** or direct. If one set of data increases as the other decreases, the relationship is **negative** or inverse. If there is no relationship between the two sets of data the relationship is said to be **zero linear correlation**.

Correlation can be shown by plotting a *line of best fit on a *scatter diagram. The steeper the line drawn, whether positive or negative, the stronger the correlation.

correspondence in mathematics, the relation between two sets of data where an operation on the members of one set maps some or all of them onto one or more members of the other. For example, if A is the set of members of a family and B is the set of months in the year, A and B are in correspondence if the operation is: '...has a birthday in the month of...'.

corresponding angles a pair of equal angles lying on the same side of a transversal (a line that cuts through two or more lines in the same plane), and making an interior and exterior angle with the intersected lines.

corridor, wildlife route linking areas of similar habitat, or between sanctuaries. For example there is a corridor linking the Masai Mara reserve in Kenya and the Serengeti in Tanzania. On a smaller scale, disused railways provide corridors into urban areas for foxes.

corrosion in earth science, an alternative name for *solution, the process by which water dissolves rocks such as limestone.

corrosion eating away and eventual destruction of metals and alloys by

chemical attack. The rusting of ordinary iron and steel is the most common form of corrosion. Rusting takes place in moist air, when the iron combines with oxygen and water to form a brown-orange deposit of *rust (hydrated iron oxide). The rate of corrosion is increased where the atmosphere is polluted with sulphur dioxide. Salty road and air conditions accelerate the rusting of car bodies.

Corrosion is largely an electrochemical process, and acidic and salty conditions favour the establishment of electrolytic cells on the metal, which cause it to be eaten away. Other examples of corrosion include the green deposit that forms on copper and bronze, called verdigris, a basic copper carbonate. The tarnish on silver is a corrosion product, a film of silver sulphide.

corruption of data introduction or presence of errors in data. Most computers use a range of verification and *validation routines to prevent corrupt data from entering the computer system or detect corrupt data that are already present.

cortex in biology, the outer part of a structure such as the brain, kidney, or adrenal gland. In botany the cortex includes nonspecialized cells lying just beneath the surface cells of the root and stem.

corticosteroid any of several steroid hormones secreted by the cortex of the *adrenal glands; also synthetic forms with similar properties. Corticosteroids have anti-inflammatory and immunosuppressive effects and may be used to treat a number of conditions, including rheumatoid arthritis, severe allergies, asthma, some skin diseases, and some cancers. Side effects can be serious, and therapy must be withdrawn very gradually.

corticotrophin-releasing hormone CRH, hormone produced by the *hypothalamus that stimulates the adrenal glands to produce the steroid cortisol, essential for normal *metabolism. CRH is also produced by the *placenta and a surge in CRH may trigger the beginning of labour.

cortisol hormone produced by the *adrenal glands. It plays a role in helping the body combat stress and is at its highest level in the blood at dawn.

cortisone natural corticosteroid produced by the *adrenal gland, now synthesized for its anti-inflammatory qualities and used in the treatment of rheumatoid arthritis.

cosecant in trigonometry, a *function of an angle in a right-angled triangle found by dividing the length of the hypotenuse (the longest side) by the length of the side opposite the angle. Thus the cosecant of an angle A, usually shortened to cosec A, is always greater than (or equal to) 1. It is the reciprocal of the sine of the angle, that is, cosec $A = 1/\sin A$.

cosine **cos**, in trigonometry, a *function of an angle in a right-angled *triangle found by dividing the length of the side adjacent to the angle by the length of the hypotenuse (the longest side). This function can be used to find either angles or sides in a right-angled triangle.

cosine rule in trigonometry, a rule that relates the sides and angles of triangles. The rule has the formula:

$$a^2 = b^2 + c^2 - 2bc \cos A$$

where a, b, and c are the lengths of the sides of the triangle, and A is the angle opposite a.

cosmic background radiation or **3° radiation**, electromagnetic radiation left over from the original formation of the universe in the *Big Bang between 10 and 20 billion years ago. It corresponds to an overall background temperature of 2.73 K ($-270.4°C/-454.7°F$), or 3°C above

absolute zero. In 1992 the US Cosmic Background Explorer satellite detected slight 'ripples' in the strength of cosmic background radiation that are believed to mark the first stage in the formation of galaxies.

cosmic radiation streams of high-energy particles and elctromagnetic radiation from outer space, consisting of electrons, protons, alpha particles, light nuclei, and gamma rays, which collide with atomic nuclei in the Earth's atmosphere and produce secondary nuclear particles (chiefly *mesons, such as pions and muons) that shower the Earth. Space shuttles carry dosimeter instruments to measure the levels of cosmic radiation.

cosmology branch of astronomy that deals with the structure and evolution of the universe as an ordered whole. Cosmologists construct 'model universes' mathematically and compare their large-scale properties with those of the observed universe.

cotangent in trigonometry, a *function of an angle in a right-angled triangle found by dividing the length of the side adjacent to the angle by the length of the side opposite it. It is usually written as cotan, or cot and it is the reciprocal of the tangent of the angle, so that cot $A = 1/\tan A$, where A is the angle in question.

cotyledon structure in the embryo of a seed plant that may form a 'leaf' after germination and is commonly known as a seed leaf. The number of cotyledons present in an embryo is an important character in the classification of flowering plants (*angiosperms).

coulomb symbol C, SI unit of electrical charge. One coulomb is the amount of charge transferred by a current of one *ampere in one second. The unit is named after French scientist Charles Coulomb (1736–1806).

Coulomb's law in physics, law that states that the *force between two charged bodies varies directly as the product of the two charges, and inversely as the square of the distance between them. It was determined in 1787 by French physicist Charles Coulomb (1736–1806) in 1787.

count rate in physics, number of particles emitted per unit time by a radioactive source. It is measured by a counter, such as a *Geiger counter, or ratemeter.

couple in mechanics, a pair of forces acting on an object that are equal in magnitude and opposite in direction, but do not act along the same straight line. The two forces produce a turning effect or moment that tends to rotate the object; however, no single resultant (unbalanced) force is produced and so the object is not moved from one position to another.

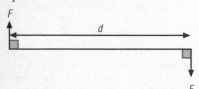

couple Two equal but opposite forces (F) will produce a turning effect on a rigid body, provided that they do not act through the same straight line. The turning effect, or moment, is equal to the magnitude of one of the turning forces multiplied by the perpendicular distance (d) between those two forces.

covalence in chemistry, a form of valence in which two atoms unite by sharing electrons in pairs, so that each atom provides half the shared electrons (see also *covalent bond).

covalent bond chemical *bond produced when two atoms share one or more pairs of electrons (usually each atom contributes an electron). The bond is often represented by a single line drawn between the two atoms. Covalently bonded substances include hydrogen (H_2), water (H_2O), and most organic substances.

Coxsackie virus one of a family of about 30 RNA-containing *viruses, including enteroviruses (those that enter through the gastrointestinal tract). Coxsackie viruses can cause severe throat infections, meningitis, and inflammation of the heart (Keshan disease).

CP/M abbreviation for control program/monitor or control program for microcomputers, one of the earliest *operating systems for microcomputers. It was written by Gary Kildall, who founded Digital Research. In the 1970s

it became a standard for microcomputers based on the Intel 8080 and Zilog Z80 8-bit microprocessors. In the 1980s it was superseded by Microsoft's MS-DOS, written for Intel's 16-bit 8086/88 microprocessors.

CPR abbreviation for **cardiopulmonary resuscitation**.

CPSR abbreviation for Computer Professionals for Social Responsibility.

CPU in computing, abbreviation for *central processing unit.

two hydrogen atoms

or H$\overset{x}{\cdot}$H, H–H
a molecule of hydrogen
sharing an electron pair

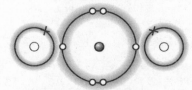

two hydrogen atoms and one
oxygen atom

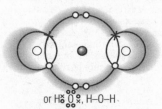

or H$\overset{\circ\circ}{\underset{\circ\circ}{\times}}O\overset{\circ\circ}{\underset{\circ\circ}{x}}$, H–O–H

a molecule of water
showing the two covalent bonds

covalent bond The formation of a covalent bond between two hydrogen atoms to form a hydrogen molecule (H_2), and between two hydrogen atoms and an oxygen atom to form a molecule of water (H_2O). The sharing means that each atom has a more stable arrangement of electrons (its outer electron shells are full).

cracking chemical reaction in which a large *alkane molecule is broken down by heat into a smaller alkane and a small *alkene molecule. The reaction is carried out at a high temperature (600°C/100°F or higher) and often in the presence of a catalyst. Cracking is a commonly used process in the *petrochemical industry. It is the main method of preparation of alkenes and is also used to manufacture petrol from the higher-boiling-point *fractions that are obtained by the *fractional distillation of crude oil (unrefined *petroleum).

ethane   ethylene

heat

cracking Cracking occurs when a larger molecule is broken down into a smaller molecule by the heating or the addition of a catalyst. When ethane is heated ethylene (ethene) is formed.

cramp painful contraction of one or more muscles. It may be due to fatigue, stress or poor posture, or, more rarely, to an imbalance of salts.

cranium the dome-shaped area of the vertebrate skull that protects the brain. It consists of eight bony plates fused together by sutures (immovable joints). Fossil remains of the human cranium have aided the development of theories concerning human evolution.

crater bowl-shaped depression in the ground, usually round and with steep sides. Craters are formed by explosive events such as the eruption of a volcano or the impact of a meteorite.

crawler on the World Wide Web, automated indexing software that scours the Web for new or updated sites. See also *'bot and *spider.

creep in civil and mechanical engineering, the property of a solid, typically a metal, under continuous stress that causes it to deform below its *yield point (the point at which any elastic solid normally stretches without any increase in load or stress). Lead, tin, and zinc, for example, exhibit creep at ordinary temperatures, as seen in the movement of the lead sheeting on the roofs of old buildings.

Cretaceous period of geological time approximately 143–65 million years ago. It is the last period of the Mesozoic era, during which angiosperm (seed-bearing) plants evolved, and dinosaurs reached a peak. The end of the Cretaceous period is marked by a mass extinction of many lifeforms, most notably the dinosaurs. The north European chalk, which forms the white cliffs of Dover, was deposited during the latter half of the Cretaceous, hence the name Cretaceous, which comes from the Latin *creta*, 'chalk'.

cretinism mental retardation resulting from congenital hypothyroidism; it is associated with poor physical development. It can be prevented by early treatment with thyroxine.

Creutzfeldt–Jakob disease CJD, rare brain disease that causes progressive physical and mental deterioration, leading to death. It claims one person in every million and is universally fatal. There have been occurrences in people treated for growth problems with pituitary hormones derived from human brains.

The most common variant, vCJD has been linked with *bovine spongiform encephalopathy (BSE),

critical angle in optics, for a ray of light passing from a denser to a less dense medium (such as from glass to air), the smallest angle of incidence at which the emergent ray grazes the surface of the denser medium – at an angle of refraction of 90°. (See page 156.)

critical density in cosmology, the minimum average density that the universe must have in order for it to stop expanding at some point in the future.

critical mass in nuclear physics, the minimum mass of fissile material that can undergo a continuous *chain

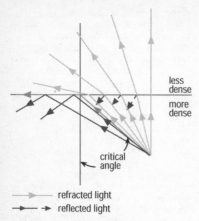

less dense

more dense

critical angle

→ refracted light

→ → reflected light

critical angle The critical angle is the angle at which light from within a transparent medium just grazes the surface of the medium.

reaction. Below this mass, too many *neutrons escape from the surface for a chain reaction to carry on; above the critical mass, the reaction may accelerate into a nuclear explosion.

critical reaction in a nuclear reactor, a self-sustaining chain reaction in which the number of neutrons being released by the fission of uranium-235 nuclei and the number of neutrons being absorbed by uranium-238 nuclei and by control rods are balanced. If balance is not achieved the reaction will either slow down and cease to generate enough power, or will build up and go out of control, as in a nuclear explosion. Control rods are used to adjust the rate of reaction and maintain balance.

critical temperature in physics, temperature above which a particular gas cannot be converted into a liquid by pressure alone. It is also the temperature at which a magnetic material loses its magnetism (the Curie temperature or point).

CRO abbreviation for *cathode-ray oscilloscope.

Crohn's disease or **regional ileitis**, chronic inflammatory bowel disease. It tends to flare up for a few days at a time, causing diarrhoea, abdominal cramps, loss of appetite, weight loss, and mild fever. The cause of Crohn's disease is unknown. However, research teams in Europe and the USA found

that a number of cases of Crohn's disease can be attributed to a fault in the gene *Nod2*, which partly controls the way the immune system responds to gut microbes. Stress may also be a factor.

crop to cut away unwanted portions of a picture. The term comes from traditional manual methods of layout and paste-up; in computing, cropping is an option made available via photo-finishing and graphics software.

crossing over in biology, a process that occurs during *meiosis. While homologous chromosomes are lying alongside each other in pairs, each partner may twist around the other and exchange corresponding chromosomal segments. It is a form of genetic *recombination, which increases variation and thus allows offspring to vary from their parents.

cross-linking in chemistry, lateral linking between two or more long-chain molecules in a *polymer. Cross-linking gives the polymer a higher melting point and makes it harder. Examples of cross-linked polymers include Bakelite and vulcanized rubber.

cross-posting on Usenet, the practice of sending a message to more than one newsgroup. A small amount of cross-posting is acceptable if the message is on a topic that is relevant to more than one newsgroup. For example, a message about top tennis player André Agassi's personal life might be posted to both rec.sport.tennis and alt.showbiz.gossip.

cross-section the surface formed when a solid is cut through by a plane perpendicular to its axis.

croup inflammation of the larynx in small children, with harsh, difficult breathing and hoarse coughing. Croup is most often associated with viral infection of the respiratory tract.

CRT abbreviation for *cathode-ray tube.

crude oil unrefined form of *petroleum.

crust rocky outer layer of the Earth, consisting of two distinct parts – the oceanic crust and the continental crust. The **oceanic** crust is on average about 10 km/6 mi thick and consists mostly of basaltic rock overlain by muddy sediments. By contrast, the **continental** crust is largely of granitic composition and has a more complex structure. Because it is continually recycled back into the mantle by the process of subduction, the oceanic crust is in no place older than about 200 million years. However, parts of the continental crust are over 3.5 billion years old.

Beneath a layer of surface sediment, the oceanic crust is made up of a layer of *basalt, followed by a layer of gabbro. The continental crust varies in thickness from about 40 km/25 mi to 70 km/45 mi, being deepest beneath mountain ranges, and thinnest above continental rift valleys. Whereas the oceanic crust is composed almost exclusively of basaltic igneous rocks and sediments, the continental crust is made of a wide variety of *sedimentary rocks, *igneous rocks, and *metamorphic rocks.

cryogenics science of very low temperatures (approaching *absolute zero), including the production of very low temperatures and the exploitation of special properties associated with them, such as the disappearance of electrical resistance (*superconductivity).

cryptography science of creating and reading codes, for example, those produced by the German Enigma machine used in World War II, those used in the secure transmission of credit card details over the Internet, and those used to ensure the privacy of e-mail messages. Unencoded text (known as plaintext) is converted to an unreadable form (known as cyphertext) by the process of encryption. The recipient must then decrypt the message before it can be read. The breaking of such codes is known as cryptanalysis. No encryption method is completely unbreakable, but cryptanalysis of a strongly encrypted message can be so time-consuming and complex as to be almost impossible.

crystal regular-shaped solid that reflects light. Examples include diamonds, grains of salt, and sugar. Particles forming a crystal are packed in an exact and ordered pattern. When this pattern is repeated many millions of times, the crystal is formed. Such an arrangement of particles, that is regular and repeating, is called a *giant molecular structure.

In *ionic compounds, such as sodium chloride (NaCl), the ions are arranged in a giant ionic lattice, with alternate positive and negative ions in a three-dimensional arrangement. The natural shape of the crystal is the same as the arrangement of ions in the lattice. In sodium chloride the ions form a cubic lattice. Hence sodium chloride crystals are cubic. In diamond there is a giant atomic structure made up of carbon atoms *covalently bonded to each other in a regular, repeating arrangement throughout the whole of the structure. In metals the atoms are also packed tightly together in a regular pattern. This gives metals like copper a crystalline structure. *Metal crystals are called grains.

A mineral can often be identified by the shape of its crystals and the system of crystallization determined. For example, extrusive *igneous rock such as basalt contains very small crystals

sodium chloride

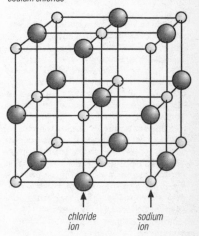

chloride   sodium
ion   ion

crystal The sodium chloride, or common salt, crystal is a regular cubic array of charged atoms (ions) – positive sodium ions and negative chloride ions. Repetition of this structure builds up into cubic salt crystals.

compared with an intrusive igneous rock such as granite. A single crystal can vary in size from a submicroscopic particle to a structure some 30 m/100 ft in length. Crystals fall into seven crystal systems or groups, classified on the basis of the relationship of three or four imaginary axes that intersect at the centre of any perfect, undistorted crystal.

crystallization formation of crystals from a liquid, gas, or solution.

crystallography scientific study of crystals. In 1912 it was found that the shape and size of the repeating atomic patterns (unit cells) in a crystal could be determined by passing X-rays through a sample. This method, known as *X-ray diffraction, opened up an entirely new way of 'seeing' atoms. It has been found that many substances have a unit cell that exhibits all the symmetry of the whole crystal; in table salt (sodium chloride, NaCl), for instance, the unit cell is an exact cube.

Many materials were not even suspected of being crystals until they were examined by X-ray crystallography. It has been shown that purified biomolecules, such as proteins and DNA, can form crystals, and such compounds may now be studied by this method. Other applications include the study of metals and their alloys, and of rocks and soils.

crystal system any of the seven crystal systems defined by symmetry, into which all known crystalline susbtances crystallize. The elements of symmetry used for this purpose are: (1) planes of **mirror symmetry**, across which a mirror image is seen, and (2) axes of **rotational symmetry**, about which, in a 360° rotation of the crystal, equivalent faces are seen two, three, four, or six times. To be assigned to a particular crystal system, a mineral must possess a certain minimum symmetry, but it may also possess additional symmetry elements. Since crystal symmetry is related to internal structure, a given mineral will always crystallize in the same system, although the crystals may not always grow into precisely the same shape. In cases where two minerals have the same chemical composition but different internal structures (for example graphite and diamond, or quartz and cristobalite), they will generally have different crystal systems.

CSCW abbreviation for computer-supported collaborative work.

CSF abbreviation for *cerebrospinal fluid.

CSH abbreviation for *corticotrophin-releasing hormone.

CSS abbreviation for Cascading Style Sheets.

CT-scan another version of *CAT scan or CAT scanner.

CT scanner medical device used to obtain detailed X-ray pictures of the inside of a patient's body. See *CAT scan.

CTS/RTS abbreviation for clear to send/ready to send, in computing, hardware handshaking used in high-speed modems. In most communications software this is an option that can be *toggled on or off. The alternative, software handshaking, is considered less reliable at high speeds.

cu abbreviation for **cubic** (measure).

CUA abbreviation for common user access, standard designed by Microsoft to ensure that identical actions, such as saving a file or accessing help, can be carried out using the same keystrokes in any piece of software. For example, in programs written to the CUA standard, help is always summoned by pressing the F1 function key. New programs should be easier to use because users will not have to learn new commands to perform standard tasks.

cube in geometry, a solid shape whose faces are all *squares. It has 6 equal-area faces and 12 equal-length edges. If the length of one edge is l, the *volume (V) of the cube is given by:

$$V = l^3$$

Its surface *area (A) is calculated by finding the area of one square:

$$l \times l = l^2;$$

and multiplying it by 6:

$$6 \times l^2$$

$$A = 6l^2$$

cube in arithmetic, to multiply a number by itself and then by itself again. For example, 5 cubed $= 5^3 = 5 \times 5 \times 5 = 125$. Alternatively, the *cube root of 125 is 5. The term also refers to a number formed by cubing; for example, 1, 8, 27, 64 are the first four cubes.

cube root number that, multiplied by itself, and then by the product, produces the *cube. For example, $3 \times 3 \times 3 = 27$, 3 being the cube root of 27, which is the cube of 3.

cubic centimetre (or metre) the metric measure of volume, corresponding to the volume of a cube whose edges are all 1 cm (or 1 metre) in length.

cubic decimetre symbol dm^3, metric measure of volume corresponding to the volume of a cube whose edges are all 1 dm (10 cm) long; it is equivalent to a capacity of one litre.

cubic equation any equation in which the largest power of x is x^3. For example, $x^3 + 3x^2y + 4y^2 = 0$ is a cubic equation.

cubic measure measure of volume, indicated either by the word 'cubic' followed by a linear measure, as in 'cubic foot', or the word 'cubed' after a linear measure, as in 'metre cubed'.

cubit earliest known unit of length, which originated between 2800 and 2300 BC. It is approximately = 50.5 cm/20.6 in long, which is about the length of the human forearm measured from the tip of the middle finger to the elbow.

cuboid six-sided three-dimensional prism whose faces are all *rectangles. A brick is a cuboid.

cuesta alternative name for *escarpment.

culture in biology, the growing of living cells and tissues in laboratory conditions.

cumulative frequency in statistics, the total frequency of a given value up to and including a certain point in a set of data. It is calculated by adding together the frequencies to give a running total and used to draw the cumulative frequency curve, the ogive. To plot a cumulative frequency diagram, the cumulative frequency is always plotted along the vertical axis and data are always plotted at the top range of the interval: for example,

frequency for weekly salary ranges £100–150 and £151–200 would be plotted at 150 and 200.

cupronickel copper alloy (75% copper and 25% nickel), used in hardware products and for coinage.

curare black, resinous poison extracted from the bark and juices of various South American trees and plants. Originally used on arrowheads by Amazonian hunters to paralyse prey, it blocks nerve stimulation of the muscles. Alkaloid derivatives (called curarines) are used in medicine as muscle relaxants during surgery.

curie symbol Ci, former unit of radioactivity, equal to 3.7×10^{10} *becquerels. One gram of radium has a radioactivity of about one curie. It was named after French physicist Pierre Curie.

curium chemical symbol Cm, synthesized, radioactive, metallic element of the actinide series, atomic number 96, relative atomic mass 247. It is produced by bombarding plutonium or americium with neutrons. Its longest-lived isotope has a half-life of 1.7×10^7 years.

current directory in a computer's file system, the *directory in which the user is positioned. As users move around a computer system, opening, reading, writing, and storing files, they navigate through that computer's directory structure. Most file commands are assumed to apply to the files in the current directory.

current, electric see *electric current.

cursor on a computer screen, the symbol that indicates the current entry position (where the next character will appear).

curve in geometry, the *locus of a point moving according to specified conditions. The circle is the locus of all points equidistant from a given point (the centre). Other common geometrical curves are the *ellipse, *parabola, and *hyperbola, which are also produced when a cone is cut by a plane at different angles.

Many curves have been invented for the solution of special problems in geometry and mechanics – for example, the cissoid (the inverse of a parabola) and the cycloid.

cusp point where two branches of a curve meet and the tangents to each branch coincide.

customer relationship management CRM, computer systems covering the consolidation of sales force, call centre, field service, help desk, and marketing automation applications. CRM systems allow companies to measure the value of their customers.

CWIS in computing, abbreviation for campus-wide information service.

cwm Welsh name for a corrie.

cyanide CN⁻, ion derived from hydrogen cyanide (HCN), and any salt containing this ion (produced when hydrogen cyanide is neutralized by alkalis), such as potassium cyanide (KCN). The principal cyanides are potassium, sodium, calcium, mercury, gold, and copper. Most cyanides are poisons.

Organic compounds containing a CN group are sometimes called cyanides but are more properly known as nitrites.

cyanobacteria singular **cyanobacterium**, alternative name for *blue-green algae. These organisms are actually not algae but bacteria. The ancestors of modern cyanobacteria generated the oxygen that caused a transformation some 2 billion years ago of the Earth's atmosphere.

cyanocobalamin chemical name for vitamin B_{12}, which is normally produced by micro-organisms in the gut. The richest sources are liver, fish, and eggs. It is essential to the replacement of cells, the maintenance of the myelin sheath which insulates nerve fibres, and the efficient use of folic acid, another vitamin in the B complex. Deficiency can result in pernicious anaemia (defective production of red blood cells), and possible degeneration of the nervous system.

cybernetics (Greek *kubernan* 'to steer') science concerned with how systems organize, regulate, and reproduce themselves, and also how they evolve and learn. In the laboratory, inanimate objects are created that behave like living systems. Applications range from the creation of electronic artificial limbs to the running of the fully automated factory where decision-making machines operate up to managerial level.

cyberspace imaginary, interactive 'worlds' created by networked computers; often used interchangeably with 'virtual world'. The invention of the word 'cyberspace' is generally credited to US science fiction writer William Gibson in his novel *Neuromancer* (1984).

cyclamate derivative of cyclohexysulphamic acid, formerly used as an artificial sweetener, 30 times sweeter than sugar. It was first synthesized in 1937. Its use in foods was banned in the USA and the UK from 1970, when studies showed that massive doses caused cancer in rats.

cycle in physics, a sequence of changes that moves a system away from, and then back to, its original state. An example is a vibration that moves a particle first in one direction and then in the opposite direction, with the particle returning to its original position at the end of the vibration.

cyclic AMP cAMP, cyclic 3,5-adenosine monophosphate, a major communicator molecule in living cells. It is formed by the action of the enzyme adenyl cyclase on *ATP and has many physiological functions. It plays a part in increasing the level of blood *glucose. *Adrenaline, which increases the level of blood sugar, does so by producing cAMP. Cyclic AMP is also involved in the synthesis of flagellar proteins in *bacteria, and in hormone release, urine regulation, and sensory and nerve processes. In some organisms, cAMP seems to be active in the breakdown of *lipids for the production of energy, when glucose is unavailable.

cyclic compound any of a group of organic chemicals that have rings of atoms in their molecules, giving them a closed-chain structure.

cyclic patterns patterns in which simple ideas are repeated to form more complex designs. Some mathematical *functions show cyclic patterns; for example, mapping round a circle.

cyclic polygon in geometry, a polygon in which each vertex (corner) lies on the circumference of a circle.

cyclic quadrilateral four-sided figure with all four of its vertices lying on the circumference of a circle. The opposite angles of cyclic quadrilaterals add up to 180°, and are therefore said to be supplementary; each external angle (formed by extending a side of the quadrilateral) is equal to the opposite interior angle.

cyclone alternative name for a *depression, an area of low atmospheric pressure with winds blowing in a anticlockwise direction in the northern hemisphere and in a clockwise direction in the southern hemisphere. A severe cyclone that forms in the tropics is called a tropical cyclone or *hurricane.

cyclosporin immunosuppressive drug derived from a fungus (*Tolypocladium inflatum*). In use by 1978, it revolutionized transplant surgery by reducing the incidence and severity of rejection of donor organs.

cyclotron circular type of particle *accelerator.

cylinder in geometry, a prism with a circular *cross-section. In everyday use, the term applies to a **right cylinder**, in which the curved surface is perpendicular to the base.

cystic fibrosis hereditary disease involving defects of various tissues, including the sweat glands, the mucous glands of the bronchi (air passages), and the pancreas. The sufferer experiences repeated chest infections and digestive disorders and generally fails to thrive. In 1989 a gene for cystic fibrosis was identified by teams of researchers in Michigan, USA, and Toronto, Canada. This discovery enabled the development of a screening test for carriers; the disease can also be detected in the unborn child.

cytochrome protein responsible for part of the process of *respiration by which food molecules are broken down in *aerobic organisms. Cytochromes are part of the electron transport chain, which uses energized electrons to reduce molecular oxygen (O_2) to oxygen ions (O^{2-}). These combine with hydrogen ions (H^+) to form water (H_2O), the end product of aerobic respiration. As electrons are passed from one cytochrome to another, energy is released and used to make *ATP.

cytokine in biology, chemical messenger that carries information from one cell to another, for example the *lymphokines.

cytology study of the structure of *cells and their functions. Major advances have been made possible in this field by the development of *electron microscopes.

cytoplasm part of plant and animal cells outside the *nucleus (and outside the large vacuole of plant cells). Strictly speaking, this includes all the *organelles (mitochondria, chloroplasts, and so on) and is the area in which most cell activities take place. However, cytoplasm is often used to refer to the jellylike matter in which the organelles are embedded (correctly termed the cytosol). Most of the activities in the cytoplasm are chemical reactions (metabolism), for example, protein synthesis.

In many cells, the cytoplasm is made up of two parts: the **ectoplasm** (or plasmagel), a dense gelatinous outer layer concerned with cell movement, and the **endoplasm** (or plasmasol), a more fluid inner part where most of the organelles are found.

cytosine in biochemistry, one of the nucleotide bases that attach to a pentose sugar (ribose or deoxyribose) forming a nucleotide. Nucleic acids (RNA, DNA) are formed from the condensation of nucleotides forming long chains. Cytosine is a pyrimidine and always bonds with the purine *guanine.

cytotoxic drug any drug used to kill the cells of a malignant tumour; it may also damage healthy cells. Side effects include nausea, vomiting, hair loss, and bone-marrow damage. Some cytotoxic drugs are also used to treat other diseases and to suppress rejection in transplant patients.

D

DAC abbreviation for *digital-to-analogue converter.

daisywheel printing head in a computer printer or typewriter that consists of a small plastic or metal disc made up of many spokes (like the petals of a daisy). At the end of each spoke is a character in relief. The daisywheel is rotated until the spoke bearing the required character is facing an inked ribbon, then a hammer strikes the spoke against the ribbon, leaving the impression of the character on the paper beneath.

dalton see *atomic mass unit.

darcy symbol D, c.g.s. unit of permeability, used mainly in geology to describe the permeability of rock (for example, to oil, gas, or water).

dark cloud in astronomy, cloud of cold dust and gas seen in silhouette against background stars or an HII region (region of hot ionized hydrogen).

dark matter matter that, according to certain modern theories of *cosmology, makes up 90–99% of the mass of the universe but so far remains undetected. Dark matter, if shown to exist, would account for many currently unexplained gravitational effects in the movement of galaxies. Theories of the composition of dark matter include unknown atomic particles (cold dark matter) or fast-moving neutrinos (hot dark matter) or a combination of both.

In 1993 astronomers identified part of the dark matter in the form of stray planets and *brown dwarfs, and, possibly, stars that have failed to ignite. These objects are known as massive astrophysical compact halo objects and may make up approximately half of the dark matter in the Milky Way's halo.

dark reaction series of reactions in *photosynthesis that do not require light. During the dark reaction, carbon dioxide is incorporated into three-carbon sugar phosphate molecules; this reaction is dependant on the *light reaction which does require light.

data compression in computing, techniques for reducing the amount of storage needed for a given amount of data. They include word tokenization (in which frequently used words are stored as shorter codes), variable bit lengths (in which common characters are represented by fewer *bits than less common ones), and run-length encoding (in which a repeated value is stored once along with a count).

Data Encryption Standard DES, in computing, widely used US government standard for encryption, adopted in 1977 and recertified for five more years in 1993. DES was developed by IBM and adopted as a government standard by the National Security Agency. It is a private-key system, so that the sender and recipient encrypt and decrypt the message using the same key.

data logging in computing, the process, usually automatic, of capturing and recording a sequence of values for later processing and analysis by computer. For example, the level in a water-storage tank might be automatically logged every hour over a seven-day period, so that a computer can produce an analysis of water use. The monitoring is carried out through *sensors or similar instruments, connected to the computer via an *interface.

data mining analysis of computer data to determine trends. It is used by retailers to find out those items often purchased together. For example, one supermarket chain found that purchases of nappies and beer were linked, and so increased sales by putting beer next to nappies. Store 'loyalty cards' enable retailers to profile customers' week-by-week shopping against their age, sex, and address.

data security in computing, precautions taken to prevent the loss or misuse of data, whether accidental or deliberate. These include measures that ensure that only authorized personnel can gain entry to a computer system or file, and regular procedures for storing and 'backing up' data, which enable files to be retrieved or recreated in the event of loss, theft, or damage.

data terminator or **rogue value**, in computing, a special value used to mark the end of a list of input data items. The computer must be able to detect that the data terminator is different from the input data in some way – for instance, a negative number might be used to signal the end of a list of positive numbers, or 'XXX' might be used to terminate the entry of a list of names. This is now considered to be very bad programming technique. A lot of old programs used data terminators, and many programmers used 00 as a year data terminator, leading to problems in the year 2000.

daughterboard in computing, small printed circuit board that plugs into a *motherboard to give it new capabilities.

dBASE family of microcomputer programs used for manipulating large quantities of data; also, a related *fourth-generation language. The first version, dBASE II, was published by Ashton-Tate in 1981; it has since become the basis for a recognized standard for database applications, known as xBase.

DBS in computing, abbreviation for *direct broadcast system.

DCE abbreviation for data communications equipment, in computing, another name for a *modem.

DCOM abbreviation for **Distributed Component Object Model**, the network version of Microsoft's *Component Object Model.

DDE in computing, abbreviation for *dynamic data exchange, a form of communication between processes used in Microsoft Windows.

DDT abbreviation for dichloro-diphenyl-trichloroethane; $(ClC_6H_5)_2CHC(HCl_2)$, insecticide discovered in 1939 by Swiss chemist Paul Müller. It is useful in the control of insects that spread malaria, but resistant strains develop. DDT is highly toxic and persists in the environment and in living tissue. Despite this and its subsequent danger to wildlife, it has evaded a worldwide ban because it remains one of the most effective ways of controlling malaria. China and India were the biggest DDT users in 1999.

deadly nightshade another name for *belladonna, a poisonous plant.

deamination removal of the amino group ($-NH_2$) from an unwanted *amino acid. This is the nitrogen-containing part, and it is converted into ammonia, uric acid, or urea (depending on the type of animal) to be excreted in the urine.

In vertebrates, deamination occurs in the *liver.

debugging finding and removing errors, or *bugs, from a computer program or system.

decagon in geometry, a ten-sided *polygon.

decay, radioactive see *radioactive decay.

decibel symbol dB, unit of measure used originally to compare sound intensities and subsequently electrical or electronic power outputs; now also used to compare voltages. A whisper has a sound intensity of 20 dB; 140 dB (a jet aircraft taking off nearby) is the threshold of pain.

deciduous describing trees and shrubs, that shed their leaves at the end of the growing season or during a dry season to reduce *transpiration (the loss of water by evaporation).

decimal fraction in mathematics, *fraction in which the denominator is any higher power of 10. Thus $3/10$, $51/100$, and $23/1,000$ are decimal fractions and are normally expressed as 0.3, 0.51, and 0.023. The use of *decimals greatly simplifies addition and multiplication of fractions, though not all fractions can be expressed exactly as decimal fractions.

decimal number system or **denary number system**, most commonly used number system, to the base ten.

Decimal numbers do not necessarily contain a decimal point; 563, 5.63, and –563 are all decimal numbers. Other systems are mainly used in computing and include the *binary number system, octal number system, and hexadecimal number system.

decimal point dot dividing a decimal number's whole part from its fractional part (the digits to the left of the point are unit digits) first used by Scottish mathematician John Napier (1550–1617). It is usually printed on the line but hand written above the line, for example 3·5. Some European countries use a comma to denote the decimal point, for example 3,56.

decision table in computing, a method of describing a procedure for a program to follow, based on comparing possible decisions and their consequences. It is often used as an aid in systems design.

declarative programming computer programming that does not describe how to solve a problem, but rather describes the logical structure of the problem. It is used in the programming language PROLOG. Running such a program is more like proving an assertion than following a *procedure.

decoder in computing, an electronic circuit used to select one of several possible data pathways. Decoders are, for example, used to direct data to individual memory locations within a computer's immediate access memory.

decomposer in biology, any organism that breaks down dead matter. Decomposers play a vital role in the *ecosystem by freeing important chemical substances, such as nitrogen compounds, locked up in dead organisms or excrement. They feed on some of the released organic matter, but leave the rest to filter back into the soil as dissolved nutrients, or pass in gas form into the atmosphere, for example as nitrogen and carbon dioxide.

The principal decomposers are bacteria and fungi, but earthworms and many other invertebrates are often included in this group. The *nitrogen cycle relies on the actions of decomposers.

decomposition chemical change in which one substance is broken down into two or more simpler substances. In biology, decomposition is the result of the action of decomposer organisms, such as bacteria and fungi. The decomposer organisms obtain food from dead organisms, such as carbon compounds, which are energy-rich. These organisms have an important role in the cycling of carbon compounds as part of the *carbon cycle. The *respiration of the organisms releases *carbon dioxide back into the atmosphere. Other organisms feed on the decomposers and they are part of the decomposer food chain. The decomposition of dead plants and animals allows chemicals to be washed out of the decaying remains into the soil. Many of these are important nutrients that plants can use.

decompression sickness illness brought about by a sudden and substantial change in atmospheric pressure. It is caused by a too rapid release of nitrogen that has been dissolved into the bloodstream under pressure; when the nitrogen forms bubbles it causes the *bends. The condition causes breathing difficulties, joint and muscle pain, and cramps, and is experienced mostly by deep-sea divers who surface too quickly.

decontamination factor in radiological protection, a measure of the effectiveness of a decontamination process. It is the ratio of the original contamination to the remaining radiation after decontamination: 1,000 and above is excellent; 10 and below is poor.

dedicated computer computer built into another device for the purpose of controlling or supplying information to it. Its use has increased dramatically since the advent of the microprocessor: washing machines, digital watches, cars, and video recorders all now have their own processors.

deep freezing method of preserving food by lowering its temperature to –18°C/0°F or below. It stops almost all spoilage processes, although there may be some residual enzyme activity in uncooked vegetables, which is why these are blanched (dipped in hot water to destroy the enzymes) before freezing. Micro-organisms cannot grow or divide while frozen, but most remain alive and can resume activity once defrosted.

Commercial freezing is usually done by one of the following methods: blast, the circulation of air at –40°C/–40°F; contact, in which a refrigerant is circulated through hollow shelves; immersion, for example, fruit in a solution of sugar and glycerol; or cryogenic means, for example, by liquid nitrogen spray.

Accelerated freeze-drying (AFD) involves rapid freezing followed by heat drying in a vacuum, for example, prawns for later rehydration. The product does not have to be stored in frozen conditions.

deep-sea trench another term for *ocean trench.

default in computing, a factory setting for user-configurable options. Default settings appear in all areas of computing, from the on-screen colour scheme in a *graphical user interface to the directories where software programs store data.

defibrillation use of electrical stimulation to restore a chaotic heartbeat to a rhythmical pattern. In fibrillation, which may occur in most kinds of heart disease, the heart muscle contracts irregularly; the heart is no longer working as an efficient pump. Paddles are applied to the chest wall, and one or more electric shocks are delivered to normalize the beat.

definition in mathematics, statement that describes a mathematical expression. If the definition is complete all the properties of the expression can be deduced from it.

deforestation destruction of forest for timber, fuel, charcoal burning, and clearing for agriculture and extractive industries, such as mining, without planting new trees to replace those lost (reforestation) or working on a cycle that allows the natural forest to regenerate.

defragmentation program or **disk optimiser**, in computing, a program that rearranges data on a disk so that files are not scattered in many small sections. See also *fragmentation.

degaussing removal of magnetism from a device. The term is usually used in reference to colour monitors and other display devices that use a *cathode-ray tube (CRT). External magnetic forces, such as the Earth's natural magnetism or a magnet placed close to the monitor, can cause distorted images and colours. Degaussing works by re-aligning the magnetic fields inside the CRT to compensate for the external magnetism.

degree symbol °, in mathematics, a unit of measurement of an angle or arc. A circle or complete rotation is divided into 360°. A degree may be subdivided into 60 minutes (symbol '), and each minute may be subdivided in turn into 60 seconds (symbol ").

Temperature is also measured in degrees, which are divided on a decimal scale. See also *Celsius, and *Fahrenheit scale.

dehydration in chemistry, the removal of water from a substance to give a product with a new chemical formula; it is not the same as drying.

There are two types of dehydration. For substances such as hydrated copper sulphate ($CuSO_4.5H_2O$) that contain *water of crystallization, dehydration means removing this water to leave the anhydrous substance. This may be achieved by heating, and is reversible.

Some substances, such as ethanol, contain the elements of water (hydrogen and oxygen) joined in a different form. **Dehydrating agents** such as concentrated sulphuric acid will remove these elements in the ratio 2:1.

delete remove or erase. In computing, the deletion of a character removes it from the file; the deletion of a file normally means removing its directory entry, rather than actually deleting it from the disk. Many systems now have an undelete or undo facility that allows the restoration of deleted material in an office program. Deleted files may be recovered by specialist software, providing they have not been overwritten. Windows provides a second chance via its Recycle Bin, where deleted files may be restored.

deliquescence phenomenon of a substance absorbing so much moisture from the air that it ultimately dissolves in it to form a solution.

Deliquescent substances make very good drying agents and are used in the bottom chambers of *desiccators. Calcium chloride ($CaCl_2$) is one of the commonest.

delta river sediments deposited when a river flows into a standing body of water with no strong currents, such as a lake, lagoon, sea, or ocean. A delta is the result of fluvial and marine processes. *Deposition is enhanced when water is saline because salty water causes small clay particles to adhere together. Other factors influencing deposition include the type of sediment, local geology, sea-level changes, plant growth, and human impact. Some examples of large deltas are those of the Mississippi, Ganges and Brahmaputra, Rhône, Po, Danube, and Nile rivers. The shape of the Nile delta is like the Greek letter *delta* or D, and gave rise to the name.

deltoid thick muscle swathing the shoulder joint from the collar bone (*clavicle) to its attachment part way down the upper arm bone (*humerus).

dementia mental deterioration as a result of physical changes in the brain. It may be due to degenerative change, circulatory disease, infection, injury, or chronic poisoning. **Senile dementia**, a progressive loss of mental faculties such as memory and orientation, is typically a disease process of old age, and can be accompanied by *depression.

demodulation in radio, the technique of separating a transmitted audio frequency signal from its modulated radio carrier wave. At the transmitter, the audio frequency signal (representing speech or music, for example) may be made to modulate the amplitude (AM broadcasting) or frequency (FM broadcasting) of a continuously transmitted radio-frequency carrier wave. At the receiver, the signal from the aerial is demodulated to extract the required speech or sound component. In early radio systems, this process was called detection. See *modulation.

demyelination pathological process in which the *myelin sheath insulating some nerve fibres is damaged. It leads to weakness and loss of function in the parts of the body served by damaged nerves. It is the central feature of *multiple sclerosis and also occurs in severe *diabetes, and in some other conditions involving neurological disease or damage.

denaturation in biology, irreversible changes occurring in the structure of proteins such as enzymes, usually caused by changes in pH or temperature, by radiation, or by chemical treatments. An example is the heating of egg albumen resulting in solid egg white.

The enzymes associated with digestion and metabolism become inactive under abnormal conditions. Heat will damage their complex structure so that the usual interactions between enzyme and substrate can no longer occur.

dendrite part of a *nerve cell or neuron. The dendrites are slender filaments projecting from the cell body. They receive incoming messages from many other nerve cells and pass them on to the cell body.

If the combined effect of these messages is strong enough, the cell body will send an electrical impulse along the axon (the threadlike extension of a nerve cell). The tip of the axon passes its message to the dendrites of other nerve cells.

dendrochronology or **tree-ring dating**, analysis of the *annual rings of trees to date past events by determining the age of timber. Since annual rings are formed by variations in the water-conducting cells produced by the plant during different seasons of the year, they also provide a means of establishing past climatic conditions in a given area.

denitrification process occurring naturally in soil, where bacteria break down *nitrates to give nitrogen gas, which returns to the atmosphere.

denominator in mathematics, the bottom number of a fraction, so called because it names the family of the fraction. The top number, or numerator, specifies how many unit fractions are to be taken.

density measure of the compactness of a substance; it is equal to its mass per unit volume and is measured, for example, in kg per cubic metre or lb per cubic foot. Density is a *scalar quantity. The average density D of a mass m occupying a volume V is given by the formula: $D = m/V$. *Relative density is the ratio of the density of a substance to that of water at 4°C/39.2°F.

density wave in astrophysics, concept proposed to account for the existence of spiral arms in *galaxies. In the density wave theory, stars in a spiral galaxy move in elliptical orbits in such a way that they crowd together in waves of temporarily enhanced density that appear as spiral arms. The idea was first proposed by Swedish astronomer Bertil Lindblad in the 1920s and developed by US astronomers Chia Lin and Frank Shu in the 1960s.

dental caries in medicine, another name for *caries.

dental formula way of showing the number of teeth in an animal's mouth. The dental formula consists of eight numbers separated by a line into two rows. The four above the line represent the teeth on one side of the upper jaw, starting at the front. If this reads 2 1 2 3 (as for humans) it means two incisors, one canine, two premolars, and three molars (see *tooth). The numbers below the line represent the lower jaw. The total number of teeth can be calculated by adding up all the numbers and multiplying by two.

dentistry care and treatment of the teeth and gums. **Orthodontics** deals with the straightening of the teeth for aesthetic and clinical reasons, and **periodontics** with care of the supporting tissue (bone and gums).

dentition type and number of teeth in a species. Different kinds of teeth have different functions; a grass-eating animal will have large molars for grinding its food, whereas a meat-eater will need powerful canines for catching and killing its prey. The teeth that are less useful to an animal's lifestyle may be reduced in size or missing altogether. An animal's dentition is represented diagramatically by a *dental formula.

denudation natural loss of soil and rock debris, blown away by wind or washed away by running water, laying bare the rock below. Over millions of years, denudation causes a general lowering of the landscape.

deoxyribonucleic acid full name of *DNA.

depopulation the decline of the population of a given area, usually caused by people moving to other areas for economic reasons, rather than an increase in the death rate or decrease in birth rate.

deposition in earth science, the dumping of the load carried by a river, glacier, or the sea. Deposition occurs when the river, glacier, or sea is no longer able to carry its load for some reason, for example a shallowing of gradient, decreasing speed, decreasing energy, decrease in the volume of water in the channel, or an increase in the friction between water and channel. *Glacial deposition occurs when ice melts.

depression in medicine, an emotional state characterized by sadness, unhappy thoughts, apathy, and dejection. Sadness is a normal response to major losses such as bereavement or unemployment. After childbirth, postnatal depression is common. Clinical depression, which is prolonged or unduly severe, often requires treatment, such as *antidepressant medication, cognitive therapy, or, in very rare cases, *electroconvulsive therapy (ECT), in which an electrical current is passed through the brain.

Periods of depression may alternate with periods of high optimism, over-enthusiasm, and confidence. This is the manic phase in a disorder known as **manic depression** or **bipolar disorder**. A manic depressive state is one in which a person switches repeatedly from one extreme to the other. Each mood can last for weeks or months. Typically, the depressive state lasts longer than the manic phase.

depression or **cyclone** or **low**, in meteorology, a region of relatively low atmospheric pressure. In mid-latitudes a depression forms as warm, moist air from the tropics mixes with cold, dry polar air, producing warm and cold boundaries (*fronts) and unstable weather – low cloud and drizzle, showers, or fierce storms. The warm air, being less dense, rises above the cold air to produce the area of low pressure on the ground. Air spirals in towards the centre of the depression in an anticlockwise direction in the northern hemisphere, clockwise in the southern hemisphere, generating winds up to gale force. Depressions tend to travel eastwards and can remain active for several days.

derivative or **differential coefficient**, in mathematics, the limit of the gradient of a chord linking two points on a curve as the distance between the points tends to zero; for a function of a single variable, $y = f(x)$, it is denoted by $f'(x)$, $Df(x)$, or dy/dx, and is equal to the gradient of the curve.

dermatitis inflammation of the skin (see *eczema), usually related to allergy. **Dermatosis** refers to any skin disorder and may be caused by contact or systemic problems.

dermatology medical speciality concerned with diagnosis and treatment of skin disorders.

DES in computing, abbreviation for *Data Encryption Standard.

desalination removal of salt, usually from sea water, to produce fresh water for irrigation or drinking. Distillation has usually been the method adopted, but in the 1970s a cheaper process, using certain polymer materials that filter the molecules of salt from the water by reverse osmosis, was developed.

Desalination plants have been built along the shores of Middle Eastern countries where fresh water is in short supply.

desert arid area with sparse vegetation (or, in rare cases, almost no vegetation). Soils are poor, and many deserts include areas of shifting sands. Deserts can be either hot or cold. Almost 33% of the Earth's land surface is desert, and this proportion is increasing. Arid land is defined as receiving less than 250 mm/9.75 in rain per year.

desertification spread of deserts by changes in climate, or by human-aided processes. Desertification can sometimes be reversed by special planting (marram grass, trees) and by the use of water-absorbent plastic grains, which, added to the soil, enable crops to be grown. About 30% of land worldwide is affected by desertification (1998), including 1 million hectares in Africa and 1.4 million hectares in Asia. The most rapid desertification is in developed countries such as the USA, Australia, and Spain.

desiccator airtight vessel, traditionally made of glass, in which materials may be stored either to dry them or to prevent them, once dried, from reabsorbing moisture.

The base of the desiccator is a chamber in which is placed a substance with a strong affinity for water (such as calcium chloride or silica gel), which removes water vapour from the desiccator atmosphere and from substances placed in it.

detergent surface-active cleansing agent. The common detergents are made from *fats (hydrocarbons) and sulphuric acid, and their long-chain molecules have a type of structure similar to that of *soap molecules: a salt group at one end attached to a long hydrocarbon 'tail'. They have the advantage over soap in that they do not produce scum by forming insoluble salts with the calcium and magnesium ions present in hard water.

To remove dirt, which is generally attached to materials by means of oil or grease, the hydrocarbon 'tails' (soluble in oil or grease) penetrate the oil or grease drops, while the 'heads' (soluble in water but insoluble in grease) remain in the water and, being salts, become ionized. Consequently the oil drops become negatively charged and tend to repel one another; thus they remain in suspension and are washed away with the dirt.

Detergents were first developed from coal tar in Germany during World War I, and synthetic organic detergents were increasingly used after World War II.

Domestic powder detergents for use in hot water have alkyl benzene as their main base, and may also include bleaches and fluorescers as whiteners, perborates to free stain-removing oxygen, and water softeners. Environment-friendly detergents contain no phosphates or bleaches.

Liquid detergents for washing dishes are based on epoxyethane (ethylene oxide). Cold-water detergents consist of a mixture of various alcohols, plus an ingredient for breaking down the surface tension of the water, so enabling the liquid to penetrate fibres and remove the dirt. When these surface-active agents (surfactants) escape the normal processing of sewage, they cause troublesome foam in rivers; phosphates in some detergents can also cause the excessive enrichment (eutrophication) of rivers and lakes.

determinant in mathematics, an array of elements written as a square, and denoted by two vertical lines enclosing the array. For a 2 × 2 matrix, the determinant is given by the difference between the products of the diagonal terms. Determinants are used to solve sets of *simultaneous equations by matrix methods.

detritus in biology, the organic debris produced during the *decomposition of animals and plants.

deuterium or **heavy hydrogen**, naturally occurring heavy isotope of hydrogen, mass number 2 (one proton and one neutron), discovered by US chemist Harold Urey in 1932. It is sometimes given the symbol D. In nature, about one in every 6,500 hydrogen atoms is deuterium. Combined with oxygen, it produces 'heavy water' (D_2O), used in the nuclear industry.

deuteron nucleus of an atom of deuterium (heavy hydrogen). It consists of one proton and one neutron, and is used in the bombardment of chemical elements to synthesize other elements.

development in biology, the process whereby a living thing transforms itself from a single cell into a vastly complicated multicellular organism, with structures, such as limbs, and functions, such as respiration, all able to work correctly in relation to each other. Most of the details of this process remain unknown, although some of the central features are becoming understood.

developmental psychology study of development of cognition and behaviour from birth to adulthood.

device driver in computing, small piece of software required to tell the operating system how to interact with a particular input or output device or peripheral.

Devonian period period of geological time 408–360 million years ago, the fourth period of the Palaeozoic era. Many desert sandstones from North America and Europe date from this time. The first land plants flourished in the Devonian period, corals were abundant in the seas, amphibians evolved from air-breathing fish, and insects developed on land. The name comes from the county of Devon in southwest England, where Devonian rocks were first studied.

dew point temperature at which the air becomes saturated with water vapour. At temperatures below the dew point, the water vapour condenses out of the air as droplets. If the droplets are large they become deposited on plants and the ground as dew; if small they remain in suspension in the air and form mist or fog.

diabase alternative name for *dolerite (a form of basalt that contains very little silica), especially dolerite that has metamorphosed.

diabetes disease that can be caused by reduced production of the *hormone *insulin, or a reduced response of the liver, muscle, and fat cells to insulin. This affects the body's ability to use and regulate sugars effectively. Diabetes mellitus is a disorder of the *islets of Langerhans in the *pancreas that prevents the production of insulin. Treatment is by strict dietary control and oral or injected insulin, depending on the type of diabetes.

diagenesis in geology, the physical, chemical, and biological processes by which a sediment becomes a *sedimentary rock. The main processes involved include compaction of the grains, and the cementing of the grains together by the growth of new minerals deposited by percolating groundwater. As a whole, diagenesis is actually a poorly understood process.

dialler in computing, element of an Internet software package that makes the connection to the online service or *Internet Service Provider. In Windows systems, this is usually the WINSOCK.DLL file, with or without a front end (part of the program that interacts with the user) to make configuration easier.

dialog box in *graphical user interfaces, a small on-screen window with blanks for user input.

dial-up connection in computing, connection to an online system or *Internet Service Provider made by dialling via a *modem over a telephone line.

diamagnetic material material that is weakly repelled by a magnet. All substances are diamagnetic but the behaviour is often masked by stronger forms of magnetism such as *paramagnetic or *ferromagnetic behaviour.

diameter straight line joining two points on the circumference of a circle that passes through the centre of that circle. It divides a circle into two equal halves.

diamond generally colourless, transparent mineral, an *allotrope of carbon. It is regarded as a precious gemstone, and is the hardest substance known (10 on the *Mohs scale). Industrial diamonds, which may be natural or synthetic, are used for cutting, grinding, and polishing.

Diamond crystallizes in the cubic system as octahedral crystals, some with curved faces and striations. The high refractive index of 2.42 and the high dispersion of light, or 'fire', account for the spectral displays seen in polished diamonds.

Diamonds may be found as alluvial diamonds on or close to the Earth's surface in riverbeds or dried watercourses; on the sea bottom (off southwest Africa); or, more commonly, in diamond-bearing volcanic pipes composed of 'blue ground', kimberlite or lamproite, where the original matrix has penetrated the Earth's crust from great depths. They are sorted from the residue of crushed ground by X-ray and other recovery methods.

There are four chief varieties of diamond: well-crystallized transparent stones, colourless or only slightly tinted, valued as gems; **boart**, poorly crystallized or inferior

diamonds; **balas**, an industrial variety, extremely hard and tough; and **carbonado**, or industrial diamond, also called black diamond or carbon, which is opaque, black or grey, and very tough. Industrial diamonds are also produced synthetically from graphite. Some synthetic diamonds conduct heat 50% more efficiently than natural diamonds and are five times greater in strength. This is a great advantage in their use to disperse heat in electronic and telecommunication devices and in the production of laser components.

Rough diamonds are often dull or greasy before being polished; around 50% are considered 'cuttable' (all or part of the diamond may be set into jewellery). Gem diamonds are valued by weight (*carat), cut (highlighting the stone's optical properties), colour, and clarity (on a scale from internally flawless to having a large inclusion clearly visible to the naked eye). They are sawn and polished using a mixture of oil and diamond powder. The two most popular cuts are the brilliant, for thicker stones, and the marquise, for shallower ones. India is the world's chief cutting centre.

Noted rough diamonds include the Cullinan, or Star of Africa (3,106 carats, over 500 g/17.5 oz before cutting, South Africa, 1905); Excelsior (995.2 carats, South Africa, 1893); and Star of Sierra Leone (968.9 carats, Yengema, 1972).

experiments Using a device known as a diamond-anvil cell, a moderate force applied to the small tips of two opposing diamonds can be used to attain extreme pressures of millions of atmospheres or more, allowing scientists to subject small amounts of material to conditions that exist deep within planet interiors. In 1999 US scientists turned a diamond into a metal using a very powerful laser to compress it.

diapause period of suspended development that occurs in some species of insects and other invertebrates, characterized by greatly reduced metabolism. Periods of diapause are often timed to coincide with the winter months, and improve the animal's chances of surviving adverse conditions.

diaphragm in mammals, a thin muscular sheet separating the thorax from the abdomen. It is attached by way of the ribs at either side and the breastbone and backbone, and a central tendon. Arching upwards against the heart and lungs, the diaphragm is important in the mechanics of breathing. It contracts at each inhalation, moving downwards to increase the volume of the chest cavity, and relaxes at exhalation.

diapirism geological process in which a particularly light rock, such as rock salt, punches upwards through the heavier layers above. The resulting structure is called a salt dome, and oil is often trapped in the curled-up rocks at each side.

diastole in biology, the relaxation of a hollow organ. In particular, the term is used to indicate the resting period between beats of the heart when blood is flowing into it.

diastolic pressure in medicine, measurement due to the pressure of blood against the arterial wall during diastole (relaxation of the heart). It is the lowest *blood pressure during the cardiac cycle. The average diastolic pressure in healthy young adults is about 80 mmHg. The variation of diastolic pressure due to changes in body position and mood is greater than that of systolic pressure. Diastolic pressure is also a more accurate predictor of hypertension (high blood pressure).

diatomic molecule molecule composed of two atoms joined together. In the case of an element such as oxygen (O_2), the atoms are identical.

dibasic acid acid containing two replaceable hydrogen atoms, such as sulphuric acid (H_2SO_4). The acid can form two series of salts, the normal salt (sulphate, SO_4^{2-}) and the acid salt (hydrogensulphate HSO_4^-).

dichloro-diphenyl-trichloroethane full name of the insecticide *DDT.

dichotomous key in biology, a practical method used to identify species. The key is written so that identification is done in steps. At each step, features of the organism are used to identify which one of two routes through the rest of the key is appropriate to the organism being identified. Dichotomous keys assume a good knowledge of the subject under investigation.

dicotyledon major subdivision of the *angiosperms, containing the great majority of flowering plants. Dicotyledons are characterized by the presence of two seed leaves, or *cotyledons, in the embryo, which is usually surrounded by the *endosperm. They generally have broad leaves with netlike veins. Dicotyledons may be small plants such as daisies and buttercups, shrubs, or trees such as oak and beech. The other subdivision of the angiosperms is the *monocotyledons.

dielectric insulator or nonconductor of electricity, such as rubber, glass, and paraffin wax. An electric field in a dielectric material gives rise to no net flow of electricity. However, the applied field causes electrons within the material to be displaced, creating an electric charge on the surface of the material. This reduces the field strength within the material by a factor known as the dielectric constant (or relative permittivity) of the material. Dielectrics are used in capacitors, to reduce dangerously strong electric fields, and have optical applications.

diesel oil lightweight fuel oil used in diesel engines. Like petrol, it is a petroleum product. When used in vehicle engines, it is also known as **derv** (diesel-engine road vehicle).

diet range of foods eaten by an animal each day; it is also a particular selection of food, or the total amount and choice of food for a specific person or group of people. Most animals require seven kinds of food in their diet: proteins, carbohydrates, fats, vitamins, minerals, water, and roughage. A diet that contains all of these things in the correct amounts and proportions is termed a balanced diet. The amounts and proportions required varies with different animals, according to their size, age, and lifestyle. The *digestive systems of animals have evolved to meet particular needs; they have also adapted to cope with the foods available in the surroundings in which they live. The necessity of finding and processing an appropriate diet is a very basic drive in animal evolution. **Dietetics** is the science of feeding individuals or groups; a dietician is a specialist in this science.

dietetics specialized branch of human nutrition, dealing with the promotion of health through the proper kinds and quantities of food.

difference in mathematics, the result obtained when subtracting one number from another. Also, those elements of one *set that are not elements of another.

difference engine mechanical calculating machine designed (and partly built in 1822) by the English mathematician Charles Babbage (1792–1871) to produce reliable tables of life expectancy. A precursor of the *analytical engine, it was to calculate mathematical functions by solving the differences between values given to *variables within equations. Babbage designed the calculator so that once the initial values for the variables were set it would produce the next few thousand values without error.

differential calculus branch of *calculus involving applications such as the determination of maximum and minimum points and rates of change.

differentiation in embryology, the process by which cells become increasingly different and specialized, giving rise to more complex structures that have particular functions in the adult organism. For instance, embryonic cells may develop into nerve, muscle, or bone cells.

differentiation in mathematics, a procedure for determining the *derivative or gradient of the tangent to a curve $f(x)$ at any point x.

Diffie-Hellman key exchange system in computing, the basis of *public-key cryptography, proposed by researchers Whitfield Diffie and Martin Hellman in 1976.

diffraction the spreading out of waves when they pass through a small gap or around a small object, resulting in some change in the direction of the waves. In order for this effect to be observed, the size of the object or gap must be comparable to or smaller than the *wavelength of the waves. Diffraction occurs with all forms of progressive waves – electromagnetic, sound, and water waves – and explains such phenomena as the ability of long-wave radio waves to bend around hills more easily than short-wave radio waves.

The wavelength of visible light ranges from about 400 to 700 nanometres, several orders of magnitude smaller than radio waves. The gap through which light travels must be extremely small for diffraction to be observed. The slight spreading of a light beam through a narrow slit causes the different wavelengths of light to interfere with each other, producing a pattern of light and dark bands. A **diffraction grating** is a plate of glass or metal ruled with close (some diffraction gratings have from 1,000 to 4,000 lines per cm), equidistant, parallel lines used for

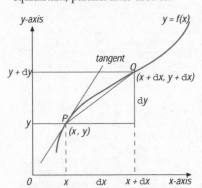

differentiation A mathematical procedure for determining the gradient, or slope, of the tangent to any curve f(x) at any point x. Part of a branch of mathematics called differential calculus, it is used to solve problems involving continuously varying quantities (such as the change in velocity or altitude of a rocket), to find the rates at which these variations occur and to obtain maximum and minimum values for the quantities.

separating a wave train such as a beam of incident light into its component frequencies. White light passing through a grating will be separated into its constituent colours. Red light is diffracted more as it has a longer wavelength; blue light is diffracted less as it has a shorter wavelength.

The wavelength of sound is between 0.5 m/1.6 ft and 2.0 m/6.6 ft. When sound waves travel through doorways or between buildings they are diffracted significantly, so that the sound is heard round corners.

diffusion net spontaneous and random movement of molecules or particles in a fluid (gas or liquid) from a region in which they are at a high concentration to a region of lower concentration, until a uniform concentration is achieved throughout. The difference in concentration between two such regions is called the **concentration gradient**. No mechanical mixing or stirring is involved. For instance, a drop of ink added to water will diffuse down the concentration gradient until evenly mixed.

Diffusion occurs because particles in a liquid or gas are free to move according to the *kinetic theory of matter. The molecules move randomly, but there is more chance that they will move out of the ink drop than into it, so the molecules diffuse until their colour becomes evenly distributed throughout. Diffusion occurs more rapidly across a higher concentration gradient and at higher temperatures.

Diffusion is quite different from the movement of molecules when a fluid is flowing. In this case movement is not random; all molecules are moving together and in the same direction.

In biological systems, diffusion plays an essential role in the transport, over short distances (for example across *cell membranes), of molecules such as nutrients, respiratory gases (*carbon dioxide and *oxygen), and neurotransmitters. It provides the means by which small molecules pass into and out of individual cells and micro-organisms, such as an amoeba, that possess no circulatory system. Plant and animal organs whose function depends on diffusion – such as the lung – have a large surface area. Diffusion of water across a semi-permeable membrane is termed *osmosis. Some important processes which involve diffusion are: the uptake of the products of *digestion from the *gut, gas exchange in the lungs, and gas exchange in the leaves of plants.

One application of diffusion is the separation of isotopes, particularly those of uranium. When uranium hexafluoride diffuses through a porous plate, the ratio of the 235 and 238 isotopes is changed slightly. With sufficient number of passes, the separation is nearly complete. There are large plants in the USA and UK for obtaining enriched fuel for fast nuclear reactors and the fissile uranium-235, originally required for the first atom

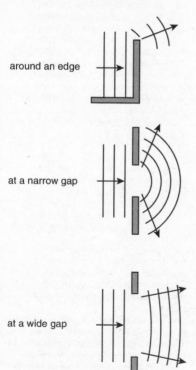

around an edge

at a narrow gap

at a wide gap

diffraction When waves pass around a barrier or through a gap, they spread out. The effect, known as diffraction, will be more pronounced at a narrow gap than at a wider gap.

bombs. Another application is the diffusion pump, used extensively in vacuum work, in which the gas to be evacuated diffuses into a chamber from which it is carried away by the vapour of a suitable medium, usually oil or mercury.

digestion process by which food eaten by an animal is broken down mechanically, and chemically by *enzymes, mostly in the *stomach and *intestines, to make the nutrients available for absorption and cell metabolism. In digestion large molecules of food are broken into smaller, soluble molecules, which are absorbed through the wall of the *gut into the bloodstream and carried to individual cells. The first stage of this may involve just the mixing of the food with water and the crushing and chopping of pieces of food by teeth or the mixing of food as it is squeezed along the gut. The second stage is the breakdown of large molecules by enzymes. The uptake of digested foods is mainly by absorption.

digestive system in the body, all the organs and tissues involved in the digestion of food. In animals, these consist of the mouth, stomach, intestines, and their associated glands. The process of digestion breaks down the food by physical and chemical means into the different elements that are needed by the body for energy and tissue building and repair. Digestion begins in the mouth and continues in the *stomach; from there most nutrients enter the small intestine from where they pass through the intestinal wall into the bloodstream; what remains is stored and concentrated into faeces in the large intestine. Birds have two additional digestive organs – the crop and gizzard. In smaller, simpler animals such as jellyfish, the digestive system is simply a cavity (coelenteron or enteric cavity) with a 'mouth' into which food is taken; the digestible portion is dissolved and absorbed in this cavity, and the remains are ejected back through the mouth.

The digestive system of humans consists primarily of the *alimentary canal, a tube which starts at the mouth, continues with the pharynx,

oesophagus (or gullet), stomach, large and small intestines, and rectum, and ends at the anus. The food moves through this canal by *peristalsis whereby waves of involuntary muscular contraction and relaxation produced by the muscles in the wall of the gut cause the food to be ground and mixed with various digestive juices. Most of these juices contain digestive enzymes, chemicals that speed up reactions involved in the breakdown of food. Other digestive juices empty into the alimentary canal from the salivary glands, gall bladder, and pancreas, which are also part of the digestive system.

The fats, proteins, and carbohydrates (starches and sugars) in foods contain very complex molecules that are broken down (see *diet) for absorption into the bloodstream: starches and complex sugars are converted to simple sugars; fats are converted to fatty acids and glycerol; and proteins are converted to amino acids and peptides. Foods such as vitamins, minerals, and water do not need to undergo digestion prior to absorption into the bloodstream. The small intestine, which is the main site of digestion and absorption, is subdivided into the duodenum, jejunum, and ileum. Covering the surface of its mucous membrane lining are a large number of small prominences called villi which increase the surface for absorption and allow the digested nutrients to diffuse into small blood-vessels lying immediately under the epithelium.

digital city in computing, area in *cyberspace, either text-based or graphical, that uses the model of a city to make it easy for visitors and residents to find specific types of information.

digital composition or **compositing**, in computing, computerized film editing. Some film special effects require shots to be cut together – composited. A sequence showing an actor hanging off the edge of a skyscraper, for example, may be put together out of footage of the actor in a safe location inserted into a shot looking down the side of the skyscraper, which may itself be a model. Traditional techniques for creating such a shot involved photographing the foreground shot with the background shot playing behind it, with an inevitable degradation of quality in the background material. In digital compositing, the same footage is digitized, and the work of merging the two sequences is done by manipulating computer files. The composite image is then transferred back onto film with no loss of quality.

digital computer computing device that operates on a two-state system, using symbols that are internally coded as binary numbers (numbers made up of combinations of the digits 0 and 1).

digital data transmission in computing, a way of sending data by converting all signals (whether

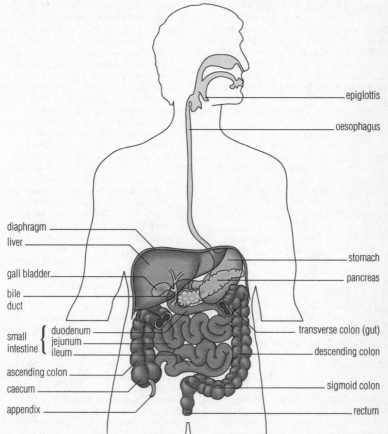

digestive system The human digestive system. When food is swallowed, it is moved down the oesophagus by the action of muscles (peristalsis) into the stomach. Digestion starts in the mouth and continues in the stomach as the food is mixed with enzymes and strong acid. After several hours, the food passes to the small intestine. Here more enzymes are added and digestion is completed. After all nutrients have been absorbed, the indigestible parts pass into the large intestine and thence to the rectum. The liver has many functions, such as storing minerals and vitamins and making bile, which is stored in the gall bladder until needed for the digestion of fats. The pancreas supplies enzymes. The appendix appears to have no function in human beings.

pictures, sounds, or words) into numeric (normally binary) codes before transmission, then reconverting them on receipt. This virtually eliminates any distortion or degradation of the signal during transmission, storage, or processing.

digitalis drug that increases the efficiency of the heart by strengthening its muscle contractions and slowing its rate. It is derived from the leaves of the common European woodland plant *Digitalis purpurea* (foxglove).

digital monitor in computing, display *monitor using standard cathode-ray tube technology that converts a digital signal from the computer into an analogue signal for display.

Digital Nervous System Microsoft's way of describing how information technology – particularly its own software products – can help companies to operate more efficiently. Digital Nervous System is used to market Microsoft's ideas to people considered not technical enough to understand DNA, its Distributed interNetwork Applications architecture.

digital recording technique whereby the pressure of sound waves is sampled more than 30,000 times a second and the values converted by computer into precise numerical values. These are recorded and, during playback, are reconverted to sound waves.

digital root a digit formed by adding the digits of a number. If this leads to a total of 10 or more, the resultant digits are added. For example, finding the digital root of 365 involves the following computations: $3 + 6 + 5 = 14$, followed by $1 + 4 = 5$. The digital root of 365 is therefore 5.

digital signal processor DSP, in computing, special-purpose integrated circuit that handles voice. DSPs are used in voice modems, which add answering machine facilities to a personal computer, and also in computer dictation systems.

digital signature in computing, method of using encryption to certify the source and integrity of a particular electronic document. Because all *ASCII characters look the same no matter who types them, methods have to be found to certify the origins of particular messages if they are to be legally binding for electronic commerce or other transactions. One type of digital signature commonly seen on the Net is generated by the program Pretty Good Privacy (PGP), which adds a digest of the message to the signature. Digital signatures play an essential part in authenticating electronic commerce transactions.

digital-to-analogue converter electronic circuit that converts a digital signal into an analogue (continuously varying) signal. Such a circuit is used to convert the digital output from a computer into the analogue voltage required to produce sound from a conventional loudspeaker.

digitize in computing, to turn analogue signals into the binary data a computer can read. Any type of analogue signal can be digitized, including pictures, sound, video, or film. The result is files that can be manipulated, stored, or transmitted by computers. See *analogue-to-digital converter.

digoxin drug derived from *digitalis that increases the force of the heart's contraction and is commonly used to treat *heart failure. It is readily effective, whereas the related drug, digitoxin, is slow-acting. Side effects include loss of appetite, nausea, vomiting, diarrhoea and, sometimes, cardiac *arrhythmia.

dihybrid inheritance in genetics, a pattern of inheritance observed when two characteristics are studied in succeeding generations. The first experiments of this type, as well as in *monohybrid inheritance, were carried out by Austrian biologist Gregor Mendel using pea plants.

dilution process of reducing the concentration of a solution by the addition of a solvent.

The extent of a dilution normally indicates the final volume of solution required. A fivefold dilution would mean the addition of sufficient solvent to make the final volume five times the original.

dimension in science, any directly measurable physical quantity such as mass (M), length (L), and time (T), and the derived units obtainable by multiplication or division from such quantities. For example, acceleration (the rate of change of velocity) has dimensions (LT^{-2}), and is expressed in such units as km s^{-2}. A quantity that is a ratio, such as relative density or humidity, is dimensionless.

dinitrogen oxide alternative name for *nitrous oxide, or 'laughing gas', one of the nitrogen oxides.

diode combination of a cold anode and a heated cathode, or the semiconductor equivalent, which incorporates a p–n junction; see *semiconductor diode. Either device allows the passage of direct current in one direction only, and so is commonly used in a *rectifier to convert alternating current (AC) to direct current (DC).

dioxin any of a family of over 200 organic chemicals, all of which are heterocyclic hydrocarbons (see *cyclic compounds). The term is commonly applied, however, to only one member of the family, 2,3,7,8-tetrachlorodibenzo-p-dioxin (2,3,7,8-TCDD), a highly toxic chemical that occurs, for example, as an impurity in the defoliant Agent Orange, used in the Vietnam War, and sometimes in the weedkiller 2,4,5-T. It has been associated with chloracne (a disfiguring skin complaint), birth defects, miscarriages, and cancer.

Disasters involving accidental release of large amounts of dioxin into the environment have occurred at Seveso, Italy, and Times Beach, Missouri, USA. Small amounts of dioxins are released by the burning of a wide range of chlorinated materials (treated wood, exhaust fumes from fuels treated with chlorinated additives, and plastics) and as a side effect of some techniques of paper-making. Dioxin may also be produced as a by-product in the manufacture of the bactericide hexachlorophene.

The possibility of food becoming contaminated by dioxins in the environment has led the European Union (EU) to decrease significantly the allowed levels of dioxin emissions from incinerators.

DIP abbreviation for *document image processing.

diphtheria acute infectious disease in which a membrane forms in the throat (threatening death by *asphyxia), along with the production of a powerful toxin that damages the heart and nerves. The organism responsible is a bacterium (*Corynebacterium diphtheriae*). It is treated with antitoxin and antibiotics.

diploblastic in biology, having a body wall composed of two layers. The outer layer is the **ectoderm**, the inner layer is the **endoderm**. This pattern of development is shown by coelenterates.

diploid having paired *chromosomes in each cell. In sexually reproducing species, one set is derived from each parent, the *gametes, or sex cells, of each parent being *haploid (having only one set of chromosomes) due to *meiosis (reduction cell division).

dip, magnetic angle at a particular point on the Earth's surface between the direction of the Earth's magnetic field and the horizontal. It is measured using a **dip circle**, which has a magnetized needle suspended so that it can turn freely in the vertical plane of the magnetic field. In the northern hemisphere the needle dips below the horizontal, pointing along the line of the magnetic field towards its north pole. At the magnetic north and south poles, the needle dips vertically and the angle of dip is 90°. See also *angle of declination.

dipole uneven distribution of magnetic or electrical characteristics within a molecule or substance so that it behaves as though it possesses two equal but opposite poles or charges, a finite distance apart.

The uneven distribution of electrons within a molecule composed of atoms of different *electronegativities may result in an apparent concentration of electrons towards one end of the molecule and a deficiency towards the other, so that it forms a dipole consisting of apparently separated but equal positive and negative charges. The product of one charge and the distance between them is the **dipole moment**. A bar magnet has a magnetic dipole and behaves as though its magnetism were concentrated in separate north and south magnetic poles.

dipole, magnetic see *magnetic dipole.

DIP switch acronym for Dual In-line Package switch, in computing, tiny switch that controls settings on devices such as printers and modems. The owner's manual will usually specify how DIP switches should be set.

direct access or **random access**, type of file access. A direct-access file contains records that can be accessed by the computer directly because each record has its own address on the storage disk. Direct access storage media include CD-ROMs and magnetic disks (such as floppy disks).

direct broadcast system DBS, in computing, combination of a small satellite dish and receiver which allows consumers to receive television and radio broadcasts from a satellite rather than via terrestrial broadcasting towers and repeaters.

direct connection in computing, connection between two computers via cable to transfer files without the intermediary of a network or online service. Each computer must be running communications software using the same protocols for file transfers. If the computers are in the same room, they can be connected using a special type of serial cable known as a null modem cable; if they are connected via telephone lines each must have a modem so that one can dial the other.

direct current DC, electric current where the electrons (negative charge) flow in one direction, and that does not reverse its flow as *alternating current does. The electricity produced by a battery is direct current. Electromagnets and electric trains use direct current.

If in one second one coulomb of electrons passes a given point, then the current flow is 1 amp DC.

A cathode ray oscilloscope (CRO) is used to display the waveforms that show the pattern of how voltage and current vary over a period of time. The waveforms for DC are straight lines, as the voltage and current do not vary over a period of time. If a resistor is connected in a circuit and

measurements of voltage across the resistor and current flowing through it are taken over a period of time, then there will be no change in the voltage and current over this period. A plot of voltage or current against time produces a straight line.

directed number in mathematics, number with a positive (+) or negative (–) sign attached, for example +5 or –5. On a graph, a positive sign shows a movement to the right or upwards; a negative sign indicates movement downwards or to the left.

direct memory access DMA, in computing, a technique used for transferring data to and from external devices without going through the *central processing unit (CPU) and thus speeding up transfer rates. DMA is used for devices such as scanners.

directory in computing, a list of file names, together with information that enables a computer to retrieve those files from backing storage. The computer operating system will usually store and update a directory on the backing storage to which it refers. So, for example, on each *disk used by a computer a directory file will be created listing the disk's contents.

The term is also used to refer to the area on a disk where files are stored; the main area, the **root** directory, is at the top-most level, and may contain several separate **sub-directories**.

directory tree in computing, collective name for a *directory and all its subdirectories.

disaccharide sugar made up of two monosaccharides or simple sugars. Sucrose, $C_{12}H_{22}O_{11}$, or table sugar, is a disaccharide.

discharge tube device in which a gas conducting an electric current emits visible light. It is usually a glass tube from which virtually all the air has been removed (so that it 'contains' a near vacuum), with electrodes at each end. When a high-voltage current is passed between the electrodes, the few remaining gas atoms in the tube (or some deliberately introduced ones) ionize and emit coloured light as they conduct the current along the tube. The light originates as electrons change energy levels in the ionized atoms.

discrete data in mathematics, data that can take only whole-number or fractional values, that is, distinct values. The opposite is *continuous data, which can take all in-between values. Examples of discrete data include *frequency and *population data. However, measurements of time and other dimensions can give rise to continuous data.

disease condition that disturbs or impairs the normal state of an organism. Diseases can occur in all living things, and normally affect the functioning of cells, tissues, organs, or systems. Diseases are usually characterized by specific symptoms and signs, and can be mild and short-lasting – such as the common cold – or severe enough to decimate a whole species – such as Dutch elm disease. Diseases can be classified as infectious or noninfectious. Infectious diseases are caused by micro-organisms, such as bacteria and viruses, invading the body; they can be spread across a species, or transmitted between one or more species. All other diseases can be grouped together as noninfectious diseases. These can have many causes: they may be inherited (*congenital diseases); they may be caused by the ingestion or absorption of harmful substances, such as toxins; they can result from poor nutrition or hygiene; or they may arise from injury or ageing. The causes of some diseases are still unknown.

disinfectant agent that kills, or prevents the growth of, bacteria and other micro-organisms. Chemical disinfectants include carbolic acid (phenol, used by English surgeon Joseph Lister (1827–1912) in surgery in the 1870s), ethanol, methanol, chlorine, and iodine.

disk in computing, a common medium for storing large volumes of data (an alternative is *magnetic tape). A **magnetic disk** is rotated at high speed in a disk-drive unit as a read/write (playback or record) head passes over its surfaces to record or read the magnetic variations that encode the data. **Optical disks**, such as *CD-ROM (compact-disc read-only memory) and *WORM (write once, read many times), are also used to store computer data. Data are recorded on the disk surface as etched microscopic pits and are read by a laser-scanning device.

disk compression technique, based on *data compression, that makes hard disks and floppy disks appear to have more storage capacity than is normally available. If the data stored on a disk can be compressed to occupy half the original amount of disk space, it will appear that the disk is twice its original size. The processes of compression (to store data) and decompression (so that data can be used) are hidden from the user by the software.

disk drive mechanical device that reads data from a *disk. Many types can also write data to the disk.

dispersion in chemistry, the distribution of the microscopic particles of a *colloid. In colloidal sulphur the dispersion is the tiny particles of sulphur in the aqueous system.

dispersion in physics, a particular property of *refraction in which the angle and velocity of waves passing through a dispersive medium depends upon their frequency. When visible white light passes through a prism it is split into a spectrum (see *electromagnetic waves). This occurs because each component frequency of light, which corresponds to a colour, is refracted by a slightly different angle, and so the light is split into its component frequencies (colours). A rainbow is formed when sunlight is dispersed by raindrops.

A white light beam splits up into the colours of which it consists (red, orange, yellow, green, blue, indigo, and violet). Red light has the lowest frequency (longest wavelength) and is the least refracted; violet light has the highest frequency (shortest wavelength) and is the most refracted. The relationship between wavelength and frequency is given by the formula: speed of light = frequency × wavelength (or wavelength = speed of light/frequency).

dispersion in statistics, the extent to which data are spread around a central point (typically the *arithmetic mean).

displacement in psychoanalysis, the transference of an emotion from the original idea with which it is associated to other ideas. It is usually thought to be indicative of *repression in that the emotional content of an unacceptable idea may be expressed without the idea itself becoming conscious.

displacement reaction chemical reaction in which a less reactive element is replaced in a compound by a more reactive one.

For example, the addition of powdered zinc to a solution of copper(II) sulphate displaces copper metal, which can be detected by its characteristic colour:

$$Zn_{(s)} + CuSO_{4(aq)} \rightarrow ZnSO_{4(aq)} + Cu_{(s)}$$

The copper is taken out of the solution and is deposited as a solid (s).

displacement vector vector that describes how an object has been moved from one position to another.

dissociation in chemistry, the process whereby a single compound splits into two or more smaller products, which may be capable of recombining to form the reactant.

Where dissociation is incomplete (not all the compound's molecules dissociate), a *chemical equilibrium exists between the compound and its dissociation products. The extent of incomplete dissociation is defined by a numerical value (dissociation constant).

distance the space between two points. Distance is normally measured using centimetres, metres, and kilometres.

Straight-line distances on a map can be easily measured with a ruler or straight edge. Simply convert the ruler measurement to distance, using the scale provided on the map. Twisting routes can be measured with a length of string, which is then straightened and measured in the normal way. For an object travelling at a constant speed the distance travelled can be calculated using the equation:

distance = speed × time.

distance ratio in a machine, the distance moved by the input force, or effort, divided by the distance moved by the output force, or load. The ratio indicates the movement magnification achieved, and is equivalent to the machine's velocity ratio.

distance–time graph graph used to describe the motion of a body by illustrating the relationship between the distance that it travels and the time taken. Plotting distance (on the vertical axis) against time (on the horizontal axis) produces a graph, the gradient of which at any point is the body's speed at that point (called the instantaneous speed). If the gradient is constant (the graph is a straight line), the body has uniform or constant speed; if the gradient varies (the graph is curved), then so does the speed, and the body may be said to be accelerating or decelerating. The shape of the graph allows the distance travelled, the speed at any point, and the average speed to be worked out.

distillation technique used to purify liquids or to separate mixtures of liquids possessing different boiling points. **Simple distillation** is used in the purification of liquids or the separation of substances in solution

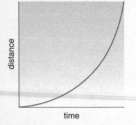
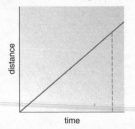
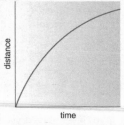

distance–time graph In a distance–time graph distance travelled by an object is plotted against the time it has taken for the object to travel that distance. If the line plotted is straight then the object is travelling at a uniform rate. The line curving upward shows acceleration; a curve downwards plots deceleration.

from their solvents – for example, in the production of pure water from a salt solution or the recovery of sodium chloride (table salt) from sea water.

The solution is boiled and the vapours of the solvent rise into a separate piece of apparatus (the condenser) where they are cooled and condensed. The liquid produced (the distillate) is the pure solvent; the substances (the solutes), now in solid form, remain in the distillation vessel. Mixtures of liquids, such as crude oil (unrefined *petroleum) or aqueous ethanol, are separated by ***fractional distillation**, or fractionation, in a *fractionating column. When the mixture is boiled, the vapours of its most volatile component (the component with the lowest boiling point) rise into the vertical column where they condense to liquid form. As they descend they are reheated to boiling point by the hotter rising vapours of the next component to reach boiling point. This boiling–condensing process occurs repeatedly inside the column. As the column is ascended, progressive enrichment by the lower-boiling-point components occurs; there is thus a temperature gradient inside the column. The vapours of the more volatile components reach the top of the column and enter the condenser for collection before those of the less volatile components. In the fractional distillation of petroleum, groups of compounds possessing similar relative molecular masses and boiling points (the fractions) are tapped off at different points on the column.

The earliest-known reference to the process is to the distillation of wine in the 12th century by Adelard of Bath. The chemical retort used for distillation was invented by Muslims, and was first seen in the West in about 1570.

distributed processing computer processing that uses more than one computer to run an application. *Local area networks, *client–server architecture, and *parallel processing involve distributed processing.

distribution in statistics, the pattern of *frequency for a set of data. This pattern can be displayed as a frequency diagram.

distributor device in the ignition system of a piston engine that distributes pulses of high-voltage electricity to the spark plugs in the cylinders. The electricity is passed to the plug leads by the tip of a rotor arm, driven by the engine camshaft, and current is fed to the rotor arm from the ignition coil. The distributor also houses the contact point or breaker, which opens and closes to interrupt the battery current to the coil, thus triggering the high-voltage pulses. With electronic ignition the distributor is absent.

disulfiram another name for *Antabuse.

diuretic any drug that increases the output of urine by the kidneys. It may be used in the treatment of high blood pressure and to relieve oedema associated with heart, lung, kidney or liver disease, and some endocrine disorders.

divergent margin in *plate tectonics, the boundary, or active zone between two lithospheric plates that are moving apart. Divergent margins are characterized by extensive normal faulting, low-magnitude earthquakes, and, in most cases, volcanic activity.

division one of the four basic operations of arithmetic, usually written in the form $a \div b$ or a/b. It is the inverse of *multiplication.

divisor in mathematics, any number that is to be divided into another number. For example in the computation $100 \div 25 = 4$, 25 is the divisor.

dizygotic twin in medicine, one of a pair of twins born to the same parents at the same time following the fertilization of two separate ova. They may be of different sexes and they are no more likely to resemble each other than any other sibling pair.

DLL in computing, the abbreviation for *dynamic link library.

DMA channel abbreviation for direct memory access channel, in computing, type of channel used for the fast transfer of data between a computer and peripherals such as CD-ROM drives. Most ISA (industry standard architecture) personal computers (PCs) have eight DMA channels, of which typically six are available for use by add-on peripherals, most of which require dedicated channels.

DNA abbreviation for deoxyribonucleic acid, molecular basis of heredity. It is a complex giant molecule that contains, in chemically coded form, the information needed for a cell to make *proteins. In other words it determines the order in which amino acids are joined to make a specific protein in a cell. DNA is a ladder-like double-stranded *nucleic acid, which forms the basis of genetic inheritance in all organisms, except for a few viruses that have only *RNA. DNA is organized into *chromosomes and, in organisms other than bacteria, it is found only in the cell nucleus.

The British scientists Francis Crick (1916–) and Maurice Wilkins and the US scientist James Watson (1928–) were awarded a Nobel Prize for Physiology or Medicine in 1962 for the discovery of the double-helical structure of DNA and determining the significance of this structure in the replication and transfer of genetic information

The two halves (the ladder sides) of the DNA molecule are formed of chains of *nucleotide subunits. Each nucleotide contains a deoxyribose sugar, a phosphate, and a base. A set of three bases – known as a **codon** – acts as a blueprint for the manufacture of a particular *amino acid, the subunit of a protein molecule. The two halves are joined together by the bases – a purine (adenine or guanine) or pyrimidine (cytosine or thymine) – forming pairs (the rungs). The bases form into two specific *base pairs: adenine with thymine and guanine with cytosine. The sequence of base pairs along the DNA acts as a code carrying information about the sequence of amino acids in proteins. Three base pairs in sequence (triplet) name an amino acid and the next three name the next amino acid that needs to be joined and so on, to make a specific protein. The specific way in which the base sequence is preserved from generation to generation. Hereditary information is stored as a specific sequence of bases.

It is important that inherited information is passed on correctly. In the process of DNA replication, which takes place before any cell divides, the two halves of DNA separate and new halves are made. Because of specific base pairing, the inherited information is copied exactly. Despite this, a mistake sometimes occurs and the sequences of bases is altered. This changes the sequence of amino acids in a protein. This is *mutation. Ionizing radiation increases the risk of mutation. In plants and animals DNA is organized into *chromosomes and is found in the *nucleus of cells. DNA in bacteria is organized differently. Bacteria have one large circular DNA molecule carrying most of their inherited information. Some bacteria also have small circular molecules of DNA, known as plasmids. These may be used in *genetic engineering to transfer genes from one organism to another.

protein synthesis Protein synthesis is quite complex. The stages are the coding by triplets of DNA bases to produce mRNA (messenger RNA), the linking of mRNA to tRNA (transfer RNA) at *ribosomes, and the linking of amino acids to form protein.

Since DNA is in the nucleus of the cell and the amino acids are joined in the *cytoplasm the information on the DNA has to be copied so that it can be taken into the cytoplasm. Sections of DNA that code for one protein are copied, making mRNA molecules. The mRNA moves out of the nucleus into the cytoplasm. In the cytoplasm, mRNA joins a ribosome. Then tRNA molecules bring amino acids. The amino acids are arranged in sequence following the code on the mRNA, and then they are joined to make protein.

codons Geneticists identify the codons by the initial letters of the constituent bases – for example, the base sequence of codon CAG is cytosine–adenine–guanine. The meaning of each of the codons in the *genetic code has been worked out by molecular geneticists. There are four different bases, which means that there must be $4 \times 4 \times 4 = 64$ different codons. Proteins are usually made up of only 20 different amino acids, so many amino acids have more than one codon (for example, GGT, GGC, GGA, and GGG all code for the same amino acid, glycine). The first chromosome to have its DNA sequenced by geneticists was chromosome 22 (one of the smallest human chromosomes, linked to schizophrenia), which had its estimated 800 genes and 33.5 million bases sequenced in 1999, leaving only a few gaps.

blueprint for the organism The information encoded by the codons is transcribed (see *transcription) by messenger RNA and is then translated into amino acids in the ribosomes and cytoplasm. The sequence of codons determines the precise order in which amino acids are linked up during manufacture and, therefore, the kind of protein that is to be produced. Because proteins are the chief structural molecules of living matter and, as enzymes, regulate all aspects of metabolism, it may be seen that the genetic code is effectively responsible for building and controlling the whole organism.

laboratory techniques The sequence of bases along the length of DNA can be determined by cutting the molecule into small portions, using restriction enzymes. This technique can also be used for transferring specific sequences of DNA from one organism to another.

ancient DNA DNA is surprisingly stable – the DNA of a nemonychid weevil trapped in amber from the Cretaceous period, and estimated as being 120–135 million years old has been extracted and analysed. US researchers extracted DNA from human hair 10,000 years old in 1994. Dinosaur DNA was also extracted in 1994 from unfossilized dinosaur bones found in a coal mine in Utah.

dimensions of DNA If the DNA of one human cell was unwound it would be almost 2 m/6.6 ft in length. The cell nucleus that contains the DNA is only 10 micrometres in diameter.

DNA fingerprinting or **DNA profiling**, another name for *genetic fingerprinting.

DNS in computing, abbreviation for *domain name server.

document image processing DIP, scanning documents for storage on *CD-ROM. The scanned images are indexed electronically, which provides much faster access than is possible with either paper or microform. See also *optical character recognition.

Document Object Model DOM, specification of a standard framework for creating object-oriented documents devised by the *W3C. DOM allows a Web page – a document – to be treated as a *container* for holding a number of *objects*. Each object is seen as a node in a tree, with each document having a *parent* or root node and any number of *child* nodes. An object might be a comment, a processing instruction (PI), a piece of text, or any other item. In an *HTML or *XML version of this dictionary, for example, the book would be the parent, headwords are child nodes, and definitions are child nodes of headwords. A processing instruction might be used to display all headwords in bold type. The hope is that DOM will provide a structure for the Web, which is, at the moment, just tens of millions of words of barely-connected text. In any event, it is much easier to use *scripting languages and intelligent software to manipulate containers of named objects than to do everything using *tags.

dodecahedron regular solid with 12 pentagonal faces and 12 vertices. It is one of the five regular *polyhedra, or Platonic solids.

dolerite igneous rock formed below the Earth's surface, a form of basalt, containing relatively little silica (mafic in composition).

dolomite in mineralogy, white mineral with a rhombohedral structure, calcium magnesium carbonate ($CaMg(CO_3)_2$). Dolomites are common in geological successions of all ages and are often formed when *limestone is changed by the replacement of the mineral calcite with the mineral dolomite.

dolomite in sedimentology, type of limestone rock where the calcite content is replaced by the mineral *dolomite. Dolomite rock may be white, grey, brown, or reddish in colour, commonly crystalline. It is used as a building material. The region of the Alps known as the Dolomites is a fine example of dolomite formation.

DOM acronym for *Document Object Model.

domain on the Internet, segment of an address that specifies an organization, its type, or its country of origin. All countries except the USA use a final two-letter code such as ca for Canada and uk for the UK. A us suffix exists for the USA, but is not widely used. US addresses usually end in one of seven 'top-level' domains, which specify the type of organization: com (commercial), mil (military), org (usually a nonprofit-making organization), info (for information) and so on. In July 1999 there were 6 million domain names registered on the Internet. By the start of 2002, there were 36,276,252 websites, a slight decline in figures from the previous year.

domain in physics, small area in a magnetic material that behaves like a tiny magnet. The magnetism of the material is due to the movement of electrons in the atoms of the domain. In an unmagnetized sample of material, the domains point in random directions, or form closed loops, so that there is no overall magnetization of the sample. In a magnetized sample, the domains are aligned so that their magnetic effects combine to produce a strong overall magnetism.

domain in mathematics, base set of numbers on which a *function works, mapping it on to a second set (the *range).

domain name server DNS, on the Internet, a type of *server that resolves an alphanumeric *domain name to its numerical *IP address (Internet Protocol address) and vice versa.

dominance in genetics, the masking of one *allele (an alternative form of a gene) by another allele. For example, if a *heterozygous person has one allele for blue eyes and one for brown eyes, his or her eye colour will be brown. The allele for blue eyes is described as *recessive and the allele for brown eyes as dominant.

For every characteristic of a plant or animal that is inherited, there are always two genes present in the cells which determine the characteristic. By

'characteristic' is meant 'height' or 'eye colour' or 'ability to make a particular enzyme'. If the two genes are identical (homozygous state) the characteristic seen in the organism is determined by either one of the two genes. However, one gene may be different from the other (heterozygous state). If so, the two genes are alleles – contrasting genes for a characteristic. In this case it is possible that one of them determines the characteristic seen and the other does not. The characteristic seen in this case is said to be dominant. The other allele, which does not contribute to the appearance of the organism in this case, will only be expressed and contribute to appearance when no dominant gene is present. This characteristic is said to be recessive.

dopa in medicine, a naturally occurring amino acid from which dopamine, a compound involved in the transmission of nervous impulses in the brain, is made. Dopamine deficiency is one of the causative factors of *Parkinson's disease.

dopamine neurotransmitter, hydroxytyramine $C_8H_{11}NO_2$, an intermediate in the formation of adrenaline. There are special nerve cells (neurons) in the brain that use dopamine for the transmission of nervous impulses. One such area of dopamine neurons lies in the basal ganglia, a region that controls movement. Patients suffering from the tremors of *Parkinson's disease show nerve degeneration in this region. Another dopamine area lies in the limbic system, a region closely involved with emotional responses. It has been found that schizophrenic patients respond well to drugs that limit dopamine excess in this area.

Doppler effect change in the observed frequency (or wavelength) of waves due to relative motion between the wave source and the observer. The Doppler effect is responsible for the perceived change in pitch of a siren as it approaches and then recedes, and for the *red shift of light from distant galaxies. It is named after the Austrian physicist Christian Doppler (1803–1853).

dormancy in botany, a phase of reduced physiological activity exhibited by certain buds, seeds, and spores. Dormancy can help a plant to survive unfavourable conditions, as in annual plants that pass the cold winter season as dormant seeds, and plants that form dormant buds.

dorsal in vertebrates, the surface of the animal closest to the backbone. For most vertebrates and invertebrates this is the upper surface, or the surface furthest from the ground. For bipedal primates such as humans, where the dorsal surface faces backwards, then the word is 'back'.

DOS acronym for Disk Operating System, computer *operating system specifically designed for use with disk storage; also used as an alternative name for a particular operating system, MS-DOS.

double blind trial in medicine, method often used in clinical trials of new medicines. Patients are placed in groups using a randomization code and each group receives a different treatment or a placebo. Neither the investigator nor the patient knows which treatment is being administered during the trial. The randomization code is broken at the end of the trial and the results are analysed statistically. Double blind trials eliminate bias but they must be rigorously controlled using recognized ethical standards.

double bond two covalent bonds between adjacent atoms, as in the *alkenes (–C=C–) and *ketones (>C=O).

double circulation in mammals, the characteristic *blood circulatory system. The *heart is divided into two halves, each half working as a separate pump. One side of the heart pumps blood to the lungs and back to the heart. The other side of the heart pumps blood to all other parts of the body and back to the heart.

double decomposition reaction between two chemical substances (usually *salts in solution) that results in the exchange of a constituent from each compound to create two different compounds.

For example, if silver nitrate solution is added to a solution of sodium chloride, there is an exchange of ions yielding sodium nitrate and silver chloride (which is precipitated).

double refraction or **birefringence**, in a crystal, when an unpolarized ray of light entering the crystal is split into two *polarized rays, one of which does obey *Snell's law of refraction, and the other does not. Calcite is such a crystal.

Down's syndrome condition caused by a chromosomal abnormality (the presence of an extra copy of chromosome 21), which in humans produces mental retardation; a flattened face; coarse, straight hair; and a fold of skin at the inner edge of the eye (hence the former name 'mongolism'). The condition can be detected by prenatal testing.

downtime in computing, time when a computer system is unavailable for use, due to maintenance or a system crash. Some downtime is inevitable on almost all systems.

dpi abbreviation for **dots per inch**, measure of the *resolution of images produced by computer screens and printers. The dot pitch limits the maximum number of dots per inch (the density of illuminated points). This is usually greater than the screen resolution, which is given in terms of pixels (picture elements) such as 800 x 600. The number of dots per inch is dependent on the screen size. Because of the common confusion between dpi and pixels, the term pixels per inch is often preferred as a measure of image sharpness on display screens.

drag and drop in computing, in *graphical user interfaces, feature that allows users to select a file name or icon using a mouse and move it to the name or icon representing a program so that the computer runs the program using that file as input data.

DRAM acronym for Dynamic Random-Access Memory, computer memory device in the form of a silicon chip commonly used to provide the *immediate-access memory of microcomputers. DRAM loses its contents unless they are read and rewritten every 2 milliseconds or so. This process is called **refreshing** the memory. DRAM is slower but cheaper than *SRAM, an alternative form of silicon-chip memory.

drift net long straight net suspended from the water surface and used by commercial fishermen. They are controversial as they are indiscriminate in what they catch. Dolphins, sharks, turtles, and other marine animals can drown as a consequence of becoming entangled.

driver in computing, a program that controls a peripheral device. Every device connected to the computer needs a driver program.

The driver ensures that communication between the computer and the device is successful.

drop-down list in computing, in a *graphical user interface, a list of options that hangs down from a blank space in a *dialog box or other on-screen form when a computer awaits user input.

dry ice solid carbon dioxide (CO_2), used as a refrigerant. At temperatures above −79°C/−110.2°F, it sublimes (turns into vapour without passing through a liquid stage) to gaseous carbon dioxide.

DSL in computing, abbreviation for Digital Subscriber Loop (or Line). Examples include *ADSL, ZDSL and, generically, xDSL.

DSP in computing, abbreviation for *digital signal processor.

DTP abbreviation for desktop publishing.

Dual In-line Memory Module DIMM, small printed circuit board carrying several *memory chips. The chips are located on both sides of the board, with independent electrical contacts on either side.

dubnium chemical symbol Db, synthesized, radioactive, metallic element of the *transactinide series, atomic number 105, relative atomic mass 262. Six isotopes have been synthesized, each with very short (fractions of a second) half-lives. Two institutions claim to have been the first to produce it: the Joint Institute for Nuclear Research in Dubna, Russia, in 1967; and the University of California at Berkeley, USA, who disputed the Soviet claim in 1970.

Duchenne muscular dystrophy commonest form of *muscular dystrophy described by French physician Guillaume Duchenne (1806–1875).

ductile describing a material that can sustain large deformations beyond the limits of its *elasticity, without fracture. Metals are very ductile, and may be pulled out into wires, or hammered or rolled into thin sheets without breaking.

ductless gland alternative name for an *endocrine gland.

dumb terminal in computing, a *terminal that has no processing capacity of its own. It works purely as a means of access to a main *central processing unit. Compare with a *personal computer used as an intelligent terminal – for example in *client–server architecture.

dump in computing, the process of rapidly transferring data to external memory or to a printer. It is usually done to help with debugging (see *bug) or as part of an error-recovery procedure designed to provide *data security. A screen dump makes a printed copy of the current screen display.

duodenal ulcer see *ulcer.

duodenum in vertebrates, a short length of *alimentary canal found between the stomach and the small intestine. Its role is in digesting carbohydrates, fats, and proteins. The smaller molecules formed are then absorbed, either by the duodenum or the ileum.

duplex or **echo**, in printing, the ability to print on both sides of the page; in computer communications, setting that *toggles the ability to send and receive signals simultaneously. **Full duplex** means two-way communication is enabled; **half duplex** means it is disabled.

DVI abbreviation for digital video interactive, in computing, a powerful compression and decompression system for digital video and audio. DVI enables 72 minutes of full-screen, full-motion video and its audio track to be stored on a CD-ROM. Originally developed by the US firm RCA, DVI is now owned by Intel and has active support from IBM and Microsoft. It can be used on the hard disk of a PC as well as on a CD-ROM.

Dvorak keyboard alternative keyboard layout to the normal typewriter keyboard layout (QWERTY). In the Dvorak layout the most commonly used keys are situated in the centre, so that keying is faster.

DWANGO acronym for Dial-up Wide Area Network Game Operation, in computing, server that enables computer users with modems to play each other at action games such as *Doom*, *Duke Nuken 3D*, and *Monster Truck Madness* without the variable delays involved in moving data over the Internet.

dye substance that, applied in solution to fabrics, stains with a permanent colour. Different types of dye are needed for different types of fibres. **Direct dyes** combine with cellulose-based fabrics like cotton, linen, and rayon, to colour the fibres. **Indirect dyes** require the presence of another substance (a mordant), with which the fabric must first be treated, to ensure that the dye will remain 'fast' during washing. **Vat dyes** are colourless soluble substances that on exposure to air yield an insoluble coloured compound that is resistant to water.

Naturally occurring dyes include indigo and madder (alizarin), which are extracted from plants; logwood, produced from wood; and cochineal, a red dye made from crushed insects. Industrial dyes are usually synthetic and are derived from coal or petroleum. English chemist William Perkin developed the first synthetic dye, mauve, in 1856 and by the early 20th century a wide range of synthetic dyes was available.

dynamic data exchange DDE, in computing, a form of interprocess communication used in Microsoft Windows, providing exchange of commands and data between two applications. DDE was used principally to include live data from one application in another – for example, spreadsheet data in a word-processed report. After Windows 3.1 DDE was replaced by *object linking and embedding.

Dynamic Host Configuration Protocol method for supplying computers with certain network *configuration information when they start up. It is commonly used for supplying computers with the configuration information they need to connect to the Internet.

Dynamic HTML in computing, the fourth version of hypertext markup language (*HTML), the language used to create Web pages. It is called Dynamic HTML because it enables dynamic effects to be incorporated in pages without the delays involved in downloading Java applets and without referring back to the server.

dynamic IP address in computing, a temporary *IP address assigned from a pool of available addresses by an *Internet Service Provider when a customer logs on to begin an online session.

dynamic link library DLL, in computing, files of executable functions that can be loaded on demand in Microsoft Windows and linked at run time. Windows itself uses DLL files for handling international keyboards, for example, and Windows word-processing programs use DLL files for functions such as spelling and hyphenation checks, and thesaurus.

dynamics or **kinetics**, in mechanics, the mathematical and physical study of the behaviour of bodies under the action of forces that produce changes of motion in them.

dynamite explosive consisting of a mixture of nitroglycerine and diatomaceous earth (diatomite, an absorbent, chalklike material). It was first devised by Swedish chemisst and engineer Alfred Nobel (1833–1896) in 1867.

dynamo in physics, a simple *generator or machine for transforming mechanical energy into electrical energy. A dynamo in basic form consists of a powerful field magnet between the poles of which a suitable conductor, usually in the form of a coil (armature), is rotated. The magnetic lines of force are cut by the rotating wire coil, which induces a current to flow through the wire. The mechanical energy of rotation is thus converted into an electric current in the armature.

dyne symbol dyn, c.g.s. unit of force. 10^5 dynes make one newton. The dyne is defined as the force that will accelerate a mass of one gram by one centimetre per second per second.

dysentery infection of the large intestine causing abdominal cramps and painful diarrhoea with blood. There are two kinds of dysentery: **amoebic** (caused by a protozoan), common in the tropics, which may lead to liver damage; and **bacterial**, the kind most often seen in the temperate zones.

dyslexia (Greek 'bad', 'pertaining to words') malfunction in the brain's synthesis and interpretation of written information, popularly known as 'word blindness'.

Dyslexia may be described as specific or developmental to distinguish it from reading or writing difficulties which are acquired. It results in poor ability in reading and writing, though the person may excel in other areas, for example, in mathematics. A similar disability with figures is called **dyscalculia**. **Acquired dyslexia** may occur as a result of brain injury or disease.

dysphagia difficulty in swallowing. It may be due to infection, obstruction, spasm in the throat or oesophagus (gullet), or neurological disease or damage.

dyspnoea difficulty in breathing, or shortness of breath disproportionate to effort. It is mostly caused by circulatory or respiratory diseases.

dysprosium chemical symbol Dy, (Greek *dusprositos* 'difficult to get near') silver-white, metallic element of the *lanthanide series, atomic number 66, relative atomic mass 162.50. It is among the most magnetic of all known substances and has a great capacity to absorb neutrons.

It was discovered in 1886 by French chemist Paul Lecoq de Boisbaudran (1838–1912).

E abbreviation for **east**.

ear organ of hearing in animals. It responds to the vibrations that constitute sound, which are translated into nerve signals and passed to the brain. A mammal's ear consists of three parts: outer ear, middle ear, and inner ear. The **outer ear** is a funnel that collects sound, directing it down a tube to the **eardrum** (tympanic membrane), which separates the outer and **middle ear**. Sounds vibrate this membrane, the mechanical movement of which is transferred to a smaller membrane leading to the **inner ear** by three small bones, the auditory ossicles. Vibrations of the inner ear membrane move fluid contained in the spiral-shaped cochlea, which vibrates hair cells that stimulate the auditory nerve connected to the brain. There are approximately 30,000 sensory hair cells (**stereocilia**). Exposure to loud noise and the process of ageing damages the stereocilia, resulting in hearing loss. Three fluid-filled canals of the inner ear detect changes of position; this mechanism, with other sensory inputs, is responsible for the sense of balance.

When a loud *noise occurs, muscles behind the eardrum contract automatically, suppressing the noise to enhance perception of sound and prevent injury.

Earth third planet from the Sun. It is almost spherical, flattened slightly at the poles, and is composed of five concentric layers: inner *core, outer core, *mantle, *crust, and atmosphere. About 70% of the surface (including

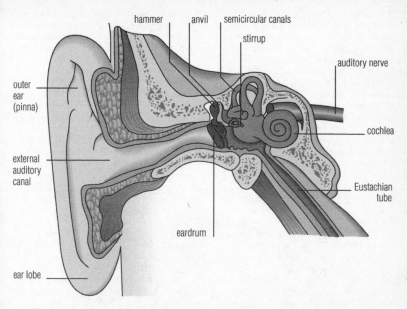

labels: hammer, anvil, semicircular canals, stirrup, auditory nerve, cochlea, Eustachian tube, eardrum, outer ear (pinna), external auditory canal, ear lobe

ear The structure of the ear. The three bones of the middle ear – hammer, anvil and stirrup – vibrate in unison and magnify sounds about 20 times. The spiral-shaped cochlea is the organ of hearing. As sound waves pass down the spiral tube, they vibrate fine hairs lining the tube, which activate the auditory nerve connected to the brain. The semicircular canals are the organs of balance, detecting movements of the head.

the north and south polar ice caps) is covered with water. The Earth is surrounded by a life-supporting atmosphere and is the only planet on which life is known to exist.

mean distance from the Sun: 149,500,000 km/92,860,000 mi

equatorial diameter: 12,755 km/ 7,920 mi

circumference: 40,070 km/24,900 mi

rotation period: 23 hours 56 minutes 4.1 seconds

year: (complete orbit, or sidereal period) 365 days 5 hours 48 minutes 46 seconds. The Earth's average speed around the Sun is 30 kps/18.5 mps. The plane of its orbit is inclined to its equatorial plane at an angle of 23.5°; this is the reason for the changing seasons

atmosphere: nitrogen 78.09%; oxygen 20.95%; argon 0.93%; carbon dioxide 0.03%; and less than 0.0001% neon, helium, krypton, hydrogen, xenon, ozone, and radon

surface: land surface 150,000,000 sq km/57,500,000 sq mi (greatest height above sea level 8,872 m/29,118 ft Mount Everest); water surface 361,000,000 sq km/139,400,000 sq mi (greatest depth 11,034 m/36,201 ft *Mariana Trench in the Pacific). The interior is thought to be an inner core about 2,600 km/1,600 mi in diameter, of solid iron and nickel; an outer core about 2,250 km/ 1,400 mi thick, of molten iron and nickel; and a mantle of mostly solid rock about 2,900 km/1,800 mi thick. The crust and the uppermost layer of the mantle form about twelve major moving plates, some of which carry the continents. The plates are in constant, slow motion, called tectonic drift

satellite: the *Moon

age: 4.6 billion years. The Earth was formed with the rest of the *Solar System by consolidation of interstellar dust. Life began 3.5–4 billion years ago. (See diagram, page 190.)

earth electrical connection between an appliance and the ground. In the event of a fault in an electrical appliance, for example, involving connection between the live part of the circuit and the outer casing, the current flows to earth, causing no harm to the user.

earthquake abrupt motion of the Earth's surface. Earthquakes are caused by the sudden release in rocks of strain accumulated over time as a result of *plate tectonics. The study of earthquakes is called *seismology. Most earthquakes occur along *faults (fractures or breaks) and Benioff zones. As two plates move past each other

they can become jammed. When sufficient strain has accumulated, the rock breaks, releasing a series of elastic waves (*seismic waves) as the plates spring free. The force of earthquakes (magnitude) is measured on the *Richter scale, and their effect (intensity) on the *Mercalli scale. The point at which an earthquake originates is the **focus** or **hypocentre**; the point on the Earth's surface directly above this is the ***epicentre**. The Alaskan (USA) earthquake of 27 March 1964 ranks as one of the greatest

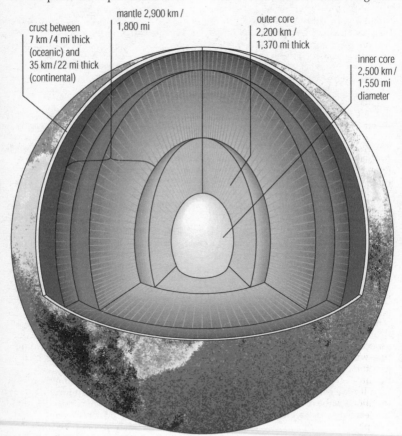

crust between 7 km / 4 mi thick (oceanic) and 35 km / 22 mi thick (continental)

mantle 2,900 km / 1,800 mi

outer core 2,200 km / 1,370 mi thick

inner core 2,500 km / 1,550 mi diameter

Earth Inside the Earth. The surface of the Earth is a thin crust about 6 km/4 mi thick under the sea and 40 km/25 mi thick under the continents. Under the crust lies the mantle, about 2,900 km/1,800 mi thick and with a temperature of 1,500–3,000°C/2,700–5,400°F. The outer core is about 2,250 km/1,400 mi thick, of molten iron and nickel. The inner core is probably solid iron and nickel, at about 5,000°C/9,000°F.

ever recorded, measuring 8.3–8.8 on the Richter scale. The 1906 San Francisco earthquake is among the most famous in history. Its magnitude was 8.3 on the Richter scale. The deadliest, most destructive earthquake in historical times is thought to have been in China in 1556. In 1987 a Californian earthquake was successfully predicted by measurement of underground pressure waves; prediction attempts have also involved the study of such phenomena as the change in gases issuing from the *crust, the level of water in wells, slight deformation of the rock surface, a sequence of minor tremors, and the behaviour of animals. The possibility of earthquake prevention is remote. However, rock slippage might be slowed at movement points, or promoted at stoppage points, by the extraction or injection of large quantities of water underground, since water serves as a lubricant. This would ease overall pressure. Human activity can create earthquakes. Mining, water extraction, and oil extraction can cause subsidence that helps generate earthquakes, while the building of large dams and underground nuclear testing have also been linked with earthquakes.

earth science   scientific study of the planet Earth as a whole. The mining and extraction of minerals and gems, the prediction of weather and earthquakes, the pollution of the atmosphere, and the forces that shape the physical world all fall within its scope of study. The emergence of the discipline reflects scientists' concern that an understanding of the global aspects of the Earth's structure and its past will hold the key to how humans affect its future, ensuring that its resources are used in a sustainable way. It is a synthesis of several traditional subjects such as *geology, *meteorology, *oceanography, *geophysics, *geochemistry, and *palaeontology.

Earth Summit   or **United Nations Conference on Environment and Development**, international meetings aiming at drawing up measures towards environmental protection of the world. The first summit took place in Rio de Janeiro, Brazil, in June 1992. Treaties were made to combat global warming and protect wildlife ('biodiversity') (the latter was not signed by the USA). The second Earth Summit was held in New York in June 1997 to review progress on the environment. The third Earth Summit was held in Johannesburg, South Africa, from 26 August to 4 September 2002.

In 1993, the Clinton administration overturned certain decisions made by George Bush at the Earth Summit. The USA, which had failed to ratify the Convention of Biological Diversity pact along with other nations, came under renewed pressure to sign it in April 1995 after India threatened to prevent US pharmaceutical and cosmetic companies from gaining access to its natural resources.

By 1996 most wealthy nations estimated that they would exceed their emissions targets, including Spain by 24%, Australia by 25%, and the USA by 3%. Britain and Germany were expected to meet their targets.

The second summit (1997) failed to agree a new deal to address the world's growing environmental crisis. Dramatic falls in aid to countries of the developing world, which the 1992 Earth Summit promised to increase, were at the heart of the breakdown. British prime minister Tony Blair condemned the USA, Japan, Canada, and Australia for failing to deliver on commitments to stabilize rising emissions of climate-changing greenhouse gases. The European Community as a whole was on target to meet its stabilization commitment because of cuts in emissions in Germany and the UK.

Deforestation was the main problem tackled at the second summit. The World Bank and the World Wide Fund for Nature signed an agreement aimed at protecting 250 million hectares/617 million acres of forest (10% of the world's forests). The importance of the issue was highlighted by the fact that deforestation progressed rapidly in developing countries since the first summit.

The Kyoto Protocol of 1997 committed the world's industrialized countries to cutting their annual emissions of harmful gases. However, in June 2001 US president George W Bush announced that the USA would not ratify the protocol.

EBCDIC abbreviation for extended binary-coded decimal interchange code, in computing, a code used for storing and communicating alphabetic and numeric characters. It is an 8-bit code, capable of holding 256 different characters, although only 85 of these are defined in the standard version. It is still used in many mainframe computers, but almost all mini- and microcomputers now use *ASCII code.

Ebola virus disease severe haemorrhagic fever similar to Marburg disease. Caused by a *filovirus, it spreads through contact with bodily fluids and has an incubation period of about 21 days. It is fatal in up to 90% of cases. There is no known cure, although a vaccine proved successful in animal trials in 1998.

The virus was first identified in 1976 when it broke out in the Ebola River region in the Sudan and the Democratic Republic of Congo (formerly Zaire). It usually affects monkeys, but in 1995 there was an outbreak in the human population of Kikwit, Democratic Republic of Congo, only 200 km/125 mi from the capital Kinshasa. The World Health Organization declared the outbreak officially over in August 1995; 244 of the 315 people who had contracted the virus had died. Further outbreaks were reported in Gabon in February 1996. By December 2000, there were 405 confirmed cases and 160 deaths following an outbreak in Uganda in September.

e-business contraction of electronic business.

eccentricity in geometry, a property of a *conic section (circle, ellipse, parabola, or hyperbola). It is the distance of any point on the curve from a fixed point (the focus) divided by the distance of that point from a fixed line (the directrix). A circle has an eccentricity of zero; for an ellipse it is less than one; for a parabola it is equal to one; and for a hyperbola it is greater than one.

ECG abbreviation for *electrocardiogram.

echinoderm marine invertebrate of the phylum Echinodermata ('spiny-skinned'), characterized by a five-radial symmetry. Echinoderms have a water-vascular system which transports substances around the body. They include starfishes (or sea stars), brittle-stars, sea lilies, sea urchins, and sea cucumbers. The skeleton is external, made of a series of limy plates. Echinoderms generally move by using tube-feet, small water-filled sacs that can be protruded or pulled back to the body.

echo in computing, user input that is printed to the screen so the user can read it.

echo repetition of a sound wave, or of a *radar or *sonar signal, by reflection from a hard surface such as a wall or building. By accurately measuring the time taken for an echo to return to the transmitter, and by knowing the speed of a radar signal (the speed of light) or a sonar signal (the speed of sound in water), it is possible to calculate the range of the object causing the echo (*echolocation).

echolocation or **biosonar**, method used by certain animals, notably bats, whales, and dolphins, to detect the positions of objects by using sound. The animal emits a stream of high-pitched sounds, generally at ultrasonic frequencies (beyond the range of human hearing), and listens for the returning echoes reflected off objects to determine their exact location.

eclipse passage of one astronomical body through the shadow of another. The term is usually used for solar and lunar eclipses. A **solar eclipse** occurs when the Moon is between the Earth and the Sun (which can happen only at new Moon), the Moon blocking the Sun's rays and casting a shadow on the Earth's surface. A **lunar eclipse** occurs when the Earth is between the Moon and the Sun (which can happen only at full Moon), the Earth blocking the Sun's rays and casting a shadow on the Moon's surface.

During a total solar eclipse the Moon appears to cover the Sun's disc completely and day turns into night.

This is known as the umbra. A total solar eclipse can last up to 7.5 minutes, and the Sun's *corona can be seen. Between two and five solar eclipses occur each year but each is visible only from a specific area. During a partial solar eclipse sunlight reaches the Earth from around the edge of the Moon. This is known as the pre-umbra. Lunar eclipses can also be partial or total. Total lunar eclipses last for up to 100 minutes; the maximum number each year is three.

ecliptic path, against the background of stars, that the Sun appears to follow each year as it is orbited by the Earth. It can be thought of as the plane of the Earth's orbit projected onto the *celestial sphere.

E. coli abbreviation for **Escherichia coli*.

ecology (Greek *oikos* 'house') study of the relationship among organisms and the environments in which they live, including all living and nonliving components. The chief environmental factors governing the distribution of plants and animals are temperature, humidity, soil, light intensity, day length, food supply, and interaction with other organisms. The term ecology was coined by the biologist Ernst Haeckel in 1866.

e-commerce contraction of *electronic commerce.

ecosystem in ecology, a unit consisting of living organisms and the environment that they live in. A simple example of an ecosystem is a pond. The pond ecosystem includes all the pond plants and animals and also the water and other substances that make up the pond itself. Individual organisms interact with each other and with their environment in a variety of relationships, such as two organisms in competition, predator and prey, or as a food source for other organisms in a *food chain. These relationships are usually complex and finely balanced, and in natural ecosystems should be self-sustaining. However, major changes to an ecosystem, such as climate change, overpopulation, or the removal of a species, may threaten the system's sustainability and result in its eventual destruction. For instance, the removal of a major carnivore predator can result in the destruction of an ecosystem through overgrazing by herbivores. Ecosystems can be large, such as the global ecosystem (the ecosphere), or small, such as the pools that collect water in the branch of a tree, and they can contain smaller systems.

ecotourism growing trend in tourism to visit sites that are of ecological interest, for example the Galapagos Islands, or Costa Rica. Ecotourism can bring about employment and income for local people, encouraging conservation, and is far less environmentally damaging than mass tourism.

One of the ideas behind ecotourism is that it is the practice of using money raised through tourism to pay for conservation and community projects, and putting this spending power in the hands of local people. However, if carried out unscrupulously it can lead to damage of environmentally-sensitive sites. In the late 1990s, ecotourism became increasingly popular, with many 'green' travel groups attempting to make entire holiday packages ecologically sound.

ECT abbreviation for *electroconvulsive therapy.

ectoparasite *parasite that lives on the outer surface of its host.

ectopic in medicine, term applied to an anatomical feature that is displaced or found in an abnormal position. An ectopic pregnancy is one occurring outside the womb, usually in a Fallopian tube.

ectoplasm outer layer of a cell's *cytoplasm.

ectotherm 'cold-blooded' animal (see *poikilothermy), such as a lizard, that relies on external warmth (ultimately from the Sun) to raise its body temperature so that it can become active. To cool the body, ectotherms seek out a cooler environment.

eczema inflammatory skin condition, a form of dermatitis, marked by dryness, rashes, itching, the formation of blisters, and the exudation of fluid. It may be allergic in origin and is sometimes complicated by infection.

eddy current electric current induced, in accordance with *Faraday's laws of electromagnetic induction, in a conductor located in a changing magnetic field. Eddy currents can cause much wasted energy in the cores of transformers and other electrical machines.

edge in topology, link on a topological map. Where edges meet and intersect, *nodes are formed.

edge connector in computing, an electrical connection formed by taking some of the metallic tracks on a *printed circuit board to the edge of the board and using them to plug directly into a matching socket.

EDI in computing, abbreviation for *electronic data interchange.

EDIFACT acronym for Electronic Data Interchange For Administration, Commerce, and Trade, in computing, the ISO and ANSI standard system for handling EDI transactions.

EDO RAM abbreviation for extended data out random-access memory, in computing, faster type of *RAM introduced in the mid-1990s.

EDP in computing, abbreviation for **electronic data processing**.

educational psychology the work of psychologists primarily in schools, including the assessment of children with achievement problems and advising on problem behaviour in the classroom.

EEG abbreviation for *electroencephalogram.

EEPROM acronym for Electrically Erasable Programmable Read-Only Memory, computer memory that can record data and retain it indefinitely. The data can be erased with an electrical charge and new data recorded.

efficiency in physics, a general term indicating the degree to which a process or device can convert energy from one form to another without loss, or how effectively energy is used, and wasted energy, such as heat and sound, minimized. It is normally expressed as a fraction or a percentage, where 100% indicates conversion with no loss. The efficiency of a machine, for example, is the ratio of the energy output to the energy input; in practice it is always less than 100% because of frictional heat losses.

For example, 75% of the electrical energy of an electric light bulb is converted into heat and only 25% is converted into light. Therefore, an electric light bulb is not an efficient energy converter. Certain electrical machines with no moving parts, such as transformers, can approach 100% efficiency.

efflorescence loss of water or crystallization of crystals exposed to air, resulting in a dry powdery surface.

effluent liquid discharge of waste from an industrial process, usually into rivers or the sea. Effluent is often toxic but is difficult to control and hard to trace.

In some cases, as at Minamata, Japan, where 43 people died of mercury poisoning, effluent can be lethal but usually its toxic effects remain unclear, because it quickly dilutes in the aquatic ecosystem.

egg in animals, the ovum, or female *gamete (reproductive cell).

After fertilization by a sperm cell, it begins to divide to form an embryo. Eggs may be deposited by the female (*ovipary) or they may develop within her body (*vivipary and *ovovivipary). In the oviparous reptiles and birds, the egg is protected by a shell, and well supplied with nutrients in the form of yolk.

ego (Latin 'I') in psychology, the processes concerned with the self and a person's conception of himself or herself, encompassing values and attitudes. In Freudian psychology, the term refers specifically to the element of the human mind that represents the conscious processes concerned with reality, in conflict with the *id (the instinctual element) and the *superego (the ethically aware element).

einsteinium chemical symbol Es, synthesized, radioactive, metallic element of the actinide series, atomic number 99, relative atomic mass 254.09.

It was produced by the first thermonuclear explosion, in 1952, and discovered in fallout debris in the form of the isotope Es-253 (half-life 20 days). Its longest-lived isotope, Es-254, with a half-life of 276 days, allowed the element to be studied at length. It is now synthesized by bombarding lower-numbered *transuranic elements in particle accelerators. It was first identified by US chemist Albert Ghiorso and his team who named it in 1955 after German-born US physicist Albert Einstein(1879–1955), in honour of his theoretical studies of mass and energy.

elastic collision in physics, a collision between two or more bodies in which the total *kinetic energy of the bodies is conserved (remains constant); none is converted into any other form of

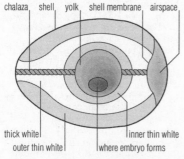

chalaza | shell | yolk | shell membrane | airspace

thick white | inner thin white
outer thin white | where embryo forms

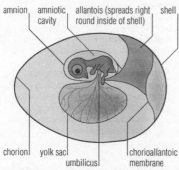

amnion | amniotic cavity | allantois (spreads right round inside of shell) | shell

chorion | yolk sac | chorioallantoic membrane
umbilicus

egg Section through a fertilized bird egg. Inside a bird's egg is a complex structure of liquids and membranes designed to meet the needs of the growing embryo. The yolk, which is rich in fat, is gradually absorbed by the embryo. The white of the egg provides protein and water. The chalaza is a twisted band of protein which holds the yolk in place and acts as a shock absorber. The airspace allows gases to be exchanged through the shell. The allantois contains many blood vessels which carry gases between the embryo and the outside.

energy. The molecules of a gas may be considered to collide elastically, but large objects may not because some of their kinetic energy will be converted on collision to heat and sound, or used to deform the object.

elasticity in physics, the ability of a solid to recover its shape once deforming forces are removed. An elastic material obeys *Hooke's law, which states that its deformation is proportional to the applied stress up to a certain point, called the **elastic limit**; beyond this point additional stresses will deform it permanently. Elastic materials include metals and rubber; however, all materials have some degree of elasticity.

elastomer any material with rubbery properties that stretches easily and then quickly returns to its original length when released.

Natural and synthetic rubbers and such materials as polychloroprene and butadiene copolymers are elastomers. The convoluted molecular chains making up these materials are uncoiled by a stretching force, but return to their original position when released because there are relatively few crosslinks between the chains.

E-layer formerly **Kennelly-Heaviside layer**, lower regions (90–120 km/56–75 mi) of the *ionosphere, which reflect radio waves, allowing their reception around the surface of the Earth. The E-layer approaches the Earth by day and recedes from it at night. Its existence was proved by the British physicist Edward Victor Appleton (1892–1965).

electrical energy form of energy carried by an electric current. It may be converted into other forms of energy such as heat, light, and motion.

Electrical appliances, such as washing machines, vacuum cleaners, radios, and televisions, use electricity in order to do work. Electric motors convert electrical energy into mechanical energy.

electric arc a continuous electric discharge of high current between two electrodes, giving out a brilliant light and heat. The phenomenon is exploited in the carbon-arc lamp, once widely used in film projectors. In the electric-arc furnace an arc struck between very large carbon electrodes and the metal charge provides the heating. In arc welding an electric arc provides the heat to fuse the metal. The discharges in low-pressure gases, as in neon and sodium lights, can also be broadly considered as electric arcs.

electric cell device in which chemical energy is converted into electrical energy; see *cell, electrical.

electric charge property of some bodies that causes them to exert forces on each other. Two bodies both with positive or both with negative charges repel each other, whereas bodies with opposite or 'unlike' charges attract each other. *Electrons possess a negative charge, and *protons an equal positive charge. The *SI unit of electric charge is the coulomb (symbol C).

A body can be charged by friction, induction, or chemical change, and the charge shows itself as an accumulation of electrons (negative charge) or a loss of electrons (positive charge) on an atom or body. Atoms generally have zero net charge but can gain electrons to become negative ions or lose them to become positive ions. So-called *static electricity, seen in such phenomena as the charging of nylon shirts when they are pulled on or off, or in brushing hair, is in fact the gain or loss of electrons from the surface atoms. A flow of charge (such as electrons through a copper wire) constitutes an electric current; the flow of current is measured in amperes (symbol A).

electric current flow of electrically charged particles through a conducting circuit due to the presence of a *potential difference. The current at any point in a circuit is the amount of charge flowing per second; its SI unit is the ampere (coulomb per second).

Current carries electrical energy from a power supply, such as a battery of electrical cells, to the components of the circuit, where it is converted into other forms of energy, such as heat, light, or motion. It may be either *direct current or *alternating current.

heating effect When current flows in a component possessing resistance, electrical energy is converted into heat energy. If the resistance of the component is R ohms and the current through it is I amperes, then the heat energy W (in joules) generated in a time t seconds is given by the formula: $W = I^2Rt$.

magnetic effect A magnetic field is created around all conductors that carry a current. When a current-bearing conductor is made into a coil it forms an *electromagnet with a *magnetic field that is similar to that of a bar magnet, but which disappears as soon as the current is switched off. The strength of the magnetic field is directly proportional to the current in the conductor – a property that allows a small electromagnet to be used to produce a pattern of magnetism on recording tape that accurately represents the sound or data stored. The direction of the field created around a conducting wire may be predicted by using *Maxwell's screw rule.

motor effect A conductor carrying current in a magnetic field experiences a force, and is impelled to move in a direction perpendicular to both the direction of the current and the direction of the magnetic field. The direction of motion may be predicted by Fleming's left-hand rule (see *Fleming's rules). The magnitude of the force experienced depends on the

direct current (for example, the current from batteries) flows in one direction and does not change in size

alternating current (for example, mains electricity in the UK) is constantly changing direction

electric current The patterns produced by a direct current and an alternating current on the screen of an oscilloscope.

length of the conductor and on the strengths of the current and the magnetic field, and is greatest when the conductor is at right angles to the field. A conductor wound into a coil that can rotate between the poles of a magnet forms the basis of an *electric motor.

electric field in physics, a region in which a particle possessing electric charge experiences a force owing to the presence of another electric charge. The strength of an electric field, E, is measured in volts per metre (V m^{-1}). It is a type of *electromagnetic field.

An electric field is formed between two metal plates that are parallel to each other and connected to a voltage supply. An electron beam moving through these plates will be deflected slightly towards the positive plate.

electricity all phenomena caused by *electric charge. There are two types of electricity: static and current. Electric charge is caused by an excess or deficit of electrons in a substance, and an electric current is the movement of charge through a material. Materials having equal numbers of positive and negative charges are termed neutral, as the charges balance out. Substances may be electrical conductors, such as metals, which allow the passage of electricity through them readily, or insulators, such as rubber, which are extremely poor conductors. Substances with relatively poor conductivities that increase with a rise in temperature or when light falls on the material are known as *semiconductors. Electric currents also flow through the nerves of organisms. For example, the optic nerve in humans carries electric signals from the eye to the brain. Electricity cannot be seen, but the effects it produces can be clearly seen; for example, a flash of lightning, or the small sparks produced by rubbing a nylon garment.

Electricity is essential to modern society. Electrical devices are used in the home, office, and industry; this is called *mains electricity, and is measured in *kilowatt hours. Electrical energy is used to power devices used for communications such as fax machines, telephones, computers, and satellites.

electric motor machine that converts electrical energy into mechanical energy. There are various types, including direct-current and induction motors, most of which produce rotary motion. A linear induction motor produces linear (in a straight line) rather than rotary motion. Electric motors and generators have the same components; they differ in the way they are used.

A simple **direct-current motor** consists of a horseshoe-shaped permanent *magnet with a wire-wound coil (*armature) mounted so that it can rotate between the poles of the magnet. The ends of a *commutator. This reverses the current (from a battery) fed to the coil on each half turn, which rotates because of the mechanical force exerted on a conductor carrying a current in a magnetic field.

electric power rate at which an electrical machine uses electrical *energy or converts it into other forms of energy – for example, light, heat, or mechanical energy. Usually measured in watts (equivalent to joules per second), it is equal to the product of the voltage and the current flowing.

In a closed circuit the potential difference (voltage) causes electrons to flow towards the positive potential. Work is done in moving the electrons, and the rate at which this work is done is called the electrical power. If the voltage is V volts and the current is I amps, then the power, P watts, is given by $P = V \times I$

An electric lamp that passes a current of 0.4 amperes at 250 volts uses 100 watts of electrical power and converts it into light and heat – in ordinary terms it is a 100-watt lamp. An electric motor that requires 6 amperes at the same voltage consumes 1,500 watts (1.5 kilowatts), equivalent to delivering about 2 horsepower of mechanical power.

electrocardiogram ECG, graphic recording of the electrical activity of the heart, as detected by electrodes placed on the skin. Electrocardiography is used in the diagnosis of heart disease.

electrochemical series or **electromotive series**, list of chemical elements arranged in descending order of the ease with which they can lose electrons to form cations (positive ions). An element can be displaced (*displacement reaction) from a compound by any element above it in the series.

electrochemistry branch of science that studies chemical reactions involving electricity. The use of electricity to produce chemical effects, *electrolysis, is employed in many industrial processes, such as electroplating, the manufacture of chlorine, and the extraction of aluminium. The use of chemical reactions to produce electricity is the basis of electrical *cells, such as the dry cell and the Leclanché cell.

Since all chemical reactions involve changes to the electronic structure of atoms, all reactions are now recognized as electrochemical in nature. Oxidation, for example, was once defined as a process in which oxygen was combined with a substance, or hydrogen was removed from a compound; it is now defined as a process in which electrons are lost.

Electrochemistry is also the basis of new methods of destroying toxic organic pollutants. For example, electrochemical cells that operate with supercritical water (a type of supercritical *fluid) have been developed to combust organic waste materials.

electroconvulsive therapy ECT; or **electroshock therapy**, treatment mainly for severe *depression, given under anaesthesia and with a muscle relaxant. An electric current is passed through one or both sides of the brain to induce alterations in its electrical activity. The treatment can cause distress and loss of concentration and memory, and so there is much controversy about its use and effectiveness.

electrode any terminal by which an electric current passes in or out of a conducting substance; for example, the anode or *cathode in an electrolytic cell. The anode is the positive electrode and the cathode is the negative electrode.

The terminals that emit and collect the flow of electrons in thermionic

*valves (electron tubes) are also called electrodes: for example, cathodes, plates (anodes), and grids.

electrodynamics branch of physics dealing with electric charges, electric currents, and associated forces. *Quantum electrodynamics (QED) studies the interaction between charged particles and their emission and absorption of electromagnetic radiation. This subject combines quantum theory and relativity theory, making accurate predictions about subatomic processes involving charged particles such as electrons and protons.

electroencephalogram EEG, graphic record of the electrical discharges of the brain, as detected by electrodes placed on the scalp. The pattern of electrical activity revealed by electroencephalography is helpful in the diagnosis of some brain disorders, in particular epilepsy.

electroencephalography in medicine, diagnostic technique that monitors the brain's electrical discharge to determine the presence of brain disease. An *electroencephalogram (EEG) is a graphical record of these changes.

electrolysis in chemistry, the production of chemical changes by passing an electric current through a solution or molten salt (the electrolyte), resulting in the migration of ions to the electrodes: positive ions (*cations) to the negative electrode (*cathode) and negative ions (*anions) to the positive electrode (*anode).

During electrolysis, the ions react with the electrode, either receiving or giving up electrons. The resultant atoms may be liberated as a gas, or deposited as a solid on the electrode, in amounts that are proportional to the amount of current passed, as discovered by English scientist Michael Faraday (1791–1867). For instance, when acidified water is electrolysed, hydrogen ions (H^+) at the cathode receive electrons to form hydrogen gas; hydroxide ions (OH^-) at the anode give up electrons to form oxygen gas and water.

One application of electrolysis is **electroplating**, in which a solution of a salt, such as silver nitrate ($AgNO_3$), is used and the object to be plated acts as the negative electrode, thus attracting silver ions (Ag^+). Electrolysis is used in many industrial processes, such as coating metals for vehicles and ships, refining *bauxite into aluminium, and the chlor-alkali industry, in which *brine (sodium chloride solution) is electrolysed to produce *chlorine, hydrogen, and *sodium hydroxide (caustic soda); it also forms the basis of a number of electrochemical analytical techniques, such as polarography.

electrolyte solution or molten substance in which an electric current is made to flow by the movement and discharge of ions in accordance with Faraday's laws of *electrolysis.

The term 'electrolyte' is frequently used to denote a substance that, when dissolved in a specified solvent, usually water, dissociates into *ions to produce an electrically conducting medium.

In medicine the term is often used for the ion itself (sodium or potassium, for example). Electrolyte balance may be severely disrupted in illness or injury.

electromagnet coil of wire wound around a soft iron core that acts as a magnet when an electric current flows through the wire. Electromagnets have many uses: in switches, electric bells, *solenoids, and metal-lifting cranes.

The strength of the electromagnet can be increased by increasing the current through the wire, changing the material of the core, or by increasing the number of turns in the wire coil. At the north pole of the electromagnet current flows anticlockwise; at the south pole flow is clockwise.

electromagnetic field in physics, region in which a particle with an *electric charge experiences a force. If it does so only when moving, it is in a pure **magnetic field**; if it does so when stationary, it is in an **electric field**. Both can be present simultaneously. For example, a light wave consists of an electric field and a magnetic field travelling simultaneously at right angles to each other.

electromagnetic force one of the four fundamental *forces of nature, the other three being the gravitational

force (gravity), the weak nuclear force, and the strong nuclear force. The particle that is the carrier for the electromagnetic force is the *photon.

electromagnetic induction in electronics, the production of an *electromotive force (emf) in a circuit by a change of magnetic flux through the circuit or by relative motion of the circuit and the magnetic flux. As a magnet is moved in and out of a coil of wire in a closed circuit an *induced current will be produced. All dynamos and generators produce electricity using this effect. When magnetic tape is driven past the playback head (a small coil) of a tape recorder, the moving magnetic field induces an emf in the head, which is then amplified to reproduce the recorded sounds.

Electromagnetic induction takes place when the magnetic field around a conductor changes. If the magnetic field is made to change quickly, the size of the current induced is larger. A galvanometer can be used to measure the direction of the current. As a magnet is pushed into a coil, the needle on the galvanometer moves in one direction. As the magnet is removed from the coil, the needle moves in the opposite direction.

If the change of magnetic flux is due to a variation in the current flowing in the same circuit, the phenomenon is known as self-induction; if it is due to a change of current flowing in another circuit it is known as mutual induction.

Lenz's law The direction of an electromagnetically-induced current (generated by moving a magnet near a wire or by moving a wire in a magnetic field) will be such as to oppose the motion producing it. This law is named after the Russian physicist Heinrich Friedrich Lenz (1804–1865), who announced it in 1833.

Faraday's laws English scientist Michael Faraday (1791–1867) proposed three laws of electromagnetic induction: (1) a changing magnetic field induces an electromagnetic force in a conductor; (2) the electromagnetic force is proportional to the rate of change of the field; and (3) the direction of the induced electromagnetic force depends on the orientation of the field.

electromagnetic pollution   the electric and magnetic fields set up by high-tension power cables, local electric sub-stations, and domestic items such as electric blankets. There have been claims that these electromagnetic fields are linked to increased levels of cancer, especially leukaemia, and to headaches, nausea, dizziness, and depression.

Although the issue has failed to receive official recognition in the UK, physicists there linked electromagnetic pollution with *radon gas as a cause of cancer in 1996. The electric fields were found to attract the radioactive decay products of radon and cause them to vibrate, making them more likely to adhere to skin and mucous membranes. However, in December 1999 British researchers announced that the world's largest study into the safety of electromagnetic fields had not found a link between electromagnetic fields and childhood cancers.

electromagnetic radiation   transfer of energy in the form of *electromagnetic waves.

electromagnetic spectrum   complete range, over all wavelengths and frequencies, of *electromagnetic waves. These include (in order of decreasing wavelength) radio and television waves, microwaves, infrared radiation, visible light, ultraviolet light, X-rays, and gamma radiation.

The colour of sunlight is made up of a whole range of colours. A glass prism can be used to split white light into separate colours that are sensitive to the human eye, ranging from red (longer wavelength) to violet (shorter wavelength). The human eye cannot detect electromagnetic radiation outside this range. Some animals, such as bees, are able to detect ultraviolet light.

electromagnetic system of units former system of absolute electromagnetic units (emu) based on the * c.g.s. system and having, as its primary electrical unit, the unit magnetic pole. It was replaced by *SI units.

electromagnetic waves   oscillating electric and magnetic fields travelling together through space at a speed of nearly 300,000 kps/186,000 mps.

Visible light is composed of electromagnetic waves. The **electromagnetic spectrum** is a family of waves that includes radio waves, infrared radiation, visible light, ultraviolet radiation, X-rays, and gamma rays. All electromagnetic waves are *transverse waves. They can be reflected, refracted, diffracted, and polarized.

Radio and television waves lie at the **long wavelength–low frequency** end of the spectrum, with wavelengths longer than 10^{-4} m. Infrared radiation has wavelengths between 10^{-4} m and 7×10^{-7} m. Visible light has yet shorter wavelengths from 7×10^{-7} m to 4×10^{-7} m. Ultraviolet radiation is near the **short wavelength–high frequency** end of the spectrum, with wavelengths between 4×10^{-7} m and 10^{-8} m. X-rays have wavelengths from 10^{-8} m to 10^{-12} m. Gamma radiation has the shortest wavelengths (less than 10^{-10} m). The different wavelengths and frequencies lend specific properties to electromagnetic waves. While visible light is diffracted by a diffraction grating, X-rays can only be diffracted by crystals. Radio waves are refracted by the atmosphere; visible light is refracted by glass or water.

electromagnetism   in physics, the properties and interactions of *magnetism and *electric currents. A current flowing through a conductor produces a magnetic field around the conductor; a varying magnetic field interacting with a conductor produces a current in the conductor. This is known as the electromagnetic effect.

electromotive force   emf, loosely, the voltage produced by an electric battery or generator in an electrical circuit or, more precisely, the energy supplied by a source of electric power in driving a unit charge around the circuit. The unit is the *volt.

A difference in charge between two points in a material can be created by an external energy source such as a battery. This causes electrons to move so that there is an excess of electrons at one point and a deficiency of electrons at a second point. This difference in charge is stored as electrical potential energy known as emf. It is the emf that

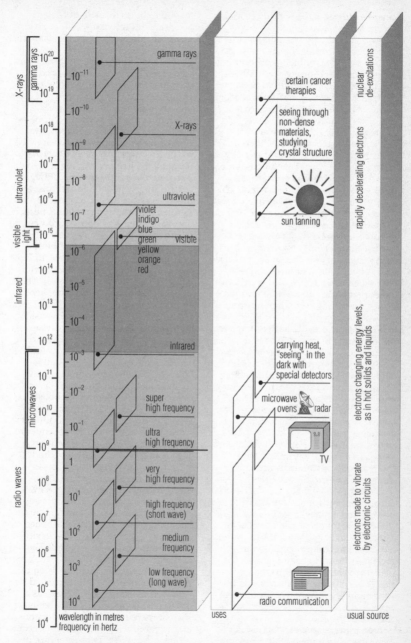

electromagnetic waves Radio waves have the lowest frequency. Infrared radiation, visible light, ultraviolet radiation, X-rays, and gamma rays have progressively higher frequencies.

causes a current to flow through a circuit.

electron negatively charged particle with negligible mass. Electrons form the outer portion of all atoms, orbiting the nucleus in groupings called shells. The first shell can hold up to two electrons; the second and third shells can hold up to eight electrons each. The electron arrangement of an element is called its **electronic configuration**; for example, the electronic configuration of the sodium atom is $Na_{(2,8,1)}$. In a neutral atom the number of electrons is equal to the number of protons in the nucleus. This electron structure is responsible for the chemical properties of the atom (see *atomic structure). Electrons are a member of the class of elementary particles known as *leptons.

Electrons carry a charge of 1.602177×10^{-19} coulombs and have a mass of 9.109×10^{-31} kg, which is $1/_{1836}$ times the mass of a *proton. Energy can be used to remove electrons from their outer orbits in conducting materials and they can be made to flow through these materials under a potential difference. This flow is known as an electric current or electricity. A beam of electrons will undergo *diffraction (scattering) and produce interference patterns in the same way as *electromagnetic waves such as light; hence they may be regarded as waves as well as particles.

electronegativity ease with which an atom can attract electrons to itself. Electronegative elements attract electrons, so forming negative ions.

US chemist Linus Pauling (1901–1994) devised an electronegativity scale to indicate the relative power of attraction of elements for electrons. Fluorine, the most non-metallic element, has a value of 4.0 on this scale; oxygen, the next most non-metallic, has a value of 3.5.

In a covalent bond between two atoms of different electronegativities, the bonding electrons will be located close to the more electronegative atom, creating a *dipole.

electron gun series of *electrodes, including a cathode for producing an electron beam. It plays an essential role in many electronic devices, including

*cathode-ray tubes (television tubes) and *electron microscopes.

electronic commerce or **e-commerce**, conducting business online. It is generally accepted that electronic commerce is a class of electronic business, and refers to the transaction of business, and more narrowly the sale and purchase of goods and services, using networked computers. This usually means via the Internet, although electronic commerce may be conducted over private networks or intranets.

electronic data interchange EDI, in computing, system for managing business-to-business transactions such as invoicing and ordering to eliminate the wastefulness of paper-based transaction systems.

electronic flash discharge tube that produces a high-intensity flash of light, used for photography in dim conditions. The tube contains a noble gas (rare gas) such as krypton. The flash lasts only a few thousandths of a second.

electronic mail see *e-mail.

electronic publishing distribution of information using computer-based media such as *multimedia and *hypertext in the creation of electronic 'books'. Critical technologies in the development of electronic publishing were *CD-ROM, with its massive yet compact storage capabilities, and the advent of computer networking with its ability to deliver information instantaneously anywhere in the world.

electronics branch of science that deals with the emission of *electrons from conductors and *semiconductors, with the subsequent manipulation of these electrons, and with the construction of electronic devices. The first electronic device was the thermionic valve, or vacuum tube, in which electrons moved in a vacuum, and led to such inventions as *radio, *television, *radar, and the digital computer. Replacement of valves with the comparatively tiny and reliable *transistor from 1948 revolutionized electronic development. Modern electronic devices are based on minute *integrated circuits (silicon chips), wafer-thin crystal slices holding tens of thousands of electronic components.

By using solid-state devices such as integrated circuits, extremely complex electronic circuits can be constructed, leading to the development of digital watches, pocket calculators, powerful microcomputers, and word processors.

electron microscope instrument that produces a magnified image by using a beam of *electrons instead of light rays, as in an optical *microscope. An **electron lens** is an arrangement of electromagnetic coils that control and focus the beam. Electrons are not visible to the eye, so instead of an eyepiece there is a fluorescent screen or a photographic plate on which the electrons form an image. The wavelength of the electron beam is much shorter than that of light, so much greater magnification and resolution (ability to distinguish detail) can be achieved. The development of the electron microscope has made possible the observation of very minute organisms, viruses, and even large molecules.

A *transmission electron microscope passes the electron beam through a very thin slice of a specimen. A *scanning electron microscope looks at the exterior of a specimen. A scanning transmission electron microscope (STEM) can produce a magnification of 90 million times. See also *atomic force microscope.

electron probe microanalyser modified *electron microscope in which the target emits X-rays when bombarded by electrons. Varying X-ray intensities indicate the presence of different chemical elements. The composition of a specimen can be mapped without the specimen being destroyed.

electrons, delocalized electrons that are not associated with individual atoms or identifiable chemical bonds, but are shared collectively by all the constituent atoms or ions of some chemical substances (such as metals, graphite, and *aromatic compounds).

A metallic solid consists of a three-dimensional arrangement of metal ions through which the delocalized electrons are free to travel. Aromatic compounds are characterized by the sharing of delocalized electrons by several atoms within the molecule.

electrons, localized pair of electrons in a *covalent bond that are located in the vicinity of the nuclei of the two contributing atoms. Such electrons cannot move beyond this area.

electron transport chain arrangement of substances within the living cell that takes part in energy production. Electron transport is the means by which reduced flavoproteins (FP) and the coenzyme NAD are oxidized in steps, such that energy from certain steps can be used for the manufacture of *ATP by oxidative phosphorylation.

The electron transport chain is a collection of *cytochromes and flavoproteins situated close to each other on the inner membrane of the *mitochondria. This allows for the proper flow of *electrons from one cytochrome to another. A compound such as $NADH_2$ or FPH_2 loses hydrogen atoms at the beginning of the chain, returning to its original form, NAD or FP. The electrons pass down the chain of cytochromes, alternately reducing and oxidizing the cytochrome as the electron arrives and moves on to the next one, until finally oxygen is reached and water produced. Hence the process is aerobic, that is, requiring oxygen.

electron volt symbol eV, unit for measuring the energy of a charged particle (*ion or *electron) in terms of the energy of motion an electron would gain from a potential difference of one volt. Because it is so small, more usual units are mega-(million) and giga- (billion) electron volts (MeV and GeV).

electrophoresis the *diffusion of charged particles through a fluid under the influence of an electric field. It can be used in the biological sciences to separate *molecules of different sizes, which diffuse at different rates. In industry, electrophoresis is used in paint-dipping operations to ensure that paint reaches awkward corners.

electroplating deposition of metals upon metallic surfaces by *electrolysis for decorative and/or protective purposes. It is used in the preparation of printing plates, 'master' audio discs, and in many other processes.

A current is passed through a bath containing a solution of a salt of the plating metal, the object to be plated being the cathode (negative terminal); the *anode (positive terminal) is either an inert substance or the plating metal. Among the metals most commonly used for plating are zinc, nickel, chromium, cadmium, copper, silver, and gold.

In **electropolishing**, the object to be polished is made the anode in an electrolytic solution and by carefully controlling conditions the high spots on the surface are dissolved away, leaving a high-quality stain-free surface. This technique is useful in polishing irregular stainless-steel articles.

electropositivity in chemistry, a measure of the ability of elements (mainly metals) to donate electrons to form positive ions. The greater the metallic character, the more electropositive the element.

electrorheological fluid another name for *smart fluid, a liquid suspension that gels when an electric field is applied across it.

electroscope apparatus for detecting *electric charge. The simple gold-leaf electroscope consists of a vertical conducting (metal) rod ending in a pair of rectangular pieces of gold foil, mounted inside and insulated from an earthed metal case or glass jar. An electric charge applied to the end of the metal rod makes the gold leaves diverge, because they each receive a similar charge (positive or negative) via the rod and so repel each other.

The polarity of the charge can be found by bringing up another charge of known polarity and applying it to the metal rod. A like charge has no effect on the gold leaves since they remain repelled. An opposite charge neutralizes the charge on the leaves and causes them to collapse.

electrostatic precipitator device that removes dust or other particles from air and other gases by electrostatic means. An electric discharge is passed through the gas, giving the impurities a negative electric charge. Positively

charged plates are then used to attract the charged particles and remove them from the gas flow. Such devices are attached to the chimneys of coal-burning power stations to remove ash particles.

electrostatics study of stationary electric charges and their fields (not currents). See *static electricity.

Charged materials behave differently if they are brought near to each other. Materials of like charge, that is both positive or both negative, will move away from each other (repelled by electrostatic force). Materials of opposite charge, that is one positive and one negative, brought close together will move towards each other (attracted by electrostatic force).

An *electroscope is an instrument that detects charges on objects that have been charged by static electricity.

electrovalent bond another name for an *ionic bond, a chemical bond in which the combining atoms lose or gain electrons to form ions.

element substance that cannot be split chemically into simpler substances. The atoms of a particular element all have the same number of protons in their nuclei (their proton or *atomic number). Elements are classified in the *periodic table of the elements. Of the known elements, 92 are known to occur naturally on Earth (those with atomic numbers 1–92). Those elements with atomic numbers above 96 do not occur in nature and must be synthesized in particle accelerators. Of the elements, 81 are stable; all the others, which include atomic numbers 43, 61, and from 84 up, are radioactive.

Elements are classified as metals, nonmetals, or metalloids (weakly metallic elements) depending on a combination of their physical and chemical properties; about 75% are metallic. Some elements occur abundantly (oxygen, aluminium); others occur moderately or rarely (chromium, neon); some, in particular the radioactive ones, are found in minute (neptunium, plutonium) or very minute (astatine, technetium) amounts. Symbols (devised by Swedish chemist Jöns Berzelius) are used to denote the elements; the symbol is usually the first letter or letters of the English or Latin name (for example, C for carbon, Ca for calcium, Fe for iron, from the Latin *ferrum*). The symbol represents one atom of the element. Two or more elements bonded together form a **compound** from which they cannot be separated by physical means. Compounds are held together by ionic or covalent bonds. The number of atoms of an element that combine to form a molecule is it **atomicity**. A molecule of oxygen (O_2) has atomicity 2; sulphur (S_8) has atomicity 8.

According to current theories, hydrogen and helium were produced in the *Big Bang at the beginning of the universe. Of the other elements, those up to atomic number 26 (iron) are made by nuclear fusion within the stars. The heavier elements, such as lead and uranium, are produced when an old star explodes; as its centre collapses, the gravitational energy squashes nuclei together to make new elements.

element in mathematics, a member of a *set.

elementary particle in physics, a subatomic particle that is not known to be made up of smaller particles, and so can be considered one of the fundamental units of matter. There are three groups of elementary particles: *quarks, *leptons, and *gauge bosons.

Quarks combine in groups of three to produce heavy particles called *baryons, and in groups of two to produce intermediate-mass particles called mesons (see *hadron). They and their composite particles are influenced by the strong nuclear force.

Leptons – the electron, muon, tau and their neutrinos – are particles that do not interact via the strong nuclear force, they are influenced by the weak nuclear force, as well as by gravitation and electromagnetism.

Gauge bosons carry forces between other particles. There are four types: gluon, photon, intermediate vector bosons (W^+, W^-, and Z), and graviton. The gluon carries the strong nuclear force, the photon the electromagnetic force, W^+, W^-, and Z the weak nuclear force, and the graviton, as yet

unobserved, the force of gravity (see *forces, fundamental).

elevation a drawing to scale of one side of an object or building.

elevation, angle of an upward angle made with the horizontal.

elevation of boiling point raising of the boiling point of a liquid above that of the pure solvent, caused by a substance being dissolved in it. The phenomenon is observed when salt is added to boiling water; the water ceases to boil because its boiling point has been elevated.

How much the boiling point is raised depends on the number of molecules of substance dissolved. For a single solvent, such as pure water, all substances in the same molecular concentration produce the same elevation of boiling point. The elevation e produced by the presence of a solute of molar concentration C is given by the equation $e = KC$, where K is a constant (called the ebullioscopic constant) for the solvent concerned.

ellipse curve joining all points (loci) around two fixed points (foci) such that the sum of the distances from those points is always constant. The diameter passing through the foci is the major axis, and the diameter bisecting this at right angles is the minor axis. An ellipse is one of a series of curves known as *conic sections. A slice across a cone that is not made parallel to, and does not pass through, the base will produce an ellipse.

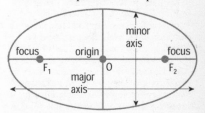

ellipse Technical terms used to describe an ellipse; for all points on the ellipse, the sum of the distances from the two foci, F_1 and F_2, is the same.

elliptical galaxy in astronomy, one of the main classes of *galaxy in the *Hubble classification, characterized by a featureless elliptical profile. Unlike spiral galaxies, elliptical galaxies have very little gas or dust and no stars have recently formed within them. They range greatly in size from giant ellipticals, which are often found at the centres of clusters of galaxies and may be strong radio sources, to tiny dwarf ellipticals, containing about a million stars, which are the most common galaxies of any type. More than 60% of known galaxies are elliptical.

e-mail contraction of electronic mail, messages sent electronically from computer to computer via network connections such as *Ethernet or the *Internet, or via telephone lines to a host system. Messages once sent are stored on the network or by the host system until the recipient picks them up. As well as text, messages may contain enclosed text files, artwork, or multimedia clips.

embolism blockage of a blood vessel by an obstruction called an embolus (usually a blood clot, fat particle, or bubble of air).

embryo early developmental stage of an animal or a plant following fertilization of an ovum (egg cell), or activation of an ovum by *parthenogenesis. In humans, the term embryo describes the fertilized egg during its first seven weeks of existence; from the eighth week onwards it is referred to as a fetus.

In animals the embryo exists either within an egg (where it is nourished by food contained in the yolk), or in mammals, in the *uterus of the mother. In mammals (except marsupials) the embryo is fed through the *placenta. The plant embryo is found within the seed in higher plants. It sometimes consists of only a few cells, but usually includes a root, a shoot (or primary bud), and one or two *cotyledons, which nourish the growing seedling.

embryology study of the changes undergone by an organism from its conception as a fertilized ovum (egg) to its emergence into the world at hatching or birth. It is mainly concerned with the changes in cell organization in the embryo and the way in which these lead to the structures and organs of the adult (the process of *differentiation).

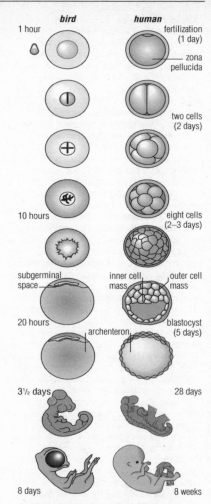

embryo The development of a bird and a human embryo. In the human, division of the fertilized egg, or ovum, begins within hours of conception. Within a week, a hollow, fluid-containing ball – a blastocyte – with a mass of cells at one end has developed. After the third week, the embryo has changed from a mass of cells into a recognizable shape. At four weeks, the embryo is 3 mm/0.1 in long, with a large bulge for the heart and small pits for the ears. At six weeks, the embryo is 1.5 cm/0.6 in long with a pulsating heart and ear flaps. By the eighth week, the embryo (now technically a fetus) is 2.5 cm/1 in long and recognizably human, with eyelids and small fingers and toes.

Applications of embryology include embryo transplants, both commercial (for example, in building up a prize dairy-cow herd quickly at low cost) and in obstetric medicine (as a method for helping couples with fertility problems to have children).

embryo research the study of human embryos at an early stage, in order to detect hereditary disease and genetic defects, and to develop cures for diseases.

As part of in vitro fertilization (IVF) treatment, eggs undergo fertilization and are allowed to grow to the eight-cell stage. One or two cells are then removed for analysis. Diseases that can be tested for include cystic fibrosis, *Duchenne muscular dystrophy, Lesch-Nyhan syndrome, Tay-Sachs disease, and haemophilia A.

embryo sac large cell within the ovule of flowering plants that represents the female *gametophyte when fully developed. It typically contains eight nuclei. Fertilization occurs when one of these nuclei, the egg nucleus, fuses with a male *gamete.

emerald clear, green gemstone variety of the mineral *beryl. It occurs naturally in Colombia, the Ural Mountains in Russia, Zimbabwe, and Australia. The green colour is caused by the presence of the element chromium in the beryl.

emf in physics, abbreviation for *electromotive force.

EMG abbreviation for **electromyography**.

emission spectroscopy in analytical chemistry, a technique for determining the identity or amount present of a chemical substance by measuring the amount of electromagnetic radiation it emits at specific wavelengths; see *spectroscopy.

emoticon contraction of emotion and icon, in computing, symbol composed of punctuation marks designed to express some form of emotion in the form of a human face. Emoticons were invented by *e-mail users to overcome the fact that communication using text only cannot convey non-verbal information (body language or vocal intonation) used in ordinary speech.

emphysema incurable lung condition characterized by disabling breathlessness. Progressive loss of the thin walls dividing the air spaces (alveoli) in the lungs reduces the area available for the exchange of oxygen and carbon dioxide, causing the lung tissue to expand. The term 'emphysema' can also refer to the presence of air in other body tissues.

empty set symbol ø, in mathematics, *set with no elements.

EMS in computing, abbreviation for *expanded memory specification.

emulator in computing, an item of software or firmware that allows one device to imitate the functioning of another. Emulator software is commonly used to allow one make of computer to run programs written for a different make of computer. This allows a user to select from a wider range of applications programs, and perhaps to save money by running programs designed for an expensive computer on a cheaper model.

emulsifier food additive used to keep oils dispersed and in suspension, in products such as mayonnaise and peanut butter. Egg yolk is a naturally occurring emulsifier, but most of the emulsifiers in commercial use today are synthetic chemicals.

emulsify in biology, to disperse *fat as millions of microscopic globules suspended in water. Emulsification is an important stage prior to *digestion of fat in the human gut.

emulsion stable dispersion of a liquid in another liquid – for example, oil and water in some cosmetic lotions.

encapsulate in computing, term used to describe the technique that uses one *protocol as an envelope for another for transmission across a network.

encephalin naturally occurring chemical produced by nerve cells in the brain that has the same effect as morphine or other derivatives of opium, acting as a natural painkiller. Unlike morphine, encephalins are quickly degraded by the body, so there is no build-up of tolerance to them, and hence no addiction. Encephalins are a variety of *peptides, as are *endorphins, which have similar effects.

encephalitis inflammation of the brain, nearly always due to viral infection but it may also occur in bacterial and other infections. It varies widely in severity, from shortlived, relatively slight effects of headache, drowsiness, and fever to paralysis, coma, and death.

encephalomyelitis in medicine, inflammation of the brain and the spinal cord. It is caused by bacteria, viruses, fungi, malignant cells, and blood following subarachnoid haemorrhage. The disease is serious and requires urgent treatment in hospital.

encryption encoding a message so that it can only be read by the intended recipient. The mathematical calculation of encryption formulae is called *cryptography.

endangered species plant or animal species whose numbers are so few that it is at risk of becoming extinct. Officially designated endangered species are listed by the World Conservation Union (or IUCN). The members of IUCN agreed to the formation of CITES (the Convention on International Trade in Endangered Species of Wild Fauna and Flora), which came into force on 1 July 1975.

endocrine gland gland that secretes hormones into the bloodstream to regulate body processes. Endocrine glands are most highly developed in vertebrates, but are also found in other animals, notably insects. In humans the main endocrine glands are the pituitary, thyroid, parathyroid, adrenal, pancreas, ovary, and testis.

endocrinology medical speciality devoted to the diagnosis and treatment of hormone disorders.

endogenous in medicine, arising within the body, for example the endogenous amines, serotonin, and noradrenaline, in the brain that are implicated in *depression.

endolymph fluid found in the inner *ear, filling the central passage of the cochlea as well as the semicircular canals.

endometriosis common gynaecological complaint in which patches of endometrium (the lining of the womb) are found outside the uterus.

endoparasite *parasite that lives inside the body of its host.

endoplasm inner, liquid part of a cell's *cytoplasm.

endoplasmic reticulum ER, a membranous system of tubes, channels, and flattened sacs that form compartments within *eukaryotic cells. It stores and transports proteins within cells and also carries various enzymes needed for the synthesis of *fats. The *ribosomes, or the organelles that carry out protein synthesis, are sometimes attached to parts of the ER.

Under the electron microscope, ER looks like a series of channels and vesicles, but it is in fact a large, sealed, baglike structure crumpled and folded into a convoluted mass. The interior of the 'bag', the ER lumen, stores various proteins needed elsewhere in the cell, then organizes them into transport vesicles formed by a small piece of ER membrane budding from the main membrane.

endorphin natural substance (a polypeptide) that modifies the action of nerve cells. Endorphins are

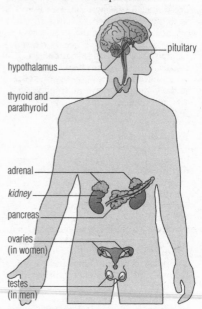

endocrine gland The main human endocrine glands. These glands produce hormones – chemical messengers – which travel in the bloodstream to stimulate certain cells.

produced by the pituitary gland and hypothalamus of vertebrates. They lower the perception of pain by reducing the transmission of signals between nerve cells.

endoskeleton internal supporting structure of vertebrates, made up of cartilage or bone. It provides support, and acts as a system of levers to which muscles are attached to provide movement. Certain parts of the skeleton (the skull and ribs) give protection to vital body organs.

endosperm nutritive tissue in the seeds of most flowering plants. It surrounds the embryo and is produced by an unusual process that parallels the *fertilization of the ovum by a male gamete. A second male gamete from the pollen grain fuses with two female nuclei within the *embryo sac. Thus endosperm cells are triploid (having three sets of chromosomes); they contain food reserves such as starch, fat, and protein that are utilized by the developing seedling.

endothermic reaction chemical reaction that requires an input of energy in the form of heat for it to proceed; the energy is absorbed from the surroundings by the reactants. A sign that this is happening is if the container holding the reactants feels cold and the temperature of the reactants falls. The energy absorbed is represented by the symbol $+\Delta H$.

The dissolving of sodium chloride in water and the process of photosynthesis are both endothermic changes. In photosynthesis the energy absorbed is in the form of light energy. In an endothermic reaction the energy needed to break the chemical bonds in the reactants is greater than the energy released when bonds are formed in the products. See *energy of reaction.

endotoxin in biology, heat stable complex of protein and lipopolysaccharide that is produced following the death of certain bacteria. Endotoxins are typically produced by the Gram negative bacteria and can cause fever. They can also cause shock by rendering the walls of the blood vessels permeable so that fluid leaks into the tissues and blood pressure falls sharply.

end user user of a computer program; in particular, someone who uses a program to perform a task (such as word processing or playing a computer game), rather than someone who writes programs (a programmer).

energy capacity for doing *work. This work may be as simple as reading a book, using a computer, or driving a car. Without energy no activity is possible. Energy can exist in many different forms For example, potential energy (PE) is energy deriving from position; thus a stretched spring has elastic PE, and an object raised to a height above the Earth's surface, or the water in an elevated reservoir, has gravitational PE. Moving bodies possess kinetic energy (KE). Energy can be converted from one form to another, but the total quantity in a system stays the same (in accordance with the *conservation of energy principle). Energy cannot be created or destroyed (although see *mass–energy equation). For example, as an apple falls it loses gravitational PE but gains KE. Although energy is never lost, after a number of conversions it tends to finish up as the kinetic energy of random motion of molecules (of the air, for example) at relatively low temperatures. This is 'degraded' energy that is difficult to convert back to other forms.

energy conservation and efficiency
All forms of energy (see *energy sources) are interconvertible by appropriate processes. Energy is transferred from one form to another, but the sum total of the energy after the conversion is always the same as the initial energy. This is the principle of conservation of energy. This principle can be illustrated by the use of energy flow diagrams, called Sankey diagrams, which show the energy transformations that take place. When a petrol engine is used to power a car, about 75% of the energy from the fuel is wasted. The total energy input equals the total energy output, but a lot of energy is wasted as heat so that the engine is only about 25% efficient. The combustion of the petrol–air mixture produces heat energy as well as kinetic energy. All forms of energy

tend to be transformed into heat and cannot then readily be converted into other, useful forms of energy.

heat transfer A difference in temperature between two objects in thermal contact leads to the transfer of energy as *heat. Heat is energy transferred due to a temperature difference. Heat is transferred by the movement of particles (that possess kinetic energy) by conduction, convection, and radiation. *Conduction involves the movement of heat through a solid material by the movement of free electrons. For example, thermal energy is lost from a house by conduction through the walls and windows. *Convection involves the transfer of energy by the movement of fluid particles. All objects radiate heat in the form of radiation of electromagnetic waves. Hotter objects emit more energy than cooler objects. Methods of reducing energy transfer as heat through the use of *insulation are important because the world's fuel reserves are limited and heating homes costs a lot of money in fuel bills. Heat transfer from the home can be reduced by a variety of methods, such as loft insulation, cavity wall insulation, and double glazing. The efficiencies of insulating materials in the building industry are compared by measuring their heat-conducting properties, represented by a U-value. A low U-value indicates a good insulating material.

energy in biology, the basis for conducting living processes. Much of life involves energy transfer. Energy is transferred from the surroundings of an organism into its body, and is also transferred within an organism's body. Energy is used by organisms to do things, such as growing or moving. When they do these things energy is transferred from one substance to another or from one place to another.

energy, alternative energy from sources that are renewable and ecologically safe, as opposed to sources that are non-renewable with toxic by-products, such as coal, oil, or gas (fossil fuels), and uranium (for nuclear power). The most important alternative energy source is flowing

water, harnessed as *hydroelectric power. Other sources include the oceans' tides and waves (see *wave power), *wind power (harnessed by windmills and wind turbines), the Sun (*solar energy), and the heat trapped in the Earth's crust (*geothermal energy) (see also *cold fusion).

The Centre for Alternative Technology, near Machynlleth in mid-Wales, was established 1975 to research and demonstrate methods of harnessing wind, water, and solar energy.

energy conservation methods of reducing *energy use through insulation, increasing energy efficiency, and changes in patterns of use. Profligate energy use by industrialized countries contributes greatly to air pollution and the *greenhouse effect when it draws on non-renewable energy sources.

It has been calculated that increasing energy efficiency alone could reduce carbon dioxide emissions in several high-income countries by 1–2% a year. The average annual decrease in energy consumption in relation to gross national product 1973–87 was 1.2% in France, 2% in the UK, 2.1% in the USA, and 2.8% in Japan.

energy level permitted energy that an electron can have in any particular atom. Energy levels can be calculated using *quantum theory. The permitted energy levels depend mainly on the distance of the electron from the nucleus. See *orbital, atomic.

energy of reaction or **enthalpy of reaction** or **heat of reaction**, energy released or absorbed during a chemical reaction, part of the energy transfer that takes place. In a chemical equation it may be represented by the symbol ΔH. In a chemical reaction, the energy stored in the reacting molecules is rarely the same as that stored in the product molecules. Depending on which is the greater, energy is either released (an *exothermic reaction) or absorbed (an *endothermic reaction) from the surroundings. The amount of energy released or absorbed by the quantities of substances represented by the chemical equation is the energy of reaction. The principle that the total amount of energy in a given chemical reaction stays the same is known as *conservation of energy.

Examples of chemical reactions that release large amounts of heat energy include the *combustion of *fuels and explosives. For example, when a mixture of hydrogen gas and oxygen gas in a glass jar is ignited by a spark, rapid and large amounts of heat energy are released shattering the glass jar in the process. The spark supplies the *activation energy required for bond breaking to occur in the reactants, enabling the reaction to begin.

$$2H_2 + O_2 \rightarrow 2H_2O + 115.6 \text{ kJ of energy}$$

Some oxidation reactions take place very slowly such as the *rusting of iron. The heat of the chemical reaction is immediately dissipated, and the activation energy required for rusting to take place is very large. The chemical equation describing the rusting of iron is:

$$4Fe + 3O_2 \rightarrow 2Fe_2O_3 + 380 \text{ kJ of energy}$$

energy sources There are two main sources of energy: the Sun, the ultimate source; and decay of radioactive elements in the Earth. Plants use the Sun's energy and convert it into food and oxygen. The remains of plants and animals that lived millions of years ago have been converted into *fossil fuels such as coal, oil, and natural gas.

energy resources So-called energy resources are stores of convertible energy. **Non-renewable resources** include the fossil fuels (coal, oil, and gas) and nuclear-fission 'fuels' – for example, uranium-235. The term 'fuel' is used for any material from which energy can be obtained. We use up fuel reserves such as coal and oil, and convert the energy they contain into other, useful forms. The chemical energy released by burning fuels can be used to do work. **Renewable resources**, such as wind, tidal, and geothermal power, have so far been less exploited. Hydroelectric projects are well established, and wind turbines and tidal systems are being developed.

engine device for converting stored energy into useful work or movement. Most engines use a fuel as their energy store. The fuel is burnt to produce heat energy – hence the name 'heat engine' – which is then converted into movement. Heat engines can be classified according to the fuel they use (*petrol engine or diesel engine), or according to whether the fuel is burnt inside (*internal combustion engine) or outside (*steam engine) the engine, or according to whether they produce a reciprocating or a rotary motion (*turbine or Wankel engine).

engine in computing, core piece of software around which other features and functions are built. A database *search engine, for example, accepts user input and handles the processing necessary to find matches between the user input and the database records.

In a computer game, the term 'engine' is also used to refer to the core software that allows users to move around the game's levels and pick up weapons and treasure.

enlargement mathematical *transformation that changes the size of a shape by multiplying all its proportions by the same scale factor. An enlargement may be either smaller or larger.

enteritis inflammation of the small intestine, usually resulting in diarrhoea. It may be caused by infection or by exposure to some form of radiation, including X-rays or radioactive isotopes.

enterovirus in medicine, one of a family of small polyhedral RNA-containing viruses that enter the body through the gut and are able to penetrate the central nervous system. They include the *polio virus; *Coxsackie viruses that cause diseases such as herpangina and epidemic myalgia; and ECHO viruses that cause aseptic meningitis.

enthalpy alternative term for *energy of reaction, the heat energy associated with a chemical change.

entropy in *thermodynamics, a parameter representing the state of disorder of a system at the atomic, ionic, or molecular level; the greater the disorder, the higher the entropy. Thus the fast-moving disordered molecules of water vapour have higher entropy than those of more ordered liquid water, which in turn have more entropy than the molecules in solid crystalline ice.

In a closed system undergoing change, entropy is a measure of the amount of energy unavailable for useful work. At *absolute zero ($-273.15°C/-459.67°F/0$ K), when all molecular motion ceases and order is assumed to be complete, entropy is zero.

enucleation in medicine, surgical removal of a complete organ or tumour; for example, the eye from its socket.

envelope in geometry, a curve that touches all the members of a family of lines or curves. For example, a family of three equal circles all touching each other and forming a triangular pattern (like a clover leaf) has two envelopes: a small circle that fits in the space in the middle, and a large circle that encompasses all three circles.

environment in ecology, the sum of conditions affecting a particular organism, including physical surroundings, climate, and influences of other living organisms. Areas affected by *environmental issues include the biosphere and *habitat. In biology, the environment includes everything outside body cells and fluid surrounding the cells. This means that materials enclosed by part of the body surface that is 'folded in' are, in fact, part of the environment and not part of the organism. So the air spaces in human lungs and the contents of the stomach are all part of the environment and not the organism, using these terms correctly. Ecology is the study of the way organisms and their environment interact with each other. Important processes in biology involve the transfer of material between an organism and its environment in exchanges of gases and food, for example during nutrition, *photosynthesis, or *respiration.

environmental audit another name for *green audit, the inspection of a company to assess its environmental impact.

environmental impact assessment **EIA**, in the UK, a process by which the potential environmental impacts of human activities, such as the construction of a power station, dam, or major housing development, are evaluated. The results of an EIA are published and discussed by different levels of government, non-governmental organizations, and the general public before a decision is made on whether or not the project can proceed.

environmental issues matters relating to the damaging effects of human activity on the biosphere, their causes, and the search for possible solutions. The political movement that supports protection of the environment is the green movement. Since the Industrial Revolution, the demands made by both the industrialized and developing nations on the Earth's natural resources are increasingly affecting the balance of the Earth's resources. Over a period of time, some of these resources are renewable – trees can be replanted, soil nutrients can be replenished – but many resources, such as minerals and fossil fuels (coal, oil, and natural gas), are *non-renewable and in danger of eventual exhaustion. In addition, humans are creating many other problems that may endanger not only their own survival, but also that of other species. For instance, *deforestation and *air pollution are not only damaging and radically altering many natural environments, they are also affecting the Earth's climate by adding to the *greenhouse effect and *global warming, while *water pollution is seriously affecting aquatic life, including fish populations, as well as human health.

Environmental pollution is normally taken to mean harm done to the natural environment by human activity. In fact, some environmental pollution can have natural sources, for example volcanic activity, which can cause major air pollution or water pollution and destroy flora and fauna. In terms of environmental issues, however, environmental pollution relates to human actions, especially in connection with energy resources. The demands of the industrialized nations for energy to power machines, provide light, heat, and so on are constantly increasing. The most versatile form of energy is electricity, which can be produced from a wide variety of other energy sources, such as the fossil fuels and nuclear power (produced from uranium). These are all non-renewable resources and, in addition, their extraction, transportation, utilization, and waste products all give rise to pollutants of one form or another. The effects of these pollutants can have consequences not only for the local environment, but also at a global level.

environment–heredity controversy see *nature–nurture controversy.

environmental lapse rate   rate at which temperature changes with altitude. Usually, because the atmosphere is heated from below, temperatures decrease as altitude increases at an average rate of −6.4°/11.5°F per 1,000 m/3,280 ft. The rate varies from place to place and according to the time of day.

enzyme   biological *catalyst produced in cells, and capable of speeding up the chemical reactions necessary for life. They are large, complex *proteins, usually soluble, and are highly specific, each chemical reaction requiring its own particular enzyme. The enzyme's specificity arises from its active site, an area with a shape corresponding to part of the molecule with which it reacts (the substrate). The shape of the enzyme where the chemical binds only allows the binding of that particular chemical, rather like a specific key only working a specific lock (the lock and key hypothesis). The enzyme and the substrate slot together forming an enzyme–substrate complex that allows the reaction to take place, after which the enzyme falls away unaltered. (See diagram, page 214.)

The activity and efficiency of enzymes are influenced by various factors, including temperature and acidity (pH). Temperatures above 60°C/140°F damage (denature) the intricate structure of enzymes, inactivating them and causing reactions to stop. Each enzyme operates best – at its maximum rate –

within a specific pH range and temperature, and is denatured by excessive acidity or alkalinity or extremes of temperature.

In *digestion, digestive enzymes include amylases (which digest starch), lipases (which digest fats), and proteases (which digest protein). Other enzymes play a part in the conversion of food energy into *ATP, the manufacture of all the molecular components of the body, the replication of *DNA when a cell divides, the production of hormones, and the control of movement of substances into and out of cells.

Enzymes have many uses in *biotechnology, from washing powders to drug production, and as research tools in molecular biology. They are involved in the making of beer, bread, cheese, and yogurt. They can be extracted from bacteria and fungi and *genetic engineering now makes it possible to tailor an enzyme for a specific purpose. The most abundant enzyme is ribulose biphosphate carboxylase. It is found in chloroplasts and is associated with photosynthesis.

enzyme-linked immunosorbant assay ELISA, in medicine, technique for accurately determining the quantity of a substance in a sample, for example hormone levels in a blood sample. It involves the preparation of an antibody to the substance and the enzyme to which the antibody binds. The presence of a substance can be measured accurately as a result of colour changes that occur following reaction between the antibody and the enzyme in the presence of the substance. The ELISA technique is used in the detection of altered hormone levels in the urine in pregnancy testing kits.

Eocene epoch   second epoch of the Tertiary period of geological time, roughly 56.5–35.5 million years ago. Originally considered the earliest division of the Tertiary, the name means 'early recent', referring to the early forms of mammals evolving at the time, following the extinction of the dinosaurs.

eon   or **aeon**, in earth science, large amount of geological time consisting of several eras. The term is also used to mean a thousand million years (10^9 y).

eotvos unit symbol E, unit for measuring small changes in the intensity of the Earth's *gravity with horizontal distance.

ephedrine drug that acts like adrenaline on the sympathetic

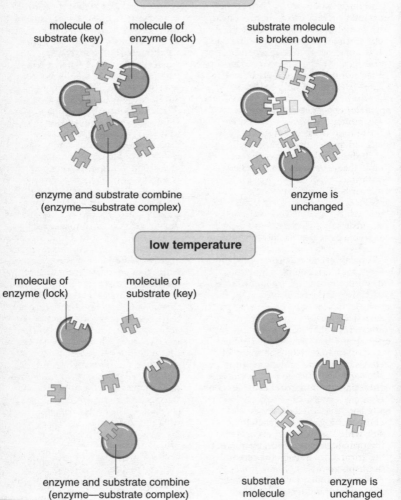

enzyme Enzymes are catalysts that can help break larger molecules into smaller molecules while remaining unchanged themselves. Like a key for a lock, each enzyme is specific to one molecule. Most also function best within a narrow temperature and pH range. As the temperature rises enzymes catalyse more molecules but beyond a certain temperature most enzymes become denatured.

*nervous system (sympathomimetic). Once used to relieve bronchospasm in *asthma, it has been superseded by safer, more specific drugs. It is contained in some cold remedies as a decongestant. Side effects include rapid heartbeat, tremor, dry mouth, and anxiety.

epicentre point on the Earth's surface immediately above the seismic focus of an *earthquake. Most building damage takes place at an earthquake's epicentre. The term is also sometimes used to refer to a point directly above or below a nuclear explosion ('at ground zero').

epicycloid in geometry, a curve resembling a series of arches traced out by a point on the circumference of a circle that rolls around another circle of a different diameter. If the two circles have the same diameter, the curve is a *cardioid.

epidermis outermost layer of *cells on an organism's body. In plants and many invertebrates such as insects, it consists of a single layer of cells. In vertebrates, it consists of several layers of cells.

The epidermis of plants and invertebrates often has an outer noncellular cuticle that protects the organism from desiccation.

epigeal describing seed germination in which the *cotyledons (seed leaves) are borne above the soil.

epiglottis small flap located behind the root of the tongue in mammals. It closes off the end of the windpipe during swallowing to prevent food from passing into it and causing choking.

epilepsy medical disorder characterized by a tendency to develop fits, which are convulsions or abnormal feelings caused by abnormal electrical discharges in the cerebral hemispheres of the *brain. Epilepsy can be controlled with a number of *anticonvulsant drugs.

epistaxis medical term for a nosebleed.

epithelium in animals, tissue of closely packed cells that forms a surface or lines a cavity or tube. Epithelial cells line the inside surfaces of fluid or air-filled tubes and spaces within the body. Epithelium may be protective (as in the skin) or secretory (as in the cells lining the wall of the gut). Epithelial cells join each other side to side to make epithelial *tissue. Epithelial cells are specialized cells. They protect cells below them and may also carry out special functions. For example, in the human lung epithelial cells line the alveoli (air spaces). These cells are very thin and form a large surface area in order to facilitate the absorption of oxygen and the loss of carbon dioxide during the process of gas exchange. They also make sure the alveoli remain moist at all times to keep them healthy.

epoch subdivision of a geological period in the geological time scale. Epochs are sometimes given their own names (such as the Palaeocene, Eocene, Oligocene, Miocene, and Pliocene epochs comprising the Tertiary period), or they are referred to as the late, early, or middle portions of a given period (as the Late Cretaceous or the Middle Triassic epoch).

EPROM acronym for Erasable Programmable Read-Only Memory, computer memory device in the form of an *integrated circuit (chip) that can record data and retain it indefinitely. The data can be erased by exposure to ultraviolet light, and new data recorded. Other kinds of computer memory chips are *ROM (read-only memory), *PROM (programmable read-only memory), and *RAM (random-access memory).

Epstein–Barr virus virus that causes *glandular fever. It is also implicated in nasopharyngeal cancer (rare in Europe but a scourge in Asia), Burkitt's lymphoma, and possibly *Hodgkin's disease. Epstein–Barr virus (EBV) was discovered in 1964 by English microbiologist Michael Epstein and his assistant Y.M. Barr. EBV is carried by 90% of people, and experimental vaccines are being developed against it.

equation in chemistry, representation of a chemical reaction by symbols and numbers; see *chemical equation. For example, the reaction of sodium hydroxide (NaOH) with hydrochloric acid (HCl) to give sodium chloride and water may be represented by:

$$NaOH + HCl \rightarrow NaCl + H_2O$$

EQUATION 216

equation in mathematics, an *expression that represents the equality of two expressions involving constants and/or variables, and thus usually includes an equals (=) sign. For example, the equation $A = \pi r^2$ equates the area A of a circle of radius r to the product πr^2. This is also known as the *formula for the area of a circle. The algebraic equation $y = mx + c$ is the general one in coordinate geometry for a straight line and is known as a *linear equation. See also *algebra, *quadratic equations, *simultaneous equations, inequations or *inequalities, and *graphs.

solving an equation To solve an equation means to find the value or values of the unknown quantity that satisfy the equation; for example, $x + 4 = 7$ is true when x is 3 The values of the unknown that make an equation true are called its solutions or roots.

In general, solving an equation depends on transforming it into a simple standard form. This can be achieved by using the following processes:

(i) adding the same quantity to each side of the equation

(ii) subtracting the same quantity from each side of the equation

(iii) multiplying each side of the equation by the same quantity (so long as it is not zero)

These processes can be used to change an equation into a simpler form but they will not alter its solution. For example, to solve the equation

$7x - 4 = 3x + 8$:

subtract $3x$ from each side in order to collect the xs on the left-hand side:

$7x - 4 - 3x = 3x + 8 - 3x$ so $4x - 4 = 8$

add 4 to each side in order to collect the numbers on the right-hand side:

$4x - 4 + 4 = 8 + 4$ so $4x = 12$

divide both sides by 4 to obtain the solution:

$4x \div 4 = 12 \div 4$ so $x = 3$

polynomials A type of equation that has been studied particularly intensively is where there is one unknown and the expression involving it is a *polynomial. A polynomial equation has the form:

$$f(x) = a_n x^n + a_{n-1} x^{n-1} + \ldots + a_2 x^2 + a_1 x + a_0$$

where $a_n, a_{n-1}, \ldots, a_0$ are all constants, n is a positive integer, and $a_n \neq 0$.

indeterminate equations An indeterminate equation is an equation for which there is an infinite set of solutions – for example, $2x = y$. A **diophantine equation** is an indeterminate equation in which both the solution and the terms must be whole numbers (after Diophantus of Alexandria, c. AD 250).

identity An equation that is true for all values of the unknown is called an identity, for example $x + x = 2x$. It is denoted by ≡.

Thus $(x + y)^2 \equiv x^2 + 2xy + y^2$ for all real numbers x, y.

equation of motion mathematical equation that gives the position and velocity of a moving object at any time. Given the mass of an object, the forces acting on it, and its initial position and velocity, an equation of motion is used to calculate its position and velocity at any later time. The equation must be based on *Newton's laws of motion or, if speeds near that of light are involved, on the theory of *relativity.

The common equations are:

$v = u + at$

$s = \frac{1}{2}(u + v)t$

$s = ut + \frac{1}{2}at^2$

$v^2 = u^2 + 2as$

in which a is the object's constant acceleration, u is its initial velocity, v is its velocity after a time t, and s is the distance travelled by it in that time.

Equator or **terrestrial equator**, great circle whose plane is perpendicular to the Earth's axis (the line joining the poles). Its length is 40,092 km/24,902 mi, divided into 360 degrees of longitude. The Equator encircles the broadest part of the Earth, and represents 0° latitude. It divides the Earth into two halves, called the northern and the southern hemispheres.

The **celestial equator** is the circle in which the plane of the Earth's Equator intersects the *celestial sphere.

equatorial coordinates   in astronomy, system for measuring the position of astronomical objects on the *celestial sphere with reference to the plane of the Earth's Equator.

equilateral   describing a geo metrical figure, having all sides of equal length.

For example, a square and a rhombus are both equilateral four-sided figures. An equilateral triangle, to which the term is most often applied, has all three sides equal and all three angles equal (at 60°).

equilibrium   in physics, an unchanging condition in which an undisturbed system can remain indefinitely in a state of balance. In a **static equilibrium**, such as an object resting on the floor, there is no motion. In a **dynamic equilibrium**, in contrast, a steady state is maintained by constant, though opposing, changes. For example, in a sealed bottle half-full of water, the constancy of the water level is a result of molecules evaporating from the surface and condensing on to it at the same rate.

equinox   time when the Sun is directly overhead at the Earth's *Equator and consequently day and night are of equal length at all latitudes. This happens twice a year: 21 March is the spring, or vernal, equinox and 23 September is the autumn equinox.

equivalent   having a different appearance but the same value. $3/5$ and $6/10$ are equivalent fractions. They both have the value 0.6.

era   any of the major divisions of geological time that includes several periods but is part of an eon. The eras of the current Phanerozoic in chronological order are the Palaeozoic, Mesozoic, and Cenozoic. We are living in the Recent epoch of the Quaternary period of the Cenozoic era.

erbium   chemical symbol Er, soft, lustrous, greyish, metallic element of the *lanthanide series, atomic number 68, relative atomic mass 167.26. It occurs with the element yttrium or as a minute part of various minerals. It was discovered in 1843 by Carl Mosander (1797–1858), and named after the town of Ytterby, Sweden, near which the lanthanides (rare-earth elements) were first found.

Erbium has been used since 1987 to amplify data pulses in optical fibres, enabling faster transmission. Erbium ions in the fibreglass, charged with infrared light, emit energy by amplifying the data pulse as it moves along the fibre.

erosion   wearing away of the Earth's surface by a moving agent, caused by the breakdown and transport of particles of rock or soil. Agents of erosion include the sea, rivers, glaciers, and wind. By contrast, *weathering does not involve transportation.

erratic   in geology, a displaced rock that has been transported by a glacier or some other natural force to a site of different geological composition.

error   in mathematics, incorrect answer to a calculation. Also, the amount by which an incorrect answer differs from the correct one. This is usually denoted by dx for a figure whose correct value is x.

error message   message produced by a computer to inform the user that an error has occurred.

erythroblast   in biology, a series of nucleated cells that go through various stages of development in the bone marrow until they form red blood cells (erythrocytes). This process is known as erythropoiesis. Erythroblasts can appear in the blood of people with blood cancers.

erythrocyte   another name for *red blood cell.

erythrocyte sedimentation rate   **ESR**, in medicine, measure of the rate at which red blood cells (erythrocytes) settle out of suspension in blood plasma. This is affected by the amount of protein in the plasma which, in turn, is affected by the presence of disease. Determination of the ESR is useful in the diagnosis of certain diseases, such as infection and malignancy.

erythromycin   an antibiotic with the chemical formula $C_{37}H_{67}NO_{13}$, isolated from a red-pigmented soil bacterium, *Streptomyces erythreus*. It is used in the treatment of a wide range of bacterial infections and is a useful alternative for those patients who are allergic to penicillin.

erythropoietin   EPO, in biology, a naturally occurring hormone, secreted mainly by the kidneys in adults and the liver in children, that stimulates production of red blood cells, which carry oxygen around the body. It is released in response to a lowered percentage of oxygen in the blood reaching the kidneys, such as in anaemic subjects. Recombinant human erythropoietin is used therapeutically to treat the anaemia associated with chronic kidney failure. A synthetic version is sometimes used illegally by athletes in endurance sports as it increases the oxygen-carrying capacity of the blood.

escape sequence in computing, string of characters sent to a *modem to switch it from sending data to a state in which it can accept and act upon commands.

escape velocity minimum velocity required for a spacecraft or other object to escape from the gravitational pull of a planetary body. In the case of the Earth, the escape velocity is 11.2 kps/6.9 mps; the Moon, 2.4 kps/1.5 mps; Mars, 5 kps/3.1 mps; and Jupiter, 59.6 kps/37 mps.

escarpment or **cuesta**, large ridge created by the erosion of dipping sedimentary rocks. It has one steep side (scarp) and one gently sloping side (dip). Escarpments are common features of chalk landscapes, such as the Chiltern Hills and the North Downs in England. Certain features are associated with chalk escarpments, including dry valleys (formed on the dip slope), combes (steep-sided valleys on the scarp slope), and springs.

Escherichia coli or **colon bacillus**, rod-shaped Gram-negative bacterium (see *bacteria) that lives, usually harmlessly, in the colon of most warm-blooded animals. It is the commonest cause of urinary tract infections in humans. It is sometimes found in water or meat where faecal contamination has occurred and can cause severe gastric problems.

The mapping of the genome of *E. coli*, consisting of 4,403 genes, was completed in 1997. It is probably the organism about which most molecular genetics is known, and is of pre-eminent importance in recombinant DNA research.

classification *Escherichia coli* is the only species in the bacterial family Enterobacteriaceae.

essential amino acid water-soluble organic molecule vital to a healthy diet because it cannot be synthesized from other food molecules; see *amino acid.

essential fatty acid organic compound consisting of a hydrocarbon chain and important in the diet; see *fatty acid.

ester organic compound formed by the reaction between an alcohol and an acid, with the elimination of water. Unlike *salts, esters are covalent compounds.

ester Molecular model of the ester ethyl ethanoate (ethyl acetate) $CH_3CO_2CH_2CH_3$.

estradiol alternative spelling of oestradiol, a type of *oestrogen (female sex hormone).

estuary river mouth widening into the sea, where fresh water mixes with salt water and tidal effects are felt.

ethanal common name **acetaldehyde**; CH_3CHO, one of the chief members of the group of organic compounds known as *aldehydes. It is a colourless flammable liquid boiling at 20.8°C/69.6°F. Ethanal is formed by the oxidation of ethanol or ethene and is used to make many other organic chemical compounds.

ethanal trimer common name **paraldehyde**; $(CH_3CHO)_3$, colourless liquid formed from ethanal (acetaldehyde). It is soluble in water.

ethanamide CH_3CONH_2, solid, crystalline compound produced by distilling ammonium ethanoate in a stream of dry ammonia. As usually prepared it has a strong odour suggestive of mice, but this is due to impurities. It is soluble in water and alcohol, melts at 82°C/179.6°F, and boils at 222°C/431.6°F.

ethane CH_3CH_3, colourless, odourless gas, the second member of the *alkane series of hydrocarbons (paraffins).

ethane-1,2-diol technical name for
*glycol.

ethanoate common name **acetate**;
$CH_3CO_2^-$, negative ion derived from
ethanoic (acetic) acid; any salt
containing this ion. In photography,
acetate film is a non-flammable film
made of cellulose ethanoate. In textiles,
it is known as acetate, and is most
commonly known in the form of a
*synthetic fibre which can be woven or
knitted to produce a variety of
different fabrics including satin, moire,
and taffeta. Fabrics made from acetate
absorb moisture, do not shrink, and are
cheaper to produce than natural fibres,
but they also tend to attract dirt and
can be damaged by heat. Acetate
filaments are also used to make filters
for cigarettes.

To produce acetate, *cellulose is
treated with acetic acid in a process
called acetylation. This forms a thick
solution which is forced through small
holes and solidified, and then, through
the process of spinning, continuous
filaments of acetate are produced.

ethanoic acid common name **acetic
acid**; CH_3CO_2H, one of the simplest
carboxylic acids (fatty acids). In the
pure state it is a colourless liquid with
an unpleasant pungent odour; it
solidifies to an icelike mass of crystals
at 16.7°C/62.4°F, and hence is often
called glacial ethanoic acid. In a dilute
form, mixed with water, it is the acid
found in vinegar. Vinegar contains 5%
or more ethanoic acid, produced by
fermentation.

Ethanoic acid is produced by the
oxidation of *ethanol. It belongs to a
homologous series, or family, of
organic compounds. Some other
members of the series are methanoic
acid, propanoic acid, and butanoic
acid. They are all weak acids.

Cellulose (derived from wood or
other sources) may be treated with
ethanoic acid to produce a cellulose
ethanoate (acetate) solution, which
can be used to make plastic items by
injection moulding or extruded to
form synthetic textile fibres.

ethanioc anhydride or **acetic
anhydride**; $(CH_3CO)_2O$, colourless
liquid, boiling point 137°C/278.6°F,
with a sharp, pungent odour

resembling that of ethanoic acid, into
which it is slowly decomposed by
water. It can be used to substitute the
ethanoyl (or acetyl) group (CH_3CO^-)
into organic compounds, without any
complicating side reactions.

ethanol or **ethyl alcohol**; C_2H_5OH,
alcohol found in beer, wine, cider,
spirits, and other alcoholic drinks.
When pure, it is a colourless liquid
with a pleasant odour, miscible with
water or ether; it burns in air with a
pale blue flame. The vapour forms an
explosive mixture with air and may be
used in high-compression internal
combustion engines.

It is produced naturally by the
fermentation of carbohydrates by yeast
cells. Industrially, it can be made by
absorption of *ethene and subsequent
reaction with water, or by the reduction
of ethanal (acetaldehyde) in the
presence of a catalyst, and is widely
used as a solvent.

Ethanol is used as a raw material in
the manufacture of ethoxyethane
(ether), trichloroethanol (chloral), and
triiodomethane (iodoform). It can also
be added to petrol, where it improves
the performance of the engine, or be
used as a fuel in its own right (as in
Brazil). In August 2001 it was
discovered that when ethanol is added
to petrol it boosts the oxygen content,
reducing the emissions of air-polluting
gases. Crops such as sugar cane may
be grown to provide ethanol (by
fermentation) for this purpose.

ethene common name **ethylene**; C_2H_4,
colourless, flammable gas, the first
member of the *alkene series of
hydrocarbons. It is the most widely
used synthetic organic chemical and is
used to produce the plastics *polythene
(polyethylene), polychloroethene, and
polyvinyl chloride (PVC). It is obtained
from natural gas or coal gas, or by the
dehydration of ethanol.

Ethene is produced during plant
metabolism and is classified as a plant
hormone. It is important in the
ripening of fruit and in *abscission.
Small amounts of ethene are often
added to the air surrounding fruit to
artificially promote ripening. Tomato
and marigold plants show distorted
growth in concentrations as low as 0.01

parts per million. Plants also release ethene when they are under stress. German physicists invented a device in 1997 that measures stress levels in plants by measuring surrounding ethene levels.

ether any of a series of organic chemical compounds having an oxygen atom linking the carbon atoms of two hydrocarbon radical groups (general formula R–O–R'); also the common name for ethoxyethane $C_2H_5OC_2H_5$ (also called diethyl ether). This is used as an anaesthetic and as an external cleansing agent before surgical operations. It is also used as a solvent, and in the extraction of oils, fats, waxes, resins, and alkaloids.

Ethoxyethane is a colourless, volatile, inflammable liquid, slightly soluble in water, and miscible with ethanol. It is prepared by treatment of ethanol with excess concentrated sulphuric acid at 140°C/284°F.

ether (or aether) in the history of science, a hypothetical medium permeating all of space. The concept originated with the Greeks, and has been revived on several occasions to explain the properties and propagation of light. It was supposed that light and other electromagnetic radiation – even in outer space – needed a medium, the ether, in which to travel. The idea was abandoned with the acceptance of *relativity.

Ethernet in computing, the most popular protocol for *local area networks. Ethernet was developed principally by the Xerox Corporation, but can now be used on most computers. It normally allows data transfer at rates up to 10 Mbps, but 100-Mbps Fast Ethernet – often called 100Base-T – is already in widespread use while the first product versions are now available of a 10 Gigabit Ethernet.

Ethernet, Fast Ethernet, and Gigabit Ethernet are all IEEE standards.

ethyl acetate common name **acetic ester**; $CH_3COOC_2H_5$, colourless liquid prepared by adding a mixture of alcohol and ethanoic acid to a mixture of alcohol and strong sulphuric acid, the whole being heated to 140°C/284°F. Ethyl acetate is characterized by a pleasant fruity odour and was formerly used for flavouring sweets and wines, and in perfumes. It is also a useful solvent. It is also known as ethyl ethanoate.

ethyl alcohol common name for *ethanol.

ethylene common name for *ethene.

ethylene glycol alternative name for *glycol.

ethyne common name **acetylene**; CHCH, colourless inflammable gas produced by mixing calcium carbide and water. It is the simplest member of the *alkyne series of hydrocarbons. It is used in the manufacture of the synthetic rubber neoprene, and in oxyacetylene welding and cutting.

Ethyne was discovered by English chemist Edmund Davy in 1836 and was used in early gas lamps, where it was produced by the reaction between water and calcium carbide. Its combustion provides more heat, relatively, than almost any other fuel known (its calorific value is five times that of hydrogen). This means that the gas gives an intensely hot flame; hence its use in oxyacetylene torches.

etiolation in botany, a form of growth seen in plants receiving insufficient light. It is characterized by long, weak stems, small leaves, and a pale yellowish colour (chlorosis) due to a lack of chlorophyll. The rapid increase in height enables a plant that is surrounded by others to quickly reach a source of light, after which a return to normal growth usually occurs.

eugenics (Greek *eugenes* 'well-born') study of ways in which the physical and mental characteristics of the human race may be improved. The eugenic principle was abused by the Nazi Party in Germany during the 1930s and early 1940s to justify attempted extermination of entire social and ethnic groups and the establishment of selective breeding programmes. Modern eugenics is concerned mainly with the elimination of genetic disease.

eukaryote in biology, one of the two major groupings (superkingdoms) into which all organisms are divided. Included are all organisms, except bacteria and cyanobacteria (*blue-green algae), which belong to the *prokaryote grouping.

The cells of eukaryotes possess a clearly defined nucleus, bounded by a membrane, within which DNA is formed into distinct chromosomes. Eukaryotic cells also contain mitochondria, chloroplasts, and other structures (organelles) that, together with a defined nucleus, are lacking in the cells of prokaryotes.

europium chemical symbol Eu, soft, greyish, metallic element of the *lanthanide series, atomic number 63, relative atomic mass 151.97. It is used in lasers and as the red phosphor in colour televisions; its compounds are used to make control rods for nuclear reactors. It was named in 1901 by French chemist Eugène Demarçay (1852–1904) after the continent of Europe, where it was first found.

Eustachian tube small air-filled canal connecting the middle *ear with the back of the throat. It is found in all land vertebrates and equalizes the pressure on both sides of the eardrum.

evaporation process in which a liquid turns into a vapour without its temperature reaching boiling point. Evaporation is the *change of state that occurs when a *liquid turns into a *gas. In a liquid the particles are close together, with forces holding them together, yet able to move about. Some particles in a liquid have more energy than others. Even when a liquid is below its boiling point, some particles have enough energy to escape and form a gas. Evaporation is greater when temperatures and wind speeds are high, and the air is dry. It is why puddles dry up in the sun, and clothes dry faster in dry, windy weather.

A liquid left to stand in a saucer eventually evaporates because, at any time, a proportion of its molecules will be fast enough (have enough kinetic energy) to escape from the attractive intermolecular forces at the liquid surface into the atmosphere. The temperature of the liquid tends to fall because the evaporating molecules remove energy from the liquid. The rate of evaporation rises with increased temperature because as the mean kinetic energy of the liquid's molecules

rises, so will the number possessing enough energy to escape.

A fall in the temperature of the liquid, known as the **cooling effect** (see *latent heat), accompanies evaporation because as the faster molecules escape from the surface, the mean energy of the remaining molecules falls. The effect may be noticed when sweat evaporates from the skin. It plays a part in temperature control of the human body.

evaporite sedimentary deposit precipitated on evaporation of salt water.

With a progressive evaporation of seawater, the most common salts are deposited in a definite sequence: calcite (calcium carbonate), gypsum (hydrous calcium sulphate), halite (sodium chloride), and finally salts of potassium and magnesium.

Calcite precipitates when seawater is reduced to half its original volume, gypsum precipitates when the seawater body is reduced to one-fifth, and halite when the volume is reduced to one-tenth. Thus the natural occurrence of chemically precipitated calcium carbonate is common, of gypsum fairly common, and of halite less common.

Because of the concentrations of different dissolved salts in seawater, halite accounts for about 95% of the chlorides precipitated if evaporation is complete. More unusual evaporite minerals include borates (for example borax, hydrous sodium borate) and sulphates (for example glauberite, a combined sulphate of sodium and calcium).

even number any number divisible by 2, hence the digits 0, 2, 4, 6, 8 are even numbers, as is any number ending in these digits, for example 1736. Any whole number which is not even is odd.

event in statistics, any happening to which a *probability can be attached. Events may be single, such as getting a score of six when one die is thrown, or they may be combined, such as the scores obtained when two dice are thrown. In some cases, such as throwing an unbiased die, each event has the same probability of occurring.

event-driven in computing, computer system that does not do anything until events are detected, such as mouse-clicks. Microsoft Windows is an event-driven operating environment.

evergreen in botany, a plant such as pine, spruce, or holly, that bears its leaves all year round. Most conifers are evergreen. Plants that shed their leaves in autumn or during a dry season are described as *deciduous.

evolution slow gradual process of change from one form to another, as in the evolution of the universe from its formation to its present state, or in the evolution of life on Earth. In biology, it is the process by which life has developed by stages from single-celled organisms into the multiplicity of animal and plant life, extinct and existing, that inhabits the Earth. The development of the concept of evolution is usually associated with the English naturalist Charles Darwin (1809–1882) who attributed the main role in evolutionary change to *natural selection acting on randomly occurring variations. These variations in species are now known to be *adaptations produced by spontaneous changes or *mutations in the genetic material of organisms. In short, evolution is the change in the genetic makeup of a population of organisms from one generation to another. Evidence shows that many species of organisms do not stay the same over generations. The most dramatic evidence of this comes from fossils.

Evolution occurs via the following processes of natural selection: individual organisms within a particular species may show a wide range of *variation because of differences in their *genes; predation, disease, and *competition cause individuals to die; individuals with characteristics most suited to the *environment are more likely to survive and breed successfully; and the genes that have enabled these individuals to survive are then passed on to the next generation, and if the environment is changing, the result is that some genes are more abundant in the next generation and the organism has evolved.

Evolutionary change can be slow, as shown in part of the fossil record.

However, it can be quite fast. If a population is reduced to a very small number, evolutionary changes can be seen over a few generations. Because micro-organisms have very short life cycles, evolutionary change in micro-organisms can be rapid. Micro-organisms can evolve resistance to a new antibiotic only a few years after the drug is first used. As a result of evolution from common ancestors, we are able to use *classification of organisms to suggest evolutionary origins.

evolutionary stable strategy ESS, in *sociobiology, an assemblage of behavioural or physical characters (collectively termed a 'strategy') of a population that is resistant to replacement by any forms bearing new traits, because the new traits will not be capable of successful reproduction.

evolutionary toxicology study of the effects of pollution on evolution. A polluted habitat may cause organisms to select for certain traits, as in **industrial melanism** for example, where some insects, such as the peppered moth, evolve to be darker in polluted areas, and therefore better camouflaged against predation.

exchange transfusion in medicine, blood *transfusion procedure carried out on a newborn infant. Blood is removed from the baby using the umbilical vein and is replaced with the same quantity of blood obtained from a donor. The procedure is repeated several times to remove damaged red blood cells while maintaining the blood volume and red blood cell count of the infant. Haemolytic disease of the newborn and severe jaundice may require an exchange transfusion.

exclusion principle in physics, a principle of atomic structure originated by Austrian-born US physicist Wolfgang Pauli (1900–1958). It states that no two electrons in a single atom may have the same set of *quantum numbers.

Hence, it is impossible to pack together certain elementary particles, such as electrons, beyond a certain critical density, otherwise they would share the same location and quantum number. A white dwarf star, which consists of electrons and other elementary particles, is thus prevented from contracting further by the exclusion principle and never collapses. Elementary particles in the class fermions obey the exclusion principle whilst those in the class bosons do not.

executable file in computing, a file – always a program of some kind – that can be run by the computer directly. The file will have been generated from a *source program by an assembler or *compiler. It will therefore not be coded in *ASCII and will not be readable as text. On MS-DOS systems executable files have an .EXE or .COM extension.

exocrine gland gland that discharges secretions, usually through a tube or a duct, onto a surface. Examples include sweat glands which release sweat on to the skin, and digestive glands which release digestive juices onto the walls of the intestine. Some animals also have *endocrine glands (ductless glands) that release hormones directly into the bloodstream.

exogenous in medicine, generated from outside the body. Drugs used to treat diseases may be described as exogenous. For example, exogenous insulin is used in the treatment of *diabetes mellitus.

exon in genetics, a sequence of bases in *DNA that codes for a protein. Exons make up only 2% of the body's total DNA. The remainder is made up of *introns. During RNA processing the introns are cut out of the molecule and the exons spliced together.

exoskeleton hardened external skeleton of insects, spiders, crabs, and other arthropods. It provides attachment for muscles and protection for the internal organs, as well as support. To permit growth it is periodically shed in a process called ecdysis.

exosphere uppermost layer of the *atmosphere. It is an ill-defined zone above the thermosphere, beginning at about 700 km/435 mi and fading off into the vacuum of space. The gases are extremely thin, with hydrogen as the main constituent.

exothermic reaction chemical reaction during which heat is given out

(see *energy of reaction). Burning sulphur in air to give sulphur dioxide is an exothermic reaction.

$$2S + O_2 \rightarrow SO_2$$

expand in algebra, to multiply out. For example, $ac + bc + ad + bd$ is the expanded form of $(a+b)(c+ d)$.

expanded memory in computing, additional memory in an MS-DOS-based computer, usually installed on an expanded-memory board. Expanded memory requires an expanded-memory manager, which gives access to a limited amount of memory at any one time, and is slower to use than extended memory. Software is available under both MS-DOS and Windows to simulate expanded memory for those applications that require it.

expansion in physics, the increase in size of a constant mass of substance caused by, for example, increasing its temperature (*thermal expansion) or its internal pressure. The **expansivity**, or coefficient of thermal expansion, of a material is its expansion (per unit volume, area, or length) per degree rise in temperature.

expansion board or **expansion card**, printed circuit board that can be inserted into a computer in order to enhance its capabilities (for example, to increase its memory) or to add facilities (such as graphics).

expectation in statistics, the theoretical *probability of a certain result.

experimental psychology application of scientific methods to the study of mental processes and behaviour.

expert system computer program for giving advice (such as diagnosing an illness or interpreting the law) that incorporates knowledge derived from human expertise. A kind of *knowledge-based system, it contains rules that can be applied to find the solution to a problem. It is a form of *artificial intelligence.

explosive any material capable of a sudden release of energy and the rapid formation of a large volume of gas, leading, when compressed, to the development of a high-pressure wave (blast).

Combustion and explosion differ essentially only in rate of reaction, and many explosives (called **low explosives**) are capable of undergoing relatively slow combustion under suitable conditions. **High explosives** produce uncontrollable blasts. The first low explosive was *gunpowder; the first high explosive was *nitroglycerine.

exponen or **index**, in mathematics, a superscript number that indicates the number of times a term is multiplied by itself; for example $x^2 = x \times x$, $4^3 = 4 \times 4 \times 4$.

exponential in mathematics, descriptive of a *function in which the variable quantity is an exponent (a number indicating the power to which another number or expression is raised).

Exponential functions and series involve the constant $e = 2.71828....$. Scottish mathematician John Napier (1550–1617) devised natural *logarithms in 1614 with e as the base.

Exponential functions are basic mathematical functions, written as e^x or exp x. The expression e^x has five definitions, two of which are: (i) e^x is the solution of the differential equation $dx/dt = x$ ($x = 1$ if $t = 0$); (ii) e^x is the limiting sum of the infinite series $1 + x + (x^2/2!) + (x^3/3!) + ... + (x^n/n!)$.

Curves of the form $y = Ae^{-ax}$, where $a > 0$, are known as decay functions; those of the form $y = Be^{bx}$, where $b > 0$, are growth functions. **Exponential growth** is not constant. It applies, for example, to population growth, where the population doubles in a short time period. A graph of population number against time produces a curve that is characteristically rather flat at first but then shoots almost directly upwards.

export file in computing, a file stored by the computer in a standard format so that it can be accessed by other programs, possibly running on different makes of computer.

expression in mathematics, a phrase written in symbols. For example, $2x^2 + 3x + 5$ is a quadratic (containing a term or terms raised to the second power but no higher) expression. Equations

consist of expressions written around an equals sign.

extend in mathematics, to draw a continuation of a line, usually in order to deduce some property of a figure.

extended character set in PC-based computing, the set of 254 characters stored in *ROM. Besides the 128 *ASCII characters, the set includes block graphics and foreign language characters.

extensor a muscle that straightens a limb.

exterior angle one of the four external angles formed when a straight line or transveral cuts through a pair of (usually parallel) lines. Also, an angle formed by extending a side of a polygon.

external modem in computing, a *modem that is a self-contained unit sitting outside a personal computer (PC) and connected to it by a cable. There are two main types of external modem: mains-powered desktop modems and credit-card sized modems that fit the PCMCIA slots in notebook and handheld computers.

extracellular matrix strong material naturally occurring in animals and plants, made up of protein and long-chain sugars (polysaccharides) in which cells are embedded. It is often called a 'biological glue', and forms part of *connective tissues such as bone and skin.

The cell walls of plants and bacteria, and the *exoskeletons of insects and other arthropods, are also formed by types of extracellular matrix.

extranet in computing, corporate *intranet that has been extended beyond the usual company boundaries to include customers, suppliers, and other third parties. An intranet is primarily a network protected from the external world by a *firewall and other security measures. The Internet is a network that exists beyond the firewall, accessible to everyone. An extranet is a hybrid whereby an intranet is extended beyond the firewall, often using the Internet although private networks can also be used.

extrasensory perception ESP, any form of perception beyond and distinct from the known sensory processes. The main forms of ESP are clairvoyance (intuitive perception or vision of events and situations without using the senses); precognition (the ability to foresee events); and telepathy or thought transference (communication between people without using any known visible, tangible, or audible medium). Verification by scientific study has yet to be achieved.

extroversion or **extraversion**, personality dimension described by the psychologists Carl Jung (1875–1961) and, later, German-born British psychologist Hans Eysenck (1916–1997). The typical extrovert is sociable, impulsive, and carefree. The opposite of extroversion is *introversion.

extrusion common method of shaping metals, plastics, and other materials. The materials, usually hot, are forced through a hole in a metal die and take its cross-sectional shape. Rods, tubes, and sheets may be made in this way.

extrusive rock or **volcanic rock**, *igneous rock formed on the surface of the Earth by volcanic activity (as opposed to intrusive, or plutonic, rocks that solidify below the Earth's surface). Magma (molten rock) erupted from volcanoes cools and solidifies quickly on the surface. The crystals that form do not have time to grow very large, so most extrusive rocks are finely grained. The term includes fine-grained crystalline or glassy rocks formed from hot lava quenched at or near Earth's surface, and those made of welded fragments of ash and glass ejected into the air during a volcanic eruption. The formation of extrusive igneous rock is part of the rock cycle.

Large amounts of extrusive rock called *basalt form at the Earth's *ocean ridges from lava that fills the void formed when two tectonic plates spread apart. Explosive volcanoes that deposit pyroclastics generally occur where one tectonic plate descends beneath another. Andesite is often formed by explosive volcanoes. Magmas that give rise to pyroclastic extrusive rocks are explosive because they are viscous. The island of Montserrat, West Indies, is an example of an explosive volcano that spews

pyroclastics of andesite composition. Magmas that produce crystalline or glassy volcanic rocks upon cooling are less viscous. The low viscosity allows the extruding lava to flow easily. Fluid-like lavas that flow from the volcanoes of the Hawaiian Islands have low viscosity and cool to form basalt.

eye organ of vision. In the human eye, the light is focused by the combined action of the curved cornea, the internal fluids, and the lens. The insect eye is compound – made up of many separate facets, known as ommatidia, each of which collects light and directs it separately to a receptor to build up an image. Invertebrates have much simpler eyes, with no lenses. Among molluscs, cephalopods have complex eyes similar to those of vertebrates. The mantis shrimp's eyes contain ten colour pigments with which to perceive colour; some flies and fish have five, while the human eye has only three. (See diagram, page 226.)

The human eye is a roughly spherical structure contained in a bony socket. Light enters it through the cornea, a transparent region at the front of the tough outer sclera and passes through the circular opening (pupil) in the iris (the coloured part of the eye). The muscular iris controls the size of the pupil and hence the amount of light entering the eye. Light then passes through the lens, which is held in position by suspensory ligaments and ciliary muscles. The ciliary muscles act on the lens (the rounded transparent structure behind the iris) to change its shape, allowing images of objects nearby and at a distance to be focused on the *retina at the back of the eye. The retina is packed with light-sensitive cells (rods and cones). Rods work well in conditions of low light but are unable to sense colour. Cones work well in bright light and are responsible for colour vision. The part of the retina on which lies the precise point at which the image is focused contains mainly cones. The rods and cones in the retina send impulses to the brain along sensory neurons in the optic nerve.

eye, defects of the abnormalities of the eye that impair vision. Glass or plastic lenses, in the form of spectacles or contact lenses, are the usual means of correction. Common optical defects are *short-sightedness or myopia; farsightedness or hypermetropia; lack of *accommodation or presbyopia; and *astigmatism. Other eye defects include *colour blindness.

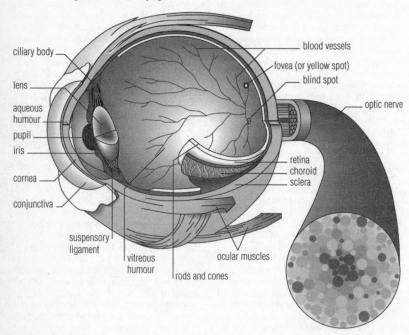

eye The human eye. The retina of the eye contains about 137 million light-sensitive cells in an area of about 650 sq mm/1 sq in. There are 130 million rod cells for black and white vision and 7 million cone cells for colour vision. The optic nerve contains about 1 million nerve fibres. The focusing muscles of the eye adjust about 100,000 times a day. To exercise the leg muscles to the same extent would need an 80 km/50 mi walk.

°F symbol for degrees *Fahrenheit.

F in physics, symbol for *farad, the SI unit of capacitance equal to that of a capacitor with a potential difference of 1 volt between plates carrying a charge of 1 coulomb.

face in geometry, a plane surface of a solid enclosed by edges. A cube has six square faces, a cuboid has six rectangular faces, and a tetrahedron has four triangular faces.

facies body of rock strata possessing unifying characteristics usually indicative of the environment in which the rocks were formed. The term is also used to describe the environment of formation itself or unifying features of the rocks that comprise the facies.

factor in algebra, certain kinds of *polynomials (expressions consisting of several or many terms) can be factorized using their common *factors. Brackets are put into an expression, and the common factor is sought. For example, the factors of $2a^2 + 6ab$ are $2a$ and $a + 3b$, since $2a^2 + 6ab = 2a(a + 3b)$. This rearrangement is called **factorization**.

factor number that divides into another *number exactly. It is also known as a *divisor. For example, the factors of 24 are 1, 2, 4, 8, 12, and 24; and the factors of 64 are 1, 2, 4, 8, 16, 32, and 64. The highest factor of both 24 and 64 is 8. This is known as the *highest common factor (HCF) of the two numbers.

factorial of a positive number, the product of all the whole numbers (integers) inclusive between 1 and the number itself. A factorial is indicated by the symbol '!'. Thus $6! = 1 \times 2 \times 3 \times 4 \times 5 \times 6 = 720$. Factorial zero, 0!, is defined as 1.

Fahrenheit scale temperature scale invented in 1714 by Polish-born Dutch physicist Gabriel Fahrenheit (1686–1736) that was commonly used in English-speaking countries until the 1970s, after which the *Celsius scale was generally adopted, in line with the rest of the world. In the Fahrenheit scale, intervals are measured in degrees (°F); °F = (°C $\times$ $^9/_5$) + 32.

Fallopian tube or **oviduct**, in mammals, one of two tubes that carry eggs from the ovary to the uterus. An egg is fertilized by sperm in the Fallopian tubes, which are lined with cells whose *cilia move the egg towards the uterus.

false of a statement, untrue. Falseness is used in proving propositions by considering the negative of the proposition to be true, then making deductions until a contradiction is reached which proves the negative to be false and the proposition to be true.

false-colour imagery graphic technique that displays images in false (not true-to-life) colours so as to enhance certain features. It is widely used in displaying electronic images taken by spacecraft; for example,

Earth-survey satellites such as *Landsat*. Any colours can be selected by a computer processing the received data.

false-memory syndrome syndrome occurring in some patients undergoing *psychotherapy or hypnosis, in which the individual 'remembers' events that never happened. For example, the patient may recall having been sexually abused as a child. As it is very difficult to establish whether memories are genuine the syndrome is surrounded by controversy.

family in biological classification, a group of related genera (see *genus). Family names are not printed in italic (unlike genus and species names), and by convention they all have the ending -idae (animals) or -aceae (plants and fungi). For example, the genera of hummingbirds are grouped in the hummingbird family, Trochilidae. Related families are grouped together in an *order.

FAQ abbreviation for frequently asked questions, in computing, file of answers to commonly asked questions on any topic. They were first used on *Usenet, where regular posters to newsgroups wrote FAQs to avoid answering repeatedly the same questions from new users. By 1996 FAQ was a common term for any information file, online or offline.

farad symbol F, SI unit of electrical capacitance (how much electric charge a *capacitor can store for a given voltage). One farad is a capacitance of one *coulomb per volt. For practical purposes the microfarad (one millionth of a farad, symbol mF) is more commonly used.

faraday unit of electrical charge equal to the charge on one mole of electrons, named after English physicist Michael Faraday (1791–1867). Its value is 9.648×10^4 coulombs.

Faraday's laws three laws of electromagnetic induction, and two laws of electrolysis, all proposed originally by English physicist Michael Faraday (1791–1867). The laws of induction are: (1) a changing magnetic field induces an electromagnetic force in a conductor; (2) the electromagnetic force is proportional to the rate of change of the field; and (3) the direction of the induced electromagnetic force depends on the orientation of the field. The laws of electrolysis are: (1) the amount of chemical change during electrolysis is proportional to the charge passing through the liquid; and (2) the amount of chemical change produced in a substance by a given amount of electricity is proportional to the electrochemical equivalent of that substance.

far point farthest point that a person can see clearly. The eye is unable to focus a sharp image on the retina of an object beyond this point. The far point for a normal eye should be at infinity; any eye that has a far point nearer than this is short-sighted (see *short-sightedness).

fast breeder or **breeder reactor**, alternative name for *fast reactor, a type of nuclear reactor.

fast reactor or **fast breeder reactor**, *nuclear reactor that makes use of fast neutrons to bring about fission. Unlike other reactors used by the nuclear-power industry, it has little or no moderator, to slow down neutrons. The reactor core is surrounded by a 'blanket' of uranium carbide. During operation, some of this uranium is converted into plutonium, which can be extracted and later used as fuel.

fat in the broadest sense, a mixture of *lipids – chiefly triglycerides (lipids containing three *fatty acid molecules linked to a molecule of glycerol). More specifically, the term refers to a lipid mixture that is solid at room temperature (20°C/68°F); lipid mixtures that are liquid at room temperature are called oils. The higher the proportion of saturated fatty acids in a mixture, the harder the fat. Fats and oils (lipids) are compounds made up of glycerol and fatty acids. Fats are insoluble in *water. Boiling fats in strong alkali forms soaps (saponification).

Fats are essential constituents of food for many animals, with a calorific value twice that of carbohydrates. However, eating too much fat, especially fat of animal origin, has been linked with heart disease in humans, where excess fat is deposited in the walls of arteries and may cause heart attacks. In many animals and plants, excess carbohydrates and proteins are converted into fats for storage. Mammals and other vertebrates store fats in specialized connective tissues (*adipose tissues), which not only act as energy reserves but also insulate the body and cushion its organs.

As a nutrient, fat serves five purposes: it is a source of energy (9 kcal/g); makes the diet palatable; provides basic building blocks for cell structure; provides essential fatty acids (linoleic and linolenic); and acts as a carrier for fat-soluble vitamins (A, D, E, and K). Indeed, fatty tissue in the body stores fat-soluble vitamins. Foods rich in fat are butter, lard, and margarine. Products high in monounsaturated or polyunsaturated fats are thought to be less likely to contribute to cardiovascular disease.

fathom (Anglo-Saxon *faethm* 'to embrace') in mining, seafaring, and handling timber, a unit of depth measurement (1.83 m/6 ft) used prior to metrication; it approximates to the distance between an adult man's hands when the arms are outstretched.

fatigue in muscle, reduced response brought about by the accumulation of lactic acid in muscle tissue due to excessive cellular activity.

fatty acid or **carboxylic acid**, organic compound consisting of a hydrocarbon chain of an even number of carbon atoms, with a carboxyl group (–COOH) at one end. The covalent bonds between the carbon atoms may be single or double; where a double bond occurs the carbon atoms concerned carry one instead of two hydrogen atoms. Chains with only single bonds have all the hydrogen they can carry, so they are said to be saturated with hydrogen. Chains with one or more double bonds are said to be unsaturated (see *polyunsaturate). Fatty acids are produced in the small intestine when fat is digested.

Saturated fatty acids include palmitic and stearic acids; unsaturated fatty acids include oleic (one double bond), linoleic (two double bonds), and linolenic (three double bonds).

Linoleic acid accounts for more than one third of some margarines. Supermarket brands that say they are high in polyunsaturates may contain as much as 39%. Fatty acids are generally found combined with glycerol in *lipids such as triglycerides.

fault fracture in the Earth's crust, on either side of which rocks have moved past each other. Faults may occur where rocks are being pushed together (compression) or pulled apart (tension) by *plate tectonics, movements of the *plates of the Earth's crust. When large forces build up quickly in rocks, they become brittle and break; *folds result from a more gradual compression. Faults involve displacements, or offsets, ranging from the microscopic scale to hundreds of kilometres. Large offsets along a fault are the result of the accumulation of many small movements (metres or less) over long periods of time. Large movements cause detectable *earthquakes, such as those experienced along the San Andreas Fault in California, USA.

Faults produce lines of weakness on the Earth's surface (along their strike) that are often exploited by processes of *weathering and *erosion. Coastal caves and geos (narrow inlets) often form along faults and, on a larger scale, rivers may follow the line of a fault. The Great Glen Fault in Scotland is an excellent example of a fault, and Loch Ness on this fault is an example of a fault-line feature.

fault gouge soft, uncemented, pulverized clay-like material found between the walls of a fault. It is created by grinding action (cataclasis) of the fault and subsequent chemical alteration of tiny mineral grains brought about by fluids that flow along the fault.

FDDI abbreviation for fibre-optic digital device interface, in computing, a series of network protocols, developed by the *American National Standards Institute, concerned with high-speed networks using *fibre optic cable.

feedback in biology, another term for *biofeedback: the influence of the outcome of one process upon the functioning of another process.

feedback general principle whereby the results produced in an ongoing reaction become factors in modifying or changing the reaction; it is the principle used in self-regulating control systems, from a simple *thermostat and steam-engine governor to automatic computer-controlled machine tools. A fully computerized control system, in which there is no operator intervention, is called a **closed-loop feedback** system. A system that also responds to control signals from an operator is called an **open-loop feedback** system.

In self-regulating systems, information about what *is* happening in a system (such as level of temperature, engine speed, or size of workpiece) is fed back to a controlling device, which compares it with what *should* be happening. If the two are different, the device takes suitable action (such as switching on a heater, allowing more steam to the engine, or resetting the tools). The idea that the Earth is a self-regulating system, with feedback operating to keep nature in balance, is a central feature of the *Gaia hypothesis.

Fehling's test chemical test to determine whether an organic substance is a reducing agent (substance that donates electrons to other substances in a chemical reaction). It is usually used to detect reducing sugars (monosaccharides, such as glucose, and the disaccharides maltose and lactose) and aldehydes.

If the test substance is heated with a freshly prepared solution containing copper(II) sulphate, sodium hydroxide and sodium potassium tartrate, the production of a brick-red precipitate indicates the presence of a reducing agent.

The test was devised by German chemist Herman von Fehling.

felsic rock a plutonic rock composed chiefly of light-coloured minerals, such as quartz, feldspar, and mica. It is derived from feldspar, lenad (meaning feldspathoid), and silica. The term felsic also applies to light-coloured minerals as a group, especially quartz, feldspar, and feldspathoids.

femtosecond *SI unit of time. It is 10^{-15} seconds (one millionth of a billionth of a second).

femur or **thigh-bone**, also the upper bone in the hind limb of a four-limbed vertebrate.

Fermat's last theorem in mathematics, states that equations of the form $x^n + y^n = z^n$ where x, y, z, and n are all *integers have no solutions if $n > 2$. French mathematician Pierre de Fermat (1601–1665) scribbled the theorem in the margin of a mathematics textbook and noted that he could have shown it to be true had he enough space in which to write the proof. The theorem remained unproven for 300 years (and therefore, strictly speaking, constituted a conjecture rather than a theorem). In 1993, Andrew Wiles, the English mathematician of Princeton University, USA, announced a proof; this turned out to be premature, but he put forward a revised proof in 1994. Fermat's last theorem was finally laid to rest in June 1997 when Wiles collected the Wolfskehl prize (the legacy bequeathed in the 19th century for the problem's solution).

Fermat's principle in physics, the principle that a ray of light, or other radiation, moves between two points along the path that takes the minimum time. The principle is named after French mathematician Pierre de Fermat (1601–1665), who used it to deduce the laws of *reflection and *refraction.

fermentation breakdown of sugars by bacteria and yeasts using a method of respiration without oxygen (*anaerobic). The enzymes in yeast break down glucose to give two products: *ethanol (alcohol) and *carbon dioxide. Fermentation processes have long been utilized in baking bread, making beer and wine, and producing cheese, yogurt, soy sauce, and many other foodstuffs.

In bread baking, the bubbles of carbon dioxide trapped in the dough make the bread rise. In the brewing process, complex sugars, such as sucrose, are first broken down by yeast into simple sugars, such as glucose. Glucose is then further decomposed into ethanol and carbon dioxide. The ethanol produced is the alcohol in alcoholic drinks, such as beer and wine; the carbon dioxide puts bubbles into beers and champagne.

The word equation for fermentation is: glucose → ethanol + carbon dioxide. The symbol equation is:

$$C_6H_{12}O_6 \rightarrow 2C_2H_5OH + 2CO_2$$

Many antibiotics are produced by fermentation; it is one of the processes that can cause food spoilage.

fermi unit of length equal to 10^{-15} m, used in atomic and nuclear physics. The unit is named after Italian-born US physicist Enrico Fermi (1901–1954).

Fermilab acronym for FERMI National Accelerator LABoratory, US centre for *particle physics at Batavia, Illinois, near Chicago. It is named after Italian-born US physicist Enrico Fermi (1901–1954). Fermilab was opened in 1972, and is the home of the Tevatron, the world's most powerful particle *accelerator. It is capable of boosting protons and antiprotons to speeds near that of light (to energies of 20 TeV).

A new main injector was built in 1998 to enable the Tevatron to function with a higher beam intensity.

fermion in physics, a subatomic particle whose spin can only take values that are half-odd-integers, such as $\frac{1}{2}$ or $\frac{3}{2}$. Fermions may be classified as *leptons, such as the *electron, and *hadrons, such as the *proton, *neutron, *mesons, and so on. All elementary particles are either fermions or *bosons. Named after Italian-born US physicist Enrico Fermi (1901–1954).

fermium chemical symbol Fm, synthesized, radioactive, metallic element of the *actinide series, atomic number 100, relative atomic mass 257. Ten isotopes are known, the longest-lived of which, Fm-257, has a half-life of 80 days. Fermium has been produced only in minute quantities in particle accelerators.

It was discovered in 1952 in the debris of the first thermonuclear explosion. The element was named in 1955 in honour of US physicist Enrico Fermi (1901–1954).

ferrimagnetism form of *magnetism in which adjacent molecular magnets are aligned anti-parallel, but have unequal strength, producing a strong overall magnetization. Ferrimagnetism is found in certain inorganic substances, such as *ferrites.

ferrite ceramic ferrimagnetic material. Ferrites are iron oxides to which small quantities of *transition metal oxides (such as cobalt and nickel oxides) have been added. They are used in transformer cores, radio antennae, and, formerly, in computer memories.

ferroelectric material ceramic dielectric material that, like ferromagnetic materials, has a *domain structure that makes it exhibit magnetism and usually the *piezoelectric effect. An example is Rochelle salt (potassium sodium tartrate tetrahydrate, $KNaC_4H_4O_6.4H_2O$).

ferromagnetism form of *magnetism that can be acquired in an external magnetic field and usually retained in its absence, so that ferromagnetic materials are used to make permanent magnets. A ferromagnetic material may therefore be said to have a high magnetic *permeability and susceptibility (which depends upon temperature). Examples are iron, cobalt, nickel, and their alloys.

fertilization in *sexual reproduction, the union of two *gametes (sex cells, often called egg or ovum, and sperm) to produce a *zygote, which combines the genetic material contributed by each parent. In self-fertilization the male and female gametes come from the same plant; in cross-fertilization they come from different plants. Self-fertilization rarely occurs in animals; usually even *hermaphrodite animals cross-fertilize each other.

The fusion of gametes combines the genetic material contributed by each parent. To avoid doubling the amount of inherited information every generation, each gamete contains only half the amount of inherited information – it is haploid. This is achieved by halving the number of *chromosomes when gametes are being produced. When the gametes fuse the full amount of information is restored (diploid state). Gametes are therefore produced by a specialized form of cell division, known as meiosis, which is only used for this purpose. This type of cell division has ways of mixing genes before the gametes are produced. This promotes variation (see *natural selection) by producing gametes with different combinations of genes.

In terrestrial insects, mammals, reptiles, and birds, fertilization occurs within the female's body. In human reproduction, the male gametes (sperm cells) are introduced into the vagina of the female. A female gamete (ovum) is produced and released from the ovary into the oviduct. Fusion of the male and female gametes usually occurs in the oviduct (the *Fallopian tube). Rarely, but more commonly when fertility drugs are used, two or more female

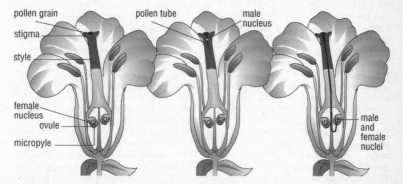

fertilization In a flowering plant pollen grains land on the surface of the stigma, and if conditions are acceptable the pollen grain germinates, forming a pollen tube, through which the male gametes pass, entering the ovule via the micropyle in order to reach the female egg.

gametes are released in humans at the same time. This results in twins or other multiple births. Just as rare is the division of a zygote into two cells by mitosis where each cell then develops to produce a new individual. These twins will be genetically identical.

In the majority of fish and amphibians, and most aquatic invertebrates, fertilization occurs externally, when both sexes release their gametes into the water. In most fungi, gametes are not released, but the hyphae of the two parents grow towards each other and fuse to achieve fertilization. In higher plants, *pollination precedes fertilization.

fetus or **foetus**, stage in mammalian *embryo development after fusion of gametes produces a zygote cell. The human embryo is usually termed a fetus after the eighth week of development, when the limbs and external features of the head are recognizable. The stage ends at birth.

fever condition of raised body temperature, usually due to infection.

fibre channel high-speed serial communications system designed to enable computers to be connected to other computers or to high-capacity storage devices using fibreoptic cable. The most common version, Fibre Channel Arbitrated Loop (FC-AL), is expected to replace *SCSI. Fibre channel has also been adapted for use over copper cables.

fibre optics branch of physics dealing with the transmission of light and images through glass or plastic fibres known as *optical fibres. Such fibres are now commonly used in both communications technology and medicine.

fibrin insoluble protein involved in blood clotting. When an injury occurs fibrin is deposited around the wound in the form of a mesh, which dries and hardens, so that bleeding stops. Fibrin is developed in the blood from a soluble protein, fibrinogen.

fibroid benign (noncancerous) growth developing in the wall of the uterus. It may give no trouble or it may grow to be enormous, causing pain and excessively heavy periods. Treatment is by surgical removal of the fibroid itself

(myomectomy) or of the uterus (hysterectomy). Fibroids occur in about 30% of women over the age of 30.

fibula rear lower bone in the hind leg of a vertebrate. It is paired and often fused with a smaller front bone, the tibia.

field in computing, a specific item of data. A field is usually part of a **record**, which in turn is part of a file.

field in physics, region of space in which an object exerts a force on a separate object because of certain properties they both possess. For example, there is a force of attraction between any two objects that have mass when one is within the gravitational field of the other.

field-length check *validation check in which the characters in an input field are counted to ensure that the correct number of characters have been entered. For example, a six-figure date field may be checked to ensure that it does contain exactly six digits.

field of view angle over which an image may be seen in a mirror or an optical instrument such as a telescope. A wide field of view allows a greater area to be surveyed without moving the instrument, but has the disadvantage that each of the objects seen is smaller. A *convex mirror gives a larger field of view than a plane or flat mirror. The field of view of an eye is called its **field of vision** or visual field.

field studies study of ecology, geography, geology, history, archaeology, and allied subjects, in the natural environment as opposed to the laboratory.

file extension in computing, the last three letters of a file name in DOS or Windows, which indicate the type of data the file contains. Extensions in common use include .TXT for 'text', .GIF for 'graphics interchange format', and .EXE for 'executable'.

file format in computing, specific way data are stored in a file. Most computer programs use proprietary file formats that cannot be read by other programs. As this is inconvenient for users, software publishers have developed filters that convert older file formats into the ones the program in use can read.

file librarian or **media librarian**, job classification for computer personnel. A file librarian stores and issues the data files used by the computer department.

film, photographic strip of transparent material (usually cellulose acetate) coated with a light-sensitive emulsion, used in cameras to take pictures. The emulsion contains a mixture of light-sensitive silver halide salts (for example, bromide or iodide) in gelatin. When the emulsion is exposed to light, the silver salts are invisibly altered, giving a latent image, which is then made visible by the process of developing. Films differ in their sensitivities to light, this being indicated by their speeds. Colour film consists of several layers of emulsion, each of which records a different colour in the light falling on it.

In **colour film** the front emulsion records blue light, then comes a yellow filter, followed by layers that record green and red light respectively. In the developing process the various images in the layers are dyed yellow, magenta (red), and cyan (blue), respectively.

When they are viewed, either as a transparency or as a colour print, the colours merge to produce the true colour of the original scene photographed.

filovirus virus causing the haemorrhagic fevers *Ebola virus disease and Marburg disease. Filoviruses have a filamentous structure and are about 1,500 nanometres in length. They reduce the number of platelets in the blood and so interfere with clotting. The incubation period is less than a week.

filter in chemistry, a porous substance, such as blotting paper, through which a mixture can be passed to separate out its solid constituents.

filter in electronics, a circuit that transmits a signal of some frequencies better than others. A low-pass filter transmits signals of low frequency and also direct current; a high-pass filter transmits high-frequency signals; a band-pass filter transmits signals in a band of frequencies.

filter in optics, a device that absorbs some parts of the visible *spectrum

and transmits others. A beam of white light can be made into a beam of coloured light by placing a transparent colour filter in the path of the beam. For example, a green filter will absorb or block all colours of the spectrum except green, which it allows to pass through. A yellow filter absorbs only light at the blue and violet end of the spectrum, transmitting red, orange, green, and yellow light.

filtrate liquid or solution that has passed through a filter.

filtration technique by which suspended solid particles in a fluid are removed by passing the mixture through a filter, usually porous paper, plastic, or cloth. The particles are retained by the filter to form a residue and the fluid passes through to make up the filtrate. For example, soot may be filtered from air, and suspended solids from water.

finite having a countable number of elements, the opposite of infinite.

finsen unit FU, unit for measuring the intensity of ultraviolet (UV) light; for instance, ultraviolet light of 2 FUs causes sunburn in 15 minutes.

fiord alternative spelling of *fjord.

firewall in computing, security system built to block access to a particular computer or network while still allowing some types of data to flow in and out onto the Internet. The firewall is often the first line of defence against hackers.

fission in physics, the splitting of a heavy atomic nucleus into two or more major fragments. It is accompanied by the emission of two or three neutrons and the release of large amounts of *nuclear energy.

Fission occurs spontaneously in nuclei of uranium-235, the main fuel used in nuclear reactors. However, the process can also be induced by bombarding nuclei with neutrons because a nucleus that has absorbed a neutron becomes unstable and soon splits. For example:

$$^{235}_{92}U + ^{1}_{0}n \rightarrow ^{236}_{92}U \rightarrow 2 \text{ nuclei} + 2\text{–}3$$
neutrons + energy

The neutrons released spontaneously by the fission of uranium nuclei may therefore be used in turn to induce further fissions, setting up a *chain reaction that must be controlled if it is not to result in a nuclear explosion. In a nuclear power station, heat energy released from the chain reaction is used to boil water to produce steam. The steam is used to drive the turbine of a generator to produce electricity. An atomic bomb uses pure uranium-235 to start an uncontrolled nuclear reaction, producing large amounts of heat energy. The minimum amount of fissile material that can undergo a continuous chain reaction is referred to as the *critical mass.

fit   in medicine, popular term for convulsion.

fitness   in genetic theory, a measure of the success with which a genetically determined character can spread in future generations. By convention, the normal character is assigned a fitness of one and variants (determined by other *alleles) are then assigned fitness values relative to this. Those with fitness greater than one will spread more rapidly and will ultimately replace the normal allele; those with fitness less than one will gradually die out.

fixed point   temperature that can be accurately reproduced and used as the basis of a temperature scale. In the Celsius scale, the fixed points are the temperature of melting ice, defined to be 0°C (32°F), and the temperature of boiling water (at standard atmospheric pressure), defined to be 100°C (212°F).

fixed-point notation   system in which numbers are represented using a set of digits with the decimal point always in its correct position. For very large and very small numbers this requires a lot of digits. In computing, the size of the numbers that can be handled in this way is limited by the capacity of the computer, and so the slower *floating-point notation is often preferred.

fixed star   description sometimes applied to a star, especially in emphasizing its difference from a planet (or 'wandering star'). Though most stars are in rapid motion through space, their great distance from Earth ensures that their relative configurations do not change appreciably over a thousand years.

fjord   or **fiord**, narrow sea inlet enclosed by high cliffs. Fjords are found in Norway, New Zealand, Southern Chile, and western parts of Scotland. They are formed when an overdeepened U-shaped glacial valley is drowned by a rise in sea-level. At the mouth of the fjord there is a characteristic lip causing a shallowing of the water. This is due to reduced glacial erosion and the deposition of *moraine at this point.

flaccidity   in botany, the loss of rigidity (turgor) in plant cells, caused by loss of water from the central vacuole so that the cytoplasm no longer pushes against the cellulose cell wall. If this condition occurs throughout the plant then wilting is seen.

flagellum   (plural flagella) small hairlike organ on the surface of certain cells. Flagella are the motile organs of certain protozoa and single-celled algae, and of the sperm cells of higher animals. Unlike *cilia, flagella usually occur singly or in pairs; they are also longer and have a more complex whiplike action.

flame test   in chemistry, the use of a flame to identify metal *cations present in a solid.

A nichrome or platinum wire is moistened with acid, dipped in a compound of the element, either powdered or in solution, and then held in a hot non-luminous flame. The colour produced in the flame is characteristic of metals present; for example, sodium burns with an orange-yellow flame, and potassium with a lilac one.

flash flood   flood of water in a normally arid area brought on by a sudden downpour of rain. Flash floods are rare and usually occur in mountainous areas. They may travel many kilometres from the site of the rainfall.

flash point   in physics, the lowest temperature at which a liquid or volatile solid heated under standard conditions gives off sufficient vapour to ignite on the application of a small flame.

flavone   sap-soluble substance that gives the ivory to deep yellow colouring of plants.

Fleming's rules memory aids used to recall the relative directions of the magnetic field, current, and motion in an electric generator or motor, using one's fingers. The three directions are represented by the thumb (for motion), forefinger (for field), and second finger (for *conventional current), all held at right angles to each other. The right hand is used for generators and the left for motors. The rules were devised by the English physicist John Fleming.

flexor any muscle that bends a limb. Flexors usually work in opposition to other muscles, the extensors, an arrangement known as antagonistic.

flint compact, hard, brittle mineral (a variety of chert), brown, black, or grey in colour, found as nodules in limestone or shale deposits. It consists of cryptocrystalline (grains too small to be visible even under a light microscope) *silica, SiO_2, principally in the crystalline form of *quartz. Implements fashioned from flint were widely used in prehistory.

The best flint, used for Neolithic tools, is **floorstone**, a shiny black flint that occurs deep within chalk.

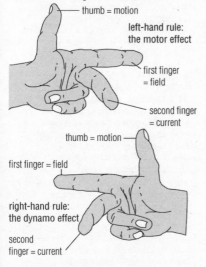

Fleming's rules give the direction of the magnetic field, motion, and current in electrical machines. The left hand is used for motors, and the right hand for generators and dynamos.

Because of their hardness (7 on the *Mohs scale), flint splinters are used for abrasive purposes and, when ground into powder, added to clay during pottery manufacture. Flints have been used for making fire by striking the flint against steel, which produces a spark, and for discharging guns. Flints in cigarette lighters are made from cerium alloy.

flip-flop in computing, another name for a *bistable circuit.

floating state of equilibrium in which a body rests on or is suspended in the surface of a fluid (liquid or gas). According to *Archimedes' principle, a body wholly or partly immersed in a fluid will be subjected to an upward force, or upthrust, equal in magnitude to the weight of the fluid it has displaced.

floating-point notation system in which numbers are represented by means of a decimal fraction and an exponent. For example, in floating-point notation, 123,000,000,000 would be represented as 0.123×10^{12}, where 0.123 is the fraction, or mantissa, and 12 the exponent. The exponent is the power of 10 by which the fraction must be multiplied in order to obtain the true value of the number.

flood plain area of periodic flooding along the course of a river valley. When river discharge exceeds the capacity of the channel, water rises over the channel banks and floods the surrounding low-lying lands. As water spills out of the channel some *alluvium (silty material) will be deposited on the banks to form levees (raised river banks). This water will slowly seep onto the flood plain, depositing a new layer of rich fertile alluvium as it does so. Many important flood plains, such as the inner Nile delta in Egypt, are in arid areas where their exceptional productivity is very important to the local economy.

A flood plain is a natural feature, flooded at regular intervals. By plotting floods that have occurred we can speak of the size of flood we would expect once every 10 years, 100 years, 500 years, and so on.

Even the most energetic flood-control plans (such as dams, dredging, and channel modification) sometimes fail, and if towns and villages are built on a flood plain there is always some risk. It is wiser to use flood plains in ways that are compatible with flooding, such as for agriculture or parks.

Flood-plain features include *meanders and oxbow lakes.

floppy disk in computing, a storage device consisting of a light, flexible disk enclosed in a cardboard or plastic jacket. The disk is placed in a disk drive, where it rotates at high speed. Data are recorded magnetically on one or both surfaces.

FLOPS acronym for FLoating point Operations Per Second, measure of the speed at which a computer program can be run.

flotation, law of law stating that a floating object displaces its own weight of the fluid in which it floats. See *Archimedes principle.

flower reproductive unit of an angiosperm (flowering plant), typically consisting of four whorls of modified leaves: *sepals, *petals, *stamens, and *carpels. These are borne on a central axis or *receptacle. The many variations in size, colour, number, and arrangement of parts are closely related to the method of pollination. Flowers adapted for wind pollination

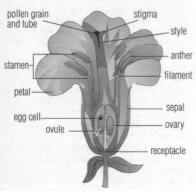

flower Cross section of a typical flower showing its basic components: sepals, petals, stamens (anthers and filaments), and carpel (ovary and stigma). Flowers vary greatly in the size, shape, colour, and arrangement of these components.

typically have reduced or absent petals and sepals and long, feathery *stigmas that hang outside the flower to trap airborne pollen. In contrast, the petals of insect-pollinated flowers are usually conspicuous and brightly coloured.

structure The sepals and petals form the **calyx** and **corolla** respectively and together comprise the **perianth** with the function of protecting the reproductive organs and attracting pollinators.

The stamens lie within the corolla, each having a slender stalk, or filament, bearing the pollen-containing anther at the top. Collectively they are known as the **androecium** (male organs). The inner whorl of the flower comprises the carpels, each usually consisting of an ovary in which are borne the *ovules, and a stigma borne at the top of a slender stalk, or style. Collectively the carpels are known as the **gynoecium** (female organs).

types of flower In size, flowers range from the tiny blooms of duckweeds scarcely visible to the naked eye to the gigantic flowers of the Malaysian *Rafflesia*, which can reach over 1 m/3 ft across. Flowers may grow either individually or in groups called *inflorescences. The stalk of the whole inflorescence is termed a **peduncle**, and the stalk of an individual flower is termed a **pedicel**. A flower is termed hermaphrodite when it contains both male and female reproductive organs. When male and female organs are carried in separate flowers, they are termed **monoecious**; when male and female flowers are on separate plants, the term **dioecious** is used.

flowering plant term generally used for *angiosperms, which bear flowers with various parts, including sepals, petals, stamens, and carpels. Sometimes the term is used more broadly, to include both angiosperms and *gymnosperms, in which case the cones of conifers and cycads are referred to as 'flowers'. Usually, however, the angiosperms and gymnosperms are referred to collectively as seed plants, or spermatophytes.

flue-gas desulphurization process of removing harmful sulphur pollution from gases emerging from a boiler. Sulphur compounds such as sulphur dioxide are commonly produced by burning *fossil fuels, especially coal in power stations, and are the main cause of *acid rain.

fluid   any substance, either liquid or gas, in which the molecules are relatively mobile and can 'flow'. A fluid can be 'pushed' by applying a force that is transmitted to different parts of the fluid. The *pressure of a fluid increases with the depth of the fluid. In water, an upward force called the upthrust is the liquid pressure that balances the weight of an object in the water. As liquids can transmit pressure, they are used in hydraulic systems.

fluid mechanics   study of the behaviour of fluids (liquids and gases) in motion. Fluid mechanics is important in the study of the weather, the design of aircraft and road vehicles, and in industries, such as the chemical industry, which deal with flowing liquids or gases.

fluid, supercritical   fluid brought by a combination of heat and pressure to the point at which, as a near vapour, it combines the properties of a gas and a liquid. Supercritical fluids are used as solvents in chemical processes, such as the extraction of lubricating oil from refinery residues or the decaffeination of coffee, because they avoid the energy-expensive need for phase changes (from liquid to gas and back again) required in conventional distillation processes.

fluorescence   short-lived *luminescence (a glow not caused by high temperature). *Phosphorescence lasts a little longer. Fluorescence is used in strip and other lighting, and was developed rapidly during World War II because it was a more efficient means of illumination than the incandescent lamp. Recently, small bulb-size fluorescence lamps have reached the market. It is claimed that, if widely used, their greater efficiency could reduce demand for electricity. Other important applications are in fluorescent screens for television and cathode-ray tubes.

fluorescence microscopy   technique for examining samples under a *microscope without slicing them into thin sections. Instead, fluorescent dyes are introduced into the tissue and used as a light source for imaging purposes. Fluorescent dyes can also be bonded to monoclonal antibodies and used to highlight areas where particular cell proteins occur.

fluoridation   addition of small amounts of fluoride salts to drinking water by certain water authorities to help prevent tooth decay. Experiments in Britain, the USA, and elsewhere have indicated that a concentration of fluoride of 1 part per million in tap water retards the decay of children's teeth by more than 50%. Much concern has been expressed about the risks of medicating the population at large by the addition of fluoride to the water supply, but the medical evidence demonstrates conclusively that there is no risk to the general health from additions of 1 part per million of fluoride to drinking water.

fluoride   F⁻, negative ion formed when hydrogen fluoride dissolves in water; compound formed between fluorine and another element in which the fluorine is the more electronegative element (see *electronegativity). In parts of India, the natural level of fluoride in water is 10 parts per million. This causes fluorosis, or chronic fluoride poisoning, mottling teeth and deforming bones.

fluorine   chemical symbol F, pale yellow, gaseous, non-metallic element, atomic number 9, relative atomic mass 19. It is the first member of the halogen group of elements, and is pungent, poisonous, and highly reactive, uniting directly with nearly all the elements. It occurs naturally as the minerals fluorite (CaF_2) and cryolite (Na_3AlF_6). Hydrogen fluoride is used in etching glass, and the freons, which all contain fluorine, are widely used as refrigerants.

Fluorine was discovered by the Swedish chemist Karl Scheele in 1771 and isolated by the French chemist Henri Moissan in 1886. Combined with uranium as UF_6, it is used in the separation of uranium isotopes.

The Infrared Space Observatory detected hydrogen fluoride molecules in an interstellar gas cloud in the constellation Sagittarius in 1997. It was the first time fluorine had been detected in space.

fluorite or **fluorspar**, glassy, brittle halide mineral, calcium fluoride CaF_2, forming cubes and octahedra; colourless when pure, otherwise violet, blue, yellow, brown, or green.

fluorocarbon compound formed by replacing the hydrogen atoms of a hydrocarbon with fluorine. Fluorocarbons are used as inert coatings, refrigerants, synthetic resins, and as propellants in aerosols. There is concern that the release of fluorocarbons – particularly those containing chlorine (chlorofluorocarbons, CFCs) – depletes the *ozone layer, allowing more ultraviolet light from the Sun to penetrate the Earth's atmosphere, and increasing the incidence of skin cancer in humans.

fluoroscopy in medicine, a technique in which X-rays are rendered visible after they have passed through the body by projecting them on to a film or screen containing a fluorescent material using a fluoroscope. The technique allows the beating of the heart or the movements in the intestine after a barium meal to be observed and it can assist diagnosis of disease in these organs.

fluvioglacial of a process or landform associated with glacial meltwater. Meltwater, flowing beneath or ahead of a glacier, is capable of transporting rocky material and creating a variety of landscape features, including eskers, kames, and outwash plains.

flux in smelting, a substance that combines with the unwanted components of the ore to produce a fusible slag, which can be separated from the molten metal. For example, the mineral fluorite, CaF_2, is used as a flux in iron smelting; it has a low melting point and will form a fusible mixture with substances of higher melting point such as silicates and oxides.

flux in soldering, a substance that improves the bonding properties of solder by removing contamination from metal surfaces and preventing their oxidation, and by reducing the surface tension of the molten solder alloy. For example, with solder made of lead–tin alloys, the flux may be resin, borax, or zinc chloride.

flythrough in *virtual reality, animation allowing users to view a model of a proposed or actual site as if they were inside it and moving through it.

FM in physics, abbreviation for *frequency modulation.

FM synthesizer abbreviation for frequency modulation synthesizer, in computing, method for generating synthetic sounds based on techniques used to transmit FM radio signals.

FMV abbreviation for full-motion video.

focal length or **focal distance**, distance from the centre of a lens or curved mirror to the focal point. For a concave mirror or convex lens, it is the distance at which rays of light parallel to the principal axis of the mirror or lens are brought to a focus (for a mirror, this is half the radius of curvature). For a convex mirror or concave lens, it is the distance from the centre to the point from which rays of light originally parallel to the principal axis of the mirror or lens diverge after being reflected or refracted.

focus in earth science, the point within the Earth's crust at which an *earthquake originates. The point on the surface that is immediately above the focus is called the epicentre.

focus or **focal point**, in optics, the point at which light rays converge, or from which they appear to diverge. Other electromagnetic rays, such as microwaves, and sound waves may also be brought together at a focus. Rays parallel to the principal axis of a lens or mirror are converged at, or appear to diverge from, the principal focus.

foetus stage in mammalian embryo development; see *fetus.

fog cloud that collects at the surface of the Earth, composed of water vapour that has condensed on particles of dust in the atmosphere. Cloud and fog are both caused by the air temperature falling below *dew point. The thickness of fog depends on the number of water particles it contains. Officially, fog refers to a condition when visibility is reduced to 1 km/0.6 mi or less, and

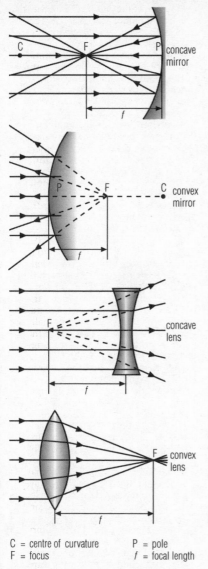

C = centre of curvature   P = pole
F = focus f = focal length

focal length The distance from the pole (P), or optical centre, of a lens or spherical mirror to its principal focus (F). The focal length of a spherical mirror is equal to half the radius of curvature ($f = CP/2$). The focal length of a lens is inversely proportional to the power of that lens (the greater the power the shorter the focal length).

mist or haze to that giving a visibility of 1–2 km/0.6–1.2 mi.

fold in geology, a deformation (bend) in beds or layers of rock. Folds are caused by pressures within the Earth's crust resulting from *plate-tectonic activity. Rocks are slowly pushed and compressed together, forming folds. Such deformation usually occurs in *sedimentary layers that are softer and more flexible. If the force is more sudden, and the rock more brittle, then a *fault forms instead of a fold.

Folding can result in gentle slopes or mountain chains such as the Rocky Mountains and the Alps. They can be eroded to form escarpments, giving rise to an undulating *topography. If the bend of the fold is arched up in the middle it is called an **anticline**; if it sags downwards in the middle it is called a **syncline**. The line along which a bed of rock folds is called its axis. The axial plane is the plane joining the axes of successive beds.

folic acid *vitamin of the B complex. It is found in liver, legumes and green leafy vegetables, and whole grain foods, and is also synthesized by the intestinal bacteria. It is essential for growth, and plays many other roles in the body. Lack of folic acid causes anaemia because it is necessary for the synthesis of nucleic acids and the formation of red blood cells.

follicle-stimulating hormone FSH, *hormone produced by the pituitary gland. It affects the ovaries in women, stimulating the production of an egg cell. Luteinizing hormone is needed to complete the process. In men, FSH stimulates the testes to produce sperm. It is used to treat some forms of infertility.

fontanelle gap (or 'soft spot') in a baby's skull before the bony plates have come together. There are two fontanelles: one in the centre front (anterior), the other towards the rear (posterior).

food chain in ecology, a sequence showing the feeding relationships between organisms in a *habitat or *ecosystem. It shows who eats whom. An organism in one food chain can belong to other food chains. This can be shown in a diagram called a **food web**.

food poisoning   any acute illness characterized by vomiting and diarrhoea and caused by eating food contaminated with harmful bacteria (for example, *listeriosis), poisonous food (for example, certain mushrooms or puffer fish), or poisoned food (such as lead or arsenic introduced accidentally during processing). A frequent cause of food poisoning is *Salmonella* bacteria. Salmonella comes in many forms, and strains are found in cattle, pigs, poultry, and eggs. Some people are more susceptible to food poisoning than others, and extra care is taken with food products designed for babies, pregnant women, or elderly people. Many types of bacteria can cause food poisoning and the incubation periods, symptoms, and methods of control vary.

food supply   availability of food, usually for human consumption. Food supply can be studied at scales ranging

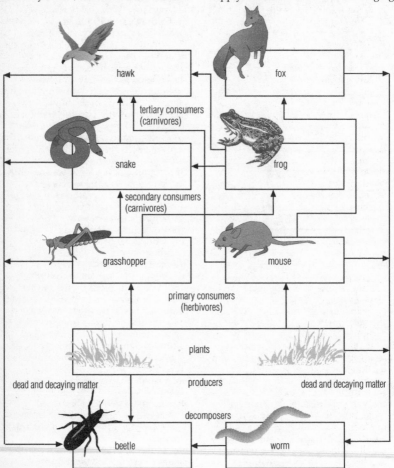

food web The complex interrelationships between animals and plants in a food web. A food web shows how different food chains are linked in an ecosystem. Note that the arrows indicate movement of energy through the web. For example, an arrow shows that energy moves from plants to the grasshopper, which eats the plants.

from individual households to global patterns. Since the 1940s the industrial and agricultural aspects of food supply have become increasingly globalized. New farming, packaging, and distribution techniques mean that the seasonal aspect of food supply has been reduced in wealthier nations such as the USA and the UK. In some less developed countries there are often problems of food scarcity and distribution caused by climate-related crop failure.

foot in geometry, point where a line meets a second line to which it is perpendicular.

foot symbol ft, imperial unit of length, equivalent to 0.3048 m, in use in Britain since Anglo-Saxon times. It originally represented the length of a human foot. One foot contains 12 inches and is one-third of a yard.

foot-and-mouth disease contagious eruptive viral disease of cloven-hoofed mammals, characterized by blisters in the mouth and around the hooves. In cattle it causes deterioration of milk yield and abortions. It is an airborne virus, which makes its eradication extremely difficult. In the UK, affected herds are destroyed; inoculation is practised in Europe and the USA. The existing vaccine for the disease leaves vaccinated animals as carriers that can infect those animals that remain unvaccinated, and a more effective vaccine is under development in the USA.

footprint in computing, the area on the desk or floor required by a computer or other peripheral device.

force any influence that tends to change the state of rest or the uniform motion in a straight line of a body. The action of an unbalanced or resultant force results in the *acceleration of a body in the direction of action of the force, or it may, if the body is unable to move freely, result in its deformation (see *Hooke's law). A force is a push or a pull on an object. A force will cause an object to move if it is stationary, change direction, slow down, or speed up if it is moving. Force is a vector quantity, possessing both magnitude and direction; the *SI unit is the newton.

Forces that make contact with an object are called contact forces. Examples are normal push/pull force, surface tension, air resistance, and frictional forces. Forces that are able to cause a push/pull without making contact with an object are called non-contact forces. Examples are magnetic force, *gravity, and electric force. A newton is used to measure how much pull the force of gravity has on an object; this corresponds to the *weight of an object. A mass of 1 kg = 9.8 newtons (N). A newton is defined as the amount of force needed to move an object of 1 kg so that it accelerates at 1 metre per second per second.

forces and motion The Italian mathematician, astronomer and physicist Galileo (1564–1642) discovered that a body moving on a perfectly smooth horizontal surface would neither speed up nor slow down. All moving bodies continue moving with the same *velocity unless a force is applied to cause an acceleration. The reason we appear to have to push something to keep it moving with constant velocity is because of frictional forces acting on all moving objects on Earth. *Friction occurs when two solid surfaces rub on each other; for example, a car tyre in contact with the ground. Friction opposes the relative motion of the two objects in contact and acts to slow the velocity of the moving object. A force is required to push the moving object and to cancel out the frictional force. If the forces combine to give a net force of zero, the object will not accelerate but will continue moving at constant velocity. A *resultant force is a single force acting on a particle or body whose effect is equivalent to the combined effects of two or more separate forces.

English physicist and mathematician Isaac Newton (1642–1727) developed this work further. According to Newton's second law of motion , the magnitude of a resultant force is equal to the rate of change of *momentum of the body on which it acts; the force F producing an acceleration a metres per second per second on a body of mass m kilograms is therefore given by:

$F = ma$. Thus Newton's second law states that change of momentum is proportional to the size of the external force and takes place in the direction in which the force acts. Momentum is a function both of the mass of a body and of its velocity. This agrees with our experience, because the idea of force is derived from muscular effort, and we know that we have to exert more strength to stop the motion of a heavy body than a light one, just as we have to exert more strength to stop a rapidly moving body than a slowly moving one. Force, then, is measured by change of momentum, momentum being equal to mass multiplied by velocity. (See also *Newton's laws of motion.) Newton's third law of motion states that if a body A exerts a force on a body B, then body B exerts an equal force on body A but in the opposite direction. This equal and opposite force is called a *reaction force.

force multiplier machine designed to multiply a small effort in order to move a larger load. The ratio by which a machine multiplies the effort is called its *mechanical advantage. Examples of a force multiplier include crowbar, wheelbarrow, nutcrackers, and bottle opener.

force ratio magnification of a force by a machine; see *mechanical advantage.

forces, fundamental in physics, four fundamental interactions currently known to be at work in the physical universe. There are two long-range forces: the **gravitational force**, or **gravity**, which keeps the planets in orbit around the Sun and acts between all particles that have mass; and the **electromagnetic force**, which stops solids from falling apart and acts between all particles with *electric charge. There are two very short-range forces, which operate over distances comparable with the size of the atomic nucleus: the **weak nuclear force**, responsible for the reactions that fuel the Sun and for the emission of *beta particles by some particles; and the **strong nuclear force**, which binds together the protons and neutrons in the nuclei of atoms. The relative strengths of the four forces are: strong, 1; electromagnetic, 10^{-2}; weak, 10^{-6}; gravitational, 10^{-40}. By 1971, the US physicists Steven Weinberg and Sheldon Glashow, the Pakistani physicist Abdus Salam, and others had developed a theory that suggested that the weak and electromagnetic forces were aspects of a single force called the **electroweak force**; experimental support came from observation at the European particle-physics laboratory *CERN in the 1980s. Physicists are now working on theories to unify all four forces.

forensic entomology branch of *forensic science, involving the study of insects on and around the corpse. Insects rapidly infest a corpse, and do so in an accepted sequence beginning with flies laying eggs. Further insects follow to feed on the decomposing flesh and fly maggots. Forensic entomologists are able to determine time of death by analysing insect colonization. They can also tell whether or not a corpse has been moved by examining the faunal community in the 'seepage area' beneath the body.

forensic medicine in medicine, branch of medicine concerned with the resolution of crimes. Examples of forensic medicine include the determination of the cause of death in suspicious circumstances or the identification of a criminal by examining tissue found at the scene of a crime. Forensic psychology involves the establishment of a psychological profile of a criminal that can assist in identification.

forensic science use of scientific techniques to solve criminal cases. A multidisciplinary field embracing chemistry, physics, botany, zoology, and medicine, forensic science includes the identification of human bodies or traces. Ballistics (the study of projectiles, such as bullets), another traditional forensic field, makes use of such tools as the comparison microscope and the electron microscope.

forging one of the main methods of shaping metals, which involves hammering or a more gradual application of pressure. A blacksmith hammers red-hot metal into shape on an anvil, and the traditional place of work is called a forge. The blacksmith's mechanical equivalent is the drop forge. The metal is shaped by the blows from a falling hammer or ram, which is usually accelerated by steam or air pressure. Hydraulic presses forge by applying pressure gradually in a squeezing action.

formaldehyde   common name for *methanal.

formalin   aqueous solution of formaldehyde (methanal) used to preserve animal specimens.

formatting   in computing, short for disk formatting (laying down a structure for organizing, saving, and retrieving data), or text formatting (changing its appearance). Modern office programs and desktop publishing packages also allow the formatting of objects such as pictures and frames.

formic acid   common name for *methanoic acid.

formula   in chemistry, a representation of a molecule, radical, or ion, in which the component chemical elements are represented by their symbols. For example, the formula for carbon dioxide is CO_2, showing that a molecule of carbon dioxide consists of one atom of carbon (C) and two atoms of oxygen (O_2). An **empirical formula** indicates the simplest ratio of the elements in a compound, without indicating how many of them there are or how they are combined. A **molecular formula** gives the number of each type of element present in one molecule. A **structural formula** shows the relative positions of the atoms and the bonds between them. For example, for ethanoic (acetic) acid, the empirical formula is CH_2O, the molecular formula is $C_2H_4O_2$, and the structural formula is CH_3COOH.

formula   in computing, an expression entered into a *cell on a *spreadsheet. The formula is displayed in the data entry line of the spreadsheet and the result is displayed in the cell. Formulae normally start with an equals (=) sign, which is followed by the expression. For example =SUM(A1:A6) would add up the contents of cells A1 to A6. Parentheses are also used to control the order of the calculation. For example =A1 + (B3 − C3).

formula   in mathematics, a set of symbols and numbers that expresses a fact or rule. For example, $A = pr^2$ is the formula for calculating the area of a circle. $E = mc^2$ is the famous formula relating energy and mass developed by German-born US physicist Einstein (1879–1955). Other common formulae exist for *density, *mass, *volume, and *area.

FORTRAN   or fortran; contraction of formula translation, high-level computer-programming language suited to mathematical and scientific computations. Developed by John Backus at IBM in 1956, it is one of the earliest computer languages still in use. A recent version, Fortran 90, is now being used on advanced parallel computers. *BASIC was strongly influenced by FORTRAN and is similar in many ways. Fortran 2000 is being developed and is expected to become an ISO standard in 2002.

fossil   (Latin *fossilis* 'dug up') cast, impression, or the actual remains of an animal or plant preserved in rock. Dead animals and plant remains that fell to the bottom of the sea bed or an inland lake were gradually buried under the accumulation of layers of sediment. Over millions of years, the sediment became *sedimentary rock and the remains preserved within the rock became fossilized. Fossils may include footprints, an internal cast, or external impression. A few fossils are preserved intact, as with mammoths fossilized in Siberian ice, or insects trapped in tree resin that is today amber. The study of fossils is called *palaeontology. Palaeontologists are able to deduce much of the geological history of a region from fossil remains. The existence of fossils is key evidence that organisms have changed with time, that is, evolved (see *evolution). About 250,000 fossil species have been discovered – a figure that is believed to represent less than 1 in 20,000 of the species that ever lived. **Microfossils** are so small they can only be seen with a microscope. They include the fossils of pollen, bone fragments, bacteria, and the remains of microscopic marine animals and plants, such as foraminifera and diatoms.

fossil fuel combustible material, such as coal, lignite, oil, *peat, and *natural gas, formed from the fossilized remains of plants that lived hundreds of millions of years ago. Such fuels are *non-renewable resources – once they are burnt, they cannot be replaced.

Fossil fuels are hydrocarbons (they contain atoms of carbon and hydrogen). They generate large quantities of heat when they burn in air, a process known as combustion. In this process carbon and hydrogen combine with oxygen in the air to form carbon dioxide, water vapour, and heat.

four-stroke cycle engine-operating cycle of most petrol and diesel engines. The 'stroke' is an upward or downward movement of a piston in a cylinder. In a petrol engine the cycle begins with the induction of a fuel mixture as the piston goes down on its first stroke. On the second stroke (up) the piston compresses the mixture in the top of the cylinder. An electric spark then ignites the mixture, and the gases produced force the piston down on its third, power, stroke. On the fourth stroke (up) the piston expels the burned gases from the cylinder into the exhaust.

fourth-generation language in computing, a type of programming language designed for the rapid programming of applications but often lacking the ability to control the individual parts of the computer. Such a language typically provides easy ways of designing screens and reports, and of using databases. Other 'generations' (the term implies a class of language rather than a chronological sequence) are *machine code (first generation); *assembly languages, or low-level languages (second); and conventional high-level languages such as *BASIC and *Pascal (third).

foxglove any of a group of flowering plants found in Europe and the Mediterranean region. They have showy spikes of bell-like flowers, and grow up to 1.5 m/5 ft high. (Genus *Digitalis*, family Scrophulariaceae.)

fractal (from Latin *fractus* 'broken') irregular shape or surface produced by a procedure of repeated subdivision. Generated on a computer screen, fractals are used in creating models of geographical or biological processes

induction stroke   compression stroke   power stroke   exhaust stroke

induction stroke   compression stroke   power stroke   exhaust stroke

four-stroke cycle Two different types of engine that function on exactly the same principle of four clearly definable strokes. However, they differ in the way these four stages occur. The Otto engine uses a mixture of fuel and air ignited by a spark, and the Wankel uses a fuel-air mixture, but a rotary arm rather than a two-way piston. The Wankel engine has fewer moving parts.

(for example, the creation of a coastline by erosion or accretion, or the growth of plants).

Sets of curves with such discordant properties were developed in the 19th century in Germany by Georg Cantor and Karl Weierstrass. The name was coined by the French mathematician Benoit Mandelbrot. Fractals are also used for computer art.

fraction in chemistry, a group of similar compounds, the boiling points of which fall within a particular range and which are separated during *fractional distillation (fractionation).

fraction (from Latin *fractus* 'broken') in mathematics, a number that indicates one or more equal parts of a whole. Usually, the number of equal parts into which the unit is divided (*denominator) is written below a horizontal or diagonal line, and the number of parts comprising the fraction (*numerator) is written above; for example, $2/3$ has numerator 2 and denominator 3. Such fractions are called *vulgar fractions or **simple fractions**. The denominator can never be zero.

fractional distillation or **fractionation**, process used to split complex *mixtures (such as *petroleum) into their components, usually by repeated heating, boiling, and condensation; see *distillation. In the laboratory it is carried out using a *fractionating column.

Fractional distillation is used to separate mixtures of miscible liquids, such as ethanol and water. The process depends on the components of the mixture having different boiling points. The liquid is heated so that it turns into a gas. The vapours pass up a fractionating column where they are gradually cooled. As each of the components of the mixture cools to its boiling point, it turns back into a liquid. The different components of the mixture condense at different levels in the fractionating column and thus may be separated.

In industry, fractional distillation is used to separate the compounds in crude oil (unrefined petroleum) into useful fractions, each fraction containing compounds with similar

boiling points. Air is also separated by fractional distillation. This is done by cooling air until it condenses and then allowing the temperature of the liquid air to rise. Each gas will distill off at its own boiling point.

fractionating column device in which many separate *distillations can occur so that a liquid mixture can be separated into its components. The technique is known as *fractional distillation, or fractionation.

Various designs exist but the primary aim is to allow maximum contact between the hot rising vapours and the cooling descending liquid. As the mixture of vapours ascends the column it becomes progressively enriched in the lower-boiling-point components, so these separate out first.

fracture break in the continuity of a bone, with or without displacement of any fragments. It may be pathological, the result of a relatively mild injury to an already diseased bone (as in osteomalacia and *Paget's disease), or, more often, it is the result of an injury to healthy bone.

fragile X syndrome the commonest inherited cause of mental retardation. It is inherited as an X-linked recessive condition and is so named because sufferers have a fragile site on one of their X chromosomes. Women carry two X chromosomes and are therefore more commonly affected than men, but less seriously. The gene was located 1991.

fragmentation in computing, the breaking up of files into many smaller sections stored on different parts of a disk. The computer *operating system stores files in this way so that maximum use can be made of disk space. Each section contains a pointer to where the next section is stored. The file allocation table keeps a record of this.

frame single photograph in a sequence representing motion, or movement, on film; in a *network, a unit of data; in word processing or desktop publishing, a marked-out area on a page that can contain text or graphics.

frame buffer in computing, a *buffer used to store a screen image.

francium chemical symbol Fr, radioactive metallic element, atomic number 87, relative atomic mass 223. It is one of the alkali metals and occurs in nature in small amounts as a decay product of actinium. Its longest-lived isotope has a half-life of only 21 minutes. Francium was discovered and named in 1939 by Marguérite Perey, to honour her country.

Frasch process process used to extract underground deposits of sulphur. Superheated steam is piped into the sulphur deposit and melts it. Compressed air is then pumped down to force the molten sulphur to the surface. The process was developed in the USA in 1891 by German-born Herman Frasch (1851–1914).

free fall the state in which a body is falling freely under the influence of *gravity, as in freefall parachuting (skydiving). In a vacuum, a freely falling body accelerates at a rate of 9.806 m sec^{-2}/32.174 ft sec^{-2}; the value varies slightly at different latitudes and altitudes. A body falling through air accelerates until it reaches a maximum speed called the *terminal velocity; thereafter, there is no further acceleration.

free radical in chemistry, an atom or molecule that has an unpaired electron and is therefore highly reactive. Most free radicals are very short-lived. They are by-products of normal cell chemistry and rapidly oxidize other molecules they encounter. Free radicals are thought to do considerable damage. They are neutralized by protective enzymes. Free radicals are often produced by high temperatures and are found in flames and explosions.

freeze–thaw form of physical *weathering, common in mountains and glacial environments, caused by the expansion of water as it freezes. Water in a crack freezes and expands in volume by 9% as it turns to ice. This expansion exerts great pressure on the rock, causing the crack to enlarge. After many cycles of freeze–thaw, rock fragments may break off to form scree slopes.

freezing *change of state from liquid to solid, as when water becomes ice. For a given substance, freezing occurs at a definite temperature, known as the

*freezing point, that is invariable under similar conditions of pressure, and the temperature remains at this point until all the liquid is frozen; the freezing point and melting point of the substance are the same temperature. By measuring the temperature of a liquid against time as it cools a cooling curve can be plotted; on the cooling curve the temperature levels out at the freezing point.

The amount of heat per unit mass that has to be removed to freeze a substance is a constant for any given substance, and is known as the latent heat of fusion.

freezing point for any given liquid, the temperature at which the liquid changes state from a liquid to a solid. The temperature remains at this point until all the liquid has solidified. It is invariable under similar conditions of pressure – for example, the freezing point of water under standard atmospheric pressure is 0°C/32°F. For a given liquid under similar conditions, the freezing point and melting point are the same temperature.

freezing point, depression of lowering of a solution's freezing point below that of the pure solvent; it depends on the number of molecules of solute dissolved in it. For a single solvent, such as pure water, all solute substances in the same molar concentration produce the same lowering of freezing point. The depression d produced by the presence of a solute of molar concentration C is given by the equation $d = KC$, where K is a constant (called the cryoscopic constant) for the solvent concerned.

Antifreeze mixtures for car radiators and the use of salt to melt ice on roads are common applications of this principle. Animals in arctic conditions, for example insects or fish, cope with the extreme cold either by manufacturing natural 'antifreeze' and staying active, or by allowing themselves to freeze in a controlled fashion, that is, they manufacture proteins to act as nuclei for the formation of ice crystals in areas that will not produce cellular damage, and so enable themselves to thaw out and become active again. Measurement of freezing-point depression is a useful method of determining the molecular weights of solutes. It is also used to detect the illicit addition of water to milk.

frequency in physics, number of periodic oscillations, vibrations, or waves occurring per unit of time. The SI unit of frequency is the hertz (Hz), one hertz being equivalent to one cycle per second. Frequency is related to wavelength and velocity by the equation:

$$f = \frac{v}{\lambda}$$

where f is frequency, v is velocity, and λ is wavelength. Frequency is the reciprocal of the *period T:

$$f = \frac{1}{T}$$

Most alternating current electrical equipment in the UK has an optimum operating frequency of 50 Hz. Frequencies lower than this will cause noticeable voltage variations in electrical appliances. This is because each time the current changes direction, the voltage falls to zero before increasing again as the current moves in the opposite direction. For example, in a television set this will be seen as flickering of the screen. At 50 Hz, the current will switch 100 times a second. This is detected as a continuous image as the human eye cannot react fast enough to detect these changes.

At one end of the electromagnetic spectrum are long radio waves with a frequency in the range 10^4–10^5 Hz and at the other extreme are gamma rays with a frequency in the range 10^{19}–10^{22} Hz.

Human beings can hear sounds from objects vibrating in the range 20–15,000 Hz. Ultrasonic frequencies well above 15,000 Hz can be detected by such mammals as bats. Infrasound (low-frequency sound) can be detected by some mammals and birds. Pigeons can detect sounds as low as 0.1 Hz; elephants communicate using sounds as low as 1 Hz. **Frequency modulation** (FM) is a method of transmitting radio signals in which the frequency of the **carrier wave** is changed and then decoded.

frequency in statistics, the number of times an event occurs. For example, in a survey carried out to find out a group of children's favourite colour of the rainbow, the colour red is chosen 26 times. This gives the colour red a frequency of 26. A table of the raw data collected, including the frequencies, is called a **frequency distribution**. It is usually presented in a **frequency table** or tally chart. The frequencies can also be shown diagrammatically using a **frequency polygon**.

frequency modulation FM, method by which radio waves are altered for the transmission of broadcasting signals. FM varies the frequency of the carrier wave in accordance with the signal being transmitted. Its advantage over AM (*amplitude modulation) is its better signal-to-noise ratio. It was invented by the US engineer Edwin Armstrong.

frequently asked questions in computing, expansion of the abbreviation *FAQ.

friction in physics, the force that opposes the movement of two bodies in contact as they move relative to each other. The **coefficient of friction** is the ratio of the force required to achieve this relative motion to the force pressing the two bodies together.

Two materials with rough surfaces rubbing together will change kinetic energy into heat and sound energy.

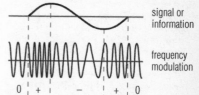

frequency modulation In FM radio transmission, the frequency of the carrier wave is modulated, rather than its amplitude (as in AM broadcasts). The FM system is not affected by the many types of interference which change the amplitude of the carrier wave, and so provides better quality reception than AM broadcasts.

Friction is greatly reduced by the use of lubricants such as oil, grease, and graphite. A layer of lubricant between two materials reduces the contact, allowing them to slide over each other smoothly. For example, engine oil used in cars reduces friction between metal parts as they move against each other. Air bearings are now used to minimize friction in high-speed rotational machinery. In joints in the human body, such as the knee, synovial fluid plays a key role as a lubricant. In other instances friction is deliberately increased by making the surfaces rough – for example, brake linings, driving belts, soles of shoes, and tyres.

Friction is also used to generate static electric charges on different materials by rubbing the materials together.

frond large leaf or leaflike structure; in ferns it is often pinnately divided. The term is also applied to the leaves of palms and less commonly to the plant bodies of certain seaweeds, liverworts, and lichens.

front in meteorology, the boundary between two air masses of different temperature or humidity. A **cold front** marks the line of advance of a cold air mass from below, as it displaces a warm air mass; a **warm front** marks the advance of a warm air mass as it rises up over a cold one. Frontal systems define the weather of the mid-latitudes, where warm tropical air is continually meeting cold air from the poles.

front-end processor small computer used to coordinate and control the communications between a large mainframe computer and its input and output devices.

frost condition of the weather that occurs when the air temperature is below freezing, 0°C/32°F. Water in the atmosphere is deposited as ice crystals on the ground or on exposed objects. As cold air is heavier than warm air and sinks to the ground, ground frost is more common than hoar (air) frost, which is formed by the condensation of water particles in the air.

fructose $C_6H_{12}O_6$, sugar that occurs naturally in honey, the nectar of flowers, and many sweet fruits; it is commercially prepared from glucose.

Fructose is a monosaccharide, whereas the more familiar cane or beet sugar is a disaccharide, made up of two monosaccharide units: fructose and glucose. It is sweeter than cane sugar and can be used to sweeten foods for people with diabetes.

fruit (from Latin *frui* 'to enjoy') in botany, the ripened ovary in flowering plants that develops from one or more seeds or carpels and encloses one or more seeds. Its function is to protect the seeds during their development and to aid in their dispersal. Fruits are often edible, sweet, juicy, and colourful. When eaten they provide vitamins, minerals, and enzymes, but little protein. Most fruits are borne by perennial plants.

FSF in computing, abbreviation for the Free Software Foundation.

FSH abbreviation for *follicle-stimulating hormone.

ft symbol for *foot, a measure of distance.

FTP abbreviation for file transfer protocol, in computing, rules for transferring files between computers on the *Internet. The use of FTP avoids incompatibility between individual computers. To use FTP over the Internet, a user must have an Internet connection, an FTP client or World Wide Web *browser, and an account on the system holding the files. Many commercial and non-commercial systems allow anonymous FTP either to distribute new versions of software products or as a public service.

FTPmail in computing, an *FTP server that can be operated by e-mail. This service is useful for people with only limited access to the Internet.

fuel any source of heat or energy, embracing the entire range of materials that burn in air (combustibles). A fuel is a substance that gives out energy when it burns. A **nuclear fuel** is any material that produces energy by nuclear fission in a nuclear reactor. *Fossil fuels are formed from the fossilized remains of plants and animals.

Crude oil (unrefined *petroleum) is purified at an oil refinery by *fractional distillation into fuels such as gasoline and kerosine. The burning of fossil fuels for energy production contributes to environmental problems such as *acid rain and the *greenhouse effect.

fuel cell cell converting chemical energy directly to electrical energy. It works on the same principle as a battery but is continually fed with fuel, usually hydrogen and oxygen. Fuel cells are silent and reliable (no moving parts) but expensive to produce. They are an example of a renewable energy source.

Hydrogen is passed over an *electrode (usually nickel or platinum) containing a *catalyst, which splits the hydrogen into electrons and protons. The electrons pass through an external circuit while the protons pass through a polymer *electrolyte membrane to another electrode, over which oxygen is passed. Water is formed at this electrode (as a by-product) in a chemical reaction involving electrons, protons, and oxygen atoms. A current is generated between the electrodes. If the spare heat also produced is used for hot water and space heating, 80% efficiency in fuel is achieved.

Fuel cells can be used to power cars, replacing the internal combustion engine, and to produce electricity on spacecraft.

fulcrum in physics, the point of support, or pivot, of a *lever.

full duplex in computing, modem setting which means that two-way communication is enabled, so that everything you type is echoed back to the screen. See *duplex.

function in computing, a small part of a program that supplies a specific value – for example, the square root of a specified number, or the current date. Most programming languages incorporate a number of built-in functions; some allow programmers to write their own. A function may have one or more arguments (the values on which the function operates). A **function key** on a keyboard is one that, when pressed, performs a designated task, such as ending a program.

function in mathematics, a function f is a non-empty set of ordered pairs

$(x, f(x))$ of which no two can have the same first element. Hence, if $f(x) = x^2$ two ordered pairs are $(-2, 4)$ and $(2, 4)$. The set of all first elements in a function's ordered pairs is called the **domain**; the set of all second elements is the **range**. Functions are used in all branches of mathematics, physics, and science generally. For example, in the *equation $y = 2x + 1$, y is a function of the symbol x. This can be written as $y = f(x)$.

In another example, the *formula $t = 2\pi\sqrt{(l/g)}$ shows that for a simple pendulum, the time of swing t is a function of its length l and of no other variable quantity (π and g, the acceleration due to gravity, are *constants).

functional group in chemistry, a small number of atoms in an arrangement that determines the chemical properties of the group and of the molecule to which it is attached (for example, the carboxyl group COOH, or the amine group NH_2). Organic compounds can be considered as structural skeletons, with a high carbon content, with functional groups attached.

functional programming computer programming based largely on the definition of *functions. There are very few functional programming languages, HOPE and ML being the most widely used, though many more conventional languages (for example, C) make extensive use of functions.

function key key on a keyboard that, when pressed, performs a designated task, such as ending a computer program.

fundamental constant physical quantity that is constant in all circumstances throughout the whole universe. Examples are the electric charge of an electron, the speed of light, *Planck's constant, and the gravitational constant.

fundamental forces see *forces, fundamental.

fundamental particle another term for *elementary particle.

fundamental vibration standing wave of the longest wavelength that can be established on a vibrating object such as a stretched string or air column. The sound produced by the fundamental vibration is the lowest-pitched (usually dominant) note heard.

The fundamental vibration of a string has a stationary *node at each end and a single *antinode at the centre where the amplitude of vibration is greatest.

fungicide any chemical *pesticide used to prevent fungus diseases in plants and animals. Inorganic and organic compounds containing sulphur are widely used.

fungus plural **fungi**, any of a unique group of organisms that includes moulds, yeasts, rusts, smuts, mildews, mushrooms, and toadstools. There are around 70,000 species of fungi known to science, although there may be as many as 1.5 million actually in existence. They are not considered to be plants for three main reasons: they have no leaves or roots; they contain no chlorophyll (green colouring) and are therefore unable to make their own food by *photosynthesis; and they reproduce by *spores. Some fungi are edible but many are highly poisonous; they often cause damage and sometimes disease to the organic matter on which they live and feed, but some fungi are exploited in the production of food and drink (for example, yeasts in baking and brewing) and in medicine (for example, penicillin).

furlong unit of measurement, originating in Anglo-Saxon England, equivalent to 220 yd (201.168 m).

fuse in electricity, a wire or strip of metal designed to melt (thus breaking the circuit) when excessive current passes through. It is a safety device that halts surges of current that would otherwise damage equipment and cause fires. In explosives, a fuse is a cord impregnated with chemicals so that it burns slowly at a predetermined rate. It is used to set off a main explosive charge, sufficient length of fuse being left to allow the person lighting it to get away to safety.

fusion in physics, the fusing of the nuclei of light elements, such as hydrogen, into those of a heavier element, such as helium. The resultant loss in their combined mass is converted into energy. Stars and thermonuclear weapons are powered by nuclear fusion.

Nuclear fusion takes place in the Sun, where hydrogen nuclei fuse at temperatures of about 10 million °C/18 million °F producing huge amounts of heat and light energy.

fuzzy logic in mathematics and computing, a form of knowledge representation suitable for notions (such as 'hot' or 'loud') that cannot be defined precisely but depend on their context. For example, a jug of water may be described as too hot or too cold, depending on whether it is to be used to wash one's face or to make tea.

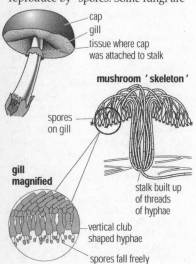

cap
gill
tissue where cap was attached to stalk

mushroom 'skeleton'

spores on gill

gill magnified

stalk built up of threads of hyphae

vertical club shaped hyphae

spores fall freely

fungus Fungi grow from spores as fine threads, or hyphae. These have no distinct cellular structure. Mushrooms and toadstools are the fruiting bodies formed by the hyphae. Gills beneath the caps of these aerial structures produce masses of spores.

G

g symbol for *gram.

GABA acronym for Gamma AminoButyric Acid, in medicine, an amino acid that occurs in the central nervous system, mainly in the tissue of the brain. It is involved in the transmission of inhibitory impulses from nerve to nerve and between nerves and tissues in the brain. Imbalances in GABA concentrations in the brain may be implicated in a variety of disorders, including *anxiety and *epilepsy.

gadolinium chemical symbol Gd, silvery-white metallic element of the lanthanide series, atomic number 64, relative atomic mass 157.25. It is found in the products of nuclear fission and used in electronic components, alloys, and products needing to withstand high temperatures.

Gaia hypothesis theory that the Earth's living and nonliving systems form an inseparable whole that is regulated and kept adapted for life by living organisms themselves. The planet therefore functions as a single organism, or a giant cell. The hypothesis was elaborated by British scientist James Lovelock (1919–) and first published in 1968.

gain in electronics, the ratio of the amplitude of the output signal produced by an amplifier to that of the input signal. In a *voltage amplifier the voltage gain is the ratio of the output voltage to the input voltage; in an inverting *operational amplifier (op-amp) it is equal to the ratio of the resistance of the feedback resistor to that of the input resistor.

gal symbol for *gallon, *galileo.

galactic coordinates in astronomy, system for measuring the position of astronomical objects on the *celestial sphere with reference to the galactic equator (or great circle).

galactose $C_6H_{12}O_6$, one of the hexose sugars, an *isomer of *glucose and *fructose.

galaxy grouping of millions or billions of stars, held together by gravity. It is believed that there are billions of galaxies in the *universe. There are different types, including spiral, barred spiral, and elliptical galaxies. Our own Galaxy, the *Milky Way, is about 100,000 light years across (a light year is the distance light travels in a year, about 9.5 billion km/6 billion mi), and contains at least 100 billion stars.

gale strong wind, usually between force seven and ten on the *Beaufort scale (measuring 45–90 km/28–56 mi per hour).

galileo symbol gal, unit of acceleration, used in geological surveying. One galileo is 10^{-2} metres per second per second. The Earth's gravitational field often differs by several milligals (thousandths of galileos) in different places, because of the varying densities of the rocks beneath the surface. Named after Italian mathematician, astronomer, and physicist Galileo (1564–1642).

Galileo spacecraft launched from the space shuttle *Atlantis* on 18 October 1989 to explore the planet Jupiter, named after Italian mathematician, astronomer, and physicist Galileo (1564–1642).Galileo's probe entered the atmosphere of Jupiter in December 1995. It radioed information back to the orbiter for 57 minutes before the craft was destroyed by atmospheric pressure. The first pictures of Jupiter were transmitted in 1996. In 1997 Galileo completed two fly-bys of Jupiter's fourth-largest and icy moon Europa, and in February 2000 it passed within 200 km/125 mi of Jupiter's third-largest moon Io.

The spacecraft flew past the planet Venus in February 1990 and passed within 970 km/600 mi of Earth in December 1990 and December 1992, using the gravitational fields of these two planets to increase its velocity. It flew past the asteroids Gaspra in 1991 and Ida in 1993, taking close-up photographs.

gall bladder small muscular sac, part of the digestive system of most, but not all, vertebrates. In humans, it is situated on the underside of the liver and connected to the small intestine by the bile duct. It stores bile from the liver.

gallium chemical symbol Ga, grey metallic element, atomic number 31, relative atomic mass 69.72. It is liquid at room temperature. Gallium arsenide (GaAs) crystals are used for semiconductors in microelectronics, since electrons travel a thousand times faster through them than through silicon. The element was discovered in 1875 by Lecoq de Boisbaudran (1838–1912).

gallium arsenide compound of gallium and arsenic, formula GaAs, used in lasers, photocells, and microwave generators. Its semiconducting properties make it a possible rival to *silicon for use in microprocessors. Chips made from gallium arsenide require less electric power and process data faster than those made from silicon.

gallon symbol gal, imperial liquid or dry measure, equal to 4.546 litres, and subdivided into four quarts or eight pints. The US gallon is equivalent to 3.785 litres.

gallstone pebblelike, insoluble accretion formed in the human gall bladder or bile ducts from cholesterol or calcium salts present in bile. Gallstones may be symptomless or they may cause pain, indigestion, or jaundice. They can be dissolved with medication or removed, either by means of an endoscope or, along with the gall bladder, in an operation known as cholecystectomy.

galvanizing process for rendering iron rust-proof, by plunging it into molten zinc (the dipping method), or by electroplating it with zinc.

galvanometer instrument for detecting small electric currents by their magnetic effect.

gamete cell that functions in sexual reproduction by merging with another gamete to form a *zygote. Examples of gametes include sperm and egg cells. In most organisms, the gametes are haploid (they contain half the number of chromosomes of the parent), owing to reduction division or *meiosis.

In higher organisms, gametes are of two distinct types: large immobile ones known as eggs or egg cells (see *ovum) and small ones known as *sperm. They come together at *fertilization. In some lower organisms the gametes are all the same, or they may belong to different mating strains but have no obvious differences in size or appearance.

gametophyte the *haploid generation in the life cycle of a plant that produces gametes; see *alternation of generations.

gamma radiation very high-frequency, high-energy electromagnetic radiation, similar in nature to X-rays but of shorter wavelength, emitted by the nuclei of radioactive substances during decay or by the interactions of high-energy electrons with matter, discovered by New-Zealand-born British physicist Ernest Rutherford (1871–1937). Cosmic gamma rays have been identified as coming from pulsars, radio galaxies, and quasars, although they cannot penetrate the Earth's atmosphere.

Gamma rays are stopped only by direct collision with an atom and are therefore very penetrating. They can, however, be stopped by about 4 cm/1.5 in of lead or by a very thick concrete shield. They are less ionizing in their effect than alpha and beta particles, but are dangerous nevertheless because they can penetrate deeply into body tissues such as bone marrow. They are not deflected by magnetic or electric fields. Gamma radiation is used to kill bacteria and other micro-organisms, sterilize medical devices, and change the molecular structure of plastics to modify their properties (for example, to improve their resistance to heat and abrasion).

ganglion plural **ganglia**, solid cluster of nervous tissue containing many cell bodies and *synapses, usually enclosed in a tissue sheath; found in invertebrates and vertebrates. In many invertebrates, the central nervous system consists mainly of ganglia connected by nerve cords. The ganglia in the head (cerebral ganglia) are usually well developed and are analogous to the brain in vertebrates. In vertebrates, most ganglia occur outside the central nervous system.

ganglion in medicine, harmless swelling or cyst developing within a tendon sheath, usually at the wrist.

gangrene death and decay of body tissue (often of a limb) due to bacterial action; the affected part gradually turns black and causes blood poisoning.

garlic perennial Asian plant belonging to the lily family, whose strong-smelling and sharp-tasting bulb, made up of several small segments, or cloves, is used in cooking. The plant has white flowers. It is widely cultivated and has been used successfully as a fungicide in the cereal grass sorghum. It also has antibacterial properties. (*Allium sativum*, family Liliaceae.)

garnet group of *silicate minerals with the formula $X_3Y_3(SiO_4)_3$, where X is calcium, magnesium, iron, or manganese, and Y is usually aluminium or sometimes iron or chromium. Garnets are used as semi-

precious gems (usually pink to deep red) and as abrasives. They occur in metamorphic rocks such as gneiss and schist.

gas form of matter, such as air, in which the molecules move randomly in otherwise empty space, filling any size or shape of container into which the gas is put. A sugar-lump sized cube of air at room temperature contains 30 trillion molecules moving at an average speed of 500 metres per second (1,800 kph/1,200 mph). Gases can be liquefied by cooling, which lowers the speed of the molecules and enables attractive forces between them to bind them together.

gas constant in physics, the constant R that appears in the equation $PV = nRT$, which describes how the pressure P, volume V, and temperature T of an ideal gas are related (n is the amount of gas in moles). This equation combines *Boyle's law and *Charles's law. R has a value of 8.3145 joules per kelvin per mole.

gas exchange movement of gases between an organism and the atmosphere, principally oxygen and carbon dioxide. All aerobic organisms (most animals and plants) take in oxygen in order to burn food and manufacture *ATP (adenosine

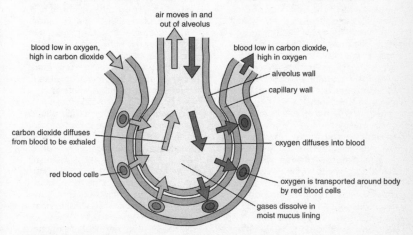

gas exchange Gas exchange in the alveolus. Blood enters the lungs low in oxygen and high in carbon dioxide. The carbon dioxide diffuses from the blood to the alveolus, where carbon dioxide levels are low. Oxygen diffuses from the alveolus, where oxygen levels are high, into the blood where the level is low, so blood leaves the lungs high in oxygen and low in carbon dioxide.

triphosphate, which provides the energy for all cell reactions). The resultant oxidation reactions release carbon dioxide as a waste product to be passed out into the *environment. Green plants also absorb carbon dioxide during *photosynthesis, and release oxygen as a waste product.

It is a characteristic of living organisms that they exchange materials with the environment. As well as the exchange of oxygen and carbon dioxide during *aerobic respiration, nutrients must be taken up from the environment to feed the organism. In humans, these include *carbohydrates, *proteins, minerals (see *mineral salts), and *vitamins. *Water is another important substance that has to be absorbed. However, in plants, only carbon dioxide, water, and minerals are required. The carbon dioxide is used in photosynthesis, which then releases oxygen as a waste product. All oxygen gas in the air on Earth is believed to have been produced by photosynthesis.

Specialized respiratory surfaces have evolved during evolution to make gas exchange more efficient. In humans and other tetrapods (four-limbed vertebrates), gas exchange occurs in the *lungs, aided by the breathing movements of the ribs. Many adult amphibia and terrestrial invertebrates can absorb oxygen directly through the skin. The bodies of insects and some spiders contain a system of air-filled tubes known as *tracheae. Fish have gills as their main respiratory surface. In plants, gas exchange generally takes place via the *stomata and the air-filled spaces between the cells in the interior of the leaf.

gas giant in astronomy, any of the four large outer planets of the Solar System, *Jupiter, Saturn, *Uranus, and *Neptune, which consist largely of gas and have no solid surface.

gas laws physical laws concerning the behaviour of gases. They include *Boyle's law and *Charles's law, which are concerned with the relationships between the pressure (P), temperature (T), and volume (V) of an ideal (hypothetical) gas. These two laws can be combined to give the **general** or

universal gas law, which may be expressed as: $PV/T = $ constant. The laws state that gas pressure depends on the temperature and volume of the gas. If the volume of a gas is kept constant, its pressure increases with temperature. When a gas is squeezed into a smaller volume, its pressure increases. If the pressure of a gas is constant, its volume increases with temperature. Such behaviour of gases depends on the *kinetic theory of matter, and particle theory, which states that all matter is composed of particles. See *gas constant.

gasohol motor fuel that is 90% petrol and 10% ethanol (alcohol). The ethanol is usually obtained by fermentation, followed by distillation, using maize, wheat, potatoes, or sugar cane. It was used in early cars before petrol became economical, and its use was revived during the 1940s war shortage and the energy shortage of the 1970s, for example in Brazil.

gasoline US term for *petrol.

gastroenteritis inflammation of the stomach and intestines, giving rise to abdominal pain, vomiting, and diarrhoea. It may be caused by food or other poisoning, allergy, or infection. Dehydration may be severe and it is a particular risk in infants.

gastrolith stone that was once part of the digestive system of a dinosaur or other extinct animal. Rock fragments were swallowed to assist in the grinding process in the dinosaur digestive tract, much as some birds now swallow grit and pebbles to grind food in their crop. Once the animal has decayed, smooth round stones remain – often the only clue to their past use is the fact that they are geologically different from their surrounding strata.

gate, logic in electronics, see *logic gate.

gauge boson or **field particle**, any of the particles that carry the four fundamental forces of nature (see *forces, fundamental). Gauge bosons are *elementary particles that cannot be subdivided, and include the photon, the graviton, the gluons, and the intermediate vector bosons (W^+, W^-, and Z). The gluon carries the strong nuclear force, the photon the electromagnetic force, W+, W−, and Z the weak nuclear force, and the graviton, as yet unobserved, the force of gravity.

gauss   symbol Gs, centimetre-gram-second (c.g.s.) unit of magnetic induction or magnetic flux density, replaced by the SI unit, the *tesla, but still commonly used. It is equal to one line of magnetic flux per square centimetre. The Earth's magnetic field is about 0.5 Gs, and changes to it over time are measured in gammas (one gamma equals 10^{-5} gauss).

gear   toothed wheel that transmits the turning movement of one shaft to another shaft. Gear wheels may be used in pairs, or in threes if both shafts are to turn in the same direction. The gear ratio – the ratio of the number of teeth on the two wheels – determines the torque ratio, the turning force on the output shaft compared with the turning force on the input shaft. The ratio of the angular velocities of the shafts is the inverse of the gear ratio.

Geiger counter   or **Geiger Müller counter**, any of a number of devices used for detecting nuclear radiation and measuring its intensity by counting the number of ionizing particles produced (see *radioactivity). It detects the momentary current that passes between *electrodes (anode and cathode) in a suitable gas (such as argon) when radiation causes the ionization of the gas. The electrodes are connected to electronic devices that enable the number of particles passing to be measured. The increased frequency of measured particles indicates the intensity of radiation. The device is named after the German physicist Hans Geiger (1882–1945).

Geissler tube   high-voltage *discharge tube in which traces of gas are ionized and conduct electricity. Since the electrified gas takes on a luminous colour characteristic of the gas, the instrument is also used in *spectroscopy. It was developed in 1858 by the German physicist Heinrich Geissler.

gel   solid produced by the formation of a three-dimensional cage structure, commonly of linked large-molecular-mass polymers, in which a liquid is trapped. It is a form of *colloid. A gel may be a jellylike mass (pectin, gelatin) or have a more rigid structure (silica gel).

gelignite   type of *dynamite, invented by Swedish chemist and engineer Alfred Nobel (1833–1896) in 1875.

gem   mineral valuable by virtue of its durability (hardness), rarity, and beauty, cut and polished for ornamental use, or engraved. Of 120 minerals known to have been used as gemstones, only about 25 are in common use in jewellery today; of these, the diamond, emerald, ruby, and sapphire are classified as precious, and all the others semi-precious; for example, the topaz, amethyst, opal, and aquamarine.

Among the synthetic precious stones to have been successfully produced are rubies, sapphires, meralds, and diamonds (first produced by General Electric in the USA in 1955). Pearls are not technically gems.

gender differences   differences between the sexes that are not anatomical or biological but are due to the influences of culture and society.

gene   basic unit of inherited material, encoded by a strand of *DNA and transcribed by *RNA. In higher organisms, genes are located on the *chromosomes. A gene consistently affects a particular character in an individual – for example, the gene for eye colour. Also termed a Mendelian gene, after Austrian biologist Gregor Mendel (1822–1884), it occurs at a particular point, or *locus, on a particular chromosome and may have several variants, or *alleles, each specifying a particular form of that character – for example, the alleles for blue or brown eyes. Some alleles show *dominance. These mask the effect of other alleles, known as *recessive. Genes can be manipulated using the techniques of *genetic engineering (gene technology).

The inheritance of genes and the way genes work is studied in *genetics. One gene carries the information that describes how one particular *protein is made. This information is stored as a chemical code on a DNA molecule and the genes are found in sequence from one end of the molecule to the other.

Each protein that is made helps to determine part of the characteristics of an organism. Between them, all the proteins determine all the inherited characteristics of an organism, though some of these characteristics can be modified by the *environment. The DNA is located in the chromosomes in the *nucleus of a cell. Many thousands of genes are present on each chromosome. The total number of genes in a human, according to estimates published in 2001 by the *Human Genome Project, is thought to be between 27,000 and 40,000, distributed between the 46 chromosomes in each human cell. Occasionally, a gene or a larger part of a chromosome or the number of chromosomes becomes accidentally altered. Such a change is a *mutation. Mutations can cause an individual to have a disease or disorder, such as Huntington's disease, cystic fibrosis, or sickle-cell anaemia. Gregor Mendel was the first to understand the mechanism of inheritance by genes, as a result of the study of plant breeding. He did not, however, know about the existence of DNA.

gene amplification   technique by which selected DNA from a single cell can be duplicated indefinitely until there is a sufficient amount to analyse by conventional genetic techniques.

   Gene amplification uses a procedure called the polymerase chain reaction. The sample of DNA is mixed with a solution of enzymes called polymerases, which enable it to replicate, and with a plentiful supply of nucleotides, the building blocks of DNA. The mixture is repeatedly heated and cooled. At each warming, the double-stranded DNA present separates into two single strands, and with each cooling the polymerase assembles a new paired strand for each single strand. Each cycle takes approximately 30 minutes to complete, so that after 10 hours there is one million times more DNA present than at the start.

   The technique can be used to test for genetic defects in a single cell taken from an embryo, before the embryo is reimplanted in in vitro fertilization.

gene bank   collection of seeds or other forms of genetic material, such as tubers, spores, bacterial or yeast cultures, live animals and plants, frozen sperm and eggs, or frozen embryos. These are stored for possible future use in agriculture, plant and animal breeding, or in medicine, genetic engineering, or the restocking of wild habitats where species have become extinct. Gene banks may be increasingly used as the rate of extinction increases, depleting the Earth's genetic variety (biodiversity).

gene imprinting   genetic phenomenon whereby a small number of genes function differently depending on whether they were inherited from the father or the mother. If two copies of an imprinted gene are inherited from one parent and none from the other, a genetic abnormality results, whereas no abnormality occurs if, as is normal, a copy is inherited from both parents. Gene imprinting is known to play a part in a number of genetic disorders and childhood diseases, for example, the Prader–Willi syndrome (characterized by mild mental retardation and compulsive eating).

gene pool   total sum of *alleles (variants of *genes) possessed by all the members of a given population or species alive at a particular time.

generalize   in mathematics, to extend a number of results to form a rule. For example, the computations $3 + 5 = 5 + 3$ and $1.5 + 2.7 = 2.7 + 1.5$ could be generalized to $a + b = b + a$.

general MIDI   GM; acronym for general Musical Instrument Digital Interface, standard set of 96 instrument and percussion 'voices' that can be used to encode musical tracks which can be reproduced on any GM-compatible synthesizer, or *MIDI.

general protection fault   computing error message; see *GPF.

generator   machine that produces *electrical energy from mechanical energy, as opposed to an *electric motor, which does the opposite. A simple generator (known as a dynamo in the UK) consists of a wire-wound coil (*armature) that is rotated between the poles of a permanent magnet. As the coil rotates it cuts across the magnetic field lines and a current is generated. A dynamo on a bicycle is an example of a simple generator.

The current generated can be increased by using a stronger magnet, by rotating the armature faster, or by winding more turns of wire onto a larger armature.

gene shears   technique in *genetic engineering which may have practical applications in the future. The gene shears are pieces of messenger *RNA that can bind to other pieces of messenger RNA, recognizing specific sequences, and cut them at that point. If a piece of *DNA which codes for the shears can be inserted in the chromosomes of a plant or animal cell, that cell will then destroy all messenger RNA of a particular type. Genetic shears may be used to protect plants against viruses which infect them and cause disease. They might also be useful against *AIDS.

gene-splicing   see *genetic engineering.

gene therapy   medical technique for curing or alleviating inherited diseases or defects that are due to a gene malfunction; certain infections, and several kinds of cancer in which affected cells from a sufferer are removed from the body, the *DNA repaired in the laboratory (*genetic engineering), and the normal functioning cells reintroduced. In 1990 a genetically engineered gene was used for the first time to treat a patient. The first human to undergo gene therapy, in 1990, was one of the so-called 'bubble babies' – a four-year-old American girl suffering from a rare enzyme (ADA) deficiency that cripples the immune system. Unable to fight off infection, such children are nursed in a germ-free bubble; they usually die in early childhood.

Cystic fibrosis is the most common inherited disorder and the one most keenly targeted by genetic engineers; the treatment has been pioneered in patients in the USA and UK. Gene therapy is not the final answer to inherited disease; it may cure the patient but it cannot prevent him or her from passing on the genetic defect to any children. However, it does hold out the promise of a cure for various

other conditions, including some forms of heart disease and some cancers; US researchers have successfully used a gene gun to target specific tumour cells.

genetic code   way in which instructions for building proteins, the basic structural molecules of living matter, are 'written' in the genetic material *DNA. This relationship between the sequence of bases (the subunits in a DNA molecule) and the sequence of *amino acids (the subunits of a protein molecule) is the basis of heredity. The code employs *codons of three bases each; it is the same in almost all organisms, except for a few minor differences recently discovered in some protozoa.

Only 2% of DNA is made up of base sequences, called **exons**, that code for proteins. The remaining DNA is known as 'junk' DNA or **introns**.

genetic counselling   in medicine, the establishment of a detailed family history of any genetic disorders for individuals who are at risk of developing them, or concerned that they may pass them on to their children, and discussion as to likely incidence and available options. Prenatal diagnosis may be available to those at risk of having a child affected by a particular disorder. Genetic counselling of individuals at risk of developing a genetic disorder in adult life can result in the individual taking action to prevent the disorder developing; for example, removal of the ovaries in women at high risk of developing ovarian cancer.

genetic disease   any disorder caused at least partly by defective genes or chromosomes. In humans there are some 3,000 genetic diseases, including cystic fibrosis, Down's syndrome, haemophilia, Huntington's chorea, some forms of anaemia, spina bifida, and Tay-Sachs disease.

genetic engineering   all-inclusive term that describes the deliberate manipulation of genetic material by biochemical techniques. It is often achieved by the introduction of new *DNA, usually by means of a virus or *plasmid. This can be for pure research, *gene therapy, or to breed functionally specific plants, animals, or bacteria. These organisms with a foreign gene added are said to be transgenic (see *transgenic organism) and the new DNA formed by this process is said to be recombinant. In most current cases the transgenic organism is a micro-organism or a plant, because ethical and safety issues are limiting its use in mammals. At the beginning of 1995 more than 60 plant species had been genetically engineered, and nearly 3,000 transgenic crops had been field-tested.

The process of genetic engineering involves several steps: the formation of DNA fragments, the insertion of DNA fragments into a vector plasmid, cloning of the plasmid, use of the plasmid to introduce the DNA into the organism, and expression of the gene.

One example of genetic engineering that has been very helpful to humans is the production of bacteria that make human insulin – the bacteria were engineered to contain the human gene for insulin. The bacteria are cultivated in fermenters to produce large amounts of insulin, which is then used to treat diabetic patients (see *diabetes). Prior to this, people with diabetes were treated with insulin from other animals. This new procedure removes the need to kill animals for insulin, and in addition the engineered insulin works better.

genetic fingerprinting or **genetic profiling**, technique developed in the UK by Professor Alec Jeffreys (1950–), and now allowed as a means of legal identification. It determines the pattern of certain parts of the genetic material *DNA that is unique to each individual. Like conventional fingerprinting, it can accurately distinguish humans from one another, with the exception of identical siblings

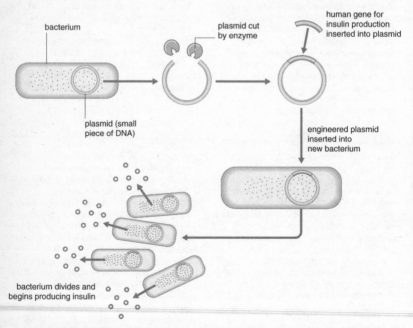

genetic engineering The genetic modification of a bacterium to produce insulin. The human gene for the production of insulin is collected from a donor chromosome and spliced into a vector plasmid (DNA found in bacteria but separate from the bacterial chromosomes). The plasmids and recipient bacteria are mixed together, during which process the bacteria absorb the plasmids. The plasmids replicate as the bacteria divide asexually (producing clones) and begin to produce insulin.

from multiple births. It can be applied to as little material as a single cell.

Genetic fingerprinting involves isolating DNA from cells, then comparing and contrasting the sequences of component chemicals between individuals. The DNA pattern can be ascertained from a sample of skin, hair, blood, or semen. Although differences are minimal (only 0.1% between unrelated people), certain regions of DNA, known as **hypervariable regions**, are unique to individuals.

genetics branch of biology concerned with the study of *heredity and variation–inheritance. It attempts to explain how characteristics of living organisms are passed on from one generation to the next. The science of genetics is based on the work of Austrian biologist Gregor Mendel (1822–1884) whose experiments with the cross-breeding (hybridization) of peas showed that the inheritance of characteristics and traits takes place by means of discrete 'particles', now known as *genes. These are present in the cells of all organisms, and are now recognized as being the basic units of heredity. All organisms possess *genotypes (sets of variable genes) and *phenotypes (characteristics produced by certain genes). Modern geneticists investigate the structure, function, and transmission of genes.

Before the publication of Mendel's work in 1865, it had been assumed that the characteristics of both parents were blended during inheritance, but Mendel showed that the genes remain intact, although their combinations change. As a result of his experiments with the cultivation of the common garden pea, Mendel introduced the concept of hybridization (see *monohybrid inheritance). Since Mendel, the study of genetics has advanced greatly, first through breeding experiments and light-microscope observations (classical genetics), later by means of biochemical and electron microscope studies (molecular genetics).

In 1944 Canadian-born bacteriologist Oswald Avery, together with his colleagues at the Rockefeller Institute,

Colin McLeod and Maclyn McCarty, showed that the genetic material was deoxyribonucleic acid (*DNA), and not protein as was previously thought. A further breakthrough was made in 1953 when US biologist James Watson (1928–) and English molecular biologist Francis Crick (1916–) published their molecular model for the structure of DNA, the double helix, based on X-ray diffraction photographs. The following decade saw the cracking of the *genetic code. The genetic code is said to be universal since the same code applies to all organisms from bacteria and viruses to higher plants and animals, including humans. Today the deliberate manipulation of genes by biochemical techniques, or *genetic engineering, takes place.

genetic screening in medicine, the determination of the genetic make-up of an individual to determine if he or she is at risk of developing a hereditary disease later in life. Genetic screening can also be used to determine if an individual is a carrier for a particular genetic disease and, hence, can pass the disease on to any children. Genetic counselling should be undertaken at the same time as genetic screening of affected individuals. Diseases that can be screened for include cystic fibrosis, Huntington's chorea, and certain forms of cancer.

genitalia reproductive organs of sexually reproducing animals, particularly the external/visible organs of mammals: in males, the penis and the scrotum, which contains the testes, and in females, the clitoris and vulva.

genome full complement of *genes carried by a single (haploid) set of *chromosomes. The term may be applied to the genetic information carried by an individual or to the range of genes found in a given species. The human genome is made up of between 27,000 and 40,000 genes, according to a rough draft of the sequenced genome completed by the *Human Genome Project in June 2000.

Genomes have been identified for many other organisms, including the bacteria *Haemofilus influenzae* and

Escherichia coli, and the mycoplasmas *Mycoplasma genitalium* and *Mycoplasma pneumoniae*. Scientists also completed a genetic blueprint for *Saccharomyces cerevisiae*, common brewer's yeast, which shares a high number of genetic sequences with humans.

genotype particular set of *alleles (variants of genes) possessed by a given organism. The term is usually used in conjunction with *phenotype, which is the product of the genotype and all environmental effects.

genus plural **genera**, group of one or more *species with many characteristics in common. Thus all doglike species (including dogs, wolves, and jackals) belong to the genus *Canis* (Latin 'dog'). Species of the same genus are thought to be descended from a common ancestor species. Related genera are grouped into *families.

geochemistry science of chemistry as it applies to geology. It deals with the relative and absolute abundances of the chemical elements and their *isotopes in the Earth, and also with the chemical changes that accompany geologic processes.

geochronology branch of geology that deals with the dating of rocks, minerals, and fossils in order to create an accurate and precise geological history of the Earth. The *geological time scale is a result of these studies. It puts stratigraphic units in chronological order and assigns actual dates, in millions of years, to those units.

geode in geology, a subspherical cavity into which crystals have grown from the outer wall into the centre. Geodes often contain very well-formed crystals of quartz (including amethyst), calcite, or other minerals.

geographical information system GIS, computer software that makes possible the visualization and manipulation of spatial data, and links such data with other information such as customer records.

geography study of the Earth's surface; its topography, climate, and physical conditions, and how these factors affect people and society. It is usually divided into **physical geography**, dealing with landforms and climates,

and **human geography**, dealing with the distribution and activities of peoples on Earth.

geological time time scale embracing the history of the Earth from its physical origin to the present day. Geological time is traditionally divided into eons (Archaean or Archaeozoic, Proterozoic, and Phanerozoic in ascending chronological order), which in turn are subdivided into eras, periods, epochs, ages, and finally chrons.

geology science of the Earth, its origin, composition, structure, and history. It is divided into several branches, inlcuding **mineralogy** (the minerals of Earth), **petrology** (rocks), **stratigraphy** (the deposition of successive beds of sedimentary rocks), **palaeontology** (fossils) and **tectonics** (the deformation and movement of the Earth's crust), **geophysics** (using physics to study the Earth's surface, interior, and atmosphere), and **geochemistry** (the science of chemistry as it applies to biology).

geometric mean in mathematics, the nth root of the product of n positive numbers. The geometric mean m of two numbers p and q is such that $m = \div(p \times q)$. For example, the geometric mean of 2 and 8 is $\div(2 \times 8) = \div16 = 4$.

geometric progression or **geometric sequence**, in mathematics, a sequence of terms (progression) in which each term is a constant multiple (called the **common ratio**) of the one preceding it. For example, 3, 12, 48, 192, 768, ... is a geometric progression with a common ratio 4, since each term is equal to the previous term multiplied by 4. Compare *arithmetic progression. The sum of n terms of a **geometric series**

$$1 + r + r^2 + r^3 + ... + r^{n-1}$$

is given by the formula

$$S_n = (1 - r^n)/(1 - r) \text{ for all } r \pi 1.$$

For mod $r < 1$ (see *absolute value), the geometric series can be summed to infinity:

$$S_\infty = 1/(1 - r).$$

In nature, many single-celled organisms reproduce by splitting in

EON	ERA	PERIOD	EPOCH	TIME (my)
PHANEROZOIC	CENOZOIC *Age of mammals*	QUATERNARY *Age of man*	HOLOCENE	0.01
			PLEISTOCENE	1.64
		TERTIARY	PLIOCENE	5.20
			MIOCENE	23.5
			OLIGOCENE	35.5
			EOCENE	56.6
			PALAEOCENE	65.0
	MESOZOIC	CRETACEOUS		146
		JURASSIC *Age of Cycads*		208
		TRIASSIC		245
	PALAEOZOIC	PERMIAN *Age of Amphibians*		290
		CARBONIFEROUS *Age of Coal* *Age of Amphibians*		363
		DEVONIAN *Age of Fishes*		409
		SILURIAN *Age of Fishes*		439
		ORDOVICIAN *Age of Marine Invertebrates*		510
		CAMBRIAN *Age of Marine Invertebrates*		570
PROTEROZOIC				2500
ARCHAEOZOIC				4600

geological timescale The time column shows millions of years ago.

two so that one cell gives rise to 2, then 4, then 8 cells, and so on, forming a geometric sequence 1, 2, 4, 8, 16, 32, ..., in which the common ratio is 2.

geometry branch of mathematics concerned with the properties of space, usually in terms of plane (two-dimensional, or 2D) and solid (three-dimensional, or 3D) figures. The subject is usually divided into **pure geometry**, which embraces roughly the plane and solid geometry dealt with in Greek mathematician Euclid's *Stoicheia/Elements* (*c* 300 BC), and **analytical** or *coordinate geometry, in which problems are solved using algebraic methods. A third, quite distinct, type includes the non-Euclidean geometries.

geomorphology branch of geology developed in the late 19th century, dealing with the morphology, or form, of the Earth's surface; nowadays it is also considered to be an integral part of physical geography. Geomorphological studies investigate the nature and origin of surface landforms, such as mountains, valleys, plains, and plateaux, and the processes that influence them. These processes include the effects of tectonic forces, *weathering, running water, waves, glacial ice, and wind, which result in the *erosion, mass movement (landslides, rockslides, mudslides), transportation, and deposition of *rocks and *soils. In addition to the natural processes that mould landforms, human activity can produce changes, either directly or indirectly, and cause the erosion, transportation, and deposition of rocks and soils, for example by poor land management practices and techniques in farming and forestry, and in the mining and construction industries.

geophysics branch of earth science using physics (for instance gravity, seismicity, and magnetism) to study the Earth's surface, interior, and atmosphere. Geophysics includes several sub-fields such as seismology, paleomagnetism, and remote sensing.

geothermal energy energy extracted for heating and electricity generation from natural steam, hot water, or hot dry rocks in the Earth's crust. It is a form of *renewable energy. Water is pumped down through an injection well where it passes through joints in the hot rocks. It rises to the surface through a recovery well and may be converted to steam or run through a heat exchanger. Steam may be directed through *turbines to produce *electrical energy. It is an important source of energy in volcanically-active areas such as Iceland and New Zealand.

The heat is produced by volcanic activity and, possibly, by the decay of the Earth's natural radioactive substances.

germ colloquial term for a micro-organism that causes disease, such as certain *bacteria and *viruses. Formerly, it was also used to mean something capable of developing into a complete organism (such as a fertilized egg, or the *embryo of a seed).

germanium chemical symbol Ge, brittle, grey-white, weakly metallic (*metalloid) element, atomic number 32, relative atomic mass 72.64. It belongs to the silicon group, and has chemical and physical properties between those of silicon and tin. Germanium is a semiconductor material and is used in the manufacture of transistors and integrated circuits. The oxide is transparent to infrared radiation, and is used in military applications. It was discovered in 1886 by German chemist Clemens Winkler (1838–1904).

In parts of Asia, germanium and plants containing it are used to treat a variety of diseases, and it is sold in the West as a food supplement despite fears that it may cause kidney damage.

German measles or **rubella**, mild, communicable virus disease, usually caught by children. It is marked by a sore throat, pinkish rash, and slight fever, and has an incubation period of two to three weeks. If a woman contracts it in the first three months of pregnancy, it may cause serious damage to the unborn child.

germination in botany, the initial stages of growth in a seed, spore, or pollen grain. Seeds germinate when they are exposed to favourable external conditions of moisture, light, and temperature, and when any factors causing dormancy have been removed.

The process begins with the uptake of water by the seed. The embryonic root, or radicle, is normally the first organ to emerge, followed by the embryonic shoot, or plumule. Food reserves, either within the *endosperm or from the *cotyledons, are broken down to nourish the rapidly growing seedling. Germination is considered to have ended with the production of the first true leaves.

germ-line therapy hypothetical application of *gene therapy to sperm and egg cells to remove the risk of an inherited disease being passed to offspring. It is controversial because of the fear it will be used to produce 'designer babies', and may result in unforseen side effects.

Gestalt (German 'form') concept of a unified whole that is greater than, or different from, the sum of its parts; that is, a complete structure whose nature is not explained simply by analysing its constituent elements. A chair, for example, will generally be recognized as a chair despite great variations between individual chairs in such attributes as size, shape, and colour.

gestation in all mammals except the monotremes (platypus and spiny anteaters), the period from the time of implantation of the embryo in the uterus to birth. This period varies among species; in humans it is about 266 days, in elephants 18–22 months, in cats about 60 days, and in some species of marsupial (such as opossum) as short as 12 days.

geyser natural spring that intermittently discharges an explosive column of steam and hot water into the air due to the build-up of steam in underground chambers. One of the most remarkable geysers is Old Faithful, in Yellowstone National Park, Wyoming, USA. Geysers also occur in New Zealand and Iceland.

g-force force experienced by a pilot or astronaut when the craft in which he or she is travelling accelerates or decelerates rapidly. The unit g denotes the acceleration due to gravity, where 1 g is the ordinary pull of gravity.

Early astronauts were subjected to launch and re-entry forces of up to 6 g or more; in the space shuttle, more than 3 g is experienced on lift-off. Pilots and astronauts wear g-suits that prevent their blood pooling too much under severe g-forces, which can lead to unconsciousness.

giant molecular structure or **giant covalent structure** or **macromolecular structure**, solid structure made up of many similar atoms joined by covalent bonds in one dimension (long chains, such as *polymers), two dimensions (flat sheets as in graphite, where each layer is a giant molecule), or in three dimensions (such as in diamond and silica).

Giant molecules do not conduct electricity as they have no free electrons as solids and no free ions when molten. They are some of the

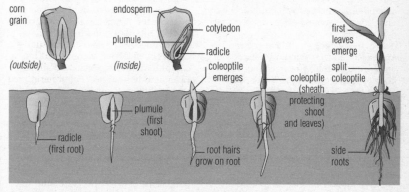

corn grain

endosperm

plumule

(outside)   (inside)

cotyledon

radicle

coleoptile emerges

plumule (first shoot)

radicle (first root)

coleoptile (sheath protecting shoot and leaves)

root hairs grow on root

first leaves emerge

split coleoptile

coleoptile

side roots

germination The germination of a corn grain. The plumule and radicle emerge from the seed coat and begin to grow into a new plant. The coleoptile protects the emerging bud and the first leaves.

hardest structures known because of the strong bonds between their atoms. These bonds also give them high melting and boiling points. *Graphite, an allotrope of carbon, is an anomaly here as free electrons between its layers enable it to conduct electricity.

giant star member of a class of stars located at the top right – of the Hertzsprung-Russell diagram, chacterized by great size and luminosity. Giants have exhausted their supply of hydrogen fuel and derive their energy from the fusion of helium and heavier elements. They are roughly 10–300 times bigger than the Sun with 30–1,000 times the luminosity. The cooler giants are known as *red giants.

gibberellin plant growth substance (see also *auxin) that promotes stem growth and may also affect the breaking of dormancy in certain buds and seeds, and the induction of flowering. Application of gibberellin can stimulate the stems of dwarf plants to additional growth, delay the ageing process in leaves, and promote the production of seedless fruit (parthenocarpy).

Gibbs' function in *thermodynamics, an expression representing part of the energy content of a system that is available to do external work, also known as the free energy G. In an equilibrium system at constant temperature and pressure, $G = H - TS$, where H is the enthalpy (heat content), T the temperature, and S the *entropy (decrease in energy availability). The function was named after US physicist Josiah Willard Gibbs.

Gibbs's free energy or **free energy**, in chemistry, a state function of a system in chemical equilibrium. Changes in the value of free energy can be used to determine if a reaction is thermodynamically favourable. See *Gibbs' function.

GIF acronym for Graphics Interchange Format, in computing, popular and economical picture file format developed by CompuServe. GIF (pronounced with a hard 'g') is one of the two most commonly used file formats for pictures on the World Wide Web (the other is *JPEG) because

pictures saved in this format take up a relatively small amount of space. The term is often used simply to mean 'pictures'. However, GIF uses a data compression system called *LZW compression, for which Unisys holds the copyright, and this has prompted moves to replace it with a different format such as *Portable Network Graphics (PNG).

giga- prefix signifying multiplication by 10^9 (1,000,000,000 or 1 billion), as in gigahertz, a unit of frequency equivalent to 1 billion hertz.

gigabyte in computing, a measure of *memory capacity, equal to 1,024 *megabytes. It is also used, less precisely, to mean 1,000 billion *bytes.

gill imperial unit of volume for liquid measure, equal to one-quarter of a pint or five fluid ounces (0.142 litre), traditionally used in selling alcoholic drinks.

Gilles de la Tourette syndrome in medicine, another name for *Tourette syndrome.

ginger southeast Asian reedlike perennial plant; the hot-tasting spicy underground root is used as a food flavouring and in preserves. (*Zingiber officinale*, family Zingiberaceae.)

gingivitis infection of the gums causing inflammation and bleeding. The gums retract from the teeth, which loosen and eventually fall out. It is caused by bacteria in the mouth.

GIS in computing, abbreviation for *geographical information system.

glacial deposition laying-down (*deposition) of *sediment once carried by a glacier. When ice melts, it deposits the material that it has been carrying. The material deposited by a glacier is called till, or in Britain **boulder clay**. It comprises angular particles of all sizes from boulders to clay that are unsorted and lacking in stratification (layering).

Unstratified till can be moulded by ice to form drumlins (egg-shaped hills). At the snout of the glacier, till piles up to form a ridge called a terminal *moraine. Glacial deposits occur in many different locations – beneath the ice (subglacial), inside it (englacial), on top of it (supraglacial), at the side of it (marginal), and in front of it (proglacial).

Stratified till that has been deposited by meltwater is termed **fluvioglacial**, because it is essentially deposited by running water. Meltwater flowing away from a glacier will carry some of the till many kilometres away. This sediment will become rounded (by the water) and, when deposited, will form a gently sloping area called an **outwash plain**. Several landforms owe their existence to meltwater (**fluvioglacial landforms**) and include the long ridges called eskers, which form parallel to the direction of the ice flow. Meltwater may fill depressions eroded by the ice to form **ribbon lakes**. Small depositional landforms may also result from glacial deposition, such as **kames** (small mounds) and **kettle holes** (small depressions, often filled with water).

In *periglacial environments on the margins of an icesheet, freeze–thaw weathering (the alternate freezing and thawing of ice in cracks in the rock) etches the outlines of rock outcrops, exploiting joints and areas of weakness, and results in aprons of scree.

glacial erosion   wearing-down and removal of rocks and soil by a *glacier. Glacial erosion forms impressive landscape features, including *glacial troughs (U-shaped valleys), arêtes (steep ridges), corries (enlarged hollows), and pyramidal peaks (high mountain peaks with three or more arêtes).

Erosional landforms result from abrasion and plucking of the underlying bedrock. Abrasion is caused by the rock debris carried by a glacier, wearing away the bedrock. The action is similar to that of sandpaper attached to a block of wood. The results include the polishing and scratching of rock surfaces to form powdered rock flour, and scratches or *striations which indicate the direction of ice movement. Plucking is a form of glacial erosion restricted to the lifting and removal of blocks of bedrock already loosened by *freeze–thaw activity.

The most extensive period of recent glacial erosion was the *Pleistocene epoch (1.6 million to 10,000 years ago) in the *Quaternary period (last 2

million years) when, over a period of 2 to 3 million years, the polar icecaps repeatedly advanced and retreated. More ancient glacial episodes are also preserved in the geological record, the earliest being in the middle *Precambrian era (4.6 billion to 570 million years ago) and the most extensive in Permo-Carboniferous times.

glacial trough   or **U-shaped valley**, steep-sided, flat-bottomed valley formed by a glacier. The erosive action of the glacier and of the debris carried by it results in the formation not only of the trough itself but also of a number of associated features, such as hanging valleys (smaller glacial valleys that enter the trough at a higher level than the trough floor). Features characteristic of glacial deposition, such as drumlins, are commonly found on the floor of the trough, together with long lakes called ribbon lakes.

glacier   body of ice, originating in mountains in snowfields above the snowline, that moves slowly downhill and is constantly built up from its source. The geographic features produced by the erosive action of glaciers (*erosion) are characteristic and include *glacial troughs (U-shaped valleys), corries, and arêtes. In lowlands, the laying down of debris carried by glaciers (*glacial deposition) produces a variety of landscape features, such as *moraines and drumlins.

Glaciers form where annual snowfall exceeds annual melting and drainage (see *glacier budget). The area at the top of the glacier is called the zone of **accumulation**. The lower area of the glacier is called the *ablation zone. In the zone of accumulation, the snow compacts to ice under the weight of the layers above and moves downhill under the force of gravity. As the ice moves, it changes its shape and structure. Partial melting of ice at the base of the glacier also produces sliding of the glacier, as the ice travels over the bedrock. In the ablation zone, melting occurs and glacial till is deposited.

glacier budget   in a glacier, the balance between *accumulation (the addition of snow and ice to the glacier) and *ablation (the loss of snow and ice by melting and evaporation). If accumulation exceeds ablation the glacier will advance; if ablation exceeds accumulation it will probably retreat.

gland specialized organ of the body that manufactures and secretes enzymes, hormones, or other chemicals. In animals, glands vary in size from small (for example, tear glands) to large (for example, the pancreas), but in plants they are always small, and may consist of a single cell. Some glands discharge their products internally, *endocrine glands, and others externally, *exocrine glands. Lymph nodes are sometimes wrongly called glands.

glandular fever or **infectious mononucleosis**, viral disease characterized at onset by fever and painfully swollen lymph nodes (in the neck); there may also be digestive upset, sore throat, and skin rashes. Lassitude persists for months and even years, and recovery can be slow. It is caused by the Epstein–Barr virus.

glass transparent or translucent substance that is physically neither a solid nor a liquid. Although glass is easily shattered, it is one of the strongest substances known. It is made by fusing certain types of sand (silica); this fusion occurs naturally in volcanic glass.

In the industrial production of common types of glass, the type of sand used, the particular chemicals added to it (for example, lead, potassium, barium), and refinements of technique determine the type of glass produced. Types of glass include: soda glass; flint glass, used in cut-crystal ware; optical glass; stained glass; heat-resistant glass; and glasses that exclude certain ranges of the light spectrum. Blown glass is either blown individually from molten glass (using a tube up to 1.5 m/4.5 ft long), as in the making of expensive crafted glass, or blown automatically into a mould – for example, in the manufacture of light bulbs and bottles; pressed glass is simply pressed into moulds, for jam jars, cheap vases, and light fittings;

while sheet glass, for windows, is made by putting the molten glass through rollers to form a 'ribbon', or by floating molten glass on molten tin in the 'float glass' process; fibreglass is made from fine glass fibres. Metallic glass is produced by treating alloys so that they take on the properties of glass while retaining the malleability and conductivity characteristic of metals.

Glauber's salt crystalline sodium sulphate decahydrate $Na_2SO_4.10H_2O$, produced by the action of sulphuric acid on common salt. It melts at $31°C/87.8°F$; the latent heat stored as it solidifies makes it a convenient thermal energy store. It is used in medicine as a laxative.

glaucoma condition in which pressure inside the eye (intraocular pressure) is raised abnormally as excess fluid accumulates. It occurs when the normal outflow of fluid within the chamber of the eye (aqueous humour) is interrupted. As pressure rises, the optic nerve suffers irreversible damage, leading to a reduction in the field of vision and, ultimately, loss of eyesight.

glia or **neuroglia**, connective tissue of the *central nervous system, composed of cells that 'service' the neurons (*nerve cells) with supportive and nutritive activities. Glial cells are far more numerous than neurons.

globalization process by which different parts of the globe become interconnected by economic, social, cultural, and political means. Globalization has become increasingly rapid since the 1970s and 1980s as a result of developments in technology, communications, and trade liberalization. Critics of globalization fear the increasing power of unelected multinational corporations, financial markets, and non-government organizations (NGOs), whose decisions can have direct and rapid effects on ordinary citizens' lives. This has led to growing antiglobalization and anticapitalist protests in the 1990s and early 21st century, which have disrupted international trade talks and meetings of international finance ministers. Supporters of globalization point to the economic benefits of growing international trade and specialization.

global variable in computing, a *variable that can be accessed by any program instruction. See also *local variable.

global warming increase in average global temperature of approximately 0.5°C/0.9°F over the past century. Much of this is thought to be related to human activity. Global temperature has been highly variable in Earth history and many fluctuations in global temperature have occurred in historical times, but this most recent episode of warming coincides with the spread of industrialization, prompting the suggestion that it is the result of an accelerated *greenhouse effect caused by atmospheric pollutants, especially *carbon dioxide gas. The melting and collapse of the Larsen Ice Shelf, Antarctica, since 1995, is a consequence of global warming. Melting of ice is expected to raise the sea level in the coming decades.

Natural, perhaps chaotic, climatic variations have not been ruled out as the cause of the current global rise in temperature, and scientists are still assessing the influence of anthropogenic (human-made) pollutants. In 1988 the World Meteorological Organization (WMO) and the United Nations (UN) set up the Intergovernmental Panel on Climate Change (IPCC), a body of more than 2,000 scientists, to investigate the causes of and issue predictions regarding climate change. In June 1996 the IPCC confirmed that global warming was taking place and that human activities were probably to blame.

Assessing the impact of humankind on the global climate is complicated by the natural variability on both geological and human time scales. The present episode of global warming has thus far still left England approximately 1°C/1.8°F cooler than during the peak of the so-called Medieval Warm Period (1000 to 1400 AD). The latter was part of a purely natural climatic fluctuation on a global scale. The interval between this period

and the recent rise in temperatures was unusually cold throughout the world, relative to historical temperatures. Scientists predict that a doubling of carbon dioxide concentrations, expected before the end of the 21st century, will increase the average global temperature by 1.4–5.8°C/2.5–10.4°F.

In addition to a rise in average global temperature, global warming has caused seasonal variations to be more pronounced in recent decades. Examples are the most severe winter on record in the eastern USA 1976–77, and the record heat waves in the Netherlands and Denmark the following year. Mountain glaciers have shrunk, late summer Arctic sea-ice has thinned by 40%, and sea levels have risen by 10–20 cm/4–8 in. Scientists have predicted a greater number of extreme weather events and sea levels are expected to rise by 9–88 cm/4–35 in by 2100. 1998 was the warmest year globally of the last millennium, according to US researchers who used tree rings and ice cores to determine temperatures over the past 1,000 years.

A 1995 UN summit in Berlin, Germany, agreed to take action to reduce gas emissions harmful to the environment. Delegates at the summit, from more than 120 countries, approved a two-year negotiating process aimed at setting specific targets and timetables for reducing nations' emissions of carbon dioxide and other greenhouse gases after the year 2000. The Kyoto Protocol of 1997 committed the world's industrialized countries to cutting their annual emissions of harmful gases. However, in June 2001 US president George W Bush announced that the USA would not ratify the protocol.

glucagon in biology, a hormone secreted by the *pancreas, which increases the concentration of glucose in the blood by promoting the breakdown of glycogen in the liver. Secretion occurs in response to a lowering of blood glucose concentrations.

glucose or **dextrose** or **grape sugar**; $C_6H_{12}O_6$, simple sugar present in the blood and manufactured by green plants during *photosynthesis. It belongs to the group of chemicals known as *carbohydrates. The *respiration reactions inside cells involves the oxidation of glucose to produce *ATP, the 'energy molecule' used to drive many of the body's biochemical reactions.

As well as being used in respiration to release useful *energy glucose may also be transported to specific parts of the body for other uses. The liver and muscles can turn it into glycogen, which acts as an energy store. It can also be turned into *fat and stored in fatty tissue. In humans and other vertebrates optimum blood glucose levels are maintained by the *hormone *insulin. An inability to control glucose levels may be caused by the disease *diabetes.

Green plants make glucose during photosynthesis. They can turn it into starch, which acts as an energy store and this starch may form an important part of the human *diet.

Glucose is prepared in syrup form by the hydrolysis of cane sugar or starch, and may be purified to a white crystalline powder. Glucose is a monosaccharide sugar (made up of a single sugar unit), unlike the more familiar sucrose (cane or beet sugar), which is a disaccharide (made up of two sugar units: glucose and fructose).

glucose tolerance test in medicine, a method of assessing the efficiency of the body to metabolize glucose, which is used in the diagnosis of *diabetes. It involves the starvation of the subject for several hours before giving a standard amount of glucose by mouth. Concentrations of glucose are measured in the blood and the urine over several hours.

gluon in physics, a *gauge boson that carries the *strong nuclear force, responsible for binding quarks together to form the strongly interacting subatomic particles known as *hadrons. There are eight kinds of gluon.

gluten protein found in cereal grains, especially wheat and rye. Gluten enables dough to expand during rising. Sensitivity to gliadin, a type of gluten, gives rise to coeliac disease.

glycerine another name for *glycerol.

glycerol or **glycerine** or **propan-1,2,3-triol**; $HOCH_2CH(OH)CH_2OH$, thick, colourless, odourless, sweetish liquid. It is obtained from vegetable and animal oils and fats (by treatment with acid, alkali, superheated steam, or an enzyme), or by fermentation of glucose, and is used in the manufacture of high explosives, in antifreeze solutions, to maintain moist conditions in fruits and tobacco, and in cosmetics.

glycine or **aminoethanoic acid**; $CH_2(NH_2)COOH$, simplest amino acid, and one of the main components of proteins. When purified, it is a sweet, colourless crystalline compound. Glycine was found in space in 1994 in the star-forming region Sagittarius B2. The discovery is important because of its bearing on the origins of life on Earth.

glycogen polymer (a polysaccharide) of the sugar *glucose made and retained in the liver as a carbohydrate store, for which reason it is sometimes called animal starch. It is a source of energy when needed by muscles, where it is converted back into glucose by the hormone *insulin and metabolized.

glycol or **ethylene glycol** or **ethane-1,2-diol**; $HOCH_2CH_2OH$, thick, colourless, odourless, sweetish liquid. It is used in antifreeze solutions, in the preparation of ethers and esters (used for explosives), as a solvent, and as a substitute for glycerol.

glycolysis conversion of glucose to *lactic acid. It takes place in the *cytoplasm of cells as part of the process of cellular respiration.

glycoside in biology, compound containing a sugar and a non-sugar unit. Many glycosides occur naturally, for example, *digitalis is a preparation of dried and powdered foxglove leaves that contains a mixture of cardiac glycosides. One of its constituents, digoxin, is used in the treatment of congestive heart failure and cardiac arrhythmias.

GM synthesizer in computing, synthesizer standard; see *general MIDI.

GMT abbreviation for *Greenwich Mean Time.

GNU in computing, suite of free Unix-like software distributed by the Free

Software Foundation. The software includes operating systems, compilers, text editors (such as EMACS), and other useful utilities.

goblet cell in biology, cup-shaped cell present in the epithelium of the respiratory and gastrointestinal tracts. Goblet cells secrete mucin, the main constituent of mucous, which lubricates the mucous membranes of these tracts.

goitre enlargement of the thyroid gland seen as a swelling on the neck. It is most pronounced in simple goitre, which is caused by iodine deficiency. More common is toxic goitre or hyperthyroidism, caused by overactivity of the thyroid gland.

gold chemical symbol Au, heavy, precious, yellow, metallic element, atomic number 79, relative atomic mass 196.97. Its symbol comes from the Latin *aurum* meaning 'gold'. It occurs in nature frequently as a free metal (see *native metal) and is highly resistant to acids, tarnishing, and corrosion. Pure gold is the most malleable of all metals and is used as gold leaf or powder, where small amounts cover vast surfaces, such as gilded domes and statues.

The elemental form is so soft that it is alloyed for strength with a number of other metals, such as silver, copper, and platinum. Its purity is then measured in *carats on a scale of 24 (24 carats is pure gold). It is used mainly for decorative purposes (jewellery, gilding) but also for coinage, dentistry, and conductivity in electronic devices. Gold has been known and worked from ancient times, and currency systems were based on it in Western civilization, where mining it became an economic and imperialistic goal. In 1990 the three leading gold-producing countries were South Africa, 605.4 tonnes; USA, 295 tonnes; and Russia, 260 tonnes. In 1989 gold deposits were found in Greenland with an estimated yield of 12 tonnes per year.

gold rush large influx of gold prospectors to an area where gold deposits have recently been discovered. The result is a dramatic increase in population. Cities such as Johannesburg, Melbourne, and San Francisco either originated or were considerably enlarged by gold rushes. Melbourne's population trebled from 77,000 to some 200,000 between 1851 and 1853, while San Francisco boomed from a small coastal village of a few hundred people to the largest city in the western USA during the California gold rush of 1848–56.

Golgi apparatus or **Golgi body**, stack of flattened membranous sacs found in the cells of *eukaryotes. Many molecules travel through the Golgi apparatus on their way to other organelles or to the endoplasmic reticulum. Some are modified or assembled inside the sacs. The Golgi apparatus is named after the Italian cell biologist Camillo Golgi (1843–1926).

It produces the membranes that surround the cell vesicles or *lysosomes.

gonad the part of an animal's body that produces the sperm or egg cells (ova) required for sexual reproduction. The sperm-producing gonad is called a *testis, and the egg-producing gonad is called an *ovary.

gonadotrophin any hormone that supports and stimulates the function of the gonads (sex glands); some gonadotrophins are used as fertility drugs.

gonorrhoea common sexually transmitted disease arising from infection with the bacterium *Neisseria gonorrhoeae*, which causes inflammation of the genito-urinary tract. After an incubation period of two to ten days, infected men experience pain while urinating and a discharge from the penis; infected women often have no external symptoms.

gorge narrow steep-sided valley or canyon that may or may not have a river at the bottom. A gorge may be formed as a *waterfall retreats upstream, eroding away the rock at the base of a river valley; or it may be caused by rejuvenation, when a river begins to cut downwards into its channel for some reason – for example, in response to a fall in sea level. Gorges are common in limestone country, where they may be formed by the collapse of the roofs of underground caverns.

gout hereditary form of *arthritis, marked by an excess of uric acid crystals in the tissues, causing pain and inflammation in one or more joints (usually of the feet or hands). Acute attacks are treated with anti-inflammatories.

GPF abbreviation for general protection fault, in Windows 3.1, error message returned by a computer when it crashes. A GPF is the same as a UAE (unexpected application error) in Windows 3.0. It often indicates that one application has tried to use memory reserved for another.

GPRS abbreviation for general packet radio service, implementation of *packet switching within *GSM. GPRS will bring Internet Protocol (IP) connectivity to the GSM network, and is sometimes referred to as $2^1/_2$G, in contrast to *UMTS, which is the third-generation standard for mobile cellular networks.

gradient on a graph, the slope of a straight or curved line. The slope of a curve at any given point is represented by the slope of the *tangent at that point.

gram symbol g, metric unit of mass; one-thousandth of a kilogram.

grand unified theory (GUT) in physics, sought-for theory that would combine the theory of the strong nuclear force (called *quantum chromodynamics) with the theory of the weak nuclear and electromagnetic forces (see *forces, fundamental). The search for the grand unified theory is part of a larger programme seeking a *unified field theory, which would combine all the forces of nature (including gravity) within one framework.

granite coarse-grained intrusive *igneous rock, typically consisting of the minerals quartz, feldspar, and biotite mica. It may be pink or grey, depending on the composition of the feldspar. Granites are chiefly used as building materials.

Granite is formed when magma (molten rock) is forced between other rocks in the Earth's crust. It cools and crystallizes deep underground. As it cools slowly large crystals are formed. Granites often form large intrusions in the core of mountain ranges, and they are usually surrounded by zones of *metamorphic rock (rock that has been altered by heat or pressure). Granite areas have characteristic moorland scenery. In exposed areas the bedrock may be weathered along joints and cracks to produce a tor, consisting of rounded blocks that appear to have been stacked upon one another.

graph pictorial representation of numerical data, such as statistical data, or a method of showing the mathematical relationship between two or more variables by drawing a diagram.

graphical user interface GUI, in computing, a type of *user interface in which programs and files appear as icons (small pictures), user options are selected from pull-down menus, and data are displayed in windows

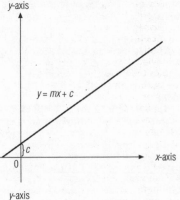

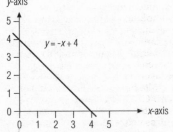

graph The equation of the straight-line graph takes the form $y = mx + c$, where m is the gradient (slope) of the line, and c is the y-intercept (the value of y where the line cuts the y-axis) for example, a graph of the equation $-y = x + 4$ will have a gradient of -1 and will cut the y-axis at $y = 4$.

(rectangular areas), which the operator can manipulate in various ways. The operator uses a pointing device, typically a *mouse, to make selections and initiate actions. It is also known as **WIMP** (Windows, Icons, Menus, Pointing device).

graphic file format format in which computer graphics are stored and transmitted. There are two main types: *raster graphics in which the image is stored as a *bit map (arrangement of dots), and *vector graphics, in which the image is stored using geometric formulae. There are many different file formats, some of which are used by specific computers, operating systems, or applications. Some formats use file compression, particularly those that are able to handle more than one colour.

graphics card in computing, a peripheral device that processes and displays graphics.

Graphics Interchange Format in computing, picture file format usually abbreviated to *GIF.

graphics tablet or **bit pad**, in computing, an input device in which a stylus or cursor is moved, by hand, over a flat surface. The computer can keep track of the position of the stylus, so enabling the operator to input drawings or diagrams into the computer.

graphite blackish-grey, soft, flaky, crystalline form of *carbon. It is used as a lubricant and as the active component of pencil lead.

Graphite, like *diamond and fullerene, is an allotrope of carbon. The carbon atoms are strongly bonded together in sheets, but the bonds between the sheets are weak, allowing the layers to slide over one another. Graphite has a very high melting point (3,500°C/6,332°F), and is a good conductor of heat and electricity. It absorbs neutrons and is therefore used to moderate the chain reaction in nuclear reactors.

gravimetric analysis in chemistry, a technique for determining, by weighing, the amount of a particular substance present in a sample. It usually involves the conversion of the test substance into a compound of known molecular weight that can be easily isolated and purified.

gravitational field region around a body in which other bodies experience a force due to its gravitational attraction. The gravitational field of a massive object such as the Earth is very strong and easily recognized as the force of gravity, whereas that of an object of much smaller mass is very weak and difficult to detect. Gravitational fields produce only attractive forces.

gravitational field strength symbol g, strength of the Earth's gravitational field at a particular point. It is defined as the gravitational force in newtons that acts on a mass of one kilogram. The value of g on the Earth's surface is taken to be 9.806 N kg^{-1}.

The symbol g is also used to represent the acceleration of a freely falling object in the Earth's gravitational field. Near the Earth's surface and in the absence of friction due to the air, all objects fall with an acceleration of 9.8 m s^{-2}/32 ft s^{-2}.

gravitational force or **gravity**, one of the four fundamental *forces of nature, the other three being the electromagnetic force, the weak nuclear force, and the strong nuclear force. The gravitational force is the weakest of the four forces, but acts over great distances. The particle that is postulated as the carrier of the gravitational force is the *graviton.

gravitational potential energy energy possessed by an object when it is placed in a position from which, if it were free to do so, it would fall under the influence of *gravity. If the object is free to fall, then the gravitational potential energy is converted into kinetic (motion) energy. The gravitational potential energy E_p of an object of mass m kilograms placed at a height h metres above the ground is given by the formula:

$$E_p = mgh$$

where g is the gravitational field strength (in newtons per kilogram) of the Earth at that place. See also *potential energy.

In a *hydroelectric power station, gravitational potential energy of water held in a high-level reservoir is used to drive turbines to produce electricity.

graviton in physics, the *gauge boson that is the postulated carrier of the gravitational force.

gravity force of attraction that arises between objects by virtue of their masses. The larger the mass of an object the more strongly it attracts other objects. On Earth, gravity causes objects to have weight; it accelerates objects (at 9.806 metres per second per second/32.174 ft per second per second) towards the centre of the Earth, the ground preventing them falling further.

The Earth's gravity also attracts the *Moon towards the Earth, keeping the Moon in orbit around the Earth. The Moon's gravity is one-sixth that of Earth's, so objects on the Moon weigh less than on Earth. The Sun contains 99.8% of the mass of the Solar System, and the resulting large force of gravity keeps the planets of the Solar System in orbit around the Sun.

A spacecraft launched from Earth must overcome the force of gravity before entering space. This is achieved by using rocket boosters at various stages of the launch. The spacecraft needs an acceleration of three times that of gravity (3 g). The speed required to escape the Earth's gravitational field is 40,000 kph/25,000 mph.

green audit inspection of a company to assess the total environmental impact of its activities or of a particular product or process.

greenhouse effect phenomenon of the Earth's atmosphere by which solar radiation, trapped by the Earth and re-emitted from the surface as long-wave infrared radiation, is prevented from escaping by various gases (the 'greenhouse gases') in the air. These gases trap heat because they readily absorb infrared radiation. As the energy cannot escape, it warms up the Earth, causing an increase in the Earth's temperature (*global warming). The main greenhouse gases are *carbon dioxide, *methane, and *chlorofluorocarbons (CFCs) as well as water vapour. Fossil-fuel consumption and forest fires are the principal causes

of carbon dioxide build-up; methane is a by-product of agriculture (rice, cattle, sheep).

The United Nations Environment Programme estimates that by 2025, average world temperatures will have risen by 1.5°C/2.7°F with a consequent rise of 20 cm/7.9 in in sea level. Low-lying areas and entire countries would be threatened by flooding and crops would be affected by the change in climate. However, predictions about global warming and its possible climatic effects are tentative and often conflict with each other.

At the 1992 Earth Summit it was agreed that by 2000 countries would stabilize carbon dioxide emissions at 1990 levels, but to halt the acceleration of global warming, emissions would probably need to be cut by 60%. Any increases in carbon dioxide emissions are expected to come from transport. The Berlin Mandate, agreed unanimously at the climate conference in Berlin in 1995, committed industrial nations to the continuing reduction of greenhouse gas emissions after 2000, when the existing pact to stabilize emissions runs out. The stabilization of carbon dioxide emissions at 1990 levels by 2000 would not be achieved by a number of developed countries, including Spain, Australia, and the USA, according to 1997 estimates. Australia is in favour of different targets for different nations, and refused to sign a communiqué at the South Pacific Forum meeting in the Cook Islands in 1997 which insisted on legally-binding reductions in greenhouse gas emissions. The United Nations Framework Convention on Climate Change (UNFCCC) adopted the Kyoto Protocol in 1997, committing the world's industrialized countries to cutting their annual emissions of harmful gases. By July 2001 the Protocol had been signed by 84 parties and ratified by 37; the USA announced its refusal to ratify the Protocol in June 2001.

Dubbed the 'greenhouse effect' by Swedish scientist Svante Arrhenius (1859–1927), it was first predicted in 1827 by French mathematician Joseph Fourier.

green movement collective term for the individuals and organizations involved in efforts to protect the environment. The movement includes political parties such as the Green Party and organizations like Friends of the Earth and Greenpeace. See also *environmental issues.

Greenwich Mean Time GMT, local time on the zero line of longitude (the **Greenwich meridian**), which passes through the Old Royal Observatory at Greenwich, London. It was replaced in 1986 by coordinated universal time (UTC), but continued to be used to measure longitudes and the world's standard time zones.

grep acronym for Global Regular Expression Print, in computing, Unix command that allows full-text searching within files. On the Net, grep is sometimes used as an all-purpose synonym for 'search'.

grey scales method of representing continuous tone images on a screen or printer. Each dot in the *bit map is represented by a number of bits and can have a different shade of grey. Compare with dithering when shades are simulated by altering the density and the pattern of black dots on a white background.

grid network of crossing parallel lines. **Rectangular grids** are used for drawing graphs. **Isometric grids** are used for drawing representations of solids in two dimensions in which lengths in the drawing match the lengths of the object.

grid reference numbering system used to specify a location on a map. The numbers representing grid lines at the bottom of the map (eastings) are given before those at the side (northings). Successive decimal digits refine the location within the grid system.

groundwater water present underground in porous rock strata and soils; it emerges at the surface as springs and streams. The groundwater's upper level is called the *water table. Rock strata that are filled with groundwater that can be extracted are called **aquifers**. Aquifers must be both porous (filled with holes) and permeable (full of holes that are interconnected so that the water is able to flow).

group in chemistry, a vertical column of elements in the *periodic table. Elements in a group have similar physical and chemical properties; for example, the group I elements (the *alkali metals: lithium, sodium, potassium, rubidium, caesium, and francium) are all highly reactive metals that form univalent ions. There is a gradation of properties down any group: in group I, melting and boiling points decrease, and density and reactivity increase.
Group 0 consists of the *noble gases (rare gases) and group II consists of the *alkaline-earth metals. Those elements placed between group II and III are the *transition metals and group VII contains the *halogens.

group in mathematics, a finite or infinite set of elements that can be combined by an operation; formally, a group must satisfy certain conditions. For example, the set of all integers (positive or negative whole numbers) forms a group with regard to addition because: (1) addition is associative, that is, the sum of two or more integers is the same regardless of the order in which the integers are added; (2) adding two integers gives another integer; (3) the set includes an identity element 0, which has no effect on any integer to which it is added (for example, $0 + 3 = 3$); and (4) each integer has an inverse (for instance, 7 has the inverse -7), such that the sum of an integer and its inverse is 0.
Group theory is the study of the properties of groups.

group psychology area of *social psychology dealing with the formation and cohesion of groups, competition and conflict among group members and between groups, the influence of the group on the individual and vice versa, communication within the group and between groups, and many other aspects of social interaction.

growth in biology, increase in size and mass during the development of an organism over a period of time. Growth is often measured as an increase in *biomass (mass of organic material, excluding water) and is associated with cell division by *mitosis, subsequent increases in cell size, and with the differentiation of cells to perform specific functions, for example red blood cells in mammals and root cells in plants. All organisms grow, although the rate of growth varies over a lifetime. Typically, growth in an organism follows an S-shaped curve, in which growth is at first slow, then fast, then, towards the end of life, non-existent. Growth may even be negative during the period before death, with decay occurring faster than cellular replacement. In humans, there is a short period of rapid growth at puberty. Growth is affected by genetic factors, which dictate the eventual size and appearance of an organism. It is dependent upon an adequate supply of water and *mineral salts (*diet), and, particularly in plants, appropriate conditions of light and temperature, which determine the rate of *photosynthesis.

growth and decay curve graph showing exponential change (growth where the increment itself grows at the same rate) as occurs with compound interest and populations.

growth hormone GH; or **somatotrophin**, hormone from the anterior *pituitary gland that promotes growth of long bones and increases protein synthesis. If it is lacking in childhood, dwarfism results; excess GH causes gigantism. Levels of GH are high in the fetus and decline throughout early childhood; there is another spurt at adolescence and then production declines throughout adulthood. After the age of about 60 only tiny amounts of GH are produced, and this 'deficiency' is responsible for some of the symptoms of aging. Growth hormone release is controlled by growth hormone releasing hormone (GHRH) and the peptide somatostatin, both produced in the hypothalamus.

growth ring another name for *annual ring.

g-scale scale for measuring force by comparing it with the force due to *gravity (g), often called *g-force.

GSM originally an abbreviation for Group Système Mobile (French) but now said to stand for Global System for Mobile (communications).

guanine in biochemistry, one of the nucleotide bases. It is part of the genetic code of DNA and RNA, where it pairs with *cytosine. Guanine is a component of molecules of guanosine and is also found in chromatophores.

guard cell in plants, a specialized cell on the undersurface of leaves for controlling gas exchange and water loss. Guard cells occur in pairs and are shaped so that a pore, or *stomata, exists between them. They can change shape with the result that the pore disappears. During warm weather, when a plant is in danger of losing excessive water, the guard cells close, cutting down evaporation from the interior of the leaf.

GUI in computing, abbreviation for *graphical user interface.

guiltware or **nagware**, in computing, variety of *shareware software that attempts to make the user register (and pay for) the software by exploiting the user's sense of guilt.

gulf any large sea inlet.

Gulf Stream warm ocean current that flows north from the warm waters of the Gulf of Mexico along the east coast of America, from which it is separated by a channel of cold water originating in the southerly Labrador current. Off Newfoundland, part of the current is diverted east across the Atlantic, where it is known as the **North Atlantic Drift**, dividing to flow north and south, and warming what would otherwise be a colder climate in the British Isles and northwest Europe.

Gulf War Syndrome mystery illness suffered by soldiers who fought in the 1991 Gulf War. Symptoms include headaches, memory loss, listlessness, depression, respiratory problems, lethargy, muscle weakness, nausea, and pain. It may be a form of *shell shock, or the symptoms could have been caused by a combination of vaccinations (to tropical diseases and diseases likely to be used in biological weapons), nerve gas, anti-nerve gas drugs, and organophosphate (OP) insecticides. In addition, troops were also exposed to Iraqi chemical weapons and smoke from burning oil wells.

gullet　another name for the
*oesophagus.

gully　long, narrow, steep-sided valley
with a flat floor. Gullies are formed by
water erosion and are more common in
unconsolidated rock and soils that are
easily eroded. They may be formed
very rapidly during periods of heavy
rainfall and are common in arid areas
that have periods of heavy rain. Gully
formation in more temperate areas is
common where the vegetation cover
has been destroyed or reduced, for
example as a result of fire or
agricultural clearance.

gum　in mammals, the soft tissues
surrounding the base of the teeth.
Gums are liable to inflammation
(gingivitis) or to infection by microbes
from food deposits (periodontal
disease).

gunpowder　or **black powder**, oldest
known *explosive, a mixture of 75%
potassium nitrate (saltpetre), 15%
charcoal, and 10% sulphur. Sulphur
ignites at a low temperature, charcoal
burns readily, and the potassium
nitrate provides oxygen for the
explosion. As gunpowder produces
lots of smoke and burns quite slowly,
it has progressively been replaced
since the late 19th century by high
explosives, although it is still widely
used for quarry blasting, fuses, and
fireworks. Gunpowder has high
*activation energy; a gun based on
gunpowder alone requires igniting by
a flint or a match.

　It was probably first invented in
China, where it was chiefly used for
fireworks. It is possible that knowledge
of it was transmitted from the Middle
East to Europe. The writings of the
English monk Roger Bacon show that
he was experimenting with
gunpowder in 1249. His mixture
contained saltpetre, charcoal, and
sulphur. The development of effective
gunpowder was essential for the
growing significance of cannons and
handguns in the late medieval period.
The Arabs produced the first known
working gun, in 1304. Gunpowder was
used in warfare from the 14th century
but it was not generally adapted to
civil purposes until the 17th century,
when it began to be used in mining.

gut　or **alimentary canal**, in the
*digestive system, the part of an
animal responsible for processing food
and preparing it for entry into the
blood.

　The gut consists of a tube divided
into segments specialized to perform
different functions. The front end (the
mouth) is adapted for food intake and
for the first stages of digestion. The
stomach is a storage area, although
digestion of protein by the enzyme
pepsin starts here; in many
herbivorous mammals this is also the
site of cellulose digestion. The small
intestine follows the stomach and is
specialized for digestion and for
absorption. The large intestine,
consisting of the colon, caecum, and
rectum, has a variety of functions,
including cellulose digestion, water
absorption, and storage of faeces. From
the gut nutrients are carried to the
liver via the hepatic portal vein, ready
for assimilation by the cells.

GVHD　abbreviation for graft-versus-
host disease.

Gy　symbol for gray, the SI unit of
absorbed ionizing radiation dose,
equal to an absorption of 1 joule per
kilogram of irradiated matter.

gymnosperm　(Greek 'naked seed') in
botany, any plant whose seeds are
exposed, as opposed to the structurally
more advanced *angiosperms, where
they are inside an ovary. The group
includes conifers and related plants
such as cycads and ginkgos, whose
seeds develop in cones. Fossil
gymnosperms have been found in
rocks about 350 million years old.

gynaecology　medical speciality
concerned with disorders of the female
reproductive system.

gynoecium　or **gynaecium**, collective
term for the female reproductive
organs of a flower, consisting of one or
more *carpels, either free or fused
together.

gypsum　common sulphate *mineral,
composed of hydrous calcium
sulphate, $CaSO_4.2H_2O$. It ranks 2 on
the *Mohs scale of hardness. Gypsum
is used for making casts and moulds,
and for blackboard chalk.

gyroscope　mechanical instrument,
invented by French physicist Léon

Foucault in 1852, used as a stabilizing device and consisting, in its simplest form, of a heavy wheel mounted on an axis fixed in a ring that can be rotated about another axis, which is also fixed in a ring capable of rotation about a third axis. Applications of the gyroscope principle include the gyrocompass, the gyropilot for automatic steering, and gyro-directed torpedoes.

GZip in computing, compression software, properly called *GNU Zip, commonly used on the Internet. Files compressed using GZip can be recognized by the file extension '.GZ'. The software is published by the Free Software Foundation and was originally developed for Unix, although a DOS version is readily available.

H symbol for *henry, the SI unit of inductance, equal to the inductance of a closed circuit with a magnetic flux of 1 weber per ampere of current.

ha symbol for *hectare.

Haber process or Haber–Bosch process, industrial process by which *ammonia is manufactured by direct combination of its elements, nitrogen and hydrogen. It is named after named after German chemist Fritz Haber (1868–1934), whose conversion of atmospheric nitrogen to ammonia was developed into an industrial process by the German scientist Carl Bosch. As the method is a *reversible reaction which reaches *chemical equilibrium, the manufacturing conditions have to be chosen carefully in order to achieve the best yield of ammonia. The reaction is carried out at 400–500°C/752–932°F and at 200 atmospheres pressure. The two gases, in the proportions of 1:3 by volume, are passed over a *catalyst of finely divided iron.

Around 10% of the reactants combine, and the unused gases are recycled. The ammonia is separated either by being dissolved in water or by being cooled to liquid form.

$$N_2 + 3H_2 \leftrightarrow 2NH_3$$

habitat in ecology, the localized *environment in which an organism lives, and which provides for all (or almost all) of its needs. The diversity of habitats found within the Earth's ecosystem is enormous, and they are changing all the time. They may vary through the year or over many years. Many can be considered inorganic or physical; for example, the Arctic icecap, a cave, or a cliff face. Others are more complex; for instance, a woodland, or a forest floor. Some habitats are so precise that they are called **microhabitats**, such as the area under a stone where a particular type of insect lives. Most habitats provide a home for many species, which form a community.

Each species is specially adapted to life in its habitat. For example, an animal is adapted to eat other members of a *food chain or food web found in the same habitat. Some species may be found in different habitats. They may be found to have different patterns of behaviour or structure in these different habitats. For example, a plant such as the blackberry may grow in an open habitat, such as a field, or in a shaded one, such as woodland. Its leaves differ in the two habitats.

hacking unauthorized access to computer systems, either for fun or for malicious or fraudulent purposes. Hackers generally use computers and telephone lines to obtain access. In computing, the term is used in a wider sense to mean using software for enjoyment or self-education, not necessarily involving unauthorized access. The most destructive form of hacking is the introduction of a computer *virus. In recognition of the potential cost to business that hacking can cause, many jurisdictions have made hacking illegal. One of the most celebrated hacking cases was that of Kevin Mitnick, who spent five years in jail 1995–2000 on 25 counts of computer and wire fraud, and one charge of cracking. Mitnick had hacked into computers at the University of Southern California and tampered with data.

Some hackers call themselves ethical or white hat hackers. Often poachers turned gamekeepers, ethical hackers probe Web sites for holes in the security system, either fixing the holes themselves or alerting the company in question. There is some dispute over the correct use of the term hacker. Some believe hacker should be used to describe someone who develops computer software, with the term cracker being used for a person who breaks into a computer system.

hadal zone   deepest level of the ocean, below the abyssal zone, at depths of greater than 6,000 m/19,500 ft. The ocean trenches are in the hadal zone. There is no light in this zone and pressure is over 600 times greater than atmospheric pressure.

hadron   in physics, a subatomic particle that experiences the strong nuclear force. Each is made up of two or three indivisible particles called *quarks. The hadrons are grouped into the *baryons (*protons, *neutrons, and *hyperons), with half-integral spins consisting, of three quarks, and the *mesons, with whole-number or zero spins, consisting of two quarks.

Hadrons were found in the 1950s and 1960s. It was shown in the early 1960s that if hadrons of the same spin are represented as points on suitable charts, simple patterns are formed. This symmetry enabled a hitherto unknown baryon, the omega-minus, to be predicted from a gap in one of the patterns; it duly turned up in experiments. See *fermion.

Hadron–Electron Ring Accelerator   HERA, particle *accelerator built under the streets of Hamburg, Germany, occupying a tunnel 6.3 km/3.9 mi in length. It is the world's most powerful collider of protons and electrons, designed to accelerate protons to energies of 820 GeV (billion electron volts), and electrons to 30 GeV. HERA began operating in 1992.

haematite   or **hematite**, principal ore of iron, consisting mainly of iron(III) oxide, Fe_2O_3. It occurs as **specular haematite** (dark, metallic lustre), **kidney ore** (reddish radiating fibres terminating in smooth, rounded surfaces), and a red earthy deposit.

haematology   medical speciality concerned with disorders of the blood.

haematoma   accumulation of blood in the tissues, causing a solid swelling. It may be due to injury, disease, or a blood clotting disorder such as *haemophilia.

haemoglobin   protein used by all vertebrates and some invertebrates for oxygen transport because the two substances combine reversibly. In vertebrates it occurs in red blood cells (erythrocytes), giving them their colour.

In the lungs or gills where the concentration of oxygen is high, oxygen attaches to haemoglobin to form **oxyhaemoglobin**. This process effectively increases the amount of oxygen that can be carried in the bloodstream. The oxygen is later released in the body tissues where it is at a low concentration, and the deoxygenated blood returned to the lungs or gills. Haemoglobin will combine also with carbon monoxide to form carboxyhaemoglobin, which has the effect of reducing the amount of oxygen that can be carried in the blood.

haemolymph   circulatory fluid of those molluscs and insects that have an 'open' circulatory system. Haemolymph contains water, amino acids, sugars, salts, and white cells like those of blood. Circulated by a pulsating heart, its main functions are to transport digestive and excretory products around the body. In molluscs, it also transports oxygen and carbon dioxide.

haemophilia   any of several inherited diseases in which normal blood clotting is impaired. The sufferer experiences prolonged bleeding from the slightest wound, as well as painful internal bleeding without apparent cause.

Haemophilias are nearly always sex-linked, transmitted through the female line only to male infants; they have afflicted a number of European royal households. Males affected by the most common form are unable to synthesize Factor VIII, a protein involved in the clotting of blood. Treatment is primarily with Factor VIII (now mass-produced by recombinant techniques), but the haemophiliac remains at risk from the slightest incident of bleeding. The disease is a painful one that causes deformities of joints.

haemorrhage   loss of blood from the circulatory system. It is 'manifest' when the blood can be seen, as when it flows from a wound, and 'occult' when the bleeding is internal, as from an ulcer or internal injury.

haemorrhoids   distended blood vessels (varicose veins) in the area of the anus, popularly called **piles**.

haemostasis   natural or surgical stoppage of bleeding. In the natural mechanism, the damaged vessel contracts, restricting the flow, and blood *platelets plug the opening, releasing chemicals essential to clotting.

hafnium   chemical symbol Hf, (Latin *Hafnia* 'Copenhagen') silvery, metallic element, atomic number 72, relative atomic mass 178.49. It occurs in nature in ores of zirconium, the properties of which it resembles. Hafnium absorbs neutrons better than most metals, so it is used in the control rods of nuclear reactors; it is also used for light-bulb filaments.

It was named in 1923 by Dutch physicist Dirk Coster (1889–1950) and Hungarian chemist Georg von Hevesy after the city of Copenhagen, where the element was discovered.

hail   precipitation in the form of pellets of ice (hailstones). Water droplets freeze as they are carried upwards. As the circulation continues, layers of ice are deposited around the droplets until they become too heavy to be supported by the air current and they

fall as a hailstorm. It is caused by the circulation of moisture in strong convection currents, usually within cumulonimbus *clouds.

hair   fine filament growing from mammalian skin. Each hair grows from a pit-shaped follicle embedded in the second layer of the skin, the dermis. It consists of dead cells impregnated with the protein *keratin.

hair analysis   diagnostic technique for ascertaining deficiencies or excesses of mineral resources in the body, using a sophisticated analytical procedure called atomic-emission spectroscopy.

Hale-Bopp   large and exceptionally active comet, which in March 1997 made its closest fly-by to Earth since 2000 BC, coming within 190 million km/118 million mi. It has a nucleus of approximately 40 km/25 mi and an extensive gas coma (when close to the Sun Hale-Bopp released 10 tonnes of gas every second). Unusually, Hale-Bopp has three tails: one consisting of dust particles, one of charged particles, and a third of sodium particles.

Hale-Bopp was discovered independently in July 1995 by two US amateur astronomers, Alan Hale and Thomas Bopp.

half duplex   in computing, a *modem setting which controls whether or not characters echo to (appear on) the screen. See *full duplex.

half-life   during *radioactive decay, the time in which the activity of a radioactive source decays to half its original value (the time taken for half the atoms to decay). In theory, the decay process is never complete and there is always some residual radioactivity. For this reason, the half-life of a radioactive isotope is measured, rather than the total decay time. It may vary from millionths of a second to billions of years.

half-life   in medicine, the time taken for the peak plasma concentration of a drug to decline by half. This is due to redistribution of the drug from the plasma to the tissues and to metabolism and excretion of the drug. The half-life of the drug is one of the factors that determines the frequency of administration required to achieve optimal therapeutic effects.

halide any compound produced by the combination of a *halogen, such as chlorine or iodine, with a less electronegative element (see *electronegativity). Halides may be formed by *ionic bonds or by *covalent bonds.

In organic chemistry, alkyl halides consist of a halogen and an alkyl group, such as methyl chloride (chloromethane).

halite mineral form of sodium chloride, NaCl. Common *salt is the mineral halite. When pure it is colourless and transparent, but it is often pink, red, or yellow. It is soft and has a low density.

halitosis bad breath. It may be caused by poor oral hygiene; disease of the mouth, throat, nose, or lungs; or disturbance of the digestion.

Hall effect production of a voltage across a conductor or semiconductor carrying a current at a right angle to a surrounding magnetic field. It was discovered in 1897 by the US physicist Edwin Hall (1855–1938). It is used in the **Hall probe** for measuring the strengths of magnetic fields and in magnetic switches.

Halley's Comet comet that orbits the Sun roughly every 75 years, named after English astronomer Edmond Halley (1656–1742), who calculated its orbit. It is the brightest and most conspicuous of the periodic comets, and recorded sightings go back over 2,000 years. The comet travels around the Sun in the opposite direction to the planets. Its orbit is inclined at almost 20° to the main plane of the Solar System and ranges between the orbits of Venus and Neptune. It will next reappear in 2061.

hallucinogen any substance that acts on the *central nervous system to produce changes in perception and mood and often hallucinations. Hallucinogens include LSD, peyote, and mescaline. Their effects are unpredictable and they are illegal in most countries.

halogen any of a group of five non-metallic elements with similar chemical bonding properties: *fluorine, *chlorine, *bromine, *iodine, and *astatine. They form a linked group (Group 7) in the *periodic table, descending from fluorine, the most reactive, to astatine, the least reactive. They all have coloured vapours and are poisonous. Melting points and boiling points increase going down the group. They combine directly with most metals to form salts, such as common salt (NaCl). Each halogen has seven electrons in its valence shell, which accounts for the chemical similarities displayed by the group.

Fluorine and chlorine are gases, bromine is a liquid, and astatine is a solid. The colour of the element darkens going down the group from fluorine, which is pale yellow, to astatine, which is black.

halon organic chemical compound containing one or two carbon atoms, together with *bromine and other *halogens. The most commonly used are halon 1211 (bromochlorodifluoromethane) and halon 1301 (bromotrifluoromethane). The halons are gases and are widely used in fire extinguishers. As destroyers of the *ozone layer, they are up to ten times more effective than *chlorofluorocarbons (CFCs), to which they are chemically related. Levels in the atmosphere are rising by about 25% each year, mainly through the testing of fire-fighting equipment. The use of halons in fire extinguishers was banned in 1994.

halophyte plant adapted to live where there is a high concentration of salt in the soil, for example in salt marshes and mud flats.

handwriting recognition ability of a computer to accept handwritten input and turn it into digital data that can be processed and displayed or stored as *ASCII characters on the computer screen.

hantavirus spherical virus causing various haemorrhagic diseases with renal complications. The virus was first identified in 1976 in Korea (and named after Hantaan River), where field mice are its main carriers. In China the hantavirus affects around 100,000 people annually. It also circulates in Japan and Russia, and a nonfatal form exists in parts of Europe, surviving in bank voles and yellow-necked field mice. Its incubation period is 12–21 days.

haploid having a single set of *chromosomes in each cell. Most higher organisms are *diploid – that is, they have two sets – but their gametes (sex cells) are haploid. Some plants, such as mosses, liverworts, and many seaweeds, are haploid, and male honey bees are haploid because they develop from eggs that have not been fertilized. See also *meiosis.

hard disk in computing, a storage device usually consisting of a rigid metal *disk coated with a magnetic material. Data are read from and written to the disk by means of a disk drive. The hard disk may be permanently fixed into the drive or in the form of a disk pack that can be removed and exchanged with a different pack. Hard disks vary from large units with capacities of more than 80,000 megabytes, intended for use with mainframe computers, to small units with capacities as low as 4,000 megabytes, intended for use with microcomputers. Smaller sizes than this do exist and originally hard drives only stored around 20 megabytes.

hardness the resistance of a material to indentation by various means, such as scratching, abrasion, wear, and drilling. Methods of heat treatment can increase the hardness of metals. A scale of hardness, *Mohs scale, was devised by German–Austrian mineralogist Friedrich Mohs (1773–1839) in the 1800s, based upon the hardness of certain minerals from soft talc (Mohs hardness 1) to diamond (10), the hardest of all materials.

hardware mechanical, electrical, and electronic components of a computer system, as opposed to the various programs, which constitute *software.

hard water water that does not lather easily with soap, and produces a deposit or *scale (limescale) in kettles. It is caused by the presence of certain salts of calcium and magnesium.

   Temporary hardness is caused by the presence of dissolved hydrogencarbonates (bicarbonates); when the water is boiled, they are converted to insoluble carbonates that precipitate as 'scale'. **Permanent hardness** is caused by sulphates and silicates, which are not affected by boiling. Water can be softened by *distillation, *ion exchange (the principle underlying commercial water softeners), targeting with low frequency magnetic waves (this alters the crystal structure of calcium salts so that they remain in suspension), addition of sodium carbonate or of large amounts of soap, or boiling (to remove temporary hardness).

Hare's apparatus in physics, a specific kind of *hydrometer used to compare the relative densities of two liquids, or to find the density of one if the other is known. It was invented by US chemist Robert Hare (1781–1858).

Harwell main research establishment of the United Kingdom Atomic Energy Authority, situated near the village of Harwell in Oxfordshire.

hassium chemical symbol Hs, synthesized, radioactive element of the *transactinide series, atomic number 108, relative atomic mass 277. It was first synthesized in 1984 by the Laboratory for Heavy-Ion Research in Darmstadt, Germany.

hay fever allergic reaction to pollen, causing sneezing, with inflammation of the nasal membranes and conjunctiva of the eyes. Symptoms are due to the release of *histamine. Treatment is by antihistamine drugs. An estimated 25% of Britons, 33% of Americans, and 40% of Australians suffer from hayfever.

hazardous waste waste substance, usually generated by industry, that represents a hazard to the environment or to people living or working nearby. Examples include radioactive wastes, acidic resins, arsenic residues, residual hardening salts, lead from car exhausts, mercury, non-ferrous sludges, organic solvents, asbestos, chlorinated solvents, and pesticides. The cumulative effects of toxic waste can take some time to become apparent (anything from a few hours to many years), and pose a serious threat to the ecological stability of the planet; its economic disposal or recycling is the subject of research.

heart muscular organ that rhythmically contracts to force blood around the body of an animal with a circulatory system. Annelid worms and some other invertebrates have simple hearts consisting of thickened sections of main blood vessels that pulse regularly. An earthworm has ten such hearts. Vertebrates have one heart. A fish heart has two chambers – the thin-walled atrium (once called the auricle) that expands to receive blood, and the thick-walled ventricle that pumps it out. Amphibians and most reptiles have two atria and one ventricle; birds and mammals have two atria and two ventricles. The beating of the heart is controlled by the autonomic nervous system and an internal control centre or pacemaker, the sinoatrial node.

The human heart is more or less conical in shape and is positioned within the chest, behind the breast bone, above the diaphragm, and between the two lungs. It has flattened back and front surfaces and is, in health, the size of a person's closed fist. However, it varies in size with the person's weight, age, sex, and state of health. Its capacity is about 20 cm³ in the newborn, reaching 150–160 cm³ in the mid-teens. The female heart has a smaller capacity and is lighter than the male. Mammals have a *double circulatory system. In this kind of

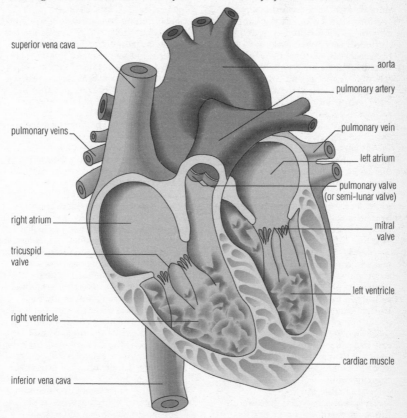

heart The structure of the human heart. During an average lifetime, the human heart beats more than 2,000 million times and pumps 500 million l/110 million gal of blood. The average pulse rate is 70–72 beats per minute at rest for adult males, and 78–82 beats per minute for adult females.

system the heart is divided into two halves, each half working as a separate pump. One side of the heart pumps blood to the lungs and back to the heart. The other side of the heart pumps blood to all other parts of the body and back to the heart.

heart attack or **myocardial infarction**, sudden onset of gripping central chest pain, often accompanied by sweating and vomiting, caused by death of a portion of the heart muscle following obstruction of a coronary artery by thrombosis (formation of a blood clot). Half of all heart attacks result in death within the first two hours, but in the remainder survival has improved following the widespread use of thrombolytic (clot-buster) drugs.

After a heart attack, most people remain in hospital for seven to ten days, and may make a gradual return to normal activity over the following months. How soon a patient is able to return to work depends on the physical and mental demands of their job. Despite widespread fears to the contrary, it is safe to return to normal sexual activity within about a month of the attack.

Women are less likely to suffer heart attack than men. However, as heart disease in women tends to be given lower priority because of lower incidence, they tend to be more likely to die as a result. The best predictor of heart attacks among older people may be an echogram (sonogram) of carotid arteries.

heartbeat regular contraction and relaxation of the heart, and the accompanying sounds. As blood passes through the heart a double beat is heard. The first is produced by the sudden closure of the valves between the atria and the ventricles. The second, slightly delayed sound, is caused by the closure of the valves found at the entrance to the major arteries leaving the heart. Diseased valves may make unusual sounds, known as heart murmurs.

heart disease disorder affecting the heart; for example, *ischaemic heart disease, in which the blood supply through the coronary arteries is reduced by *atherosclerosis; valvular

heart disease, in which a heart valve is damaged; and cardiomyopathy, where the heart muscle itself is diseased. See also *coronary artery disease.

heart failure condition arising when the heart's pumping mechanism is inadequate. It results in back pressure of blood, causing congestion of the liver and lungs, failure of the peripheral blood supply and oedema. It may be a consequence of *hypertension or various types of heart disease. Treatment is with a *diuretic and heart drugs, especially *digoxin.

heat form of energy possessed by a substance by virtue of the vibrational movement (kinetic energy) of its molecules or atoms. Heat energy is transferred by conduction, convection, and radiation. It always flows from a region of higher *temperature (heat intensity) to one of lower temperature. Its effect on a substance may be simply to raise its temperature, or to cause it to expand, melt (if a solid), vaporize (if a liquid), or increase its pressure (if a confined gas).

heat capacity in physics, the quantity of heat required to raise the temperature of an object by one degree. The **specific heat capacity** of a substance is the heat capacity per unit of mass, measured in joules per kilogram per kelvin ($J kg^{-1} K^{-1}$).

heat of reaction alternative term for *energy of reaction.

heat pump machine, run by electricity or another power source, that cools the interior of a building by removing heat from interior air and pumping it out or, conversely, heats the inside by extracting energy from the atmosphere or from a hot-water source and pumping it in.

heat treatment in industry, the subjection of metals and alloys to controlled heating and cooling after fabrication to relieve internal stresses and improve their physical properties. Methods include *annealing, *quenching, and *tempering.

heavy water or **deuterium oxide**; D_2O, water containing the isotope deuterium instead of hydrogen (relative molecular mass 20 as opposed to 18 for ordinary water).

Its chemical properties are identical with those of ordinary water, but its physical properties differ slightly. It occurs in ordinary water in the ratio of about one part by mass of deuterium to 5,000 parts by mass of hydrogen, and can be concentrated by electrolysis, the ordinary water being more readily decomposed by this means than the heavy water. It has been used in the nuclear industry because it can slow down fast neutrons, thereby controlling the chain reaction.

hectare symbol ha, metric unit of area equal to 100 ares or 10,000 square metres (2.47 acres).

height of a plane figure or solid, the perpendicular distance from the vertex to the base; see *altitude.

Heimlich manoeuvre technique to relieve choking due to a blockage in the *larynx (windpipe). The manoeuvre consists of the following steps: Stand behind the patient and wrap your arms around the waist. Making a fist with your hand, place it slightly above the patient's navel. Grip the fist with your other hand, and press hard into the abdomen – a quick, upward thrust. Repeat this movement until the blockage is dislodged. If the patient is pregnant, the fist should be in a higher position, at the base of the breastbone.

Helicobacter pylori spiral-shaped swimming bacterium that causes gastritis and stomach *ulcers when it colonizes the stomach lining. Without antibiotic treatment, infection can be permanent. *H. pylori* may also contribute towards stomach cancer.

helium chemical symbol He, (Greek *helios* 'Sun') colourless, odourless, gaseous, non-metallic element, atomic number 2, relative atomic mass 4.0026. It is grouped with the *noble gases (rare gases) in Group 0 of the *periodic table. Helium is nonreactive because of its full outer shell of *electrons and forms no compounds. It is the second most abundant element (after hydrogen) in the universe, and has the lowest boiling ($-268.9°C/-452°F$) and melting points ($-272.2°C/-458°F$) of all the elements. It is present in small quantities in the Earth's atmosphere from gases issuing from radioactive

elements (from *alpha decay) in the Earth's crust; after hydrogen it is the second-lightest element.

Helium is a component of most stars, including the Sun, where the nuclear-fusion process converts hydrogen into helium with the production of heat and light. It is obtained by the compression and fractionation of naturally occurring gases. It is used for inflating balloons and as a dilutant for oxygen in deep-sea breathing systems. Liquid helium is used extensively in low-temperature physics (cryogenics).

Helium was originally discovered in 1868 in the spectrum of the Sun. It was found on Earth in 1895.

helix in mathematics, a three-dimensional curve resembling a spring, corkscrew, or screw thread. It is generated by a line that encircles a cylinder or cone at a constant angle.

helper application in computing, in Web *browsers, an external application that adds the ability to display certain types of files. Common helper applications include RealAudio, which allows browsers to play live sound tracks such as radio broadcasts or recorded lectures; Acrobat; and mIRC, which allows access to Internet Relay Chat via the World Wide Web.

hemisphere half a sphere, produced when a sphere is sliced along a great circle.

henry symbol H, SI unit of *inductance (the reaction of an electric current against the magnetic field that surrounds it). One henry is the inductance of a circuit that produces an opposing voltage of one volt when the current changes at one ampere per second.

hepatic of or pertaining to the liver.

hepatitis any inflammatory disease of the liver, usually caused by a virus. Other causes include alcohol, drugs, gallstones, *lupus erythematous, and amoebic *dysentery. Symptoms include weakness, nausea, and jaundice.

herbaceous plant plant with very little or no wood, dying back at the end of every summer. The herbaceous perennials survive winters as underground storage organs such as bulbs and tubers.

herbalism   in alternative medicine, the prescription and use of plants and their derivatives for medication. Herbal products are favoured by alternative practitioners as 'natural medicine', as opposed to modern synthesized medicines and drugs, which are regarded with suspicion because of the dangers of side effects and dependence.

herbivore   animal that feeds on green plants (or photosynthetic single-celled organisms) or their products, including seeds, fruit, and nectar. The most numerous type of herbivore is thought to be the zooplankton, tiny invertebrates in the surface waters of the oceans that feed on small photosynthetic algae. Herbivores are more numerous than other animals because their food is the most abundant. They form a vital link in the food chain between plants and *carnivores.

heredity   in biology, the transmission of traits from parent to offspring. See also *genetics.

hermaphrodite   organism that has both male and female sex organs. Hermaphroditism is the norm in such species as earthworms and snails, and is common in flowering plants. Cross-fertilization is common among hermaphrodites, with the parents functioning as male and female simultaneously, or as one or the other sex at different stages in their development. Human hermaphrodites are extremely rare.

hernia   or **rupture**, protrusion of part of an internal organ through a weakness in the surrounding muscular wall, usually in the groin. The appearance is that of a rounded soft lump or swelling.

heroin   or **diamorphine**, powerful *opiate analgesic, an acetyl derivative of *morphine. It is more addictive than morphine but causes less nausea.

herpes   any of several infectious diseases caused by viruses of the herpes group. **Herpes simplex I** is the causative agent of a common inflammation, the cold sore. **Herpes simplex II** is responsible for genital herpes, a highly contagious, sexually transmitted disease characterized by painful blisters in the genital area. It can be transmitted in the birth canal from mother to newborn. **Herpes zoster** causes *shingles; another herpes virus causes chickenpox.

herpes genitalis   or **genital herpes**, in medicine, an infection of the genital organs caused by the *Herpes simplex* virus. It is transmitted by sexual intercourse. Infection is marked by the formation of blisters on the genitals and pain. Treatment is with antiviral drugs, such as aciclovir, used orally or as creams. It can be passed from mother to child during birth, causing life-threatening illnesses and blindness.

hertz   symbol Hz, SI unit of frequency (the number of repetitions of a regular occurrence in one second). A wave source has a frequency of 1 Hz if it produces one wave each second. Human beings have an audible range from approximately 20 Hz to 20,000 Hz. Radio waves are often measured in megahertz (MHz), millions of hertz, and the clock rate of a computer is usually measured in megahertz. The unit is named after German physicist Heinrich Hertz (1857–1894).

heterogeneous reaction   in chemistry, a reaction where there is an interface between the different components or reactants. Examples of heterogeneous reactions are those between a gas and a solid, a gas and a liquid, two immiscible liquids, or two different solids.

heterotroph   any living organism that obtains its energy from organic substances produced by other organisms. All animals and fungi are heterotrophs, and they include *herbivores, *carnivores, and *saprotrophs (those that feed on dead animal and plant material).

heterozygous   in a living organism, having two different *alleles for a given trait. In *homozygous organisms, by contrast, both chromosomes carry the same allele. In an outbreeding population an individual organism will generally be heterozygous for some genes but homozygous for others.

For example, in humans, alleles for both blue- and brown-pigmented eyes exist, but the 'blue' allele is *recessive to the dominant 'brown' allele.

Only individuals with blue eyes are predictably homozygous for this trait; brown-eyed people can be either homozygous or heterozygous.

hexagon six-sided *polygon. A regular hexagon has all six sides of equal length, and all six angles of equal size, 120°.

hexanedioic acid or **adipic acid**; $(CH_2)_4(COOH)_2$, crystalline solid acid, obtained by the oxidation of certain fatty or waxy bodies. It is a dibasic acid, akin to oxalic and succinic acids, with the typical reactions of carboxylic acids, and is used in the manufacture of *nylon 66.

HF in physics, abbreviation for high *frequency. HF radio waves have frequencies in the range 3–30 MHz.

hibernation state of dormancy in which certain animals spend the winter. It is associated with a dramatic reduction in all metabolic processes, including body temperature, breathing, and heart rate.

The body temperature of the Arctic ground squirrel falls to below 0°C/32°F during hibernation. Hibernating bats may breathe only once every 45 minutes, and can go for up to 2 hours without taking a breath.

hidden file computer file in an MS-DOS system that is not normally displayed when the directory listing command is given. Hidden files include certain system files, principally so that there is less chance of modifying or deleting them by accident, but any file can be made hidden if required.

hierarchical storage management in computing, the organization of data storage so that information which is used most often is stored using the fastest access technology, and information used less often is stored on slower and less expensive storage devices.

Higgs boson or **Higgs particle**, postulated *elementary particle whose existence would explain why particles have mass. The current theory of elementary particles, called the *standard model, cannot explain how mass arises. To overcome this difficulty, English theoretical physicist Peter Higgs (1929–) of the University of Edinburgh and Thomas Kibble of Imperial College, London, proposed in 1964 a new particle that binds to other particles and gives them their mass.

highest common factor HCF, in a set of numbers, the highest number that will divide every member of the set without leaving a remainder. For example, 6 is the highest common factor of 36, 48, and 72.

high-level language in computing, a programming language designed to suit the requirements of the programmer; it is independent of the internal machine code of any particular computer. High-level languages are used to solve problems and are often described as **problem-oriented languages**; for example, *BASIC was designed to be easily learnt by first-time programmers; *COBOL is used to write programs solving business problems; and *FORTRAN is used for programs solving scientific and mathematical problems. With the increasing popularity of windows-based systems, the next generation of programming languages was designed to facilitate the development of GUI interfaces; for example, *Visual Basic wraps the BASIC language in a graphical programming environment. Support for *object-oriented programming has also become more common, for example in *C++ and *Java. In contrast, **low-level languages**, such as *assembly languages, closely reflect the machine codes of specific computers, and are therefore described as **machine-oriented languages**.

Hill reaction Production of oxygen from water during photosynthesis, named after British biochemist Robert Hill (1899–1991).

hinge joint in vertebrates, a joint where movement occurs in one plane only. Examples are the elbow and knee, which are controlled by pairs of muscles, the *flexors and *extensors.

Hippocratic oath oath laid down by the Greek physician Hippocrates (c.460–c.370 BC), who is regarded as the father of medicine. Taken by all doctors on qualifying, it lays down moral and ethical guidelines for the practice of medicine.

hirsutism excessive growth of hair of masculine type and distribution in women. It may be caused by hormone imbalance, disease of the ovaries or some drugs. Unwanted hair on the face and legs is treated by electrolysis.

histamine inflammatory substance normally released in damaged tissues, which also accounts for many of the symptoms of *allergy. It is an amine, $C_5H_9N_3$. Substances that neutralize its activity are known as *antihistamines. Histamine was first described in 1911 by British physiologist Henry Dale (1875–1968).

histogram in statistics, a graph showing frequency of data, in which the horizontal axis details discrete units or class boundaries, and the vertical axis represents the frequency. Blocks are drawn such that their areas (rather than their height as in a *bar chart) are proportional to the frequencies within a class or across several class boundaries. There are no spaces between blocks.

histology study of plant and animal tissue by visual examination, usually with a *microscope.

histology in medicine, the laboratory study of cells and tissues.

hit in computing, request sent to a file server.

HIV abbreviation for human immunodeficiency virus, infectious agent that is believed to cause *AIDS. It was first discovered in 1983 by Luc Montagnier of the Pasteur Institute in Paris, who called it lymphocyte-associated virus (LAV). Independently, US scientist Robert Gallo of the National Cancer Institute in Bethesda, Maryland, claimed its discovery in 1984 and named it human T-lymphocytotrophic virus 3 (HTLV-III).

HLA system abbreviation for human leucocyte antigen system, a series of gene groups (termed A, B, C, and D) that code for proteins on the surface of cells. If two people have the same HLA types, they are histocompatible (having compatible tissue types). HLA types feature in tissue testing before a transplant.

Hodgkin's disease or **lymphadenoma**, rare form of cancer mainly affecting the lymph nodes and spleen, named after English physician Thomas Hodgkin (1798–1866) who first

recognised the disease. It undermines the immune system, leaving the sufferer susceptible to infection. However, it responds well to radiotherapy and *cytotoxic drugs, and long-term survival is usual.

holistic medicine umbrella term for an approach that virtually all alternative therapies profess, which considers the overall health and lifestyle profile of a patient, and treats specific ailments not primarily as conditions to be alleviated but rather as symptoms of more fundamental disease.

holmium chemical symbol Ho, (Latin *Holmia* 'Stockholm') silvery, metallic element of the *lanthanide series, atomic number 67, relative atomic mass 164.93. It occurs in combination with other rare-earth metals and in various minerals such as gadolinite. Its compounds are highly magnetic.

The element was discovered in 1878, spectroscopically, by the Swiss chemists Jacques-Louis Soret and Marc Delafontaine, and independently in 1879 by Swedish chemist Per Cleve (1840–1905), who named it after Stockholm, near which it was found.

Holocene epoch period of geological time that began 10,000 years ago, and continues into the present. During this epoch the climate became warmer, the glaciers retreated, and human civilizations developed significantly. It is the second and current epoch of the Quaternary period.

hologram three-dimensional image produced by *holography. Small, inexpensive holograms appear on credit cards and software licences to guarantee their authenticity.

holography method of producing three-dimensional (3-D) images, called *holograms, by means of *laser light. Holography uses a photographic technique (involving the splitting of a laser beam into two beams) to produce a picture, or hologram, that contains 3-D information about the object photographed. Some holograms show meaningless patterns in ordinary light and produce a 3-D image only when laser light is projected through them, but reflection holograms produce images when ordinary light is reflected from them (as found on credit cards).

Although the possibility of holography was suggested as early as 1947 (by Hungarian-born British physicist Dennis Gabor), it could not be demonstrated until a pure, coherent light source, the laser, became available in 1963. The first laser-recorded holograms were created by Emmett Leith and Juris Upatnieks at the University of Michigan, USA, and Yuri Denisyuk in the Soviet Union.

The technique of holography is also applicable to sound, and bats may navigate by ultrasonic holography. Holographic techniques also have applications in storing dental records, detecting stresses and strains in construction and in retail goods, detecting forged paintings and documents, and producing three-dimensional body scans. The technique of detecting strains is of widespread application. It involves making two different holograms of an object on one plate, the object being stressed between exposures. If the object has distorted during stressing, the hologram will be greatly changed, and the distortion readily apparent.

Using holography, digital data can be recorded page by page in a crystal. As many as 10,000 pages (100 megabytes) of digital data can be stored in an iron-doped lithium niobate crystal of 1 cubic cm/0.06 cubic in.

homeopathy or **homoeopathy**, system of alternative medicine based on the principle that symptoms of disease are part of the body's self-healing processes, and on the practice of administering extremely diluted doses of natural substances found to produce in a healthy person the symptoms manifest in the illness being treated. Developed by the German physician Samuel Hahnemann (1755–1843), the system is widely practised today as an alternative to allopathic (orthodox) medicine, and many controlled tests and achieved cures testify its efficacy.

homeostasis maintenance of a constant environment around living cells, particularly with regard to pH, salt concentration, temperature, and blood sugar levels. Stable conditions are important for the efficient

functioning of the *enzyme reactions within the cells. In humans, homeostasis in the blood (which provides fluid for all tissues) is ensured by several organs. The *kidneys regulate pH, urea, and water concentration. The lungs regulate oxygen and carbon dioxide (see *breathing). Temperature is regulated by the liver and the skin (see *temperature regulation). Glucose levels in the blood are regulated by the *liver and the pancreas. (See diagram, page 288.)

homeothermy maintenance of a constant body temperature in endothermic (warm-blooded) animals, by the use of chemical processes to compensate for heat loss or gain when external temperatures change. Such processes include generation of heat by the breakdown of food and the contraction of muscles, and loss of heat by sweating, panting, and other means.

home page in computing, opening page on a particular site on the World Wide Web. The term is also used for the page that loads automatically when a user opens a Web *browser, and for a user's own personal Web pages.

homoeopathy variant spelling of *homeopathy.

Homo erectus see *human species origins of

homogeneous reaction in chemistry, a reaction where there is no interface between the components. The term applies to all reactions where only gases are involved or where all the components are in solution.

Homo habilis see *human species origins of

homologous in biology, a term describing an organ or structure possessed by members of different taxonomic groups (for example, species, genera, families, orders) that originally derived from the same structure in a common ancestor. The wing of a bat, the arm of a monkey, and the flipper of a seal are homologous because they all derive from the forelimb of an ancestral mammal.

homologous series in chemistry, any of a number of series of organic compounds with similar chemical properties in which members differ by a constant relative molecular mass.

Alkanes (paraffins), alkenes (olefins), and alkynes (acetylenes) form such series in which members differ in mass by 14, 12, and 10 atomic mass units respectively. For example, the alkane homologous series begins with methane (CH_4), ethane (C_2H_6), propane (C_3H_8), butane (C_4H_{10}), and pentane (C_5H_{12}), each member differing from the previous one by a CH_2 group (or 14 atomic mass units).

Homo sapiens see *human species origins of

homozygous in a living organism, having two identical *alleles for a given trait. Individuals homozygous for a trait always breed true; that is, they produce offspring that resemble them in appearance when bred with a genetically similar individual; inbred varieties or species are homozygous for almost all traits. *Recessive alleles are only expressed in the homozygous condition. *Heterozygous organisms have two different alleles for a given trait.

Hooke's law law stating that the deformation of a body is proportional to the magnitude of the deforming

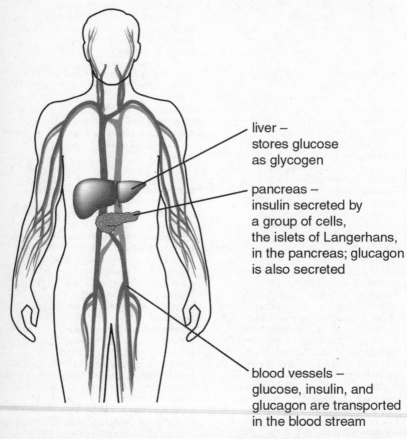

liver –
stores glucose
as glycogen

pancreas –
insulin secreted by
a group of cells,
the islets of Langerhans,
in the pancreas; glucagon
is also secreted

blood vessels –
glucose, insulin, and
glucagon are transported
in the blood stream

homeostasis How blood-sugar levels are maintained in the body. When blood-sugar levels rise, insulin is secreted by the islets of Langerhans in the pancreas to facilitate the storage of glucose as glycogen. When blood-sugar levels fall too low, glucagon is secreted from the pancreas to facilitate the breakdown of glycogen back into glucose.

Alkane	Alcohol	Aldehyde	Ketone	Carboxylic acid	Alkene
CH_4 methane	CH_3OH methanol	HCHO methanal	——	HCO_2H methanoic acid	——
CH_3CH_3 ethane	CH_3CH_2OH ethanol	CH_3CHO ethanal	——	CH_3CO_2H ethanoic acid	CH_2CH_2 ethene
$CH_3CH_2CH_3$ propane	$CH_3CH_2CH_2OH$ propanol	CH_3CH_2CHO propanal	CH_3COCH_3 propanone	$CH_3CH_2CO_2H$ propanoic acid	CH_2CHCH_3 propene
methane	methanol	methanal	propanone	methanoic acid	ethene

homologous series Six different types of homologous series: organic chemicals with similar chemical properties in which members differ by a constant relative atomic mass. For example, all the members of the alkane series differ by a relative atomic mass of 14.

force, provided that the body's elastic limit (see *elasticity) is not exceeded. If the elastic limit is not reached, the body will return to its original size once the force is removed. The law was discovered by English physicist Robert Hooke (1635–1703) in 1676.

A force applied to a spring by adding a weight to it will cause the spring to stretch in proportion to the applied force. If the force is removed and the spring returns to its original size, the material of the spring is said to be elastic. If the spring does not return to its original size, the material is said to be non-elastic. The elastic limit is the point at which a certain amount of force will deform the material so it becomes non-elastic.

For example, if a spring is stretched by 2 cm by a weight of 1 N, it will be stretched by 4 cm by a weight of 2 N, and so on. However, once the load exceeds the elastic limit for the spring, Hooke's law will no longer be obeyed and each successive increase in weight will result in a greater extension until finally the spring breaks.

Hope's apparatus in physics, an apparatus used to demonstrate the temperature at which water has its maximum density. It is named after Thomas Charles Hope (1766–1844).

horizon in astronomy, the great circle dividing the visible part of the sky from the part hidden by the Earth.

horizon limit to which one can see across the surface of the sea or a level plain, that is, about 5 km/3 mi at 1.5 m/5 ft above sea level, and about 65 km/40 mi at 300 m/1,000 ft.

horizontal having the same direction as the horizon, flat and level. A marble would not roll off a horizontal table without being pushed. On paper, a horizontal line is shown by a line parallel to the top of the page. In practice, horizontal planes are checked with a spirit level.

hormone in biology, a chemical secretion of the ductless *endocrine glands and specialized nerve cells (see *neurohormones) concerned with control of body functions. Hormones act as chemical messengers and are transported to all parts of the body by the *bloodstream where they affect target organs. The major glands are the thyroid, parathyroid, pituitary, adrenal, pancreas, ovary, and testis. There are also hormone-secreting cells in the kidney, liver, gastrointestinal tract, thymus (in the neck), pineal (in the brain), and placenta. Hormones bring about changes in the functions of various organs according to the body's requirements. The *hypothalamus, which adjoins the pituitary gland at the base of the brain, is a control centre for overall coordination of hormone secretion; the thyroid hormones determine the rate of general body chemistry; the adrenal hormones prepare the organism during stress for 'fight or flight'; and the sexual hormones such as oestrogen and testosterone govern reproductive functions. Plants produce chemicals that affect growth and development. These chemicals can also be called hormones (see *plant hormone).

Hormones often bring about a slower response than messages sent by the *nerve cells, but the response to hormones may be longer lasting.

Examples of hormones include: insulin and glucagon (produced in the pancreas), which are involved in glucose regulation; ADH (produced in the pituitary gland), which regulates the concentration of urine produced by the *kidney; oestrogen and progesterone (produced in the ovaries), which regulate the growth and functioning of sex organs for *sexual reproduction; testosterone (produced in the testes), which regulates male sexual development; follicle-stimulating hormone (FSH) and luteinizing hormone (LH) (from the pituitary gland), which regulate the menstrual cycle in females; and adrenaline (released from the adrenal glands), which prepares the body for increased activity.

hormone-replacement therapy HRT, use of *oestrogen and progesterone to help limit the unpleasant effects of the menopause in women. The treatment was first used in the 1970s.

horsepower abbreviation hp, imperial unit of power, now replaced by the *watt. It was first used by the engineer James Watt, who employed it to compare the power of steam engines with that of horses.

host in biology, an organism that is parasitized by another. In commensalism, the partner that does not benefit may also be called the host.

host or **host computer**, large computer that supports a number of smaller computers or terminals that are connected to it via a network. Hosts may be mainframe computers that service large number of terminals or green screens, or, for example, Internet hosts that serve Web pages and files to personal computers attached via the Internet.

hostname alternative name given to a computer when it is connected to the *Internet. The hostname is used instead of the unique Internet address which is in the form of four numbers separated by dots. The hostname is an alias, or nickname, and is easier to remember than the number.

hot key in computing, a key stroke (or sequence of key strokes) that triggers a memory-resident program. Such programs are called *terminate and stay resident. Hot keys should be chosen so that they do not conflict with key sequences in commonly used applications.

hot spot in earth science, area where a strong current or 'plume' of *magma rises upwards below the Earth's crust. The magma spreads horizontally in all directions, and may break through where the crust is thin. Hot spots occur within, rather than on the edges of, lithospheric *plates. However, the magma usually reaches the surface at *plate margins. Examples of hot spots include Hawaii, Iceland, and Yellowstone National Park, Wyoming, USA.

hour period of time comprising 60 minutes; 24 hours make one calendar day.

hp abbreviation for *horsepower.

HPGL abbreviation for Hewlett-Packard Graphics Language, file format used in *vector graphics. HPGL is often generated by *CAD systems.

HSCSD abbreviation for high speed circuit switched data, wireless service introduced in 2000 by UK mobile phone operator Orange, allowing users of *laptop computers and *personal digital assistants (PDAs) to transmit data at a speed of 28.8 kilobytes per second (kbps). HSCSD is a rival of *GPRS, which requires the use of a mobile phone to connect laptops and PDAs.

ht abbreviation for **height**.

HTML abbreviation for hypertext markup language, standard for structuring and describing a document on the *World Wide Web. The HTML standard provides labels for constituent parts of a document (for example headings and paragraphs) and permits the inclusion of images, sounds, and 'hyperlinks' to other documents. A *browser program is then used to convert this information into a graphical document on-screen. The specifications for HTML version 4, called *Dynamic HTML, were adopted at the end of 1997.

HTML extension   in computing, any proprietary addition to the standard specification of HTML (hypertext markup language).

HTTP   abbreviation for hypertext transfer protocol, in computing, the *protocol used for communications between client (the Web *browser) and *server on the World Wide Web.

Hubble classification   in astronomy, scheme for classifying *galaxies according to their shapes, originally devised by US astronomer Edwin Hubble (1889–1953) in the 1920s.

Hubble constant   in astronomy, measure of the rate at which the universe is expanding, named after US astronomer Edwin Hubble (1889–1953). Observations suggest that galaxies are moving apart at a rate of 50–100 kps/30–60 mps for every million *parsecs of distance. This means that the universe, which began at one point according to the *Big Bang theory, is between 10 billion and 20 billion years old. Observations by the Hubble Space Telescope in 1996 produced a figure for the constant of 73 kps/45 mps.

Hubble's law   law that relates a galaxy's distance from us to its speed of recession as the universe expands, formulated in 1929 by US astronomer Edwin Hubble (1889–1953). He found that galaxies are moving apart at speeds that increase in direct proportion to their distance apart. The rate of expansion is known as the Hubble constant.

Hubble Space Telescope   HST, space-based astronomical observing facility, orbiting the Earth at an altitude of 610 km/380 mi. It consists of a 2.4 m/94 in telescope and four complimentary scientific instruments, is roughly cylindrical, 13 m/43 ft long, and 4 m/13 ft in diameter, with two large solar panels. HST produces a wealth of scientific data, and allows astronomers to observe the birth of stars, find planets around neighbouring stars, follow the expanding remnants of exploding stars, and search for black holes in the centre of galaxies. HST is a cooperative programme between the European Space Agency (ESA) and the US agency NASA, and is the first spacecraft specifically designed to be serviced in orbit as a permanent space-based observatory. It was launched in 1990.

Human Genome Project   HGP, research scheme to map the complete nucleotide (see *nucleic acid) sequence of human *DNA. It was begun in 1990 and a working draft of the genome (a mapping of 97% of the genome, sequencing of 85%, and completion of 24% of the human genome) was achieved in June 2000, with the results being published in February 2001. The publicly-funded Human Genome Organization (HUGO) coordinated the US$300 million project (the largest research project ever undertaken in the life sciences), which took place in over 20 centres around the world. Sequencing was also carried out commercially by US biotechnology company Celera Genomics. The completed detailed mapping of the genome is scheduled for 2003.

human papillomavirus   HPV, virus responsible for the common wart. It is one of a group of viruses implicated in tumour formation and is believed to play an important part in triggering cervical cancer that is the second most common cancer worldwide; it is also linked with cancer of the anus, which is very rare. Vaccines are being developed against HPV.

human reproduction   production of offspring by humans. Human reproduction is an example of *sexual reproduction, where the male produces sperm and the female eggs. These gametes contain only half the normal number of chromosomes, 23 instead of 46, so that on fertilization the resulting cell has the correct genetic complement. Fertilization is internal, which increases the chances of conception; unusually for mammals, pregnancy can occur at any time of the year. Human beings are also remarkable for the length of childhood and for the highly complex systems of parental care found in society. The use of contraception and the development of laboratory methods of insemination and fertilization are issues that make human reproduction more than a merely biological phenomenon.

human species, origins of evolution of humans from ancestral primates. The African apes (gorilla and chimpanzee) are shown by anatomical and molecular comparisons to be the closest living relatives of humans. The oldest known hominids (of the human group) had been the australopithecines, found in Africa, dating from 3.5–4.4 million years ago. But in December 2000, scientists unearthed the fossilized remains of a hominid dating back 6 million years. The first hominids to use tools appeared 2 million years ago, and hominids first used fire and moved out of Africa 1.7 million years ago. Modern humans are all believed to descend from one African female of 200,000 years ago, although there is a rival theory that humans evolved in different parts of the world simultaneously.

Miocene apes Genetic studies indicate that the last common ancestor between chimpanzees and humans lived 5 to 10 million years ago. There are only fragmentary remains of ape and hominid fossils from this period. Dispute continues over the hominid status of *Ramapithecus*, the jaws and teeth of which have been found in India and Kenya in late Miocene deposits, dating from between 14 and 10 million years ago. The lower jaw of a fossil ape found in the Otavi Mountains, Namibia, comes from deposits dated between 10 and 15 million years ago, and is similar to finds from East Africa and Turkey. It is thought to be close to the initial divergence of the great apes and humans.

australopithecines *Australopithecus afarensis*, found in Ethiopia and Kenya, date from 3.9 to 4.4 million years ago.

These hominids walked upright and they were either direct ancestors or an offshoot of the line that led to modern humans. They may have been the ancestors of *Homo habilis* (considered by some to be a species of *Australopithecus*), who appeared about 2 million years later, had slightly larger bodies and brains, and were probably the first to use stone tools. Also living in Africa at the same time was *A. africanus*, a gracile hominid thought to be a meat-eater, and *A. robustus*, a hominid with robust bones, large teeth, heavy jaws, and thought to be a vegetarian. They are not generally considered to be our ancestors.

A new species of *Australopithecus* was discovered in Ethiopia in 1999. Named *A. garhi*, the fossils date from 2.5 million years ago and also share anatomical features with *Homo* species. The most complete australopithecine skeleton to date was found in South Africa in April 2000. It is about 1.8 million years old and from a female *A. robustus*.

The skull of an unknown hominid species, *Kenyanthropus platyops*, was discovered in Kenya in March 2001. Approximately 3.5 million years old, it is contemporary with the australopithecines, previously the oldest known hominids, leading to the suggestion that humans are descended from *K. platyops*, rather than the australopithecines as has been thought.

Homo erectus Over 1.7 million years ago, *Homo erectus*, believed by some to be descended from *H. habilis*, appeared in Africa. *H. erectus* had prominent brow ridges, a flattened cranium, with the widest part of the skull low down, and jaws with a rounded tooth row, but the chin, characteristic of modern humans, is lacking. They also had much larger brains (900–1,200 cu cm), and were probably the first to use fire and the first to move out of Africa. Their remains are found as far afield as China, West Asia, Spain, and southern Britain. Modern human *H. sapiens sapiens* and the Neanderthals *H. sapiens neanderthalensis* are probably descended from *H. erectus*.

Australian palaeontologists announced the discovery of stone tools dated at about 800,000 to 900,000 years old and belonging to *H. erectus* on Flores, an island near Bali, in 1998. The discovery provided strong evidence that *H. erectus* were seafarers and had the language abilities and social structure to organize the movements of large groups to colonize new islands. In 2000 Japanese archaeologists discovered that *H. erectus* were probably building hut-like shelters around 500,000 years ago, the oldest known artificial structures.

Neanderthals Neanderthals were large-brained and heavily built, probably adapted to the cold conditions of the ice ages. They lived in Europe and the Middle East, and disappeared about 40,000 years ago, leaving *H. sapiens sapiens* as the only remaining species of the hominid group. Possible intermediate forms between Neanderthals and *H. sapiens sapiens* have been found at Mount Carmel in Israel and at Broken Hill in Zambia, but it seems that *H. sapiens sapiens* appeared in Europe quite rapidly and either wiped out the Neanderthals or interbred with them.

modern humans There are currently two major views of human evolution: the 'out of Africa' model, according to which *H. sapiens* emerged from *H. erectus*, or a descendant species, in Africa and then spread throughout the world; and the multiregional model, according to which selection pressures led to the emergence of similar advanced types of *H. sapiens* from *H. erectus* in different parts of the world at around the same time. Analysis of DNA in recent human populations suggests that *H. sapiens* originated about 200,000 years ago in Africa from a single female ancestor, 'Eve'. The oldest known fossils of *H. sapiens* also come from Africa, dating from 100,000–150,000 years ago. Separation of human populations occurred later, with separation of Asian, European, and Australian populations taking place between 100,000 and 50,000 years ago.

The human genome consists of between 27,000 to 40,000 genes. Of these only about 1.5% differ between humans and the great apes.

Humans are distinguished from apes by the complexity of their brain and its size relative to body size; by their small jaw, which is situated under the face and is correlated with reduction in the size of the anterior teeth, especially the canines, which no longer project beyond the tooth row; by their bipedalism, which affects the position of the head on the vertebral column, the lumbar and cervical curvature of the vertebral column, and the structure of the pelvis, knee joint, and foot; by their complex language; and by their elaborate culture.

The broad characteristics of human behaviour are a continuation of primate behaviour rather than a departure from it. For example, tool use, once a criterion for human status, has been found regularly in gorillas, orang-utans, and chimpanzees, and sporadically in baboons and macaques. Chimpanzees even make tools. In hominid evolution, manual dexterity has increased so that more precise tools can be made. Cooperation in hunting, also once thought to be a unique human characteristic, has been found in chimpanzees, and some gorillas and chimpanzees have been taught to use sign language to communicate.

humerus   upper bone of the forelimb of tetrapods. In humans, the humerus is the bone above the elbow.

humidity   quantity of water vapour in a given volume of the atmosphere (absolute humidity), or the ratio of the amount of water vapour in the atmosphere to the saturation value at the same temperature (relative humidity). At *dew point the relative humidity is 100% and the air is said to be saturated. Condensation (the conversion of vapour to liquid) may then occur. Relative humidity is measured by various types of hygrometer.

humour   in biology, any fluid or semi-fluid tissue in the body, such as the vitreous and aqueous humours of the *eye.

humours, theory of   theory prevalent in the West in classical and medieval times that the human body was composed of four kinds of fluid:

phlegm, blood, choler or yellow bile, and melancholy or black bile. Physical and mental characteristics were explained by different proportions of humours in individuals.

humus component of *soil consisting of decomposed or partly decomposed organic matter, dark in colour and usually richer towards the surface. It has a higher carbon content than the original material and a lower nitrogen content, and is an important source of minerals in soil fertility.

hundredweight symbol cwt, imperial unit of mass, equal to 112 lb (50.8 kg). It is sometimes called the long hundredweight, to distinguish it from the short hundredweight or **cental**, equal to 100 lb (45.4 kg).

Huntington's chorea rare hereditary disease of the nervous system that usually begins in middle age. It is characterized by involuntary movements (*chorea), emotional disturbances, and rapid mental degeneration progressing to *dementia. There is no known cure but the genetic mutation giving rise to the disease was located in 1993, making it easier to test individuals for the disease and increasing the chances of developing a cure.

hurricane or **tropical cyclone** or **typhoon**, a severe *depression (region of very low atmospheric pressure) in tropical regions, called **typhoon** in the North Pacific. It is a revolving storm originating at latitudes between 5° and 20° north or south of the Equator, when the surface temperature of the ocean is above 27°C/80°F. A central calm area, called the eye, is surrounded by inwardly spiralling winds (anticlockwise in the northern hemisphere and clockwise in the southern hemisphere) of up to 320 kph/200 mph. A hurricane is accompanied by lightning and torrential rain, and can cause extensive damage. In meteorology, a hurricane is a wind of force 12 or more on the *Beaufort scale.

HWM abbreviation for **high water mark**.

hybrid offspring from a cross between individuals of two different species, or two inbred lines within a species. In most cases, hybrids between species are infertile and unable to reproduce sexually. In plants, however, doubling of the chromosomes (see *polyploid) can restore the fertility of such hybrids.

hybridization production of a *hybrid.

hydrate chemical compound that has discrete water molecules combined with it. The water is known as **water of crystallization** and the number of water molecules associated with one molecule of the compound is denoted in both its name and chemical formula: for example, $CuSO_4.5H_2O$ is copper(II) sulphate pentahydrate.

hydration in chemistry, the combination of water and another substance to produce a single product. It is the opposite of *dehydration.

hydration in earth science, a form of *chemical weathering caused by the expansion of certain minerals as they absorb water. The expansion weakens the parent rock and may cause it to break up.

hydraulic action in earth science, the erosive force exerted by water (as distinct from erosion by the rock particles that are carried by water). It can wear away the banks of a river, particularly at the outer curve of a *meander (bend in the river), where the current flows most strongly.

hydraulic radius measure of a river's channel efficiency (its ability to move water and sediment), used by water engineers to assess the likelihood of flooding. The hydraulic radius of a channel is defined as the ratio of its cross-sectional area to its wetted perimeter (the part of the cross-section – bed and bank – that is in contact with the water).

hydraulics field of study concerned with utilizing the properties of water and other liquids, in particular the way they flow and transmit pressure, and with the application of these properties in engineering. It applies the principles of *hydrostatics and hydrodynamics. The oldest type of hydraulic machine is the **hydraulic press**, invented by Joseph Bramah in England in 1795. The hydraulic principle of fluid pressure transmitting a small force over a small area in order to produce a larger force over a larger area is commonly used in vehicle braking systems, the forging press, and the hydraulic systems of aircraft and earth-moving machinery.

hydride chemical compound containing hydrogen and one other element, and in which the hydrogen is the more electronegative element (see *electronegativity).

Hydrides of the more reactive metals may be ionic compounds containing a hydride anion (H^-).

hydrocarbon any of a class of chemical compounds containing only hydrogen and carbon (for example, the alkanes and alkenes). Hydrocarbons are obtained industrially principally from petroleum and coal tar.

Unsaturated hydrocarbons contain at least one double or triple carbon–carbon *bond, whereas saturated hydrocarbons contain only single bonds.

hydrocele accumulation of clear fluid in a sac, usually in the *testis. It causes swelling of the scrotum. It is treated by draining or by excision of the sac.

hydrocephalus potentially serious increase in the volume of cerebrospinal fluid (CSF) within the ventricles of the brain. In infants, since their skull plates have not fused, it causes enlargement of the head, and there is a risk of brain damage from CSF pressure on the developing brain.

hydrochloric acid HCl, highly corrosive solution of hydrogen chloride (a colourless, acidic gas) in water. The concentrated acid is about 35% hydrogen chloride. The acid is a typical strong, monobasic acid forming only one series of salts, the chlorides. It has many industrial uses, including recovery of zinc from galvanized scrap iron and the production of chlorine. It is also produced in the stomachs of animals for the purposes of digestion.

hydrocyanic acid or **prussic acid**, solution of hydrogen cyanide gas (HCN) in water. It is a colourless, highly poisonous, volatile liquid, smelling of bitter almonds.

hydrodynamics branch of physics dealing with fluids (liquids and gases) in motion.

hydroelectric power electricity generated by the motion (*kinetic energy) of water. In a typical scheme, the potential energy of water stored in a reservoir, often created by damming a river, is converted into kinetic energy as it is piped into water *turbines. The turbines are coupled to *generators to produce electricity. Hydroelectric power provides about one-fifth of the world's electricity, supplying more than a billion people. Hydroelectricity is a non-polluting, renewable energy resource, produced from water that can be recycled.

hydrofoil wing that develops lift in the water in much the same way that an aeroplane wing develops lift in the air. A hydrofoil boat is one whose hull rises out of the water owing to the lift, and the boat skims along on the hydrofoils. The first hydrofoil was fitted to a boat in 1906. The first commercial hydrofoil went into operation in 1956. One of the most advanced hydrofoil boats is the Boeing jetfoil. Hydrofoils are now widely used for fast island ferries in calm seas.

hydrogen chemical symbol H, (Greek *hydro* + *gen* 'water generator') colourless, odourless, gaseous, non-metallic element, atomic number 1, relative atomic mass 1.00797. It is the lightest of all the elements and occurs on Earth, chiefly in combination with oxygen, as water. Hydrogen is the most abundant element in the universe, where it accounts for 93% of the total number of atoms and 76% of the total mass. It is a component of most stars, including the Sun, whose heat and light are produced through the nuclear-fusion process that converts hydrogen into helium. When subjected to a pressure 500,000 times greater than that of the Earth's atmosphere, hydrogen becomes a solid with metallic properties, as in one of the inner zones of Jupiter. Hydrogen's common and industrial uses include the hardening of oils and fats by hydrogenation, the creation of high-temperature flames for welding, and as rocket fuel. It has been proposed as a fuel for road vehicles.

Its isotopes *deuterium and *tritium (half-life 12.5 years) are used in nuclear weapons, and deuterons (deuterium nuclei) are used in synthesizing elements. The element's name refers to the generation of water by the combustion of hydrogen, and was coined in 1787 by French chemist Louis Guyton de Morveau (1737–1816), after having been discovered in 1766 by English physicist and chemist Henry Cavendish (1731–1810).

hydrogenation addition of hydrogen to an unsaturated organic molecule (one that contains *double bonds or *triple bonds). It is widely used in the manufacture of margarine and low-fat spreads by the addition of hydrogen to vegetable oils.

Vegetable oils contain double carbon-to-carbon bonds and are therefore examples of unsaturated compounds. When hydrogen is added to these double bonds, the oils become saturated and more solid in consistency.

hydrogen bond weak electrostatic bond that forms between covalently bonded hydrogen atoms and a strongly *electronegative atom with a lone electron pair (for example, oxygen, nitrogen, and fluorine). Hydrogen bonds (denoted by a dashed line) are of great importance in biochemical processes, particularly in the N–H group which enables proteins and nucleic acids to form the three-dimensional structures necessary for their biological activity.

hydrogencarbonate or **bicarbonate**, compound containing the ion HCO_3^-, an acid salt of carbonic acid (solution of carbon dioxide in water). When heated or treated with dilute acids, it gives off carbon dioxide. The most important compounds are *sodium hydrogencarbonate (bicarbonate of soda), and *calcium hydrogencarbonate.

hydrogen cyanide HCN, poisonous gas formed by the reaction of sodium cyanide with dilute sulphuric acid; it is used for fumigation.

The salts formed from it are cyanides – for example sodium cyanide, used in hardening steel and extracting gold and silver from their ores. If dissolved in water, hydrogen cyanide gives hydrocyanic acid.

hydrogen peroxide H_2O_2, colourless syrupy liquid used, in diluted form, as an antiseptic. Oxygen is released when hydrogen peroxide is added to water and the froth helps to discharge dead tissue from wounds and ulcers. It is also used as a mouthwash and as a bleach and as a rocket propellant.

hydrogen sulphate HSO_4^-, compound containing the hydrogen sulphate ion. Hydrogen sulphates are *acid salts.

hydrogen sulphide H_2S, poisonous gas with the smell of rotten eggs. It is found in certain types of crude oil where it is formed by decomposition of sulphur compounds. It is removed from the oil at the refinery and converted to elemental sulphur.

hydrogen trioxide H_2O_3, relatively stable compound of hydrogen and oxygen present in the atmosphere and possibly also in living tissue. It was first synthesized in 1994; previously it had been assumed to be too unstable.

It is produced in a reaction similar to that used for the commercial production of hydrogen peroxide (H_2O_2) but ozone (O_3) is used instead of oxygen. Hydrogen trioxide is stable at low temperatures but begins to decompose slowly at $-40°C/-40°F$ forming the high-energy form of oxygen, singlet oxygen.

hydrological cycle also known as the *water cycle, by which water is circulated between the Earth's surface and its atmosphere.

hydrology study of the location and movement of inland water, both frozen and liquid, above and below ground. It is applied to major civil engineering projects such as irrigation schemes, dams, and hydroelectric power, and in planning water supply. Hydrologic studies are also undertaken to assess drinking water supplies, to track water underground, and to understand the role of water in geological processes such as fault movement and mineral deposition.

hydrolysis chemical reaction in which the action of water or its ions breaks down a substance into smaller molecules. Hydrolysis occurs in certain inorganic salts in solution, in nearly all non-metallic chlorides, in esters, and in other organic substances. It is one of the mechanisms for the breakdown of food by the body, as in the conversion of starch to glucose.

hydrometer instrument used to measure the relative density of liquids (the density compared with that of water). A hydrometer consists of a thin glass tube ending in a sphere that leads into a smaller sphere, the latter being weighted so that the hydrometer floats upright, sinking deeper into less dense liquids than into denser liquids. Hydrometers are used in brewing and to test the strength of acid in car batteries.

hydrophilic (Greek 'water-loving') describing *functional groups with a strong affinity for water, such as the carboxyl group (–COOH).

    If a molecule contains both a hydrophilic and a *hydrophobic group (a group that repels water), it may have an affinity for both aqueous and nonaqueous molecules. Such compounds are used to stabilize *emulsions or as *detergents.

hydrophobia another name for the disease *rabies.

hydrophobic (Greek 'water-hating') describing *functional groups that repel water (the opposite of *hydrophilic).

hydrophyte plant adapted to live in water, or in waterlogged soil.

hydroponics cultivation of plants without soil, using specially prepared solutions of mineral salts. Beginning in the 1930s, large crops were grown by hydroponic methods, at first in California but since then in many other parts of the world.

hydrostatics branch of *statics dealing with fluids in equilibrium – that is, in a static condition. Practical applications include shipbuilding and dam design.

hydrothermal in geology, pertaining to a fluid whose principal component is hot water, or to a mineral deposit believed to be precipitated from such a fluid.

hydrothermal vein crack in rock filled with minerals precipitated through the action of circulating high-temperature fluids. Igneous activity often gives rise to the circulation of heated fluids that migrate outwards and move through the surrounding rock. When such solutions carry metallic ions, ore-mineral deposition occurs in the new surroundings on cooling.

hydrothermal vent or **smoker**, crack in the ocean floor, commonly associated with an *ocean ridge, through which hot, mineral-rich water flows into the cold ocean water, forming thick clouds of suspended material. The clouds may be dark or light, depending on the mineral content, thus producing 'white smokers' or 'black smokers'. In some cases the water is clear.

hydrotropism the directional growth of a plant in response to water; see *tropism.

hydroxide any inorganic chemical compound containing one or more hydroxyl (OH) groups and generally combined with a metal. Hydroxides include sodium hydroxide (caustic soda, NaOH), potassium hydroxide (caustic potash, KOH), and calcium hydroxide (slaked lime, $Ca(OH)_2$).

hydroxyl group atom of hydrogen and an atom of oxygen bonded together and covalently bonded to an organic molecule. Common compounds containing hydroxyl groups are alcohols and phenols. In chemical reactions, the hydroxyl group (–OH) frequently behaves as a single entity.

hydroxypropanoic acid technical name for *lactic acid.

hyperactivity condition of excessive activity in young children, combined with restlessness, inability to concentrate, and difficulty in learning. There are various causes, ranging from temperamental predisposition to brain disease. In some cases food additives have come under suspicion; in such instances modification of the diet may help. Mostly there is improvement at puberty, but symptoms may persist in the small proportion diagnosed as having *attention-deficit hyperactivity disorder (ADHD).

hyperalgesia in medicine, excessive sensitivity to pain.

hyperbola in geometry, a curve formed by cutting a right circular cone with a plane so that the angle between the plane and the base is greater than the angle between the base and the side of the cone. All hyperbolae are bounded by two asymptotes (straight lines to which the hyperbola moves closer and closer to but never reaches). A hyperbola is a member of the family of curves known as *conic sections.

A hyperbola can also be defined as a path traced by a point that moves such that the ratio of its distance from a fixed point (focus) and a fixed straight line (directrix) is a constant and greater than 1; that is, it has an *eccentricity greater than 1.

hypercharge in physics, a property of certain *elementary particles, analogous to electric charge, that accounts for the absence of some expected behaviour (such as certain decays).

hyperlink in computing, link from one document to another or, within the same document, from one place to another. It can be activated by clicking on the link with a *mouse. The link is usually highlighted in some way, for example by the inclusion of a small graphic. Documents linked in this way are described as *hypertext. Examples of programs that use hypertext and hyperlinks are Windows help files, Acrobat, and Mosaic.

hypermedia in computing, system that uses links to lead users to related graphics, audio, animation, or video files in the same way that *hypertext systems link related pieces of text. The World Wide Web is an example of a hypermedia system, as is Hypercard.

hypermetropia or **long-sightedness**, defect of vision in which a person is able to focus on objects in the distance, but not on close objects. It is caused by the failure of the lens to return to its normal rounded shape, or by the eyeball being too short, with the result that the image is focused on a point behind the retina. Hypermetropia is corrected by wearing glasses fitted with *converging lenses, each of which acts like a magnifying glass.

hyperon in physics, any of a group of highly unstable *elementary particles that includes all the baryons with the exception of protons and neutrons. They are all composed of three quarks. The lambda, xi, sigma, and omega particles are hyperons.

hypertension abnormally high *blood pressure due to a variety of causes, leading to excessive contraction of the smooth muscle cells of the walls of the arteries. It increases the risk of kidney disease, stroke, and heart attack.

hypertext in computing, a method of forming connections between different files (including office documents, graphics, and Web pages) so that the user can click a 'link' with the *mouse to jump between them. For example, a software program might display a map of a country; if the user clicks on a particular city the program displays information about that city. The linked files do not need to be on the same computer, or even in the same country, for a hyperlink to be created.

hyperventilation in medicine, an abnormally rapid resting respiratory rate. It occurs in patients with chest and heart diseases, such as severe chronic obstructive lung disease, because of a lack of oxygen or an increase in carbon dioxide in the blood.

hypha (plural **hyphae**) delicate, usually branching filament, many of which collectively form the mycelium and fruiting bodies of a *fungus. Food molecules and other substances are transported along hyphae by the movement of the cytoplasm, known as 'cytoplasmic streaming'.

hypnosis artificially induced state of relaxation or altered attention characterized by heightened suggestibility. There is evidence that, with susceptible persons, the sense of pain may be diminished, memory of past events enhanced, and illusions or hallucinations experienced. Post-hypnotic amnesia (forgetting what happened during hypnosis) and post-hypnotic suggestion (performing an action after hypnosis that had been suggested during it) have also been demonstrated.

hypnotic any substance (such as *barbiturate, benzodiazepine, alcohol) that depresses brain function, inducing sleep. Prolonged use may lead to physical or psychological addiction.

hypogeal describing seed germination in which the *cotyledons remain below ground. It can refer to fruits that develop underground, such as peanuts *Arachis hypogea*.

hypoglycaemia condition of abnormally low level of sugar (glucose) in the blood, which starves the brain. It causes weakness, sweating, and mental confusion, sometimes fainting.

hypotension in medicine, abnormally low blood pressure. Postural hypotension refers to the sudden fall in blood pressure that can occur on standing up suddenly. The patient experiences dizziness and may fall. It is most common in the elderly.

hypotenuse longest side of a right-angled triangle, opposite the right angle. It is of particular application in *Pythagoras' theorem (the square of the hypotenuse equals the sum of the squares of the other two sides), and in trigonometry where the ratios *sine and *cosine are defined as the ratios opposite/hypotenuse and adjacent/hypotenuse respectively.

hypothalamus region of the brain below the *cerebrum which regulates rhythmic activity and physiological stability within the body, including water balance and temperature. It regulates the production of the pituitary gland's hormones and controls that part of the *nervous system governing the involuntary muscles.

hypothermia condition in which the deep (core) temperature of the body falls below 35°C. If it is not discovered coma and death ensue. Most at risk are the aged and babies (particularly if premature).

hypothesis in science, an idea concerning an event and its possible

explanation. The term is one favoured by the followers of the philosopher Karl Popper, who argue that the merit of a scientific hypothesis lies in its ability to make testable predictions.

hypothesis an unproven idea. It is part of the process of solving problems: to form a hypothesis and then test it by experiment.

hypoventilation in medicine, an abnormally slow resting respiratory rate characterized by shallow breathing. It can result in a lack of oxygen in the blood and can lead to organ damage and death if it is severe and remains untreated. Injury to the respiratory centre and some drugs, such as opioid analgesics, can cause hypoventilation.

hypoxia shortage of oxygen in the tissues. See also *anoxia.

hysteresis phenomenon seen in the elastic and electromagnetic behaviour of materials, in which a lag occurs between the application or removal of a force or field and its effect.

hysteria according to the work of Austrian physician Sigmund Freud (1856–1939), the conversion of a psychological conflict or anxiety feeling into a physical symptom, such as paralysis, blindness, recurrent cough, vomiting, and general malaise. The term is little used today in diagnosis, *psychosomatic illness being the modern term.

Hz in physics, the symbol for *hertz.

IAB in computing, abbreviation for Internet Architecture Board.

IAEA abbreviation for International Atomic Energy Agency.

iatrogenic caused by medical treatment; the term 'iatrogenic disease' may be applied to any pathological condition or complication that is

caused by the treatment, the facilities, or the staff.

IBM abbreviation for International Business Machines, multinational company, the largest manufacturer of computers in the world. The company is a descendant of the Tabulating Machine Company, formed in 1896 by

US inventor Herman Hollerith to exploit his punched-card machines. It adopted its present name in 1924.

IC abbreviation for *integrated circuit.

ICANN acronym for Internet Corporation for Assigned Names and Numbers, not-for-profit organization set up by the US government in 1999 to oversee the issuing of top-level Internet *domain names. The private company Network Solutions had previously held a monopoly on this. ICANN has licensed many other companies to issue top-level domain names.

ICE acronym for Information Content Exchange, protocol developed in 1998 for the transfer of data to partnering Web sites. Applications based on ICE allow companies to construct syndicated publishing networks, Web superstores, and online reseller channels.

ice solid formed by water when it freezes. It is colourless and its crystals are hexagonal. The water molecules are held together by hydrogen bonds.

The freezing point of ice, used as a standard for measuring temperature, is 0° for the Celsius and Réaumur scales and 32° for the Fahrenheit scale. Ice expands in the act of freezing (hence burst pipes), becoming less dense than water (0.9175 at 5°C/41°F). In 1998 US geologists succeeded in creating ice made up of irregular crystals. The new type of ice was created by exerting an enormous pressure (6,500 atmospheres) on water molecules, until their hydrogen atoms were squeezed into a disorderly state.

ice age any period of extensive glaciation (in which icesheets and icecaps expand over the Earth) occurring in the Earth's history, but particularly that in the *Pleistocene epoch (last 2 million years), immediately preceding historic times. On the North American continent, *glaciers reached as far south as the Great Lakes, and an icesheet spread over northern Europe, leaving its remains as far south as Switzerland. In Britain ice reached as far south as a line from Bristol to Banbury to Exeter. There were several glacial advances separated by interglacial (warm)

stages, during which the ice melted and temperatures were higher than today. We are currently in an interglacial phase of an ice age.

Other ice ages have occurred throughout geological time: there were four in the Precambrian era, one in the Ordovician, and one at the end of the Carboniferous and beginning of the Permian. The occurrence of an ice age is governed by a combination of factors (the **Milankovitch hypothesis**): (1) the Earth's change of attitude in relation to the Sun – that is, the way it tilts in a 41,000-year cycle and at the same time wobbles on its axis in a 22,000-year cycle, making the time of its closest approach to the Sun come at different seasons; and (2) the 92,000-year cycle of eccentricity in its orbit around the Sun, changing it from an elliptical to a near circular orbit, the severest period of an ice age coinciding with the approach to circularity. There is a possibility that the Pleistocene ice age is not yet over. It may reach another maximum in another 60,000 years.

Ice Age, Little period of particularly severe winters that gripped northern Europe between the 13th and 17th centuries. Contemporary writings and paintings show that Alpine glaciers were much more extensive than at present, and rivers such as the Thames, which do not ice over today, were so frozen that festivals could be held on them.

iceberg floating mass of ice, about 80% of which is submerged, rising sometimes to 100 m/300 ft above sea level. Glaciers that reach the coast become extended into a broad foot; as this enters the sea, masses break off and drift towards temperate latitudes, becoming a danger to shipping.

icecap body of ice that is larger than a glacier but smaller than an ice sheet. Such ice masses cover mountain ranges, such as the Alps, or small islands. Glaciers often originate from icecaps.

Iceland spar form of *calcite, $CaCO_3$, originally found in Iceland. In its pure form Iceland spar is transparent and exhibits the peculiar phenomenon of producing two images of anything seen through it (birefringence or *double refraction). It is used in optical instruments. The crystals cleave into perfect rhombohedra.

ice sheet body of ice that covers a large land mass or continent; it is larger than an ice cap. During the last *ice age, ice sheets spread over large parts of Europe and North America. Today there are two ice sheets, covering much of Antarctica and Greenland. About 96% of all present-day ice is in the form of ice sheets. The ice sheet covering western Greenland has increased in thickness by 2 m/ 6.5 ft 1981–93; this increase is the equivalent of a 10% rise in global sea levels.

icon in computing, a small picture on the computer screen, or *VDU, representing an object or function that the user may manipulate or otherwise use. It is a feature of *graphical user interface (GUI) systems. Icons make computers easier to use by allowing the user to point to and click with a *mouse on pictures, rather than type commands.

icosahedron plural **icosahedra**, regular solid with 20 equilateral (equal-sided) triangular faces. It is one of the five regular *polyhedra, or Platonic solids.

id in Freudian psychology, the mass of motivational and instinctual elements of the human mind, whose activity is largely governed by the arousal of specific needs. It is regarded as the *unconscious element of the human psyche, and is said to be in conflict with the *ego and the *superego.

IDE abbreviation for intelligent drive electronics or integrated drive electronics, interface for mass-storage devices where the controller is integrated into the *disk drive (either a hard disk, a high-capacity removable disk drive, or a CD-ROM drive). It is the most popular interface used in modern hard disks and is most commonly used to refer to the advanced technology attachment (ATA) specification.

IDEA acronym for International Data Encryption Algorithm, in computing, an encryption *algorithm, developed in 1990 in Zürich, Switzerland. For reasons of speed, it is used in the encryption program Pretty Good Privacy (PGP) along with *RSA.

identity in mathematics, a number or operation that leaves others unchanged when combined with them. Zero is the identity for addition; one is the identity for multiplication. For example:

$$7 + 0 = 7$$

$$7 \times 1 = 7$$

identity the distinct and recognizable nature of an individual, which results from a unique combination of characteristics and qualities. In philosophy, identity is the sameness of a person, which may continue in spite of changes in bodily appearance, personality, intellectual abilities, memory, and so on. In psychology, identity refers to one's conception of oneself and sense of continuous being, particularly as an individual distinguishable from, but interacting, with others.

id Software computer software company that publishes popular games such as *Doom* and *Quake*, based in Texas, USA. An entire subculture has built up around id's games because of its habit of releasing *source code to enable fans to write their own additional game levels using settings of their own choice.

igneous rock rock formed from the cooling and solidification of molten rock called *magma. The acidic nature of this rock type means that areas with underlying igneous rock are particularly susceptible to the effects of acid rain. Igneous rocks that crystallize slowly from magma below the Earth's surface have large crystals. Examples include dolerite and granite.

Igneous rocks that crystallize from magma below the Earth's surface are called **plutonic** or **intrusive**, depending on the depth of formation. They have large crystals produced by slow cooling; examples include dolerite and *granite. Those extruded at the surface from *lava are called **extrusive** or **volcanic**. Rapid cooling results in small crystals; *basalt is an example.

ignis fatuus another name for *will-o'-the-wisp.

ignition temperature or **fire point**, minimum temperature to which a substance must be heated before it will spontaneously burn independently of the source of heat; for example, ethanol has an ignition temperature of 425°C/798°F and a *flash point of 12°C/54°F.

ileum part of the small intestine of the *digestive system, between the duodenum and the colon, that absorbs digested food.

illumination or **illuminance**, the brightness or intensity of light falling on a surface. It depends upon the brightness and distance of light sources and the angle at which the light strikes the surface. The SI unit is the *lux.

image in mathematics, a point or number that is produced as the result of a *transformation or mapping.

image picture or appearance of a real object, formed by light that passes through a lens or is reflected from a mirror. If rays of light actually pass through an image, it is called a **real image**. Real images, such as those produced by a camera or projector lens, can be projected onto a screen. An image that cannot be projected onto a screen, such as that seen in a flat mirror, is known as a **virtual image**.

In a pinhole camera, rays of light travelling from the object through the pinhole of the camera form a real image on the screen of the camera. The image is upside down, or inverted.

image compression in computing, one of a number of methods used to reduce the amount of information required to represent an image, so that it takes up less computer memory and can be transmitted more rapidly and economically via telecommunications systems. It plays a major role in fax transmission and in videophone and multimedia systems.

image intensifier electronic device that brightens a dark image. Image intensifiers are used for seeing at night; for example, in military situations.

The intensifier first forms an image on a photocathode, which emits electrons in proportion to the intensity of the light falling on it. The electron flow is increased by one or more amplifiers. Finally, a fluorescent screen converts the electrons back into visible light, now bright enough to see.

imaginary number term often used to describe the non-real element of a *complex number. For the complex number $(a + ib)$, ib is the imaginary number where $i = \sqrt{(-1)}$, and b any real number.

immediate access memory in computing, *memory provided in the *central processing unit to store the programs and data in current use.

immiscible describing liquids that will not mix with each other, such as oil and water. When two immiscible liquids are shaken together, an emulsion is produced. This normally forms separate layers on being left to stand. Immiscible liquids may be separated using a separating funnel.

immunity protection that organisms have against foreign micro-organisms, such as bacteria and viruses, and against cancerous cells (see *cancer). The cells that provide this protection are called white blood cells, or leucocytes, and make up the immune system. They include neutrophils and *macrophages, which can engulf invading organisms and other unwanted material, and natural killer cells that destroy cells infected by viruses and cancerous cells. Some of the most important immune cells are the *B cells and *T cells. Immune cells coordinate their activities by means of chemical messengers or *lymphokines, including the antiviral messenger *interferon. The lymph nodes play a major role in organizing the immune response.

Immunity is also provided by a range of physical barriers such as the skin, tear fluid, acid in the stomach, and mucus in the airways. *AIDS is one of many viral diseases in which the immune system is affected.

immunization process of conferring immunity to infectious disease by artificial methods, in other words making someone not liable to catch a disease. The most widely used technique is *vaccination.

Immunization is an important public

health measure. If most of the population has been immunized against a particular disease, it is impossible for an epidemic to take hold.

In vaccination people can be immunized against a disease by introducing small quantities of dead or inactive forms of the disease-causing micro-organism into the body. In the vaccine are chemicals which act as antigens. These stimulate the white blood cells to produce antibodies. Antibodies are capable of binding to disease-causing bacteria, resulting in their destruction. If the body is effectively 'warned' about the antigen by this means, the body is then able to produce enough of the appropriate antibody whenever the living disease-causing organism enters the body. The micro-organism can then be killed before it harms the body. The person contacted by the disease will probably feel well all the time and will be unaware that he or she has been in contact with the disease. This is called active immunity.

Vaccination can be used to protect against diseases caused by viruses. It is, in fact, the best way to deal with virus diseases, because antibiotics are not effective against them. An example is the MMR vaccine used to protect children against measles, mumps, and rubella. A study of thousands of children in Finland who have had the MMR vaccine has shown that there is only a very low risk of damage being caused to a child as a result of having the vaccine. The risk is much lower than the damage caused by catching one of the diseases.

If vaccination covers a large proportion of the population at risk, a disease can become very rare, or even die out. Smallpox was eliminated in this way. The worries over MMR vaccine have reduced the numbers of children being vaccinated, which increases the risk of measles, mumps, or rubella breaking out in an epidemic.

immunoglobulin   human globulin *protein that can be separated from blood and administered to confer immediate immunity on the recipient. It participates in the immune reaction as the antibody for a specific *antigen (disease-causing agent).

Normal immunoglobulin (gamma globulin) is the fraction of the blood serum that, in general, contains the most antibodies, and is obtained from

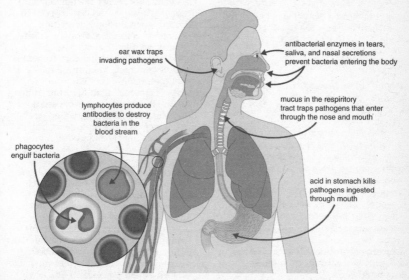

ear wax traps invading pathogens

antibacterial enzymes in tears, saliva, and nasal secretions prevent bacteria entering the body

lymphocytes produce antibodies to destroy bacteria in the blood stream

mucus in the respiratory tract traps pathogens that enter through the nose and mouth

phagocytes engulf bacteria

acid in stomach kills pathogens ingested through mouth

immunity How the body's defence mechanisms keep out invading pathogens or destroy those that succeed in entering the body.

plasma pooled from about a thousand donors. It is given for short-term (two to three months) protection when a person is at risk, mainly from hepatitis A (infectious hepatitis), or when a pregnant woman, not immunized against *German measles, is exposed to the rubella virus.

Specific immunoglobulins are injected when a susceptible (nonimmunized) person is at risk of infection from a potentially fatal disease, such as hepatitis B (serum hepatitis), rabies, or tetanus. These immunoglobulins are prepared from blood pooled from donors convalescing from the disease.

impedance symbol Z, total opposition of a circuit to the passage of alternating electric current. It includes the resistance R and the reactance X (caused by *capacitance or *inductance); the impedance can then be found using the equation $Z^2 = R^2 + X^2$.

imperial system traditional system of units developed in the UK, based largely on the foot, pound, and second (f.p.s.) system.

impermeable rock rock that does not allow water to pass through it – for example, clay, shale, and slate. Unlike *permeable rocks, which absorb water, impermeable rocks can support rivers. They therefore experience considerable erosion (unless, like slate, they are very hard) and commonly form lowland areas.

imply in mathematics, to lead logically to. For example, if $2x = 10$, then $x = 5$. The second statement follows from the first.

import file in computing, a file that can be read by a program even though it was produced as an *export file by a different program or make of computer.

imprinting in ethology, the process whereby a young animal learns to recognize both specific individuals (for example, its mother) and its own species.

improper fraction *fraction whose numerator is larger than its denominator.

impulse in mechanics, the product of a force and the time over which it acts.

An impulse applied to a body causes its *momentum to change and is equal to that change in momentum. It is measured in newton seconds (N s). For example, the impulse J given to a football when it is kicked is given by:

$$J = Ft$$

where F is the kick force in newtons and t is the time in seconds for which the boot is in contact with the ball.

impulse in biology, a means of carrying information between different parts of the body. Nerve impulses travel at speed along nerve cells (neurones). A sequence of impulses carries information.

in abbreviation for *inch, a measure of distance.

inbreeding in *genetics, the mating of closely related individuals. It is considered undesirable because it increases the risk that offspring will inherit copies of rare deleterious *recessive alleles (genes) from both parents and so suffer from disabilities.

incandescence emission of light from a substance in consequence of its high temperature. The colour of the emitted light from liquids or solids depends on their temperature, and for solids generally the higher the temperature the whiter the light. Gases may become incandescent through *ionizing radiation, as in the glowing vacuum *discharge tube.

inch imperial unit of linear measure, a twelfth of a foot, equal to 2.54 centimetres.

incisor sharp tooth at the front of the mammalian mouth. Incisors are used for biting or nibbling, as when a rabbit or a sheep eats grass. Rodents, such as rats and squirrels, have large continually-growing incisors, adapted for gnawing. The elephant tusk is a greatly enlarged incisor. In humans, the incisors are the four teeth at the front centre of each jaw.

inclusive fitness in *genetics, the success with which a given variant (or allele) of a *gene is passed on to future generations by a particular individual, after additional copies of the allele in the individual's relatives and their offspring have been taken into account.

incubation period in infectious disease, the time-lag between catching the *pathogen and developing symptoms of the condition.

independent variable the variable that does not depend on another variable for its values. The symbol x is usually employed to denote the independent variable, while y is used for the dependent variable. Time is always an independent variable.

indeterminacy principle alternative name for *uncertainty principle.

index plural **indices**, (Latin 'sign', 'indicator') in mathematics, another term for *exponent, the number that indicates the power to which a term should be raised.

indexed sequential file in computing, a type of file access in which an index is used to obtain the address of the *block containing the required record.

index notation plural **indices notation**, the index, or *power of a number indicates how many times the number is to be multiplied by itself. For example, 4^2 (four squared) $= 4 \times 4 = 16$; 2^3 (two cubed) $= 2 \times 2 \times 2 = 8$; and 5^4 (five to the power four) $= 5 \times 5 \times 5 \times 5 = 625$. In these examples, 4^2, 2^3, and 5^4 are all written in index notation. There are three important rules for the use of indices:

(i) $a^m \times a^n = a^{m+n}$

(ii) $a^m \div a^n = a^{m-n}$

(iii) $(a^m)^n = a^{mn}$

In all these three rules, m and n are assumed to be positive whole numbers and in rule (ii), a is not zero and $m > n$. For example:

$3^3 \times 3^4 = 3^{3+4} = 3^7$;
$5^4 \div 5^2 = 5^{4-2} = 5^2$;
$(2^3)^4 = 2^{3 \times 4} = 2^{12}$

When using fractional indices, $x^{1/n} = \sqrt[n]{x}$, so $x^{1/2} = \sqrt[2]{x}$ and $x^{1/4} = \sqrt[4]{x}$; also $x^{m/n} = (\sqrt[n]{x})^m$ or $\sqrt[n]{(x^m)}$. The same rules apply when using fractional indices as for whole number indices.

indicator in chemistry, a compound that changes its structure and colour in response to its environment. The commonest chemical indicators detect changes in *pH (for example, litmus and *universal indicator) or in the oxidation state of a system (redox indicators).

indium chemical symbol In, (Latin *indicum* 'indigo') soft, ductile, silver-white, metallic element, atomic number 49, relative atomic mass 114.82. It occurs in nature in some zinc ores, is resistant to abrasion, and is used as a coating on metal parts. It was discovered in 1863 by German metallurgists Ferdinand Reich (1799–1882) and Hieronymus Richter (1824–1898), who named it after the two indigo lines of its spectrum.

induced current electric current that appears in a closed circuit when there is relative movement of its conductor in a magnetic field. The effect is known as the **dynamo effect**, and is used in all *dynamos and generators to produce electricity. See *electromagnetic induction. The magnitude of the induced current depends upon the rate at which the magnetic flux is cut by the conductor, and its direction is given by Fleming's right-hand rule (see *Fleming's rules).

inductance in physics, phenomenon in which a changing current in a circuit builds up a magnetic field which induces an *electromotive force either in the same circuit and opposing the current (self-inductance) or in another circuit (mutual inductance). The SI unit of inductance is the henry (symbol H).

A component designed to introduce inductance into a circuit is called an *inductor (sometimes inductance) and is usually in the form of a coil of wire. The energy stored in the magnetic field of the coil is proportional to its inductance and the current flowing through it. See *electromagnetic induction.

induction in physics, an alteration in the physical properties of a body that is brought about by the influence of a field. See *electromagnetic induction and *magnetic induction.

induction coil type of electrical transformer, similar to an ignition coil, that produces an intermittent high-voltage alternating current from a low-voltage direct current supply. It has a primary coil consisting of a few turns of thick wire wound around an iron core and subjected to a low voltage (usually from a battery). Wound on top of this is a secondary coil made up of many turns of thin wire. An iron armature and make-and-break mechanism (similar to that in an electric bell) repeatedly interrupts the current to the primary coil, producing a high-voltage, rapidly alternating current in the secondary circuit.

An **induction motor** employs *alternating current. It comprises a stationary current-carrying coil (stator) surrounding another coil (rotor), which rotates because of the current induced in it by the magnetic field created by the stator; it thus requires no commutator.

inductor device included in an electrical circuit because of its inductance.

inequality in mathematics, a statement that one quantity is larger or smaller than another, employing the symbols < and >. Inequalities may be solved by finding sets of numbers that satisfy them. For example, to find the solution set to the inequality $2x + 3 < 19$, the inequality can be treated like an equation:

$$2x + 3 < 19, \quad \text{so} \quad 2x < 19 - 3,$$
$$2x < 16, \quad \text{and so} \quad x < 8.$$

Therefore, the solution set consists of all values of x less than 8 on the number line. Inequality relationships involving variables are sometimes called **inequations**.

inert gas alternative name for *noble gas or rare gas.

inertia in physics, the tendency of an object to remain in a state of rest or uniform motion until an external force is applied, as described by Isaac Newton's first law of motion (see *Newton's laws of motion).

inferiority complex in psychology, a complex or cluster of repressed fears, described by the Austrian psychologist Alfred Adler (1870–1937), based on physical inferiority. The term is popularly used to describe general feelings of inferiority and the overcompensation that often ensues.

inferno in astrophysics, a unit for describing the temperature inside a star. One inferno is 1 billion K, or approximately 1 billion °C.

infinite series in mathematics, a series of numbers consisting of a denumerably infinite sequence of terms. The sequence $n, n^2, n^3, \ldots$ gives the series $n + n^2 + n^3 + \ldots$. For example, $1 + 2 + 3 + \ldots$ is a divergent infinite arithmetic series, and $8 + 4 + 2 + 1 + 1/2 + \ldots$ is a convergent infinite geometric series that has a sum to infinity of 16.

infinity symbol ∞, mathematical quantity that is larger than any fixed assignable quantity. By convention, the result of dividing any number by zero is regarded as infinity.

inflorescence in plants, a branch, or system of branches, bearing two or more individual flowers. Inflorescences can be divided into two main types: cymose (or definite) and racemose (or indefinite). In a **cymose inflorescence**, the tip of the main axis produces a single flower and subsequent flowers arise on lower side branches, as in forget-me-not *Myosotis* and chickweed *Stellaria*; the oldest flowers are, therefore, found at the tip. A **racemose inflorescence** has an active growing region at the tip of its main axis, and bears flowers along its length, as in hyacinth *Hyacinthus*; the oldest flowers are found near the base or, in cases where the inflorescence is flattened, towards the outside. The stalk of the inflorescence is called a peduncle; the stalk of each individual flower is called a pedicel.

influenza any of various viral infections primarily affecting the air passages, accompanied by *systemic effects such as fever, chills, headache, joint and muscle pains, and lassitude. Treatment is with bed rest and analgesic drugs such as aspirin or paracetamol.

information superhighway popular collective name for the *Internet and other related large-scale computer networks. The term was first used in 1993 by US vice-president Al Gore in a speech outlining plans to build a high-speed national data communications network.

information technology IT, collective term for the various technologies involved in processing and transmitting information. They include computing, telecommunications, and microelectronics. The term became popular in the UK after the Government's 'Information Technology Year' in 1982.

infrared radiation i.r., electromagnetic *radiation of wavelength between about 700 nanometres and 1 millimetre – that is, between the limit of the red end of the visible spectrum and the shortest microwaves. All bodies above the *absolute zero of temperature absorb and radiate infrared radiation. Infrared radiation is used in medical photography and treatment, and in industry, astronomy, and criminology.

The human eye cannot detect infrared, but its effects can be demonstrated. For example, an electric hob operates at high temperatures and only the visible light it gives out can be seen. As it cools down, the visible light is no longer seen. However, the heat (infrared radiation) that continues to be given out can be felt. Infrared absorption spectra are used in chemical analysis, particularly for organic compounds. Objects that radiate infrared radiation can be photographed or made visible in the dark on specially sensitized emulsions. This is important for military purposes and in detecting people buried under rubble. The strong absorption by many substances of infrared radiation is a useful method of applying heat.

infrastructure on the Internet, the underlying structure of telephone links, leased lines, and computer programs that makes communication possible.

ingestion process of taking food into the mouth. The method of food capture varies but may involve biting, sucking, or filtering. Many single-celled organisms have a region of their cell wall that acts as a mouth. In these cases surrounding tiny hairs (cilia) sweep food particles together, ready for ingestion.

inheritance in biology, the passing of characteristics from parents to offspring. Characteristics that can be passed on in this way are determined by *genes (see also *allele).

The study of inheritance is called genetics and was founded by the Austrian biologist Gregor Mendel (1822–1884). He found that crossing two parents having one contrasting characteristic (for example, one tall and the other dwarf) showed that inheritance was the consequence of passing on 'particles' or, as we now know them, genes, which determine the characteristics. This kind of cross is called a monohybrid cross.

In *asexual reproduction the offspring inherit the same characteristics as the parent. In *sexual reproduction, the offspring inherit a mix of the characteristics of the two parents.

inherited disease disease resulting from the effect of having defective *genes. These diseases are inherited because genetic material is accurately copied before being passed onto the offspring from a parent. However, they can also arise as a result of a sudden change in the DNA known as a *mutation. Common examples of inherited diseases in humans are cystic fibrosis, sickle-cell anaemia, polydactyly, and Huntingdon's disease. Downs syndrome is an example of a disease arising from mutation, but which is not inherited.

Cystic fibrosis, sickle-cell anaemia, polydactyly, and Huntingdon's disease are all each the result of a single defective gene. Down's syndrome, however, is usually the result of having an extra chromosome containing thousands of genes. In polydactyly (extra fingers) and Huntingdon's disease (physical and mental degeneration) the diseases show dominance – this means that inheriting a single defective gene will result in the disease. Cystic fibrosis (viscous secretions, especially in the lungs) is recessive, which means that two defective genes have to be inherited for the disease to occur.

In sickle-cell anaemia people who inherit a single defective gene are normal most of the time, but can show anaemia from time to time. However, if a person inherits two defective genes, they show severe illness. This defective gene exhibits incomplete dominance. The gene for sickle-cell anaemia is more common than would be expected considering the anaemia it causes. It is more common than expected because inheritance of a single defective gene makes the person partly immune to malaria, so *natural selection favours these individuals where malaria is common.

inhibition, neural in biology, the process in which activity in one *nerve cell suppresses activity in another. Neural inhibition in networks of nerve cells leading from sensory organs, or to muscles, plays an important role in allowing an animal to make fine sensory discriminations and to exercise fine control over movements.

inoculation injection into the body of dead or weakened disease-carrying organisms or their toxins (*vaccine) to produce immunity by inducing a mild form of a disease.

inorganic chemistry branch of chemistry dealing with the chemical properties of the elements and their compounds, excluding the more complex covalent compounds of carbon, which are considered in *organic chemistry.

The origins of inorganic chemistry lay in observing the characteristics and experimenting with the uses of the substances (compounds and elements) that could be extracted from mineral ores. These could be classified according to their chemical properties: elements could be classified as metals or non-metals; compounds as acids or bases, oxidizing or reducing agents, ionic compounds (such as salts), or covalent compounds (such as gases). The arrangement of elements into groups possessing similar properties led to Mendeleyev's *periodic table of the elements, which prompted chemists to predict the properties of undiscovered elements that might occupy gaps in the table. This, in turn, led to the discovery of new elements, including a number of highly radioactive elements that do not occur naturally.

inorganic compound chemical compound that does not contain carbon and is not manufactured by living organisms. Water, sodium chloride, and potassium are inorganic compounds because they are widely found outside living cells. However, carbon dioxide is considered inorganic,

contains carbon, and is manufactured by organisms during respiration. Carbonates and carbon monoxide are also regarded as inorganic compounds. See *organic compound.

input device device for entering information into a computer. Input devices include keyboards, joysticks, mice, light pens, touch-sensitive screens, scanners, graphics tablets, speech-recognition devices, and vision systems. The input into an electronic system is usually through *switches or *sensors. Compare with an output device.

insemination, artificial see *artificial insemination.

insulation process or material that prevents or reduces the flow of electricity, heat, or sound from one place to another.

Materials that are poor conductors of heat, such as glass, brick, water, or air, are good insulators. They play a vital role, for example, in keeping homes and people warm.

insulator any poor *conductor of heat, sound, or electricity. Most substances lacking free (mobile) *electrons, such as non-metals, are electrical or thermal insulators that resist the flow of electricity or heat through them. Plastics and rubber are good insulators. Usually, devices of glass or porcelain, called insulators, are used for insulating and supporting overhead wires.

insulin protein *hormone, produced by specialized cells in the *islets of Langerhans in the pancreas. Insulin regulates the metabolism (rate of activity) of glucose, fats, and proteins. In this way it helps to regulate the concentration of *glucose in the *blood of mammals. If the blood glucose concentration is too high, the pancreas releases insulin into the blood. This causes blood glucose levels to fall. This is partly due to increased uptake of glucose into most body cells (except liver and brain) but also because the liver converts glucose into insoluble glycogen and stores it instead of making glucose. In *diabetes a person's blood glucose may rise to such a high concentration that it can kill. This may be because the pancreas does not make enough insulin, or because the body cells respond less to the insulin.

Normally, insulin is secreted in response to rising blood sugar levels (after a meal, for example), stimulating the body's cells to store the excess. Failure of this regulatory mechanism in diabetes mellitus requires treatment with insulin injections or capsules taken by mouth. Types vary from pig and beef insulins to synthetic and bioengineered ones. They may be combined with other substances to make them longer- or shorter-acting. Implanted, battery-powered insulin pumps deliver the hormone at a preset rate, to eliminate the unnatural rises and falls that result from conventional, subcutaneous (under the skin) delivery. Human insulin has now been produced from bacteria by *genetic engineering techniques. In 1990 the Medical College of Ohio developed gelatin capsules and an aspirin-like drug which helps the insulin pass into the bloodstream.

Insulin was discovered by Canadian physician Frederick Banting (1891–1941) and Canadian physiologist Charles Best in 1922, who pioneered its use in treating diabetes. Its structure was determined by English biochemist Frederick Sanger (1918–) in 1955.

integer any whole number. Integers may be positive or negative; 0 is an integer, and is often considered positive. Formally, integers are members of the set

$$Z = \{\dots -3, -2, -1, 0, 1, 2, 3, \dots \}$$

This is the integer set of the number line.

*Fractions, such as $1/2$ and the *decimal 0.35, are known as non-integral numbers ('not integers').

The *addition (their sum) and *subtraction (their *difference) of two integers will always result in an integer. The *multiplication of two integers (their product) will always result in an integer; however, the *division of integers may result in a non-integer.

integral calculus branch of mathematics using the process of *integration. It is concerned with finding volumes and areas and

summing infinitesimally small quantities.

integrated circuit IC; or **silicon chip**, miniaturized electronic circuit produced on a single crystal, or chip, of a semiconducting material – usually silicon. It may contain many millions of components and yet measure only 5 mm/0.2 in square and 1 mm/0.04 in thick. The IC is encapsulated within a plastic or ceramic case, and linked via gold wires to metal pins with which it is connected to a *printed circuit board and the other components that make up such electronic devices as computers and calculators.

integrated pest management IPM, the use of a coordinated array of methods to control pests, including biological control, chemical *pesticides, crop rotation, and avoiding monoculture. By cutting back on the level of chemicals used the system can be both economical and beneficial to health and the environment.

Integrated Services Digital Network ISDN, internationally developed telecommunications network for sending signals in digital format that offers faster data transfer rates than traditional analogue telephone circuits. It involves converting the 'local loop' – the link between the user's telephone (or private automatic branch exchange) and the digital telephone exchange – from an analogue system into a digital system, thereby greatly increasing the amount of information that can be carried. The words 'integrated services' refer to the capacity for ISDN to make two connections simultaneously, in any combination of data (voice, video, and fax), over a single line. The data is sent and received in 'digital' format and ISDN is described as a 'network' because it extends from the local telephone exchange to the remote user. The first large-scale use of ISDN began in Japan in 1988.

integration in mathematics, a method in *calculus of determining the solutions of definite or indefinite integrals. An example of a definite integral can be thought of as finding the area under a curve (as represented by an algebraic expression or function) between particular values of the function's variable. In practice, integral calculus provides scientists with a powerful tool for doing calculations that involve a continually varying quantity (such as determining the position at any given instant of a space rocket that is accelerating away from Earth). Its basic principles were discovered in the late 1660s independently by the German philosopher Leibniz (1646–1716) and the English physicist and mathematician Isaac Newton (1642–1727).

integument in seed-producing plants, the protective coat surrounding the ovule. In flowering plants there are two, in gymnosperms only one. A small hole at one end, the micropyle, allows a pollen tube to penetrate through to the egg during fertilization.

intellectual property rights right of control over the copying, distribution, and sale of intellectual property which is codified in the copyright laws.

intelligence in psychology, a general concept that summarizes the abilities of an individual in reasoning and problem solving, particularly in novel situations. These consist of a wide range of verbal and nonverbal skills and therefore some psychologists dispute a unitary concept of intelligence.

intelligent terminal in computing, a *terminal with its own processor that can take some of the processing load away from the main computer. The amount of processing done by the terminal varies. It may be concerned mainly with formatting the input and output data exchanged with the main computer, as on a thin client, or it may be more extensive, as when a *personal computer is used as a terminal.

intensity in physics, power (or energy per second) per unit area carried by a form of radiation or wave motion. It is an indication of the concentration of energy present and, if measured at varying distances from the source, of the effect of distance on this. For example, the intensity of light is a measure of its brightness, and may be shown to diminish with distance from its source in accordance with the

*inverse square law (its intensity is inversely proportional to the square of the distance).

interactive computing in computing, a system for processing data in which the operator is in direct communication with the computer, receiving immediate responses to input data. In *batch processing, by contrast, the necessary data and instructions are prepared in advance and processed by the computer with little or no intervention from the operator.

Interactive Multimedia Association IMA, organization founded in 1987 to promote the growth of the multimedia industry. Based in Anapolis, Maryland, USA, the IMA runs special interest groups, summit meetings, conferences, and trade shows for its member companies.

interactive video IV, computer-mediated system that enables the user to interact with and control information (including text, recorded speech, or moving images) stored on video disk. IV is most commonly used for training purposes, using analogue video disks, but has wider applications with digital video systems such as CD-I (Compact Disc Interactive, from Philips and Sony) which are based on the CD-ROM format derived from audio compact discs.

intercept in geometry, point at which a line or curve cuts across a given axis. Also, the segment cut out of a transversal by the pair of lines it cuts across. If a set of parallel lines make equal intercepts on one transversal, it will make equal intercepts on any other.

interface in computing, the point of contact between two programs or pieces of equipment. The term is most often used for the physical connection between the computer and a peripheral device, which is used to compensate for differences in such operating characteristics as speed, data coding, voltage, and power consumption. For example, a printer interface is the cabling and circuitry used to transfer data from a computer to a printer, and to compensate for differences in speed and coding.

interference in physics, the phenomenon of two or more wave motions interacting and combining to produce a resultant wave of larger or smaller amplitude (depending on whether the combining waves are in or out of *phase with each other).

Interference of white light (multiwavelength) results in spectral coloured fringes; for example, the iridescent colours of oil films seen on water or soap bubbles (demonstrated by *Newton's rings). Interference of sound waves of similar frequency produces the phenomenon of beats, often used by musicians when tuning an instrument. With monochromatic light (of a single wavelength), interference produces patterns of light and dark bands. This is the basis of *holography, for example. Interferometry can also be applied to radio waves, and is a powerful tool in modern astronomy.

interferometer in physics, a device that splits a beam of light into two parts, the parts being recombined after travelling different paths to form an interference pattern of light and dark bands. Interferometers are used in many branches of science and industry where accurate measurements of distances and angles are needed.

In the Michelson interferometer, a light beam is split into two by a semisilvered mirror. The two beams are then reflected off fully silvered mirrors and recombined. The pattern of dark and light bands is sensitive to small alterations in the placing of the mirrors, so the interferometer can detect changes in their position to within one ten-millionth of a metre. Using lasers, compact devices of this kind can be built to measure distances, for example to check the accuracy of machine tools.

In radio astronomy, interferometers consist of separate radio telescopes, each observing the same distant object, such as a galaxy, in the sky. The signal received by each telescope is fed into a computer. Because the telescopes are in different places, the distance travelled by the signal to reach each differs and the overall signal is akin to the interference pattern in the Michelson interferometer. Computer analysis of the overall signal can build up a

detailed picture of the source of the radio waves.

In space technology, interferometers are used in radio and radar systems. These include space-vehicle guidance systems, in which the position of the spacecraft is determined by combining the signals received by two precisely spaced antennae mounted on it.

interferon or **IFN**, naturally occurring cellular protein that makes up part of mammalian defences against viral disease. Three types (alpha, beta, and gamma) are produced by infected cells and enter the bloodstream and uninfected cells, making them immune to virus attack.

Interferon was discovered in 1957 by Scottish virologist Alick Isaacs. Interferons are cytokines, small molecules that carry signals from one cell to another. They can be divided into two main types: **type I** (alpha, beta, tau, and omega) interferons are more effective at bolstering cells' ability to resist infection; **type II** (gamma) interferon is more important to the normal functioning of the immune system. Alpha interferon may be used to treat some cancers; interferon beta 1b has been found useful in the treatment of *multiple sclerosis.

interior angle one of the four internal angles formed when a transversal cuts two or more (usually parallel) lines. Also, one of the angles inside a *polygon.

interlacing technique for increasing resolution on computer graphic displays. The electron beam traces alternate lines on each pass, providing twice the number of lines of a noninterlaced screen. However, screen refresh is slower and screen flicker may be increased over that seen on an equivalent noninterlaced screen.

interleukin in medicine, polypeptide produced by an activated lymphocyte and involved in signalling between cells in the immune system as part of the immune response and in the formation of blood. There are several different interleukins with different activities and some have applications in the treatment of diseases, such as cancer.

intermediate vector boson member of a group of elementary particles, W^+, W^-, and Z, which mediate the *weak nuclear force. This force is responsible for, among other things, beta decay.

intermercial contraction of interstitial commercial, advertising screen interposed between one Web page, where the user mouse-clicks on a link, and the destination page that the user is trying to reach.

intermolecular force or **van der Waals' force**, force of attraction between molecules. Intermolecular forces are relatively weak; hence simple molecular compounds are gases, liquids, or low-melting-point solids.

internal-combustion engine heat engine in which fuel is burned inside the engine, contrasting with an external-combustion engine (such as the steam engine) in which fuel is burned in a separate unit. The diesel engine and *petrol engine are both internal-combustion engines. Gas *turbines and *jet and *rocket engines are also considered to be internal-combustion engines because they burn their fuel inside their combustion chambers.

internal modem in computing, a *modem that fits into a slot inside a personal computer. On older PCs, an internal modem may prove a better choice for high-speed data communications than an external modem, as it may have built-in features which make up for features missing in older computers. Internal modems are generally also cheaper, except for the small-sized PCMCIA types.

internal resistance or **source resistance**, the resistance inside a power supply, such as a battery of cells, that limits the current that it can supply to a circuit.

International Date Line IDL, imaginary line that approximately follows the 180° line of longitude. The date is put forward a day when crossing the line going west, and back a day when going east. The IDL was chosen at the International Meridian Conference in 1884.

International Organization for Standardization or ISO, (Greek *isos* 'equal') global federation of national standards bodies from over 130 countries, with one body for each country. The ISO is a non-governmental organization, established in 1947, and its mission is to promote globally the development of standardization of technical terms, specifications, and units, and other related activities. The work of the ISO results in international agreements, which are published as International Standards. Its headquarters are in Geneva, Switzerland.

International Union for the Conservation of Nature IUCN, alternative name for the World Conservation Union.

internet global public computer network that provides the communication infrastructure for applications such as *e-mail, the *World Wide Web, and *FTP. The Internet is not one individual network, but an interconnected system of smaller networks using common *protocols to pass *packets of information from one computer to another.

internet service provider ISP, organization that provides Internet services, including access to the Internet. Several types of company provide Internet access, including online information services, electronic conferencing systems, and local bulletin board systems (BBSs). Most of the more recently-founded ISPs offer only direct access to the Internet without the burden of running services of their own just for their members. ISPs vary in the way they charge for services: some charge a flat monthly or quarterly fee; others do not charge for Internet provision instead getting revenue from advertising and *electronic commerce; others obtain revenue through complex arrangements with the companies that provide physical delivery such as telecommunications and cable companies. ISPs serve individuals as well as companies and act as gatekeepers to the Internet. They are linked to each other through Network Access Points (NAPs) which are major Internet connection points.

Internet worm in computing, a virus; see *worm.

InterNIC in computing, service that administers *domain names and maintains a number of Internet user directories. Users interested in registering a particular domain name can use the InterNIC's resources to check if the domain name or a similar one is already in use. See *ICANN.

interplanetary matter gas, dust, and charged particles from a variety of sources that occupies the space between the planets. Dust left over from the formation of the Solar System, or from disintegrating comets, orbits the Sun in or near the ecliptic plane and is responsible for scattering sunlight to cause the zodiacal light (mainly the smaller particles) and for *meteor showers in the Earth's atmosphere (larger particles). The charged particles mostly originate in the *solar wind, but some are cosmic rays from deep space.

interpolation estimate of a value lying between two known values. For example, it is known that 13^2 is 169 and 14^2 is 196. Thus, the square of 13.5 can be interpolated as halfway between 169 and 196, i.e. 182.5 (the exact value is 182.25).

interpolation mathematical technique for using two values to calculate intermediate values. It is used in computer graphics to create smooth shadings.

interpreter computer program that translates and executes a program written in a high-level language. Unlike a *compiler, which produces a complete machine-code translation of the high-level program in one operation, an interpreter translates the source program, instruction by instruction, each time that program is run.

interquartile range in statistics, a measure of the range of data, equalling the difference in value between the upper and lower *quartiles. It provides information on the spread of data in the middle 50% of a distribution and is not affected by freak extreme values.

intersection on a graph, the point where two lines or curves meet. The intersections of graphs provide the graphical solutions of equations.

intersection in set theory, the set of elements that belong to both set A and set B. Intersections in *Venn diagrams are where the circles overlap.

interstellar cirrus in astronomy, wispy cloud-like structures discovered in the mid-1980s by the Infrared Astronomy Satellite and believed to be the remains of dust shells blown into space from cool giant or supergiant stars.

interstellar matter medium of electrons, ions, atoms, molecules, and dust grains that fills the space between stars in our own and other galaxies. Over 100 different types of molecule exist in gas clouds in the Earth's Galaxy. Most have been detected by their radio emissions, but some have been found by the absorption lines they produce in the spectra of starlight. The most complex molecules, many of them based on *carbon, are found in the dense clouds where stars are forming. They may be significant for the origin of life elsewhere in space.

intertropical convergence zone ITCZ, area of heavy rainfall found in the tropics and formed as the trade winds converge and rise to form cloud and rain. It moves a few degrees northwards during the northern summer and a few degrees southwards during the southern summer, following the apparent movement of the Sun. The ITCZ is responsible for most of the rain that falls in Africa. The doldrums are also associated with this zone.

interval in statistics, the difference between the smallest and largest measurement in an *class interval.

intestine in vertebrates, the digestive tract from the stomach outlet to the anus. The human **small intestine** is 6 m/20 ft long, 4 cm/1.5 in in diameter, and consists of the duodenum, jejunum, and ileum; the **large intestine** is 1.5 m/5 ft long, 6 cm/2.5 in in diameter, and includes the caecum, colon, and rectum. The contents of the intestine are passed along slowly by *peristalsis (waves of involuntary muscular action). The term intestine is also applied to the lower digestive tract of invertebrates.

intranet in computing, an organization's computer network based on the same *TCP/IP *protocols as the Internet. The intranet resembles an Internet Web site to the user and is effectively an internal Internet. The difference between the Internet and an intranet is that access to an intranet is limited to those with authorization from the organization. This authorization usually extends to all employees. A *firewall and other security precautions prevent unwanted external access.

intron or **junk DNA**, in genetics, a sequence of bases in *DNA that carries no genetic information. Introns, discovered in 1977, make up approximately 98% of DNA (the rest is made up of *exons). Introns may be present within genes but are removed during translation. Their function is unknown.

About 10% of the human genome is made up of one base sequence, *Alu*, that occurs in about 1 million separate locations. It is made up of 283 nucleotides, has no determinable function (though some do have an effect on nearby genes), and is a *transposon ('jumping gene').

introspection observing or examining the contents of one's own mind or consciousness. For example, 'looking' at and describing a 'picture' or image in the 'mind's eye', or trying to examine what is happening when one performs mental arithmetic.

introversion in psychology, preoccupation with the self, generally coupled with a lack of sociability. The opposite of introversion is *extroversion.

inverse function *function that exactly reverses the transformation produced by a function f; it is usually written as f^{-1}. For example, to find the inverse of a function $y = 3x + 2$, x is made the subject of the equation: $3x = y - 2$, so $x = (y - 2)/3$. So $3x + 2$ and $(x - 2)/3$ are mutually inverse functions. Multiplication and division are inverse operations (see *reciprocals).

An inverse function is clearly demonstrated on a calculator by entering any number, pressing x^2, then pressing $\sqrt{x}$ to get the inverse. The functions on a scientific calculator can be inversed in a similar way.

inverse multiplexing in computing, technique for combining individual low-bandwidth channels into a single high-bandwidth channel. It is used to create high-speed telephone links for applications such as videoconferencing which require the transmission of huge quantities of data.

inverse square law statement that the magnitude of an effect (usually a force) at a point is inversely proportional to the square of the distance between that point and the object exerting the force.

Light, sound, electrostatic force (Coulomb's law), and gravitational force (Newton's law) all obey the inverse square law.

invertebrate animal without a backbone. The invertebrates form all of the major divisions of the animal kingdom called phyla, with the exception of vertebrates. Invertebrates include the sponges, coelenterates, flatworms, nematodes, annelids, arthropods, molluscs, and echinoderms.

Oxford zoologists estimated in 1996 that two or more British invertebrate species become extinct each year, a rate of 1% per century.

Primitive aquatic chordates such as sea squirts and lancelets, which only have notochords and do not possess a vertebral column of cartilage or bone, are sometimes called invertebrate chordates, but this is misleading, since the notochord is the precursor of the backbone in advanced chordates.

in vitro process biological experiment or technique carried out in a laboratory, outside the body of a living organism (literally 'in glass', for example in a test tube). By contrast, an in vivo process takes place within the body of an organism.

in vivo process biological experiment or technique carried out within a living organism; by contrast, an in vitro process takes place outside the organism, in an artificial environment such as a laboratory.

involuntary action behaviour not under conscious control, for example the contractions of the gut during peristalsis or the secretion of adrenaline by the adrenal glands. Breathing and urination reflexes are involuntary, although both can be controlled voluntarily to some extent. These processes are regulated by the *autonomic nervous system.

I/O abbreviation for input/output, see *input devices and output devices. The term is also used to describe transfer to and from disk – that is, disk I/O.

iodide compound formed between iodine and another element in which the iodine is the more electronegative element (see *electronegativity, *halide).

iodine chemical symbol I, (Greek *iodes* 'violet') greyish-black non-metallic element, atomic number 53, relative atomic mass 126.9044. It is a member of the *halogen group (Group 7 of the *periodic table). Its crystals sublime, giving off, when heated, a violet vapour with an irritating odour resembling that of chlorine. It occurs only in combination with other elements. Its salts are known as **iodides** and are found in sea water. As a mineral nutrient it is vital to the proper functioning of the thyroid gland, where it occurs in trace amounts as part of the hormone thyroxine. Absence of iodine from the diet leads to *goitre. Iodine is used in photography, in medicine as an antiseptic, and in making dyes.

Its radioactive isotope iodine-131 (half-life of eight days) is a dangerous fission product from nuclear explosions and from the nuclear reactors in power plants, since, if ingested, it can be taken up by the thyroid and damage it. It was discovered in 1811 by French chemist B Courtois (1777–1838).

iodoform or **triiodomethane**; CHI_3, antiseptic that crystallizes into yellow hexagonal plates. It is soluble in ether, alcohol, and chloroform, but not in water.

ion atom, or group of atoms, that is either positively charged (*cation) or negatively charged (*anion), as a result of the loss or gain of electrons during chemical reactions or exposure to certain forms of radiation. In solution or in the molten state, ionic compounds such as salts, acids, alkalis, and metal oxides conduct electricity. These compounds are known as *electrolytes.

Ions are produced during *electrolysis, for example the salt zinc chloride ($ZnCl_2$) dissociates into the positively-charged Zn^{2+} and negatively-charged Cl^- when electrolysed.

ion exchange process whereby an ion in one compound is replaced by a different ion, of the same charge, from another compound. Ion exchange is used in commercial water softeners to exchange the dissolved ions responsible for the water's hardness with others that do not have this effect. For example, when hard water is passed over an ion-exchange resin, the dissolved calcium and magnesium ions are replaced by either sodium or hydrogen ions, so the hardness is removed. The addition of *washing soda to hard water is also an example of ion exchange:

$$Na_2CO_{3(aq)} + CaSO_{4(aq)} \rightarrow CaCO_{3(s)} + Na_2SO_{4(aq)}$$

Ion exchange is the basis of a type of *chromatography in which the components of a mixture of ions in solution are separated according to the ease with which they will replace the ions on the polymer matrix through which they flow. The exchange of positively charged ions is called cation exchange; that of negatively charged ions is called anion exchange.

ion half equation equation that describes the reactions occurring at the electrodes of a chemical cell or in electrolysis. It indicates which ion is losing electrons (oxidation) or gaining electrons (reduction).

Examples are given from the electrolysis of dilute hydrochloric acid (HCl):

$$2Cl^- - 2e^- \rightarrow Cl_2 \text{ (positive electrode)}$$

$$2H^+ + 2e^- \rightarrow H_2 \text{ (negative electrode)}$$

ionic bond or **electrovalent bond**, bond produced when atoms of one element donate electrons to atoms of another element, forming positively and negatively charged *ions respectively. The attraction between the oppositely charged ions constitutes the bond. Sodium chloride (Na^+Cl^-) is a typical ionic compound.

Each ion has the electronic structure of a *noble gas (rare gas; see *noble gas

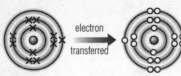

electronic arrangement, 2.8.1 of a sodium atom

electronic arrangement, 2.8.7 of a chlorine atom

becomes a sodium ion, Na$^+$, with an electron arrangement 2.8

becomes a chloride ion, Cl$^-$, with an electron arrangement 2.8.8

ionic bond The formation of an ionic bond between a sodium atom and a chlorine atom to form a molecule of sodium chloride. The sodium atom transfers an electron from its outer electron shell (becoming the positive ion Na+) to the chlorine atom (which becomes the negative chloride ion Cl$^-$). The opposite charges mean that the ions are strongly attracted to each other. The formation of the bond means that each atom becomes more stable, having a full quota of electrons in its outer shell.

structure). The maximum number of electrons that can be gained is usually two.

ionic compound substance composed of oppositely charged ions. All salts, most bases, and some acids are examples of ionic compounds. They possess the following general properties: they are crystalline solids with a high melting point; are soluble in water and insoluble in organic solvents; and always conduct electricity when molten or in aqueous solution. A typical ionic compound is sodium chloride (Na$^+$Cl$^-$).

ionic equation equation showing only those ions in a chemical reaction that actually undergo a change, either by combining together to form an insoluble salt or by combining together to form one or more molecular compounds. Examples are the

precipitation of insoluble barium sulphate when barium and sulphate ions are combined in solution, and the production of ammonia and water from ammonium hydroxide.

$$Ba^{2+}_{(aq)} + SO_4^{2-}_{(aq)} \rightarrow BaSO_{4(s)}$$

$$NH_4^+_{(aq)} + OH^-_{(aq)} \rightarrow NH_{3(g)} + H_2O_{(l)}$$

The other ions in the mixtures do not take part and are called *spectator ions.

ionization process of ion formation. It can be achieved in two ways. The first way is by the loss or gain of electrons by atoms to form positive or negative ions.

$$Na - e^- \rightarrow Na^+$$

$$\tfrac{1}{2}Cl_2 + e^- \rightarrow Cl^-$$

In the second mechanism, ions are formed when a covalent bond breaks, as when hydrogen chloride gas is dissolved in water. One portion of the molecule retains both electrons, forming a negative ion, and the other portion becomes positively charged. This bond-fission process is sometimes called dissociation.

$$HCl_{(g)} + aq \leftrightarrow H^+_{(aq)} + Cl^-_{(aq)}$$

ionization chamber device for measuring *ionizing radiation. The radiation ionizes the gas in the chamber and the ions formed are collected and measured as an electric charge. Ionization chambers are used for determining the intensity of X-rays or the disintegration rate of radioactive materials.

ionization potential measure of the energy required to remove an *electron from an *atom. Elements with a low ionization potential readily lose electrons to form *cations.

ionizing radiation radiation that removes electrons from atoms during its passage, thereby leaving ions in its path. Alpha and beta particles are far more ionizing in their effect than are neutrons or gamma radiation.

ionosphere ionized layer of Earth's outer *atmosphere (60–1,000 km/ 38–620 mi) that contains sufficient free electrons to modify the way in which radio waves are propagated, for instance by reflecting them back to

Earth. The ionosphere is thought to be produced by absorption of the Sun's ultraviolet radiation. The British Antarctic Survey estimates that the ionosphere is decreasing at a rate of 1 km/0.6 mi every five years, based on an analysis of data from 1960 to 1998. Global warming is the probable cause.

ion plating method of applying corrosion-resistant metal coatings. The article is placed in argon gas, together with some coating metal, which vaporizes on heating and becomes ionized (acquires charged atoms) as it diffuses through the gas to form the coating. It has important applications in the aerospace industry.

IP address abbreviation for Internet protocol address, in computing, numbered address assigned to an Internet *host. Traditionally, IP addresses are 32-bit, which means that numbered addresses have four sections separated by dots, each a decimal number between 0 and 255. This is also called IPv4, and can provide a maximum of 4,294,967,296 different IP addresses. A newer specification called IPv6 is being developed to create the huge number of extra IP addresses that will be needed in the future. IPv6 uses six sections, each with numbers between 0 and 255, providing 281,474,976,710,656 IP addresses (65,536 times more than IPv4).

IR or **ir**, in physics, abbreviation for **infrared**.

IRC in computing, abbreviation for Internet Relay Chat.

IrDA abbreviation for **Infrared Data Association**, which sets the standard by which infrared signals may be used to **beam** data between two devices: for example, between two computers, or a computer and a printer. Users of handheld computers frequently exchange electronic business cards or *vCards using built-in IrDA ports.

iridium chemical symbol Ir, (Latin *iridis* 'rainbow') hard, brittle, silver-white, metallic element, atomic number 77, relative atomic mass 192.22. It is resistant to tarnish and corrosion. Iridium is one of the so-called platinum group of metals; it occurs in platinum ores and as a free metal (*native metal)

with osmium in osmiridium, a natural alloy that includes platinum, ruthenium, and rhodium.

It is alloyed with platinum for jewellery and used for watch bearings and in scientific instruments. It was named in 1804 by English chemist Smithson Tennant (1761–1815) for the iridescence of its salts in solution.

iris in anatomy, the coloured muscular diaphragm that controls the size of the pupil in the vertebrate eye. It contains radial muscle that increases the pupil diameter and circular muscle that constricts the pupil diameter. Both types of muscle respond involuntarily to light intensity.

iron chemical symbol Fe, (Germanic *eis* 'strong') hard, malleable and ductile, silver-grey, metallic element, atomic number 26, relative atomic mass 55.847. It chemical symbol comes from the Latin *ferrum*. It is the fourth most abundant element in the Earth's crust. Iron occurs in concentrated deposits as the ores haematite (Fe_2O_3), spathic ore ($FeCO_3$), and magnetite (Fe_3O_4). It sometimes occurs as a free metal, occasionally as fragments of iron or iron–nickel meteorites.

Iron is extracted from iron ore in a *blast furnace. The chemical *reactions of iron with oxygen, water, acids, and other substances, can be explained by its position in the middle of the *reactivity series of metals. Iron is the most useful of all metals; it is strongly magnetic and is the basis for *steel (an alloy with carbon and other elements) and *cast iron. Steel is used for buildings, bridges, ships, car bodies, and tools. Stainless steel is used for car parts, kitchen sinks, and cutlery. In electrical equipment iron is used in permanent magnets and electromagnets, and forms the cores of transformers and magnetic amplifiers. The *corrosion of iron is called *rusting and is an example of an *oxidation reaction. In the human body, iron is an essential component of haemoglobin, the molecule in red blood cells that transports oxygen to all parts of the body. A deficiency in the diet causes a form of anaemia.

iron ore any mineral from which iron is extracted. The chief iron ores are *magnetite, a black oxide; *haematite, or kidney ore, a reddish oxide; limonite, brown, impure oxyhydroxides of iron; and siderite, a brownish carbonate.

Iron ores are found in a number of different forms, including distinct layers in igneous intrusions, as components of contact metamorphic rocks, and as sedimentary beds. Much of the world's iron is extracted in Russia, Kazakhstan, and the Ukraine. Other important producers are the USA, Australia, France, Brazil, and Canada; over 40 countries produce significant quantities of ore.

iron pyrites or **pyrite**; FeS_2, common iron ore. Brassy yellow, and occurring in cubic crystals, it is often called 'fool's gold', since only those who have never seen gold would mistake it.

IRQ abbreviation for interrupt request, in computing, a request for an interrupt to be generated to allow programs and devices to function even while the computer is performing another task. A PC-compatible computer uses a series of interrupt requests numbered from IRQ0 to IRQ9, which are assigned to devices such as sound cards, mice, and serial communications ports.

irrational number number that cannot be expressed as an exact *fraction. Irrational numbers include some square roots (for example, $\sqrt{2}$, $\sqrt{3}$, and $\sqrt{5}$ are irrational); numbers such as π (for circles), which is approximately equal to the *decimal 3.14159; and e (the base of *natural logarithms, approximately 2.71828). If an irrational number is expressed as a decimal it would go on for ever without repeating. An irrational number multiplied by itself gives a *rational number.

irregular galaxy class of *galaxy with little structure, which does not conform to any of the standard shapes in the *Hubble classification. The two satellite galaxies of the *Milky Way, the Magellanic Clouds, are both irregulars. Some galaxies previously classified as irregulars are now known to be normal galaxies distorted by tidal effects or undergoing bursts of star formation.

ISA bus abbreviation for industry standard architecture bus, in computing, 16-bit data *bus introduced in 1984 with the IBM PC AT and still in common use in PCs, alongside the superior *PCI bus. PC hardware and software manufacturers would like to get rid of the ISA bus as it is not compatible with *Plug and Play.

ischaemia reduction of blood supply to any part of the body.

ischaemic heart disease IHD, disorder caused by reduced perfusion of the coronary arteries due to *atherosclerosis. It is the commonest cause of death in the Western world, leading to more than a million deaths each year in the USA and about 160,000 in the UK. See also *coronary artery disease.

Early symptoms of IHD include *angina or palpitations, but sometimes a heart attack is the first indication that a person is affected.

ISDN abbreviation for *Integrated Services Digital Network, a telecommunications system.

island area of land surrounded entirely by water. Australia is classed as a continent rather than an island, because of its size. Islands can be formed in many ways. **Continental islands** were once part of the mainland, but became isolated (by tectonic movement, erosion, or a rise in sea level, for example). **Volcanic islands**, such as Japan, were formed by the explosion of underwater volcanoes. **Coral islands** consist mainly of coral, built up over many years. An **atoll** is a circular coral reef surrounding a lagoon; atolls were formed when a coral reef grew up around a volcanic island that subsequently sank or was submerged by a rise in sea level. **Barrier islands** are found by the shore in shallow water, and are formed by the deposition of sediment eroded from the shoreline.

islets of Langerhans groups of cells within the pancreas responsible for the secretion of the hormone insulin, first described by German anatomist Paul Langerhans (1847–1888). They are sensitive to blood sugar levels, producing more hormone when glucose levels rise.

ISO abbreviation for Infrared Space Observatory.

ISO alternative name for the *International Organization for Standardization, derived from the Greek word *isos* meaning 'equal'.

ISO 9660 in computing, standard file format for *CD-ROM disks, synonymous with High-Sierra format. This format is compatible with most systems, so the same disk can contain both Apple Macintosh and PC versions.

isobar line drawn on maps and weather charts linking all places with the same atmospheric pressure (usually measured in millibars). When used in weather forecasting, the distance between the isobars is an indication of the barometric gradient (the rate of change in pressure).

Where the isobars are close together, cyclonic weather is indicated, bringing strong winds and a depression, and where far apart anticyclonic, bringing calmer, settled conditions.

ISOC in computing, abbreviation for Internet Society.

isomer chemical compound having the same *molecular formula but with different molecular structure. For example, the organic compounds butane ($CH_3(CH_2)_2CH_3$) and methyl propane ($CH_3CH(CH_3)CH_3$) are isomers, each possessing four carbon atoms and ten hydrogen atoms but differing in the way that these are arranged with respect to each other.

Structural isomers have obviously different constructions, but **geometrical** and **optical isomers** must be drawn or modelled in order to appreciate the difference in their three-dimensional arrangement. Geometrical isomers have a plane of symmetry and arise because of the restricted rotation of atoms around a bond; optical isomers are mirror images of each other. For instance, 1,1-dichloroethene ($CH_2=CCl_2$) and 1,2-dichloroethene ($CHCl=CHCl$) are structural isomers, but there are two possible geometric isomers of the latter (depending on whether the chlorine atoms are on the same side or on opposite sides of the plane of the carbon–carbon double bond).

isometric transformation in mathematics, *transformation in which length is preserved.

isomorphism existence of substances of different chemical composition but with similar crystalline form.

isoprene technical name **methylbutadiene**; $CH_2CHC(CH_3)CH_2$, colourless, volatile liquid obtained from petroleum and coal, used to make synthetic rubber.

isosceles triangle *triangle with two sides equal, hence its base angles are also equal. The triangle has an axis of symmetry which is an *altitude of the triangle.

isotherm line on a map, linking all places having the same temperature at a given time.

isotonic solution in medicine, a solution that can be mixed with body fluids without affecting their constituents. Solutions which are isotonic with blood, such as sodium chloride 0.9%, have the same osmotic pressure as serum and they do not affect the membranes of the red blood cells.

butane $CH_3(CH_2)_2CH_3$

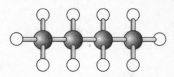

methyl propane $CH_3CH(CH_3)CH_3$

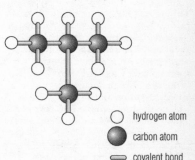

○ hydrogen atom

● carbon atom

⬭ covalent bond

isomer The chemicals butane and methyl propane are isomers. Each has the molecular formula C_4H_{10}, but with different spatial arrangements of atoms in their molecules.

isotope one of two or more atoms that have the same atomic number (same number of protons), but which contain a different number of neutrons, thus differing in their *relative atomic mass. They may be stable or radioactive (as a *radioisotope), naturally occurring, or synthesized. For example, hydrogen has the isotopes ^2H (*deuterium) and ^3H (*tritium). The term was coined by English chemist Frederick Soddy, a pioneer researcher in atomic disintegration.

Elements at the lower end of the periodic table have atoms with roughly the same number of protons as neutrons. These elements are called **stable isotopes**. The stable isotopes of oxygen include ^{16}O, ^{17}O, and ^{18}O; those of carbon include ^{12}C and ^{14}C. Elements with high atomic mass numbers have many more neutrons than protons and are therefore less stable. It is these isotopes that are more prone to *radioactive decay. Examples are ^{238}U (uranium-238) and ^{60}Co (cobalt-60).

ISP in computing, abbreviation for *Internet Service Provider.

IT abbreviation for *information technology.

iteration in computing, a method of solving a problem by performing the same steps repeatedly until a certain condition is satisfied. For example, in one method of *sorting, adjacent items are repeatedly exchanged until the data are in the required sequence.

ITU abbreviation for **International Telecommunications Union**, the standards-setting body for the communications industry.

IUCN abbreviation for International Union for the Conservation of Nature, an alternative name for the World Conservation Union.

IUPAC abbreviation for International Union of Pure and Applied Chemistry, organization that recommends the nomenclature to be used for naming substances, the units to be used, and which conventions are to be adopted when describing particular changes.

J in physics, the symbol for *joule, the SI unit of energy.

jacinth or **hyacinth**, red or yellowish-red gem, a variety of zircon, $ZrSiO_4$.

jade semi-precious stone consisting of either jadeite, $NaAlSi_2O_6$ (a pyroxene), or nephrite, $Ca_2(Mg,Fe)_5Si_8O_{22}(OH,F)_2$ (an amphibole), ranging from colourless through shades of green to black according to the iron content. Jade ranks 5.5–6.5 on the *Mohs scale of hardness.

James–Lange theory in psychology, a theory that sensory impressions resulting from exposure to an emotional situation are caused by physiological reactions. In this way emotions of pleasure, fear, or amusement may be seen as the result of changes of bodily activity; fear is the sensation of sweating and trembling rather than the emotional reaction to a terrifying situation.

JANET acronym for Joint Academic NETwork, in computing, network linking academic and research institutes in the UK. JANET is composed of many interconnected smaller networks, controlled by a committee called the Joint Network Team. Sites that subscribe to the network can communicate with each other, with other academic networks, and with external services.

jasper hard, compact variety of chalcedony SiO_2, usually coloured red, brown, or yellow. Jasper can be used as a gem.

jaundice   yellow discoloration of the skin and whites of the eyes caused by an excess of bile pigment in the bloodstream. Approximately 60% of newborn babies exhibit some degree of jaundice, which is treated by bathing in white, blue, or green light that converts the bile pigment bilirubin into a water-soluble compound that can be excreted in urine. A serious form of jaundice occurs in rhesus disease (see *rhesus factor).

Java   in computing, programming language much like C developed by James Gosling at *Sun Microsystems in 1995. Java has been adopted as a multipurpose, cross-platform lingua franca for network computing, including the *World Wide Web. When users connect to a server that uses Java, they download a small program called an applet onto their computers. The applet then runs on the computer's own processor via a *Java Virtual Machine program or JVM. In 2000 Java 2 Micro Edition was released, designed for use in handheld devices, telephones, and pagers.

Java 3D   in computing, a set of special 3-dimensional graphics functions built into the latest version of the *Java programming language. Java 3D allows programmers to develop 3D models and *virtual worlds. Because Java programs are portable, and can be run using a Web *browser, Java 3D is becoming a serious alternative to another computing language, *VRML, for developing Web-based virtual worlds.

Java Beans   specification devised by *Sun Microsystems, according to which *object-oriented programs can be created in the *Java language. A Java Bean is similar to an *ActiveX control in Microsoft's software architecture.

JavaScript   *scripting language commonly used to add interactive elements to Web pages. JavaScript was developed by Netscape Communications as LiveScript (it was not derived from Java) and has been standardized by ECMA (European Computer Manufacturers' Association) as ECMAScript.

Java Virtual Machine   JVM, in computing, a program that sits on top of a computer's usual operating system and runs Java applets.

JET   abbreviation for Joint European Torus, research facility at Culham, near Abingdon, Oxfordshire, UK, that conducts experiments on nuclear fusion. It is the focus of the European effort to produce a safe and environmentally sound fusion-power reactor. On 9 November 1991 the JET *tokamak, operating with a mixture of deuterium and tritium, produced a 1.7 megawatt pulse of power in an experiment that lasted two seconds. In 1997 isotopes of deuterium and tritium were fused to produce a record 21 megajoule of nuclear fusion power. JET has tested the first large-scale plant of the type needed to process and supply tritium in a future fusion power station.

jet   in earth science, hard, black variety of lignite, a type of coal. It is cut and polished for use in jewellery and ornaments. Articles made of jet have been found in Bronze Age tombs.

jet propulsion   method of propulsion in which an object is propelled in one direction by a jet, or stream of gases, moving in the other. This follows from Isaac Newton's third law of motion: 'To every action, there is an equal and opposite reaction.' (see *Newton's laws of motion) The most widespread application of the jet principle is in the jet (gas turbine) engine, the most common kind of aircraft engine.

Joint European Torus   experimental nuclear-fusion machine, known as *JET.

Josephson junction   device used in 'superchips' (large and complex integrated circuits) to speed the passage of signals by a phenomenon called 'electron tunnelling'. Although these superchips respond a thousand times faster than the *silicon chip, they have the disadvantage that the components of the Josephson junctions operate only at temperatures close to *absolute zero. They are named after English theoretical physicist Brian Josephson.

joule   symbol J, SI unit of work and energy (such as *potential energy, *kinetic energy, or *electrical energy).

Named after English physicist James Joule (1818–1889).

Joule–Kelvin effect or **Joule–Thomson effect**, in physics, the fall in temperature of a gas as it expands adiabatically (without loss or gain of heat to the system) through a narrow jet. It can be felt when, for example, compressed air escapes through the valve of an inflated bicycle tyre. It is the basic principle of some refrigerators.

JPEG acronym for Joint Photographic Experts Group, used to describe a compression standard set up by that group and now widely accepted for the storage and transmission of colour images. The JPEG compression standard reduces the size of image files considerably, although it is a *lossy compression format. Repeating editing and resaving JPEGs will adversely affect their appearance.

jugular vein one of two veins in the necks of vertebrates; they return blood from the head to the superior (or anterior) *vena cava and thence to the heart.

jump in computing, a programming instruction that causes the computer to branch to a different part of a program, rather than execute the next instruction in the program sequence. Unconditional jumps are always executed; conditional jumps are only executed if a particular condition is satisfied.

jumper in computing, rectangular plug used to make connections on a circuit board. By pushing a jumper onto a particular set of pins on the board, or removing another, users can adjust the configuration of their computer's circuitry. Most home users, however, prefer to leave the insides of their machines with all the factory settings intact.

jungle popular name for *rainforest.

junk DNA another name for *intron, a region of DNA that contains no genetic information.

Jupiter fifth planet from the Sun and the largest in the Solar System, with a mass equal to 70% of all the other planets combined and 318 larger than that of the Earth. Its main feature is the Great Red Spot, a cloud of rising gases,

14,000 km/8,500 mi wide and 30,000 km/20,000 mi long, revolving anticlockwise.

mean distance from the Sun: 778 million km/484 million mi

equatorial diameter: 142,800 km/88,700 mi

rotation period: 9 hours 51 minutes

year: 11.86 Earth years

atmosphere: consists of clouds of white ammonia crystals, drawn out into belts by the planet's high speed of rotation (the fastest of any planet). Darker orange and brown clouds at lower levels may contain sulphur, as well as simple organic compounds. Temperatures range from –140°C/ –220°F in the upper atmosphere to as much as 24,000°C/43,000°F near the core. This is the result of heat left over from Jupiter's formation, and it is this that drives the turbulent weather patterns of the planet. The Great Red Spot was first observed in 1664. Its top is higher than the surrounding clouds; its colour is thought to be due to red phosphorus. The Southern Equatorial Belt in which the Great Red Spot occurs is subject to unexplained fluctuation. In 1989 it sustained a dramatic and sudden fading. Jupiter's strong magnetic field gives rise to a large surrounding magnetic 'shell', or magnetosphere, from which bursts of radio waves are detected. Jupiter's faint rings are made up of dust from its moons, particularly the four inner moons

surface: largely composed of hydrogen and helium, which under the high pressure and temperature of the interior behave not as a gas but as a supercritical fluid. Under even more extreme conditions, at a depth of 30,000 km/18,000 mi, hydrogen transforms into a metallic liquid. Jupiter probably has a molten rock core whose mass is 15 to 20 times greater than that of the Earth

In 1995, the *Galileo probe revealed Jupiter's atmosphere to consist of 0.2% water, less than previously estimated.

satellites: Jupiter has 28 known moons. The four largest moons, Io, Europa (which is the size of the Moon), Ganymede, and Callisto, are the Galilean satellites, discovered in 1610 by the Italian mathematician, astronomer and physicist Galileo Galilei (1564–1642) (Ganymede, which is larger than Mercury, is the largest moon in the Solar System). Three small moons were discovered in 1979 by the US Voyager probes, as was a faint ring of dust around Jupiter's equator 55,000 km/34,000 mi above the cloud tops. One of Jupiter's small inner moons, Almathea (diameter 250 km/155 mi), was shown by pictures from the Galileo probe in April 2000 to have a long, narrow, bright region, as yet unidentified. A new moon was first observed in orbiting Jupiter in October 1999 by US researchers at the Kitt Peak Observatory, Arizona. It was thought to be an asteroid and named S/1999J1, but was confirmed to be a moon in July 2000. The moon is only 5 km/3 mi in diameter and orbits Jupiter once every two years at a distance of 24 million km/15 million mi.

Ten previously unobserved moons were discovered orbiting Jupiter in November and December 2000. These moons are all believed to be less than 5 km/3.1 mi in diameter, and were observed by astronomers at the Mauna Kea observatory, Hawaii.

Jurassic period   period of geological time 208–146 million years ago; the middle period of the Mesozoic era. Climates worldwide were equable, creating forests of conifers and ferns; dinosaurs were abundant, birds evolved, and limestones and iron ores were deposited.

The name comes from the Jura Mountains in France and Switzerland, where the rocks formed during this period were first studied.

justification   in printing and word processing, the arrangement of text so that it is aligned with either the left or right margin, or both.

k   symbol for **kilo-**, as in kg (kilogram) and km (kilometre).

K   abbreviation for thousand, as in a salary of £30K.

K   symbol for **kelvin**, a scale of temperature.

Kagoshima Space Centre   launch site of Japan's Institute of Space and Astronautical Science (ISAS), situated on South Kyushu Island. Japan's first satellite was launched from Kagoshima in 1970. More than 300 rockets had been launched by 2001.

kaolin   or **china clay**, group of clay minerals, such as *kaolinite, $Al_2Si_2O_5(OH)_4$, derived from the alteration of aluminium silicate minerals, such as feldspars and *mica. It is used in medicine to treat digestive upsets, and in poultices.

kaolinite   white or greyish *clay mineral, hydrated aluminium silicate,

$Al_2Si_2O_5(OH)_4$, formed mainly by the decomposition of feldspar in granite. It is made up of platelike crystals, the atoms of which are bonded together in two-dimensional sheets, between which the bonds are weak, so that they are able to slip over one another, a process made more easy by a layer of water. China clay (kaolin) is derived from it. It is mined in France, the UK, Germany, China, and the USA.

Kaposi's sarcoma   malignant (cancerous) tumour developing from blood vessels in the skin. Previously rare in the Western world, it is now seen in some patients with *AIDS.

karat   or **carat**, the unit of purity in gold in the USA. Pure gold is 24-karat; 22-karat (the purest used in jewelry) is 22 parts gold and two parts alloy (to give greater strength); 18-karat is 75% gold.

karyotype in biology, the set of *chromosomes characteristic of a given species. It is described as the number, shape, and size of the chromosomes in a single cell of an organism. In humans, for example, the karyotype consists of 46 chromosomes, in mice 40, crayfish 200, and in fruit flies 8.

The diagrammatic representation of a complete chromosome set is called a **karyogram**.

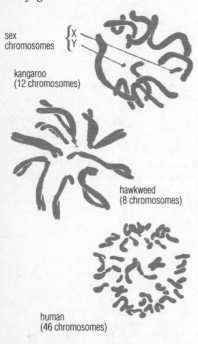

sex chromosomes {X Y}

kangaroo (12 chromosomes)

hawkweed (8 chromosomes)

human (46 chromosomes)

karyotype The characteristics, or karyotype, of the chromosomes vary according to species. The kangaroo has 12 chromosomes, the hawkweed has 8, and a human being has 46.

katabatic wind cool wind that blows down a valley on calm clear nights. (By contrast, an anabatic wind is warm and moves up a valley in the early morning.) When the sky is clear, heat escapes rapidly from ground surfaces, and the air above the ground becomes chilled. The cold dense air moves downhill, forming a wind that tends to blow most strongly just before dawn.

kcal symbol for **kilocalorie** (see *calorie).

Kelvin scale temperature scale used by scientists. It begins at *absolute zero (−273.15°C) and increases in kelvins, the same degree intervals as the Celsius scale; that is, 0°C is the same as 273.15 K and 100°C is 373.15 K. It is named after the Irish physicist William Thomson, 1st Baron Kelvin (1824–1907).

Kennedy Space Center KSC, NASA launch site on Merritt Island, near Cape Canaveral, Florida, used for *Apollo* project and *space shuttle launches. The first *Apollo* flight to land on the Moon (1969) and *Skylab*, the first orbiting laboratory (1973), were launched from the site.

Kepler's laws in astronomy, three laws of planetary motion formulated in 1609 and 1619 by German mathematician and astronomer Johannes Kepler (1571–1630): (1) the orbit of each planet is an ellipse with the Sun at one of the foci; (2) the radius vector of each planet sweeps out equal areas in equal times; (3) the squares of the periods of the planets are proportional to the cubes of their mean distances from the Sun.

keratin fibrous protein found in the *skin of vertebrates and also in hair, nails, claws, hooves, feathers, and the outer coating of horns.

If pressure is put on some parts of the skin, more keratin is produced, forming thick calluses that protect the layers of skin beneath.

keratitis in medicine, eye disease characterized by inflammation of the cornea. This may be the result of mechanical trauma, for example being hit in the eye, contact with chemicals, or infection with bacteria or viruses. The affected eye is red and painful and those affected should avoid exposure to light due to sensitivity. Treatment varies depending on the cause of the disorder.

kernel the inner, softer part of a nut, or of a seed within a hard shell.

kerosene thin oil obtained from the *fractional distillation of crude oil (unrefined *petroleum; a highly refined form is used in jet aircraft fuel. Kerosene is a mixture of hydrocarbons of the *alkane series, consisting mainly of hydrocarbons with 11 or 12 carbon atoms. Boiling points range from 160°C/320°F to 250°C/480°F. Crude oil contains approximately 10–15% kerosene.

ketone member of the group of organic compounds containing the *carbonyl group (>C=O) bonded to two atoms of carbon (instead of one carbon and one hydrogen as in *aldehydes). Ketones are liquids or low-melting-point solids, slightly soluble in water. An example is propanone (acetone, CH_3COCH_3), used as a solvent.

key in cryptography, the password needed to both encode and decipher a file. The key performs a sequence of operations on the original data. The recipient of the encoded file will need to apply another key in order to reverse all the operations in the correct order. Current encryption techniques such as Pretty Good Privacy (PGP) make use of a *public key and a secret one.

key-to-disk system or **key-to-tape system**, in computing, a system that enables large amounts of data to be entered at a keyboard and transferred directly onto computer-readable disks or tapes.

kg symbol for *kilogram.

kidney in vertebrates, one of a pair of organs responsible for fluid regulation, excretion of waste products, and maintaining the ionic composition of the blood – in other words the regulation of the concentrations of some chemicals in the blood. The kidneys are situated on the rear wall of the abdomen. Each one consists of a number of long tubules (see *nephron) – the outer parts filter the aqueous components of blood, and the inner parts selectively reabsorb vital salts, leaving waste products in the remaining fluid (urine), which is passed through the ureter to the bladder.

The kidneys tasks can be broken down into a number of stages: filtering of the blood; re-absorbance of all sugar; re-absorbance of the dissolved ions needed by the body; re-absorbance of as much water as the body needs; and release of urea, excess ions, and excess water as urine.

The action of the kidneys is vital and so kidney failure is serious. However, if one kidney fails, the other enlarges to take over its function. A patient with two defective kidneys may still continue near-normal life with the aid of dialysis using a kidney machine or continuous ambulatory peritoneal dialysis (CAPD), or by a kidney transplant. Other diseases of the kidney can include the formation of kidney stones. These hard stones can build up as a result of high levels of blood calcium or high levels of uric acid, and can cause intense pain as they travel down the ureter, as well as causing bleeding in the tissues of the urinary tract.

killer application in computing, term originally coined to describe a computer program so good or so compelling to certain potential users that they buy the computer that the program runs on for no other reason than to be able to use that program. The term's meaning has since been extended to describe something, such as an idea or a product, which is extremely compelling.

killfile in computing, file specifying material that you do not wish to see when accessing a newsgroup. By entering names, subjects or phrases into a killfile, users can make *Usenet a more pleasant experience, filtering out tedious threads, offensive subject headings, *spamming or contributions from other irritating subscribers.

kilo- prefix denoting multiplication by 1,000, as in kilohertz, a unit of frequency equal to 1,000 hertz.

kilobyte **K** or **KB**, in computing, a unit of memory equal to 1,024 *bytes. It is sometimes used, less precisely, to mean 1,000 bytes.

kilogram symbol kg, SI unit of mass equal to 1,000 grams (2.24 lb). It is defined as a mass equal to that of the international prototype, a platinum-iridium cylinder held at the International Bureau of Weights and Measures in Sèvres, France.

kilometre symbol km, unit of length equal to 1,000 metres, equivalent to 3,280.89 ft or 0.6214 (about $5/8$) of a mile.

kilostream link in computing, British Telecom-leased line that can transfer data at 64 kilobits per second.

kilowatt symbol kW, unit of power equal to 1,000 watts or about 1.34 horsepower. If an electrical appliance has a power rating of 1 kW, it will change 1000 J of electrical energy into other forms of energy every second; for example, a 1 kW electric heater changes 1000 J of electrical energy into 1000 J of heat energy every second it is turned on.

kilowatt-hour symbol kWh, commercial unit of electrical energy, defined as the work done by a power of 1,000 watts in one hour and equal to 3.6 megajoules. It is used to calculate the cost of electrical energy taken from the domestic supply.

kinesis (plural **kineses**) in biology, a nondirectional movement in response to a stimulus; for example, woodlice move faster in drier surroundings. **Taxis** is a similar pattern of behaviour, but there the response is directional.

kinetic energy the energy of a body resulting from motion.

A moving body has kinetic energy. This energy is equal to the work that would have to be done in bringing the body to rest, and is dependent on both the body's mass and speed. The kinetic energy in joules of a mass m kilograms travelling with speed v metres per second is given by the formula:

$$KE = \frac{1}{2}mv^2$$

If a moving object collides with another object, then work is done. For example, if a moving car collides with a stationary car, it will cause the stationary car to move. The force from the moving object is used to move the stationary object by a certain distance.

kinetics branch of chemistry that investigates the rates of chemical reactions.

kinetics branch of *dynamics dealing with the action of forces producing or changing the motion of a body; **kinematics** deals with motion without reference to force or mass.

kinetic theory theory describing the physical properties of matter in terms of the behaviour – principally movement – of its component atoms or molecules. It states that all matter is made up of very small particles that are in constant motion, and can be used to explain the properties of solids, liquids, and gases, as well as changes of state. In a solid, the particles are arranged close together in a regular pattern and vibrate on the spot. In a liquid, the particles are still close together but in an irregular arrangement, and the particles are moving a little faster and are able to slide past one another. In a gas, the particles are far apart and moving rapidly, bouncing off the walls of their container. The temperature of a substance is dependent on the velocity of movement of its constituent particles, increased temperature being accompanied by increased movement.

A gas consists of rapidly moving atoms or molecules and, according to kinetic theory, it is their continual impact on the walls of the containing vessel that accounts for the pressure of the gas. The slowing of molecular motion as temperature falls, according to kinetic theory, accounts for the physical properties of liquids and solids, culminating in the concept of no molecular motion at *absolute zero (0 K/–273.15°C).

kingdom primary division in biological *classification. At one time, only two kingdoms were recognized: animals and plants. Today most biologists prefer a five-kingdom system, even though it still involves grouping together organisms that are probably unrelated. One widely accepted scheme is as follows: **Kingdom Animalia** (all multicellular animals); **Kingdom Plantae** (all plants, including seaweeds and other algae, except blue-green); **Kingdom Fungi** (all fungi, including the unicellular yeasts, but not slime moulds); **Kingdom Protista** or **Protoctista** (protozoa, diatoms, dinoflagellates, slime moulds, and various other lower organisms with eukaryotic cells); and **Kingdom Monera** (all prokaryotes – the bacteria and cyanobacteria, or *blue-green algae). The first four of these kingdoms make up the eukaryotes.

kin selection   in biology, the idea that *altruism shown to genetic relatives can be worthwhile, because those relatives share some genes with the individual that is behaving altruistically, and may continue to reproduce. See *inclusive fitness.

kiosk   in computing, any computer that has been set up to act as an information centre in a public place. Users navigate the display using keyboards or touch screens, but are never allowed to access the computer's operating system. A kiosk in a museum might show an interactive multimedia display, or one in a library might give readers access to catalogues.

Kirchhoff's laws   two laws governing electric circuits devised by the German physicist Gustav Kirchhoff (1824–1887). **Kirchhoff's first law** states that the total current entering any junction in a circuit is the same as the total current leaving it. This is an expression of the conservation of electric charge. **Kirchhoff's second law** states that the sum of the potential drops across each resistance in any closed loop in a circuit is equal to the total electromotive force acting in that loop. The laws are equally applicable to DC and AC circuits.

kiss of life   or **artificial ventilation**, in first aid, another name for *artificial respiration.

kite   quadrilateral with two pairs of adjacent equal sides. The geometry of this figure follows from the fact that it has one axis of symmetry.

kleptomania   (Greek *kleptes* 'thief') behavioural disorder characterized by an overpowering desire to possess articles for which one has no need. In kleptomania, as opposed to ordinary theft, there is no obvious need or use for what is stolen and sometimes the sufferer has no memory of the theft.

kleptoparasitism   habitual stealing of food from another organism. Skuas kleptoparasitize other seabirds, forcing them to relinquish their catches midflight. The Spanish slug *Deroceras hilbrandi* takes prey from the insect-eating plant *Pinguicula vallisneriifolia* whilst leaving the edible plant unharmed. Many small spiders are

kleptoparasites living on the webs of bigger spiders.

km   symbol for *kilometre.

knot   in navigation, unit by which a ship's speed is measured, equivalent to one *nautical mile per hour (one knot equals about 1.15 miles per hour). It is also sometimes used in aviation.

knowledge-based system   KBS, computer program that uses an encoding of human knowledge to help solve problems. It was discovered during research into *artificial intelligence that adding heuristics (rules of thumb) enabled programs to tackle problems that were otherwise difficult to solve by the usual techniques of computer science.

knowledge management   process by which an organization gathers, organizes, shares, and analyses its knowledge. Data entered via a keyboard, a scanner, or voice input can be catalogued, indexed, filtered, and linked. The information can then be refined, for example through *data mining, in preparation for dissemination.

Korsakoff's syndrome   organic disorder in which the brain fails to cope with new information, though distant memory is preserved. It is usually due to chronic alcoholism.

kph   or **km/h**, symbol for **kilometres per hour**.

Krebs cycle   or **citric acid cycle** or **tricarboxylic acid cycle**, discovered by German-born British biochemist Hans Adolf Krebs (1900–1981). It is the final part of the chain of biochemical reactions by which organisms break down food using oxygen to release energy (respiration). It takes place within structures called *mitochondria in the body's cells, and breaks down food molecules in a series of small steps, producing energy-rich molecules of *ATP.

krypton   chemical symbol Kr, (Greek *kryptos* 'hidden') colourless, odourless, gaseous, non-metallic element, atomic number 36, relative atomic mass 83.80. It is grouped with the *noble gases (rare gases) in Group 0 of the *periodic table, and was long believed not to enter into reactions, but it is now known to combine with fluorine under certain conditions; it remains inert to all other reagents. As with the other noble gases, krypton's lack of reactivity is due to its full outer shell of electrons. It is present in very small quantities in the air (about 114 parts per million). It is used chiefly in fluorescent lamps, lasers, and gas-filled electronic valves.

Krypton was discovered in 1898 in the residue from liquid air by British chemists William Ramsay (1852–1916) and Morris Travers (1872–1961); the name refers to their difficulty in isolating it.

kW symbol for *****kilowatt**.

l symbol for *****litre**, a measure of liquid volume.

labelled compound or **tagged compound**, chemical compound in which a radioactive isotope is substituted for a stable one. The path taken by such a compound through a system can be followed, for example by measuring the radiation emitted. This powerful and sensitive technique is used in medicine, chemistry, biochemistry, and industry.

lactic acid or **2-hydroxypropanoic acid**; $CH_3CHOHCOOH$, organic acid, a colourless, almost odourless liquid, produced by certain bacteria during fermentation and by active muscle cells when they are exercised hard and are experiencing *oxygen debt. An accumulation of lactic acid in the muscles may cause cramp. It occurs in yogurt, buttermilk, sour cream, poor wine, and certain plant extracts, and is used in food preservation and in the preparation of pharmaceuticals.

lactose white sugar, found in solution in milk; it forms 5% of cow's milk. It is commercially prepared from the whey obtained from cheese-making. Like table sugar (sucrose), it is a disaccharide, consisting of two basic sugar units (monosaccharides), in this case, glucose and galactose. Unlike sucrose, it is tasteless.

laevulose or **L-fructose** or **fruit sugar**; $CH_2OH(CHOH)_3COCH_2OH$, ketose (simple sugar) contained in most sweet fruits, honey, and starches, together with dextrose.

lagoon coastal body of shallow salt water, usually with limited access to the sea. The term is normally used to describe the shallow sea area cut off by a coral reef or barrier islands.

lake body of still water lying in depressed ground without direct communication with the sea. Lakes are common in formerly glaciated regions, along the courses of slow rivers, and in low land near the sea. The main classifications are by origin: **glacial lakes**, formed by glacial scouring; **barrier lakes**, formed by *landslides and glacial *moraines; **crater lakes**, found in *volcanoes; and **tectonic lakes**, occurring in natural fissures.

Crater lakes form in the calderas of extinct volcanoes, for example Crater Lake, Oregon, USA. Subsidence of the roofs of limestone caves in karst landscape exposes the subterranean stream network and provides a cavity in which a lake can develop. Tectonic lakes form during tectonic movement, as when a *rift valley is formed. Lake Tanganyika was created in conjunction with the East African Great Rift Valley. Glaciers produce several distinct types of lake, such as the lochs of Scotland and the Great Lakes of North America.

Lakes are mainly freshwater, but salt and bitter lakes are found in areas of low annual rainfall and little surface run-off, so that the rate of evaporation exceeds the rate of inflow, allowing mineral salts to accumulate. The Dead Sea has a salinity of about 250 parts per 1,000 and the Great Salt Lake,

Utah, about 220 parts per 1,000. Salinity can also be caused by volcanic gases or fluids, for example Lake Natron, Tanzania.

In the 20th century large artificial lakes have been created in connection with hydroelectric and other works. Some lakes have become polluted as a result of human activity. Sometimes eutrophication (a state of overnourishment) occurs, when agricultural fertilizers leaching into lakes cause an explosion of aquatic life, which then depletes the lake's oxygen supply until it is no longer able to support life.

Lamarckism theory of evolution, now discredited, advocated during the early 19th century by French naturalist Jean Baptiste Lamarck. Lamarckism is the theory that acquired characteristics, such as the increased body mass of an athlete, were inherited. It differs from the Darwinian theory of evolution.

LAN in computing, abbreviation for *local area network.

landfill site large holes in the ground used for dumping household and commercial waste. Landfill disposal has been the preferred option in the UK and the USA for many years, with up to 85% of household waste being dumped in this fashion. However, the sites can be dangerous, releasing toxins and other leachates (see *leaching) into the soil and the policy is itself wasteful both in terms of the materials dumped and land usage.

landform the shape, size, form, nature, and characteristics of a specific feature of the land's surface, whether above ground or underwater. The term *geomorphology is used to describe the study of landforms.

land reclamation conversion of derelict or otherwise unusable areas into productive land. For example, where industrial or agricultural activities, such as sand and gravel extraction or open-cast mining, have created large areas of derelict or waste ground, the companies involved are usually required to improve the land so that it can be used.

landslide sudden downward movement of a mass of soil or rocks from a cliff or steep slope. Landslides happen when a slope becomes unstable, usually because the base has been undercut or because materials within the mass have become wet and slippery.

lanthanide any of a series of 15 metallic elements (also known as rare earths) with atomic numbers 57 (lanthanum) to 71 (lutetium). One of its members, promethium, is radioactive. All occur in nature. Lanthanides are grouped because of their chemical similarities (most are trivalent, but some can be divalent or tetravalent), their properties differing only slightly with atomic number.

Lanthanides were called rare earths originally because they were not widespread and were difficult to identify and separate from their ores by their discoverers. The series is set out in a band in the periodic table of the elements, as are the *actinides.

lanthanum chemical symbol La, (Greek *lanthanein* 'to be hidden') soft, silvery, ductile and malleable, metallic element, atomic number 57, relative atomic mass 138.91, the first of the lanthanide series. It is used in making alloys. It was named in 1839 by Swedish chemist Carl Mosander (1797–1858).

lapis lazuli rock containing the blue mineral lazurite in a matrix of white calcite with small amounts of other minerals. It occurs in silica-poor igneous rocks and metamorphic limestones found in Afghanistan, Siberia, Iran, and Chile. Lapis lazuli was a valuable pigment of the Middle Ages, also used as a gemstone and in inlaying and ornamental work.

Large Electron Positron Collider LEP, world's largest particle *accelerator, in operation 1989–2000 at the CERN laboratories near Geneva in Switzerland. It occupies a tunnel 3.8 m/ 12.5 ft wide and 27 km/16.7 mi long, which is buried 180 m/590 ft underground and forms a ring consisting of eight curved and eight straight sections. In June 1996, LEP resumed operation after a £210 million upgrade. The upgraded machine, known as LEP2, generated collision energy of 161 gigaelectron volts.

larva stage between hatching and adulthood in those species in which the young have a different appearance and way of life from the adults. Examples include tadpoles (frogs) and caterpillars (butterflies and moths). Larvae are typical of the invertebrates, some of which (for example, shrimps) have two or more distinct larval stages. Among vertebrates, it is only the amphibians and some fishes that have a larval stage.

larynx in mammals, a cavity at the upper end of the trachea (windpipe) containing the vocal cords. It is stiffened with cartilage and lined with mucous membrane. Amphibians and reptiles have much simpler larynxes, with no vocal cords. Birds have a similar cavity, called the **syrinx**, found lower down the trachea, where it branches to form the bronchi. It is very complex, with well-developed vocal cords.

laser acronym for **light amplification by stimulated emission of radiation**, device for producing a narrow beam of light, capable of travelling over vast distances without dispersion, and of being focused to give enormous power densities (10^8 watts per cm^2 for high-energy lasers). The laser operates on a principle similar to that of the *maser (a high-frequency microwave amplifier or oscillator). The uses of lasers include communications (a laser beam can carry much more information than can radio waves), cutting, drilling, welding, satellite tracking, medical and biological research, and surgery. Sound wave vibrations from the window glass of a room can be picked up by a

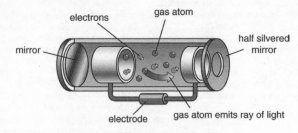

electrons   gas atom

mirror

half silvered mirror

electrode   gas atom emits ray of light

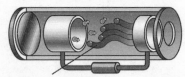

light ray hits another energised atom causing more light to be emitted

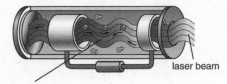

laser beam

light rays bounce between the mirrors causing a build up of light

laser In a gas laser, electrons moving between the electrodes pass energy to gas atoms. An energized atom emits a ray of light. The ray hits another energized atom causing it to emit a further light ray. The rays bounce between mirrors at each end causing a build-up of light. Eventually it becomes strong enough to pass through the half-silvered mirror at one end, producing a laser beam.

reflected laser beam. Lasers are also used as entertainment in theatres, concerts, and light shows.

lat. abbreviation for *latitude.

latency time needed for an electronic message to travel across a communications system and for the remote system to respond. Latency is inherent in all computer systems because nothing happens instantly, but usually things happen so quickly that the delay does not matter. However, latency is a critical factor in real-time systems where something is being controlled: a robot arm, for example, or an aircraft wing. It particularly concerns users when they want things to happen quickly – as when playing fast action games over a network – or when distance means the latency (delay) is very large. For example, a control message could take hours or days to reach a spacecraft.

latent heat in physics, the heat absorbed or released by a substance as it changes state (for example, from solid to liquid) at constant temperature and pressure.

lateral inversion the reversal experienced by an image formed in a plane (flat) mirror. Although the image is the correct way up, its left and right sides are transposed. The impression that most people have of their own face is based on the image that they see in a mirror; however, this is quite different from the face that other people see – for example, a hair parting that appears to be on the left when viewed in a mirror is seen by everyone else to be on the right.

lateral moraine linear ridge of rocky debris deposited near the edge of a *glacier. Much of the debris is material that has fallen from the valley side onto the glacier's edge, having been weathered by *freeze-thaw (the alternate freezing and thawing of ice in cracks); it will, therefore, tend to be angular in nature. Where two glaciers merge, two lateral moraines may join together to form a **medial moraine** running along the centre of the merged glacier.

latitude and longitude imaginary lines used to locate position on the globe. Lines of latitude are drawn parallel to the Equator, with 0° at the Equator and 90° at the north and south poles. Lines of longitude are drawn at right-angles to these, with 0° (the Prime Meridian) passing through Greenwich, England.

The 0-degree line of latitude is defined by Earth's Equator, a characteristic definable by astronomical observation. It was determined as early as AD 150 by Egyptian astronomer Ptolemy in his world atlas. The prime meridian, or 0-degree line of longitude, is a matter of convention rather than physics. Prior to the latter half of the

Point X lies on longitude 60°W

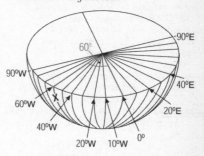

Point X lies on latitude 20°S

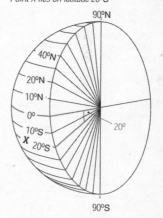

latitude and longitude Locating a point on a globe using latitude and longitude. Longitude is the angle between the terrestrial meridian through a place and the standard meridian 0° passing through Greenwich, England. Latitude is the angular distance of a place from the equator.

18th century, sailors navigated by referring to their position east or west of any arbitrary meridian. When Nevil Maskelyne (1732–1811), English astronomer and fifth Astronomer Royal, published the *Nautical Almanac* he referred all of his lunar–stellar distance tables to the Greenwich meridian. These tables were relied upon for computing longitudinal position and so the Greenwich meridian became widely accepted.

Chronometers, time-keeping devices with sufficient accuracy for longitude determination, invented by English instrument-maker John Harrison (1693–1776) and perfected in 1759, would gradually replace the lunar distance method for navigation, but reliance on the Greenwich meridian persisted because the *Nautical Almanac* was used by sailors to verify their position. The Greenwich meridian was officially adopted as the Prime Meridian by the International Meridian Conference held in Washington, DC, in 1884.

lattice network of straight lines.

lattice points points of intersection of the lines in a lattice.

laughing gas popular name for *nitrous oxide, an anaesthetic.

Laurasia northern landmass formed 200 million years ago by the splitting of the single world continent *Pangaea. (The southern landmass was Gondwanaland.) It consisted of what was to become North America, Greenland, Europe, and Asia, and is believed to have broken up about 100 million years ago with the separation of North America from Europe.

lava molten *magma that erupts from a *volcano and cools to form extrusive *igneous rock. Lava types differ in composition, temperature, gas content, and viscosity (resistance to flow). The three major lava types are basalt (dark, fluid, and relatively low silica content), rhyolite (light, viscous, high silica content), and andesite (an intermediate lava).

lawrencium chemical symbol Lr, synthesized, radioactive, metallic element, the last of the actinide series, atomic number 103, relative atomic mass 262. Its only known isotope,

Lr-257, has a half-life of 4.3 seconds and was originally synthesized at the University of California at Berkeley in 1961 by bombarding californium with boron nuclei. The original symbol, Lw, was officially changed in 1963.

The element was named after Ernest Lawrence (1901–1958), the US inventor of the cyclotron.

lb (Latin *'libra'*) symbol for *pound (weight).

LCD abbreviation for *liquid-crystal display.

LCD projector a type of projector that uses a series of colour *liquid-crystal display (LCD) panels through which a very bright light is shone. The images on the LCD screens are focused onto a screen using a lens and are controlled by a computer, either instead of, or alongside, a normal monitor.

LDAP abbreviation for lightweight directory access protocol, in computing, Internet standard that enables a client PC or workstation to look up an e-mail address on an LDAP server over a *TCP/IP network. LDAP is a simplified version of the 'heavyweight' X.500 directory access protocol in the OSI (Open Systems Interconnection) standards suite.

L-dopa chemical, normally produced by the body, which is converted by an enzyme to dopamine in the brain. It is essential for integrated movement of individual muscle groups.

LDR abbreviation for *light-dependent resistor*.

leaching process by which substances are washed through or out of the soil. Fertilizers leached out of the soil drain into rivers, lakes, and ponds and cause *water pollution. In tropical areas, leaching of the soil after the destruction of forests removes scarce nutrients and can lead to a dramatic loss of soil fertility. The leaching of soluble minerals in soils can lead to the formation of distinct soil horizons as different minerals are deposited at successively lower levels.

lead chemical symbol Pb, heavy, soft, malleable, grey, metallic element, atomic number 82, relative atomic mass 207.19. Its chemical symbol comes from the Latin *plumbum*. Usually found as an ore (most often as the sulphide galena), it occasionally occurs as a free metal (*native metal), and is the final stable product of the decay of uranium. Lead is the softest and weakest of the commonly used metals, with a low melting point; it is a poor conductor of electricity and resists acid corrosion. As a cumulative poison, lead enters the body from lead water pipes, lead-based paints, and leaded petrol. (In humans, exposure to lead shortly after birth is associated with impaired mental health between the ages of two and four.) The metal is an effective shield against radiation and is used in batteries, glass, ceramics, and alloys such as pewter and solder.

lead–acid cell   type of *accumulator (storage battery).

lead(II) nitrate $Pb(NO_3)_2$, one of only two common water-soluble compounds of lead (the other is lead ethanoate or acetate). When heated, it decrepitates and decomposes readily into oxygen, brown nitrogen(IV) oxide gas, and the red-yellow solid lead(II) oxide.

$$2Pb(NO_3)_2 \rightarrow 2PbO + 4NO_2 + O_2$$

lead ore   any of several minerals from which lead is extracted. The primary ore is galena or lead sulphite PbS. This is unstable, and on prolonged exposure to the atmosphere it oxidizes into the minerals cerussite $PbCO_3$ and anglesite $PbSO_4$. Lead ores are usually associated with other metals, particularly silver – which can be mined at the same time – and zinc, which can cause problems during smelting.

lead(II) oxide   or **lead monoxide**; PbO, yellow or red solid, an amphoteric oxide (one that reacts with both acids and bases). The other oxides of lead are the brown solid lead(IV) oxide (PbO_2) and red lead (Pb_3O_4).

leaf   lateral outgrowth on the stem of a plant, and in most species is the primary organ of *photosynthesis (the process in which the *energy from absorbed sunlight is used to combine *carbon dioxide and *water to form sugars). The chief leaf types are cotyledons (seed leaves), scale leaves (on underground stems), foliage leaves, and bracts (in the axil of which a flower is produced).

learning theory   in psychology, any theory or body of theories about how behaviour in animals and human beings is acquired or modified by experience. Two main theories are classical and operant *conditioning.

leased line   in computing, permanent dedicated digital telephone link used for round-the-clock connection within a network or between offices. For example, a bank may use leased lines to carry financial data between branches and head office. The infrastructure of the Internet is a network of leased lines that deliver guaranteed *bandwidth at a fixed cost, regardless of how much traffic they carry. The enormous economies produced by the heavy use of such lines makes the Net a very cheap method of communication.

least action, principle of   in physics, an alternative expression of *Newton's laws of motion that states that a particle moving between two points under the influence of a force will follow the path along which its total action is least. Action is a quantity related to the average difference between the kinetic energy and the potential energy of the particle along its path. The principle is only true where no energy is lost from the system; for example an object moving in free fall in a gravitational field. It is closely related to *Fermat's principle of least time which governs the path taken by a ray of light.

Le Chatelier's principle   or **Le Chatelier-Braun principle**, principle that if a change in conditions is imposed on a system in equilibrium, the system will react to counteract that change and restore the equilibrium. First stated in 1884 by French chemist Henri le Chatelier (1850–1936), it has been found to apply widely outside the field of chemistry.

lecithin   lipid (fat), containing nitrogen and phosphorus, that forms a vital part of the cell membranes of plant and animal cells. The name is from the Greek *lekithos* 'egg yolk', eggs being a major source of lecithin.

LED   abbreviation for *light-emitting diode.

left-hand rule in physics, a memory aid used to recall the relative directions of motion, magnetic field, and current in an electric motor. It was devised by English physicist John Fleming. (See *Fleming's rules*).

legacy application in computing, inherited application, usually an old one that runs on a large minicomputer or mainframe, and that may be too important to scrap or too expensive to change. 'Legacy' implies that such applications are valuable and should be looked after. Those who want to be rid of legacy applications use different metaphors, such as 'slum clearance'.

legacy system in computing, old, in-house, back-room, pre-Internet, corporate computing systems. Legacy systems are a hangover from the pre-Internet business world. Legacy systems have to be overhauled so that a company's front- and back-room operations can integrate seamlessly with the Internet and Internet applications. There is considerable corporate resistance to upgrading legacy systems, which is unsurprising when some estimates believe the investment in legacy systems to be over $4 trillion worldwide.

legionnaires' disease pneumonia-like disease, so called because it was first identified when it broke out at a convention of the American Legion in Philadelphia in 1976. Legionnaires' disease is caused by the bacterium *Legionella pneumophila*, which breeds in warm water (for example, in the cooling towers of air-conditioning systems). It is spread in minute water droplets, which may be inhaled. The disease can be treated successfully with antibiotics, though mortality can be high in elderly patients.

legume plant of the family Leguminosae, which has a pod containing dry seeds. The family includes peas, beans, lentils, clover, and alfalfa (lucerne). Legumes are important in agriculture because of their specialized roots, which have nodules containing bacteria capable of fixing nitrogen from the air and increasing the fertility of the soil. The edible seeds of legumes are called **pulses**.

leishmaniasis any of several parasitic diseases caused by microscopic protozoans of the genus *Leishmania*, identified by William Leishman (1865–1926), and transmitted by sandflies. It occurs in two main forms: **visceral** (also called kala-azar), in which various internal organs are affected, and **cutaneous**, where the disease is apparent mainly in the skin. Leishmaniasis occurs in the Mediterranean region, Africa, Asia, and Central and South America. There are 12 million cases of leishmaniasis annually. The disease kills 8,000 people a year in South America and results in hundreds of thousands more suffering permanent disfigurement and disability through skin lesions, joint pain, and swelling of the liver and spleen.

lens in optics, a piece of a transparent material, such as glass, with two polished surfaces – one concave or convex, and the other plane, concave, or convex – that modifies rays of light. A convex lens brings rays of light together; a concave lens makes the rays diverge. Lenses are essential to spectacles, microscopes, telescopes, cameras, and almost all optical instruments.

lenticel small pore on the stems of woody plants or the trunks of trees. Lenticels are a means of gas exchange between the stem interior and the atmosphere. They consist of loosely packed cells with many air spaces in between, and are easily seen on smooth-barked trees such as cherries, where they form horizontal lines on the trunk.

Lenz's law in physics, a law stating that the direction of an electromagnetically induced current (generated by moving a magnet near a wire or a wire in a magnetic field) will be such as to oppose the motion producing it. It is named after the Russian physicist Heinrich Friedrich Lenz (1804–1865), who announced it in 1833.

leprosy or **Hansen's disease**, chronic, progressive disease caused by the bacterium *Mycobacterium leprae*, closely related to that of tuberculosis. The infection attacks the skin and nerves.

Leprosy is endemic in 28 countries, and confined almost entirely to the tropics. It is controlled with drugs. Worldwide, there are 700,000 new cases of leprosy a year (2001).

lepton any of a class of *elementary particles that 'feel' the weak nuclear and electromagnetic force and are not affected by the strong nuclear force. Leptons are particles with half-integral spin and comprise the *electron, *muon, and *tau, and their *neutrinos (the electron neutrino, muon neutrino, and tau neutrino), as well as their six *antiparticles. The muon (found by the US physicist Carl Anderson in cosmic radiation in 1937) produces the muon neutrino when it decays; the tau, a surprise discovery of the 1970s, produces the tau neutrino when it decays. See *fermion.

leptoquark in physics, a hypothetical particle made up of a *quark combined with a *lepton, or a new particle created by their interaction.

leucine one of the nine essential *amino acids.

leucocyte another name for a *white blood cell.

leukaemia any one of a group of cancers of the blood cells, with widespread involvement of the bone marrow and other blood-forming tissue. The central feature of leukaemia is runaway production of white blood cells that are immature or in some way abnormal. These rogue cells, which lack the defensive capacity of healthy white cells, overwhelm the normal ones, leaving the victim vulnerable to infection. Treatment is with radiotherapy and *cytotoxic drugs to suppress replication of abnormal cells, or by bone-marrow transplantation.

lever simple machine consisting of a rigid rod pivoted at a fixed point called the fulcrum, used for shifting or raising a heavy load or applying force. Levers are classified into orders according to where the effort is applied, and the load-moving force developed, in relation to the position of the fulcrum. (See diagram, page 336.)

LF in physics, abbreviation for low *frequency. LF radio waves have frequencies in the range 30–300 kHz.

LH abbreviation for ***luteinizing hormone**.

libido in Freudian psychology, the energy of the sex instinct, which is to be found even in a newborn child. The libido develops through a number of phases, described by Austrian physician Sigmund Freud (1856–1939) in his theory of infantile sexuality. The source of the libido is the *id.

library program one of a collection, or library, of regularly used software routines, held in a computer backing store. For example, a programmer might store a routine for sorting a file into key field order, and so could incorporate it easily into any new

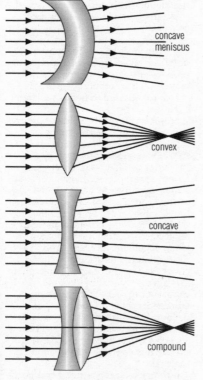

concave meniscus

convex

concave

compound

lens The passage of light through lenses. The concave lenses diverges a beam of light from a distant source. The convex and compound lenses focus light from a distant source to a point. The distance between the focus and the lens is called the focal length. The shorter the focus, the more powerful the lens.

program being developed instead of having to rewrite it.

lichen any organism of a unique group that consists of associations of a specific *fungus and a specific *alga living together in a mutually beneficial relationship. Found as coloured patches or spongelike masses on trees, rocks, and other surfaces, lichens flourish in harsh conditions. (Group Lichenes.)

life cycle in biology, the sequence of developmental stages through which members of a given species pass. Most vertebrates have a simple life cycle consisting of *fertilization of sex cells or *gametes, a period of development as an *embryo, a period of juvenile growth after hatching or birth, an adulthood including *sexual reproduction, and finally death. Invertebrate life cycles are generally more complex and may involve major reconstitution of the individual's appearance (*metamorphosis) and

first-order lever

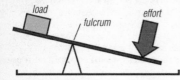

second-order lever

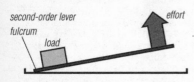

third-order lever

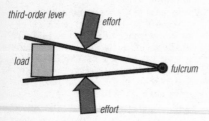

lever Types of lever. Practical applications of the first-order lever include the crowbar, seesaw, and scissors. The wheelbarrow is a second-order lever; tweezers or tongs are third-order levers.

completely different styles of life. Plants have a special type of life cycle with two distinct phases, known as *alternation of generations. Many insects such as cicadas, dragonflies, and mayflies have a long larvae or pupae phase and a short adult phase. Dragonflies live an aquatic life as larvae and an aerial life during the adult phase. In many invertebrates and protozoa there is a sequence of stages in the life cycle, and in parasites different stages often occur in different host organisms.

life-cycle analysis assessment of the environmental impact of a product, taking into account all aspects of production (including resources used), packaging, distribution and ultimate end.

This 'cradle-to-grave' approach can expose inconsistencies in many so-called 'eco-friendly' labels, applied to products such as soap powders, which may be biodegradable but which are perhaps contained in nonrecyclable containers.

life sciences scientific study of the living world as a whole, a new synthesis of several traditional scientific disciplines including *biology, *zoology, and *botany, and newer, more specialized areas of study such as biophysics and *sociobiology.

ligament strong, flexible connective tissue, made of the protein *collagen, which joins bone to bone at moveable joints and sometimes encloses the joints. Ligaments prevent bone dislocation (under normal circumstances) but allow joint flexion. The ligaments around the joints are composed of white fibrous tissue. Other ligaments are composed of yellow elastic tissue, which is adapted to support a continuous but varying stress, as in the ligament connecting the various cartilages of the *larynx (voice box).

ligand in chemistry, a group that bonds symmetrically to a central atom or ion of a metal; the result is called a **coordination complex**. An example of a neutral ligand is ammonia; the nitrosyl ion NO^+ is a charged ligand. An example of a coordination complex is hexaminocobalt chloride,

[Co(NH$_3$)$_6$]Cl$_3$, in which the central cobalt ion (Co^{3+}) is surrounded by covalent bonds with six ammonia molecules and ionic bonds with three chloride ions.

Ligands are used in medicine as an antidote to heavy metal poisoning, removing the metal ions by attaching themselves to form a harmless compound.

light *electromagnetic waves (made up of electric and magnetic components) in the visible range, having a wavelength from about 400 nanometres in the extreme violet to about 700 nanometres in the extreme red. Light is considered to exhibit particle and wave properties, and the fundamental particle, or quantum, of light is called the photon. A light wave comprises two transverse waves of electric and magnetic fields travelling at right angles to each other, and as such is a form of electromagnetic radiation. The speed of light (and of all electromagnetic radiation) in a vacuum is approximately 300,000 km/186,000 mi per second, and is a universal constant denoted by c.

Light is a form of energy that is mainly visible to the human *eye. Light is radiated by hot objects such as the Sun or an electric light bulb.

The nature of light travelling through a medium can be explained with the help of Young's double-slit experiment. A double slit is placed between a light source and a screen. Light diffracted by the double slit forms a pattern of bright and dark bands (called fringes) on the screen. The fringe pattern is due to the light waves moving through the slits and being diffracted. At a point where light waves are in phase (a crest and crest meeting at the same point) a constructive wave front is formed with a larger amplitude than that of the two waves forming the constructive wave front. This is seen as a bright band on the screen. At a point where the light waves are out of phase (a crest and a trough meeting at the same point) a destructive wave front is formed with zero amplitude (the waves effectively cancel each other out). This is seen as a dark band on the screen.

light bulb incandescent filament lamp, first demonstrated by Joseph Swan in the UK in 1878 and US scientist and inventor Thomas Edison (1847–1931) in the USA in 1879. The present-day light bulb is a thin glass bulb filled with an inert mixture of nitrogen and argon gas. It contains a filament made of fine tungsten wire. When electricity is passed through the wire, it glows white hot, producing light.

light curve in astronomy, graph showing how the brightness of an astronomical object varies with time. Analysis of the light curves of variable stars, for example, gives information about the physical processes causing the variation.

light-dependent resistor LDR, component of electronic circuits whose resistance varies with the level of illumination on its surface. Usually resistance decreases as illumination increases, as the nature of the material is altered by the presence of light. LDRs are used in light-measuring or light-sensing instruments where light intensity is converted to a digital signal (for example, in the exposure-meter circuit of an automatic camera), and in switches (such as those that switch on street lights at dusk).

LDRs are made from *semiconductors, such as cadmium sulphide.

light-emitting diode LED, electronic component that converts electrical energy into light or infrared radiation in the range of 550 nm (green light) to 1,300 nm (infrared). They are used for displaying symbols in electronic instruments and devices. An LED is a *diode made of *semiconductor material, such as gallium arsenide phosphide, that glows when electricity is passed through it. The first digital watches and calculators had LED displays, but many later models use *liquid-crystal displays.

In 1993 chemists at the University of Cambridge, England, developed LEDs from the polymer poly (p-phenylenevinyl) (PPV) that emit as much light as conventional LEDs and in a variety of colours. A new generation of LEDs that can produce light in the mid-infrared range

(300–1,000 nm) safely and cheaply have also been developed, using thin alternating layers of indium arsenide and indium arsenide antimonide.

lightning high-voltage electrical discharge between two rainclouds or between a cloud and the Earth, caused by the build-up of electrical charges. Air in the path of lightning ionizes (becomes a conductor), and expands; the accompanying noise is heard as thunder. Currents of 20,000 amperes and temperatures of 30,000°C/54,000°F are common. Lightning causes nitrogen oxides to form in the atmosphere and approximately 25% of atmospheric nitrogen oxides are formed in this way.

lightning conductor device that protects a tall building from lightning strike, by providing an easier path for current to flow to earth than through the building. It consists of a thick copper strip of very low resistance connected to the ground below. A good connection to the ground is essential and is made by burying a large metal plate deep in the damp earth. In the event of a direct lightning strike, the current in the conductor may be so great as to melt or even vaporize the metal, but the damage to the building will nevertheless be limited.

light reaction part of the *photosynthesis process in green plants that requires sunlight (as opposed to the *dark reaction, which does not). During the light reaction light energy is used to generate ATP (by the phosphorylation of ADP), which is necessary for the dark reaction.

light second unit of length, equal to the distance travelled by a beam of light in a vacuum in one second. It is equal to 2.99792458×10^8 m$/9.8357103 \times 10^8$ ft. See *light year.

light watt unit of radiant power (brightness of light). One light watt is the power required to produce a perceived brightness equal to that of light at a wavelength of 550 nanometres and 680 lumens.

light year distance travelled by a beam of light in a vacuum in one year. It is equal to approximately 9.4605×10^{12} km$/5.9128 \times 10^{12}$ mi.

lignin naturally occurring substance produced by plants to strengthen their tissues. It is difficult for *enzymes to attack lignin, so living organisms cannot digest wood, with the exception of a few specialized fungi and bacteria. Lignin is the essential ingredient of all wood and is, therefore, of great commercial importance.

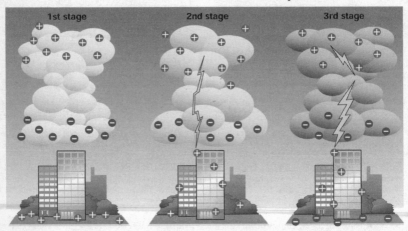

lightning The build-up of electrical charge during a thunderstorm that causes lightning. Negative charge builds up at the bottom of a cloud; positive charges rise from the ground and also within the cloud, moving to the top of it. A conducting channel forms through the cloud and a giant spark jumps between opposite charges causing lightning to strike within the cloud and from cloud to ground.

lignite type of *coal that is brown and fibrous, with a relatively low carbon content. As a fuel it is less efficient because more of it must be burned to produce the same amount of energy generated by bituminous coal. Lignite also has a high sulphur content and is more polluting. It is burned to generate power in Scandinavia and some European countries because it is the only fuel resource available without importing.

lime or **quicklime**; technical name **calcium oxide**; CaO, white powdery substance used in making mortar and cement. It is made commercially by heating calcium carbonate ($CaCO_3$), obtained from limestone or chalk, in a lime kiln. Quicklime readily absorbs water to become calcium hydroxide $Ca(OH)_2$, known as slaked lime, which is used to reduce soil acidity.

limestone sedimentary rock composed chiefly of calcium carbonate ($CaCO_3$), either derived from the shells of marine organisms or precipitated from solution, mostly in the ocean. Various types of limestone are used as building stone. Karst is a type of limestone landscape. Caves commonly occur in limestone. *Marble is metamorphosed limestone. Certain so-called marbles are not in fact marbles but fine-grained fossiliferous limestones that have been polished.

limestone pavement bare rock surface resembling a block of chocolate, found on limestone plateaus. It is formed by the weathering of limestone into individual upstanding blocks, called clints, separated from each other by joints, called grykes. The weathering process is thought to entail a combination of freeze-thaw (the alternate freezing and thawing of ice in cracks) and carbonation (the dissolving of minerals in the limestone by weakly acidic rainwater). Malham Tarn in North Yorkshire is an example of a limestone pavement.

limewater common name for a dilute solution of slaked lime (calcium hydroxide, $Ca(OH)_2$). In chemistry, it is used to detect the presence of carbon dioxide. If a gas containing carbon dioxide is bubbled through limewater, the solution turns milky owing to the formation of calcium carbonate ($CaCO_3$). Continued bubbling of the gas causes the limewater to clear again as the calcium carbonate is converted to the more soluble calcium hydrogencarbonate ($Ca(HCO_3)_2$).

limit in mathematics, in an infinite sequence, the final value towards which the sequence is tending. For example, the limit of the sequence $1/2$, $3/4$, $7/8$, $15/16$ ... is 1, although no member of the sequence will ever exactly equal 1 no matter how many terms are added together. The limit of the ratios of a Fibonacci sequence is $(\sqrt{5} + 1)/2$. This number is also the golden section.

limiting factor in biology, any factor affecting the rate of a metabolic reaction. Levels of light or of carbon dioxide are limiting factors in *photosynthesis because both are necessary for the production of carbohydrates. In experiments, photosynthesis is observed to slow down and eventually stop as the levels of light decrease. It is believed that the concentrations of carbon dioxide building up in the atmosphere through the burning of fossil fuels will allow faster plant growth.

linac contraction of *linear accelerator, a type of particle accelerator in which the particles are accelerated along a straight tube.

linear accelerator or **linac**, in physics, a type of particle *accelerator in which the particles move along a straight tube. Particles pass through a linear accelerator only once – unlike those in a cyclotron or synchrotron (ring-shaped accelerators), which make many revolutions, gaining energy each time.

The world's longest linac is the Stanford Linear Collider, in which electrons and positrons are accelerated along a straight track 3.2 km/2 mi long and then steered into a head-on collision with other particles. The first linear accelerator was built in 1928 by the Norwegian engineer Ralph Wideröe to investigate the behaviour of heavy ions (large atoms with one or more electrons removed), but devices capable of accelerating smaller particles such as protons and electrons could not be built until after World War II and the development of high-power radio- and microwave-frequency generators.

linear equation in mathematics, a relationship between two variables that, when plotted on Cartesian axes, produces a straight-line graph; the equation has the general form $y = mx + c$, where m is the slope of the line represented by the equation and c is the y-intercept, or the value of y where the line crosses the y-axis in the *Cartesian coordinate system. Sets of linear equations can be used to describe the behaviour of buildings, bridges, trusses, and other static structures.

linear motor type of electric motor, an induction motor in which the fixed stator and moving armature are straight and parallel to each other (rather than being circular and one inside the other as in an ordinary induction motor). Linear motors are used, for example, to power sliding doors. There is a magnetic force between the stator and armature; this force has been used to support a vehicle, as in the experimental maglev linear motor train.

linear programming in mathematics and economics, a set of techniques for finding the maxima or minima of certain variables governed by linear equations or inequalities. These maxima and minima are used to represent 'best' solutions in terms of goals such as maximizing profit or minimizing cost.

line of best fit on a *scatter diagram, line drawn as near as possible to the various points so as to best represent the trend being graphed. The sums of the displacements of the points on either side of the line should be equal.

line of force in physics, an imaginary line representing the direction of force at any point in a magnetic, gravitational, or electrical field.

link in computing, an image or item of text in a *World Wide Web document that acts as a route to another Web page or file on the Internet. Links are created by using *HTML to combine an on-screen 'anchor' with a hidden Hypertext Reference (HRF), usually

the *URL (Web address) of the item in question.

linkage in genetics, the association between two or more genes that tend to be inherited together because they are on the same chromosome. The closer together they are on the chromosome, the less likely they are to be separated by crossing over (one of the processes of *recombination) and they are then described as being 'tightly linked'.

linolenic acid $C_{18}H_{30}O_2$, an essential *fatty acid found in linseed oil and in most seed fats.

Linux contraction of Linus Unix, in computing, operating system based on an original core program written by Linus Torvalds, a 22-year-old student at the University of Helsinki, Finland, in 1991–92. Linux is a non-proprietary system, made up of freely available ('open source') code created over several years by *Unix enthusiasts all over the world. Each programmer retains the copyright to his or her creation, but makes it freely available on the Internet. Linux retains the flexibility and many of the advanced programming features that make Unix popular for technically minded users, but can run on an ordinary PC instead of an expensive Unix workstation.

LINX, the contraction of London Internet Exchange, in computing, Docklands-based hub for all Internet traffic within the UK, run by the main British Internet Service Providers as a joint enterprise. By making sure that all traffic between British Internet users stays in the UK, the LINX improves the speed and reliability of network connections.

lipase enzyme responsible for breaking down fats into fatty acids and glycerol. It is produced by the *pancreas and requires a slightly alkaline environment. The products of fat digestion are absorbed by the intestinal wall.

lipid any of a large number of esters of fatty acids, commonly formed by the reaction of a fatty acid with glycerol. They are soluble in alcohol but not in water. Lipids are the chief constituents of plant and animal waxes, fats, and oils.

Phospholipids are lipids that also contain a phosphate group, usually linked to an organic base; they are major components of biological cell membranes.

lipophilic (Greek 'fat-loving') in chemistry, describing *functional groups with an affinity for fats and oils.

lipophobic (Greek 'fat-hating') in chemistry, describing *functional groups that tend to repel fats and oils.

liposome in medicine, a minute droplet of oil that is separated from a medium containing water by a *phospholipid layer. Drugs, such as cytotoxic agents, can be incorporated into liposomes and given by injection or by mouth. The liposomes allow the drug to reach the site of action, such as a tumour, without being broken down in the body.

liquefaction process of converting a gas to a liquid, normally associated with low temperatures and high pressures (see *condensation).

liquefaction in earth science, the conversion of a soft deposit, such as clay, to a jellylike state by severe shaking. During an earthquake buildings and lines of communication built on materials prone to liquefaction will sink and topple. In the Alaskan earthquake of 1964 liquefaction led to the destruction of much of the city of Anchorage.

liquefied petroleum gas LPG, liquid form of butane, propane, or pentane, produced by the distillation of petroleum during oil refining. At room temperature these substances are gases, although they can be easily liquefied and stored under pressure in metal containers. They are used for heating and cooking where other fuels are not available: camping stoves and cigarette lighters, for instance, often use liquefied butane as fuel.

liquid state of matter between a *solid and a *gas. A liquid forms a level surface and assumes the shape of its container. The way that liquids behave can be explained by the *kinetic theory of matter and particle theory. Its atoms do not occupy fixed positions as in a crystalline solid, nor do they have total freedom of movement as in a gas.

Unlike a gas, a liquid is difficult to compress since pressure applied at one point is equally transmitted throughout (Pascal's principle). *Hydraulics makes use of this property.

liquid air air that has been cooled so much that it has liquefied. This happens at temperatures below about −196°C/−321°F. The various constituent gases, including nitrogen, oxygen, argon, and neon, can be separated from liquid air by the technique of *fractional distillation.

liquid-crystal display LCD, display of numbers (for example, in a calculator) or pictures (such as on a pocket television screen) produced by molecules of a substance in a semiliquid state with some crystalline properties, so that clusters of molecules align in parallel formations. The display is a blank until the application of an electric field, which 'twists' the molecules so that they reflect or transmit light falling on them. There two main types of LCD are **passive matrix** and **active matrix**. (See diagram, page 342.)

liquid-droplet model Model for the nucleus proposed in 1939 by Danish physicist Niels Bohr (1885–1962). In this model, nuclear particles are pulled together by short-range forces, similar to the way in which molecules in a drop of liquid are attracted to one another. In the case of uranium, the extra energy produced by the absorption of a neutron causes the nuclear particles to separate into two groups of approximately the same size, thus breaking the nucleus into two smaller nuclei – a process called nuclear fission. The model was vindicated when Bohr correctly predicted the differing behaviour of nuclei of uranium-235 and uranium-238 from the fact that the numbers of neutrons in the two nuclei is odd and even respectively.

LISP contraction of list processing, high-level computer-programming language designed for manipulating lists of data items. It is used primarily in research into *artificial intelligence (AI).

listeriosis disease of animals that may occasionally infect humans, caused by the bacterium *Listeria monocytogenes*.

The bacteria multiply at temperatures close to 0°C/32°F, which means they may flourish in precooked frozen meals if the cooking has not been thorough. Listeriosis causes flulike symptoms and inflammation of the brain and its surrounding membranes. It can be treated with penicillin. Named after English surgeon Joseph Lister (1827–1912), the founder of antiseptic surgery.

lithification conversion of an unconsolidated (loose) sediment into

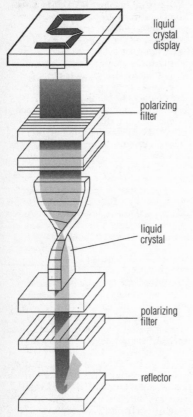

liquid crystal display

polarizing filter

liquid crystal

polarizing filter

reflector

liquid-crystal display A liquid-crystal display consists of a liquid crystal sandwiched between polarizing filters similar to polaroid sunglasses. When a segment of the seven-segment display is electrified, the liquid crystal twists the polarized light from the front filter, allowing the light to bounce off the rear reflector and illuminate the segment.

solid sedimentary rock by **compaction** of mineral grains that make up the sediment, **cementation** by crystallization of new minerals from percolating water solutions, and new growth of the original mineral grains. The term is less commonly used to refer to solidification of *magma to form igneous rock.

lithium chemical symbol Li, (Greek *lithos* 'stone') soft, ductile, silver-white, metallic element, atomic number 3, relative atomic mass 6.941. It is one of the *alkali metals (being found at the top of Group 1 of the *periodic table). It has a very low density (far less than most woods), and floats on water (relative density 0.57); it is the lightest of all metals. Although in the *reactivity series it is the least reactive of the alkali metals, it reacts readily with water and the oxygen in air, and so has to be kept under oil. Lithium is used to harden alloys, and in batteries; its compounds are used in medicine to treat *manic depression.

Lithium was named in 1818 by Swedish chemist Jöns Berzelius, having been discovered the previous year by his student Johan A Arfwedson. Berzelius named it after 'stone' because it is found in most igneous rocks and many mineral springs.

lithosphere upper rocky layer of the Earth that forms the jigsaw of plates that take part in the movements of *plate tectonics. The lithosphere comprises the *crust and a portion of the upper *mantle. It is regarded as being rigid and brittle and moves about on the more plastic and less brittle asthenosphere. The lithosphere ranges in thickness from 2–3 km/1–2 mi at mid-ocean ridges to 150 km/93 mi beneath old ocean crust, to 250 km/ 155 mi under cratons.

litre symbol l, metric unit of volume, equal to one cubic decimetre (1.76 imperial pints/2.11 US pints). It was formerly defined as the volume occupied by one kilogram of pure water at 4°C at standard pressure, but this is slightly larger than one cubic decimetre.

liver in vertebrates, large organ with many regulatory and storage functions. The human liver is situated in the upper abdomen, and weighs about 2 kg/4.5 lb. It is divided into four lobes. The liver receives the products of digestion (food absorbed from the *gut and carried to the liver by the bloodstream), converts *glucose to glycogen (a long-chain carbohydrate used for storage), and then back to glucose when needed. In this way the liver regulates the level of glucose in the blood (see *homeostasis). This is partly controlled by a *hormone, *insulin. The liver removes excess amino acids from the blood, converting them to urea, which is excreted by the kidneys. The liver also synthesizes vitamins, produces bile and blood-clotting factors, and removes damaged red cells and toxins such as alcohol from the blood.

If more *protein is eaten than is needed to make different proteins for the body, the excess is broken down. This breakdown takes place in the liver. One product of this breakdown is urea and this has to be lost from the body in the urine. The liver also stores some vitamins, produces *bile, and breaks down *red blood cells.

lm in physics, symbol for *lumen, the SI unit of luminous flux, equal to the flux emitted by a 1-candela point source in a 1-steradian solid angle.

loam type of fertile soil, a mixture of sand, silt, clay, and organic material. It is porous, which allows for good air circulation and retention of moisture.

local anaesthetic in medicine, an *anaesthetic that is applied only to the specific area under treatment.

local area network LAN, in computing, a *network restricted to a single room or building. Local area networks enable around 500 devices, usually microcomputers acting as workstations, as well as peripheral devices, such as printers, to be connected together.

local bus in computing, an extension of the central processing unit (CPU) *bus (electrical pathway), designed to

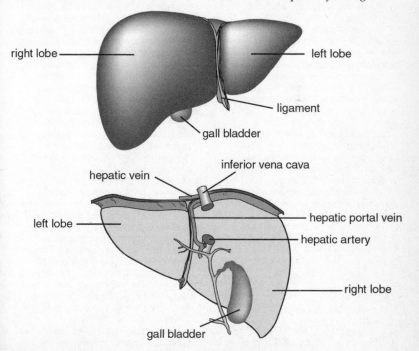

liver The structure of the human liver, front view (top) and cross-section.

speed up data transfer between the CPU, disks, graphics boards, and other devices. The *PCI bus has become the standard on PCs and more recently has been adopted for the Apple Macintosh.

local variable in computing, a *variable that can be accessed only by the instructions within a particular subroutine.

lockjaw former name for *tetanus, a type of bacterial infection.

locus in biology, the point on a *chromosome where a particular *gene occurs.

locus (Latin 'place') in mathematics, traditionally the path traced out by a moving point, but now defined as the set of all points on a curve satisfying given conditions. The locus of points a fixed distance from a fixed point is a circle. The locus of a point equidistant from two fixed points is a straight line that perpendicularly bisects the line joining them. The locus of points a fixed distance from a line is two parallel lines running either side.

log in mathematics, abbreviation for *logarithm.

log any apparatus for measuring the speed of a ship; also the daily record of events on board a ship or aircraft.

logarithm or **log**, the *exponent or index of a number to a specified base – usually 10. For example, the logarithm to the base 10 of 1,000 is 3 because 10^3 = 1,000; the logarithm of 2 is 0.3010 because $2 = 10^{0.3010}$. The whole-number part of a logarithm is called the **characteristic**; the fractional part is called the **mantissa**. Before the advent of cheap electronic calculators, multiplication and division could be simplified by being replaced with the addition and subtraction of logarithms.

For any two numbers x and y (where $x = b^a$ and $y = b^c$), $x \times y = b^a \times b^c = b^{a+c}$; hence one would add the logarithms of x and y, and look up this answer in antilogarithm tables.

Tables of logarithms and antilogarithms are available that show conversions of numbers into logarithms, and vice versa. For example, to multiply 6,560 by 980, one looks up their logarithms (3.8169 and 2.9912), adds them together (6.8081), then looks up the antilogarithm of this to get the

answer (6,428,800). **Natural** or **Napierian logarithms** are to the base e, an *irrational number equal to approximately 2.7183.

The principle of logarithms is also the basis of the slide rule. With the general availability of the electronic pocket calculator, the need for logarithms has been reduced. The first log tables (to base e) were published by the Scottish mathematician John Napier (1550–1617) in 1614. Base-ten logs were introduced by the Englishman Henry Briggs (1561–1631) and Dutch mathematician Adriaen Vlacq (1600–1667).

log file in computing, file that keeps a record of computer transactions. A log file might track the length and type of connection made to a network, or compile details of faxes sent by computer.

logic gate or **logic circuit**, in electronics, one of the basic components used in building *integrated circuits. The five basic types of gate make logical decisions based on the functions OR, AND, NOT, NOR (NOT OR) and NAND (NOT AND). With the exception of the NOT gate, each has two or more inputs.

log off or **log out**, in computing, the process by which a user identifies himself or herself to a multiuser computer and leaves the system.

log on or **log in**, in computing, the process by which a user identifies himself or herself to a multiuser computer and enters the system. Logging on usually requires the user to enter a password before access is allowed.

lone pair in chemistry, a pair of electrons in the outermost shell of an atom that are not used in bonding. In certain circumstances, they will allow the atom to bond with atoms, ions, or molecules (such as boron trifluoride, BF_3) that are deficient in electrons, forming coordinate covalent (dative) bonds in which they provide both of the bonding electrons.

long. abbreviation for **longitude**; see *latitude and longitude.

longitude see *latitude and longitude.

longitudinal wave *wave in which the displacement of the medium's particles

circuit symbols

OR gate   AND gate   NOT or inverter gate   NOR gate   NAND gate

truth tables

inputs		output
0	0	0
0	1	1
1	0	1
1	1	1

OR gate

inputs		output
0	0	0
0	1	0
1	0	0
1	1	1

AND gate

inputs	output
0	1
1	0

NOT gate

inputs		output
0	0	1
0	1	0
1	0	0
1	1	0

NOR gate

inputs		output
0	0	1
0	1	1
1	0	1
1	1	0

NAND gate

logic gate The circuit symbols for the five basic types of logic gate: OR, AND, NOT, NOR, and NAND. The truth table displays the output results of each possible combination of input signal.

is in line with or parallel to the direction of travel of the wave motion.

Various methods are used to reproduce waves, such as a ripple tank or a loosely-coiled spring, in order to understand their properties. If a loosely-coiled spring is given a slight 'pushing' force, the wave travels in the same direction as the movement of the coils. The waves consist of a series of compressions (where the coils of the spring are pressed together) and rarefactions (where the coils of the spring are more widely spaced). This is known as a longitudinal wave. The moving coils represent waves travelling along the spring. The spring remains as it was after the wave has travelled along the spring; waves carry energy from one place to another but they do not transfer matter. Sound waves are an example of longitudinal waves.

longshore drift movement of material along a beach. When a wave breaks at an angle to the beach, pebbles are carried up the beach in the direction of the wave (**swash**). The wave returns to the sea at right angles to the beach (**backwash**) because that is the steepest gradient, carrying some pebbles with it. In this way, material moves in a zigzag fashion along a beach. Longshore drift is responsible for the erosion of beaches and the formation of spits (ridges of sand or shingle projecting into the water). Attempts are often made to halt longshore drift by erecting barriers, or groynes, at right angles to the shore.

long-sightedness nontechnical term for *hypermetropia, a vision defect.

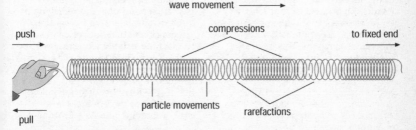

longitudinal wave The diagram illustrates the motion of a longitudinal wave. Sound, for example, travels through air in longitudinal waves: the waves vibrate back and forth in the direction of travel. In the compressions the particles are pushed together, and in the rarefactions they are pulled apart.

look-and-feel in computing, the general appearance of a user interface (usually a *graphical user interface). The concept of look-and-feel was the subject of several court cases in the USA. Apple sued Microsoft on the basis that the look-and-feel of Microsoft Windows infringed its copyright. The case was decided principally in Microsoft's favour.

look-up table in computing, a specific area on a worksheet in a *spreadsheet program that is set aside for storing a grid of information. This can then be referenced in a *formula using special look-up functions. A look-up table can be referred to directly, by cell reference, or can be used indirectly using functions to determine the position of a given value. A good example is an invoicing system that 'looks up' the price of a product.

loop in computing, short for *program loop.

loop in mathematics, part of a curve that encloses a space when the curve crosses itself. In a flow chart, a path that keeps on returning to the same question.

loopback in computing, any connection that sends an output signal to the same system's input. Loopback adaptors are used in electrical testing.

Los Alamos town in New Mexico, USA, which has had a centre for atomic and space research since 1942. In World War II the first atom (nuclear fission) bomb was built there (under US physicist Robert Oppenheimer (1904–1967)); the hydrogen bomb was also developed there.

lossless compression in computing, *data compression technique that reduces the number of *bits used to represent data in a file, thereby reducing its size while retaining all the original information. This makes it suitable for computer code and text files. Lossless compression typically achieves space savings of 30%.

lossy compression in computing, *data compression technique that dramatically reduces the size of a file by eliminating superfluous data. The lost information is either unnoticeable to the user, or can be recovered during decompression by extrapolation of the existing data. *JPEG and *MPEG are lossy methods that can reduce the size of graphics, audio, and video files by over 90%.

loudness subjective judgement of the level or power of sound reaching the ear. The human ear cannot give an absolute value to the loudness of a single sound, but can only make comparisons between two different sounds. The precise measure of the power of a sound wave at a particular point is called its *intensity. Accurate comparisons of sound levels may be made using sound-level meters, which are calibrated in units called *decibels.

loudspeaker electromechanical device that converts electrical signals into sound waves, which are radiated into the air. The most common type of loudspeaker is the **moving-coil speaker**. For example, electrical signals from a radio are fed to a coil of fine wire wound around the top of a cone. The coil is positioned between the poles of a permanent magnet. When signals pass through it, the coil becomes an electromagnet, experiencing a force at right angles to the direction of the current and magnetic field, causing the coil to move. As the signal varies, the coil and the cone vibrate, setting up sound waves. If the electrical signals have a frequency of 2,000 hertz (Hz), sound with a frequency of 2,000 Hz is produced.

lowest common denominator lcd, smallest number that is a multiple and thus divides exactly into each of the denominators of a set of fractions.

low-level language in computing, a programming language designed for a particular computer and reflecting its internal *machine code; low-level languages are therefore often described as **machine-oriented** languages. They cannot easily be converted to run on a computer with a different central processing unit, and they are relatively difficult to learn because a detailed knowledge of the internal working of the computer is required. Since they must be translated into machine code by an assembler program, low-level languages are also called *assembly languages.

LPG abbreviation for **liquefied petroleum gas**.

LSI abbreviation for large-scale integration, technology that enables whole electrical circuits to be etched into a piece of semiconducting material just a few millimetres square.

lubricant substance used between moving surfaces to reduce friction. Carbon-based (organic) lubricants, commonly called grease and oil, are recovered from petroleum distillation. Extensive research has been carried out on chemical additives to lubricants, which can reduce corrosive wear, prevent the accumulation of 'cold sludge' (often the result of stop-start driving in city traffic jams), keep pace with the higher working temperatures of aviation gas turbines, or provide radiation-resistant greases for nuclear power plants. Silicon-based spray-on lubricants are also used; they tend to attract dust and dirt less than carbon-based ones. A solid lubricant is graphite, an allotropic form of carbon, either flaked or emulsified (colloidal) in water or oil.

lumen symbol lm, SI unit of luminous flux (the amount of light passing through an area per second).

lumen in biology, the space enclosed by an organ, such as the bladder, or a tubular structure, such as the gastrointestinal tract.

luminescence emission of light from a body when its atoms are excited by means other than raising its temperature. Short-lived luminescence is called fluorescence; longer-lived luminescence is called phosphorescence.

luminous paint preparation containing a mixture of pigment, oil, and a phosphorescent sulphide, usually calcium or barium. After exposure to light it appears luminous in the dark. The luminous paint used on watch faces contains radium, is radioactive, and therefore does not require exposure to light.

Lunar Alps conspicuous mountain range on the Moon, northeast of the Sea of Showers (Mare Imbrium), cut by a valley 150 km/93 mi long. The highest peak (about 3,660 m/12,000 ft) is Mont Blanc.

lung in mammals, large cavity of the body, used for *gas exchange. Most tetrapod (four-limbed) vertebrates have a pair of lungs occupying the thorax. The lungs are essentially an infolding of the body surface – a sheet of thin, moist membrane made of a single layer of cells, which is folded so as to occupy less space while having a large surface area for the uptake of *oxygen and loss of *carbon dioxide. The folding creates tiny sacs called *alveoli. Outside the walls of the alveoli there are lots of blood *capillaries for transporting the products of gas exchange. The lung tissue, consisting of multitudes of air sacs and blood vessels, is very light and spongy, and functions by bringing inhaled air into close contact with the blood for efficient gas exchange. The efficiency of lungs is enhanced by *breathing movements, by the thinness and moistness of their surfaces, and by a constant supply of circulating blood.

The lungs inflate and deflate as a result of breathing movements (ventilation). Breathing movements are caused by movements of *muscles between the ribs and the muscles of the diaphragm. Air flows into the mouth and then along ever narrower tubes, trachea, bronchi, and tiny broncheoles. However, the last part of the journey to the alveoli is by diffusion only, as is the exchange with the blood.

Dust in the air is usually trapped by the mucus lining the tubes leading to the lungs. Cells lining the tubes are specialized cells (see *epithelium) and have hair-like structures – cilia – that sweep the trapped dust up to the mouth where it is swallowed. Some dust may reach the lungs where white blood cells may destroy it. However, the lungs can be damaged if dust is not removed. Many miners suffer from lungs damaged by the effects of coal dust, and many other forms of industrial dusts are equally dangerous.

In humans, the principal diseases of the lungs are tuberculosis, pneumonia, bronchitis, emphysema, and cancer. Bronchitis is an irritation of the airways resulting in them becoming narrower than normal so that a person cannot breathe fully. Emphysema is permanent damage to the alveolar walls resulting in too little surface for gas exchange. This too results in difficulties in breathing. The commonest cause of both bronchitis and emphysema is smoking.

lung cancer in medicine, cancer of the lung. The main risk factor is smoking, with almost nine out of ten cases attributed to it. Other risk factors include workspace exposure to carcinogenic substances such as asbestos and radiation. Warning symptoms include a persistent and otherwise unexplained cough, breathing difficulties, and pain in the chest or shoulder. Treatment is with chemotherapy, radiotherapy, and surgery.

lupus in medicine, any of various diseases characterized by lesions of the skin. One form (lupus vulgaris) is caused by the tubercle bacillus (see *tuberculosis). The organism produces ulcers that spread and eat away the

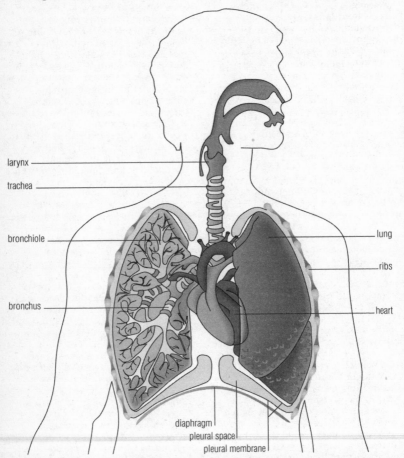

larynx
trachea
bronchiole
bronchus
lung
ribs
heart
diaphragm
pleural space
pleural membrane

lung The human lungs contain 300,000 million tiny blood vessels which would stretch for 2,400 km/1,500 mi if laid end to end. A healthy adult at rest breathes 12 times a minute; a baby breathes at twice this rate. Each breath brings 350 millilitres of fresh air into the lungs, and expels 150 millilitres of stale air from the nose and throat.

underlying tissues. Treatment is primarily with standard antituberculous drugs, but ultraviolet light may also be used.

luteinizing hormone *hormone produced by the pituitary gland. In males, it stimulates the testes to produce androgens (male sex hormones). In females, it works together with follicle-stimulating hormone to initiate production of egg cells by the ovary. If fertilization occurs, it plays a part in maintaining the pregnancy by controlling the levels of the hormones oestrogen and progesterone in the body.

lutetium chemical symbol Lu, (Latin *Lutetia* 'Paris') silver-white, metallic element, the last of the *lanthanide series, atomic number 71, relative atomic mass 174.97. It is used in the 'cracking', or breakdown, of petroleum and in other chemical processes. It was named by its discoverer, French chemist Georges Urbain, after his native city.

lux symbol lx, SI unit of illuminance or illumination (the light falling on an object). It is equivalent to one *lumen per square metre or to the illuminance

of a surface one metre distant from a point source of one *candela.

LWM abbreviation for **low water mark**.

lx in physics, symbol for *lux, the SI unit of illumination.

lycopene red carotenoid pigment found in tomatoes, rose hips, and many berries. It is the parent substance of all the natural carotenoids.

Lyme disease disease transmitted by tick bites that affects all the systems of the body. It has a 10% mortality rate. First described in 1975 following an outbreak in children living around Lyme, Connecticut, USA, it is caused by the micro-organism *Borrelia burgdorferi*, isolated by Burgdorfer and Barbour in the USA in 1982. Symptoms include a red 'bull's eye' rash around the bite, but this only occurs in about two-thirds of cases and otherwise the early stages are difficult to distinguish from flu. Untreated, the disease attacks the joints, nervous system, heart, liver, kidneys, and eyes, but responds to penicillin or tetracycline.

lymph fluid found in the lymphatic system of vertebrates.

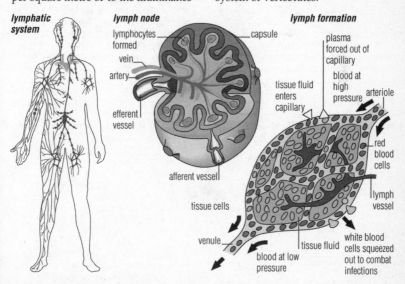

lymph Lymph is the fluid that carries nutrients and white blood cells to the tissues. Lymph enters the tissue from the capillaries (right) and is drained from the tissues by lymph vessels. The lymph vessels form a network (left) called the lymphatic system. At various points in the lymphatic system, lymph nodes (centre) filter and clean the lymph.

Lymph is drained from the tissues by lymph capillaries, which empty into larger lymph vessels (lymphatics). These lead to lymph nodes (small, round bodies chiefly situated in the neck, armpit, groin, thorax, and abdomen), which process the *lymphocytes produced by the bone marrow, and filter out harmful substances and bacteria. From the lymph nodes, vessels carry the lymph to the thoracic duct and the right lymphatic duct, which drain into the large veins in the neck. Some vertebrates, such as amphibians, have a lymph heart, which pumps lymph through the lymph vessels.

lymph nodes small masses of lymphatic tissue in the body that occur at various points along the major lymphatic vessels. Tonsils and adenoids are large lymph nodes. As the lymph passes through them it is filtered, and bacteria and other micro-organisms are engulfed by cells known as macrophages.

Lymph nodes are sometimes mistakenly called lymph 'glands', and the term 'swollen glands' refers to swelling of the lymph nodes caused by infection.

lymphocyte type of white blood cell with a large nucleus, produced in the bone marrow. Most occur in the *lymph and blood, and around sites of infection. **B lymphocytes** or *B cells are responsible for producing *antibodies. **T lymphocytes** or *T cells have several roles in the mechanism of *immunity.

lymphokines chemical messengers produced by lymphocytes that carry messages between the cells of the immune system (see *immunity). Examples include interferon, which initiates defensive reactions to viruses, and the interleukins, which activate specific immune cells.

lymphoma any malignant tumour of the *lymph nodes.

lysine one of the nine essential *amino acids.

lysis in biology, any process that destroys a cell by rupturing its membrane or cell wall (see *lysosome).

lysosome membrane-enclosed structure, or organelle, inside a *cell, principally found in animal cells. Lysosomes contain enzymes that can break down proteins and other biological substances. They play a part in digestion, and in the white blood cells known as phagocytes the lysosome enzymes attack ingested bacteria.

Lyssavirus type of *virus that includes *rabies. A new lyssavirus was identified in Australian fruitbats in 1996.

LZW compression abbreviation for Lempel–Zif–Welsh compression, data compression system used to reduce the size of data files and thus shorten the time needed to transmit them. LZW compression is the basis of the * GIF graphics interchange format. When US computer company Unisys announced in 1995 that it wanted licence fees for the use of LZW compression for commercial applications, users began to develop alternatives such as *Portable Network Graphics (PNG).

m symbol for *metre.

MA abbreviation for *mechanical advantage.

machine code in computing, a set of instructions that a computer's central processing unit (CPU) can understand and obey directly, without any translation. Each type of CPU has its own machine code. Because machine-code programs consist entirely of binary digits (bits), most programmers write their programs in an easy-to-use *high-level language. A high-level program must be translated into

machine code – by means of a
*compiler or *interpreter program –
before it can be executed by a computer.

machine-readable of data, readable
directly by a computer without the
need for retyping. The term is usually
applied to files on disk or tape, but can
also be applied to typed or printed text
that can be scanned for *optical
character recognition or bar codes.

Mach number ratio of the speed of a
body to the speed of sound in the
medium through which the body
travels, named after Austrian
philosopher and physicist Ernst Mach
(1838–1916). In the Earth's atmosphere,
Mach 1 is reached when a body (such
as an aircraft or spacecraft) 'passes the
sound barrier', at a velocity of 331
m/1,087 ft per second (1,192 kph/740
mph) at sea level. A *space shuttle
reaches Mach 15 (about 17,700
kph/11,000 mph an hour) 6.5 minutes
after launch.

macro in computer programming, a
new command created by combining a
number of existing ones. For example,
a word-processing macro might create
a letterhead or fax cover sheet,
inserting words, fonts, and logos with
a single keystroke or mouse click.
Macros are also useful to automate
computer communications – for
example, users can write a macro to
ask their computer to dial an *Internet
Service Provider (ISP), retrieve e-mail
and *Usenet articles, and then
disconnect. A **macro key** on the
keyboard combines the effects of
pressing several individual keys.

macro- prefix meaning on a very large
scale, as opposed to micro-.

macromolecule in chemistry, a very
large molecule, generally a *polymer.

macrophage type of *white blood cell,
or leucocyte, found in all vertebrate
animals. Macrophages specialize in the
removal of bacteria and other micro-
organisms, or of cell debris after injury.
Like phagocytes, they engulf foreign
matter, but they are larger than
phagocytes and have a longer lifespan.
They are found throughout the body,
but mainly in the lymph and
connective tissues, and especially the
lungs, where they ingest dust, fibres,
and other inhaled particles.

macro virus in computing, a computer
virus that hides in Microsoft Word
documents or, less commonly, Excel
files.

mad cow disease common name for
*bovine spongiform encephalopathy,
an incurable brain condition in cattle.

maelstrom whirlpool off the Lofoten
Islands, Norway, also known as the
Moskenesstraumen, which gave its
name to whirlpools in general.

mafic rock plutonic rock composed
chiefly of dark-coloured minerals such
as olivine and pyroxene that contain
abundant magnesium and iron. It is
derived from **magnesium** and **ferric**
(iron). The term **mafic** also applies to
dark-coloured minerals rich in iron
and magnesium as a group. 'Mafic
rocks' usually refers to dark-coloured
igneous rocks such as basalt, but can
also refer to their metamorphic
counterparts.

magic bullet term sometimes used for
a drug that is specifically targeted on
certain cells or tissues in the body, such
as a small collection of cancerous cells
(see *cancer) or cells that have been
invaded by a virus. Such drugs can be
made in various ways, but *monoclonal
antibodies are increasingly being
used to direct the drug to a specific
target.

magic numbers in atomic physics
certain numbers of *neutrons or
*protons (2, 8, 20, 28, 50, 82, 126) in the
nuclei of elements of outstanding
stability, such as lead and helium. Such
stability is the result of neutrons and
protons being arranged in completed
'layers' or 'shells'.

magic square in mathematics, a square
array of numbers in which the rows,
columns, and diagonals add up to the
same total. A simple example
employing the numbers 1 to 9, with a
total of 15, has a first row of 6, 7, 2, a
second row of 1, 5, 9, and a third row
of 8, 3, 4.

    A **pandiagonal magic square** is one
in which all the broken diagonals also
add up to the magic constant.

magma molten rock material that
originates in the lower part of the
Earth's crust, or *mantle, where it
reaches temperatures as high as
1,000°C/1,832°F. *Igneous rocks are

formed from magma. *Lava is magma that has extruded onto the surface.

magnesia common name for *magnesium oxide.

magnesium chemical symbol Mg, lightweight, very ductile and malleable, silver-white, metallic element, atomic number 12, relative atomic mass 24.305. It is one of the *alkaline-earth metals, and the lightest of the commonly used metals. Magnesium silicate, carbonate, and chloride are widely distributed in nature. The metal is used in alloys, flares, and flash bulbs. It is a necessary trace element in the human diet, and green plants cannot grow without it since it is an essential constituent of the photosynthetic pigment *chlorophyll ($C_{55}H_{72}MgN_4O_5$).

It was named after the ancient Greek city of Magnesia, near where it was first found. It was first recognized as an element by Scottish chemist Joseph Black (1728–1799) in 1755 and discovered in its oxide by English chemist Humphry Davy 1808. Pure magnesium was isolated in 1828 by French chemist Antoine-Alexandre-Brutus Bussy.

magnesium oxide or **magnesia**; MgO, white powder or colourless crystals, formed when magnesium is burned in air or oxygen; a typical basic oxide. It is used to treat acidity of the stomach, and in some industrial processes; for example, as a lining brick in furnaces, because it is very stable when heated (refractory oxide).

magnet any object that forms a magnetic field (displays *magnetism), either permanently or temporarily through induction, causing it to attract materials such as iron, cobalt, nickel, and alloys of these. It always has two *magnetic poles, called north and south.

magnetic dip see *dip, magnetic and *angle of dip.

magnetic dipole the pair of north and south magnetic poles, separated by a short distance, that makes up all magnets. Individual magnets are often called 'magnetic dipoles'. Single magnetic poles, or monopoles, have never been observed despite being searched for. See also magnetic *domain.

magnetic field region around a permanent magnet, or around a conductor carrying an electric current, in which a force acts on a moving charge or on a magnet placed in the field. The force cannot be seen; only the effects it produces are visible. The field can be represented by lines of force parallel at each point to the direction of a small compass needle placed on them at that point. These invisible lines of force are called the magnetic field or the flux lines. A magnetic field's magnitude is given by the *magnetic flux density (the number of flux lines per unit area), expressed in *teslas. See also *polar reversal.

magnetic flux measurement of the strength of the magnetic field around electric currents and magnets. Its SI

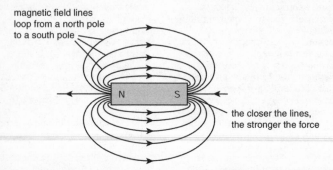

magnetic field lines loop from a north pole to a south pole

the closer the lines, the stronger the force

magnetic field A magnet is an object that forms a magnetic field. It has a north pole and a south pole. As iron is a magnetic material, iron filings shaken around a magnet will form along the lines of force and produce the pattern of the magnetic field.

unit is the *weber; one weber per square metre is equal to one tesla.

The amount of magnetic flux through an area equals the product of the area and the magnetic field strength at a point within that area.

magnetic induction production of magnetic properties in unmagnetized iron or other ferromagnetic material when it is brought close to a magnet. The material is influenced by the magnet's magnetic field. The magnetic lines of force pass through the material causing its particles to align themselves parallel to the lines of the magnetic field. The induced magnetism may be temporary, disappearing as soon as the magnet is removed, or permanent, depending on the nature of the material and the strength of the magnet.

*Electromagnets make use of temporary induced magnetism to lift sheets of steel; the magnetism induced in the steel by the approach of the electromagnet enables it to be picked up and transported. To release the sheet, the current supplying the electromagnet is temporarily switched off and the induced magnetism disappears.

magnetic-ink character recognition MICR, in computing, a technique that enables special characters printed in magnetic ink to be read and input rapidly to a computer. MICR is used extensively in banking because magnetic-ink characters are difficult to forge and are therefore ideal for marking and identifying cheques.

magnetic pole region of a magnet in which its magnetic properties are strongest. Every magnet has two poles, called north and south. The north (or north-seeking) pole is so named because a freely-suspended magnet will turn so that this pole points towards the Earth's magnetic north pole. The north pole of one magnet will be attracted to the south pole of another, but will be repelled by its north pole. So unlike poles attract, like poles repel.

magnetic resonance imaging MRI, diagnostic scanning system based on the principles of nuclear magnetic resonance (NMR). MRI yields finely detailed three-dimensional images of structures within the body without exposing the patient to harmful radiation. The technique is invaluable for imaging the soft tissues of the body, in particular the brain and the spinal cord.

magnetic storm in meteorology, a sudden disturbance affecting the Earth's magnetic field, causing anomalies in radio transmissions and magnetic compasses. It is probably caused by *sunspot activity.

magnetic strip or **magnetic stripe**, thin strip of magnetic material attached to a plastic card and used for recording data. Magnetic strips are used on credit cards, bank cards (as used at cash dispensing machines), telephone cards, and railway tickets.

magnetic tape narrow plastic ribbon coated with an easily magnetizable material on which data can be recorded. It is used in sound recording, audio-visual systems (videotape), and computing. For mass storage on commercial mainframe computers, large reel-to-reel tapes are still used, but cartridges are becoming popular. Various types of cartridge are now standard on minicomputers and PCs, while audio cassettes were used with early home computers.

magnetism phenomena associated with *magnetic fields. Magnetic fields are produced by moving charged particles. In electromagnets, electrons flow through a coil of wire connected to a battery; in permanent magnets, spinning electrons within the atoms generate the field.

magnetite black, strongly magnetic opaque mineral, Fe_3O_4, of the spinel group, an important ore of iron. Widely distributed, magnetite is found in nearly all igneous and metamorphic rocks. Some deposits, called lodestone, are permanently magnetized. Lodestone has been used as a compass since the first millennium BC. Today the orientations of magnetite grains in rocks are used in the study of the Earth's magnetic field (see *palaeomagnetism).

magnetohydrodynamics MHD, field of science concerned with the behaviour of ionized gases or liquids in a magnetic field. Systems have been developed that use MHD to generate electrical power. MHD-driven ships have been tested in Japan. In 1991 two cylindrical thrusters with electrodes and niobium–titanium superconducting coils, soaked in liquid helium, were placed under the passenger boat *Yamato 1*. The boat, 30 m/100 ft long, was designed to travel at 8 knots. An electric current passed through the electrodes accelerates water through the thrusters, like air through a jet engine, propelling the boat forward.

magnetomotive force or **magnetic potential**, work done in carrying a unit magnetic pole around a magnetic circuit. The concept is analogous to *electromotive force in an electric circuit.

magnetosphere volume of space, surrounding a planet, in which the planet's magnetic field has a significant influence. The Earth's magnetosphere extends 64,000 km/40,000 mi towards the Sun, but many times this distance on the side away from the Sun. That of Jupiter is much larger, and, if it were visible, would appear from the Earth to have roughly the same extent as the full Moon. The Russian-led space missions Coronas-I (launched in 1994) and Coronas-F (launched in 2001) were designed to investigate the magnetosphere of the Sun.

magnetron thermionic *valve (electron tube) for generating very high-frequency oscillations, used in radar and to produce microwaves in a microwave oven. The flow of electrons from the tube's cathode to one or more anodes is controlled by an applied magnetic field.

magnification measure of the enlargement or reduction of an object in an imaging optical system. **Linear magnification** is the ratio of the size (height) of the image to that of the object. **Angular magnification** is the ratio of the angle subtended at the observer's eye by the image to the angle subtended by the object when viewed directly.

magnitude in astronomy, measure of the brightness of a star or other celestial object. The larger the number denoting the magnitude, the fainter the object. Zero or first magnitude indicates some of the brightest stars. Still brighter are those of negative magnitude, such as Sirius, whose magnitude is −1.46. **Apparent magnitude** is the brightness of an object as seen from the Earth; **absolute magnitude** is the brightness at a standard distance of 10 parsecs (32.616 light years).

magnitude in mathematics, size irrespective of sign, used especially for *vectors irrespective of direction.

Magnox early type of nuclear reactor used in the UK, for example in Calder Hall, the world's first commercial nuclear power station. This type of reactor uses uranium fuel encased in tubes of magnesium alloy called Magnox. Carbon dioxide gas is used as a coolant to extract heat from the reactor core. See also *nuclear energy.

mail-bombing or **dumping**, in computing, sending large amounts of *e-mail to an individual or organization, usually in retaliation for a breach of netiquette. The aim is to completely fill the recipient's *hard disk with immense, useless files, causing at best irritation, and at worst total computer failure. While mail-bombing often achieves its aim of annoying the individual concerned, it also inconveniences systems administrators and other users.

mailbox in computing, folder in which e-mail is stored, typically divided into 'in' and 'out' trays. Users usually have two mailboxes: one on their PC, and another at their mail *server at the *internet service provider (ISP), where incoming messages await collection.

mail server software in client/server computing (see *client–server architecture), software that stores e-mail and distributes it only to the authorized recipient.

mainboard new (and more politically correct) name for a *motherboard.

mainframe large computer used for commercial data processing and other large-scale operations. Because of the general increase in computing power, the differences between the mainframe, *supercomputer, minicomputer, and microcomputer (personal computer) are becoming less marked.

mains electricity domestic electricity supply system. In the UK, electricity is supplied to houses, offices, and most factories as an *alternating current at a frequency of 50 hertz and a root-mean-square voltage of 230 volts. An advantage of having an alternating supply is that it may easily be changed, using a *transformer, to a lower and safer voltage, such as 9 volts, for operating toys and for recharging batteries.

major arc larger of the two arcs formed when a circle is divided into two unequal parts by a straight line or chord.

malachite common *copper ore, basic copper carbonate, $Cu_2CO_3(OH)_2$. It is a source of green pigment and is used as an antifungal agent in fish farming, as well as being polished for use in jewellery, ornaments, and art objects.

malaria infectious parasitic disease of the tropics transmitted by mosquitoes, marked by periodic fever and an enlarged spleen. When a female mosquito of the *Anopheles* genus bites a human who has malaria, it takes in with the human blood one of four malaria protozoa of the genus *Plasmodium*. This matures within the insect and is then transferred when the mosquito bites a new victim. Malaria affects around 300–500 million people each year, in 103 countries, and over 2 million people die of the disease annually. In sub-Saharan Africa alone between 1.5 and 2 million children die from malaria and its consequences each year.

malic acid or **hydroxysuccinic acid**; $HOOCCH_2CH(OH)COOH$, organic crystalline acid that can be extracted from apples, plums, cherries, grapes, and other fruits, but occurs in all living cells in smaller amounts, being one of the intermediates of the *Krebs cycle.

malignant term used to describe a cancerous growth. It is also applied to any condition that could kill if untreated (as in malignant *hypertension).

maltase enzyme found in plants and animals that breaks down the disaccharide maltose into glucose.

Maltese-cross tube cathode-ray tube used to demonstrate some of the properties of cathode rays. The cathode rays, or electron streams, emitted by the tube's *electron gun are directed towards a *fluorescent screen in front of which hangs a metal Maltese cross. Those electrons that hit the screen give up their kinetic energy and cause its phosphor coating to fluoresce. However, the sharply defined cross-shaped shadow cast on the screen shows that electrons are unable to pass through the Maltese cross. Cathode rays are thereby shown to travel in straight lines, and to be unable to pass through metal.

maltose $C_{12}H_{22}O_{11}$, *disaccharide sugar in which both monosaccharide units are glucose. It is produced by the enzymic hydrolysis of starch and is a major constituent of malt, produced in the early stages of beer and whisky manufacture.

mandible anatomical name for the lower jaw bone in mammals.

manganese chemical symbol Mn, hard, brittle, grey-white metallic element, atomic number 25, relative atomic mass 54.9380. It resembles iron (and rusts), but it is not magnetic and is softer. It is used chiefly in making steel alloys, also alloys with aluminium and copper. It is used in fertilizers, paints, and industrial chemicals. It is a necessary trace element in human nutrition. The name is old, deriving from the French and Latin forms of Magnesia, a mineral-rich region of Italy.

manganese ore any mineral from which manganese is produced. The main ores are the oxides, such as **pyrolusite**, MnO_2; **hausmannite**, Mn_3O_4; and **manganite**, $MnO(OH)$.

manganese(IV) oxide or **manganese dioxide**; MnO_2, brown solid that acts as a depolarizer in dry batteries by oxidizing the hydrogen gas produced to water; without this process, the performance of the battery is impaired. It acts as a *catalyst in decomposing hydrogen peroxide to obtain oxygen:

$$2H_2O_{2\ (aq)} \rightarrow 2H_2O_{\ (l)} + O_{2\ (g)}$$

It oxidizes concentrated hydrochloric acid to produce chlorine:.

$$MnO_2 + 4HCl \rightarrow MnCl_2 + Cl_2 + 2H_2O$$

mania in psychiatry, term used to describe high mood. The affected individual can appear cheerful and optimistic or irritable and angry. Sleep is reduced and the sufferer can be overactive to the point of physical exhaustion. Speech is rapid and can convey grandiose delusions. Mania often occurs as part of *manic depression. It is treated with an antipsychotic drug to control the mood and, in the long term, with *lithium.

manic depression or **bipolar disorder**, mental disorder characterized by recurring periods of either *depression or mania (inappropriate elation, agitation, and rapid thought and speech) or both. Sufferers may be genetically predisposed to the condition. Some cases have been improved by taking prescribed doses of *lithium.

manometer instrument for measuring the pressure of liquids (including human blood pressure) or gases. In its basic form, it is a U-tube partly filled with coloured liquid. Greater pressure on the liquid surface in one arm will force the level of the liquid in the other arm to rise. A difference between the pressures in the two arms is therefore registered as a difference in the heights of the liquid in the arms.

mantle intermediate zone of the Earth between the *crust and the *core, accounting for 82% of the Earth's volume. The crust, made up of separate tectonic *plates, floats on the mantle which is made of dark semi-liquid rock that is rich in magnesium and silicon. The temperature of the mantle can be as high as 3,700°C/6,692°F. Heat generated in the core causes convection currents in the semi-liquid mantle; rock rises and then slowly sinks again as it cools, causing the movements of the tectonic plates. The boundary (junction) between the mantle and the crust is called the Mohorovicic discontinuity, which lies at an average depth of 32 km/20 mi. The boundary between the mantle and the core is called the Gutenburg discontinuity, and lies at an average depth of 2,900 km/1,813 mi.

mantle keel in earth science, relatively cold slab of mantle material attached to the underside of a continental craton (core of a continent composed of old, highly deformed metamorphic rock), and protruding down into the mantle like the keel of a boat. Their presence suggests that tectonic processes may have been different at the time the cratons were formed.

MAOI abbreviation for *monoamine-oxidase inhibitor.

marble rock formed by metamorphosis of sedimentary *limestone. It takes and retains a good polish, and is used in building and sculpture. In its pure form it is white and consists almost entirely of calcite ($CaCO_3$). Mineral impurities give it various colours and patterns. Carrara, Italy, is known for white marble.

Mariana Trench lowest region on the Earth's surface; the deepest part of the sea floor. The trench is 2,400 km/1,500 mi long and is situated 300 km/200 mi east of the Mariana Islands, in the northwestern Pacific Ocean. Its deepest part is the gorge known as the Challenger Deep, which extends 11,034 m/36,210 ft below sea level.

marijuana dried leaves and flowers of the hemp plant cannabis, used as a drug; it is illegal in most countries. It is eaten or inhaled and causes euphoria, distortion of time, and heightened sensations of sight and sound. Mexico is the world's largest producer.

Markov chain in statistics, an ordered sequence of discrete states (random variables) $x_1, x_2, ..., x_i, ..., x_n$ such that the probability of x_i depends only on n and/or the state x_{i-1} which has preceded it. If independent of n, the chain is said to be homogeneous. Formulated by Russian mathematician Andrei Andreyevich Markov (1856–1922).

mark sensing in computing, a technique that enables pencil marks made in predetermined positions on specially prepared forms to be rapidly read and input to a computer. The technique makes use of the fact that pencil marks contain graphite and therefore conduct electricity. A **mark sense reader** scans the form by passing small metal brushes over the paper surface. Whenever a brush touches a pencil mark a circuit is completed and the mark is detected.

Mars fourth planet from the Sun. It is much smaller than Venus or Earth, with a mass 0.11 that of Earth. Mars is slightly pear-shaped, with a low, level northern hemisphere, which is comparatively uncratered and geologically 'young', and a heavily cratered 'ancient' southern hemisphere.

mean distance from the Sun: 227.9 million km/141.6 million mi

equatorial diameter: 6,780 km/ 4,210 mi

rotation period: 24 hours 37 minutes

year: 687 Earth days

atmosphere: 95% carbon dioxide, 3% nitrogen, 1.5% argon, and 0.15% oxygen. Red atmospheric dust from the surface whipped up by winds of up to 450 kph/280 mph accounts for the light pink sky. The surface pressure is less than 1% of the Earth's atmospheric pressure at sea level

surface: the landscape is a dusty, red, eroded lava plain. Mars has white polar caps (water ice and frozen carbon dioxide) that advance and retreat with the seasons

satellites: two small satellites: Phobos and Deimos

marsh low-lying wetland. Freshwater marshes are common wherever groundwater, surface springs, streams, or run-off cause frequent flooding, or more or less permanent shallow water. A marsh is alkaline whereas a *bog is acid. Marshes develop on inorganic silt or clay soils. Rushes are typical marsh plants. Large marshes dominated by papyrus, cattail, and reeds, with standing water throughout the year, are commonly called *swamps. Near the sea, *salt marshes may form.

marsh gas gas consisting mostly of *methane. It is produced in swamps and marshes by the action of bacteria on dead vegetation.

maser acronym for **microwave amplification by stimulated emission of radiation**, in physics, a high-frequency microwave amplifier or oscillator in which the signal to be amplified is used to stimulate excited atoms into emitting energy at the same frequency. Atoms or molecules are raised to a higher energy level and then allowed to lose this energy by

radiation emitted at a precise frequency. The principle has been extended to other parts of the electromagnetic spectrum as, for example, in the *laser.

masochism desire to subject oneself to physical or mental pain, humiliation, or punishment, for erotic pleasure, to alleviate guilt, or out of destructive impulses turned inward. The term is derived from Leopold von Sacher-Masoch.

mass in physics, quantity of matter in a body as measured by its inertia, including all the particles of which the body is made up. Mass determines the acceleration produced in a body by a given force acting on it, the acceleration being inversely proportional to the mass of the body. The mass also determines the force exerted on a body by *gravity on Earth, although this attraction varies slightly from place to place (the mass itself will remain the same). In the SI system, the base unit of mass is the kilogram.

At a given place, equal masses experience equal gravitational forces, which are known as the weights of the bodies. Masses may, therefore, be compared by comparing the weights of bodies at the same place. The standard unit of mass to which all other masses are compared is a platinum-iridium cylinder of 1 kg, which is kept at the International Bureau of Weights and Measures in Sèvres, France.

mass action, law of in chemistry, a law stating that at a given temperature the rate at which a chemical reaction takes place is proportional to the product of the active masses of the reactants. The active mass is taken to be the molar concentration of each reactant.

mass–energy equation The equation $E = mc^2$ developed by German-born US physicist Albert Einstein (1879–1955), denoting the equivalence of mass and energy, following his special theory of *relativity. In SI units, E is the energy in joules, m is the mass in kilograms, and c, the speed of light in a vacuum, is in metres per second.

The conversion of mass into energy in accordance with this equation applies universally, although it is only in nuclear reactions that the percentage change in mass is large enough to detect. This is the basis of atomic power.

mass extinction event that produces the extinction of many species at about the same time. One notable example is the boundary between the Cretaceous and Tertiary periods (known as the K-T boundary) that saw the extinction of the dinosaurs and other large reptiles, and many of the marine invertebrates as well. Mass extinctions have taken place frequently during Earth's history.

There have been five major mass extinctions, in which 75% or more of the world's species have been wiped out: End Ordovician period (440 million years ago) in which about 85% of species were destroyed (second most severe); Late Devonian period (365 million years ago) which took place in two waves a million years apart, and was the third most severe, with marine species particularly badly hit; Late Permian period (251 million years ago), the gravest mass Late Triassic (205 million years ago), in which about 76% of species were destroyed, mainly marine; Late Cretaceous period (65 million years ago), in which 75–80 of species became extinct, including dinosaurs.

mass number or **nucleon number**; symbol A, sum of the numbers of *protons and *neutrons in the nucleus of an atom. It is used along with the *atomic number (the number of protons) in *nuclear notation: in symbols that represent nuclear *isotopes, such as $^{14}_6C$, the lower number is the atomic number, and the upper number is the mass number. Since the mass of the *electrons in an atom are negligible, the total number of protons is neutrons (mass number) in an atom determines its mass.

mass spectrometer apparatus for analysing chemical composition by separating ions by their mass; the ions may be elements, isotopes, or molecular compounds. Positive ions (charged particles) of a substance are separated by an electromagnetic system, designed to focus particles of equal mass to a point where they can be detected. This permits accurate measurement of the relative

concentrations of the various ionic masses present. A mass spectrometer can be used to identify compounds or to measure the relative abundance of compounds in a sample.

mast cell in medicine, one of the histamine-containing cells involved in the production of inflammatory reactions, in response to an external trigger factor. For example, cells in the airways or in the bronchial epithelium release *histamine and other mediators in response to allergic stimuli, such as allergens, infection, stress, and exercise.

mastectomy surgical removal of a breast. Performed for cancer, it may also involve removal of lymphatic tissue in the armpit. Around 250,000 mastectomies are performed worldwide annually (1995).

mathematical induction formal method of proof in which the proposition $P(n + 1)$ is proved true on the hypothesis that the proposition $P(n)$ is true. The proposition is then shown to be true for a particular value of n, say k, and therefore by induction the proposition must be true for $n = k + 1, k + 2, k + 3, \ldots$. In many cases $k = 1$, so then the proposition is true for all positive integers.

mathematics science of relationships between numbers, between spatial configurations, and abstract structures. The main divisions of **pure mathematics** include geometry, arithmetic, algebra, calculus, and trigonometry. Mechanics, statistics, numerical analysis, computing, the mathematical theories of astronomy, electricity, optics, thermodynamics, and atomic studies come under the heading of **applied mathematics**.

matrix in biology, usually refers to the *extracellular matrix.

matrix in mathematics, a square ($n \times n$) or rectangular ($m \times n$) array of elements (numbers or algebraic variables) used to facilitate the study of problems in which the relation between the elements is important. They are a means of condensing information about mathematical systems and can be used for, among other things, solving simultaneous linear equations (see *simultaneous equations and *transformation).

The advantage of matrices is that they can be studied algebraically by assigning a single symbol to a matrix rather than considering each element separately. The symbol used is usually a bold capital letter, but often a matrix is denoted by a symbol like (a_{ij}), meaning 'the matrix with element a in row i column j'. The size of a matrix is described by stating the number of its rows and then the number of its columns so, for example, a matrix with three rows and two columns is a 3×2 matrix. A matrix with equal numbers of rows and columns is called a 'square matrix'.

Much early matrix theory was developed by the British mathematician Arthur Cayley, although the term was coined by his contemporary James Sylvester (1814–1897).

matter in physics, anything that has mass. All matter is made up of *atoms, which in turn are made up of *elementary particles; it ordinarily exists in one of three physical states: solid, liquid, or gas.

states of matter Whether matter exists as a solid, liquid, or gas depends on its temperature and the pressure on it. *Kinetic theory describes how the state of a material depends on the movement and arrangement of its atoms or molecules. In a solid, the atoms or molecules vibrate in a fixed position. In a liquid, they do not occupy fixed positions as in a solid, and yet neither do they have the freedom of random movement that occurs within a gas, so the atoms or molecules within a liquid will always follow the shape of their container. The transition between states takes place at definite temperatures, called melting point and boiling point.

conservation of matter In chemical reactions matter is conserved, so no matter is lost or gained and the sum of the mass of the reactants will always equal the sum of the end products.

maxilla anatomical name for the upper jaw.

maximum and minimum in *coordinate geometry, points at which the slope of a curve representing a *function changes from positive to negative (maximum), or from negative to positive (minimum). A tangent to the curve at a maximum or minimum has zero gradient.

maxwell symbol Mx, c.g.s. unit of magnetic flux (the strength of a *magnetic field in an area multiplied by the area). It is now replaced by the SI unit, the *weber (one maxwell equals 10^{-8} weber).

Maxwell–Boltzmann distribution in physics, a statistical equation describing the distribution of velocities among the molecules of a gas. It is named after Scottish physicist James Maxwell (1831–1879) and Austrian physicist Ludwig Boltzmann (1844–1906), who derived the equation, independently of each other, in the 1860s.

One form of the distribution is $n = N \exp(-E/RT)$, where N is the total number of molecules, n is the number of molecules with energy in excess of E, T is the absolute temperature (temperature in kelvin), R is the *gas constant, and exp is the exponential function.

Maxwell's screw rule in physics, a rule formulated by Scottish physicist James Maxwell (1831–1879) that predicts the

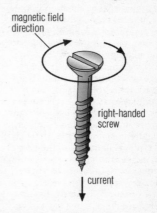

Maxwell's screw rule Maxwell's screw rule, named after the physicist James Clerk Maxwell, predicts the direction of the magnetic field produced around a wire carrying electric current. If a right-handed screw is turned so that it moves forward in the same direction as the current, its direction of rotation will give the direction of the magnetic field.

direction of the magnetic field produced around a wire carrying electric current. It states that if a right-handed screw is turned so that it moves forwards in the same direction as the conventional current, its direction of rotation will give the direction of the magnetic field.

MBONE abbreviation for multicast backbone, in computing, layer of the Internet designed to deliver *packets of multimedia data, enabling video and audio communication. It can be used for telephony and video-conferencing – however, it can deliver a maximum of only five video frames per second, as opposed to television's 30. Large rock concerts are occasionally broadcast on the MBONE.

m-commerce contraction of *mobile commerce.

MDMA (3,4-methylenedio-xymethamphetamine) psychedelic drug, also known as ecstasy.

ME abbreviation for **myalgic encephalomyelitis**, a popular name for *chronic fatigue syndrome.

mean in mathematics, a measure of the average of a number of terms or quantities. The simple *arithmetic mean* is the average value of the quantities, that is, the sum of the quantities divided by their number. The **weighted mean** takes into account the frequency of the terms that are summed; it is calculated by multiplying each term by the number of times it occurs, summing the results and dividing this total by the total number of occurrences. The **geometric mean** of n quantities is the nth root of their product. In statistics, it is a measure of central tendency of a set of data, that is one measure used to express the frequency distribution of a number of recorded events.

meander loop-shaped curve in a *river flowing sinuously across flat country. As a river flows, any curve in its course is accentuated (intensified) by the current. On the outside of the curve the velocity, and therefore the erosion, of the current is greatest. Here the river cuts into the outside bank, producing a **river cliff**. On the inside of the curve the current is slow and so it deposits any transported material,

building up a gentle slip-off slope. As each meander migrates in the direction of its outer bank, the river gradually changes its course across the flood plain.

mean deviation in statistics, a measure of the spread of a population from the *mean.

mean free path in physics, the average distance travelled by a particle, atom, or molecule between successive collisions. It is of importance in the *kinetic theory of gases.

mean life in nuclear physics, the average lifetime of a nucleus of a radioactive isotope; it is equal to 1.44 times the *half-life. See *radioactivity.

measles also known as **rubeola**, acute virus disease spread by airborne infection. Symptoms are fever, severe catarrh, small spots inside the mouth, and a raised, blotchy red rash appearing for about a week after two weeks' incubation. Prevention is by vaccination.

More than 1 million children a year die of measles. In industrialized countries measles is not usually a serious disease, but serious complications may develop, so most developed countries have a vaccination programme. Measles kills approximately 1 in 5,000 children who develop the disease in developed countries. It also has a 1 in 1,000 risk of causing encephalitis, which leaves half of those children who develop it brain-damaged.

mechanical advantage **MA**, in physics, the ratio by which the load moved by a machine is greater than the effort applied to that machine. In equation terms: MA = load/effort.

mechanical equivalent of heat in physics, a constant factor relating the calorie (the * c.g.s. unit of heat) to the joule (the unit of mechanical energy), equal to 4.1868 joules per calorie. It is redundant in the SI system of units, which measures heat and all forms of energy in *joules.

mechanical weathering in earth science, an alternative name for *physical weathering.

mechanics branch of physics dealing with the motions of bodies and the forces causing these motions, and also with the forces acting on bodies in *equilibrium. It is usually divided into *dynamics and *statics.

media singular **medium**, in computing, the collective name for materials on which data can be recorded. For example, paper is a medium that can be used to record printed data; a floppy disk is a medium for recording magnetic data.

medial moraine linear ridge of rocky debris running along the centre of a glacier. Medial moraines are commonly formed by the joining of two *lateral moraines when two glaciers merge.

median in mathematics and statistics, the middle number of an ordered group of numbers. If there is no middle number (because there is an even number of terms), the median is the *mean (average) of the two middle numbers. For example, the median of the group 2, 3, 7, 11, 12 is 7; that of 3, 4, 7, 9, 11, 13 is 8 (the mean of 7 and 9). The median together with the *mode and *arithmetic mean make up the *average of a set of data. In addition it is useful to know the *range or spread of the data.

In geometry, the term refers to a line from the vertex of a triangle to the midpoint of the opposite side.

medicine, alternative forms of medical treatment that do not use synthetic drugs or surgery in response to the symptoms of a disease, but aim to treat the patient as a whole (holism). The emphasis is on maintaining health (with diet and exercise) and on dealing with the underlying causes rather than just the symptoms of illness. It may involve the use of herbal remedies and techniques like acupuncture, *homeopathy, and chiropractic. Some alternative treatments are increasingly accepted by orthodox medicine, but the absence of enforceable standards in some fields has led to the proliferation of eccentric or untrained practitioners.

Mediterranean climate type of climate found on the west coasts of continents between 30° and 45° north and south of the Equator – that is, in Mediterranean Europe, California, parts of southern Australia, Cape Province (South Africa), and central Chile. The characteristics of the climate are hot, dry summers and warm, wet winters. Summers in southern Europe are hot with little cloud cover, and the winters are mild. Other Mediterranean climate areas are less warm in summer.

medulla central part of an organ. In the mammalian kidney, the medulla lies beneath the outer cortex and is responsible for the reabsorption of water from the filtrate. In plants, it is a region of packing tissue in the centre of the stem. In the vertebrate brain, the medulla is the posterior region responsible for the coordination of basic activities, such as breathing and temperature control.

medusa free-swimming phase in the life cycle of a coelenterate, such as a jellyfish or coral. The other phase is the sedentary **polyp**.

mega- prefix denoting multiplication by a million. For example, a megawatt (MW) is equivalent to a million watts.

megabyte **MB**, in computing, a unit of memory equal to 1,024 *kilobytes. It is sometimes used, less precisely, to mean 1 million bytes.

megaton one million (10^6) tons. Used with reference to the explosive power of a nuclear weapon, it is equivalent to the explosive force of one million tons of trinitrotoluene (TNT).

meiosis in biology, a process of cell division in which the number of *chromosomes in the cell is halved. It only occurs in *eukaryotic cells, and is part of a life cycle that involves *sexual reproduction because it allows the genes of two parents to be combined without the total number of chromosomes increasing. Cells in reproductive organs – testes and ovaries in humans – divide to form sex cells (gametes) by meiosis.

In meiosis, the nucleus of a cell divides twice. A single cell produces four cells by the end of meiosis. In sexually reproducing *diploid animals (having two sets of chromosomes per cell), meiosis occurs during formation of the *gametes (sex cells – see *sperm and *ovum), so that the gametes are *haploid (having only one set of chromosomes – half the number of chromosomes of the parent cell). When the gametes unite during *fertilization, the diploid condition is restored.

In plants, meiosis occurs just before spore formation. Thus the spores are haploid and in lower plants such as mosses they develop into a haploid plant called a gametophyte which produces the gametes (see *alternation of generations). See also *mitosis.

Meiosis gives rise to variation. This is an important part of sexual reproduction. The variation produced is inherited, which means that evolution can take place as a result of the natural selection of certain variants to suit a changing environment. The way that meiosis gives rise to variation is by recombining genes from chromosomes in new ways. When the number of chromosomes is halved, there is some randomness in the way parts of chromosomes are selected to go into the gametes.

meitnerium chemical symbol Mt, synthesized radioactive element of the *transactinide series, atomic number 109, relative atomic mass 268. It was first produced in 1982 at the Laboratory for Heavy-Ion Research in Darmstadt, Germany, by fusing bismuth and iron nuclei; it took a week to obtain a single new, fused nucleus. It was named in 1997 after the

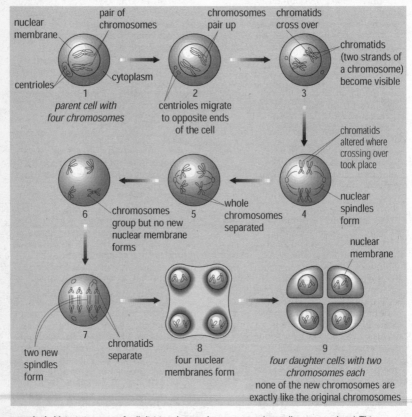

meiosis Meiosis is a type of cell division that produces gametes (sex cells, sperm and egg). This sequence shows an animal cell but only four chromosomes are present in the parent cell (1). There are two stages in the division process. In the first stage (2–6), the chromosomes come together in pairs and exchange genetic material. This is called crossing over. In the second stage (7–9), the cell divides to produce four gamete cells, each with only one copy of each chromosome from the parent cell.

Austrian-born Swedish physicist Lise Meitner. Its temporary name was unnilennium.

melamine $C_3H_6N_6$, *thermosetting *polymer based on urea–formaldehyde. It is extremely resistant to heat and is also scratch-resistant. Its uses include synthetic resins.

melanin brown pigment that gives colour to the eyes, skin, hair, feathers, and scales of many vertebrates. In humans, melanin helps protect the skin against ultraviolet radiation from sunlight. Both genetic and environmental factors determine the amount of melanin in the skin.

melanism black coloration of animal bodies caused by large amounts of the pigment melanin. Melanin is of significance in insects, because melanic ones warm more rapidly in sunshine than do pale ones, and can be more active in cool weather. A fall in temperature may stimulate such insects to produce more melanin. In industrial areas, dark insects and pigeons match sooty backgrounds and escape predation, but they are at a disadvantage in rural areas where they do not match their backgrounds. This is known as **industrial melanism**.

melatonin *hormone secreted by the *pineal body that influences the body's *circadian rhythms (metabolic activities centred approximately on a 24-hour cycle). Melatonin secretion is inhibited by bright light and the variations in secretion have been linked to *seasonal affective disorder (SAD).

melting point temperature at which a substance melts, or changes from solid to liquid form. A pure substance under standard conditions of pressure (usually one atmosphere) has a definite melting point. If heat is supplied to a solid at its melting point, the temperature does not change until the melting process is complete. The melting point of ice is 0°C or 32°F.

meltwater water produced by the melting of snow and ice, particularly in glaciated areas. Streams of meltwater flowing from glaciers transport rocky materials away from the ice to form outwash. Features formed by the deposition of debris carried by meltwater or by its erosive action are called **fluvioglacial features**; they include eskers, kames, and outwash plains.

member in mathematics, one of the elements belonging to a set. For example, 25 and 2,500 are both members of the set of square numbers, but 250 is not.

membrane in living things, a continuous layer, made up principally of fat molecules, that encloses a *cell or *organelles within a cell. Small molecules, such as water and sugars, can pass through the cell membrane by *diffusion. Large molecules, such as proteins, are transported across the membrane via special channels, a process often involving energy input. The *Golgi apparatus within the cell is thought to produce certain membranes.

In cell organelles, enzymes may be attached to the membrane at specific positions, often alongside other enzymes involved in the same process, like workers at a conveyor belt. Thus membranes help to make cellular processes more efficient.

memory in computing, the part of a system used to store data and programs either permanently or temporarily. There are two main types: immediate access memory and backing storage. Memory capacity is measured in *bytes or, more conveniently, in kilobytes (units of 1,024 bytes) or megabytes (units of 1,024 kilobytes).

memory ability to store and recall observations and sensations. Memory does not seem to be based in any particular part of the brain; it may depend on changes to the pathways followed by nerve impulses as they move through the brain. Memory can be improved by regular use as the connections between *nerve cells (neurons) become 'well-worn paths' in the brain. Events stored in **short-term memory** are forgotten quickly, whereas those in **long-term memory** can last for many years, enabling recall of information and recognition of people and places over long periods of time.

Short-term memory is the most likely to be impaired by illness or drugs whereas long-term memory is very resistant to such damage. Memory changes with age and otherwise healthy people may experience a natural decline after the age of about 40. Research is just beginning to uncover the biochemical and electrical bases of the human memory. Older people who are highly educated show less memory loss and fewer problems with thinking.

memory resident present in the main (*RAM) memory of the computer. For an application to be run, it has to be memory resident. Some applications are kept in memory (see *terminate and stay resident), while most are deleted from the memory when their task is complete. However, the memory is usually not large enough to hold all applications and *swapping in and out of memory is necessary. This slows down the application.

mendelevium chemical symbol Md, synthesized, radioactive metallic element of the *actinide series, atomic number 101, relative atomic mass 258. It was first produced by bombardment of Es-253 with helium nuclei. Its longest-lived isotope, Md-258, has a half-life of about two months. The element is chemically similar to thulium. It was named by the US physicists at the University of California at Berkeley who first synthesized it in 1955 after the Russian chemist Dmitri Mendeleyev, who in 1869 devised the basis for the periodic table of the elements.

Mendelism in genetics, the theory of inheritance originally outlined by Austrian biologist Gregor Mendel (1822–1884). He suggested that, in sexually reproducing species, all characteristics are inherited through indivisible 'factors' (now identified with *genes) contributed by each parent to its offspring.

Ménière's disease or **Ménière's syndrome**, recurring condition of the inner ear caused by an accumulation of fluid in the labyrinth of the ear that affects mechanisms of both hearing and balance. It usually develops in the middle or later years. Symptoms, which include deafness, ringing in the ears (*tinnitus), nausea, vertigo, and loss of balance, may be eased by drugs, but there is no cure.

meningitis inflammation of the meninges (membranes) surrounding the brain, caused by bacterial or viral infection. Bacterial meningitis, though treatable by antibiotics, is the more serious threat. Diagnosis is by lumbar puncture.

meningococcus in medicine, the causative organism of meningococcal *meningitis, *Neisseria meningitidis*.

meniscus in physics, the curved shape of the surface of a liquid in a thin tube, caused by the cohesive effects of *surface tension (capillary action). When the walls of the container are made wet by the liquid, the meniscus is concave, but with highly viscous liquids (such as mercury) the meniscus is convex. Also, a meniscus lens is a concavo-convex or convexo-concave *lens.

menstrual cycle biological cycle occurring in female mammals of reproductive age that prepares the body for pregnancy. At the beginning of the cycle, a Graafian (egg) follicle develops in the ovary, and the inner wall of the uterus forms a soft spongy lining. The egg (*ovum) is released from the ovary, and the *uterus lining (endometrium) becomes vascularized (filled with *blood vessels). At this stage fertilization can take place. If fertilization does not occur, the corpus luteum (remains of the Graafian follicle) degenerates, and the uterine lining breaks down, and is shed. This is what causes the loss of blood that marks menstruation. The cycle then begins again. Human menstruation takes place from puberty to menopause, except during pregnancy, occurring about every 28 days.

mental disability arrested or incomplete development of mental capacities. It can be very mild, but in more severe cases is associated with social problems and difficulties in living independently. A person may be born with a mental disability (for example, *Down's syndrome) or may acquire it through brain damage. Between 90 and 130 million people in the world suffer from such disabilities.

mental health well-being and soundness of mind, not only in terms

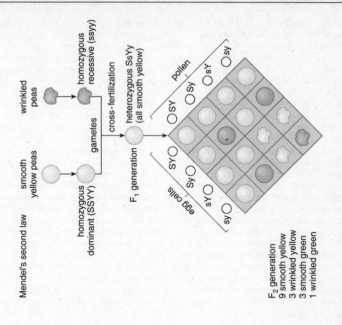

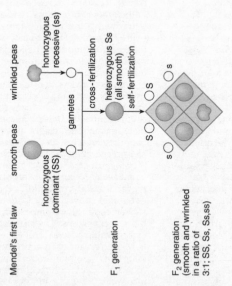

Mendelism Mendel's laws explain the proportion of offspring having various characteristics. When pea plants having smooth yellow peas are crossed with plants with wrinkled green peas, the first-generation offspring all have smooth yellow peas. The second-generation offspring, however, contain smooth yellow, wrinkled green, smooth green, and wrinkled yellow peas. This can be understood by tracing the passage of alleles Y, S, s, y throughout the generations. S and Y are dominant genes.

of intellectual abilities, but also in terms of the capability to deal with everyday problems, and the capacity to get on well with other people and to form and sustain relationships.

mental illness disordered functioning of the mind. Since normal working cannot easily be defined, the borderline between mild mental illness and normality is a matter of opinion (not to be confused with normative behaviour). It is broadly divided into two categories: *neurosis, in which the patient remains in touch with reality; and *psychosis, in which perception, thought, and belief are disordered.

mental test any of various standardized methods of assessing or measuring mental abilities and personality traits. Where selected for their reliability and validity, mental tests are indispensable aids in assessing educational attainment, in the clinical assessment of mental disorders and their treatment, and in careers guidance and job selection.

Mercalli scale qualitative scale describing the **intensity** of an *earthquake. It differs from the *Richter scale, which indicates earthquake **magnitude** and is quantitative. It is named after the Italian seismologist Giuseppe Mercalli (1850–1914).

Intensity is a subjective value, based on observed phenomena, and varies from place to place even when describing the same earthquake.

Mercator's projection System devised by Flemish mapmaker Gerardus Mercator (1512–1594) in which the parallels and meridians on maps are drawn uniformly at 90º. The projection continues to be used, in particular for navigational charts, because compass courses can be drawn as straight lines, but the true area of countries is increasingly distorted the further north or south they are from the Equator.

Mercury closest planet to the Sun. Its mass is 0.056 that of Earth. On its sunward side the surface temperature reaches over 400°C/752°F, but on the 'night' side it falls to −170°C/−274°F.

mean distance from the Sun: 58 million km/36 million mi

equatorial diameter: 4,880 km/ 3,030 mi

rotation period: 59 Earth days
year: 88 Earth days
atmosphere: Mercury's small mass and high daytime temperature mean that it is impossible for an atmosphere to be retained.

surface: composed of silicate rock often in the form of lava flows. In 1974 the US space probe Mariner 10 showed that Mercury's surface is cratered by meteorite impacts.

satellites: none

NASA's Mariner 10 probe, launched on 3 November 1973, arrived at Mercury on 29 March 1974, and provided the first close-up images of the planet. NASA plans a US$286 million mission to launch the Messenger probe in 2004 to orbit Mercury in 2008, to photograph the planet's surface, analyse its atmospheric composition, and map its magnetic field.

mercury chemical symbol Hg; or **quicksilver**, (Latin *mercurius*) heavy, silver-grey, metallic element, atomic number 80, relative atomic mass 200.59. Its symbol comes from the Latin *hydrargyrum*. It is a dense, mobile liquid with a low melting point (−38.87°C/−37.96°F). Its chief source is the mineral cinnabar, HgS, but it sometimes occurs in nature as a free metal.

uses Its alloys with other metals are called amalgams (a silver–mercury amalgam is used in dentistry for filling cavities in teeth). Industrial uses include drugs and chemicals, mercury-vapour lamps, arc rectifiers, power-control switches, barometers, and thermometers.

hazards Mercury is a cumulative poison that can contaminate the food chain, and cause intestinal disturbance, kidney and brain damage, and birth defects in humans. The World Health Organization's 'safe' limit for mercury is 0.5 milligrams of mercury per kilogram of muscle tissue. The US Environmental Protection Agency recommended a maximum safe level for mercury of 0.1 mg per kilogram of body weight in January 1998 (a fifth of that recommended by WHO).The discharge into the sea by industry of organic mercury compounds such as dimethylmercury was the chief cause of mercury poisoning in the latter half of the 20th century.

Between 1953 and 1975, 684 people in the Japanese fishing village of Minamata were poisoned (115 fatally) by organic mercury wastes that had been dumped into the bay and had accumulated in the bodies of fish and shellfish.

In a landmark settlement, a British multinational chemical company in April 1997 agreed to pay £1.3 million in compensation to 20 South African workers who were poisoned by mercury. Four of the workers had died and a number of others were suffering severe brain and other neurological damage. The workers had accused Thor Chemical Holdings of adopting working practices in South Africa that would not have been allowed in Britain. The claimants had all worked at Thor's mercury plant at Cato Ridge in Natal. Thor had operated a mercury plant at Margate, in Kent, which during the 1980s was repeatedly criticized by the Health and Safety Executive (HSE) for bad working practices and over-exposure of British workers to mercury. Under pressure from the HSE, Thor closed down its mercury operations in Britain in 1987 and expanded them in South Africa.

history The element was known to the ancient Chinese and Hindus, and is found in Egyptian tombs of about 1500 BC. It was named by the alchemists after the fast-moving god, for its fluidity.

meridian half a great circle drawn on the Earth's surface passing through both poles and thus through all places with the same longitude. Terrestrial longitudes are usually measured from the Greenwich Meridian.

An astronomical meridian is a great circle passing through the celestial pole and the zenith (the point immediately overhead).

meristem region of plant tissue containing cells that are actively dividing to produce new tissues (or have the potential to do so). Meristems found in the tip of roots and stems, the apical meristems, are responsible for the growth in length of these organs.

mesmerism former term for *hypnosis, after Austrian physician Friedrich Mesmer.

meson in physics, a group of unstable subatomic particles made up of a *quark and an antiquark. It is found in cosmic radiation, and is emitted by nuclei under bombardment by very high-energy particles. See *hadron, *boson.

mesophyll tissue between the upper and lower epidermis of a leaf blade (lamina), consisting of parenchyma-like cells containing numerous *chloroplasts.

mesosphere layer in the Earth's *atmosphere above the stratosphere and below the thermosphere. It lies between about 50 km/31 mi and 80 km/50 mi above the ground.

mesothelioma malignant tumour occurring in the *pleura, peritoneum or pericardium. Pleural mesothelioma may occur as a consequence of *asbestosis.

The incidence of mesothelioma increased sixfold in Britain 1968–1991. A virus similar to the monkey virus SV40 may trigger mesothelioma.

Mesozoic era of geological time 245–65 million years ago, consisting of the Triassic, Jurassic, and Cretaceous periods. At the beginning of the era, the continents were joined together as Pangaea; dinosaurs and other giant reptiles dominated the sea and air; and ferns, horsetails, and cycads thrived in a warm climate worldwide. By the end of the Mesozoic era, the continents had begun to assume their present positions, flowering plants were dominant, and many of the large reptiles and marine fauna were becoming extinct.

message-ID in computing, special number given to every item of *e-mail as it travels across the Internet. Message-IDs are especially important for controlling traffic in *Usenet. Articles are initially offered across the network by their message-IDs, enabling news servers to check whether they have already received them and either take the rest of the message or move on to the next message-ID.

messenger RNA mRNA, single-stranded nucleic acid (made up of *nucleotides) found in *ribosomes, *mitochondria, and nucleoli of cells that carries coded information for building chains of *amino acids into polypeptides.

metabolism chemical processes of living organisms enabling them to grow and to function. It involves a constant alternation of building up complex molecules (**anabolism**) and breaking them down (**catabolism**). For example, green plants build up complex organic substances from water, carbon dioxide, and mineral salts (*photosynthesis); by digestion animals partially break down complex organic substances, ingested as food, and subsequently resynthesize them for use in their own bodies (see *digestive system). Within cells, complex molecules are broken down by the process of *respiration. The waste products of metabolism are removed by excretion.

metadata in computing, data about data. Originally a term used by the information science community, it is becoming much more common with the growth of the *Internet. On the Web, metadata describes the content of a Web page. It might record the name of the author and the date the page was created, and may contain keywords to describe the Web page to *search engines. The metadata is not displayed on the Web page itself, but makes the page more understandable to Web *crawlers and Web *robots.

metal any of a class of chemical elements with specific physical and chemical characteristics. Metallic elements compose about 75% of the 112 elements in the *periodic table of the elements. Physical properties include a sonorous tone when struck; good conduction of heat and electricity; high melting and boiling points; opacity but good reflection of light; malleability, which enables them to be cold-worked and rolled into sheets; and ductility, which permits them to be drawn into thin wires.

The majority of metals are found in nature in a combined form only, as compounds or mineral ores; about 16 of them also occur in the elemental form, as *native metals. Their chemical properties are largely determined by the extent to which their atoms can lose one or more electrons and form positive ions (cations).

All metals except mercury are solid at ordinary temperatures, and all of them will crystallize under suitable conditions. The chief chemical properties of metals also include their strong affinity for certain non-metallic elements, for example sulphur and chlorine, with which they form sulphides and chlorides. Metals will, when fused, enter into the forming of *alloys.

By comparing the reactions of metals with oxygen, water, acids, and other substances, the metals can be arranged in order of reactivity, known as the *reactivity series of metals.

metal fatigue condition in which metals fail or fracture under relatively light loads, when these loads are applied repeatedly. Structures that are subject to flexing, such as the airframes of aircraft, are prone to metal fatigue.

metallic bond force of attraction operating in a *metal that holds the atoms together in a metallic structure. In metallic bonding, metal atoms form a close-packed, regular arrangement. The atoms lose their outer-shell electrons to become positive ions. The outer electrons become a 'sea' of mobile electrons surrounding a lattice of positive ions. The lattice is held together by the strong attractive forces between the mobile electrons and the positive ions.

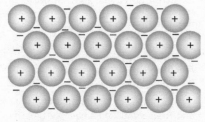

metallic bond In metallic bonding, metal ions are tightly packed with their outer shell electrons overlapping, so each electron becomes detached from its parent atom (delocalized). The metal is held together by the strong forces of attraction between the positive nuclei and the surrounding sea of delocalized electrons.

The properties of metals can be explained in terms of metallic bonding. Metals conduct electricity as the electrons are free to move. Conduction of heat occurs by vibration of the positive ions as well as via the mobile electrons. Metals are both ductile and malleable because the bonding is not broken when metals are deformed; instead, the metal ions slide over each other to new lattice positions.

metallic character chemical properties associated with those elements classed as metals. These properties, which arise from the element's ability to lose electrons, are: the displacement of hydrogen from dilute acids; the formation of basic oxides; the formation of ionic chlorides; and their reducing reaction, as in the thermite process (see *reduction).

In the periodic table of the elements, metallic character increases down any group and across a period from right to left.

metalloid or **semimetal**, any chemical element having some of but not all the properties of metals; metalloids are thus usually electrically semiconducting. They comprise the elements germanium, arsenic, antimony, and tellurium. Metalloids are found in the *periodic table of the elements between metals and non-metals.

metallurgy the science and technology of producing metals, which includes extraction, alloying, and hardening. **Extractive**, or **process, metallurgy** is concerned with the extraction of metals from their *ores and refining and adapting them for use. **Physical metallurgy** is concerned with their properties and application. **Metallography** establishes the microscopic structures that contribute to hardness, ductility, and strength.

metamorphic rock rock altered in structure, composition, and texture by pressure or heat after its original formation. (Rock that actually melts under heat is called *igneous rock upon cooling.) For example, limestone can be metamorphosed by heat into marble, and shale by pressure into slate. The term was coined in 1833 by Scottish geologist Charles Lyell.

Metamorphism is part of the rock cycle, the gradual formation, change, and re-formation of rocks over millions of years.

metamorphosis period during the life cycle of many invertebrates, most amphibians, and some fish, during which the individual's body changes from one form to another through a major reconstitution of its tissues. For example, adult frogs are produced by metamorphosis from tadpoles, and butterflies are produced from caterpillars following metamorphosis within a pupa. (See diagram, page 370.) In classical thought and literature, metamorphosis is the transformation of a living being into another shape, either living or inanimate (for example Niobe). The Roman poet Ovid wrote about this theme.

metastasis spread of a malignant tumour from its original site. During metastasis, tumour cells become detached and enter blood or lymphatic vessels, travelling to other parts of the body. The pattern of the spread can be predicted to a certain degree because of this: many cancers spread to the lungs where most deoxygenated blood is returned, whereas intestinal cancers usually spread first to the liver.

meta-tags in Web pages, *tags used to document the page and/or provide information that can be used by search engines to classify or index the page, not to control the display of information. Some Web programmers use meta-tags to try to cheat search engines – for example by repeating words frequently, or by using popular search words that have no relevance to the page – while search engine programmers make considerable efforts to negate this misuse. The battle is unseen by most Web users, who rarely use the browser's **view source** menu item to read meta-tags.

metazoa another name for animals. It reflects a former system of classification, in which there were two main divisions within the animal kingdom, the multicellular animals, or metazoa, and the single-celled 'animals' or protozoa. The *protozoa are no longer included in the animal kingdom, so only the metazoa remain.

meteor flash of light in the sky, popularly known as a **shooting** or **falling star**, caused by a particle of dust, a *meteoroid, entering the atmosphere at speeds up to 70 kps/ 45 mps and burning up by friction at a height of around 100 km/60 mi. On any clear night, several **sporadic meteors** can be seen each hour.

meteor-burst communications technique for sending messages by bouncing radio waves off the trails of *meteors. High-speed computer-controlled equipment is used to sense the presence of a meteor and to broadcast a signal during the short time that the meteor races across the sky.

meteorite piece of rock or metal from space that reaches the surface of the Earth, Moon, or other body. Most meteorites are thought to be fragments from asteroids, although some may be pieces from the heads of comets. Most are stony, although some are made of iron and a few have a mixed rock–iron composition.

meteoroid small natural object in interplanetary space. Meteoroids are smaller than *asteroids, ranging from the size of a pebble up, and move through space at high speeds. There is no official distinction between meteoroids and asteroids, although the term 'asteroid' is generally reserved for objects larger than 1.6 km/1 mi in diameter.

meteorology scientific observation and study of the *atmosphere, so that *weather can be accurately forecast.

Data from meteorological stations and weather satellites are collated by computers at central agencies, and forecast and weather maps based on current readings are issued at regular intervals. Modern analysis, employing some of the most powerful computers, can give useful forecasts for up to six days ahead.

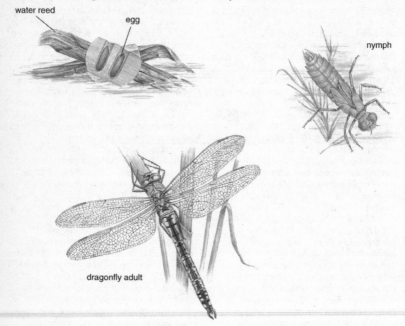

water reed

egg

nymph

dragonfly adult

metamorphosis The life cycle of the dragonfly. Eggs are usually laid under water and the larvae hatch as free-moving, water-dwelling nymphs. The carnivorous larva lives in the water for several weeks (or even years in some species) and undergoes a series of moults as it grows. It leaves the water when it is ready to undergo its final moult, during which it sheds its 'skin' to reveal the winged adult.

At meteorological stations readings are taken of the factors determining weather conditions: atmospheric pressure, temperature, humidity, wind (using the *Beaufort scale), cloud cover (measuring both type of cloud and coverage), and precipitation such as rain, snow, and hail (measured at 12-hour intervals). *Satellites are used either to relay information transmitted from the Earth-based stations, or to send pictures of cloud development, indicating wind patterns, and snow and ice cover.

history Apart from some observations included by Aristotle in his book *Meteorologia*, meteorology did not become a precise science until the end of the 16th century, when Italian mathematician, astronomer, and physicist Galileo (1564–1642) and the Florentine academicians constructed the first thermometer of any importance, and when Evangelista Torricelli in 1643 discovered the principle of the barometer. The work of the Irish chemist and physicist Robert Boyle (1627–1691) on gases, and that of his assistant, English physicist Robert Hooke (1635–1703), on barometers, advanced the physics necessary for the understanding of the weather. The invention by Polish-born Dutch physicist Gabriel Fahrenheit (1686–1736) of a superior mercury thermometer provided further means for temperature recording.

weather maps In the early 19th century a chain of meteorological stations was established in France, and weather maps were constructed from the data collected. The first weather map in England, showing the trade winds and monsoons, was made in 1688, and the first telegraphic weather report appeared on 31 August 1848. The first daily telegraphic weather map was prepared at the Great Exhibition in 1851, but the Meteorological Office was not established in London until 1855. The first regular daily collections of weather observations by telegraph and the first British daily weather reports were made in 1860, and the first daily printed maps appeared in 1868.

collecting data Observations can be collected not only from land stations, but also from weather ships, aircraft, and self-recording and automatic transmitting stations, such as the radiosonde. *Radar may be used to map clouds and storms. Satellites have played an important role in televising pictures of global cloud distribution.

methanal or **formaldehyde**; HCHO, gas at ordinary temperatures, condensing to a liquid at –21°C/–5.8°F. It has a powerful, penetrating smell. Dissolved in water, it is used as a biological preservative. It is used in the manufacture of plastics, dyes, foam (for example, urea-formaldehyde foam, used in insulation), and in medicine.

methane CH_4, simplest *hydrocarbon of the *alkane series. Colourless, odourless, and lighter than air, it burns with a bluish flame and explodes when mixed with air or oxygen. As the chief constituent of natural gas, methane's main use is as a fuel. It also occurs in the explosive firedamp of coal mines. Methane emitted by rotting vegetation forms marsh gas, which may ignite by spontaneous combustion to produce the pale flame seen over marshland and known as *will-o'-the-wisp.

Methane causes about 38% of the warming of the globe through the *greenhouse effect; weight for weight it is 60–70 times more potent than carbon dioxide at trapping solar radiation in the atmosphere and so heating the planet. The rate of increase of atmospheric methane is declining and global emissions remained relatively constant over the period 1984–96, so atmospheric levels were predicted, in 1998, to stabilize by the 2020s. An estimated 15% of all methane gas in the atmosphere is produced by cattle and other cud-chewing animals, and 20% is produced by termites that feed on soil.

methanogenic bacteria one of a group of primitive micro-organisms, the Archaea. They give off methane gas as a by-product of their metabolism, and are common in sewage treatment plants and hot springs, where the temperature is high and oxygen is absent. Archaeons were originally classified as bacteria, but were found to be unique in 1996 following the gene sequencing of the deep-sea vent *Methanococcus jannaschii*.

methanoic acid or **formic acid**; HCOOH, colourless, slightly fuming liquid that freezes at 8°C/46.4°F and boils at 101°C/213.8°F. It occurs in stinging ants, nettles, sweat, and pine needles, and is used in dyeing, tanning, and electroplating.

methanol common name **methyl alcohol**; CH₃OH, simplest of the alcohols. It can be made by the dry distillation of wood (hence it is also known as wood alcohol), but is usually made from coal or natural gas. When pure, it is a colourless, flammable liquid with a pleasant odour, and is highly poisonous.

Methanol is used to produce methanal (formaldehyde, from which resins and plastics can be made), methyl-*tert*-butyl ether (MTB, a replacement for lead as an octane-booster in petrol), vinyl acetate (largely used in paint manufacture), and petrol. In 1993 Japanese engineers built an engine, made largely of ceramics, that runs on methanol. The prototype is lighter and has a cleaner exhaust than comparable metal, petrol-powered engines.

methionine one of the nine essential *amino acids. It is also used as an antidote to paracetamol poisoning.

methyl alcohol common name for *methanol.

methylated spirit alcohol that has been rendered undrinkable, and is used for industrial purposes, as a fuel for spirit burners or a solvent.

It is nevertheless drunk by some individuals, resulting eventually in death. One of the poisonous substances in it is *methanol, or methyl alcohol, and this gives it its name. (The 'alcohol' of alcoholic drinks is ethanol.)

methyl benzene alternative name for *toluene.

methyl bromide or **bromoethane**; CH₃Br, pesticide gas used to fumigate soil. It is a major *ozone depleter. Industry produced 50,000 tonnes of methyl bromide annually in 1995. At a meeting in July 1998 the European Union (EU) proposed a total ban on usage by 2001. Some countries have failed significantly to reduce their ozone-depleting emissions, for example China, which at the end of

1999 was producing 10% of the world's methyl bromide.

methyl orange C₁₄H₁₄N₃NaO₃S, orange-yellow powder used as an acid–base indicator in chemical tests, and as a stain in the preparation of slides of biological material. Its colour changes with *pH; below pH 3.1 it is red, above pH 4.4 it is yellow.

metre symbol m, SI unit of length, equivalent to 1.093 yards or 39.37 inches. It is defined by scientists as the length of the path travelled by light in a vacuum during a time interval of 1/299,792,458 of a second.

metric system system of weights and measures developed in France in the 18th century and recognized by other countries in the 19th century.

In 1960 an international conference on weights and measures recommended the universal adoption of a revised International System (Système International d'Unités, or SI), with seven prescribed 'base units': the metre (m) for length, kilogram (kg) for mass, second (s) for time, ampere (A) for electric current, kelvin (K) for thermodynamic temperature, candela (cd) for luminous intensity, and mole (mol) for quantity of matter. See Appendix for full details.

metropolitan area network MAN, high-speed communications network within the environs of a city or metropolitan area. In the UK, MANs are starting to become more common as more schools, colleges, and public services become linked to the *Internet. MANs provide a high-speed network *backbone for all of these services, and the high cost of building the MAN can be recovered by charging commercial organisations to use it to gain faster access to the Internet.

mg symbol for **milligram**.

mho SI unit of electrical conductance, now called the *siemens; equivalent to a reciprocal ohm.

mi symbol for *mile.

mica any of a group of silicate minerals that split easily into thin flakes along lines of weakness in their crystal structure (perfect basal cleavage). They are glossy, have a pearly lustre, and are found in many igneous and metamorphic rocks. Their good thermal and electrical insulation qualities make them valuable in industry. Their chemical composition is complicated, but they are silicates with silicon–oxygen tetrahedra arranged in continuous sheets, with weak bonding between the layers, resulting in perfect cleavage. A common example of mica is muscovite (white mica), $KAl_2Si_3AlO_{10}(OH,F)_2$.

micro- symbol μ, prefix denoting a one-millionth part (10^{-6}). For example, a micrometre, μm, is one-millionth of a metre.

microbe another name for *microorganism.

microbiology study of microorganisms, mostly viruses and single-celled organisms such as bacteria, protozoa, and yeasts. The practical applications of microbiology are in medicine (since many micro-organisms cause disease); in brewing, baking, and other food and beverage processes, where the micro-organisms carry out fermentation; and in genetic engineering, which is creating increasing interest in the field of microbiology.

microchip popular name for the silicon chip, or *integrated circuit.

microclimate climate of a small area, such as a woodland, lake, or even a hedgerow. Significant differences can exist between the climates of two neighbouring areas – for example, a town is usually warmer than the surrounding countryside (forming a heat island), and a woodland cooler, darker, and less windy than an area of open land. Microclimates play a significant role in agriculture and horticulture, as different crops require different growing conditions.

micrometre symbol μm, one-millionth of a *metre.

micron obsolete name for the micrometre, one-millionth of a metre.

micro-organism or **microbe**, living organism invisible to the naked eye but visible under a microscope. Micro-organisms include *viruses and single-celled organisms such as bacteria and yeasts. Yeasts are fungi, but other fungi are often big enough to see with the naked eye and so are not micro-organisms. The study of micro-organisms is known as microbiology.

microphone primary component in a sound-reproducing system, whereby the mechanical energy of sound waves is converted into electrical signals by means of a *transducer. A diaphragm is attached to a coil of wire placed between two poles of a permanent magnet. Sound waves cause the diaphragm to vibrate, which in turn causes the coil of wire to move in the magnetic field of the permanent magnet. An induced electrical current, matching the pattern of the sound waves, flows through the coil and is fed to an amplifier. The amplified signals are either stored or sent to a loudspeaker.

microscope instrument for forming magnified images with high resolution for detail. Optical and electron microscopes are the ones chiefly in use; other types include acoustic, *scanning tunnelling, and *atomic force microscopes.

The **optical microscope** usually has two sets of glass lenses and an eyepiece. The first true compund microscope was developed in 1609 in the Netherlands by Zacharias Janssen (1580– c. 1638). **Fluorescence microscopy** makes use of fluorescent dyes to illuminate samples, or to highlight the presence of particular substances within a sample. Various illumination systems are also used to highlight details.

The *transmission electron microscope, developed from 1932, passes a beam of electrons, instead of a beam of light, through a specimen. Since electrons are not visible, the eyepiece is replaced with a fluorescent screen or photographic plate; far higher magnification and resolution are possible than with the optical microscope.

The *scanning electron microscope (SEM), developed in the mid-1960s, moves a fine beam of electrons over the surface of a specimen, the reflected electrons being collected to form the image. The specimen has to be in a vacuum chamber.

The **acoustic microscope** passes an ultrasonic (ultrahigh-frequency sound) wave through the specimen, the transmitted sound being used to form an image on a computer screen.

The **scanned-probe microscope**, developed in the late 1980s, runs a probe, with a tip so fine that it may consist only of a single atom, across the surface of the specimen. In the **scanning tunnelling microscope**, an electric current that flows through a probe is used to construct an image of the specimen. In 1988 a scanning tunnelling microscope was used to photograph a single protein molecule for the first time. In the **atomic force microscope**, the force felt by a probe is measured and used to form the image. These instruments can magnify a million times and give images of single atoms.

microtubules   tiny tubes found in almost all cells with a nucleus. They help to define the shape of a cell by forming scaffolding for cilia and they also form the fibres of mitotic spindle (see *mitosis).

microwave radiation   *electromagnetic wave with a wavelength in the range 0.3 cm to 30 cm/0.1 in to 12 in, or 300–300,000 megahertz (between radio waves and *infrared radiation). Microwaves are used in radar, in radio broadcasting, in satellite communications, and in microwave heating and cooking.

In microwave cooking, microwaves are used to transfer energy to food. The frequency of the microwaves matches the natural frequency at which the water molecules in food vibrate. The vibration energy of the water molecules results in an increase in temperature, which allows the food to be cooked.

Mid-Atlantic Ridge   *mid-ocean ridge that runs along the centre of the Atlantic Ocean, parallel to its edges, for some 14,000 km/8,800 mi – almost from the Arctic to the Antarctic. Like other ocean ridges, the Mid-Atlantic Ridge is essentially a linear, segmented volcano.

The Mid-Atlantic Ridge runs down the centre of the ocean because the ocean crust has continually grown outwards from the ridge at a steady rate during the past 200 million years.

Iceland straddles the ridge and was formed by volcanic outpourings.

MIDI   acronym for Musical Instrument Digital Interface, manufacturers' standard allowing different pieces of digital music equipment used in composing and recording to be freely connected.

mid-ocean ridge   long submarine mountain range that winds along the middle of the ocean floor. The mid-ocean ridge system is essentially a segmented, linear *shield volcano. There are a number of major ridges, including the *Mid-Atlantic Ridge, which runs down the centre of the Atlantic; the East Pacific Rise in the southeast Pacific; and the Southeast Indian Ridge. These ridges are now known to be spreading centres, or divergent margins, where two plates of oceanic lithosphere are moving away from one another (see *plate tectonics). Ocean ridges can rise thousands of metres above the surrounding seabed, and extend for up to 60,000 km/37,000 mi in length.

Ocean ridges usually have a *rift valley along their crests, indicating where the flanks are being pulled apart by the growth of the plates of the *lithosphere beneath. The crests are generally free of sediment; increasing depths of sediment are found with increasing distance down the flanks.

migraine   acute, sometimes incapacitating headache (generally only on one side), accompanied by nausea, that recurs, often with advance symptoms such as flashing lights. No cure has been discovered, but ergotamine normally relieves the symptoms. Some sufferers learn to avoid certain foods, such as chocolate, which suggests an allergic factor.

migration   movement, either seasonal or as part of a single life cycle, of certain animals, chiefly birds and fish, to distant breeding or feeding grounds.

mil   abbreviation for military, in the Internet's *domain name system (DNS), one of the top-level domains along with net, gov, org, com and info. US military organizations typically have Internet addresses of the form name.mil.

Milankovitch hypothesis combination of factors governing the occurrence of *ice ages proposed in 1930 by the Yugoslav geophysicist M Milankovitch (1879–1958). These include the variation in the angle of the Earth's axis, and the geometry of the Earth's orbit around the Sun.

mile symbol mi, imperial unit of linear measure. A statute mile is equal to 1,760 yards (1.60934 km), and an international nautical mile is equal to 2,026 yards (1,852 m).

milk teeth or **deciduous teeth**, teeth that erupt in childhood between the ages of 6 and 30 months. They are replaced by the permanent teeth, which erupt between the ages of 6 and 21 years. See also *dentition and *tooth.

Milky Way faint band of light crossing the night sky, consisting of stars in the plane of our Galaxy. The name Milky Way is often used for the Galaxy itself. It is a spiral *galaxy, 100,000 light years in diameter and 2,000 light years thick, containing at least 100 billion *stars. The Sun is in one of its spiral arms, about 25,000 light years from the centre, not far from its central plane.

milli- symbol m, prefix denoting a one-thousandth part (10^{-3}). For example, a millimetre, mm, is one thousandth of a metre.

millibar unit of pressure, equal to one-thousandth of a *bar.

millilitre one-thousandth of a litre (ml), equivalent to one cubic centimetre (cc).

millimetre of mercury symbol mmHg, unit of pressure, used in medicine for measuring blood, pressure defined as the pressure exerted by a column of mercury one millimetre high, under the action of gravity.

MIME acronym for Multipurpose Internet Mail Extension, in computing, standard for transferring multimedia *e-mail messages and *World Wide Web *hypertext documents over the Internet. Under MIME, binary files (any file not in plain text, such as graphics and audio) are translated into a form of *ASCII before transmission, and then turned back into binary form by the recipient. See also *UUencode.

mimicry imitation of one species (or group of species) by another. The most common form is **Batesian mimicry** (named after English naturalist H W Bates), where the mimic resembles a model that is poisonous or unpleasant to eat, and has aposematic, or warning, coloration; the mimic thus benefits from the fact that predators have learned to avoid the model. Hoverflies that resemble bees or wasps are an example. Appearance is usually the basis for mimicry, but calls, songs, scents, and other signals can also be mimicked.

mineral naturally formed inorganic substance with a particular chemical composition and a regularly repeating internal structure. Either in their perfect crystalline form or otherwise, minerals are the constituents of *rocks. In more general usage, a mineral is any substance economically valuable for mining (including coal and oil, despite their organic origins).

Mineral-forming processes include: melting of pre-existing rock and subsequent crystallization of a mineral to form igneous or volcanic rocks; weathering of rocks exposed at the land surface, with subsequent transport and grading by surface waters, ice, or wind to form sediments; and recrystallization through increasing temperature and pressure with depth to form *metamorphic rocks. The transformation and recycling of the minerals of the Earth's outer layers is known as the rock cycle.

Minerals are usually classified as magmatic, *sedimentary, and *metamorphic. The magmatic minerals, in *igneous rock, include the feldspars, quartz, pyroxenes, amphiboles, micas, and olivines that crystallize from silica-rich rock melts within the crust or from extruded lavas. The most commonly occurring sedimentary minerals are either pure concentrates or mixtures of sand, clay minerals, and carbonates (chiefly calcite, aragonite, and dolomite). Minerals typical of metamorphism include andalusite, cordierite, garnet, tremolite, lawsonite, pumpellyite, glaucophane, wollastonite, chlorite, micas, hornblende, staurolite, kyanite, and diopside.

mineralogy study of minerals. The classification of minerals is based chiefly on their chemical composition and the kind of chemical bonding that holds their atoms together. The mineralogist also studies their crystallographic and physical characters, occurrence, and mode of formation. The systematic study of minerals began in the 18th century, with the division of minerals into four classes: earths, metals, salts, and bituminous substances, distinguished by their reactions to heat and water.

mineral oil oil obtained from mineral sources, for example coal or petroleum, as distinct from oil obtained from vegetable or animal sources.

mineral salt in nutrition, simple inorganic chemicals that are required, as nutrients, by living organisms. Plants usually obtain their mineral salts from the soil, while animals get theirs from their food. Important mineral salts include iron salts (needed by both plants and animals), magnesium salts (needed mainly by plants, to make *chlorophyll), and calcium salts (needed by animals to make bone or shell). A *trace element is required only in tiny amounts.

minor arc the smaller of the two arcs formed when a circle is divided into two unequal parts by a straight line or *chord.

minor planet name sometimes given to the larger members of the *asteroid belt.

minute unit of time consisting of 60 seconds; also a unit of angle equal to one sixtieth of a degree.

Miocene ('middle recent') fourth epoch of the Tertiary period of geological time, 23.5–5.2 million years ago. At this time grasslands spread over the interior of continents, and hoofed mammals rapidly evolved.

Miocene apes see *human species, origins of.

mips acronym for Million Instructions Per Second, in computing, a measure of the speed of a processor. It does not equal the computer power in all cases.

mirage illusion seen in hot weather of water on the horizon, or of distant objects being enlarged. The effect is caused by the *refraction, or bending, of light.

mirror site in computing, archive site which keeps a copy of another site's files for downloading by *FTP. Software archives such as those of the University of Michigan, and the many companies that distribute software by FTP, have several mirror sites around the world, so that users can choose the nearest site.

mist low cloud caused by the condensation of water vapour in the lower part of the *atmosphere. Mist is less thick than *fog, visibility being 1–2 km/0.6–1.2 mi.

mitochondria singular **mitochondrion**, membrane-enclosed organelles within *eukaryotic cells, containing *enzymes responsible for energy production during *aerobic respiration. They are found in both plant and animal cells. Mitochondria absorb oxygen (O_2) and complete the breakdown of glucose to carbon dioxide (CO_2) and water (H_2O) to produce energy in the form of *ATP, which is used in life processes in the cell. These rodlike or spherical bodies are thought to be derived from free-living bacteria that, at a very early stage in the history of life, invaded larger cells and took up a symbiotic way of life inside them. Each still contains its own small loop of DNA called mitochondrial DNA, and new mitochondria arise by division of existing ones. Mitochondria each have 37 genes.

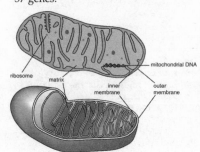

mitochondria A cross-section of a mitochondrion (above) and a 3-D cutaway. Mitochondria have an outer membrane and an inner membrane. The inner membrane is much folded to form christae. Mitochondria have their own DNA and ribosomes.

Mutations in mitochondrial genes are always inherited from the mother. These mutations have been linked to a number of disorders, mainly degenerative, including Alzheimer's disease and diabetes.

mitosis in biology, the process of cell division by which one parent cell produces two genetically identical 'daughter' cells. The genetic material of *eukaryotic cells is carried on a number of *chromosomes. During mitosis the DNA is duplicated and the chromosome number doubled – identical copies of the chromosomes are separated into the two daughter cells, which contain the same amount of DNA as the original cell. To control movements of chromosomes during cell division so that both new cells get the correct number, a system of protein tubules, known as the spindle, organizes the chromosomes into position in the middle of the cell before they replicate. The spindle then controls the movement of chromosomes as the cell goes through the stages of division: **interphase**, **prophase**, **metaphase**, **anaphase**, and **telophase**. See also *meiosis.

Mitosis is used for growth and for *asexual reproduction. *Growth is the increase in size and weight of an organism over a period of time. In biology growth is often measured as *biomass. Growth results from mitosis

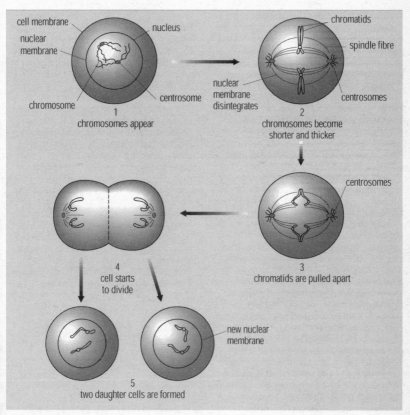

cell membrane
nucleus
nuclear membrane
chromatids
spindle fibre
chromosome
centrosome
nuclear membrane disintegrates
centrosomes
1
chromosomes appear
2
chromosomes become shorter and thicker

centrosomes
4
cell starts to divide
3
chromatids are pulled apart

new nuclear membrane
5
two daughter cells are formed

mitosis The stages of mitosis, the process of cell division that takes place when a plant or animal cell divides for growth or repair. The two 'daughter' cells each receive the same number of chromosomes as were in the original cell.

followed by the increase in size of the new cells.

mixture in chemistry, a substance containing two or more elements or compounds that still retain their separate physical and chemical properties. There is no chemical bonding between them and they can be separated from each other by physical means (compare *compound).

The separation of mixtures can be achieved by a number of methods, such as *filtration and distillation. Examples of mixtures include rocks, air, crude oil (unrefined petroleum), all solutions (such as sea water), and all metal alloys (such as steel).

m.k.s. system system of units in which the base units metre, kilogram, and second replace the centimetre, gram, and second of the * c.g.s. system. From it developed the SI system (see *SI units).

ml symbol for **millilitre**.

mm symbol for **millimetre**.

mmHg symbol for *millimetre of mercury.

MMR vaccine combined vaccine given to small children to prevent *measles, *mumps, and rubella (*German measles).

mobile commerce or **m-commerce**, transaction of commerce on the move using a wireless mobile device, such as a mobile phone or a *personal digital assistant (PDA), instead of a *personal computer connected to a fixed telephone line.

mobile ion in chemistry, an ion that is free to move; such ions are only found in the aqueous solutions or melts (molten masses) of an *electrolyte. The mobility of the ions in an electrolyte is what allows it to conduct electricity.

Möbius strip structure made by giving a half twist to a flat strip of paper and joining the ends together. It has certain remarkable properties, arising from the fact that it has only one edge and one side. If cut down the centre of the strip, instead of two new strips of paper, only one long strip is produced. It was invented by the German mathematician August Möbius (1790–1868).

mode in mathematics, the element that appears most frequently in a given set of data. For example, the mode for the data 0, 0, 9, 9, 9, 12, 87, 87 is 9.

model in computing, set of assumptions and criteria based on actual phenomena, used to conduct a computer simulation. Models are used to predict the behaviour of a system such as the movement of a hurricane or the flow of goods from a store. In industry, they are important tools for testing new products: engineers subject *virtual prototypes of aircraft or bridges to various scenarios to find out what adjustments are necessary to the design. However, a model is only as good as the assumptions that underlie it.

modem contraction of modulator/demodulator, device for transmitting computer data over telephone lines. Such a device is used to convert digital signals produced by computers to *analogue signals compatible with the telephone network, and back again.

moderator in computing, person or group of people that screens submissions to certain newsgroups and mailing lists before passing them on for wider circulation. The aim of moderation is not to censor, but to ensure that the quality of debate is maintained by filtering out *spamming, irrelevant ('off-topic'), or gratuitously offensive postings.

modulation in radio transmission, the variation of frequency, or amplitude, of a radio carrier wave, in accordance with the audio characteristics of the speaking voice, music, or other signal being transmitted.

modulus in mathematics, a number that divides exactly into the difference between two given numbers. Also, the multiplication factor used to convert a logarithm of one base to a logarithm of

Möbius strip The Möbius strip has only one side and one edge. It consists of a strip of paper connected at its ends with a half-twist in the middle.

another base. Also, another name for *absolute value.

Mohs scale   scale of hardness for minerals (in ascending order): 1 talc; 2 gypsum; 3 calcite; 4 fluorite; 5 apatite; 6 orthoclase; 7 quartz; 8 topaz; 9 corundum; 10 diamond. The scale was devised by German minerologist Friedrich Mohs (1773–1839) in 1812. The scale is useful in mineral identification because any mineral will scratch any other mineral lower on the scale than itself, and similarly it will be scratched by any other mineral higher on the scale.

mol   symbol for *mole, the SI unit of amount of substance, equal to the amount that contains as many elementary entities (such as atoms, ions, or molecules) as there are atoms in 12 grams of carbon.

molar   one of the large teeth found towards the back of the mammalian mouth. The structure of the jaw, and the relation of the muscles, allows a massive force to be applied to molars. In herbivores the molars are flat with sharp ridges of enamel and are used for grinding, an adaptation to a diet of tough plant material. Carnivores have sharp powerful molars called carnassials, which are adapted for cutting meat.

molarity   *concentration of a solution expressed as the number of *moles of solute per cubic decimetre of solution.

molar solution   solution that contains one *mole of a substance per litre of solvent.

molar volume   volume occupied by one *mole (the molecular mass in grams) of any gas at standard temperature and pressure, equal to 2.24136×10^{-2} m^3.

mole   symbol mol, unit of the amount of a substance. One mole of a substance is the mass that contains the same number of particles (atoms, molecules, ions, or electrons) as there are atoms in 12 grams of the *isotope carbon-12. One mole of a substance is 6.022045×10^{23} atoms, which is *Avogadro's number. It is obtained by weighing out the relative atomic mass (RAM) or relative molecular mass (RMM) in grams (so one mole of carbon weighs 12 g).

molecular biology   study of the molecular basis of life, including the biochemistry of molecules such as DNA, RNA, and proteins, and the molecular structure and function of the various parts of living cells.

molecular clock   use of rates of *mutation in genetic material to calculate the length of time elapsed since two related species diverged from each other during evolution. The method can be based on comparisons of the DNA or of widely occurring proteins, such as haemoglobin.

molecular cloud   in astronomy, enormous cloud of cool interstellar dust and gas containing hydrogen molecules and more complex molecular species. Giant molecular clouds (GMCs), about a million times as massive as the Sun and up to 300 light years in diameter, are regions in which stars are being born. The Orion nebula is part of a GMC.

molecular formula   in chemistry, formula indicating the actual number of atoms of each element present in a single *molecule of a chemical compound. For example, the molecular formula of carbon dioxide is CO_2, indicating that one molecule of carbon dioxide is made up of one atom of carbon and two atoms of oxygen. This is determined by two pieces of information: the empirical *formula and the *relative molecular mass, which is determined experimentally.

molecular mass   or **relative molecular mass**, mass of a molecule, calculated relative to one-twelfth the mass of an atom of carbon-12. It is found by adding the relative atomic masses of the atoms that make up the molecule.

molecular solid   in chemistry, solid composed of molecules that are held together by relatively weak *intermolecular forces. Such solids are low-melting and tend to dissolve in organic solvents. Examples of molecular solids are sulphur, ice, sucrose, and solid carbon dioxide.

molecular weight   see *relative molecular mass.

molecule   smallest configuration of an element or compound that can exist independently. One molecule is made up of a group of atoms held together by *covalent or *ionic bonds. Several non-metal elements exist as molecules. For example, hydrogen *atoms, at room temperature, do not exist independently. They are bonded in pairs to form hydrogen molecules. A molecule of a compound consists of two or more different atoms bonded together. For example, carbon dioxide is made up of molecules, each containing one carbon and two oxygen atoms bonded together. The *molecular formula is made up of the chemical symbols representing each element in the molecule and numbers showing how many atoms of each element are present. For example, the formula for hydrogen is H_2, and for carbon dioxide is CO_2. Molecules vary in size and complexity from the hydrogen molecule to the large *macromolecules of proteins. In general, elements and compounds with molecular structures have similar properties. They have low melting and boiling points, so that many molecular substances are gases or liquids at room temperatures. They are usually insoluble in water and do not conduct electricity even when melted.

Molecules may be held together by ionic bonds, in which the atoms gain or lose electrons to form *ions, or by covalent bonds, where electrons from each atom are shared to form the bond.

The existence of molecules was first inferred from the Italian physicist Amedeo *Avogadro's hypothesis in 1811. He observed that when gases combine, they do so in simple proportions. For example, exactly one volume of oxygen and two volumes of hydrogen combine to produce water. He hypothesized that equal volumes of gases at the same temperature and *pressure contain equal numbers of molecules. Avogadro's hypothesis only became generally accepted in 1860 when proposed by the Italian chemist Stanislao Cannizzaro. The movement of some molecules can be observed in a microscope. As early as 1827, Robert Brown observed that very fine pollen grains suspended in water move about in a continuously agitated manner. This continuous, random motion of particles in a fluid medium (gas or

liquid) as they are subjected to impact from the molecules of the medium is known as *Brownian movement. The spontaneous and random movement of molecules or particles in a fluid can also be observed as *diffusion occurs from a region in which they are at a high concentration to a region of lower concentration, until a uniform concentration is achieved throughout. No mechanical mixing or stirring is involved. For example, if a drop of ink is added to water, its molecules will diffuse until the colour becomes evenly distributed. See *kinetic theory.

change of state As matter is heated its temperature may rise or it may cause a *change of state. As the internal energy of matter increases the energy possessed by each particle increases too. This can be visualized as the *kinetic energy of the molecules increasing, causing them to move more quickly. This movement includes both vibration within the molecule (assuming the substance is made of more than one atom) and rotation. A solid is made of particles that are held together by forces. As a solid is heated, the particles vibrate more vigorously, taking up more space, and causing the material to expand. As the temperature of the solid increases, it reaches its *melting point and turns into a liquid. The particles in a liquid can move around more freely but there are still forces between them. As further energy is added, the particles move faster until they are able to overcome the forces between them. When the *boiling point is reached the liquid boils and becomes a gas. Gas particles move around independently of one another except when they collide. Different objects require different amounts of heat energy to change their temperatures by the same amount. The *heat capacity of an object is the quantity of heat required to raise its temperature by one degree. The *specific heat capacity of a substance is the heat capacity per unit of mass, measured in joules per kilogram per kelvin. As a substance is changing state while being heated, its temperature remains constant, provided that thermal energy is being added. For example, water boils at a constant temperature as it turns to steam. The energy required to cause the change of state is called *latent heat. This energy is used to break down the forces holding the particles together so that the change in state can occur. *Specific latent heat is the thermal energy required to change the state of a certain mass of that substance without any temperature change. *Evaporation causes cooling as a liquid vaporizes. Heat is transferred by the movement of particles (that possess kinetic energy) by conduction, convection, and radiation. *Conduction involves the movement of heat through a solid material by the movement of free electrons. *Convection involves the transfer of energy by the movement of fluid particles. *Convection currents are caused by the expansion of a liquid or gas as its temperature rises. The expanded material, being less dense, rises above colder and therefore denser material.

the size and shape of molecules
The shape of a molecule profoundly affects its chemical, physical, and biological properties. Optical *isomers (molecules that are mirror images of each other) rotate plane *polarized light in opposite directions; isomers of drug molecules may have different biological effects; and *enzyme reactions are crucially dependent on the shape of the enzyme and the substrate on which it acts. A wheel-shaped molecule containing 700 atoms and with a relative molecular mass of about 24,000 was produced by German chemists in 1995. Containing 154 *molybdenum atoms surrounded by oxygen atoms, it belongs to the class of compounds known as metal clusters.

mole fraction   in chemistry, proportion of the number of molecules of one species taking part in a chemical reaction in relation to the sum of the molecules of all the components involved.

molybdenum   chemical symbol Mo, (Greek molybdos 'lead') heavy, hard, lustrous, silver-white, metallic element, atomic number 42, relative atomic mass 95.94. The chief ore is the mineral sulphide molybdenite. The element is highly resistant to heat and conducts electricity easily. It is used in alloys, often to harden steels. It is a necessary trace element in human nutrition. It was named in 1781 by Swedish chemist Karl Scheele, after its isolation by another Swedish chemist Peter Jacob Hjelm (1746–1813), for its resemblance to lead ore.

moment of a force   in physics, measure of the turning effect, or torque, produced by a force acting on a body. It is equal to the product of the force and the perpendicular distance from its line of action to the point, or pivot, about which the body will turn. The turning force around the pivot is called the moment. Its unit is the newton metre.

moment of inertia   in physics, the sum of all the point masses of a rotating object multiplied by the squares of their respective distances from the axis of rotation.

In linear dynamics, Newton's second law of motion (see *Newton's laws of motion) states that the force F on a moving object equals the products of its mass m and acceleration a ($F = ma$); the analogous equation in rotational dynamics is $T = Ia$, where T is the torque (the turning effect of a force) that causes an angular acceleration a and I is the moment of inertia. For a given object, I depends on its shape and the position of its axis of rotation.

momentum   product of the mass of a body and its velocity. If the mass of a body is m kilograms and its velocity is v m s^{-1}, then its momentum is given by: momentum = mv. Its unit is the kilogram metre-per-second (kg m s^{-1}) or the newton second. The momentum of a body does not change unless a resultant or unbalanced force acts on that body (see *Newton's laws of motion).

According to Newton's second law of motion, the magnitude of a resultant force F equals the rate of change of momentum brought about by its action, or: $F = (mv-mu)/t$ where mu is the initial momentum of the body, mv is its final momentum, and t is the time in seconds over which the force acts. The change in momentum, or *impulse, produced can therefore be

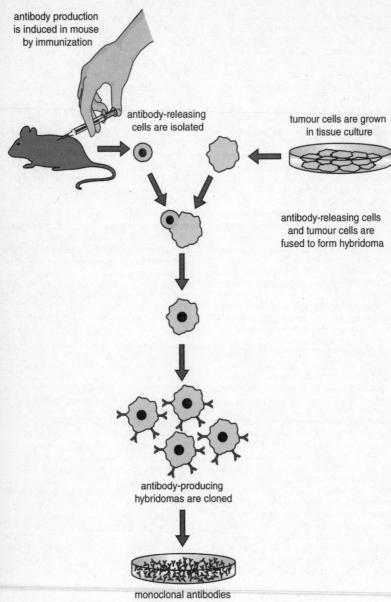

antibody production
is induced in mouse
by immunization

antibody-releasing
cells are isolated

tumour cells are grown
in tissue culture

antibody-releasing cells
and tumour cells are
fused to form hybridoma

antibody-producing
hybridomas are cloned

monoclonal antibodies
are isolated and cultivated

monoclonal antibody Monoclonal antibody production. Antibodies are collected from a mouse after immunization and are fused with tumour cells grown in tissue culture to form antibody hybridomas. These are cloned to give large numbers from which the antibodies can be harvested.

expressed as: impulse $= mv - mu = Ft$. The law of conservation of momentum is one of the fundamental concepts of classical physics. It states that the total momentum of all bodies in a closed system is constant and unaffected by processes occurring within the system. The **angular momentum** of an orbiting or rotating body of mass m travelling at a velocity v in a circular orbit of radius R is expressed as mvR. Angular momentum is conserved, and should any of the values alter (such as the radius of orbit), the other values (such as the velocity) will compensate to preserve the value of angular momentum, and that lost by one component is passed to another.

monitor or **screen**, output device on which a computer displays information for the benefit of the operator user – usually in the form of a *graphical user interface such as Windows. The commonest type is the *cathode-ray tube (CRT), which is similar to a television screen. Portable computers often use *liquid crystal display (LCD) screens. These are harder to read than CRTs, but require less power, making them suitable for battery operation.

monoamine-oxidase inhibitor (MAOI) antidepressant drug that works by curtailing the activity of the enzyme monoamine-oxidase in the brain. MAOIs interact dangerously with other drugs and some foods (including cheese), as well as having a range of unwanted side effects. For this reason their use is restricted to patients who do not respond to other, safer antidepressants.

monobasic acid acid that has only one replaceable hydrogen atom and can therefore form only one series of salts; for example, HCl.

monoclonal antibody **MAB**, antibody produced by fusing an antibody-producing lymphocyte (white blood cell) with a cancerous myeloma (bone-marrow) cell. The resulting fused cell, called a hybridoma, is immortal and can be used to produce large quantities of a single, specific antibody. By choosing antibodies that are directed against antigens found on cancer cells, and combining them with cytotoxic

drugs, it is hoped to make so-called magic bullets that will be able to pick out and kill cancers.

It is the antigens on the outer cell walls of germs entering the body that provoke the production of antibodies as a first line of defence against disease. Antibodies 'recognize' these foreign antigens, and, in locking on to them, cause the release of chemical signals in the bloodstream to alert the immune system for further action. MABs are copies of these natural antibodies, with the same ability to recognize specific antigens. Introduced into the body, they can be targeted at disease sites.

The full potential of these biological missiles, developed by César Milstein and others at Cambridge University, England, in 1975, is still under investigation. However, they are already in use in blood-grouping, in pinpointing viruses and other sources of disease, in tracing cancer sites, and in developing vaccines.

monocotyledon angiosperm (flowering plant) having an embryo with a single cotyledon, or seed leaf (as opposed to *dicotyledons, which have two). Monocotyledons usually have narrow leaves with parallel veins and smooth edges, and hollow or soft stems. Their flower parts are arranged in threes. Most are small plants such as orchids, grasses, and lilies, but some are trees such as palms.

monocyte type of white blood cell. They are found in the tissues and the lymphatic and circulatory systems where their purpose is to remove foreign particles, such as bacteria and tissue debris, by ingesting them.

monohybrid inheritance pattern of inheritance seen in simple *genetics experiments, where the two animals (or two plants) being crossed are genetically identical except for one gene, which is heterozygous. In other words, inheritance is the passing of characteristics from parents to offspring, while monohybrid inheritance is inheritance for a single characteristic. (See diagram, page 384.)

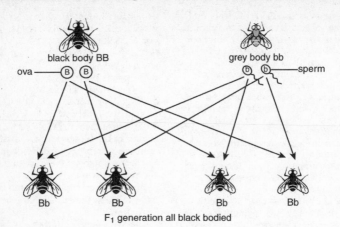

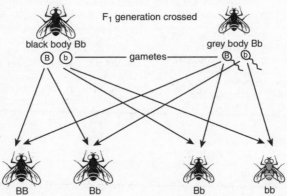

F$_2$ generation has black to grey ratio of 3:1

monohybrid inheritance A simple monohybrid cross. The allele for a black body is dominant and the allele for a grey body is recessive in fruit flies, therefore a cross between a fly with a black body and one with a grey body will produce a generation of black-bodied flies (F$_1$ generation). If two flies from the F$_1$ generation are crossed then the ratio of black to grey flies in the F$_2$ generation will be 3:1.

monomer chemical compound composed of simple molecules from which *polymers can be made. Under certain conditions the simple molecules (of the monomer) join together (polymerize) to form a very long chain molecule (macromolecule) called a polymer. For example, the polymerization of *ethene (ethylene) monomers produces the polymer *polythene (polyethylene):

$$2n\text{CH}_2{=}\text{CH}_2 \rightarrow (\text{CH}_2{-}\text{CH}_2{-}\text{CH}_2{-}\text{CH}_2)_n$$

monosaccharide or **simple sugar**, *carbohydrate that cannot be hydrolysed (split) into smaller carbohydrate units. Examples are glucose and fructose, both of which have the molecular formula $C_6H_{12}O_6$.

monosodium glutamate MSG; $NaC_5H_8NO_4$, white, crystalline powder, the sodium salt of glutamic acid (an *amino acid found in proteins that plays a role in the metabolism of plants and animals). It has no flavour of its own, but enhances the flavour of foods such as meat and fish. It is used to enhance the flavour of many packaged and 'fast foods', and in Chinese cooking. Ill effects may arise from its overconsumption, and some people are very sensitive to it, even in small amounts. It is commercially derived from vegetable protein. It occurs naturally in soybeans and seaweed.

monozygotic twin one of a set of *twins born to the same parents at the same time following the fertilization of a single ovum. They are of the same sex and virtually identical in appearance.

monsoon wind pattern that brings seasonally heavy rain to South Asia; it blows towards the sea in winter and towards the land in summer. The monsoon may cause destructive flooding all over India and Southeast Asia from April to September, leaving thousands of people homeless each year.

Moon natural satellite of Earth, 3,476 km 2,160 mi in diameter, with a mass 0.012 (approximately one-eightieth) that of Earth. Its surface gravity is only 0.16 (one-sixth) that of Earth. Its average distance from Earth is 384,400 km/238,855 mi, and it orbits in a west-to-east direction every 27.32 days (the **sidereal month**). It spins on its axis with one side permanently turned towards Earth. The Moon has no atmosphere and was thought to have no water until ice was discovered on its surface in 1998.

phases The Moon is illuminated by sunlight, and goes through a cycle of phases of shadow, waxing from **new** (dark) via **first quarter** (half Moon) to

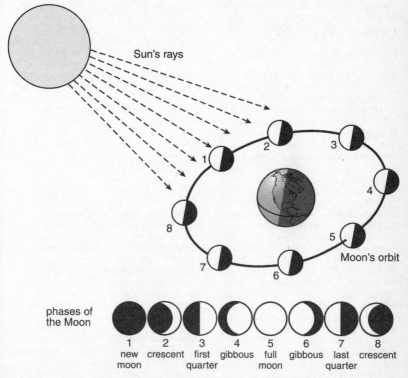

Moon The phases of the Moon as viewed from the Earth. The Moon takes a lunar month (about 29 days) to orbit the Earth. When the Earth is between the Sun and the Moon, the Moon appears to be 'full' because we can see all the light from the Sun reflected by the Moon.

full, and waning back again to new every 29.53 days (the **synodic month**, also known as a **lunation**). On its sunlit side, temperatures reach 110°C/230°F, but during the two-week lunar night the surface temperature drops to −170°C/−274°F.

origins The origin of the Moon is still open to debate. Scientists suggest the following theories: that it split from the Earth; that it was a separate body captured by Earth's gravity; that it formed in orbit around Earth; or that it was formed from debris thrown off when a body the size of Mars struck Earth.

research 70% of the far side of the Moon was photographed from the Soviet Lunik 3 in October 1959. Much of our information about the Moon has been derived from this and other photographs and measurements taken by US and Soviet Moon probes, from geological samples brought back by US *Apollo* astronauts and by Soviet Luna probes, and from experiments set up by US astronauts 1969–72. US astronaut Neil Armstrong (1930–) became the first person to set foot on the Moon on 20 July 1969. The Moon landing was part of the *Apollo* project. The US probe Lunar Prospector, launched in January 1998, examined the composition of the lunar crust, recorded gamma rays, and mapped the lunar magnetic field. It also discovered the ice on the moon in March 1998.

composition The Moon is rocky, with a surface heavily scarred by *meteorite impacts that have formed craters up to 240 km/150 mi across. Seismic observations indicate that the Moon's surface extends downwards for tens of kilometres; below this crust is a solid mantle about 1,100 km/688 mi thick, and below that a silicate core, part of which may be molten. Rocks brought back by astronauts show that the Moon is 4.6 billion years old, the same age as Earth. It is made up of the same chemical elements as Earth, but in different proportions, and differs from Earth in that most of the Moon's surface features were formed within the first billion years of its history when it was hit repeatedly by meteorites. The youngest craters are surrounded

by bright rays of ejected rock. The largest scars have been filled by dark lava to produce the lowland plains called seas, or **maria** (plural of mare). These dark patches form the so-called 'man-in-the-Moon' pattern. Inside some craters that are permanently in shadow is up to 300 million tonnes/330 million tons of ice existing as a thin layer of crystals.

One of the Moon's easiest features to observe is the mare Plato, which is about 100 km/62 mi in diameter and 2,700 m/8,860 ft deep, and at times is visible with the naked eye.

Moon probe crewless spacecraft used to investigate the Moon. Early probes flew past the Moon or crash-landed on it, but later ones achieved soft landings or went into orbit. Soviet probes included the Luna series. US probes (Ranger space probes, Surveyor, Lunar Orbiter) prepared the way for the *Apollo* project crewed flights.

moonstone translucent, pearly variety of potassium sodium feldspar, found in Sri Lanka and Myanmar, and distinguished by a blue, silvery, or red opalescent tint. It is valued as a gem.

moor in earth science, a stretch of land, usually at a height, which is characterized by a vegetation of heather, coarse grass, and bracken. A moor may be poorly drained and contain boggy hollows.

moraine rocky debris or till carried along and deposited by a *glacier. Material eroded from the side of a glaciated valley and carried along the glacier's edge is called a **lateral moraine**; that worn from the valley floor and carried along the base of the glacier is called a **ground moraine**. Rubble dropped at the snout (front end) of a melting glacier is called a **terminal moraine**.

When two glaciers converge their lateral moraines unite to form a **medial moraine**. Debris that has fallen down crevasses and becomes embedded in the ice is termed an **englacial moraine**; when this is exposed at the surface due to partial melting it becomes an **ablation moraine**.

morphine narcotic alkaloid, $C_{17}H_{19}NO_3$, derived from *opium and prescribed only to alleviate severe pain. Its use produces serious side effects, including nausea, constipation, tolerance, and addiction, but it is highly valued for the relief of the terminally ill.

morphology in biology, the study of the physical structure and form of organisms, in particular their soft tissues.

Mössbauer effect recoil-free emission of gamma rays from atomic nuclei under certain conditions. The effect was discovered in 1958 by German physicist Rudolf Mössbauer (1929–), and used in 1960 to provide the first laboratory test of the general theory of relativity developed by German-born US physicist Einstein (1879–1955).

motherboard *printed circuit board that contains the main components of a microcomputer. The power, memory capacity, and capability of the microcomputer may be enhanced by adding expansion boards to the motherboard, now more commonly called a mainboard.

motility ability to move spontaneously. The term is often restricted to those cells that are capable of independent locomotion, such as spermatozoa. Many single-celled organisms are motile, for example, the amoeba. Research has shown that cells capable of movement, including vertebrate muscle cells, have certain biochemical features in common. Filaments of the proteins actin and myosin are associated with motility, as are the metabolic processes needed for breaking down the energy-rich compound *ATP (adenosine triphosphate).

motion process of moving; an object that is moving is said to be 'in motion'. The speed at which an object is moving is often measured in metres per second (m/s) or kilometres per hour (kph). The motion of objects can be described using *Newton's laws of motion.

motor effect tendency of a wire carrying an electric current in a magnetic field to move. The direction of the movement is given by the left-hand rule (see *Fleming's rules). This effect is used in the *electric motor. It also explains why streams of electrons produced, for instance, in a television tube can be directed by electromagnets.

motor nerve in anatomy, any nerve that transmits impulses from the central nervous system to muscles or organs. Motor nerves cause voluntary and involuntary muscle contractions, and stimulate glands to secrete hormones.

motor neuron disease MND; or **amyotrophic lateral sclerosis**, chronic disease in which there is progressive degeneration of the nerve cells which instigate movement. It leads to weakness, wasting, and loss of muscle function and usually proves fatal within two to three years of onset. Motor neuron disease occurs in both familial and sporadic forms but its causes remain unclear. A gene believed to be implicated in familial cases was discovered in 1993.

mountain natural upward projection of the Earth's surface, higher and steeper than a hill. Mountains are at least 330 m/1,000 ft above the surrounding topography. The existing rock below a high mountain may be subjected to high temperatures and pressures, causing metamorphism. Plutonic activity also can accompany mountain building.

mountain sickness see *altitude sickness.

mouse in computing, an input device used to control a pointer on a computer screen. It is a feature of *graphical user interface (GUI) systems. The mouse is about the size of a pack of playing cards, is connected to the computer by a wire or infrared link, and incorporates one or more buttons that can be pressed. Moving the mouse across a flat surface causes a corresponding movement of the pointer. In this way, the operator can manipulate objects on the screen and make menu selections.

mouth cavity forming the entrance to the digestive tract (gut or alimentary canal). In land vertebrates, air from the nostrils enters the mouth cavity to pass down the trachea. The mouth in mammals is enclosed by the jaws, cheeks, and palate. It contains teeth that may have a variety of functions depending on the way of life of the mammal. They may be used to hold things, to kill other organisms, or to cut food.

moving-coil meter instrument used to detect and measure electrical current. A coil of wire pivoted between the poles of a permanent magnet is turned by the motor effect of an electric current (by which a force acts on a wire carrying a current in a magnetic field). The extent to which the coil turns can then be related to the magnitude of the current.

mp in chemistry, abbreviation for *melting point.

MP3 abbreviation for MPEG-1 Audio Layer 3, audio compression format that uses perceptual audio coding and psychoacoustic compression to strip inaudible and unnecessary information from sound signals. MP3 packs a minute of sound, at near CD quality, into about 1 megabyte. This makes delivering music directly over the Internet a viable prospect. MP3 is Layer 3 of the *MPEG-2 video compression system developed by the Moving Pictures Expert Group (MPEG).

MPC abbreviation for multimedia personal computer, standard defining the minimum specification for developing and running CD-ROM software. It has been rendered largely obsolete by the fact that most of today's PCs include multimedia features as a matter of course. The current MPC specification, MPC III, requires at least 8 MB of RAM, a 75MHz Pentium processor, a VGA monitor, and a quad-speed CD-ROM disk drive.

MPEG pronounced 'empeg'; acronym for Moving Picture Experts Group, in computing, a committee of the International Standards Organization, formed in 1988, that sets standards for digital audio and video compression; hence, any file that has been compressed using those standards. The **MPEG-1** is the standard for the digital coding of video pictures for CD recording; **MPEG-2** is a common standard for broadcast-quality video; and **MPEG-4** for Internet telephony. (There is no MPEG-3 as it was absorbed into MPEG-2.)

MRI abbreviation for *magnetic resonance imaging*.

MRSA abbreviation for methicillin-resistant *Staphylococcus* aureus, an infection that causes serious problems in hospitals.

mucous membrane thin skin lining all animal body cavities and canals that come into contact with the air (for example, eyelids, breathing and digestive passages, genital tract). It contains goblet cells that secrete mucus, a moistening, lubricating, and protective fluid. In the air passages mucus captures dust and bacteria. In the gut it helps food slip along, and protects the epithelial cells from being damaged by digestive enzymes. Mucous membranes line the air passages from the *mouth to the *lungs and to the *gut. The layer of cells next to the space in the tubes is an *epithelium. In the air passages many of these cells have hair-like projections called cilia and the epithelium is then called a ciliated epithelium.

mucus lubricating and protective fluid, secreted by mucous membranes in many different parts of the body. In the gut, mucus smooths the passage of food and keeps potentially damaging digestive enzymes away from the gut lining. In the lungs, it traps airborne particles so that they can be expelled.

multimedia computerized method of presenting information by combining audio and video components using text, sound, and graphics (still, animated, and video sequences). For example, a multimedia database of musical instruments may allow a user not only to search and retrieve text about a particular instrument but also to see pictures of it and hear it play a piece of music. Multimedia applications emphasize interactivity between the computer and the user.

multiple any number that is the product of a given number. For example, the sequences 5, 10, 15, 20, 25... are multiples of 5; and 2, 4, 6, 8, 10... are multiples of 2. These sequences form *multiplication tables. Some multiples are common to several multiplication tables. In the sequence 10, 20, 30... all are multiples of 2, 5, and 10. The lowest of this sequence is 10 and this is, therefore, known as the **lowest common multiple** (lcm) of 2, 5, and 10.

multiple proportions, law of in chemistry, the principle that if two elements combine with each other to form more than one compound, then the ratio of the masses of one of them that combine with a particular mass of the other is a small whole number.

multiple sclerosis MS; or **disseminated sclerosis**, incurable chronic disease of the central nervous system, occurring in young or middle adulthood. Most prevalent in temperate zones, it affects more women than men. It is characterized by degeneration of the myelin sheath that surrounds nerves in the brain and spinal cord.

multiplication one of the four basic operations of arithmetic, usually written in the form $a \times b$ or ab, and involving repeated addition in the sense that a is added to itself b times. Multiplication obeys the associative law (for example, $a \times b \times c = c \times b \times a$), the commutative law (a special case of the associative law where there are only two numbers involved; for example, $ab = ba$), and the distributive law (multiplication over addition; for example, $m(a + b) = ma + mb$), and every number (except 0) has a multiplicative inverse. The number 1 is the identity for multiplication.

multitasking or **multiprogramming**, in computing, a system in which one processor appears to run several different programs (or different parts of the same program) at the same time. All the programs are held in memory together and each is allowed to run for a certain period.

multi-user or **multi-access**, in computing, an operating system that enables several users to access centrally stored data and programs simultaneously over a network. Each user has a terminal, which may be local (connected directly to the computer) or remote (connected to the computer via a modem and a telephone line).

mumps or **infectious parotitis**, virus infection marked by fever, pain, and swelling of one or both parotid salivary glands (situated in front of the ears). It is usually short-lived in children, although meningitis is a possible complication. In adults the symptoms are more serious and it may cause sterility in men.

Münchhausen's syndrome emotional disorder in which a patient feigns or invents symptoms to secure medical treatment. It is the chronic form of factitious disorder, which is more common, and probably underdiagnosed. In some cases the patient will secretly ingest substances to produce real symptoms. It was named after the exaggerated tales of Baron Münchhausen. Some patients invent symptoms for their children, a phenomenon known as Münchhausen's by proxy.

muon *elementary particle similar to the electron except for its mass which is 207 times greater than that of the electron. It has a half-life of 2 millionths of a second, decaying into electrons and *neutrinos. The muon was originally thought to be a *meson but is now classified as a *lepton. See also *tau.

muscle contractile animal tissue that produces locomotion and power and maintains the movement of body substances. Muscle contains very specialized animal cells – long cells – that can contract to between one-half and one-third of their relaxed length. Muscle tissue is sometimes found in large amounts, forming muscles, that are organs. Muscle tissue enables movement. It may move the whole body, or part of it, or some material along a tube within it. Muscles can only do this work by contracting. This explains why muscles are usually found in pairs (antagonistic pairs) where the work done in the contraction of one causes the stretching of the other. (See diagram, page 390.)

Muscle is categorized into three main groups: striped (or striated) muscles are activated by *motor nerves under voluntary control – their ends are usually attached via tendons to bones; involuntary or smooth muscles are controlled by motor nerves of the *autonomic nervous system, and are located in the gut, blood vessels, iris, and various ducts; cardiac muscle occurs only in the *heart, and is also controlled by the autonomic nervous system.

muscular dystrophy any of a group of inherited chronic muscle disorders marked by weakening and wasting of muscle. Muscle fibres degenerate, to be replaced by fatty tissue, although the nerve supply remains unimpaired. Death occurs in early adult life.

mushroom fruiting body of certain fungi (see *fungus), consisting of an upright stem and a spore-producing cap with radiating gills on the undersurface. There are many edible species belonging to the genus *Agaricus*, including the field mushroom (*A. campestris*).

musical instrument digital interface manufacturer's standard for digital music equipment; see *MIDI.

mutagen any substance that increases the rate of gene *mutation. A mutagen may also act as a *carcinogen.

mutation in biology, a change in the *genes produced by a change in the *DNA that makes up the hereditary material of all living organisms. It can be a change in a single gene or a change that affects sections of *chromosomes. In the process of DNA replication, which takes place before any cell divides, the two halves of DNA separate and new halves are made. Because of specific base pairing, the inherited information is copied exactly. Despite this, rarely, a mistake occurs and the sequence of bases is altered. This changes the sequence of amino acids in a protein. This is mutation, the raw material of evolution. The consequences of mutation are varied. Due to the redundancy built into genetic code many mutations have no effect upon DNA functions. Genes describe how to make *proteins. As a result of mutation a protein may not be produced, may be produced but act abnormally, or remain fully functional. Only a few mutations improve the organism's performance and are therefore favoured by *natural selection. Mutation rates are increased by certain chemicals and by ionizing radiation.

Common mutations include the omission or insertion of a base (one of the chemical subunits of DNA); these are known as **point mutations**. Larger-scale mutations include removal of a whole segment of DNA or its inversion within the DNA strand. Not all mutations affect the organism, because there is a certain amount of redundancy in the genetic information. If a mutation is 'translated' from DNA into the protein that makes up the

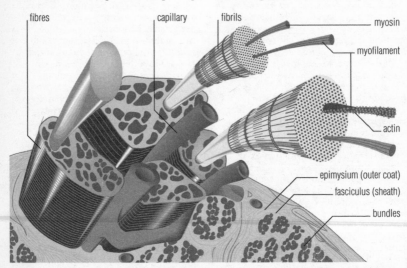

muscle Muscles make up 35–45% of our body weight; there are over 650 skeletal muscles. Muscle cells may be up to 20 cm/0.8 in long. They are arranged in bundles, fibres, fibrils, and myofilaments.

organism's structure, it may be in a non-functional part of the protein and thus have no detectable effect. This is known as a **neutral mutation**, and is of importance in *molecular clock studies because such mutations tend to accumulate gradually as time passes. Some mutations do affect genes that control protein production or functional parts of protein, and most of these are lethal to the organism.

mutism another word for dumbness. Inability to speak may be innate or acquired (due to brain damage or disease); elective mutism is where the person chooses not to speak.

mutual induction in physics, the production of an electromotive force (emf) or voltage in an electric circuit caused by a changing *magnetic flux in a neighbouring circuit. The two circuits are often coils of wire, as in a *transformer, and the size of the induced emf depends on the numbers of turns of wire in each of the coils.

mutualism association between two organisms of different species whereby both profit from the relationship; see *symbiosis.

mutually exclusive in probability theory, describing two *events that cannot happen together. For example, it is not possible to throw a 6 and a *prime number on the same die at the same time. The *probability that both events will happen is zero. The probability that either one or the other of two mutually exclusive events will happen is the sum of their separate probabilities. The addition rule is

$$P(A \text{ or } B) = P(A) + P(B)$$

myalgic encephalomyelitis ME, popular name for *chronic fatigue syndrome.

myasthenia gravis in medicine, an uncommon condition characterized by loss of muscle power, especially in the face and neck. The muscles tire rapidly and fail to respond to repeated nervous stimulation. *Autoimmunity is the cause.

mycelium interwoven mass of threadlike filaments or *hyphae,

forming the main body of most fungi. The reproductive structures, or 'fruiting bodies', grow from the mycelium.

mycology the study of fungi (see *fungus). Its most important sections are medical and agricultural mycology, since fungi cause many diseases in humans, animals, and plants. Research into fungi has led to their use as fermenters, antibiotics, and hallucinogenic drugs; vitamins, flavourings, and hormones are also obtained from them. In studying fungi, mycologists have made major contributions to the store of biological and biochemical knowledge.

mycosis any disease caused by a fungus, such as actinomycosis or *ringworm.

myelin sheath insulating layer that surrounds nerve cells in vertebrate animals. It serves to speed up the passage of nerve impulses.

myelitis in medicine, inflammation of the spinal cord. It occurs in diseases such as *meningitis.

myeloma or **multiple myeloma**, malignant disease of the bone marrow, usually occurring in older people. The symptoms include fatigue, severe bone pain and backache. It causes anaemia, vertebral collapse, clotting problems and damage to the eyes and internal organs. It is treated with *chemotherapy and *radiotherapy.

myopia or **short-sightedness**, defect of the eye in which a person can see clearly only those objects that are close up. It is caused either by the eyeball being too long or by the cornea and lens system of the eye being too powerful, both of which cause the images of distant objects to be formed in front of the retina instead of on it. Nearby objects are sharply perceived. Myopia can be corrected by suitable glasses or contact lenses.

myxomatosis contagious, usually fatal, virus infection of rabbits which causes much suffering. It has been deliberately introduced in the UK and Australia since the 1950s to reduce the rabbit population.

N

n in mathematics, variable used to denote an integer number.

N abbreviation for **north**; **newton, a unit of force; the chemical symbol for **nitrogen**.

nadir point on the celestial sphere vertically below the observer and hence diametrically opposite the **zenith**.

nagware in computing, another name for *guiltware.

name server see *domain name server.

NAND gate type of *logic gate.

nano- prefix used in *SI units of measurement, equivalent to a one-billionth part (10^{-9}). For example, a nanosecond is one-billionth of a second.

nanotechnology experimental technology using individual atoms or molecules as the components of minute machines, measured by the nanometre, or millionth of a millimetre. Nanotechnology research in the 1990s focused on testing molecular structures and refining ways to manipulate atoms using a scanning tunnelling microscope. The ultimate aim is to create very small computers and molecular machines which could perform vital engineering or medical tasks.

naphtha mixtures of hydrocarbons obtained by destructive distillation of petroleum, coal tar, and shale oil. It is a raw material for the petrochemical and plastics industries. The term was originally applied to naturally occurring liquid hydrocarbons.

naphthalene $C_{10}H_8$, solid, white, shiny, aromatic hydrocarbon obtained from coal tar. The smell of moth balls is due to their naphthalene content. It is used in making indigo and certain azo dyes, as a mild disinfectant, and as a pesticide.

Napier's bones An early mechanical calculating device for multiplication and division.

In 1617 Scottish mathematician John Napier (1550–1617) published his description of what was arguably the first mechanical calculator – a set of numbered rods, usually made of bone or ivory and therefore known as Napier's bones. Using them, multiplication became merely a process of reading off the appropriate figures and making simple additions.

narcissism in psychology, an exaggeration of normal self-respect and self-involvement which may amount to mental disorder when it precludes relationships with other people.

narcolepsy rare neurological disorder characterized by an abnormal tendency to fall asleep. Often associated with cataplexy, it is controlled by drugs.

narcotic pain-relieving and sleep-inducing drug. The term is usually applied to heroin, morphine, and other opium derivatives, but may also be used for other drugs which depress brain activity, including anaesthetic agents and *hypnotics.

NASA acronym for **National Aeronautics and Space Administration**, US government agency for space flight and aeronautical research, founded in 1958 by the National Aeronautics and Space Act. Its headquarters are in Washington, DC, and its main installations include the *Kennedy Space Center on Merritt Island in Florida, the Johnson Space Center in Houston, Texas, the Jet Propulsion Laboratory in Pasadena, California, the Goddard Space Flight Center in Beltsville, Maryland, and the Marshall Space Flight Center in Huntsville, Alabama. NASA's early planetary and lunar programmes included the Pioneer probes, from 1958, which gathered data for the later crewed missions, and the *Apollo* project, which took the first astronauts to the Moon in *Apollo 11* on 16–24 July 1969.

National Aeronautics and Space Administration full name of *NASA.

national grid the network of cables and wires, carried overhead on pylons or buried under the ground, that connects consumers of electrical energy to power stations, and interconnects the power stations. It ensures that power can be made available to all customers at any time, allowing demand to be shared by several power stations, and particular power stations to be shut down for maintenance work from time to time.

native metal or **free metal**, any of the metallic elements that occur in nature in the chemically uncombined or elemental form (in addition to any combined form). They include bismuth, cobalt, copper, gold, iridium, iron, lead, mercury, nickel, osmium, palladium, platinum, ruthenium, rhodium, tin, and silver. Some are commonly found in the free state, such as gold; others occur almost exclusively in the combined state, but under unusual conditions do occur as native metals, such as mercury. Examples of native non-metals are carbon and sulphur.

natural frequency frequency at which a mechanical system will vibrate freely. A pendulum, for example, always oscillates at the same frequency when set in motion. More complicated systems, such as bridges, also vibrate with a fixed natural frequency. If a varying force with a frequency equal to the natural frequency is applied to such an object the vibrations can become violent, a phenomenon known as *resonance.

natural gas mixture of flammable gases found in the Earth's crust (often in association with petroleum). It is one of the world's three main fossil fuels (with coal and oil).

natural hazard naturally occurring phenomenon capable of causing destruction, injury, disease, or death. Examples include earthquakes, floods, hurricanes, or famine. Natural hazards occur globally and can play an important role in shaping the landscape. The events only become hazards where people are affected. Because of this, natural hazards are usually measured in terms of the damage they cause to persons or property. Human activities can trigger natural hazards, for example skiers crossing the top of a snowpack may cause an avalanche.

natural logarithm in mathematics, the *exponent of a number expressed to base e, where e represents the *irrational number 2.71828... . Natural *logarithms are also called Napierian logarithms, after their inventor, the Scottish mathematician John Napier (1550–1617).

natural number one of the set of numbers used for counting. Natural numbers comprise all the *positive integers, excluding zero.

natural radioactivity radioactivity generated by those radioactive elements that exist in the Earth's crust. All the elements from polonium (atomic number 84) to uranium (atomic number 92) are radioactive. *Radioisotopes of some lighter elements are also found in nature (for example potassium-40).

natural selection process by which gene frequencies in a population change through certain individuals producing more descendants than others because they are better able to survive and reproduce in their environment. The accumulated effect of natural selection is to produce *adaptations such as the insulating coat of a polar bear or the spadelike forelimbs of a mole. The process is slow, relying firstly on random variation in the genes of an organism being produced by *mutation and secondly on the genetic *recombination of sexual reproduction. It was recognized by English naturalist Charles Darwin and Welsh naturalist Alfred Russel Wallace as the main process driving *evolution.

nature–nurture controversy or **environment–heredity controversy**, long-standing dispute among philosophers and psychologists over the relative importance of environment, that is, upbringing, experience, and learning ('nurture'), and heredity, that is, genetic inheritance ('nature'), in determining the make-up of an organism, as related to human personality and intelligence.

One area of contention is the reason for differences between individuals; for example, in performing intelligence tests. The environmentalist position assumes that individuals do not differ significantly in their inherited mental abilities and that subsequent differences are due to learning, or to differences in early experiences. Opponents insist that certain differences in the capacities of individuals (and hence their behaviour) can be attributed to inherited differences in their genetic make-up.

nature reserve area set aside to protect a habitat and the wildlife that lives within it, with only restricted admission for the public. A nature reserve often provides a sanctuary for rare species and rare habitats, such as marshland. The world's largest is Etosha Reserve, Namibia; area 99,520 sq km/38,415 sq mi.

nautical mile unit of distance used in navigation, an internationally agreed standard (since 1959) equalling the average length of one minute of arc on a great circle of the Earth, or 1,852 m/6,076 ft. The term formerly applied to various units of distance used in navigation.

navigate in computing, to find your way around hyperspace or a *hypertext document, especially when using a *browser.

navigation, biological ability of animals or insects to navigate. Although many animals navigate by following established routes or known landmarks, many animals can navigate without such aids; for example, birds can fly several thousand miles back to their nest site, over unknown terrain.

Such feats may be based on compass information derived from the position of the Sun, Moon, or stars, or on the characteristic patterns of Earth's magnetic field.

navigation map in computing, specialized tool to help users find their way around a Web site. Colourful graphics overlay a hidden grid, like that on a conventional map, containing *hypertext links. Users navigate by placing their cursor on an image and clicking the mouse, sending a 'map reference' back to the Web site which activates a link. Navigation maps give the designer more control over how

the page will appear on the screen, and are more attractive than conventional links.

NDA abbreviation for *nondisclosure agreement.

Neanderthals see *human species origins of.

near point the closest position to the eye to which an object may be brought and still be seen clearly. For a normal human eye the near point is about 25 cm; however, it gradually moves further away with age, particularly after the age of 40.

nebula cloud of gas and dust in space. Nebulae are the birthplaces of stars, but some nebulae are produced by gas thrown off from dying stars (see *planetary nebula; *supernova). Nebulae are classified depending on whether they emit, reflect, or absorb light.

nebular hypothesis hypothesis that the Solar System evolved from a *nebula. 18th-century Swedish mystic and scientist Emanuel Swedenborg and 18th-century German philosopher Immanuel Kant both put forward explanations of this kind, but 18th–19th-century French astronomer Pierre Laplace was the first to develop a nebular hypothesis on strictly scientific lines.

negative number number less than *zero. On a number line, any number to the left of zero is negative. Negative numbers are always written with a minus sign in front, for example, –6 is negative, 6 is positive. Scales often display negative numbers. On a temperature scale, 5°C below freezing is –5°C.

nematode any of a group of unsegmented worms that are pointed at both ends, with a tough, smooth outer skin. They include many free-living species found in soil and water, including the sea, but a large number are parasites, such as the roundworms and pinworms that live in humans, or the eelworms that attack plant roots. They differ from flatworms in that they have two openings to the gut (a mouth and an anus). The group includes *Caenorhabditis elegans* which is a model genetic organism and the first multicellular animal to have its complete genome sequenced. (Phylum Nematoda.)

neodymium chemical symbol Nd, yellowish metallic element of the *lanthanide series, atomic number 60, relative atomic mass 144.24. Its rose-coloured salts are used in colouring glass, and neodymium is used in lasers. It was named in 1885 by Austrian chemist Carl von Welsbach, who fractionated it away from didymium (originally thought to be an element but actually a mixture of rare-earth metals consisting largely of neodymium, praesodymium, and cerium).

neon chemical symbol Ne, (Greek *neos* 'new') colourless, odourless, non-metallic, gaseous element, atomic number 10, relative atomic mass 20.183. It is grouped with the *noble gases (rare gases) in Group 0 of the *periodic table. Neon is nonreactive, and forms no compounds. It occurs in small quantities in the Earth's atmosphere. Tubes containing neon are used in electric advertising signs, giving off a fiery red glow; it is also used in lasers. Neon was discovered by Scottish chemist William Ramsay (1852–1916) and English chemist Morris Travers (1872–1961).

neonate medical term for a newborn baby.

neoplasm (Greek 'new growth') any lump or tumour, which may be benign or malignant (cancerous).

neoteny in biology, the retention of some juvenile characteristics in an animal that seems otherwise mature. An example is provided by the axolotl, a salamander that can reproduce sexually although still in its larval form.

nephritis or **Bright's disease**, general term used to describe inflammation of the kidney. The degree of illness varies, and it may be acute (often following a recent streptococcal infection), or chronic, requiring a range of treatments from antibiotics to dialysis or transplant.

nephron microscopic unit in vertebrate kidneys that forms **urine**. A human kidney is composed of over a million nephrons. Each nephron consists of a knot of blood capillaries called a glomerulus, contained in the *Bowman's capsule, and a long narrow tubule

enmeshed with yet more capillaries. Waste materials and water pass from the bloodstream into the tubule, and essential minerals and some water are reabsorbed from the tubule back into the bloodstream. The remaining filtrate (urine) is passed out from the body. (See diagram, page 396.)

Neptune eighth planet in average distance from the Sun. It is a giant gas (hydrogen, helium, methane) planet, with a mass 17.2 times that of Earth. It has the fastest winds in the Solar System.

mean distance from the Sun: 4.4 billion km/2.794 billion mi

equatorial diameter: 48,600 km/30,200 mi

rotation period: 16 hours 7 minutes

year: 164.8 Earth years

atmosphere: methane in its atmosphere absorbs red light and gives the planet a blue colouring. Consists primarily of hydrogen (85%) with helium (13%) and methane (1–2%)

surface: hydrogen, helium, and methane. Its interior is believed to have a central rocky core covered by a layer of ice

satellites: of Neptune's eight moons, two (Triton and Nereid) are visible from Earth. Six more were discovered by the Voyager 2 probe in 1989, of which Proteus (diameter 415 km/260 mi) is larger than Nereid (300 km/200 mi)

rings: there are four faint rings: Galle, Le Verrier, Arago, and Adams (in order from Neptune). Galle is the widest at 1,700 km/1,060 mi. Le Verrier and Arago are divided by a wide diffuse particle band called the plateau.

neptunium chemical symbol Np, silvery, radioactive metallic element of the *actinide series, atomic number 93, relative atomic mass 237. It occurs in nature in minute amounts in *pitchblende and other uranium ores, where it is produced from the decay of neutron-bombarded uranium in these ores. The longest-lived isotope, Np-237, has a half-life of 2.2 million years. The element can be produced by bombardment of U-238 with neutrons and is chemically highly reactive.

It was first synthesized in 1940 by US physicists Edwin McMillan and

Philip Abelson (1913–), who named it after the planet Neptune (since it comes after uranium as the planet Neptune comes after Uranus).

Neptunium was the first *transuranic element to be synthesized.

Nernst effect in physics, phenomenon in which an electric potential develops across a conductor located at right angles to a magnetic field and with heat flowing through it. All three factors – potential, magnetic field, and heat flow – are at right angles to each other.

It is named after German physical chemist Hermann Nernst.

nerve bundle of nerve cells enclosed in a sheath of connective tissue and transmitting nerve impulses to and from the brain and spinal cord. A single nerve may contain both *motor and sensory nerve cells, but they function independently.

nerve cell or **neuron**, elongated cell that transmits information rapidly between different parts of the body, the basic functional unit of the *nervous system. Each nerve cell has a cell body, containing the nucleus, from which trail processes called dendrites, responsible for receiving incoming signals. The unit of information is the nerve impulse, a travelling wave of chemical and electrical changes involving the membrane of the nerve cell. The cell's longest process, the *axon, carries impulses away from the cell body. The *brain contains many nerve cells.

A reflex arc is a simple example of how nerve cells help to control and coordinate processes in the body. It involves the use of three types of nerve

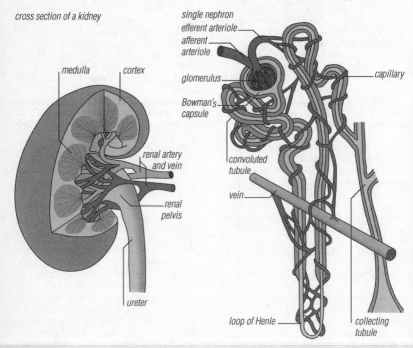

cross section of a kidney

medulla — cortex

renal artery and vein

renal pelvis

ureter

single nephron

efferent arteriole

afferent arteriole

glomerulus — capillary

Bowman's capsule

convoluted tubule

vein

loop of Henle — collecting tubule

nephron The kidney (left) contains more than a million filtering units, or nephrons (right), consisting of the glomerulus, Bowman's capsule, and the loop of Henle. Blood flows through the glomerulus – a tight knot of fine blood vessels from which water and metabolic wastes filter into the tubule. This filtrate flows through the convoluted tubule and loop of Henle where most of the water and useful molecules are reabsorbed into the blood capillaries. The waste materials are passed to the collecting tubule as urine.

cell. These are the sensory nerve cell, the intermediate (relay) nerve cell and the motor nerve cell. Where each nerve cell connects with the next there is a tiny gap called a *synapse. Impulses can cross this gap.

The impulse involves the passage of sodium and potassium ions across the nerve-cell membrane. Sequential changes in the permeability of the membrane to positive sodium (Na^+) ions and potassium (K^+) ions produce electrical signals called action potentials. Impulses are received by the cell body and passed, as a pulse of electric charge, along the axon. The axon terminates at the synapse, a

specialized area closely linked to the next cell (which may be another nerve cell or a specialized effector cell such as a muscle). On reaching the synapse, the impulse releases a chemical *neurotransmitter, which diffuses across to the neighbouring cell and there stimulates another impulse or the action of the effector cell.

Nerve impulses travel quickly – in humans, they may reach speeds of 160 m/525 ft per second, although many are much slower.

nervous system   system of interconnected *nerve cells of most invertebrates and all vertebrates. It is composed of the *central and *autonomic nervous systems. It may be as simple as the nerve net of coelenterates (for example, jellyfishes) or as complex as the mammalian nervous system, with a central nervous system comprising *brain and spinal cord and a peripheral nervous system connecting up with sensory organs, muscles, and glands.

In a nervous system, specialized cells called nerve cells or neurones carry messages as nerve impulses quickly from one part of the body to another. These impulses may be carrying information about the outside world (stimuli) which allows the body to respond quickly to them. However, much of the information being carried around is to do with organizing processes inside the body (nervous coordination).

In mammals, some examples of processes regulated by the nervous system are changes in *heart rate, changes in ventilation rate (*breathing rate), the movement of food along the *alimentary canal, and changes in the size of the iris which alter the amount of light entering the *eye.

The nervous system includes the brain, the area in which collected information is used to make decisions and from where responses are initiated. Information is collected from all over the body, including from specialized sense organs such as the eye. The rest of the nervous system includes the spinal cord and nerve cells carrying information to and from the brain. For example, responses to

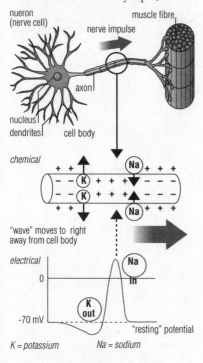

nueron (nerve cell)

muscle fibre

nerve impulse

axon

nucleus
dendrites   cell body

chemical

"wave" moves to right away from cell body

electrical

0

-70 mV

"resting" potential

K = potassium Na = sodium

nerve cell The anatomy and action of a nerve cell. The nerve cell or neuron consists of a cell body with the nucleus and projections called dendrites which pick up messages. An extension of the cell, the axon, connects one cell to the dendrites of the next. When a nerve cell is stimulated, waves of sodium (Na^+) and potassium (K^+) ions carry an electrical impulse down the axon.

stimuli involve nerve impulses which are carried along nerve cells from the brain, through part of the spinal cord, and then to *muscles or other cells of the body.

net in mathematics, a plan which can be used to make a model of a solid.

net abbreviation for network, in the Internet's *domain name system (DNS), one of the top-level domains, along with org, edu, com, gov, and mil. However, the use of the name has not been controlled and a name.net address does not necessarily identify the user as someone who works for a company involved in Internet networking.

network in computing, a method of connecting computers so that they can share data and *peripheral devices, such as printers. The main types are classified by the pattern of the connections – star or ring network, for example – or by the degree of geographical spread allowed; for example, *local area networks (LANs) for communication within a room or building, and *wide area networks (WANs) for more remote systems. The *Internet is the linking of computer networks of institutions throughout the world: there are around 500 million users.

network computer NC, simple computer consisting essentially of a microprocessor, a *RAM chip, and a monitor. NCs are designed to function as part of a network, connected to a central *server via the Internet or an *intranet, downloading software (especially *object oriented programs) as required.

network interface card NIC, in computing, item of computer hardware that allows computers to be connected to a computer network.

network operating system NOS, in computing, software designed to enable a *LAN or other network to operate. The main task of every NOS is to tell both the central file server and the *workstations connected to it how to communicate with each other. Network operating systems may also include security and back-up features, remote access facilities, and a centralized database.

Network Time Protocol NTP, *Internet standards by which computers can tell each other the time. It is important for computer systems to have the correct date and time, and especially important for online transaction processing systems such as banks. NTP *servers base their own clocks on the times transmitted from super-accurate atomic clocks. Other computers, which represent clients, periodically check their own time with that of the NTP server, to remain in accordance.

neuralgia sharp or burning pain originating in a nerve and spreading over its area of distribution. Trigeminal neuralgia, a common form, is a severe pain on one side of the face.

neural network artificial network of processors that attempts to mimic the structure of nerve cells (neurons) in the human brain. Neural networks may be electronic, optical, or simulated by computer software.

neural tube embryological structure from which the brain and spinal cord are derived. It forms when two edges of a groove in a plate of primitive nerve tissue fuse. Incomplete fusion results in *neural tube defects such as spinal bifida.

neural tube defect NTD, any of a group of abnormalities caused by incomplete closure of the *neural tube. These include: spina bifida, where part of the spinal cord and its membranes are exposed through a gap in the backbone; meningocele, where the meninges protrude through a gap in the spine; and *anencephaly that is incompatible with life. These conditions can be detected by prenatal testing.

neurasthenia obsolete term for nervous exhaustion, covering mild *depression and various symptoms of *neurosis. Formerly thought to be a bodily malfunction, it is now generally considered to be mental in origin. Dating from the mid-19th century, the term became widely used to describe the symptoms of soldiers returning from the front in World War I.

neuroblastoma malignant tumour arising from embryonic nerve cells. It develops most often in the adrenal gland; secondary growths may be widespread.

neurohormone chemical secreted by nerve cells and carried by the blood to target cells. The function of the neurohormone is to act as a messenger; for example, the neurohormone ADH (antidiuretic hormone) is secreted in the pituitary gland and carried to the kidney, where it promotes water reabsorption in the kidney tubules.

neuroleptic alternative name for *antipsychotic drug.

neuron another name for a *nerve cell.

neuropsychology branch of neurology that overlaps with psychiatry and psychology and is mainly concerned with the *cerebral cortex, specifically those disorders of perception, memory, language, and behaviour that result from brain injury or disease.

neurosis in psychology, a general term referring to emotional disorders, such as anxiety, depression, and phobias. The main disturbance tends to be one of mood; contact with reality is relatively unaffected, in contrast to *psychosis.

neuroticism personality dimension described by German-born British psychologist Hans Eysenck (1916–1997). People with high neuroticism are worriers, emotional, and moody.

neurotoxin any substance that destroys nerve tissue.

neurotransmitter chemical that diffuses across a *synapse, and thus transmits impulses between *nerve cells, or between nerve cells and effector organs (for example, muscles). Common neurotransmitters are noradrenaline (which also acts as a hormone) and acetylcholine, the latter being most frequent at junctions between nerve and muscle. Nearly 50 different neurotransmitters have been identified.

neutralization in chemistry, a process occurring when the excess acid (or excess base) in a substance is reacted with added base (or added acid) so that the resulting substance is neither acidic nor basic.

In theory neutralization involves adding acid or base as required to achieve *pH 7. When the colour of an *indicator is used to test for neutralization, the final pH may differ from pH 7 depending upon the indicator used. It will also differ from pH 7 in reactions between strong acids and weak bases and weak acids and strong bases, because the salt formed will have acidic or basic properties respectively.

neutral oxide oxide that has neither acidic nor basic properties (see *oxide). Neutral oxides are only formed by *non-metals. Examples are carbon monoxide, water, and nitrogen(I) oxide.

neutral solution solution of pH 7, in which the concentrations of $H^+_{(aq)}$ and $OH^-_{(aq)}$ ions are equal. It is therefore a solution which is neither acidic nor alkaline. An alkali may be *neutralized by an acid.

neutrino in physics, any of three uncharged *elementary particles (and their antiparticles) of the *lepton class, having a mass that is very small (possibly zero). The most familiar type, the antiparticle of the electron neutrino, is emitted in the beta decay of a nucleus. The other two are the muon and tau neutrinos.

neutron one of the three main subatomic particles, the others being the *proton and the *electron. Neutrons have about the same mass as protons but no electric charge, and occur in the nuclei of all atoms except hydrogen. They contribute to the mass of atoms but do not affect their chemistry. For instance, the *isotopes of a single element differ only in the number of neutrons in their nuclei but have identical chemical properties. Neutrons and protons have masses approximately 2,000 times those of electrons.

The neutron is a composite particle, being made up of three *quarks, and therefore belongs to the baryon group of the *hadrons. Outside a nucleus, a free neutron is unstable, decaying with a half-life of 11.6 minutes into a proton, an electron, and an antineutrino. The process by which a neutron changes into a proton is called *beta decay.

The neutron was discovered by the British chemist James Chadwick in 1932.

neutron bomb or **enhanced radiation weapon**, small hydrogen bomb for battlefield use that kills by radiation, with minimal damage to buildings and other structures.

neutron number   symbol N, number of neutrons possessed by an atomic nucleus. *Isotopes are atoms of the same element possessing different neutron numbers.

neutron star   very small, 'superdense' star composed mostly of *neutrons. They are thought to form when massive stars explode as *supernovae, during which the protons and electrons of the star's atoms merge, owing to intense gravitational collapse, to make neutrons. A neutron star has a mass two to three times that of the Sun, compressed into a globe only 20 km/12 mi in diameter.

If its mass is any greater, its gravity will be so strong that it will shrink even further to become a *black hole. Being so small, neutron stars can spin very quickly. The rapidly flashing radio stars called pulsars are believed to be neutron stars. The flashing is caused by a rotating beam of radio energy similar in behaviour to a lighthouse beam.

newton   symbol N, SI unit of *force. A newton is defined as the amount of force needed to move an object of 1 kg so that it accelerates at 1 metre per second per second. It is also used as a unit of weight. The weight of a medium size (100 g/3 oz) apple is one newton.

Newtonian physics   physics based on the concepts of the English physicist and mathematician Isaac Newton (1642–1727), before the formulation of quantum theory or relativity theory.

Newton's laws of motion   in physics, formulated by English physicist and mathematician Isaac Newton (1642–1727). Three laws that form the basis of Newtonian mechanics, describing the motion of objects. (1) Unless acted upon by an unbalanced force, a body at rest stays at rest, and a moving body continues moving at the same speed in the same straight line. (2) An unbalanced force applied to a body gives it an acceleration proportional to the force (and in the direction of the force) and inversely proportional to the mass of the body. (3) When a body A exerts a force on a body B, B exerts an equal and opposite force on A; that is, to every action there is an equal and opposite reaction.

the first law As an example, if a car is travelling at a certain speed in a certain direction, it will continue to travel at that speed in the same direction unless it is acted upon by an unbalanced force such as *friction in the brake mechanism, which will slow down the car. A person in the car will continue to move forward (in accordance with the first law) unless acted upon by a force; for example, the restraining force of a seat belt.

the second law This can be demonstrated using a ticker timer and a trolley with mass that can be varied. If the mass is kept constant and the amount of force applied in pulling the trolley varies, then the dots on the ticker timer tape become further apart; the trolley is changing velocity and therefore accelerating. The acceleration (a) is proportional to the force (F) applied. This can be expressed as acceleration/force = a constant, or a/F = a constant.

If the force applied in pulling the trolley is kept the same (constant), and the mass placed on the trolley is varied (from low to high), then the dots on the ticker timer tape become closer together as the mass gets larger; the acceleration decreases as mass increases. This can be expressed as acceleration being inversely proportional to mass (m), or $a \propto 1/m$.

The two equations can be combined to give an overall equation, expressing Newton's second law of motion. This is: force = mass × acceleration, or $F = ma$.

the third law As an example, a book placed on a table will remain at rest. The force of gravity acting on the book pulls the book towards the ground. The table opposes the force (weight) exerted on the book by gravity. The table exerts the same amount of force in an upward direction. Hence the book remains at rest on the table.

Newton's rings   in optics, observed by English physicist and mathematician Isaac Newton (1642–1727). An *interference phenomenon seen (using white light) as concentric rings of spectral colours where light passes through a thin film of transparent medium, such as the wedge of air between a large-radius convex lens and a flat glass plate. With monochromatic light (light of a single wavelength), the rings take the form of alternate light and dark bands. They are caused by interference (interaction) between light rays reflected from the plate and those reflected from the curved surface of the lens.

niche in ecology, the 'place' occupied by a species in its habitat, including all chemical, physical, and biological components, such as what it eats, the time of day at which the species feeds, temperature, moisture, the parts of the habitat that it uses (for example, trees or open grassland), the way it reproduces, and how it behaves.

It is believed that no two species can occupy exactly the same niche, because they would be in direct competition for the same resources at every stage of their life cycle.

nickel chemical symbol Ni, hard, malleable and ductile, silver-white, metallic element, atomic number 28, relative atomic mass 58.71. It occurs in igneous rocks and as a free metal (*native metal), occasionally occurring in fragments of iron–nickel meteorites. It is a component of the Earth's core, which is held to consist principally of iron with some nickel. It has a high melting point, low electrical and thermal conductivity, and can be magnetized. It does not tarnish and therefore is much used for alloys, electroplating, and for coinage.

It was discovered in 1751 by Swedish mineralogist Axel Cronstedt (1722–1765) and the name given is an abbreviated form of *kopparnickel*, Swedish 'false copper', since the ore in which it is found resembles copper but yields none.

nickel ore any mineral ore from which nickel is obtained. The main minerals are arsenides such as chloanthite ($NiAs_2$), and the sulphides millerite (NiS) and pentlandite (($Ni,Fe)_9S_8$), the commonest ore. The chief nickel-producing countries are Canada, Russia, Kazakhstan, Cuba, and Australia.

nicotinamide adenine dinucleotide NAD, naturally occurring compound that acts as a *coenzyme. It consists of two *nucleotides, adenosine mononucleotide and nicotinamide mononucleotide, joined by a phosphate bridge. A similar compound, NAD phosphate, **NADP**, is produced by phosphorylating the ribose part of the adenosine mononucleotide. NAD takes part in oxidation–reduction reactions, being converted to $NADH_2$ by accepting two hydrogen atoms. In this form it is able to enter the *electron transport chain where it is oxidized back to NAD, producing large quantities of *ATP in the process. NADP is an important oxidation–reduction factor in *fatty acid biosynthesis.

nicotine $C_{10}H_{14}N_2$, *alkaloid (nitrogenous compound) obtained from the dried leaves of the tobacco plant *Nicotiana tabacum*. A colourless oil, soluble in water, it turns brown on exposure to the air. Nicotine is found in tobacco smoke. It can be described as a recreational drug. It stimulates the human body and produces feelings that cause people to carry on smoking. However, nicotine is usually addictive. Regular smokers find that it is very difficult or impossible to give up smoking even though they may try to. This is a problem, in that other chemicals in smoke cause a wide range of diseases and increase the risk of dying early. Nicotine has also been used as an insecticide. Nicotine in its pure form is one of the most powerful poisons known. It is named after a 16th-century French diplomat, Jacques Nicot, who introduced tobacco to France.

nicotinic acid or **niacin**, water-soluble *vitamin ($C_5H_5N.COOH$) of the B complex, found in meat, fish, and cereals; it can also be formed in small amounts in the body from the essential *amino acid tryptophan. Absence of nicotinic acid from the diet leads to the disease pellagra.

niobium chemical symbol Nb, soft, grey-white, somewhat ductile and malleable, metallic element, atomic number 41, relative atomic mass 92.906. It occurs in nature with tantalum, which it resembles in chemical properties. It is used in making stainless steel and other alloys for jet engines and rockets and for making superconductor magnets.

Niobium was discovered in 1801 by the English chemist Charles Hatchett (1765–1847), who named it columbium (symbol Cb), a name that is sometimes still used in metallurgy. In 1844 it was renamed after Niobe by the German chemist Heinrich Rose (1795–1864) because of its similarity to tantalum (Niobe is the daughter of Tantalus in Greek mythology).

nitrate salt or ester of nitric acid, containing the NO_3^- ion. Nitrates are used in explosives, in the chemical and pharmaceutical industries, in curing meat (as *nitre), and as fertilizers. They are the most water-soluble salts known and play a major part in the nitrogen cycle. Nitrates in the soil, whether naturally occurring or from inorganic or organic fertilizers, can be used by plants to make proteins and nucleic acids. However, run-off from fields can result in *nitrate pollution.

nitrate pollution contamination of water by nitrates. Increased use of artificial fertilizers and land cultivation means that higher levels of nitrates are being washed from the soil into rivers, lakes, and aquifers. There they cause an excessive enrichment of the water (eutrophication), leading to a rapid growth of algae, which in turn darkens the water and reduces its oxygen content. The water is expensive to purify and many plants and animals die. High levels are now found in drinking water in arable areas. These may be harmful to newborn babies, and it is possible that they contribute to stomach cancer, although the evidence for this is unproven.

nitr or **saltpetre**, potassium nitrate, KNO_3, a mineral found on and just under the ground in desert regions; used in explosives. Nitre occurs in Bihar, India, Iran, and Cape Province, South Africa. The salt was formerly used for the manufacture of gunpowder, but the supply of nitre for explosives is today largely met by making the salt from nitratine (also called Chile saltpetre, $NaNO_3$).

Saltpetre is a preservative and is widely used for curing meats.

nitric acid or **aqua fortis**; HNO_3, fuming acid obtained by the oxidation of ammonia or the action of sulphuric acid on potassium nitrate. It is a highly corrosive acid, dissolving most metals, and a strong oxidizing agent. It is used in the nitration and esterification of organic substances, and in the making of sulphuric acid, nitrates, explosives, plastics, and dyes.

nitric oxide or **nitrogen monoxide (NO)**, colourless gas released when metallic copper reacts with nitric acid and when nitrogen and oxygen combine at high temperatures. It is oxidized to nitrogen dioxide on contact with air. Nitric oxide has a wide range of functions in the body. It is involved in the transmission of nerve impulses and the protection of nerve cells against stress. It is released by macrophages in the immune system in response to viral and bacterial infection or to the proliferation of cancer cells. It is also important in the control of blood pressure.

nitrification process that takes place in soil when bacteria oxidize ammonia, turning it into nitrates. Nitrates can be absorbed by the roots of plants, so this is a vital stage in the *nitrogen cycle.

nitrite salt or ester of nitrous acid, containing the nitrite ion (NO_2^-). Nitrites are used as preservatives (for example, to prevent the growth of botulism spores) and as colouring agents in cured meats such as bacon and sausages.

nitrocellulose alternative name for *cellulose nitrate.

nitrogen chemical symbol N, (Greek *nitron* 'native soda', sodium or potassium nitrate) colourless, odourless, tasteless, gaseous, non-metallic element, atomic number 7, relative atomic mass 14.0067. It forms almost 80% of the Earth's atmosphere by volume and is a constituent of all plant and animal tissues (in proteins and nucleic acids). Nitrogen is obtained for industrial use by the liquefaction and *fractional distillation of air. Its compounds are used in the manufacture of foods, drugs, fertilizers, dyes, and explosives.

Nitrogen has been recognized as a plant nutrient, found in manures and other organic matter, from early times, long before the complex cycle of *nitrogen fixation was understood. It was isolated in 1772 by the English chemist Daniel Rutherford (1749–1819) and named in 1790 by the French chemist Jean Chaptal (1756–1832).

nitrogen cycle   process of nitrogen passing through the ecosystem. Nitrogen, in the form of inorganic compounds (such as nitrates) in the soil, is absorbed by plants and turned into organic compounds (such as proteins) in plant tissue. A proportion of this nitrogen is eaten by *herbivores, with some of this in turn being passed on to the carnivores, which feed on the herbivores. The nitrogen is ultimately returned to the soil as excrement and when organisms die, and is converted back to inorganic forms by *decomposers.

Although about 78% of the atmosphere is nitrogen, this cannot be used directly by most organisms. However, certain bacteria and cyanobacteria (*blue-green algae) are capable of nitrogen fixation. Some nitrogen-fixing bacteria live mutually with leguminous plants (peas and beans) or other plants (for example, alder), where they form characteristic nodules on the roots. The presence of such plants increases the nitrate content, and hence the fertility, of the soil.

nitrogen fixation   process by which nitrogen in the atmosphere is converted into nitrogenous compounds by the action of micro-organisms, such as cyanobacteria (see *blue-green algae) and bacteria, in conjunction with certain legumes (see *root nodule). Several chemical processes duplicate nitrogen fixation to produce fertilizers; see *nitrogen cycle.

nitrogen oxide   any chemical compound that contains only nitrogen and oxygen. All nitrogen oxides are gases. Nitrogen monoxide and nitrogen dioxide contribute to air pollution. See also *nitrous oxide.

nitrogen monoxide   NO, or nitric oxide, is a colourless gas released when metallic copper reacts with concentrated *nitric acid. It is also

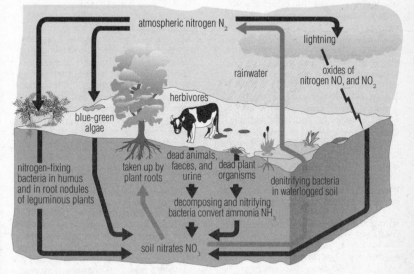

atmospheric nitrogen N_2

lightning

rainwater

oxides of nitrogen NO, and NO_2

herbivores

blue-green algae

nitrogen-fixing bacteria in humus and in root nodules of leguminous plants

taken up by plant roots

dead animals, faeces, and urine

dead plant organisms

denitrifying bacteria in waterlogged soil

decomposing and nitrifying bacteria convert ammonia NH_3

soil nitrates NO_3

nitrogen cycle The nitrogen cycle is one of a number of cycles during which the chemicals necessary for life are recycled. The carbon, sulphur, and phosphorus cycles are others. Since there is only a limited amount of these chemicals in the Earth and its atmosphere, the chemicals must be continuously recycled if life is to go on.

produced when nitrogen and oxygen combine at high temperature. On contact with air it is oxidized to nitrogen dioxide. Nitrogen monoxide was discovered during the 1980s to act as a chemical messenger in small quantities within the human body, despite being toxic at higher concentrations and its rapid reaction with oxygen. The medical condition of septic shock is linked to overproduction by the body of nitrogen monoxide. Nitrogen monoxide has an unpaired electron, which can be removed to produce the nitrosyl ion, NO^+.

nitrogen dioxide nitrogen(IV) oxide, NO_2, is a brown, acidic, pungent gas that is harmful if inhaled and contributes to the formation of *acid rain, as it dissolves in water to form nitric acid. It is the most common of the nitrogen oxides and is obtained by heating most nitrate salts (for example *lead(II) nitrate, $Pb(NO_3)_2$). If liquefied, it gives a colourless solution (N_2O_4). It has been used in rocket fuels.

nitroglycerine or **glycerol trinitrate**; $C_3H_5(ONO_2)_3$, flammable, explosive oil produced by the action of nitric and sulphuric acids on glycerol. Although poisonous, it is used in cardiac medicine. It explodes with great violence if heated in a confined space and is used in the preparation of dynamite, cordite, and other high explosives. It was synthesized by the Italian Ascanio Soberro in 1846, and is unusual among explosives in that it is a liquid. Nitroglycerine is an effective explosive because it has low *activation energy, and produces little smoke when burned. However, it was initially so reactive it was virtually unusable. The innovation made by Swedish chemist and engineer Alfred Nobel (1833–1896) was to purify nitroglycerine (using water, with which it is immiscible, to dissolve the impurities), and thereby make it more stable, and absorb it into wood flour to make dynamite.

nitrous acid HNO_2, weak acid that, in solution with water, decomposes quickly to form nitric acid and nitrogen dioxide.

nitrous oxide or **dinitrogen oxide**; N_2O, colourless, nonflammable gas that, used in conjunction with oxygen, reduces sensitivity to pain. In higher doses it is an anaesthetic. Well tolerated, it is often combined with other anaesthetic gases to enable them to be used in lower doses. It may be self-administered; for example, in childbirth. It is a greenhouse gas; about 10% of nitrous oxide released into the atmosphere comes from the manufacture of nylon. It used to be known as 'laughing gas'.

nobelium chemical symbol No, synthesized, radioactive, metallic element of the *actinide series, atomic number 102, relative atomic mass 259. It is synthesized by bombarding curium with carbon nuclei. It was named in 1957 after the Nobel Institute in Stockholm, Sweden, where it was claimed to have been first synthesized. Later evaluations determined that this was in fact not so, as the successful 1958 synthesis at the University of California at Berkeley produced a different set of data. The name was not, however, challenged. In 1992 the International Unions for Pure and Applied Chemistry and Physics (IUPAC and IUPAP) gave credit to Russian scientists in Dubna for the discovery of nobelium.

noble gas or **rare gas** or **inert gas**, any of a group of six elements (*helium, *neon, *argon, *krypton, *xenon, and *radon), originally named 'inert' because they were thought not to enter into any chemical reactions. This is now known to be incorrect: in 1962, xenon was made to combine with fluorine, and since then, compounds of argon, krypton, and radon with fluorine and/or oxygen have been described.

The extreme unreactivity of the noble gases is due to the stability of their *electronic structure. All the electron shells (*energy levels) of inert gas atoms are full and, except for helium, they all have eight electrons in their outermost (*valency) shell. The apparent stability of this electronic arrangement led to the formulation of the *octet rule to explain the different types of chemical bond found in simple compounds. The noble gases are in Group 0 of the *periodic table of the elements.

noble gas structure configuration of electrons in *noble gases (rare gases): helium, neon, argon, krypton, xenon, and radon. This is characterized by full electron shells around the nucleus of an atom, which render the element stable. Any ion, produced by the gain or loss of electrons, that achieves an electronic configuration similar to one of the noble gases is said to have a noble gas structure.

NO CARRIER in computing, error message returned by a modem when the telephone line drops unexpectedly because of line noise or other interruptions.

noctilucent cloud clouds of ice forming in the upper atmosphere at around 83 km/52 mi. They are visible on summer nights, particularly when sunspot activity is low.

node in physics, a position in a *standing wave pattern at which there is no vibration. Points at which there is maximum vibration are called **antinodes**. Stretched strings, for example, can show nodes when they vibrate. Guitarists can produce special effects (harmonics) by touching a sounding string lightly to produce a node.

node in computing, any device connected to a network, such as a *router, a *bridge, a hub, and a *server.

nodule in geology, a lump of mineral or other matter found within rocks or formed on the seabed surface; mining technology is being developed to exploit them.

noise unwanted sound. Permanent, incurable loss of hearing can be caused by prolonged exposure to high noise levels (above 85 decibels). Over 55 decibels on a daily outdoor basis is regarded as an unacceptable level. In scientific and engineering terms, a noise is any random, unpredictable signal.

nondisclosure agreement NDA, one-sided or mutually binding agreement where the parties agree to protect the confidentiality of any discussions that take place or any information that changes hands. NDAs are frequently requested by entrepreneurs attempting to raise finance although very few venture capitalists will agree to sign.

NDAs are often signed with suppliers by manufacturers, programmers, journalists, and others, in exchange for detailed information or copies of new products in advance of their public launch. For example, a manufacturer might sign an NDA to get a copy of a new operating system in order to have compatible hardware ready for the program's launch. NDAs are often required in order to participate in beta (pre-launch) tests of important pieces of software.

nonlinear video editing in computing, video editing method that processes compressed video data stored on a hard disk. This makes it much easier for editors to find their way around the material, and enables them to commence editing at any point in the tape – hence the name.

non-metal one of a set of elements (around 20 in total) with certain physical and chemical properties opposite to those of *metal elements. The division between metal and non-metal elements forms the simplest division of the *periodic table of the elements. Common physical properties are that non-metals have low electrical conductivity, are brittle when solid, or are gases or liquids. Exceptions include *graphite, a form of carbon, which is a good electrical conductor.

In structure the non-metals are very diverse. *Hydrogen (H), *oxygen (O), *nitrogen (N), *fluorine (F), *chlorine (Cl), and the noble gases (rare gases), *helium (He), *neon (Ne), *argon (Ar), *krypton (Kr), *xenon (Xe), and radioactive *radon (Rn), are gases. Only *bromine (Br) is a liquid at room temperature, and the rest are solids.

Non-metals are the chemical opposites of metals. Metals form positively charged ions or *cations; non-metals form negatively charged ions or *anions. The exceptions are the chemically unreactive noble gases (rare gases), although xenon does react under certain conditions. Non-metals are electronegative, which means that they are able to gain electrons when bonding with metals. Apart from the noble gases, the non-metal elements have incomplete outer electron shells, and so try to gain enough electrons to fill them. The noble gases do not react because they already have complete outer electron shells. The type of bonding where ions are formed is known as *ionic bonding. Non-metals may also share electrons with other non-metal elements to complete their outer shell; this type of bonding is known as *covalent bonding.

non-renewable resource natural resource, such as coal, oil, or natural gas, that takes millions of years to form naturally and therefore cannot be replaced once it is consumed; it will eventually be used up. The main energy sources used by humans are non-renewable; *renewable resources, such as solar, tidal, wind, and geothermal power, have so far been less exploited.

Fossil fuels like coal, oil, and gas generate a considerable amount of energy when they are burnt (the process of combustion). Non-renewable resources have a high carbon content because their origin lies in the photosynthetic activity of plants millions of years ago. The fuels release this carbon back into the atmosphere as carbon dioxide. The rate at which such fuels are being burnt is thus resulting in a rise in the concentration of carbon dioxide in the atmosphere, a cause of the *greenhouse effect.

nonvolatile memory in computing, *memory that does not lose its contents when the power supply to the computer is disconnected.

noradrenaline in the body, a *hormone that acts directly on specific receptors to stimulate the sympathetic nervous system. Released by nerve stimulation or by drugs, it slows the heart rate mainly by constricting arterioles (small arteries) and so raising blood pressure. It is used therapeutically to treat shock.

NOR gate in electronics, a type of *logic gate.

normal distribution in statistics, a distribution widely used to model variation in a set of data which is symmetrical about its mean value. The bell-shaped curve that results when a normal distribution is represented graphically by plotting the distribution $f(x)$ against x is called the normal distribution curve. The curve is symmetrical about the mean value.

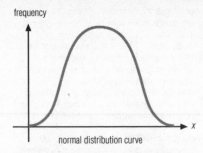

normal distribution curve

normal distribution A curve used in statistics to represent the distribution of data which is symmetrical about its mean value.

North Atlantic Drift warm *ocean current in the North Atlantic Ocean; an extension of the *Gulf Stream. It flows east across the Atlantic and has a mellowing effect on the climate of northwestern Europe, particularly the British Isles and Scandinavia.

northern lights common name for the *aurora borealis.

NOT gate or **inverter gate**, in electronics, a type of *logic gate.

notochord stiff but flexible rod that lies between the gut and the nerve cord of all embryonic and larval chordates, including the vertebrates. It forms the supporting structure of the adult lancelet, but in vertebrates it is replaced by the vertebral column, or spine.

nova plural **novae**, faint star that suddenly erupts in brightness by 10,000 times or more, remains bright for a few days, and then fades away and is not seen again for very many years, if at all. Novae are believed to occur in close binary star systems, where gas from one star flows to a companion *white dwarf. The gas ignites and is thrown off in an explosion at speeds of 1,500 kps/930 mps or more. Unlike a *supernova, the star is not completely disrupted by the outburst. After a few weeks or months it subsides to its previous state; it may erupt many more times.

nuclear energy or **atomic energy**, energy released from the inner core, or *nucleus, of the atom. Energy produced by nuclear *fission (the splitting of certain atomic nuclei) has been harnessed since the 1950s to generate electricity, and research continues into the possible controlled use of *nuclear fusion (the fusing, or combining, of atomic nuclei).

In nuclear power stations, fission of radioactive substances (see *radioactivity) takes place, releasing large amounts of heat energy. The heat is used to produce the steam that drives *turbines and *generators, producing electrical power.

Nuclear energy is produced from the transformation of matter into energy. The amount of energy can be calculated from the equation $E = mc^2$ developed by German-born US physicist Einstein (1879–1955), where E is the heat energy given out, m is the mass of radioactive substance, and c is the speed of light.

The Sun is an example of a natural nuclear reactor. Millions of atoms of hydrogen fuse together to form millions of atoms of helium, generating a continuous supply of heat and light energy. This is called a fusion reaction. A hydrogen bomb obtains its energy from a fusion reaction.

nuclear fusion process whereby two atomic nuclei are fused, with the release of a large amount of energy. Very high temperatures and pressures are required for the process. Under these conditions the atoms involved are stripped of all their electrons so that the remaining particles, which together make up a **plasma**, can come close together at very high speeds and overcome the mutual repulsion of the positive charges on the atomic nuclei. At very close range the strong nuclear force will come into play, fusing the particles to form a larger nucleus. As fusion is accompanied by the release of large amounts of energy, the process might one day be harnessed to form the basis of commercial energy production. Methods of achieving controlled fusion are therefore the subject of research around the world.

Fusion is the process by which the Sun and the other stars produce their energy.

nuclear magnetic resonance NMR, in medicine, scanning technique used to produce images of the organs in the body. It is another term for *magnetic resonance imaging (MRI).

nuclear medicine in medicine, the use of radioactive isotopes in the diagnosis, investigation, and treatment of disease. See *radioisotope scanning.

nuclear notation method used for labelling an atom according to the composition of its nucleus. The atoms or isotopes of a particular element are represented by the symbol $^A_Z X$ where A is the mass number of their nuclei, Z is their atomic number, and X is the chemical symbol for that element.

nuclear physics study of the properties of the nucleus of the *atom, including the structure of nuclei; nuclear forces; the interactions between particles and nuclei; and the study of radioactive decay. The study of elementary particles is *particle physics.

nuclear reaction reaction involving the nuclei of atoms. Atomic nuclei can undergo changes either as a result of radioactive decay, as in the decay of radium to radon (with the emission of an alpha particle):

$$^{226}_{88}Ra \rightarrow ^{222}_{86}Rn + ^4_2He$$

or as a result of particle bombardment in a machine or device, as in the production of cobalt-60 by the bombardment of cobalt-59 with neutrons:

$$^{59}_{27}Co + ^1_0n \rightarrow ^{60}_{27}Co + \gamma.$$

Nuclear *fission and nuclear *fusion are examples of nuclear reactions. The enormous amounts of energy released arise from the mass–energy relation put forward by German-born US physicist Einstein (1879–1955), stating that $E = mc^2$ (where E is energy, m is mass, and c is the velocity of light).

In nuclear reactions the sum of the masses of all the products is less than the sum of the masses of the reacting particles. This lost mass is converted to energy according to Einstein's equation.

nuclear reactor device for producing *nuclear energy in a controlled manner. There are various types of reactor in use, all using nuclear *fission. In a **gas-cooled reactor**, a circulating gas under pressure (such as carbon dioxide) removes heat from the core of the reactor, which usually contains natural uranium. The efficiency of the fission process is increased by slowing neutrons in the core by using a moderator such as carbon. The reaction is controlled with neutron-absorbing rods made of boron. An **advanced gas-cooled reactor** (AGR) generally has enriched uranium as its fuel. A **water-cooled reactor**, such as the steam-generating heavy water (deuterium oxide) reactor, has water circulating through the hot core. The water is converted to steam, which drives turbo-alternators for generating electricity. The most widely used reactor is the **pressurized-water reactor** (PWR), which contains a sealed system of pressurized water that is heated to form steam in heat exchangers in an external circuit. The **fast reactor** has no moderator and uses fast neutrons to bring about fission. It uses a mixture of plutonium and uranium oxide as fuel. When operating, uranium is converted to plutonium, which can be extracted and used later as fuel. It is also called the fast breeder or breeder reactor because it produces more plutonium than it consumes. Heat is removed from the reactor by a coolant of liquid sodium.

nuclear safety measures to avoid accidents in the operation of nuclear reactors and in the production and disposal of nuclear weapons and of *nuclear waste. There are no guarantees of the safety of any of the various methods of disposal. Nuclear safety is a controversial subject – some governments do not acknowledge the hazards of *atomic radiation and *radiation sickness.

nuclear accidents Windscale (now Sellafield), Cumbria, England. In 1957 fire destroyed the core of a reactor, releasing large quantities of radioactive fumes into the atmosphere. In 1990 a scientific study revealed an increased risk of leukaemia in children whose fathers had worked at Sellafield between 1950 and 1985.

Ticonderoga, 130 km/80 mi off the coast of Japan. In 1965 a US Navy Skyhawk jet bomber fell off the deck of this ship, sinking in 4,900 m/16,000 ft of water. It carried a one-megaton hydrogen bomb. The accident was only revealed in 1989.

Three Mile Island, Harrisburg, Pennsylvania, USA. In 1979 a combination of mechanical and electrical failure, as well as operator error, caused a pressurized water reactor to leak radioactive matter.

Church Rock, New Mexico, USA. In July 1979, 380 million litres/100 million gallons of radioactive water containing uranium leaked from a pond into the Rio Purco, causing the water to become over 6,500 times as radioactive as safety standards allow for drinking water.

Chernobyl, Ukraine. In April 1986 there was an explosive leak, caused by overheating, from a nonpressurized boiling-water reactor, one of the largest in Europe. The resulting clouds of radioactive material spread as far as the UK. Thirty-one people were killed in the explosion, and thousands of square kilometres of land were contaminated by fallout. By June 1992, seven times as many children in the Ukraine and Belarus were contracting thyroid cancer as before the accident, and the incidence of leukaemia was rising; it was estimated that more than 6,000 people had died as a result of the accident, and that the death toll in the Ukraine alone would eventually reach 40,000.

Tomsk, Siberia, Russia. In April 1993 a tank exploded at a uranium reprocessing plant, sending a cloud of radioactive particles into the air.

nuclear testing detonation of nuclear devices to verify their reliability, power, and destructive abilities. Although carried out secretly in remote regions of the world, such tests are easily detected by the seismic shock waves produced. The first tests were carried out in the atmosphere during the 1950s by the USA, the USSR, and the UK. The Comprehensive Test Ban Treaty was signed by 149 countries in 1996.

These tests produced large quantities of radioactive fallout, still active today. A test ban in 1963 prohibited all tests except those carried out underground. About 2,000 tests have been carried out since World War II, an average of one every nine days. The USA carried out more than 200 secret nuclear tests over the period 1963–90.

France conducted the first of eight planned underground tests in September 1995 at the Mururoa atoll in French Polynesia. However, after carrying out six of these tests, it announced a 'definitive end' to its nuclear-testing programme in January 1996; the head of the programme admitted that radioactive iodine had leaked into the sea around the Mururoa atoll, but in 'insignificant amounts'.

During the second half of May 1998 India and Pakistan conducted a combined total of 11 nuclear tests (five by India followed by six by Pakistan), provoking widespread international outrage, led by the 'Big Five' nuclear powers – Britain, China, France, Russia, and the USA. A campaign by the Group of Eight industrial powers, launched by forceful US sanctions against India and Pakistan, claimed some success with the announcement in early June by both Delhi and Islamabad of test moratoria. On 12 June the G-8 countries announced a freeze on all non-humanitarian loans to the two countries in protest at their recent tests. The freeze was largely symbolic given the halt to all International Monetary Fund and World Bank lending imposed late May.

nuclear waste   the radioactive and toxic by-products of the nuclear energy and nuclear weapons industries. Nuclear waste may have an active life of several thousand years. Reactor waste is of three types: **high-level** spent fuel, or the residue when nuclear fuel has been removed from a reactor and reprocessed; **intermediate**, which may be long-or short-lived; and **low-level**, but bulky, waste from reactors, which has only short-lived radioactivity. Disposal, by burial on land or at sea, has raised problems of safety, environmental pollution, and security.

nucleic acid   complex organic acid made up of a long chain of *nucleotides, present in the nucleus and sometimes the cytoplasm of the living cell. The two types, known as *DNA (deoxyribonucleic acid) and *RNA (ribonucleic acid), form the basis of heredity. The nucleotides are made up of a sugar (deoxyribose or ribose), a phosphate group, and one of four purine or pyrimidine bases. The order of the bases along the nucleic acid strand contains the genetic code.

nucleolus   in biology, an RNA-rich structure found in the nucleus of eukaryotic cells. It produces the RNA that makes up the *ribosomes, from instructions in the DNA.

nucleon   in particle physics, either a *proton or a *neutron, when present in the atomic nucleus. **Nucleon number** is an alternative name for the *mass number of an atom.

nucleon number   or **mass number** (symbol A), the sum of the numbers of protons and neutrons in the nucleus of an atom. With the *proton number, it is used in nuclear notation – for example, in the symbol $^{14}_{6}C$ representing the isotope carbon-14, the lower number is the proton number, and the upper is the nucleon number.

nucleotide   organic compound consisting of a purine (adenine or guanine) or a pyrimidine (thymine, uracil, or cytosine) base linked to a sugar (deoxyribose or ribose) and a phosphate group. *DNA and *RNA are made up of long chains of nucleotides.

nucleus   in astronomy, compact central core of a *galaxy, often containing powerful radio, X-ray, and infrared sources. Active galaxies have extremely energetic nuclei.

nucleus   in biology, the central, membrane-enclosed part of a eukaryotic cell, containing threads of *DNA. It is found in both plant and animal cells. During cell division the threads of DNA coil up to form *chromosomes. The nucleus controls the function of the cell by determining which proteins are produced within it. It is where inherited information (see *inheritance) is stored as *genes. Because proteins are the chief structural molecules of living matter and, as enzymes, regulate all aspects of metabolism, it may be seen that the genetic code within the nucleus is effectively responsible for building and controlling the whole organism.

The nucleus contains the **nucleolus**, the part of the cell where ribosomes are produced. Movement of molecules into and out of the nucleus occurs through the nuclear pores. An average mammalian nucleus has approximately 3,000 pores.

nucleus plural **nuclei**, in physics, the positively-charged central part of an *atom, which constitutes almost all its mass. Except for hydrogen nuclei, which have only one proton, nuclei are composed of both protons and neutrons. Surrounding the nucleus are electrons, of equal and opposite charge to that of the protons, thus giving the atom a neutral charge. Nuclei that are unstable may undergo *radioactive decay or nuclear *fission. In all stars, including our Sun, small nuclei join together to make more stable, larger nuclei. This process is called nuclear *fusion.

The nucleus was discovered by the New Zealand-born British physicist Ernest Rutherford (1871–1937) in 1911 as a result of experiments in firing alpha particles through very thin gold foil.

A few of the particles were deflected back, and Rutherford, astonished, reported: 'It was almost as if you fired a 15-inch shell at a piece of tissue paper and it came back and hit you!' The deflection, he deduced, was due to the positively-charged alpha particles being repelled by approaching a small but dense positively-charged nucleus.

nuclide in physics, a species of atomic nucleus characterized by the number of protons (Z) and the number of neutrons (N). Nuclides with identical *proton number but differing *neutron number are called *isotopes.

nugget piece of gold found as a lump of the *native metal. Nuggets occur in *alluvial deposits where river-borne particles of the metal have adhered to one another.

null character character with the *ASCII value 0. A null character is used by some programming languages, most notably C, to mark the end of a character string.

null-modem special cable that is used to connect the *serial interfaces of two computers, so as to allow them to exchange data.

null string in computing, a string, usually denoted by '–', containing nothing or a *null character. A null string is used in some programming languages to denote the last of a series of values.

number theory in mathematics, the abstract study of the structure of number systems and the properties of positive integers (whole numbers). For example, the theories of factors and prime numbers fall within this area, as does the work of mathematicians Giuseppe Peano (1858–1932), Pierre de Fermat (1601–1665), and Karl Gauss.

numerator number or symbol that appears above the line in a vulgar fraction. For example, the numerator of 5/6 is 5. The numerator represents the fraction's dividend and indicates how many of the equal parts indicated by the denominator (number or symbol below the line) comprise the fraction.

nutation in astronomy, slight 'nodding' of the Earth in space, caused by the varying gravitational pulls of the Sun and Moon. Nutation changes the angle of the Earth's axial tilt (average 23.5°) by about 9 seconds of arc to either side of its mean position, a complete cycle taking just over 18.5 years.

nutrient cycle transfer of nutrients from one part of an *ecosystem to another. Trees, for example, take up nutrients such as calcium and potassium from the soil through their root systems and store them in leaves. When the leaves fall they are decomposed by bacteria and the nutrients are released back into the soil where they become available for root uptake again.

nylon synthetic long-chain polymer similar in chemical structure to protein. Nylon was the first fully *synthetic fibre. It is a polyamide and is made from petroleum, natural gas, air, and water by *condensation polymerization, nylon was developed in 1935 by the US chemist W H Carothers and his associates, who worked for Du Pont. It is used in the manufacture of moulded articles, textiles, and medical sutures. Nylon fibres are stronger and more elastic than silk and are relatively insensitive to moisture and mildew.

Nylon is used for a wide range of different textiles including carpets, and can be used in knitting or weaving. It is also used for simulating other fabrics such as silks and furs.

oasis area of land made fertile by the presence of water near the surface in an otherwise arid region. The occurrence of oases affects the distribution of plants, animals, and people in the desert regions of the world.

object in computing, an item created on a page in desktop publishing and many office applications.

object linking and embedding OLE, enhancement to *dynamic data exchange, which makes it possible not only to include live data from one application in another application, but also to edit the data in the original application without leaving the application in which the data has been included. (See also *ActiveX.)

object-oriented programming OOP, computer programming based on *objects, in which each object has its own program code and data and can interact with other objects. Data items are closely linked to the procedures that operate on them. For example, a circle on the screen might be an object: it has data, such as a centre point and a radius, as well as procedures for moving it, erasing it, changing its size, and so on.

object program in computing, the *machine code translation of a program written in a *source language.

obsession persistently intruding thought, emotion, or impulse, often recognized by the sufferer as irrational, but nevertheless causing distress. It may be a brooding on destiny or death, or chronic doubts interfering with everyday life (such as fearing the gas is not turned off and repeatedly checking), or an impulse leading to repetitive action, such as continually washing one's hands.

obsessive-compulsive disorder OCD, in psychiatry, anxiety disorder that manifests itself in the need to check constantly that certain acts have been performed 'correctly'. Sufferers may, for example, feel compelled to wash themselves repeatedly or return home again and again to check that doors have been locked and appliances switched off. They may also hoard certain objects and insist in these being arranged in a precise way, or be troubled by intrusive and unpleasant thoughts. In extreme cases, normal life is disrupted through the hours devoted to compulsive actions. Treatment involves cognitive therapy and drug therapy with serotonin-blocking drugs such as *Prozac.

obstetrics medical speciality concerned with the management of pregnancy, childbirth, and the immediate postnatal period.

obtuse angle angle greater than 90° but less than 180°.

occluded front weather *front formed when a cold front catches up with a warm front. It brings clouds and rain as air is forced to rise upwards along the front, cooling and condensing as it does so.

occupational psychology study of human behaviour at work. It includes dealing with problems in organizations, advising on management difficulties, and investigating the relationship between humans and machines (as in the design of aircraft controls). Another area is psychometrics and the use of assessment to assist in selection of personnel.

ocean great mass of salt water. Geographically speaking three oceans exist – the Atlantic, Indian, and Pacific – to which the Arctic is often added. They cover approximately 70% or 363,000,000 sq km/140,000,000 sq mi of the total surface area of the Earth. According to figures released in August 2001, the total volume of the world's oceans is 1,370 million cubic km/329 million cubic mi. Water levels recorded in the world's oceans have shown an increase of 10–15 cm/4–6 in over the past 100 years.

depth (average) 3,660 m/12,000 ft, but shallow ledges (continental shelves) 180 m/600 ft run out from the continents, beyond which the continental slope reaches down to the abyssal zone, the largest area, ranging from 2,000–6,000 m/6,500–19,500 ft. Only the *deep-sea trenches go deeper, the deepest recorded being 11,034 m/36,201 ft (by the *Vityaz*, USSR) in the Mariana Trench of the western Pacific in 1957

features deep trenches (off eastern and southeastern Asia, and western South America), volcanic belts (in the western Pacific and eastern Indian Ocean), and ocean ridges (in the mid-Atlantic, eastern Pacific, and Indian Ocean).

temperature varies on the surface with latitude (–2°C/28°F to +29°C)/84°F; decreases rapidly to 370 m/1,200 ft, then more slowly to 2,200 m/7,200 ft; and hardly at all beyond that

seawater contains about 3% dissolved salts, the most abundant being *sodium chloride; salts come from the weathering of rocks on land; rainwater flowing over rocks, soils, and organic matter on land dissolves small amounts of substances, which pass into rivers to be carried to the sea. Salt concentration in the oceans remains remarkably constant as water is evaporated by the Sun and fresh water added by rivers. Positive ions present in sea water include sodium, magnesium, potassium, and calcium; negative ions include chloride, sulphate, hydrogencarbonate, and bromide

commercial extraction of minerals includes bromine, magnesium, potassium, salt (sodium chloride); those potentially recoverable include

aluminium, calcium, copper, gold, manganese, silver.

pollution Oceans have always been used as a dumping area for human waste, but as the quantity of waste increases, and land areas for dumping diminish, the problem is exacerbated. Today ocean pollutants include airborne emissions from land (33% by weight of total marine pollution); oil from both shipping and land-based sources; toxins from industrial, agricultural, and domestic uses; sewage; sediments from mining, forestry, and farming; plastic litter; and radioactive isotopes. Thermal pollution by cooling water from power plants or other industry is also a problem, killing coral and other temperature-sensitive sedentary species.

ocean current fast-flowing body of seawater forced by the wind or by variations in water density (as a result of temperature or salinity variations) between two areas. Ocean currents are partly responsible for transferring heat from the Equator to the poles and thereby evening out the global heat imbalance.

oceanography study of the oceans. Its subdivisions deal with each ocean's extent and depth, the water's evolution and composition, its physics and chemistry, the bottom topography, currents and wind, tidal ranges, biology, and the various aspects of human use. Computer simulations are widely used in oceanography to plot the possible movements of the waters, and many studies are carried out by remote sensing.

Oceanography involves the study of water movements – currents, waves, and tides – and the chemical and physical properties of the seawater. It deals with the origin and topography of the ocean floor – ocean trenches and ridges formed by *plate tectonics, and continental shelves from the submerged portions of the continents.

ocean trench submarine valley. Ocean trenches are characterized by the presence of a volcanic arc on the concave side of the trench. Trenches are now known to be related to subduction zones, places where a plate of oceanic *lithosphere dives beneath another plate of either oceanic or continental lithosphere. Ocean trenches are found around the edge of the Pacific Ocean and the northeastern Indian Ocean; minor ones occur in the Caribbean and near the Falkland Islands.

Ocean trenches represent the deepest parts of the ocean floor, the deepest being the *Mariana Trench which has a depth of 11,034 m/36,201 ft. At depths of below 6 km/3.6 mi there is no light and very high pressure; ocean trenches are inhabited by crustaceans, coelenterates (for example, sea anemones), polychaetes (a type of worm), molluscs, and echinoderms.

OCR abbreviation for *optical character recognition.

octagon *polygon with eight sides. A regular octagon has all eight sides of equal length, and all eight angles of equal size, 135°.

octahedron, regular regular solid with eight faces, each of which is an equilateral triangle. It is one of the five regular polyhedra or Platonic solids. The figure made by joining the midpoints of the faces is a perfect cube and the vertices of the octahedron are themselves the midpoints of the faces of a surrounding cube. For this reason, the cube and the octahedron are called dual solids.

octane rating numerical classification of petroleum fuels indicating their combustion characteristics.

The efficient running of an *internal combustion engine depends on the ignition of a petrol-air mixture at the correct time during the cycle of the engine. Higher-rated petrol burns faster than lower-rated fuels. The use of the correct grade must be matched to the engine.

octet rule in chemistry, rule stating that elements combine in a way that gives them the electronic structure of the nearest *noble gas (rare gas). The joining together of elements to form compounds is called bonding. All the noble gases except helium have eight electrons in their outermost shell, hence the term octet.

odd number any number not divisible by 2, thus odd numbers form the infinite sequence 1, 3, 5, 7 ... . Every square number n^2 is the sum of the first n odd numbers. For example, $49 = 7^2 = 1 + 3 + 5 + 7 + 9 + 11 + 13$.

Oedipus complex in psychology, the unconscious antagonism of a son to his father, whom he sees as a rival for his mother's affection. For a girl antagonistic to her mother as a rival for her father's affection, the term is **Electra complex**. The terms were coined by Austrian physician Sigmund Freud (1856–1939).

OEM in computing, abbreviation for original equipment manufacturer.

oersted symbol Oe, * c.g.s. unit of *magnetic field strength, now replaced by the SI unit ampere per metre. The Earth's magnetic field is about 0.5 oersted; the field near the poles of a small bar magnet is several hundred oersteds; and a powerful *electromagnet can have a field strength of 30,000 oersteds.

oesophagus muscular tube through which food travels from the mouth to the stomach. The human oesophagus is about 23 cm/9 in long. It extends downwards from the *pharynx, immediately behind the windpipe. It is lined with a mucous membrane made of epithelial cells (see *epithelium), which secretes lubricant fluid to assist the downward movement of food. In its wall is *muscle, which contracts to squeeze the food towards the stomach (*peristalsis).

oestrogen any of a group of *hormones produced by the *ovaries of vertebrates; the term is also used for various synthetic hormones that mimic their effects. The principal oestrogen in mammals is oestradiol. Oestrogens control female sexual development, promote the growth of female *secondary sexual characteristics at puberty, stimulate egg (*ovum) production, and, in mammals, prepare the lining of the uterus for pregnancy. In other words, together with another hormone, progesterone, they regulate the growth and functioning of sex organs for *sexual reproduction. Oestrogens are also used in female oral contraceptives, to inhibit the production of ova.

oestrus in mammals, the period during a female's reproductive cycle (also known as the oestrus cycle or *menstrual cycle) when mating is most likely to occur. It usually coincides with ovulation.

offline in computing, not connected, so that data cannot be transferred, for example, to a printer. The opposite of *online.

offline browsing in computing, downloading and copying Web pages onto a computer so that they can be viewed without being connected to the Internet. By taking advantage of off-peak hours, when telephone charges are low and the network responds faster, offline browsing is a thrifty way of using the *World Wide Web.

ohm symbol Ω, SI unit of electrical *resistance (the property of a conductor that restricts the flow of electrons through it). Named after German physicist Georg Simon Ohm (1789–1854).

It was originally defined with reference to the resistance of a column of mercury, but is now taken as the resistance between two points when a potential difference of one volt between them produces a current of one ampere.

Ohm's law law that states that, for many materials over a wide range of conditions, the current flowing in a conductor maintained at constant temperature is directly proportional to the potential difference (voltage) between its ends. The law was discovered by German physicist Georg Ohm (1789–1854) in 1827. He found that if the voltage across a conducting material is changed, the current flow through the material is changed proportionally (for example, if the voltage is doubled then the current also doubles).

oil flammable substance, usually insoluble in water, and composed chiefly of carbon and hydrogen. Oils may be solids (fats and waxes) or liquids. Various plants produce vegetable oils; mineral oils are based on petroleum.

The crude oil (unrefined *petroleum) found beneath the Earth's surface is formed from the remains of dead plants and animals. As these plants and animals died they were buried with mud near the sea floor. Over

millions of years, heat from the Earth's interior and pressure from overlying rocks slowly changed the dead remains into hydrocarbons (substances containing hydrogen and carbon). The hydrocarbons, being light molecules, moved upwards and became trapped beneath impermeable rocks.

Oil reservoirs are often found beneath the seabed and drilling technology is used to locate these supplies of oil. Crude oil extracted from the ground is refined in *fractional distillation columns to produce more useful products such as petrol, diesel, kerosene, plastics, and chemicals for pharmaceuticals.

oil spill oil discharged from an ocean-going tanker, pipeline, or oil installation, often as a result of damage. An oil spill kills shore life, clogging the feathers of birds and suffocating other creatures. At sea, toxic chemicals spread into the water below, poisoning sea life. Mixed with dust, the oil forms globules that sink to the seabed, poisoning sea life there as well. Oil spills are broken up by the use of detergents but such chemicals can themselves damage wildlife. The annual spillage of oil is 8 million barrels (280 million gallons) a year. At any given time tankers are carrying 500 million barrels (17.5 billion gallons).

OLE in computing, abbreviation for *object linking and embedding.

olefin common name for *alkene.

Oligocene epoch third epoch of the Tertiary period of geological time, 35.5–3.25 million years ago. The name, from Greek, means 'a little recent', referring to the presence of the remains of some modern types of animals existing at that time.

oligodendrocyte type of glial cell (see *glia) surrounding the neurons in the brain and spinal cord. They produce the insulating *myelin sheath surrounding the nerve axon.

omega symbol Ω, last letter of the Greek alphabet used in astronomy as a symbol for the mass density of the universe. If Ω is less than 1.0 the universe will expand forever; if it is more than 1.0 the gravitational pull of its mass will be strong enough to reverse its expansion and cause its eventual collapse. The value of Ω is estimated as probably being between 0.1 and 1.0. In physics Ω stands for angular frequency.

omnivore animal that feeds on both plant and animal material. Omnivores have digestive adaptations intermediate between those of *herbivores and *carnivores, with relatively unspecialized digestive systems and gut micro-organisms that can digest a variety of foodstuffs. Omnivores include humans, the chimpanzee, the cockroach, and the ant.

OMR abbreviation for *optical mark recognition.

onchocerciasis or **river blindness**, disease found in tropical Africa and Central America. It is transmitted by bloodsucking black flies, which infect the victim with parasitic filarial worms (genus *Onchocerca*), producing skin disorders and intense itching; some invade the eyes and may cause blindness.

oncogene gene carried by a virus that induces a cell to divide abnormally, giving rise to a cancer. Oncogenes arise from mutations in genes (proto-oncogenes) found in all normal cells. They are usually also found in viruses that are capable of transforming normal cells to tumour cells. Such viruses are able to insert their oncogenes into the host cell's DNA, causing it to divide uncontrollably. More than one oncogene may be necessary to transform a cell in this way.

oncology medical speciality concerned with the diagnosis and treatment of *neoplasms, especially cancer.

online in computing, connected, so that data can be transferred, for example, to a printer or from a network. The opposite of *offline.

ontogeny process of development of a living organism, including the part of development that takes place after hatching or birth. The idea that 'ontogeny recapitulates phylogeny' (the development of an organism goes through the same stages as its evolutionary history), proposed by the German scientist Ernst Heinrich Haeckel, is now discredited.

onyx semi-precious variety of chalcedonic *silica (SiO_2) in which the crystals are too fine to be detected under a microscope, a state known as cryptocrystalline. It has straight parallel bands of different colours: milk-white, black, and red.

oocyte in medicine, an immature ovum. Only a fraction of the oocytes produced in the ovary survive until puberty and not all of these undergo meiosis to become an ovum that can be fertilized by a sperm.

oosphere another name for the female gamete, or *ovum, of certain plants such as algae.

opal form of hydrous *silica ($SiO_2.nH_2O$), often occurring as stalactites and found in many types of rock. The common opal is translucent, milk-white, yellow, red, blue, or green, and lustrous. Precious opal is opalescent, the characteristic play of colours being caused by close-packed silica spheres diffracting light rays within the stone.

open cluster or **galactic cluster**, loose cluster of young stars. More than 1,200 open clusters have been catalogued, each containing between a dozen and several thousand stars. They are of interest to astronomers because they represent samples of stars that have been formed at the same time from similar material. Examples include the Pleiades and the Hyades. See also *star cluster.

open-hearth furnace method of steelmaking, now largely superseded by the *basic–oxygen process. It was developed in 1864 in England by German-born William and Friedrich Siemens, and improved by Pierre and Emile Martin in France in the same year. In the furnace, which has a wide, saucer-shaped hearth and a low roof, molten pig iron and scrap are packed into the shallow hearth and heated by overhead gas burners using preheated air.

open source computing term to describe the distribution of software with its *source code, its original instructions. Anyone may examine open source software's codes, make improvements, and pass these on to the open source community. Its use created one of the most adaptable systems for developing software, including powerful operating systems like GNU, the free UNIX-like software distributed by the Free Software Foundation, and the non-proprietary Linux operating system.

open systems systems that conform to Open Systems Interconnection or *POSIX standards. *Unix was the original basis of open systems and most nonproprietary open systems still use this *operating system.

operating system OS, in computing, a program that controls the basic operation of a computer. A typical OS controls the peripheral devices such as printers, organizes the filing system, provides a means of communicating with the operator, and runs other programs.

operation in mathematics, action on numbers, matrices, or vectors that combines them to form others. The basic operations on numbers are addition, subtraction, multiplication, and division. Matrices involve the same operations as numbers (under specific conditions), but two different types of multiplication are used for vectors: vector multiplication and scalar multiplication.

operational amplifier or **op-amp**, processor that amplifies the difference between two incoming electrical signals. It is a current gain device. Operational amplifiers have two inputs, an inverting input (–) and a noninverting input (+). The input signal from a sensor or switch, often as part of a potential divider, is compared with a similar signal at the other input. This then decides the type of output. If a positive signal is applied to the noninverting input, the output will be positive. If the input is negative, the output will be negative. If a positive signal is given to the inverting input, the output will be negative. A negative input will give a positive output. If the signals at the two inputs are the same, the output will be zero. *Feedback loops are attached from the output to the input in order to control the amplification of the output.

operon group of genes that are found next to each other on a chromosome, and are turned on and off as an integrated unit. They usually produce enzymes that control different steps in the same biochemical pathway by a single operator gene. Operons were discovered in 1961 (by the French biochemists François Jacob and Jacques Monod) in bacteria.

They are less common in higher organisms where the control of metabolism is a more complex process.

ophthalmology medical speciality concerned with diseases of the eye and its surrounding tissues.

opiate, endogenous naturally produced chemical in the body which has effects similar to morphine and other opiate drugs; a type of neurotransmitter.

Examples include *endorphins and *encephalins.

opium drug extracted from the unripe seeds of the opium poppy (*Papaver somniferum*) of southwestern Asia. An addictive *narcotic, it contains several alkaloids, including **morphine**, one of the most powerful natural painkillers and addictive narcotics known, and **codeine**, a milder painkiller.

optical aberration see *aberration, optical.

optical activity in chemistry, the ability of certain crystals, liquids, and solutions to rotate the plane of *polarized light as it passes through them. The phenomenon is related to the three-dimensional arrangement of the atoms making up the molecules concerned. Only substances that lack any form of structural symmetry exhibit optical activity.

optical character recognition OCR, in computing, a technique for inputting text to a computer by means of a document reader. First, a scanner produces a digital image of the text; then character-recognition software makes use of stored knowledge about the shapes of individual characters to convert the digital image to a set of internal codes that can be stored and processed by a computer.

optical computer computer in which both light and electrical signals are used in the *central processing unit. The technology is still not fully developed, but such a computer promises to be faster and less vulnerable to outside electrical interference than one that relies solely on electricity.

optical disk in computing, a storage medium in which laser technology is used to record and read large volumes of digital data. Types include *CD-ROM, *WORM, and floptical disk.

optical fibre very fine, optically-pure glass fibre through which light can be reflected to transmit images or data from one end to the other. Although expensive to produce and install, optical fibres can carry more data than traditional cables, and are less susceptible to interference. Standard optical fibre transmitters can send up to 10 billion bits of information per second by switching a laser beam on and off.

optical illusion scene or picture that fools the eye. An example of a natural optical illusion is that the Moon appears bigger when it is on the horizon than when it is high in the sky, owing to the *refraction of light rays by the Earth's atmosphere.

optical mark recognition OMR, in computing, a technique that enables marks made in predetermined positions on computer-input forms to be detected optically and input to a computer. An **optical mark reader** shines a light beam onto the input document and is able to detect the marks because less light is reflected back from them than from the paler, unmarked paper.

optic nerve large nerve passing from the eye to the brain, carrying visual information. In mammals, it may contain up to a million nerve fibres, connecting the sensory cells of the retina to the optical centres in the brain. Embryologically, the optic nerve develops as an outgrowth of the brain.

optics branch of physics that deals with the study of *light and vision – for example, shadows and mirror images, lenses, microscopes, telescopes, and cameras. On striking a surface, light rays are reflected or refracted with some absorption of

twisted pair cable

coaxial cable

fibre optic cable

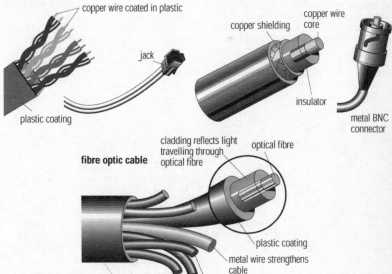

optical fibre The major differences in construction between twisted pair (telephone), coaxial, and fibre optic cable.

energy, and the study of this is known as geometrical optics.

optoelectronics branch of electronics concerned with the development of devices (based on the *semiconductor gallium arsenide) that respond not only to the *electrons of electronic data transmission, but also to *photons ('particles' of light).

orbit path of one body in space around another, such as the orbit of the Earth around the Sun or of the Moon around the Earth. When the two bodies are similar in mass, as in a binary star, both bodies move around their common centre of mass. The movement of objects in orbit follows *Kepler's laws, which apply to artificial satellites as well as to natural bodies.

orbital, atomic region around the nucleus of an atom (or, in a molecule, around several nuclei) in which an *electron is likely to be found. According to *quantum theory, the position of an electron is uncertain; it may be found at any point. However, it is more likely to be found in some places than in others, and this pattern of probabilities makes up the orbital.

An atom or molecule has numerous orbitals, each of which has a fixed size and shape. An orbital is characterized by three numbers, called *quantum numbers, representing its energy (and hence size), its angular momentum (and hence shape), and its orientation. Each orbital can be occupied by one or (if their spins are aligned in opposite directions) two electrons.

order in biological classification, a group of related *families. For example, the horse, rhinoceros, and tapir families are grouped in the order Perissodactyla, the odd-toed ungulates, because they all have either one or three toes on each foot. The names of orders are not shown in italic (unlike genus and species names) and by convention they have the ending '-formes' in birds and fish; '-a' in mammals, amphibians, reptiles, and other animals; and '-ales' in fungi and plants. Related orders are grouped together in a *class.

order to arrange with regard to size or quantity or other quality, for example alphabetical order. Putting a set of things in order is the same as mapping them on to the set of natural numbers. An infinite set can sometimes be mapped in this way in which case it is said to be countable.

ordered pair in mathematics, any pair of numbers whose order makes a difference to their meaning. Coordinates are an ordered pair because the point (2, 3) is not the same as the point (3, 2). Vulgar fractions are ordered pairs because the top number gives the quantity of parts while the bottom gives the number of parts into which the unit has been divided.

ordinal number in mathematics, one of the series first, second, third, fourth,

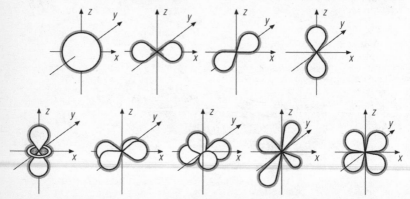

orbital, atomic The shapes of atomic orbitals. An atomic orbital is a picture of the 'electron cloud' that surrounds the nucleus of an atom. There are four basic shapes for atomic orbitals: spherical, dumbbell, clover-leaf, and complex (shown at bottom left).

.... Ordinal numbers relate to order, whereas *cardinal numbers (1, 2, 3, 4, ...) relate to quantity, or count.

ordinate in *coordinate geometry, the y coordinate of a point; that is, the vertical distance of the point from the horizontal or x-axis. For example, a point with the coordinates (3,4) has an ordinate of 4. See *abscissa.

Ordovician period period of geological time 510–439 million years ago; the second period of the *Palaeozoic era. Animal life was confined to the sea: reef-building algae and the first jawless fish are characteristic. The period is named after the Ordovices, an ancient Welsh people, because the system of rocks formed in the Ordovician period was first studied in Wales.

ore body of rock, a vein within it, or a deposit of sediment, worth mining for the economically valuable mineral it contains. The term is usually applied to sources of metals. Occasionally metals are found uncombined (native metals), but more often they occur as compounds such as carbonates, sulphides, or oxides. The ores often contain unwanted impurities that must be removed when the metal is extracted.

Commercially valuable ores include bauxite (aluminium oxide, Al_2O_3) haematite (iron(III) oxide, Fe_2O_3), zinc blende (zinc sulphide, ZnS), and rutile (titanium dioxide, TiO_2).

Hydrothermal ore deposits are formed from fluids such as saline water passing through fissures in the host rock at an elevated temperature. Examples are the 'porphyry copper' deposits of Chile and Bolivia, the submarine copper–zinc–iron sulphide deposits recently discovered on the East Pacific Rise, and the limestone lead–zinc deposits that occur in the southern USA and in the Pennines of Britain.

Other ores are concentrated by igneous processes, causing the ore metals to become segregated from a magma – for example, the chromite- and platinum-rich bands within the bushveld, South Africa. Erosion and transportation in rivers of material from an existing rock source can lead to further concentration of heavy minerals in a deposit – for example, Malaysian tin deposits. Weathering of rocks in situ can result in residual metal-rich soils, such as the nickel-bearing laterites of New Caledonia.

organ in biology, a part of a living body that has a distinctive function or set of functions. Examples include the *oesophagus, *liver, or *brain in animals, or a *leaf in plants. An organ is composed of a group of coordinated *tissues. A group of organs working together to perform a function is called an organ system, for example, the *digestive system comprises a number of organs including the oesophagus, stomach, the small *intestine, the pancreas, and the liver. The tissues of a leaf include the epidermis, palisade mesophyll, and spongy mesophyll. The tissues of the oesophagus include *muscle, and *epithelium.

organelle discrete and specialized structure in a living cell; organelles include mitochondria, chloroplasts, lysosomes, ribosomes, and the nucleus.

organic chemistry branch of chemistry that deals with the more complex covalent compounds of carbon. Organic compounds form the chemical basis of life and are more abundant than inorganic compounds. In a typical organic compound, each carbon atom forms bonds covalently with each of its neighbouring carbon atoms in a chain or ring, and additionally with other atoms, commonly hydrogen, oxygen, nitrogen, or sulphur. (See diagram, page 420.) The basis of organic chemistry is the ability of carbon to form long chains of atoms, branching chains, rings, and other complex structures. Compounds containing only carbon and hydrogen are known as **hydrocarbons**.

organic compound in chemistry, a class of compounds that contain carbon (carbonates, carbon monoxide, and carbon dioxide are excluded). The original distinction between organic and inorganic compounds was based on the belief that the molecules of living systems were unique, and could not be synthesized in the laboratory. Today it is routine to manufacture thousands of organic chemicals both in research and in the drug industry. Certain simple compounds of carbon, such as carbonates, oxides of carbon, carbon disulphide, and carbides are usually treated in *inorganic chemistry.

organophosphate insecticide insecticidal compounds that cause the irreversible inhibition of the

Formula	Name	Atomic bonding
CH_3	methyl	
CH_2CH_3	ethyl	
CC	double bond	
CHO	aldehyde	
CH_2OH	alcohol	
CO	ketone	
$COOH$	acid	
CH_2NH_2	amine	
C_6H_6	benzene ring	

organic chemistry Common organic molecule groupings. Organic chemistry is the study of carbon compounds, which make up over 90% of all chemical compounds. This diversity arises because carbon atoms can combine in many different ways with other atoms, forming a wide variety of rings and chains.

cholinesterase enzymes that break down acetylcholine. As this mechanism of action is very toxic to humans, the compounds should be used with great care. Malathion and permethrin may be used to control lice in humans and have many applications in veterinary medicine and agriculture. In 1998 organophosphates were the most widely used insecticides, with 40% of the global market.

OR gate in electronics, a type of *logic gate.

origin in mathematics, the point where the x–axis meets the y–axis. The coordinates of the origin are $(0, 0)$.

ornithology study of birds. It covers scientific aspects relating to their structure and classification, and their habits, song, flight, and value to agriculture as destroyers of insect pests. Worldwide scientific banding (or the fitting of coded rings to captured specimens) has resulted in accurate information on bird movements and distribution. There is an International Council for Bird Preservation with its headquarters at the Natural History Museum, London.

orogenesis in its original, literal sense, orogenesis means 'mountain building', but today it more specifically refers to the tectonics of mountain building (as opposed to mountain building by erosion).

Orogenesis is brought about by the movements of the rigid plates making up the Earth's crust and uppermost mantle (described by *plate tectonics). Where two plates collide at a destructive margin rocks become folded and lifted to form chains of mountains (such as the Himalayas). Processes associated with orogeny are faulting and thrusting (see *fault), folding, metamorphism, and plutonism. However, many topographical features of mountains – cirques, U-shaped valleys – are the result of *non-orogenic* processes, such as weathering, erosion, and glaciation. Isostasy (uplift due to the buoyancy of the Earth's crust) can also influence mountain physiography.

orthopaedics (Greek *orthos* 'straight'; *pais* 'child') medical speciality concerned with the correction of disease or damage in bones and joints.

oscillating universe in astronomy, theory stating that the gravitational attraction of the mass within the universe will eventually slow down and stop the expansion of the universe. The outward motions of the galaxies will then be reversed, eventually resulting in a 'Big Crunch' where all the matter in the universe will be contracted into a small volume of high density. This could undergo a further *Big Bang, thereby creating another expansion phase. The theory suggests that the universe would alternately expand and collapse through alternate Big Bangs and Big Crunches.

oscillation one complete to-and-fro movement of a vibrating object or system. For any particular vibration, the time for one oscillation is called its *period and the number of oscillations in one second is called its *frequency. The maximum displacement of the vibrating object from its rest position is called the *amplitude of the oscillation.

oscillator any device producing a desired oscillation (vibration). There are many types of oscillator for different purposes, involving various arrangements of thermionic *valves or components such as *transistors, *inductors, *capacitors, and *resistors.

oscilloscope another name for *cathode-ray oscilloscope.

osmium chemical symbol Os, (Greek *osme* 'odour') hard, heavy, bluish-white, metallic element, atomic number 76, relative atomic mass 190.32. It is the densest of the elements,

and is resistant to tarnish and corrosion. It occurs in platinum ores and as a free metal (see *native metal) with iridium in a natural alloy called osmiridium, containing traces of platinum, ruthenium, and rhodium. Its uses include pen points and light-bulb filaments; like platinum, it is a useful catalyst.

It was discovered in 1803 and named in 1804 by English chemist Smithson Tennant (1761–1815) after the irritating smell of one of its oxides.

osmoregulation process whereby the water content of living organisms is maintained at a constant level. If the water balance is disrupted, the concentration of salts will be too high or too low, and vital functions, such as nerve conduction, will be adversely affected.

osmosis movement of *water through a partially (selectively) permeable membrane separating solutions of different concentrations. Water passes by *diffusion from a **weak solution** (high water concentration) to a **strong solution** (low water concentration) until the two concentrations are equal. A membrane is partially permeable if it lets water through but not the molecules or ions dissolved in the water (the solute; for example, sugar molecules). Many cell membranes behave in this way, and osmosis is a vital mechanism in the transport of fluids in living organisms. One example is in the transport of water from soil (weak solution) into the roots of plants (stronger solution of cell sap) via the *root hair cells. Another is the

or

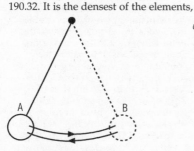

one complete oscillation or cycle is from A to B and back to A

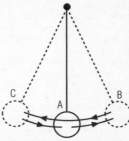

one complete oscillation or cycle is from A to B to C and back to A, moving in the same direction again

oscillation Alternate ways of defining the oscillation (one complete swing) of a pendulum.

uptake of water by the *epithelium lining the *gut in animals. There are also membranes that humans can manufacture that are partially permeable.

If solutions have different concentrations of dissolved substance, there will be a difference in the concentration of water as well. Where there is a lot of dissolved substance there is less water. Where there is less dissolved substance there is more water. Water behaves like any other substance in diffusion. It diffuses from an area where there is a high concentration of it to an area where there is a lower concentration of it. Water diffuses down the concentration gradient for water and across the membrane. This means that it will diffuse from the dilute solution into the concentrated solution.

Excessive flow of water into a cell by osmosis can burst the cell. Cells protect against this using processes of osmoregulation. If external pressure is applied to the stronger solution, osmosis is arrested. By this mechanism plant cells can osmoregulate, since the cell wall of a fully turgid cell exerts pressure on the solution within the cell. Animal cells such as the red blood cell cannot osmoregulate in this way since they have no cell wall. Instead, the correct concentration of plasma is maintained by the *kidney. This is important because if the concentration changes significantly, osmosis will take place and a lot of water will either diffuse out of cells into the blood or into cells from the blood, thereby damaging them.

Otto cycle alternative name for the *four-stroke cycle, introduced by the German engineer Nikolaus Otto (1832–1891) in 1876.

ounce unit of mass, one-sixteenth of a pound avoirdupois, equal to 437.5 grains (28.35 g); also one-twelfth of a pound troy, equal to 480 grains.

ovary in female animals, the organ that generates the *ovum. In humans, the ovaries are two whitish rounded bodies about 25 mm/1 in by 35 mm/1.5 in, located in the lower abdomen on either side of the uterus. Every month, from puberty to the onset of the menopause, an ovum is released from the ovary. This is called ovulation, and forms part of the *menstrual cycle. In botany, an ovary is the expanded basal portion of the *carpel of flowering plants, containing one or more *ovules. It is hollow with a thick wall to protect the ovules. Following fertilization of the ovum, it develops into the fruit wall or pericarp.

The ovaries of female animals secrete the hormones responsible for the secondary sexual characteristics of the female, such as smooth, hairless facial skin and enlarged breasts. An ovary in a half-grown human fetus contains 5 million eggs, and so the unborn baby already possesses the female genetic information for the next generation.

In botany, the relative position of the ovary to the other floral parts is often a distinguishing character in classification; it may be either inferior or superior, depending on whether the petals and sepals are inserted above or below.

overflow error in computing, an error that occurs if a number is outside the computer's range and is too large to deal with.

overlay in computing, set of specialized data for use with a larger database. A database of a particular country's geography, for example, that includes roads, towns, and natural features such as rivers and lakes might come with overlays that can be displayed or turned off at the user's command, such as the distribution of speed cameras.

oviparous method of animal reproduction in which eggs are laid by the female and develop outside the body, in contrast to *ovoviviparous and *viviparous. It is the most common form of reproduction.

ovoviviparous method of animal reproduction in which fertilized eggs develop within the female (unlike *oviparous), and the embryo gains no nutritional substances from the female (unlike *viviparous). It occurs in some invertebrates, fishes, and reptiles.

ovulation in female animals, the process of releasing egg cells (ova) from the *ovary. In mammals it occurs as part of the *menstrual cycle.

ovule structure found in seed plants that develops into a seed after fertilization. It consists of an *embryo sac containing the female gamete (*ovum or egg cell), surrounded by nutritive tissue, the nucellus. Outside this there are one or two coverings that provide protection, developing into the testa, or seed coat, following fertilization.

ovum plural **ova**, female gamete (sex cell) before fertilization. In mammals it is called an egg, and is produced by a special cell division called *meiosis in the ovaries during the *menstrual cycle. In plants, where it is also known as an egg cell or oosphere, the ovum is produced in an ovule. The ovum does not move by itself. It must be fertilized by a male gamete before it can develop further, except in cases of *parthenogenesis.

oxalic acid or **ethamedioic acid**; (COOH)$_2$.2H$_2$O, white, poisonous solid, soluble in water, alcohol, and ether. Oxalic acid is found in rhubarb leaves, and its salts (oxalates) occur in wood sorrel (genus *Oxalis*, family Oxalidaceae) and other plants. It also occurs naturally in human body cells. It is used in the leather and textile industries, in dyeing and bleaching, ink manufacture, metal polishes, and for removing rust and ink stains.

oxidation in chemistry, the loss of *electrons, gain of oxygen, or loss of hydrogen by an atom, ion, or molecule during a chemical reaction.

At a simple level, oxidation may be regarded as the reaction of a substance with oxygen. For example, *rusting, *respiration, and *combustion (burning) are all oxidation reactions. When carbon is burned in air it is oxidized to carbon dioxide:

$$C + O_2 \rightarrow CO_2$$

Oxidation may be brought about by reaction with another compound (oxidizing agent), which simultaneously undergoes *reduction, or electrically at the anode (positive electrode) of an electrolytic cell.

oxidation number Roman numeral often seen in a chemical name, indicating the *valency of the element immediately before the number.

Examples are lead(II) nitrate, manganese(IV) oxide (manganese dioxide), and potassium manganate(VII) (potassium permanganate).

oxide compound of oxygen and another element, frequently produced by burning the element or a compound of it in air or oxygen.

Oxides of metals are normally *bases and will react with an acid to produce a *salt in which the metal forms the cation (positive ion). Some of them will also react with a strong alkali to produce a salt in which the metal is part of a complex anion (negative ion; see *amphoteric). Most oxides of non-metals are acidic (dissolve in water to form an *acid). Some oxides display no pronounced acidic or basic properties.

oxide film thin film of oxide formed on the surface of some metals as soon as they are exposed to the air. This oxide film makes the metal much more resistant to chemical attack. The considerable lack of reactivity of aluminium to most reagents arises from such a film.

The thickness of the oxide film can be increased by *anodizing the aluminium.

oxidizing agent substance that will oxidize another substance (see *oxidation).

In a redox reaction, the oxidizing agent is the substance that is itself reduced. Common oxidizing agents include oxygen, chlorine, nitric acid, and potassium manganate(VII).

oxyfuel fuel enriched with oxygen to decrease carbon monoxide (CO) emissions. Oxygen is added in the form of chemicals such as methyl tertiary butyl ether (MTBE) and ethanol.

Cars produce CO when there is insufficient oxygen present to convert all the carbon in the petrol to CO$_2$. This occurs mostly at low temperatures, such as during the first five minutes of starting the engine and in cold weather. CO emissions are reduced by the addition of oxygen-rich chemicals. The use of oxyfuels in winter is compulsory in 35 US cities. There are fears, however, that MTBE can cause health problems, including nausea, headaches, and skin rashes.

oxygen chemical symbol O, (Greek *oxys* 'acid'; *genes* 'forming') colourless, odourless, tasteless, *non-metallic, gaseous element, atomic number 8, relative atomic mass 15.9994. It is the most abundant element in the Earth's crust (almost 50% by mass), forms about 21% by volume of the atmosphere, and is present in combined form in water and many other substances. Oxygen is a by-product of *photosynthesis and the basis for *respiration in plants and animals.

Oxygen is very reactive and combines with all other elements except the *noble gases (rare gases) and fluorine. *Combustion (burning) and *rusting are two examples of reactions involving oxygen. It is present in carbon dioxide, silicon dioxide (quartz), iron ore, calcium carbonate (limestone). In nature it exists as a molecule composed of two atoms (O_2); single atoms of oxygen are very short-lived owing to their reactivity. They can be produced in electric sparks and by the Sun's ultraviolet radiation in the upper atmosphere, where they rapidly combine with molecular oxygen to form ozone (O_3), an allotrope of oxygen.

Oxygen is obtained for industrial use by the *fractional distillation of *liquid air, by the *electrolysis of water, or by heating manganese(IV) oxide with potassium chlorate. In the laboratory it is prepared by the action of the *catalyst manganese(IV) oxide on hydrogen peroxide. The simple laboratory test for oxygen is that it relights a glowing spill. Oxygen is essential for *combustion, and is used with ethyne (acetylene) in high-temperature oxyacetylene welding and cutting torches.

The element was first identified by English chemist Joseph Priestley in 1774 and independently in the same year by Swedish chemist Karl Scheele. It was named by French chemist Antoine Lavoisier (1743–1794) in 1777.

oxygen debt physiological state produced by vigorous exercise, in which the lungs cannot supply all the oxygen that the muscles need. In other words, the lungs and *bloodstream, pumped by the *heart, cannot supply sufficient *oxygen for *aerobic respiration in the *muscles. In such a situation the muscles can continue to break down *glucose to liberate energy for a short time using *anaerobic respiration. This partial breakdown produces *lactic acid, which results in a sensation of fatigue when it reaches certain levels in the muscles and the blood. This explains why it is possible to run faster in a sprint than over longer distances. During the sprint, the muscles can respire anaerobically. Once the vigorous muscle movements cease, the body breaks down the accumulated lactic acid on top of the 'normal' breakdown of glucose in aerobic respiration, using up extra oxygen to do so. Panting after exercise is an automatic mechanism to 'pay off' the oxygen debt.

oxyhaemoglobin oxygenated form of *haemoglobin, the protein found in the red blood cells.

oxytocin hormone that stimulates the uterus in late pregnancy to initiate and sustain labour. It is secreted by the *pituitary gland. After birth, it stimulates the uterine muscles to contract, reducing bleeding at the site where the placenta was attached.

oz abbreviation for *ounce.

ozone gas consisting of three atoms of oxygen (O_3), which is therefore an allotrope of oxygen. It is formed when the molecule of the stable form of oxygen (O_2) is split by ultraviolet radiation or electrical discharge. It forms the *ozone layer in the upper atmosphere, which protects life on Earth from ultraviolet rays, a cause of skin cancer.

Ozone is a highly reactive pale-blue gas with a penetrating odour.

ozone depleter any chemical that destroys the ozone in the stratosphere. Most ozone depleters are chemically stable compounds containing chlorine or bromine, which remain unchanged for long enough to drift up to the upper atmosphere. The best known are *chlorofluorocarbons (CFCs).

Other ozone depleters include halons, used in some fire extinguishers; methyl chloroform and carbon tetrachloride, both solvents;

some CFC substitutes; and the pesticide methyl bromide.

CFCs accounted for approximately 75% of ozone depletion in 1995, whereas methyl chloroform (atmospheric concentrations of which had markedly decreased during 1990–94) accounted for an estimated 12.5%. The ozone depletion rate overall is now decreasing as international agreements to curb the use of ozone-depleting chemicals begin to take effect. In 1996 there was a decrease in ozone depleters in the lower atmosphere. This trend is expected to continue into the stratosphere over the next few years.

ozone layer thin layer of the gas *ozone in the upper atmosphere that shields the Earth from harmful ultraviolet rays. A continent-sized hole has formed over Antarctica as a result of damage to the ozone layer. This has been caused in part by *chlorofluorocarbons (CFCs), but many reactions destroy ozone in the stratosphere: nitric oxide, chlorine, and bromine atoms are implicated.

It is believed that the ozone layer is depleting at a rate of about 5% every 10 years over northern Europe, with depletion extending south to the Mediterranean and southern USA. However, ozone depletion over the polar regions is the most dramatic manifestation of a general global effect. Ozone levels over the Arctic in spring 1997 fell over 10% since 1987, despite the reduction in the concentration of CFCs and other industrial compounds which destroy the ozone when exposed to sunlight. It is thought that this may be because of an expanding vortex of cold air forming in the lower stratosphere above the Arctic, leading to increased ozone loss. It is expected that an Arctic hole as large as that over Antarctica could remain a threat to the northern hemisphere for several decades.

The size of the hole in the ozone layer in October 1998 was three times the size of the USA, larger than it had ever been before. In autumn 2000 the hole in the ozone layer was at its largest ever. Observers had hoped that its 1998 level was due to El Niño and would not be exceeded.

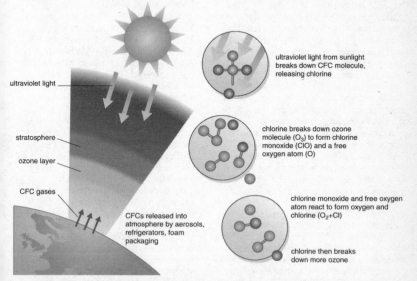

ultraviolet light from sunlight breaks down CFC molecule, releasing chlorine

ultraviolet light

chlorine breaks down ozone molecule (O_3) to form chlorine monoxide (ClO) and a free oxygen atom (O)

stratosphere

ozone layer

CFC gases

CFCs released into atmosphere by aerosols, refrigerators, foam packaging

chlorine monoxide and free oxygen atom react to form oxygen and chlorine (O_2+Cl)

chlorine then breaks down more ozone

ozone layer The destruction of the ozone layer by chlorofluorocarbons (CFCs). CFCs discharged into the atmosphere break down in sunlight releasing chlorine, which breaks down the ozone to form chlorine monoxide and a free oxygen atom. These products react together to form oxygen and chlorine, leaving the chlorine to break down another ozone molecule, and so on.

Pa in physics, symbol for *pascal, the SI unit of pressure, equal to a pressure of one newton per square metre.

packet in computing, unit of data sent across a network. As well as the actual substance of the message, every packet carries error-control information and details of its origin and its final target, enabling a *router to send it on to the intended recipient. This means that packets belonging to the same file can travel via different routes over the network, to be automatically reassembled in the correct sequence when they arrive at their destination. All traffic on the Internet consists of packets. See also *TCP/IP and *X.25.

packet radio in computing, the use of amateur (ham) radio, instead of telephones, to communicate between computers. A terminal node controller (TNC) replaces the modem, a radio transceiver takes the place of the telephone, and the phone system is replaced by radio waves. Packet radio, which works on several different frequencies, has a complete network of its own, complete with satellite links and terrestrial relays. It cannot be used to access the Internet, but it can be connected through the Internet.

packet switching method of transmitting data between computers connected in a *network. Packet switched networks do not provide a dedicated connection between two locations, as with a circuit switched network. Packet switched networks make more effective use of *bandwidth than circuit switched networks, and are more resilient to breaks in network links because there are always multiple routes from source to destination.

paediatrics medical speciality concerned with the care of children. Paediatricians treat childhood diseases such as measles, chicken pox, and mumps, and immunize children against more serious infections such as diphtheria. Their role also includes treating and identifying disorders caused by lack of proper nutrition or child abuse.

page-description language in computing, a control language used to describe the contents and layout of a complete printed page. Page-description languages are frequently used to control the operation of laser printers. The most popular page-description languages are Adobe Postscript and Hewlett-Packard Printer Control Language.

Paget's disease chronic disease of the later years, named after English surgeon James Paget (1814–1899), characterized by thickening and structural disorganization of bone. Most common sites are the pelvis, lumbar spine (lower back), skull and long bones. There may be no symptoms or the victim may suffer severe pain. Treatment is with drugs. The condition is common in the UK but rare in Africa, the Middle East and Asia. Its medical name is osteitis deformans.

painkiller agent for relieving pain. Types of painkiller include analgesics such as *aspirin and aspirin substitutes, *morphine, *codeine, paracetamol, and synthetic versions of the natural inhibitors, the encephalins and endorphins, which avoid the side effects of the others.

PAL acronym for Phase Alternation by Line, video standard used in the UK, other parts of Europe, and China. It has a higher definition and different screen format from the US NTSC standard. Running a television program written for PAL on an NTSC system can result in the bottom of the screen image being cut off.

Palaeocene epoch (Greek 'old' + 'recent') first epoch of the Tertiary period of geological time, 65–56.5 million years ago. Many types of mammals spread rapidly after the disappearance of the great reptiles of the Mesozoic. Flying mammals replaced the flying reptiles, swimming mammals replaced the swimming reptiles, and all the ecological niches vacated by the reptiles were adopted by mammals.

At the end of the Palaeocene there was a mass extinction that caused more than half of all bottom-dwelling organisms to disappear worldwide, over a period of around 1,000 years. Surface-dwelling organisms remained unaffected, as did those on land. The cause of this extinction remains unknown, though US palaeontologists have found evidence (released in 1998) that it may have been caused by the Earth releasing tonnes of methane into the oceans causing increased water temperatures.

palaeomagnetic stratigraphy use of distinctive sequences of magnetic polarity reversals to date rocks. Magnetism retained in rocks at the time of their formation are matched with known dated sequences of *polar reversals or with known patterns of secular variation.

palaeomagnetism study of the magnetic properties of rocks in order to reconstruct the Earth's ancient magnetic field and the former positions of the continents, using traces left by the Earth's magenetic field in igneous rocks before they cool. Palaeomagnetism shows that the Earth's magnetic field has reversed itself – the magnetic north pole becoming the magnetic south pole, and vice versa – at approximate half-million-year intervals, with shorter reversal periods in between the major spans.

Starting in the 1960s, this known pattern of magnetic reversals was used to demonstrate seafloor spreading or the formation of new ocean crust on either side of mid-oceanic ridges. As new material hardened on either side of a ridge, it would retain the imprint of the magnetic field, furnishing datable proof that material was spreading steadily outward. Palaeomagnetism is also used to demonstrate *continental drift by

determining the direction of the magnetic field of dated rocks from different continents.

palaeontology the study of ancient life, encompassing the structure of ancient organisms and their environment, evolution, and ecology, as revealed by their *fossils and the rocks in which those fossils are found. The practical aspects of palaeontology are based on using the presence of different fossils to date particular rock strata and to identify rocks that were laid down under particular conditions; for instance, giving rise to the formation of oil. The use of fossils to trace the age of rocks was pioneered in Germany by Johann Friedrich Blumenbach (1752–1830) at Göttingen, followed by Georges Cuvier and Alexandre Brongniart in France in 1811.

Palaeozoic era era of geological time 570–245 million years ago. It comprises the Cambrian, Ordovician, Silurian, Devonian, Carboniferous, and Permian periods. The Cambrian, Ordovician, and Silurian constitute the Lower or Early Palaeozoic; the Devonian, Carboniferous, and Permian make up the Upper or Late Palaeozoic. The era includes the evolution of hard-shelled multicellular life forms in the sea; the invasion of land by plants and animals; and the evolution of fish, amphibians, and early reptiles.

palindromic in medicine, a term used to describe symptoms or diseases that recur.

palladium chemical symbol Pd, lightweight, ductile and malleable, silver-white, metallic element, atomic number 46, relative atomic mass 106.42. It is one of the so-called platinum group of metals, and is resistant to tarnish and corrosion. It often occurs in nature as a free metal (see *native metal) in a natural alloy with platinum. Palladium is used as a catalyst, in alloys of gold (to make white gold) and silver, in electroplating, and in dentistry. It was discovered in 1803 by English physicist William Wollaston, and named after the asteroid Pallas (found in 1802).

palsy in medicine, another term for *paralysis.

pancreas in vertebrates, a gland of the digestive system located close to the duodenum. When stimulated by the hormone secretin, it releases enzymes into the duodenum that digest starches, proteins, and fats. In humans, it is about 18 cm/7 in long, and lies behind and below the stomach. It contains groups of cells called the *islets of Langerhans, which secrete the hormones insulin and glucagon that regulate the blood sugar level.

Pangaea or **Pangea**, (Greek 'all-land') single land mass, made up of all the present continents, believed to have existed between 300 and 200 million years ago; the rest of the Earth was covered by the Panthalassa ocean. Pangaea split into two land masses – *Laurasia in the north and Gondwanaland in the south – which subsequently broke up into several continents. These then moved slowly to their present positions (see *plate tectonics).

The former existence of a single 'supercontinent' was proposed by German meteorologist Alfred Wegener in 1912.

pantothenic acid water-soluble *vitamin ($C_9H_{17}NO_5$) of the B complex, found in a wide variety of foods. Its absence from the diet can lead to dermatitis, and it is known to be involved in the breakdown of fats and carbohydrates.

paper sizes standard European sizes for paper, designated by a letter (A, B, or C) and a number (0–6). The letter indicates the size of the basic sheet at manufacture; the number is how many times it has been folded. A4 is obtained by folding an A3 sheet, which is half an A2 sheet, in half, and so on.

papilloma overgrowth of cells at the surface of tissues, such as skin, bladder or intestinal lining. Papillomas are localized and usually benign. A wart is a papilloma.

papovavirus in medicine, group of viruses that include the causative agents of *papilloma (warts and verrucas) in humans.

parabola in mathematics, a curve formed by cutting a right circular cone with a plane parallel to the sloping side of the cone. A parabola is one of the family of curves known as *conic sections. The graph of $y = x^2$ is a parabola. It can also be defined as a path traced out by a point that moves in such a way that the distance from a fixed point (focus) is equal to its distance from a fixed straight line (directrix); it thus has an *eccentricity of 1.

The trajectories of missiles within the Earth's gravitational field approximate closely to parabolas (ignoring the effect of air resistance). The corresponding solid figure, the paraboloid, is formed by rotating a parabola about its axis. It is a common shape for headlight reflectors, dish-shaped microwave and radar aerials, and radiotelescopes, since a source of radiation placed at the focus of a paraboloidal reflector is propagated as a parallel beam.

paracetamol analgesic, particularly effective for musculoskeletal pain. It is as effective as aspirin in reducing fever, and less irritating to the stomach, but has little anti-inflammatory action. An overdose can cause severe, often irreversible or even fatal, liver and kidney damage.

paraffin common name for *alkane, any member of the series of hydrocarbons with the general formula C_nH_{2n+2}. The lower members are gases, such as methane (marsh or natural gas). The middle ones (mainly liquid) form the basis of petrol, kerosene, and lubricating oils, while the higher ones (paraffin waxes) are used in ointment and cosmetic bases.

paraldehyde common name for *ethanal trimer.

parallax in *virtual reality, the distance between the viewer's left and right eyes in the virtual world.

parallel circuit electrical circuit in which the components are connected side by side. The current flowing in the circuit is shared by the components.

voltage in a parallel circuit
Components are connected across the same voltage source, and therefore the voltage across each of the components is the same (provided the resistance of each component is also the same). This can be written as $V_T = V_1 = V_2$, where V_T is the total voltage, V_1 is the voltage

across component 1, and V_2 is the voltage across component 2. For example, if the voltage across component 1 is 6 volts, then the voltage across component 2 is also 6 volts. The mains supply in the UK is 240 volts. All electrical appliances are designed to operate efficiently at this voltage. Appliances are connected in parallel circuits with the same voltage applied to each appliance.

current in a parallel circuit The current divides to flow through each of the components in the parallel paths. The amount of current that flows through each of the components depends on the resistance of each component. A component with a lower resistance will allow more current to flow through. If I_1 is the current flowing through component 1 and I_2 is the current flowing through component 2, then the total current, I_T, is given by $I_T = I_1 + I_2$.

resistance in a parallel circuit The total resistance R_T of two conductors of resistances R_1 and R_2 connected in parallel is given by:

$$\frac{1}{R_T} = \frac{1}{R_1} + \frac{1}{R_2}$$

parallel circuit In a parallel circuit, the components are connected side by side, so that the current is split between two or more parallel paths or conductors.

parallel device in computing, a device that communicates binary data by sending the bits that represent each character simultaneously along a set of separate data lines, unlike a *serial device.

parallel lines and parallel planes straight lines or planes that always remain a constant distance from one another no matter how far they are extended. Examples of parallel lines found in everyday life include railway lines, the verticals of goal posts, and shelves in a book-case. Parallel lines are shown in diagrams by using single or double arrows:

parallelogram a quadrilateral (a shape with four sides) with opposite pairs of sides equal in length and parallel, and opposite angles equal. The diagonals of a parallelogram bisect each other. Its area is the product of the length (l) of one side and the perpendicular distance (height h) between this and the opposite side; the formula is $A = l \times h$. In the special case when all four sides are equal in length, the parallelogram is known as a rhombus, and when the internal angles are right angles, it is a rectangle or square.

parallelogram of forces in physics and applied mathematics, a method of calculating the resultant (combined effect) of two different forces acting together on an object. Because a force has both magnitude and direction it is a *vector quantity and can be represented by a straight line. A second force acting at the same point in a

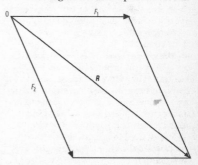

parallelogram of forces The diagram shows how the parallelogram of forces can be used to calculate the resultant (R) of two different forces (F_1 and F_2) acting together on an object. The two forces are represented by two lines drawn at an angle to each other. By completing the parallelogram (of which the two lines are sides), a diagonal may be drawn from the original angle to the opposite corner to represent the resultant force vector.

different direction can be represented by another line drawn at an angle to the first. By completing the parallelogram (of which the two lines are sides) a diagonal may be drawn from the original angle to the opposite corner to represent the resultant force vector.

parallel processing emerging computer technology that allows more than one computation at the same time. Although in the 1990s this technology enabled only a small number of computer processor units to work in parallel, in theory thousands or millions of processors could be used at the same time.

parallel running in computing, a method of implementing a new computer system in which the new system and the old system are run together for a short while. The old system is therefore available to take over from its replacement should any faults arise. An alternative method is *pilot running.

paralysis loss of voluntary movement due to failure of nerve impulses to reach the muscles involved. It may result from almost any disorder of the nervous system, including brain or spinal cord injury, poliomyelitis, stroke, and progressive conditions such as a tumour or multiple sclerosis. Paralysis may also involve loss of sensation due to sensory nerve disturbance.

paramagnetic material material that is weakly pulled towards a strong magnet. The effect is caused by unpaired electrons in the atoms of the material, which cause the atoms to act as weak magnets. A paramagnetic material has a fairly low magnetic susceptibility that is inversely proportional to temperature. See *magnetism.

parameter variable factor or characteristic. For example, length is one parameter of a rectangle; its height is another. In computing, it is frequently useful to describe a program or object with a set of variable parameters rather than fixed values.

paranoia mental disorder marked by delusions of grandeur or persecution. In popular usage, paranoia means baseless or exaggerated fear and suspicion.

paranormal not within the range of, or explicable by, established science. Paranormal phenomena include *extrasensory perception** (ESP) which takes in clairvoyance, precognition, and telepathy; **telekinesis**, the movement of objects from one position to another by human mental concentration; and **mediumship**, supposed contact with the spirits of the dead, usually via an intermediate 'guide' in the other world. *Parapsychology is the study of such phenomena.

paraplegia paralysis of the lower limbs, involving loss of both movement and sensation; it is usually due to spinal injury.

parapsychology (Greek *para* 'beyond') study of *paranormal phenomena, which are generally subdivided into two types: *extrasensory perception (ESP), or the paracognitive; and psychokinesis (PK), telekinesis, or the paraphysical – movement of an object without the use of physical force or energy.

parasite organism that lives on or in another organism (called the host) and depends on it for nutrition, often at the expense of the host's welfare. Parasites that live inside the host, such as liver flukes and tapeworms, are called **endoparasites**; those that live on the exterior, such as fleas and lice, are called **ectoparasites**.

Parasitic wasps, such as ichneumons, are more correctly **parisitoids**, as they ultimately kill their hosts.

parasympathetic nervous system in medicine, a division of the *autonomic nervous system.

parenchyma plant tissue composed of loosely packed, more or less spherical cells, with thin cellulose walls. Although parenchyma often has no specialized function, it is usually present in large amounts, forming a packing or ground tissue. It usually has many intercellular spaces.

parity of a number, the state of being either even or odd. In computing, the term refers to the number of 1s in the binary codes used to represent data. A binary representation has **even parity** if it contains an even number of 1s and **odd parity** if it contains an odd number of 1s.

parity check form of *validation of data.

Parkinson's disease or **parkinsonism** or **paralysis agitans**, first described by British neurologist James Parkinson (1755–1824) in 1817, degenerative disease of the brain characterized by a progressive loss of mobility, muscular rigidity, tremor, and speech difficulties. The condition is mainly seen in people over the age of 50.

parotid gland one of two salivary glands located on either side of the mouth just in front of the ear. Their ducts open on the inside of the cheeks.

parse in computing, software facility for breaking down a data stream into individual pieces of information that can be acted upon. On the World Wide Web, for example, data entered by a user can be sent to a database program for storage and later analysis; the ability to do this depends on being able to feed the right bit of data into the right record field.

parsec symbol pc, unit used for distances to stars and galaxies. One parsec is equal to 3.2616 *light years, 2.063×10^5 *astronomical units, and 3.857×10^{13} km.

parthenogenesis development of an ovum (egg) without any genetic contribution from a male. Parthenogenesis is the normal means of reproduction in a few plants (for example, dandelions) and animals (for example, certain fish). Some sexually reproducing species, such as aphids, show parthenogenesis at some stage in their life cycle to accelerate reproduction to take advantage of good conditions.

particle detector one of a number of instruments designed to detect subatomic particles and track their paths; they include the *cloud chamber, *bubble chamber, *spark chamber, and multiwire chamber.

The earliest particle detector was the cloud chamber, which contains a super-saturated vapour in which particles leave a trail of droplets, in much the same way that a jet aircraft leaves a trail of vapour in the sky. A bubble chamber contains a superheated liquid in which a particle leaves a trail of bubbles. A spark chamber contains a series of closely-packed parallel metal plates, each at a high voltage. As particles pass through the chamber, they leave a visible spark between the plates. A modern multiwire chamber consists of an array of fine, closely-packed wires, each at a high voltage. As a particle passes through the chamber, it produces an electrical signal in the wires. A computer analyses the signal and reconstructs the path of the particles. Multiwire detectors can be used to detect X-rays and gamma rays, and are used as detectors in *positron emission tomography (PET).

particle physics study of the particles that make up all atoms, and of their interactions. More than 300 subatomic particles have now been identified by physicists, categorized into several classes according to their mass, electric charge, spin, magnetic moment, and interaction. Subatomic particles include the *elementary particles (*quarks, *leptons, and *gauge bosons), which are indivisible, so far as is known, and so may be considered the fundamental units of matter; and the *hadrons (*baryons, such as the *proton and *neutron, and *mesons), which are composite particles, made up of two or three quarks. Quarks, protons, *electrons, and neutrinos are the only stable particles (the neutron being stable only when in the atomic nucleus). The unstable particles decay rapidly into other particles, and are known from experiments with particle accelerators and cosmic radiation. See *atomic structure.

Pioneering research took place at the Cavendish laboratory, Cambridge, England. In 1897 the English physicist J J Thomson discovered that all atoms contain identical, negatively charged particles (electrons), which can easily be freed. By 1911 the New Zealand physicist Ernest Rutherford (1871–1937) had shown that the electrons surround a very small, positively charged *nucleus. In the case of hydrogen, this was found to consist of a single positively charged particle, a proton.

The nuclei of other elements are made up of protons and uncharged particles called neutrons. 1932 saw the discovery of a particle (whose existence had been predicted by the British theoretical physicist Paul Dirac in 1928) with the mass of an electron, but an equal and opposite charge – the *positron. This was the first example of *antimatter; it is now believed that all particles have corresponding antiparticles. In 1934 the Italian-born US physicist Enrico Fermi (1901–1954) argued that a hitherto unsuspected particle, the *neutrino, must accompany electrons in beta-emission.

particles and fundamental forces
By the mid-1930s, four types of fundamental *force interacting between particles had been identified.

The *electromagnetic force (1) acts between all particles with electric charge, and is related to the exchange between these particles of *gauge bosons called *photons, packets of electromagnetic radiation.

In 1935 the Japanese physicist Hideki Yukawa suggested that the *strong nuclear force (2) (binding protons and neutrons together in the nucleus) was transmitted by the exchange of particles with a mass about one-tenth of that of a proton; these particles, called *pions (originally pi mesons), were found by the British physicist Cecil Powell in 1946. Yukawa's theory was largely superseded from 1973 by the theory of *quantum chromodynamics, which postulates that the strong nuclear force is transmitted by the exchange of gauge bosons called *gluons between the *quarks and antiquarks making up protons and neutrons.

Theoretical work on the *weak nuclear force (3) began with Enrico Fermi in the 1930s. The existence of the gauge bosons that carry this force, the W and Z particles, was confirmed in 1983 at CERN, the European nuclear research organization.

The fourth fundamental force, *gravity (4), is experienced by all matter; the postulated carrier of this force has been named the *graviton.

Leptons 'feel' the weak nuclear and electromagnetic force and are not affected by the strong nuclear force. Hadrons – mesons and baryons – 'feel' the strong nuclear force.

quarks In 1964 the US physicists Murray Gell-Mann and George Zweig suggested that all hadrons were built from three 'flavours' of a new particle with half-integral spin and a charge of magnitude either $1/3$ or $2/3$ that of an electron; Gell-Mann named the particle the **quark**. Mesons are quark–antiquark pairs (spins either add to one or cancel to zero), and baryons are quark triplets. To account for new mesons such as the psi (J) particle the number of quark flavours had risen to six by 1985.

particle size in chemistry the size of the grains that make up a powder. The grain size has an effect on certain properties of a substance. Finely divided powders have a greater surface area for contact; they therefore react more quickly, dissolve more readily, and are of increased efficiency as catalysts compared with their larger-sized counterparts.

particle, subatomic a particle that is smaller than an atom; see *particle physics.

parvovirus in medicine, one of a group of viruses responsible for epidemic nausea and vomiting. Outbreaks can affect whole communities and are more common in the winter. Symptoms include nausea, vomiting, diarrhoea, and giddiness but the disease is not serious and usually lasts for less than three days.

Pascal acronym for **Program Appliqué à la Selection et la Compilation Automatique de la Littérature**, high-level computer-programming language. Designed by Niklaus Wirth in the 1960s as an aid to teaching programming, it is still widely used as such in universities, and as a good general-purpose programming language. Most professional programmers, however, now use *C or *C++. Pascal was named after the 17th-century French mathematician Blaise Pascal (1623–1662).

pascal symbol Pa, SI unit of pressure, equal to one newton per square metre. It replaces *bars and millibars (10^5 Pa equals one bar). It is named after the

French mathematician Blaise Pascal
(1623–1662).

Pascal's principle   see *liquid.

Pascal's triangle   triangular array of
numbers (with 1 at the apex), in which
each number is the sum of the pair of
numbers above it. It is named after
French mathematician Blaise Pascal
(1623–1662), who used it in his study
of probability. When plotted at equal
distances along a horizontal axis, the
numbers in the rows give the binomial
probability distribution (with equal
probability of success and failure) of an
event, such as the result of tossing a
coin.

Pascal's triangle In Pascal's triangle, each
number is the sum of the two numbers
immediately above it, left and right – for
example, 2 is the sum of 1 and 1, and 4 is the
sum of 3 and 1. Furthermore, the sum of each
row equals a power of 2 – for example, the sum
of the 3rd row is $4 = 2^2$; the sum of the 4th
row is $8 = 2^3$.

passive matrix display or **passive
matrix LCD**, in computing, *liquid
crystal display (LCD) produced by
passing a current between an array of
electrodes set between glass plates.
Passive matrix screens lack the
transistors that enhance the
performance of *active matrix LCDs,
which makes them relatively
inexpensive, but lacking in contrast
and slow to react.

passive smoking   inhalation of tobacco
smoke from other people's cigarettes.

Pasteur effect   modification in the
process of energy production in certain
organisms. A bacterium or facultative
cell which normally produces its **ATP**
anaerobically, that is, without the use
of oxygen, does so by means of
*glycolysis. In this process the yield of
ATP per molecule of glucose is low, so
the cell uses up a great deal of its
glucose for its energy requirements. An

aerobic organism, which uses oxygen,
can employ the *Krebs cycle to
produce its energy, and the yield of
ATP per molecule of glucose is much
greater. The Pasteur effect occurs when
a facultative anaerobic cell is provided
with oxygen, so that its high rate of
glucose metabolism is slowed down. It
then produces energy from glucose far
more efficiently because Krebs cycle is
used, and in addition the production of
*lactic acid, a normal product of
anaerobic glycolysis, is automatically
stopped. Named after French chemist
and microbiologist Louis Pasteur
(1822–1895).

patch   in computing, modification or
update made to a program, consisting
of a short segment of additional code.
Developers often correct bugs or fine-
tune software by releasing a patch
which rewrites existing codes and adds
new material.

pathogen   (Greek 'disease producing')
any micro-organism that causes
disease. Most pathogens are *parasites,
and the diseases they cause are
incidental to their search for food or
shelter inside the host. Nonparasitic
organisms, such as soil bacteria or
those living in the human gut and
feeding on waste foodstuffs, can also
become pathogenic to a person whose
immune system or liver is damaged.
The larger parasites that can cause
disease, such as nematode worms, are
not usually described as pathogens.

pathology   medical speciality
concerned with the study of disease
processes and how these provoke
structural and functional changes in
the body.

Pauli exclusion principle   in physics, a
principle of atomic structure
discovered by Austrian-born physicist
Wolfgang Pauli (1900–1958). See
*exclusion principle.

payment service provider   PSP,
company which allows an *electronic
commerce Web site to accept payments
online from credit and debit cards,
without the need for the site to hold
payment details itself. The card
transactions are cleared and
cardholders' accounts debited
immediately. PSPs include Datacash,
Netbanx, Secpay, and Worldpay.

PC Card standard for 'credit card' memory and device cards used in *portable computers. As well as providing flash memory, PC Cards can provide either additional disk storage, or modem or fax functionality. PC Card was adopted as being a simpler and more accurate name than PCMCIA.

PCI abbreviation for peripheral component interconnect, form of *local bus connection between external devices and the main *central processing unit. Developed (but not owned) by Intel, it was available as 32-bit in 1993, but is now available as 64-bit.

PCL *page description language, developed by Hewlett Packard for use on Laserjet laser printers. Versions PCL 1 to PCL 4 used *raster graphics fonts; PCL 5 uses outline fonts.

PCM abbreviation for *pulse-code modulation.

PCX in computing, bitmapped *graphics file format, originally developed by Z-Soft for use with PC-Paintbrush, but now used and generated by many applications and hardware such as scanners.

pd abbreviation for *potential difference.

PDF abbreviation for portable document format, in computing, file format created by Adobe's Acrobat system that retains the entire content of an electronic document (including layouts, graphics, styled text, and navigation features) regardless of the computer system on which it is viewed. Because they are platform-independent, PDF files are often an effective way to send documents over the Internet.

peat organic matter found in bogs and formed by the incomplete decomposition of plants such as sphagnum moss. Northern Asia, Canada, Finland, Ireland, and other places have large deposits, which have been dried and used as fuel from ancient times. Peat can also be used as a soil additive.

Peat bogs began to be formed when glaciers retreated, about 9,000 years ago. They grow at the rate of only a millimetre a year, and large-scale digging can result in destruction both of the bog and of specialized plants growing there. The destruction of peat bogs is responsible for diminishing fish stocks in coastal waters; the run-off from the peatlands carries high concentrations of iron, which affects the growth of the plankton on which the fish feed.

Approximately 60% of the world's wetlands are peat. In May 1999 the Ramsar Convention on the Conservation of Wetlands approved a peatlands action plan that should have a major impact on the conservation of peat bogs.

pectoral relating to the upper area of the thorax associated with the muscles and bones used in moving the arms or forelimbs, in vertebrates. In birds, the *pectoralis major* is the very large muscle used to produce a powerful downbeat of the wing during flight.

pediment broad, gently inclined erosion surface formed at the base of a mountain as it erodes and retreats. Pediments consist of bedrock and are often covered with a thin layer of sediments, called alluvium, which have been eroded off the mountain.

peer-to-peer networking in computing, method of file sharing in which computers are linked to each other as opposed to being linked to a central file server.

pelagic of or pertaining to the open ocean, as opposed to bottom or shore areas. **Pelagic sediment** is fine-grained fragmental material that has settled from the surface waters, usually the siliceous and calcareous skeletal remains of marine organisms, such as radiolarians and foraminifera.

Peltier effect in physics, a change in temperature at the junction of two different metals produced when an electric current flows through them. The extent of the change depends on what the conducting metals are, and the nature of change (rise or fall in temperature) depends on the direction of current flow. It is the reverse of the *Seebeck effect. It is named after the French physicist Jean Charles Peltier (1785–1845) who discovered it in 1834.

pelvis in vertebrates, the lower area of the abdomen featuring the bones and muscles used to move the legs or hindlimbs. The **pelvic girdle** is a set of bones that allows movement of the legs in relation to the rest of the body and provides sites for the attachment of relevant muscles.

penicillin   any of a group of *antibiotic (bacteria killing) compounds obtained from filtrates of moulds of the genus *Penicillium* (especially *P. notatum*) or produced synthetically. Penicillin was the first antibiotic to be discovered (by Scottish bacteriologist Alexander Fleming (1881–1955)); it kills a broad spectrum of bacteria, many of which cause disease in humans.

peninsula   land surrounded on three sides by water but still attached to a larger landmass. Florida, USA, is an example.

penis   male reproductive organ containing the *urethra, the channel through which urine and *semen are voided. It transfers sperm to the female reproductive tract to fertilize the ovum. In mammals, the penis is made erect by vessels that fill with blood, and in most mammals (but not humans) is stiffened by a bone.

Pennsylvanian period   US term for the Upper or Late *Carboniferous period of geological time, 323–290 million years ago; it is named after the US state, which contains vast coral deposits.

pentadactyl limb   typical limb of the mammals, birds, reptiles, and amphibians. These vertebrates (animals with backbone) are all descended from primitive amphibians whose immediate ancestors were fleshy-finned fish. The limb which evolved in those amphibians had three parts: a 'hand/foot' with five digits (fingers/toes), a lower limb containing two bones, and an upper limb containing one bone.

pentagon   five-sided plane figure. A regular pentagon has all five sides of equal length and all five angles of equal size, 108°. It has golden section proportions between its sides and diagonals. The five-pointed star formed by drawing all the diagonals of a regular pentagon is called a **pentagram**. This star has further golden sections.

pentanol   common name **amyl alcohol**; $C_5H_{11}OH$, clear, colourless, oily liquid, usually with a characteristic choking odour. It is obtained by the fermentation of starches and from the distillation of petroleum. There are eight possible isomers.

Pentium   in computing, microprocessor produced by Intel in 1993. The Pentium followed on from the 486 processor and would have been called the 586, but Intel wanted to distinguish its processor from those of rival chip manufacturers, and was unable to register numbers as a trademark. The Pentium family was extended by the Pentium Pro in 1995 and the Pentium II in 1997, and also by the addition of MMX instructions in 1996. All members of the family are 32-bit chips with 64-bit data buses for faster access to memory and the *PCI expansion bus. The Pentium III, running at 1.0 GHz, was released in 2000, and a 2.0 GHz Pentium 4 in 2001.

pepsin   *enzyme that breaks down proteins during digestion. It is produced by the walls of the *stomach. It requires a strongly acidic environment such as that present in the stomach. It digests large protein molecules into smaller protein molecules (smaller polypeptides) and is therefore a protease – an enzyme that breaks down a protein.

peptide   molecule comprising two or more *amino acid molecules (not necessarily different) joined by **peptide bonds**, whereby the acid group of one acid is linked to the amino group of the other (–CO.NH–). The number of amino acid molecules in the peptide is indicated by referring to it as a di-, tri-, or polypeptide (two, three, or many amino acids).

Proteins are built up of interacting polypeptide chains with various types of bonds occurring between the chains. Incomplete hydrolysis (splitting up) of a protein yields a mixture of peptides, examination of which helps to determine the sequence in which the amino acids occur within the protein.

peptide bond   bond that joins two peptides together within a protein. The carboxyl (–COOH) group on one *amino acid reacts with the amino

(–NH$_2$) group on another amino acid to form a peptide bond (–CO.NH–) with the elimination of water. Peptide bonds are broken by hydrolytic enzymes called peptidases.

percentage way of representing a number as a *fraction of 100. For example, 45 percent (45%) equals $^{45}/_{100}$, and 45% of 20 is $^{45}/_{100} \times 20 = 9$.

percentile in a cumulative frequency distribution, one of the 99 values of a variable that divide its distribution into 100 parts of equal frequency. In practice, only certain of the percentiles are used. They are the *median (or 50th percentile), the lower and the upper *quartiles (respectively, the 25th and 75th percentiles), the 10th percentile that cuts off the bottom 10% of a frequency distribution, and the 90th that cuts off the top 10%. The 5th and 95th are also sometimes used.

percolation gradual movement or transfer of water thorough porous substances (such as porous rocks or soil).

pericarp wall of a *fruit. It encloses the seeds and is derived from the *ovary wall. In fruits such as the acorn, the pericarp becomes dry and hard, forming a shell around the seed. In fleshy fruits the pericarp is typically made up of three distinct layers. The **epicarp**, or **exocarp**, forms the tough outer skin of the fruit, while the **mesocarp** is often fleshy and forms the middle layers. The innermost layer or **endocarp**, which surrounds the seeds, may be membranous or thick and hard, as in the drupe (stone) of cherries, plums, and apricots.

peridot pale-green, transparent gem variety of the mineral olivine.

periglacial environment bordering a glacial area but not actually covered by ice all year round, or having similar climatic and environmental characteristics, such as in mountainous areas. Periglacial areas today include parts of Siberia, Greenland, and North America. The rock and soil in these areas is frozen to a depth of several metres (*permafrost) with only the top few centimetres thawing during the brief summer (the active layer). The vegetation is characteristic of *tundra.

During the last ice age all of

southern England was periglacial. Weathering by *freeze–thaw (the alternate freezing and thawing of ice in rock cracks) would have been severe, and solifluction (movement of soil that is saturated by water) would have taken place on a large scale, causing wet topsoil to slip from valley sides.

perimeter or **boundary**, line drawn around the edge of an area or shape. For example, the perimeter of a rectangle is the sum of the lengths of its four sides; the perimeter of a circle is known as its *circumference.

perineum part of the external floor of the *pelvis which lies between the anus and the *vulva in the female, and between the anus and *scrotum in the male. It is a tough sheet of *ligaments and *muscles.

period in chemistry, a horizontal row of elements in the *periodic table. There is a gradation of properties along each period, from metallic (Group 1, the alkali metals) to non-metallic (Group 7, the halogens).

period another name for menstruation; see *menstrual cycle.

period in physics, the time taken for one complete cycle of a repeated sequence of events. For example, the time taken for a pendulum to swing from side to side and back again is the period of the pendulum. The period is the reciprocal of the *frequency.

periodic table of the elements a table in which the *elements are arranged in order of their atomic number. There are eight groups of elements, plus a block of *transition metals in the centre. Group 1 contains the *alkali metals; Group 2 the alkaline-earth *metals; Group 7 the *halogens; and Group 0 the *noble gases (rare gases). A zigzag line through the groups separates the metals on the left-hand side of the table from the non-metals on the right. The horizontal rows in the table are called **periods**. The table summarizes the major properties of the elements and enables predictions to be made about their behaviour.

There are striking similarities in the chemical properties of the elements in each of the *groups (vertical columns), which are numbered 1–7 (I–VII) and 0 to reflect the number of electrons in the

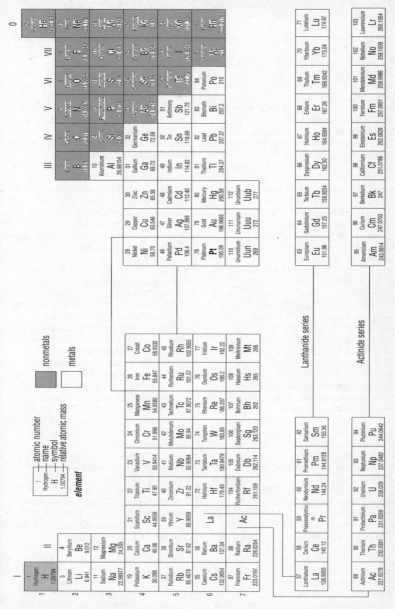

periodic table of the elements The periodic table of the elements arranges the elements into horizontal rows (called periods) and vertical columns (called groups) according to their atomic numbers. The elements in a group or column all have similar properties – for example, all the elements in the far right-hand column are noble gases.

outermost unfilled shell and hence the maximum *valency. Reactivity increases down a group. A gradation (trend) of properties may be traced along the periods. Metallic character increases across a period from right to left, and down a group. A large block of elements, between Groups 2 and 3, contains the transition elements, characterized by displaying more than one valence state.

peripheral device in computing, any item connected to a computer's *central processing unit (CPU). Typical peripherals include keyboard, mouse, monitor, and printer. Users who enjoy playing games might add a joystick or a *trackball; others might connect a *modem, scanner, or *integrated services digital network (ISDN) terminal to their machines.

peristalsis wavelike contractions, produced by the contraction of smooth muscle, that pass along tubular organs, such as the intestines. The same term describes the wavelike motion of earthworms and other invertebrates, in which part of the body contracts as another part elongates.

Perl acronym for Practical Extraction and Report Language, open source programming language developed in 1987 by US linguist Larry Wall for processing text. However, it has proved extremely popular as a way of writing *CGI scripts for Web pages. Hailed as the 'duct tape' of the Internet, Perl is used by over half of the sites on the World Wide Web.

permafrost condition in which a deep layer of soil does not thaw out during the summer. Permafrost occurs under *periglacial conditions. It is claimed that 26% of the world's land surface is permafrost.

Permafrost gives rise to a poorly drained form of grassland typical of northern Canada, Siberia, and Alaska known as *tundra.

permanent hardness hardness of water that cannot be removed by boiling (see *hard water).

permeability in physics, the degree to which the presence of a substance alters the magnetic field around it. Most substances have a small constant permeability. When the permeability is less than 1, the material is a *diamagnetic material; when it is greater than 1, it is a *paramagnetic material. *Ferrimagnetic materials have very large permeabilities. See also *magnetism.

permeable rock rock which through which water can pass either via a network of spaces between particles or along bedding planes, cracks, and fissures. Permeable rocks can become saturated. Examples of permeable rocks include limestone (which is heavily jointed) and chalk (porous).

Unlike *impermeable rocks, which do not allow water to pass through, permeable rocks rarely support rivers and are therefore subject to less erosion. As a result they commonly form upland areas (such as the chalk downs of southeastern England, and the limestone Pennines of northern England).

Permian period of geological time 290–245 million years ago, the last period of the Palaeozoic era. Its end was marked by a dramatic change in marine life – the greatest mass extinction in geological history – including the extinction of many corals and trilobites. Deserts were widespread, terrestrial amphibians and mammal-like reptiles flourished, and cone-bearing plants (gymnosperms) came to prominence. In the oceans, 49% of families and 72% of genera vanished in the late Permian. On land, 78% of reptile families and 67% of amphibian families disappeared.

permutation in mathematics, a specified arrangement of a group of objects.

It is the arrangement of a distinct objects taken b at a time in all possible orders. It is given by $a!/(a - b)!$, where '!' stands for *factorial. For example, the number of permutations of four letters taken from any group of six different letters is $6!/2! = (1 \times 2 \times 3 \times 4 \times 5 \times 6)/(1 \times 2) = 360$. The theoretical number of four-letter 'words' that can be made from an alphabet of 26 letters is $26!/22! = 358,800$.

See also *combination.

perpendicular at a right angle; also, a line at right angles to another line or to a plane. Everyday examples include lamp posts, which are perpendicular to the road, and walls, which are perpendicular to the ground.

perpetual motion idea that a machine can be designed and constructed in such a way that, once started, it will do work indefinitely without requiring any further input of energy (motive power). Such a device would contradict at least one of the two laws of thermodynamics that state that (1) energy can neither be created nor destroyed (the law of conservation of energy) and (2) heat cannot by itself flow from a cooler to a hotter object. As a result, all practical (real) machines require a continuous supply of energy, and no heat engine is able to convert all the heat into useful work.

persistent vegetative state PVS, condition arising from overwhelming damage to the *cerebral cortex. The patient lies unresponsive and, though the eyes may be open, he or she remains unaware of his or her surroundings, makes no purposeful gestures, and never speaks. The vegetative patient may survive for many years in a condition that many observers regard as a living death. However, some patients can regain consciousness, even after as long as five years in a vegetative state.

personal computer PC, another name for microcomputer. The term is also used, more specifically, to mean the IBM Personal Computer and computers compatible with it.

personal digital assistant PDA, handheld computer designed to store names, addresses, and diary information, and to send and receive faxes and e-mail. They aim to provide a more flexible and powerful alternative to the Filofax or diary.

personal identification device PID, device, such as a magnetic card, carrying machine readable identification, which provides authorization for access to a computer system. PIDs are often used in conjunction with a PIN.

personal productivity software in computing, work-oriented software such as word processing, *spreadsheets, or databases.

Perspex trade name for a clear, light-weight, tough plastic first produced in 1930. It is widely used for watch glasses, advertising signs, domestic baths, motorboat windscreens, aircraft canopies, and protective shields. it is formed by *addition polymerization and its chemical name is polymethylmethacrylate (PMMA). It is manufactured under other names: Plexiglas, Lucite, Acrylite, and Rhoplex (in the USA), and Oroglas (in Europe).

pervious rock rock that allows water to pass through via bedding planes, cracks, and fissures, but not through pores within the rock (that is, they are non-porous). See *permeable rock.

pesticide any chemical used in farming, gardening, or in the home to combat pests. Pesticides are of three main types: **insecticides** (to kill insects), **fungicides** (to kill fungal diseases), and **herbicides** (to kill plants, mainly those considered weeds). Pesticides cause a number of pollution problems through spray drift onto surrounding areas, direct contamination of users or the public, and as residues on food. The World Health Organization (WHO) estimated in 1999 that 20,000 people die annually worldwide from pesticide poisoning incidents.

The safest pesticides include those made from plants, such as the insecticides pyrethrum and derris. Pyrethrins are safe and insects do not develop resistance to them. Their impact on the environment is very small as the ingredients break down harmlessly.

More potent are synthetic products, such as chlorinated hydrocarbons. These products, including DDT and dieldrin, are highly toxic to wildlife and often to humans, so their use is now restricted by law in some areas and is declining. Safer pesticides such as malathion are based on organic phosphorus compounds, but they still present hazards to health. An international treaty to ban persistent organic pollutants (POPs), including pesticides such as DDT, was signed in Stockholm, Sweden, in May 2001. The United Nations Food and Agriculture

Organization reported in the same month that more than 500,000 tonnes of pesticide waste resides in dumps worldwide.

petaflops in computing, jargon for 1,000 trillion computations per second. This is still many years from being realized.

petal part of a flower whose function is to attract pollinators such as insects or birds. Petals are frequently large and brightly coloured and may also be scented. Some have a nectary at the base and markings on the petal surface, known as honey guides, to direct pollinators to the source of the nectar. In wind-pollinated plants, however, the petals are usually small and insignificant, and sometimes absent altogether. Petals are derived from modified leaves, and are known collectively as a corolla.

petiole in botany, the stalk attaching the leaf blade, or lamina, to the stem. Typically it is continuous with the midrib of the leaf and attached to the base of the lamina, but occasionally it is attached to the lower surface of the lamina, as in the nasturtium (a peltate leaf). Petioles that are flattened and leaflike are termed phyllodes. Leaves that lack a petiole are said to be sessile.

petrochemical chemical derived from the processing of petroleum (crude oil). **Petrochemical industries** are those that obtain their raw materials from the processing of petroleum and natural gas. Polymers, detergents, solvents, and nitrogen fertilizers are all major products of the petrochemical industries. Inorganic chemical products include carbon black, sulphur, ammonia, and hydrogen peroxide.

petrol mixture of hydrocarbons derived from *petroleum, mainly used as a fuel for internal-combustion engines. It is colourless and highly volatile. **Leaded petrol** contains antiknock (a mixture of tetra ethyl lead and dibromoethane), which improves the combustion of petrol and the performance of a car engine. The lead from the exhaust fumes enters the atmosphere, mostly as simple lead compounds. There is strong evidence that it can act as a nerve poison on young children and cause mental impairment. This prompted a gradual switch to the use of **unleaded petrol** in the UK.

The changeover from leaded petrol gained momentum from 1989 owing to a change in the tax on petrol, making it cheaper to buy unleaded fuel. Unleaded petrol contains a different mixture of hydrocarbons from leaded petrol. Leaded petrol cannot be used in cars fitted with a *catalytic converter.

In the USA, petrol is called gasoline, and unleaded petrol has been used for many years.

petrol engine the most commonly used source of power for motor vehicles, introduced by the German engineers Gottlieb Daimler and Karl Benz in 1885. The petrol engine is a complex piece of machinery made up of about 150 moving parts. It is a reciprocating piston engine, in which a number of pistons move up and down in cylinders. A mixture of petrol and air is introduced to the space above the pistons and ignited. The gases produced force the pistons down, generating power. The engine-operating cycle is repeated every four strokes (upward or downward movement) of the piston, this being known as the *four-stroke cycle. The motion of the pistons rotate a crankshaft, at the end of which is a heavy flywheel. From the flywheel the power is transferred to the car's driving wheels via the transmission system of clutch, gearbox, and final drive.

petroleum or **crude oil**, natural mineral oil, a thick greenish-brown flammable liquid found underground in permeable rocks. Petroleum consists of hydrocarbons mixed with oxygen, sulphur, nitrogen, and other elements in varying proportions. It is thought to be derived from ancient organic material that has been converted by, first, bacterial action, then heat, and pressure (but its origin may be chemical also).

From crude petroleum, various products are made by *fractional distillation and other processes; for example, fuel oil, petrol, kerosene, diesel, and lubricating oil. Petroleum products and chemicals are used in large quantities in the manufacture of detergents, artificial fibres, plastics, insecticides, fertilizers, pharmaceuticals, toiletries, and synthetic rubber.

Petroleum was formed from the remains of marine plant and animal life which existed many millions of years ago (hence it is known as a *fossil fuel). Some of these remains were deposited along with rock-forming sediments under the sea where they were decomposed anaerobically (without oxygen) by bacteria which changed the fats in the sediments into fatty acids which were then changed into an asphaltic material called kerogen. This was then converted over millions of years into petroleum by the combined action of heat and pressure. At an early stage the organic material was squeezed out of its original sedimentary mud into adjacent sandstones. Small globules of oil collected together in the pores of the rock and eventually migrated upwards through layers of porous rock by the action of the oil's own surface tension (capillary action), by the force of water movement within the rock, and by gas pressure. This migration ended either when the petroleum emerged through a fissure as a seepage of gas or oil onto the Earth's surface, or when it was trapped in porous reservoir rocks, such as sandstone or limestone, in anticlines and other traps below impervious rock layers.

pollution The burning of petroleum fuel is one cause of *air pollution. The transport of oil can lead to catastrophes – for example, the *Torrey Canyon* tanker lost off southwestern England in 1967, which led to an agreement by the international oil companies in 1968 to pay compensation for massive shore pollution. The 1989 oil spill in Alaska from the *Exxon Valdez* damaged the area's fragile environment, despite clean-up efforts. Drilling for oil involves the risks of accidental spillage and drilling-rig accidents. The problems associated with oil have led to the various alternative *energy technologies.

petrology branch of geology that deals with the study of rocks, their mineral compositions, their textures, and their origins.

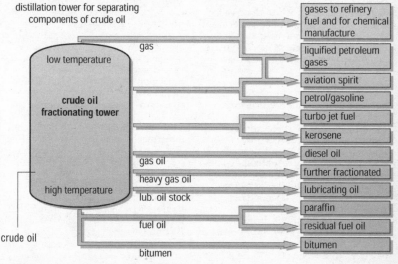

petroleum Refining petroleum using a distillation column. The crude petroleum is fed in at the bottom of the column where the temperature is high. The gases produced rise up the column, cooling as they travel. At different heights up the column, different gases condense to liquids called fractions, and are drawn off.

pewter any of various alloys of mostly tin with varying amounts of lead, copper, or antimony. Pewter has been known for centuries and was once widely used for domestic utensils but is now used mainly for ornamental ware.

PGML abbreviation for precision graphics markup language, hybrid of *PDF and *HTML under development by Adobe. PGML is particularly suitable for use on electronic book readers and *personal digital assistants. See also *ClearType.

PGP abbreviation for the encryption program Pretty Good Privacy.

pH scale from 0 to 14 for measuring acidity or alkalinity. A pH of 7 indicates neutrality, below 7 is *acid, while above 7 is alkaline. Strong acids, such as those used in car batteries, have a pH of about 2; strong *alkalis such as sodium hydroxide have a pH of about 13.

Acidic fruits such as citrus fruits are about pH 4. Fertile soils have a pH of about 6.5 to 7, while weak alkalis such as soap have a pH of about 9 to 10.

phage another name for a *bacteriophage, a virus that attacks bacteria.

phagocyte type of *white blood cell, or leucocyte, that can engulf a bacterium or other invading micro-organism. Phagocytes are found in blood, lymph, and other body tissues, where they also ingest foreign matter and dead tissue. A *macrophage differs in size and lifespan.

Phanerozoic eon (Greek *phanero* 'visible') eon in Earth history, consisting of the most recent 570 million years. It comprises the Palaeozoic, Mesozoic, and Cenozoic eras. The vast majority of fossils come from this eon, owing to the evolution of hard shells and internal skeletons. The name means 'interval of well-displayed life'.

pharmacokinetics in medicine, the relationship between the absorption, distribution, metabolism, and excretion of a drug in mathematical terms. The effects and the duration of action of the drug are also taken into account. Clinical pharmacokinetics is the application of pharmacokinetic studies to clinical practice and to the safe and effective therapeutic management of the individual patient.

pharmacology study of the properties of drugs and their effects on the human body.

pharynx muscular cavity behind the nose and mouth, extending downwards from the base of the skull. Its walls are made of muscle strengthened with a fibrous layer and lined with mucous membrane. The internal nostrils lead backwards into the pharynx, which continues downwards into the oesophagus and (through the epiglottis) into the windpipe. On each

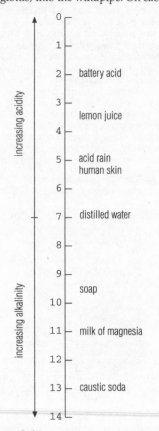

pH The pHs of some common substances. The lower the pH, the more acidic the substance; the higher the pH, the more alkaline the substance. A pH of 7 is neutral.

side, a Eustachian tube enters the pharynx from the middle ear cavity.

phase in chemistry, a physical state of matter. For example, ice and liquid water are different phases of water; a mixture of the two is termed a two-phase system.

phase in physics, a stage in an oscillatory motion, such as a wave motion: two waves are in phase when their peaks and their troughs coincide. Otherwise, there is a **phase difference**, which has consequences in *interference phenomena and *alternating current electricity.

phencyclidine hydrochloride (PCP) technical name for the drug angel dust.

phenol member of a group of aromatic chemical compounds with weakly acidic properties, which are characterized by a hydroxyl (OH) group attached directly to an aromatic ring. The simplest of the phenols, derived from benzene, is also known as phenol and has the formula C_6H_5OH. It is sometimes called **carbolic acid** and can be extracted from coal tar.

Pure phenol consists of colourless, needle-shaped crystals, which take up moisture from the atmosphere. It has a strong and characteristic smell and was once used as an antiseptic. It is, however, toxic by absorption through the skin.

phenolphthalein acid–base *indicator that is clear below pH 8 and red above

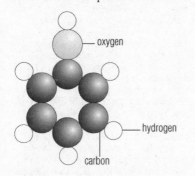

oxygen

hydrogen

carbon

phenol The phenol molecule with its ring of six carbon atoms and a hydroxyl (OH) group attached. Phenol was first extracted from coal tar in 1834. It is used to make phenolic and epoxy resins, explosives, pharmaceuticals, perfumes, and nylon.

pH 9.6. It is used in titrating (see *titration) weak acids against strong bases.

phenotype in genetics, visible traits, those actually displayed by an organism. The phenotype is not a direct reflection of the *genotype because some alleles are masked by the presence of other, dominant alleles (see *dominance). The phenotype is further modified by the effects of the environment (for example, poor nutrition stunts growth).

phenylanaline one of the nine essential amino acids. *Phenylketonuria is a rare genetic disease which results from the inability to metabolize the phenylalanine present in food.

phenylketonuria inherited metabolic condition in which the liver of a child cannot control the level of phenylalanine (an *amino acid derived from protein food) in the bloodstream. The condition must be detected promptly and a special diet started in the first few weeks of life if brain damage is to be avoided. Untreated, it causes stunted growth, epilepsy, and severe mental disability.

pheromone chemical signal (such as an odour) that is emitted by one animal and affects the behaviour of others. Pheromones are used by many animal species to attract mates.

phloem tissue found in vascular plants. Its main function is to transport sugars and other food materials such as amino acids (see *protein) from the leaves, where they are produced, to all other parts of the plant. This could be from the leaves to the roots to provide the chemicals needed for growth. However, it could be from a leaf and up to a developing fruit that is rich in sugars. The sugars are made by *photosynthesis, which occurs in green parts of plants, such as leaves (see *leaf). The amino acids are made from sugars and *minerals, such as nitrate absorbed from the soil. Phloem tissue is usually found close to the other transport tissue in plants, *xylem, which transports *water and minerals. In non-woody plants phloem and xylem are found in bundles, such as the veins of a leaf.

Phloem is composed of sieve elements and their associated companion cells, together with some sclerenchyma and *parenchyma cell types. Sieve elements are long, thin-walled cells joined end to end, forming sieve tubes; large pores in the end walls allow the continuous passage of nutrients. Phloem is usually found in association with *xylem, the water-conducting tissue, but unlike the latter it is a living tissue.

phlogiston hypothetical substance formerly believed to have been produced during combustion. The substance, originally named terra pinguis ('fatty earth') by the German chemist Johann Joachim Becher in 1669, was renamed by Georg Stahl at the beginning of the 18th century. The phlogiston theory was replaced by the theory of oxygen gain and loss, first enunciated by the French chemist Antoine Lavoisier (1743–1794).

phobia excessive irrational fear of an object or situation – for example, agoraphobia (fear of open spaces and crowded places), acrophobia (fear of heights), and claustrophobia (fear of enclosed places). *Behaviour therapy is one form of treatment.

Phong shading in computing, type of shading used in animation and 3-D graphics, based on a computerized model of how light is reflected from surfaces.

phonograph a device in which the vibrations of the human voice are engraved by a needle on a revolving cylinder coated with tin foil, invented in 1877 by US scientist and inventor Thomas Alva Edison (1847–1931). This invention began the era of recorded sound.

phosphate salt or ester of *phosphoric acid. Incomplete neutralization of phosphoric acid gives rise to acid phosphates (see *acid salt and *buffer). Phosphates are used as fertilizers, and are required for the development of healthy root systems. They are involved in many biochemical processes, often as part of complex molecules, such as *ATP.

phospholipid any *lipid consisting of a glycerol backbone, a phosphate group, and two long chains. Phospholipids are found everywhere in living systems as the basis for biological membranes.

phosphor any substance that is phosphorescent, that is, gives out visible light when it is illuminated by a beam of electrons or ultraviolet light. The television screen is coated on the inside with phosphors that glow when beams of electrons strike them. Fluorescent lamp tubes are also phosphor-coated. Phosphors are also used in Day-Glo paints, and as optical brighteners in detergents.

phosphorescence emission of light by certain substances after they have absorbed energy, whether from visible light, other electromagnetic radiation such as ultraviolet rays or X-rays, or cathode rays (a beam of electrons). When the stimulating energy is removed phosphorescence persists for more than 0.1 nanoseconds before ceasing (unlike *fluorescence, which stops immediately). See also *luminescence.

phosphoric acid acid derived from phosphorus and oxygen. Its commonest form (H_3PO_4) is also known as orthophosphoric acid, and is produced by the action of phosphorus pentoxide (P_2O_5) on water. It is used in rust removers and for rust-proofing iron and steel.

phosphorus chemical symbol P, (Greek *phosphoros* 'bearer of light') highly reactive, non-metallic element, atomic number 15, relative atomic mass 30.9738. It occurs in nature as phosphates (commonly in the form of the mineral apatite), and is essential to plant and animal life. Compounds of phosphorus are used in fertilizers, various organic chemicals, for matches and fireworks, and in glass and steel.

Phosphorus was first identified in 1674 by German alchemist Hennig Brand (born c. 1630), who prepared it from urine. The element has three allotropic forms: a black powder; a white-yellow, waxy solid that ignites spontaneously in air to form the poisonous gas phosphorus pentoxide; and a red-brown powder that neither ignites spontaneously nor is poisonous.

PhotoCD picture storage and viewing system developed by Kodak and

Philips. The aim of Kodak's PhotoCD is to allow the user to put up to 100 photos onto compact disc: images are transferred from film to a PhotoCD disk and can then be viewed by means of Kodak's own PhotoCD player, which plugs into a television set, or by using suitable software on a multimedia PC.

photocell or **photoelectric cell**, device for measuring or detecting light or other electromagnetic radiation, since its electrical state is altered by the effect of light. In a **photoemissive** cell, the radiation causes electrons to be emitted and a current to flow (*photoelectric effect); a **photovoltaic** cell causes an *electromotive force to be generated in the presence of light across the boundary of two substances. A **photoconductive** cell, which contains a semiconductor, increases its conductivity when exposed to electromagnetic radiation.

photochemical reaction any chemical reaction in which light is produced or light initiates the reaction. Light can initiate reactions by exciting atoms or molecules and making them more reactive: the light energy becomes converted to chemical energy. Many photochemical reactions set up a *chain reaction and produce *free radicals. This type of reaction is seen in the bleaching of dyes or the yellowing of paper by sunlight. It is harnessed by plants in *photosynthesis and by humans in photography.

   Chemical reactions that produce light are most commonly seen when materials are burned. Light-emitting reactions are used by living organisms in *bioluminescence. One photochemical reaction is the action of sunlight on car exhaust fumes, which results in the production of *ozone. Some large cities, such as Los Angeles, USA, and Santiago, Chile, now suffer serious pollution due to photochemical smog.

photodiode semiconductor *p–n junction diode used to detect light or measure its intensity. The photodiode is encapsulated in a transparent plastic case that allows light to fall onto the junction. When this occurs, the reverse-bias resistance (high resistance in the opposite direction to normal current

flow) drops and allows a larger reverse-biased current to flow through the device. The increase in current can then be related to the amount of light falling on the junction.

photoelectric cell alternative name for *photocell.

photoelectric effect process by which electromagnetic radiation, including visible light, incident on a material releases an electric charge. It is commonly thought of as the emission of *electrons from a substance (usually a metallic surface) when it is struck by *photons (quanta of electromagnetic radiation), usually those of visible light or ultraviolet radiation.

   The energy of the emitted electrons depends on the frequency of the incident radiation (which must exceed a characteristic threshold frequency); the higher the frequency, the greater the energy of its photons. The number of electrons emitted depends on the radiation's intensity (rate of transfer of energy per unit area).

   The effect was first observed 1887 by German physicist Heinrich Hertz (1857–1894, better known for his researches on electromagnetic waves), who showed that electric sparks occur more easily when the electrodes are illuminated with ultraviolet light (see *ultraviolet radiation). It was established in 1900 as being due to the ejection of electrons by the light. This discovery, along with the identification of the *beta particles observed in radioactivity as being electrons also, helped establish the fundamental nature of electrons as a constituent of all atoms. The theory of the photoelectric effect, a *quantum theory of radiation, was formulated by German-born US physicist Albert Einstein (1879–1955) in 1905. Einstein suggested that the kinetic energy of each electron, $1/2mv^2$, is equal to the difference in the incident light energy, hv, and the light energy needed to overcome the threshold of emission, hv^0, where h is *Planck's constant and v^0 is the threshold frequency. This can be written mathematically as:

$$\frac{1}{2}mv^2 = hv - hv^0$$

photoluminescence see
*luminescence.

photolysis chemical reaction that is
driven by light or ultraviolet radiation.
For example, the light reaction of
*photosynthesis (the process by which
green plants manufacture
carbohydrates from carbon dioxide
and water) is a photolytic reaction.

photometer instrument that measures
light, electromagnetic radiation in the
visible range, usually by comparing
relative intensities from different
sources. Bunsen's grease-spot
photometer of 1844 compares the
intensity of a light source with a
known source by each illuminating
one half of a translucent area. Modern
photometers use *photocells, as in a
photographer's exposure meter. A
photomultiplier can also be used as a
photometer.

photomultiplier instrument that
detects low levels of electromagnetic
radiation (usually visible light or
*infrared radiation) and amplifies it to
produce a detectable signal.

photon in physics, the *elementary
particle or 'package' (quantum) of
energy in which light and other forms
of electromagnetic radiation are
emitted. The photon has both particle
and wave properties; it has no charge,
is considered massless but possesses
momentum and energy. It is one of the
*gauge bosons, and is the carrier of the
*electromagnetic force, one of the
fundamental forces of nature.

According to *quantum theory the
energy of a photon is given by the
formula $E = hf$, where h is *Planck's
constant and f is the frequency of the
radiation emitted.

photoperiodism biological mechanism
that determines the timing of certain
activities by responding to changes
in day length. The flowering of
many plants is initiated in this way.
Photoperiodism in plants is regulated
by a light-sensitive pigment,
phytochrome. The breeding seasons of
many temperate-zone animals are also
triggered by increasing or declining
day length, as part of their biorhythms.

Autumn-flowering plants (for
example, chrysanthemum and
soybean) and autumn-breeding

mammals (such as goats and deer)
require days that are shorter than a
critical length; spring-flowering and
spring-breeding ones (such as radish
and lettuce, and birds) are triggered by
longer days.

photophobia intolerance of light. It
may be prevalent in some diseases,
especially migraine.

photosphere visible surface of the
Sun, which emits light and heat. About
300 km/200 mi deep, it consists of
incandescent gas at a temperature of
5,800 K (5,530°C/9,980°F).

photosynthesis process by which
green plants trap light energy from
sunlight in order to combine *carbon
dioxide and *water to make high-
energy chemicals – *glucose and other
*carbohydrates. The simple sugar
glucose provides the basic food for
both plants and animals. For photosynthesis to occur, the plant must
possess *chlorophyll (the green
chemical that absorbs the energy from
the sunlight) and must have a supply
of carbon dioxide and water.
Photosynthesis takes place inside
*chloroplasts, which contain the
enzymes and chlorophyll necessary for
the process. They are found mainly in
the leaf cells of plants. The by-product
of photosynthesis, oxygen, is of great
importance to all living organisms, and
virtually all atmospheric oxygen has
originated from photosynthesis.

Through photosynthesis, plants are
able to make some of their own food –
this is one way in which the nutrition
of plants differs from that of animals.
The glucose they produce may be
converted to starch and stored. Starch
can then be converted back to glucose
to provide energy for the plant at a
later stage, for example during the
night.

phototropism movement of part of a
plant toward or away from a source of
light. Leaves are positively phototropic,
detecting the source of light and
orientating themselves to receive the
maximum amount.

phrenology study of the shape and
protuberances of the skull, based on
the now discredited theory of the
Austrian physician Franz Josef Gall
that such features revealed

measurable psychological and intellectual traits.

phyllotaxis arrangement of leaves on a plant stem. Leaves are nearly always arranged in a regular pattern and in the majority of plants they are inserted singly, either in a **spiral** arrangement up the stem, or on **alternate** sides. Other principal forms are opposite leaves, where two arise from the same node, and whorled, where three or more arise from the same node.

phylogeny historical sequence of changes that occurs in a given species during the course of its evolution. It was once erroneously associated with ontogeny (the process of development of a living organism).

phylum plural **phyla**, major grouping in biological classification. Mammals, birds, reptiles, amphibians, fishes, and tunicates belong to the phylum Chordata; the phylum Mollusca consists of snails, slugs, mussels, clams, squid, and octopuses; the phylum Porifera contains sponges; and the phylum Echinodermata includes starfish, sea urchins, and sea cucumbers. In classifying plants (where the term 'division' often takes the place of 'phylum'), there are between four and nine phyla depending on the criteria used; all flowering plants belong to a single phylum, Angiospermata, and all conifers to another, Gymnospermata. Related phyla are grouped together in a *kingdom; phyla are subdivided into *classes.

physical change in chemistry, a type of change that does not produce a new chemical substance, but rather a change physical state (see also *chemical change). Boiling and melting are examples of physical change.

physical chemistry branch of chemistry concerned with examining the relationships between the chemical compositions of substances and the physical properties that they display. Most chemical reactions exhibit some physical phenomenon (change of state, temperature, pressure, or volume, or the use or production of electricity), and the measurement and study of such phenomena has led to many chemical theories and laws.

physical weatherin or **mechanical weathering**, form of *weathering responsible for the mechanical breakdown of rocks but involving no chemical change.

physics branch of science concerned with the laws that govern the structure of the universe, and the investigation of the properties of matter and energy and their interactions. Before the 19th century, physics was known as **natural philosophy**.

physiological psychology aspect of *experimental psychology concerned with physiological and neurological processes as the basis of experience and behaviour. It overlaps considerably with such fields as anatomy, physiology, neurology, and biochemistry.

physiology branch of biology that deals with the functioning of living organisms, as opposed to anatomy, which studies their structures.

physiotherapy treatment of injury and disease by physical means such as exercise, heat, manipulation, massage, and electrical stimulation.

phytomenadione one form of vitamin K, a fat-soluble chemical found in green vegetables. It is involved in the production of prothrombin, which is essential in blood clotting. It is given to newborns to prevent potentially fatal brain haemorrhages.

pi symbol π, ratio of the circumference of a circle to its diameter. Pi is an irrational number: it cannot be expressed as the ratio of two integers, and its expression as a decimal never terminates and never starts recurring. The value of pi is 3.1415926, correct to seven decimal places. Common approximations to pi are $^{22}/_7$ and 3.14, although the value 3 can be used as a rough estimation.

PIC device abbreviation for programmable interface controller, microcontroller that can be programmed to carry out specific tasks. The PIC contains a low-power microprocessor, together with a small amount of Flash *RAM used to store the program and a smaller amount of working RAM in which the program data is held while running. PICs also contain all the necessary circuitry to connect to the outside world, plus timers and a section of *EEPROM computer memory in which the processor's instruction set, or language, is stored.

picric acid technical name **2,4,6-trinitrophenol**; $C_6H_2(NO_2)_3OH$, strong acid that is used to dye wool and silks yellow, for the treatment of burns, and in the manufacture of explosives. It is a yellow, crystalline solid.

PICS acronym for Platform for Internet Content Selection, in computing, method of classifying data according to its content. Under PICS, the creator and the reader of a file can add descriptive electronic labels to it, making it possible for users to sort documents according to keywords on the label. The system, introduced in 1996, aims to help parents to control what their children can see on the Internet, for example by blocking access to pornographic or violent material; it also enables people to highlight subjects in which they are especially interested.

PICT in computing, object-oriented file format used on the Apple Macintosh computer. The format uses QuickDraw and is supported by almost all graphics applications on the Macintosh.

pie chart method of displaying proportional information by dividing a circle up into different-sized sectors (slices of pie). The angle of each sector is proportional to the size, expressed as a percentage, of the group of data that it represents.

piezoelectric effect property of some crystals (for example, quartz) to develop an electromotive force or voltage across opposite faces when subjected to tension or compression, and, conversely, to expand or contract in size when subjected to an electromotive force. Piezoelectric crystal *oscillators are used as frequency standards (for example, replacing balance wheels in watches), and for producing ultrasound. Crystalline quartz is a good example of a piezoelectric material.

pig iron or **cast iron**, iron produced in a *blast furnace. It contains around 4% carbon plus some other impurities.

piles popular name for *haemorrhoids.

pilot running in computing, a method of implementing a new computer system in which the work is gradually transferred from the old system to the new system over a period of time. This ensures that any faults in the new system are resolved before the old system is withdrawn. An alternative method is *parallel running.

pineal body or **pineal gland**, cone-shaped outgrowth of the vertebrate brain. In some lower vertebrates, it develops a rudimentary lens and retina, which show it to be derived from an eye, or pair of eyes, situated on the top of the head in ancestral vertebrates. In fishes that can change colour to match their background, the pineal perceives the light level and controls the colour change. In birds, the pineal detects changes in daylight and stimulates breeding behaviour as spring approaches. Mammals also have a pineal gland, but it is located deeper within the brain. It secretes a hormone, melatonin, thought to influence rhythms of activity. In humans, it is a small piece of tissue attached by a stalk to the rear wall of the third ventricle of the brain.

PING acronym for Packet INternet Groper, in computing, short message sent over a network by one computer to check whether another is correctly connected to it. By extension, one can 'ping' other people – for example, checking addresses on a mailing list by sending an *e-mail to all members requesting an acknowledgement.

pinna in botany, the primary division of a *pinnate leaf. In mammals, the pinna is the external part of the ear.

pinnate leaf leaf that is divided up into many small leaflets, arranged in rows along either side of a midrib, as in ash trees (*Fraxinus*). It is a type of compound leaf. Each leaflet is known as a **pinna**, and where the pinnae are themselves divided, the secondary divisions are known as pinnules.

pins and needles or **paresthesiae**, in medicine, numbness and disturbed sensation of the limbs. It may be experienced following long periods of immobility or as a symptom of some neurological diseases.

pint symbol pt, imperial dry or liquid measure of capacity equal to 20 fluid ounces, half a quart, one-eighth of a gallon, or 0.568 litre. In the USA, a liquid pint is equal to 0.473 litre, while a dry pint is equal to 0.550 litre.

pion or **pi meson**, in physics, a subatomic particle with a neutral form (mass 135 MeV) and a charged form (mass 139 MeV). The charged pion decays into muons and neutrinos and the neutral form decays into gamma-ray photons. They belong to the *hadron class of *elementary particles.

pipette device for the accurate measurement of a known volume of liquid, usually for transfer from one container to another, used in chemistry and biology laboratories.

A pipette may be a slender glass tube, often with an enlarged bulb, which is calibrated in one or more positions; or it may be a plastic device with an adjustable plunger, fitted with one or more disposable plastic tips.

PIPEX acronym for Public Internet Protocol EXchange, in computing, UK-based Internet provider that started operations in 1992, specializing in serving the commercial sector. The company became one of the UK's major *backbone providers, was acquired by *UUNET in 1996, and was spun off again at the end of 1998.

pitch in chemistry, a black, sticky substance, hard when cold, but liquid when hot, used for waterproofing, roofing, and paving. It is made by the destructive distillation of wood or coal tar, and has been used since antiquity for caulking wooden ships.

pitch in mechanics, the distance between the adjacent threads of a screw or bolt. When a screw is turned through one full turn it moves a distance equal to the pitch of its thread. A screw thread is a simple type of machine, acting like a rolled-up inclined plane, or ramp (as may be illustrated by rolling a long paper triangle around a pencil). A screw has a *mechanical advantage greater than one.

pitchblende or **uraninite**, brownish-black mineral, the major uranium ore, consisting mainly of uranium oxide (UO_2). It also contains some lead (the final, stable product of uranium decay) and variable amounts of most of the naturally occurring radioactive elements, which are products of either the decay or the fissioning of uranium isotopes. The uranium yield is 50–80%; it is also a source of radium, polonium, and actinium. Pitchblende was first studied by French scientists Pierre and Marie Curie, who found radium and polonium in its residues in 1898.

pitot tube instrument that measures fluid (gas or liquid) flow. It is used to measure the speed of aircraft, and works by sensing pressure differences in different directions in the airstream.

It was invented in the 1730s by the French scientist Henri Pitot (1695–1771).

pituitary gland major *endocrine gland of vertebrates, situated in the centre of the brain. It is attached to the *hypothalamus by a stalk. The pituitary consists of two lobes. The posterior lobe is an extension of the hypothalamus, and is in effect nervous tissue. It stores two hormones synthesized in the hypothalamus: *ADH and *oxytocin. The anterior lobe secretes six hormones, some of which control the activities of other glands (thyroid, gonads, and adrenal cortex); others are direct-acting hormones affecting milk secretion and controlling growth. (See diagram, page 450.)

pixel derived from picture element, single dot on a computer screen. All screen images are made up of a collection of pixels, with each pixel being either off (dark) or on (illuminated, possibly in colour). The number of pixels available determines the screen's resolution. Typical resolutions of microcomputer screens vary from 320×200 pixels to 800×600 pixels, but screens with $1,024 \times 768$ pixels or more are now common for high-quality graphic (pictorial) displays.

PKI abbreviation for *public key infrastructure.

placebo (Latin 'I will please') any harmless substance, often called a 'sugar pill', that has no active ingredient, but may nevertheless bring about improvement in the patient's condition.

placenta organ that attaches the developing *embryo or *fetus to the *uterus (womb) in placental mammals (mammals other than marsupials, platypuses, and echidnas). Composed of maternal and embryonic tissue, it links the blood supply of the embryo to the blood supply of the mother, allowing the exchange of oxygen, nutrients, and waste products. The two blood systems are not in direct contact, but are separated by thin membranes, with materials diffusing across from one system to the other. The placenta also produces hormones that maintain and regulate pregnancy. It is shed as part of the afterbirth.

plague term applied to any epidemic disease with a high mortality rate, but it usually refers to bubonic plague. This is a disease transmitted by fleas (carried by the black rat), which infect

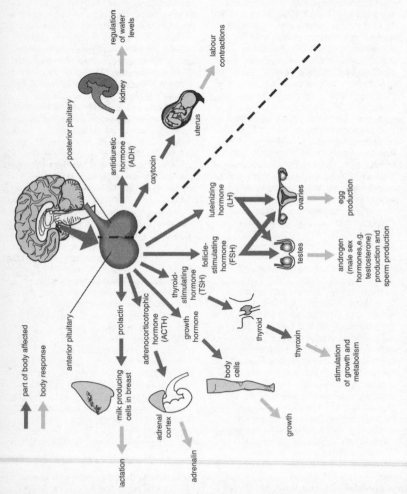

pituitary gland The hormones secreted from the pituitary and the effects they have on the body. Two hormones (ADH and oxytocin) are secreted from the posterior pituitary and six (prolactin, ACTH, growth hormone, TSH, FSH, and LH) from the anterior pituitary.

the sufferer with the bacillus *Yersinia pestis*. An early symptom is swelling of lymph nodes, usually in the armpit and groin; such swellings are called 'buboes'. It causes virulent blood poisoning and the death rate is high.

plain or **grassland**, land, usually flat, upon which grass predominates. The plains cover large areas of the Earth's surface, especially between the *deserts of the tropics and the *rainforests of the Equator, and have rain in one season only. In such regions the *climate belts move north and south during the year, bringing rainforest conditions at one time and desert conditions at another. Temperate plains include the North European Plain, the High Plains of the USA and Canada, and the Russian Plain (also known as the steppe).

plaintext in computing, another name for *cleartext.

Planck's constant symbol h, fundamental constant that relates the energy (E) of one quantum of electromagnetic radiation (a 'packet' of energy; see *quantum theory) to the frequency (f) of its radiation by $E = hf$. Its value is 6.6262×10^{-34} joule seconds. Named after German physicist Max Planck (1858–1947).

plane flat surface. Planes are either parallel or they intersect in a straight line. Vertical planes, for example the join between two walls, intersect in a vertical line. Horizontal planes do not intersect since they are all parallel.

planet (Greek 'wanderer') large celestial body in orbit around a star, composed of rock, metal, or gas. There are nine planets in the *Solar System orbiting the *Sun: Mercury, Venus, Earth, Mars, Jupiter, Saturn, Neptune, Uranus, and Pluto. The inner four, called the **terrestrial planets**, are small and rocky, and have few natural *satellites. The outer planets, with the exception of Pluto, are called the **major planets**, and have denser atmospheres consisting mainly of hydrogen and helium gases, and many natural satellites. The largest planet in the Solar System is Jupiter (about 780 million km/490 million mi from the Sun) with a diameter of 140,000 km/87,500 mi, which contains a mass greater than all the other planets combined. The smallest (and furthest from the Sun at about 5,900 million km/3,600 million mi) is Pluto with a diameter of 2,300 km/1,400 mi.

planetary nebula shell of gas thrown off by a star at the end of its life. Planetary nebulae have nothing to do with planets. They were named by German-born English astronomer William Herschel, who thought their rounded shape resembled the disc of a planet. After a star such as the Sun has expanded to become a *red giant, its outer layers are ejected into space to form a planetary nebula, leaving the core as a *white dwarf at the centre.

plankton small, often microscopic, forms of plant and animal life that live in the upper layers of fresh and salt water, and are an important source of food for larger animals. Marine plankton is concentrated in areas where rising currents bring mineral salts to the surface.

plant organism that carries out *photosynthesis, has cellulose cell walls and complex cells, and is immobile. A few parasitic plants have lost the ability to photosynthesize but are still considered to be plants. Plants are *autotrophs, that is, they make carbohydrates from water and carbon dioxide, and are the primary producers in all food chains, so that all animal life is dependent on them. They play a vital part in the carbon cycle, removing carbon dioxide from the atmosphere and generating oxygen. The study of plants is known as *botany.

plant cell *cell in a plant, which, like all cells, has a *cell membrane, *cytoplasm, and a *nucleus, but which differs from an animal cell by having a *cell wall outside the cell surface membrane and a large *vacuole. They may also have *chloroplasts in the cytoplasm.

The cell surface membrane keeps the cell together by being strong, although it is still very thin and flexible. This membrane controls what enters and leaves the cell – for example, nutrients must be allowed to enter and waste materials must be allowed to leave. The cell has some cytoplasm, but most of the space inside the cell is generally filled with a large vacuole. The cytoplasm contains the nucleus, which controls the activities of the cell, and often contains many chloroplasts, which capture sunlight and make food for the plant through *photosynthesis. Chloroplasts are green – this explains why most plants look green.

plant classification taxonomy or classification of plants. Originally the plant kingdom included bacteria, diatoms, dinoflagellates, fungi, and slime moulds, but these are not now thought of as plants. The groups that are always classified as plants are the bryophytes (mosses and liverworts), pteridophytes (ferns, horsetails, and club mosses), gymnosperms (conifers, yews, cycads, and ginkgos), and angiosperms (flowering plants). The angiosperms are split into monocotyledons (for example, orchids, grasses, lilies) and dicotyledons (for example, oak, buttercup, geranium, and daisy).

plant hormone substance produced by a plant that has a marked effect on its growth, flowering, leaf fall, fruit ripening, or some other process. Examples include *auxin, *gibberellin, *ethene, and cytokinin.

Unlike animal *hormones, these substances are not produced by a particular area of the plant body, and they may be less specific in their effects. It has therefore been suggested that they should not be described as hormones at all.

plant propagation production of plants. Botanists and horticulturalists can use a wide variety of means for propagating plants. There are the natural techniques of *vegetative reproduction, together with cuttings, grafting, and micropropagation. The range is wide because most plant tissue, unlike animal tissue, can give rise to the complete range of tissue types within a particular species.

plaque any abnormal deposit on a body surface, especially the thin, transparent film of sticky protein (called mucin) and bacteria on tooth surfaces. If not removed, this film forms tartar (calculus), promotes tooth decay, and leads to gum disease. Another form of plaque is a deposit of fatty or fibrous material in the walls of blood vessels causing atheroma.

plasma in biology, the liquid component of the *blood. It is a straw-coloured fluid, largely composed of water (around 90%), in which a number of substances are dissolved. These include a variety of proteins (around 7%) such as fibrinogen (important in *blood clotting), inorganic mineral salts such as sodium and calcium, waste products such as *urea, traces of *hormones, and *antibodies to defend against infection.

plasma in physics, ionized gas produced at extremely high temperatures, as in the Sun and other stars. It contains positive and negative charges in equal numbers. It is a good electrical conductor. In thermonuclear reactions the plasma produced is confined through the use of magnetic fields.

plasma display type of flat display, which uses an ionized gas between two panels containing grids of wires. When current flows through the wires a *pixel is charged causing it to light up. This technology has been developed to produce wall-hung flat screen televisions of up to 122 cm/ 50 in when measured diagonally.

plasmid small, mobile piece of *DNA found in bacteria that, for example, confers antibiotic resistance, used in *genetic engineering. Plasmids are separate from the bacterial chromosome but still multiply during cell growth. Their size ranges from 3% to 20% of the size of the chromosome. Some plasmids carry 'fertility genes' that enable them to move from one bacterium to another and transfer genetic information between strains. Plasmid genes determine a wide variety of bacterial properties including resistance to antibiotics and the ability to produce toxins.

plasmolysis separation of the plant cell cytoplasm from the cell wall as a result of water loss. As moisture leaves the vacuole the total volume of the cytoplasm decreases while the cell itself, being rigid, hardly changes. Plasmolysis is induced in the laboratory by immersing a plant cell in a strongly saline or sugary solution, so that water is lost by osmosis. Plasmolysis is unlikely to occur in the wild except in severe conditions.

plaster of Paris form of calcium sulphate $CaCo_3.\frac{1}{2}H_2O$, obtained from gypsum; it is mixed with water for making casts and moulds.

plastic any of the stable synthetic materials that are fluid at some stage in their manufacture, when they can be shaped, and that later set to rigid or semi-rigid solids. Plastics today are chiefly derived from petroleum. Most are polymers, made up of long chains of identical molecules and can be *thermoplastic or *thermoset plastics. Since plastics have afforded an economical replacement for ivory in the manufacture of piano keys and billiard balls, the industrial chemist may well have been responsible for the survival of the elephant.

Most plastics cannot be broken down by micro-organisms, so cannot easily be disposed of. Incineration leads to the release of toxic fumes, unless carried out at very high temperatures.

plastid general name for a cell *organelle of plants that is enclosed by a double membrane and contains a series of internal membranes and vesicles. Plastids contain *DNA and are produced by division of existing plastids. They can be classified into two main groups: the **chromoplasts**, which contain pigments such as carotenes and chlorophyll, and the **leucoplasts**, which are colourless; however, the distinction between the two is not always clear-cut.

plate or **tectonic plate** or **lithospheric plate**, one of several relatively distinct sections of the *lithosphere, approximately 100 km/60 mi thick, which together comprise the outermost layer of the Earth (like the pieces of the cracked shell of a hard-boiled egg). The plates are made up of two types of crustal material: oceanic crust (sima) and continental crust (sial), both of which are underlain by a solid layer of *mantle. Dense **oceanic crust** lies beneath Earth's oceans and consists largely of *basalt. **Continental crust**, which underlies the continents and the continental shelves, is thicker, less

dense, and consists of rocks that are rich in silica and aluminium. Due to convection in the Earth's mantle (see *plate tectonics) these pieces of lithosphere are in motion, riding on a more plastic layer of the mantle, called the asthenosphere. Mountains, volcanoes, earthquakes, and other geological features and events all come about as a result of interaction between these plates.

plateau elevated area of fairly flat land, or a mountainous region in which the peaks are at the same height. An **intermontane plateau** is one surrounded by mountains. A **piedmont plateau** is one that lies between the mountains and low-lying land. A **continental plateau** rises abruptly from low-lying lands or the sea. Examples are the Tibetan Plateau and the Massif Central in France.

platelet tiny disc-shaped structure found in the blood, which helps it to clot. Platelets are not true cells, but membrane-bound cell fragments without nuclei that bud off from large cells in the bone marrow.

plate margin or **plate boundary**, the meeting place of one *plate (plates make up the top layer of the Earth's structure) with another plate. There are four types of plate margin – destructive, constructive, collision, and conservative. A *volcano may be found along two of the types of plate margin, and an *earthquake may occur at all four plate margins.

destructive or convergent plate margins At a destructive margin an oceanic plate moves towards (and disappears into the *mantle of) a continental plate or another oceanic plate. This is the **subduction zone**. As it is forced downwards, pressure at the margins increases, and this can result in violent earthquakes. The heat produced by friction turns the crust into *magma (liquid rock). The magma tries to rise to the surface and, if it succeeds, violent volcanic eruptions occur.

constructive or divergent plate margins At a *constructive margin the Earth's crust is forced apart. Magma rises and solidifies to create a new oceanic crust and forms a mid-ocean ridge. This ridge is made from *igneous rock; such ridges usually form below sea level on the sea bed (an exception to this is in Iceland).

collision plate margins A collision margin occurs when two plates moving together are both made from continental crust. Continental crust cannot sink or be destroyed, and as a result the land between them is pushed upwards to form high 'fold' mountains like the Himalayas. Earthquakes are common along collision margins but there are no volcanic eruptions.

conservative plate margins At a conservative margin two plates try to slide past each other slowly. Quite often, the two plates stick and pressure builds up; the release of this pressure creates a severe earthquake. There are no volcanic eruptions along conservative plate margins because the crust is neither being created nor destroyed. The San Andreas Fault in California lies above the North American and Pacific plates, and is an example of a conservative plate margin.

plate tectonics theory formulated in the 1960s to explain the phenomena of *continental drift and sea-floor spreading, and the formation of the major physical features of the Earth's surface. The Earth's outermost layer, the *lithosphere, is seen as a jigsaw puzzle of rigid major and minor plates that move relative to each other, probably under the influence of convection currents in the *mantle beneath. At the margins of the plates, where they collide or move apart or slide past one another, major landforms such as *mountains, *rift valleys, *volcanoes, *ocean trenches, and **mid-ocean ridges** are created. The rate of plate movement is on average 2–3 cm/1 in per year and at most 15 cm/6 in per year.

The concept of plate tectonics brings together under one unifying theory many phenomena observed in the Earth's crust that were previously thought to be unrelated. The size of the crust plates is variable, as they are constantly changing, but six or seven large plates now cover much of the Earth's surface, the remainder being occupied by a number of smaller plates. Each large plate may include both continental and ocean lithosphere. As a result of seismic studies it is known that the lithosphere is a rigid layer extending to depths of about 50–100 km/30–60 mi, overlying the upper part of the mantle (the asthenosphere), which is composed of rocks very close to melting point. This zone of mechanical weakness allows the movement of the overlying plates. The margins of the plates are defined by major earthquake zones and belts of volcanic and tectonic activity. Almost all earthquake, volcanic, and tectonic activity is confined to the margins of plates, and shows that the plates are in constant motion (see *plate margin).

platform in computing, *operating system, together with the *hardware on which it runs.

platinum chemical symbol Pt, (Spanish *platina* 'little silver') heavy, soft, silver-white, malleable and ductile, metallic element, atomic number 78, relative atomic mass 195.08. It is the first of a group of six metallic elements (platinum, osmium, iridium, rhodium, ruthenium, and palladium) that possess similar properties, such as resistance to tarnish, corrosion, and attack by acid, and that often occur as free metals (*native metals). They often occur in natural alloys with each other, the commonest of which is osmiridium. Both pure and as an alloy, platinum is used in dentistry, jewellery, and as a catalyst.

Platonic solid in geometry, another name for a regular *polyhedron, one of five possible three-dimensional figures with all its faces the same size and shape.

Platyhelminthes invertebrate phylum consisting of the flatworms.

pleiotropy process whereby a given gene influences several different observed characteristics of an organism. For example, in the fruit fly *Drosophila* the vestigial gene reduces the size of wings, modifies the halteres, changes the number of egg strings in the ovaries, and changes the direction of certain bristles. Many human syndromes are caused by pleiotropic genes, for example Marfan's syndrome

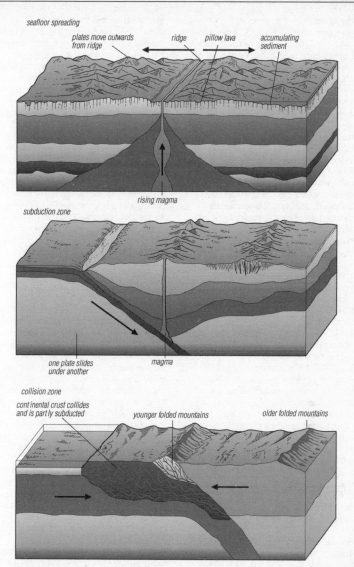

plate tectonics Constructive and destructive action in plate tectonics. (top) Seafloor spreading. The upwelling of magma forces apart the crust plates, producing new crust at the joint. Rapid extrusion of magma produces a domed ridge; more gentle spreading produces a central valley. (middle) The drawing downwards of an oceanic plate beneath a continent produces a range of volcanic fold mountains parallel to the plate edge. (bottom) Collision of continental plates produces immense fold mountains, such as the Himalayas. Younger mountains are found near the coast with older ranges inland. The plates of the Earth's lithosphere are always changing in size and shape of as material is added at constructive margins and removed at destructive margins. The process is extremely slow, but it means that the tectonic history of the Earth cannot be traced back further than about 200 million years.

where the slender physique, hypermobility of the joints, elongation of the limbs, dislocation of the lens, and susceptibility to heart disease are all caused by one gene.

Pleistocene epoch first part of the Quaternary period of geological time, beginning 1.64 million years ago and ending 10,000 years ago. The polar ice caps were extensive and glaciers were abundant during the ice age of this period, and humans evolved into modern *Homo sapiens sapiens* about 100,000 years ago.

pleura covering of the lungs and of the inner surface of the chest wall; the surfaces are lubricated by fluid to prevent friction from breathing movements.

pleurisy inflammation of the pleura, the thin, secretory membrane that covers the lungs and lines the space in which they rest. Pleurisy is nearly always due to bacterial or viral infection, but may also be a complication of other diseases.

Pliocene Epoch ('almost recent') fifth and last epoch of the Tertiary period of geological time, 5.2–1.64 million years ago. The earliest hominid, the humanlike ape *Australopithecines*, evolved in Africa.

plotter or **graph plotter**, device that draws pictures or diagrams under computer control.

plug and play in computing, item of hardware or software that configures itself and the user's system automatically when first installed. Having been thus 'plugged' in, it can be used ('played' with) immediately. In the PC industry, suppliers have adopted a form of Plug and Play (PnP) developed by Microsoft and Intel in 1993.

plug-in in computing, small add-on file which enhances the operation of an application program, often by enabling it to launch, display, or interpret a file created using another one. The first plug-ins were made for graphics programs in the 1980s, but the practice became very popular in the mid-1990s, when a range of plug-ins became available to enhance the multimedia capabilities of Netscape's Navigator *browser. Plug-ins are often created

and distributed by independent developers rather than the manufacturer of the program they extend.

plumbago alternative name for the mineral *graphite.

plumule part of a seed embryo that develops into the shoot, bearing the first true leaves of a plant. In most seeds, for example the sunflower, the plumule is a small conical structure without any leaf structure. Growth of the plumule does not occur until the *cotyledons have grown above ground. This is *epigeal germination. However, in seeds such as the broad bean, a leaf structure is visible on the plumule in the seed. These seeds develop by the plumule growing up through the soil with the cotyledons remaining below the surface. This is known as *hypogeal germination.

Pluto smallest and, usually, outermost planet of the Solar System. The existence of Pluto was predicted by calculation by US astronomer Percival Lowell and the planet was located by US astronomer Clyde Tombaugh in 1930. Its highly elliptical orbit occasionally takes it within the orbit of Neptune, as in 1979–99. Pluto has a mass about 0.002 of that of Earth.

mean distance from the Sun: 5.8 billion km/3.6 billion mi

equatorial diameter: 2,300 km/1,438 mi

rotation period: 6.39 Earth days

year: 248.5 Earth years

atmosphere: thin atmosphere with small amounts of methane gas

surface: low density, composed of rock and ice, primarily frozen methane; there is an ice cap at Pluto's north pole

satellites: one moon, Charon

plutonium chemical symbol Pu, silvery-white, radioactive, metallic element of the *actinide series, atomic number 94, relative atomic mass 244. It occurs in nature in minute quantities in *pitchblende and other ores, but is produced in quantity only synthetically. It has six allotropic forms (see *allotropy) and is one of three fissile elements (elements capable of splitting into other elements – the others are thorium and uranium).

Plutonium dioxide, PuO_2, a yellow crystalline solid, is the compound most widely used in the nuclear industry. It was believed to be inert until US researchers discovered in 1999 that it reacts very slowly with oxygen and water to form a previously unknown green crystalline compound that is soluble in water.

PM10 abbreviation for particulate matter less than 10 micrometres (10 µm) across, clusters of small particles, such as carbon particles, in the air that come mostly from vehicle exhausts. There is a link between increase in PM10 levels and a rise in death rate, increased hospital admissions, and asthma incidence. The elderly and those with chronic heart or lung disease are most at risk.

pneumonia inflammation of the lungs, generally due to bacterial or viral infection but also to particulate matter or gases. It is characterized by a build-up of fluid in the alveoli, the clustered air sacs (at the ends of the air passages) where oxygen exchange takes place.

PNG abbreviation for *Portable Network Graphics.

p–n junction diode in electronics, another name for *semiconductor diode.

poikilothermy condition in which an animal's body temperature is largely dependent on the temperature of the air or water in which it lives. It is characteristic of all animals except birds and mammals, which maintain their body temperatures by *homeothermy (they are 'warm-blooded').

point in geometry, a basic element, whose position in the Cartesian system may be determined by its *coordinates.

point and click in computing, basic method of navigating a *Web page or a multimedia CD-ROM. The user points at an object using a cursor and a mouse, and clicks to activate it.

point of sale POS, in business premises, place at which a sale is made. This could be at a retail store (such as a supermarket checkout), at home (door-to-door sales), or on the phone (phone selling). At the point of sale there will often be promotional materials and other inducements to purchase, such as notices of special offers. In conjunction with electronic funds transfer, point of sale is part of the terminology of 'cashless shopping', enabling buyers to transfer funds directly from their bank accounts to the shop's.

point-of-sale terminal or **POS terminal**, computer terminal used in shops to input and output data at the point where a sale is transacted; for example, at a supermarket checkout. The POS terminal inputs information about the identity of each item sold, retrieves the price and other details from a central computer, and prints out a fully itemized receipt for the customer. It may also input sales data for the shop's computerized stock-control system.

Point-to-Point Protocol in computing, method of connecting a computer to the Internet; usually abbreviated to PPP.

poise symbol P, c.g.s. unit of dynamic *viscosity (the property of liquids that determines how readily they flow). It is equal to one dyne-second per square centimetre. For most liquids the centipoise (one hundredth of a poise) is used. Water at 20°C/68°F has a viscosity of 1.002 centipoise.

Poiseuille's formula relationship describing the rate of flow of a fluid through a narrow tube. For a capillary (very narrow) tube of length l and radius r with a pressure difference p between its ends, and a liquid of *viscosity η, the velocity of flow expressed as the volume per second is $\pi p r^4 / 8 l \eta$. The formula was devised in 1843 by French physicist Jean Louis Poiseuille (1799–1869).

poison or **toxin**, any chemical substance that, when introduced into or applied to the body, is capable of injuring health or destroying life.

The liver removes some poisons from the blood. The majority of poisons may be divided into **corrosives**, such as sulphuric, nitric, and hydrochloric acids; **irritants**, including arsenic and copper sulphate; **narcotics**, such as opium and carbon monoxide; and **narcotico-irritants** from any substances of plant origin including phenol acid and tobacco.

Poisson's ratio The ratio of the lateral contraction of a body to its longitudinal extension. The ratio is constant for a given material. It is named after French applied mathematician and physicist Siméon-Denis Poisson (1781–1840).

polar coordinates in mathematics, a way of defining the position of a point in terms of its distance r from a fixed point (the origin) and its angle θ to a fixed line or axis. The coordinates of the point are (r, θ).

polarimetry in astronomy, any technique for measuring the degree of polarization of radiation from stars, galaxies, and other objects.

polarization in physics, restriction in the directions in which transverse waves can vibrate. Light can be polarized by passing it through certain materials (such as Polaroid) or on being reflected. Radio waves and other types of electromagnetic radiation can also be polarized.

polarized light light in which the electromagnetic vibrations take place in one particular plane. In ordinary (unpolarized) light, the electric fields vibrate in all planes perpendicular to the direction of propagation. After reflection from a polished surface or transmission through certain materials (such as Polaroid), the electric fields are confined to one direction, and the light is said to be **linearly polarized**. In **circularly polarized** and **elliptically polarized** light, the electric fields are confined to one direction, but the direction rotates as the light propagates. Polarized light is used to test the strength of sugar solutions and to measure stresses in transparent materials.

polarography electrochemical technique for the analysis of oxidizable and reducible compounds in solution. It involves the diffusion of the substance to be analysed onto the surface of a small electrode, usually a bead of mercury, where oxidation or reduction occurs at an electric potential characteristic of that substance.

polar reversal or **magnetic reversal**, change in polarity of Earth's magnetic field. Like all magnets, Earth's magnetic field has two opposing regions, or poles, positioned approximately near geographical North and South Poles. During a period of normal polarity the region of attraction corresponds with the North Pole. Today, a compass needle, like other magnetic materials, aligns itself parallel to the magnetizing force and points to the North Pole. During a period of reversed polarity, the region of attraction would change to the South Pole and the needle of a compass would point south.

Studies of the magnetism retained in rocks at the time of their formation (like small compasses frozen in time) have shown that the polarity of the magnetic field has reversed repeatedly throughout geological time.

The reason for polar reversals is not known. Although the average time between reversals over the last 10 million years has been 250,000 years, the rate of reversal has changed continuously over geological time. The most recent reversal was 780,000 years ago; scientists have no way of predicting when the next reversal will occur. The reversal process probably takes a few thousand years. Dating rocks using distinctive sequences of magnetic reversals is called *magnetic stratigraphy.

pole either of the geographic north and south points of the axis about which the Earth rotates. The geographic poles differ from the magnetic poles, which are the points towards which a freely suspended magnetic needle will point.

In 1985 the magnetic north pole was some 350 km/218 mi northwest of Resolute Bay, Northwest Territories, Canada. It moves northwards about 10 km/6 mi each year, although it can vary in a day about 80 km/50 mi from its average position. It is relocated every decade in order to update navigational charts.

It is thought that periodic changes in the Earth's core cause a reversal of the magnetic poles (see *polar reversal, *magnetic field). Many animals, including migrating birds and fish, are believed to orient themselves partly using the Earth's magnetic field. A permanent scientific base collects data at the South Pole.

pole, magnetic see *magnetic pole.

polio or **poliomyelitis**, viral infection of the central nervous system affecting nerves that activate muscles. The disease used to be known as infantile paralysis since children were most often affected. Two kinds of vaccine are available, one injected and one given by mouth. The Americas were declared to be polio-free by the Pan American Health Organization in 1994. In 1997 the World Health Organization (WHO) reported that causes of polio had dropped by nearly 90% since 1988 when the organization began its programme to eradicate the disease by the year 2000. Most remaining cases were in Africa and southeast Asia in early 2000.

pollen grains of seed plants that contain the male gametes. In *angiosperms (flowering plants) pollen is produced within *anthers; in most *gymnosperms (cone-bearing plants) it is produced in male cones. A pollen grain is typically yellow and, when mature, has a hard outer wall. Pollen of insect-pollinated plants (see *pollination) is often sticky and spiny and larger than the smooth, light grains produced by wind-pollinated species.

pollen tube outgrowth from a pollen grain that grows towards the *ovule, following germination of the grain on the *stigma. In *angiosperms (flowering plants) the pollen tube reaches the ovule by growing down through the *style, carrying the male gametes inside. The gametes are discharged into the ovule and one fertilizes the egg cell.

pollination process by which pollen is transferred from one plant to another. The male *gametes are contained in pollen grains, which must be transferred from the anther to the stigma in *angiosperms (flowering plants), and from the male cone to the female cone in *gymnosperms (cone-bearing plants). Fertilization (not the same as pollination) occurs after the growth of the pollen tube to the ovary. Self-pollination occurs when pollen is transferred to a stigma of the same flower, or to another flower on the same plant; cross-pollination occurs when pollen is transferred to another

plant. This involves external pollen-carrying agents, such as wind, water, insects, birds, bats, and other small mammals.

Animal pollinators carry the pollen on their bodies and are attracted to the flower by scent, or by the sight of the petals. Most flowers are adapted for pollination by one particular agent only. Bat-pollinated flowers tend to smell of garlic, rotting vegetation, or fungus. Those that rely on animals generally produce nectar, a sugary liquid, or surplus pollen, or both, on which the pollinator feeds. Thus the relationship between pollinator and plant is an example of mutualism, in which both benefit. However, in some

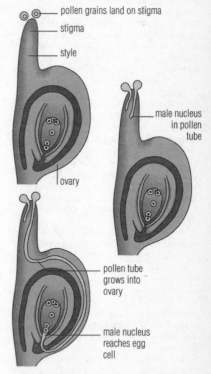

pollen grains land on stigma
stigma
style
male nucleus in pollen tube
ovary
pollen tube grows into ovary
male nucleus reaches egg cell

pollen Pollination, the process by which pollen grains transfer their male nuclei (gametes) to the ovary of a flower. The pollen grains land on the stigma and form a pollen tube that grows down into the ovary. The male nuclei travel along the pollen tube.

plants the pollinator receives no benefit (as in *pseudocopulation), while in others, nectar may be removed by animals that do not effect pollination.

polling in computing, a technique for transferring data from a terminal to the central computer of a *multiuser system. The computer automatically makes a connection with each terminal in turn, interrogates it to check whether it is holding data for transmission, and, if it is, collects the data.

pollinium group of pollen grains that is transported as a single unit during pollination. Pollinia are common in orchids.

polluter-pays principle the idea that whoever causes pollution is responsible for the cost of repairing any damage. The principle is accepted in British law but has in practice often been ignored; for example, farmers causing the death of fish through slurry pollution have not been fined the full costs of restocking the river.

polonium chemical symbol Po, radioactive, metallic element, atomic number 84, relative atomic mass 209. Polonium occurs in nature in small amounts and was isolated from *pitchblende. It is the element having the largest number of isotopes (27) and is 5,000 times as radioactive as radium, liberating considerable amounts of heat. It was the first element to have its radioactive properties recognized and investigated.

Polonium was isolated in 1898 from the pitchblende residues analysed by French scientists Pierre and Marie Curie, and named after Marie Curie's native Poland.

polyamides Synthetic polymers made by *condensation polymerisation containing the amide group. See *nylon

polychlorinated biphenyl PCB, any of a group of chlorinated isomers of biphenyl (C_6H_5)$_2$. They are dangerous industrial chemicals, valuable for their fire-resistant qualities. They constitute an environmental hazard because of their persistent toxicity. Since 1973 their use had been limited by international agreement. In December 2000, 122 nations agreed a treaty to ban the toxic chemicals known as persistent organic polluters (POPs), which include PCBs, although they are unlikely to be totally eliminated until about 2025.

polyester synthetic resin formed by the *condensation polymerization of polyhydric alcohols (alcohols containing more than one hydroxyl group) with dibasic acids (acids containing two replaceable hydrogen atoms). Polyesters are thermosetting *plastics, used for constructional plastics and, with glass fibre added as reinforcement, they are used in car bodies and boat hulls. Polyester is also a major *synthetic fibre used for knitting or weaving fabrics which are strong but lightweight, and resist creasing but can be heat-set into pleats. Polyester is often mixed with other fibres and can be found in a wide range of different textiles.

polyethene alternative (and approved systematic) name for * polythene.

polyethylene alternative term for *polythene.

polygon in geometry, a plane (two-dimensional) figure with three or more straight-line sides. **Regular polygons** have sides of the same length, and all the *exterior angles are equal. Common polygons have names that define the number of sides (for example, *triangle (3), *quadrilateral (4), *pentagon (5), *hexagon (6), **heptagon** (7), *octagon (8), and so on). These are all convex polygons, having no *interior angle greater than 180°.

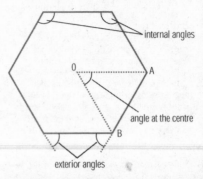

polygon The types of angles in a regular polygon are shown here. These are the angles in a regular hexagon.

| dodecahedron | icosahedron | tetrahedron | cube | octahedron |

polyhedron The five regular polyhedra or Platonic solids.

polyhedron in geometry, a solid figure with four or more plane faces. The more faces there are on a polyhedron, the more closely it approximates to a sphere. Knowledge of the properties of polyhedra is needed in crystallography and stereochemistry to determine the shapes of crystals and molecules.

polymer very long-chain molecule made up of many repeated simple units (*monomers) linked together by *polymerization. There are many polymers, both natural (cellulose, chitin, lignin, rubber) and synthetic (polyethylene and nylon, types of plastic). Synthetic polymers belong to two groups: thermosoftening and thermosetting (see *plastic).

The size of the polymer matrix is determined by the amount of monomer used; it therefore does not form a molecule of constant molecular size or mass.

polymerase chain reaction PCR, technique developed during the 1980s to clone short strands of DNA from the *genome of an organism. The aim is to produce enough of the DNA to be able to sequence and identify it. It was developed by US biochemist Kary Mullis in 1983.

polymerization chemical union of two or more (usually small) molecules of the same kind to form a new compound. **Addition polymerization** produces simple multiples of the same compound. **Condensation polymerization** joins molecules together with the elimination of water or another small molecule.

Addition polymerization uses only a single monomer (basic molecule); condensation polymerization may involve two or more different monomers (**co-polymerization**).

polymorph in biology, white blood cell that has a nucleus of irregular shape. Polymorphs constitute about 70% of the white blood cells and are involved in the immune response.

polymorphism in genetics, the coexistence of several distinctly different types in a *population (groups of animals of one species). Examples include the different blood groups in humans, different colour forms in some butterflies, and snail shell size, length, shape, colour, and stripiness.

addition of ethane molecules to form polyethene

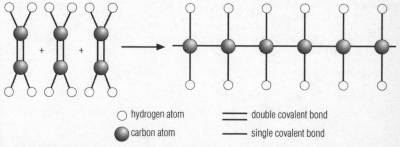

| ○ hydrogen atom | ═══ double covalent bond |
| ● carbon atom | ─── single covalent bond |

polymerization In polymerization, small molecules (monomers) join together to make large molecules (polymers). In the polymerization of ethene to polyethene (polythene), electrons are transferred from the carbon–carbon double bond of the ethene molecule, allowing the molecules to join together as a long chain of carbon–carbon single bonds.

polymorphism in mineralogy, the ability of a substance to adopt different internal structures and external forms, in response to different conditions of temperature and/or pressure. For example, diamond and graphite are both forms of the element carbon, but they have very different properties and appearance.

Silica (SiO_2) also has several polymorphs, including quartz, tridymite, cristobalite, and stishovite (the latter a very high-pressure form found in meteoritic impact craters).

polynomial in mathematics, an algebraic expression that has one or more *variables (denoted by letters). A polynomial equation has the form:

$$f(x) = a_n x^n + a_{n-1} x^{n-l} + \ldots + a_2 x^2 + a_1 x + a_0$$

where $a_n, a_{n-1}, \ldots, a_0$ are all constants, n is a positive integer, and $a_n \neq 0$.

The 'degree' of a polynomial equation is simply the degree of the polynomial involved. A polynomial of degree one, that is, whose highest power of x is 1, as in $2x + 1$, is called a linear polynomial:

$3x^2 + 2x + 1$ is quadratic;

$4x^3 + 3x^2 + 2x + 1$ is cubic.

polyp or **polypus**, small 'stalked' benign tumour, usually found on mucous membrane of the nose or bowels. Intestinal polyps are usually removed, since some have been found to be precursors of cancer.

polypeptide long-chain *peptide.

polyploid in genetics, possessing three or more sets of chromosomes in cases where the normal complement is two sets (*diploid). Polyploidy arises spontaneously and is common in plants (mainly among flowering plants), but rare in animals. Many crop plants are natural polyploids, including wheat, which has four sets of chromosomes per cell (durum wheat) or six sets (common wheat).

Plant breeders can induce the formation of polyploids by treatment with a chemical, colchicine.

polypropylene or **polypropene**, $(C_3H_6)_n$, plastic made by *addition polymerization, or linking together, of *propene molecules ($CH_2=CH–CH_3$). Polypropylene is one of the most important plastics and has a large number of different uses, including packaging, house ware, textiles, carpets, artificial turf, and ropes. It is a very light *synthetic fibre, which does not absorb moisture, and so stays dry and warm. These properties make it a good fibre for use in clothing designed for cold environments.

polysaccharide long-chain *carbohydrate made up of hundreds or thousands of linked simple sugars (monosaccharides) such as glucose and closely related molecules. The polysaccharides are natural polymers. They either act as energy-rich food stores in plants (starch) and animals (glycogen), or have structural roles in the plant cell wall (cellulose, pectin) or the tough outer skeleton of insects and similar creatures (chitin). See also *carbohydrate.

polystyrene or polyphenylethene. A type of *plastic made by *addition polymerization from phenylethene ($C_6H_5CHCH_2)_n$. Polystyrene is a used in kitchen utensils or, in an expanded form, in insulation and ceiling tiles. CFCs (*chlorofluorocarbons) are used to produce expanded polystyrene so alternatives are being sought.

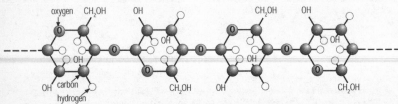

polysaccharide A typical polysaccharide molecule, glycogen (animal starch), is formed from linked glucose ($C_6H_{12}O_6$) molecules. A glycogen molecule has 100–1,000 linked glucose units.

polytetrafluoroethene PTFE, polymer made from the monomer tetrafluoroethene (CF_2CF_2) by *addition polymerization. It is a thermosetting plastic with a high melting point that is used to produce 'nonstick' surfaces on pans and to coat bearings. Its trade name is Teflon.

polythene or **polyethylene** or **polyethene**, *polymer of the gas *ethene (C_2H_4) made by *addition polymerization. It is a tough, white, translucent, waxy thermoplastic (which means it can be repeatedly softened by heating). It is used for packaging, bottles, toys, wood preservation, electric cable, pipes, and tubing.

Polythene is produced in two forms: low-density polythene, made by high-pressure *polymerization of ethene gas (the *monomer), and high-density polythene, which is made at lower pressure by using catalysts. This form, first made in 1953 by German chemist Karl Ziegler, is more rigid at low temperatures and softer at higher temperatures than the low-density type. Polythene was first made in the 1930s at very high temperatures by the British chemical company ICI.

polyunsaturate type of *fat or oil containing a high proportion of triglyceride molecules whose *fatty acid chains contain several double bonds. By contrast, the fatty-acid chains of the triglycerides in saturated fats (such as lard) contain only single bonds. Medical evidence suggests that polyunsaturated fats, used widely in margarines and cooking fats, are less likely to contribute to cardiovascular disease than saturated fats, but there is also some evidence that they may have adverse effects on health.

polyurethane polymer made from the monomer urethane. It is a thermoset *plastic, used in liquid form as a paint or varnish, and in foam form for upholstery and in lining materials (where it may be a fire hazard).

polyvinyl chloride PVC or polychloroethene, type of *plastic used for drainpipes, floor tiles, audio discs, shoes, and handbags. It is derived from vinyl chloride (CH_2=CHCl) by *addition polymerization. Swedish scientists identified a link between regular exposure to PVC and testicular cancer, in 1998, increasing the demand for a ban on PVC in commercial products.

pomeron in physics, hypothetical object with energy and momentum, but no colour or electrical charge, produced when an electron strikes a proton. Pomerons were first suggested in 1958 by Russian physicist Isaac Pomeranchuk.

pooter small device for collecting invertebrates, consisting of a jar to which two tubes are attached. A sharp suck on one of the tubes, while the other is held just above an insect, will propel the animal into the jar. A filter wrapped around the mouth tube, prevents debris or organisms from being swallowed.

PoP acronym for Point of Presence, in computing, place where users can access a network via a telephone connection. A PoP is a collection of modems and other equipment which are permanently connected to the network. Compare *vPoP.

POP3 acronym for Post Office Protocol 3, on the Internet, one of the two most common mail *protocols.

population in biology and ecology, a group of organisms of one species, living in a certain area. The organisms are able to interbreed. It also refers to the members of a given species in a *community of living things. The area can be small. For example, one can refer to the population of duckweed (a small floating plant found on the surface of ponds) on a pond. Since the pond is a *habitat, one can consider the population of duckweed in a habitat and forming part of the community of plants and animals there. However, it is also possible to use the term population for all the organisms of one species in a large geographical area, for example the elephant population in Africa. It could also be used to describe all the organisms of that species on Earth, for example the world population of humans. Population sizes in habitats change over a period of time. The timescale may be daily, seasonal, or there may be changes over the years.

population in statistics, the universal set from which a sample of data is selected. The chief object of statistics is to find out population characteristics by taking samples.

population genetics branch of genetics that studies the way in which the frequencies of different *alleles (alternative forms of a gene) in populations of organisms change, as a result of natural selection and other processes.

pop-up menu in computing, a menu that appears in a (new) window when an option is selected with a mouse or keystroke sequence in a *graphical user interface (GUI), such as Microsoft Windows. Compare with *pull-down menu.

pore in biology, a small opening in the skin that releases sweat and sebum. Sebum acts as a natural lubricant and protects the skin from the effects of moisture or excessive dryness.

porphyria group of rare genetic disorders caused by an enzyme defect. Porphyria affects the digestive tract, causing abdominal distress; the nervous system, causing psychotic disorder, epilepsy, and weakness; the circulatory system, causing high blood pressure; and the skin, causing extreme sensitivity to light. No specific treatments exist.

port in computing, a socket that enables a computer processor to communicate with an external device. It may be an **input port** (such as a

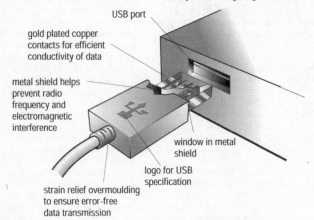

USB port

gold plated copper contacts for efficient conductivity of data

metal shield helps prevent radio frequency and electromagnetic interference

window in metal shield

logo for USB specification

strain relief overmoulding to ensure error-free data transmission

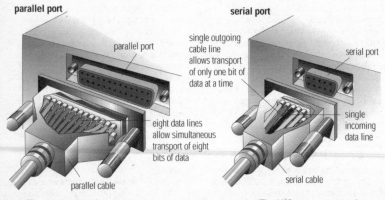

parallel port

parallel port

serial port

single outgoing cable line allows transport of only one bit of data at a time

serial port

eight data lines allow simultaneous transport of eight bits of data

single incoming data line

parallel cable

serial cable

port The three types of communications port in a microcomputer. The USB port transmits data at a faster rate than a serial or parallel port.

joystick port), or an **output port** (such as a printer port), or both (an **i/o port**).

portability in computing, the characteristic of certain programs that enables them to run on different types of computer with minimum modification. Programs written in a *high-level language can usually be run on any computer that has a compiler or interpreter for that particular language.

Portable Network Graphics PNG, file format for images created under the auspices of the *W3C. PNG offers an alternative to the *GIF format, which ran into problems over its use of *LZW compression. See also *graphic file format.

port address on the Internet, a way for a host system to specify which *server a particular application will use.

portal *Web site designed to be used as a start-up site for browsing sessions and to provide a gateway to the rest of the Web. To increase their attractiveness, portals expanded rapidly in the late 1990s to offer a wide variety of services such as personalized start-up pages, free e-mail, directories, customized news and weather reports, online calendars, games, and free Web space. The first portals were online services such as America Online (AOL), which added gateways to the Internet. Others have come from different areas: Yahoo! started as a hierarchical guide to Web sites, Excite as a search engine, and Netscape's Netcenter as the default home page for users of Netscape's Web browser. CNet's Snap! was the first Web site designed to act as a portal. Traditional portals are under threat from a number of areas. Free ISPs are capturing customers from portals, educated Web users choose to go straight to niche sites that interest them, and satellite and cable companies offer interactive television with Internet access channelled through their own Web sites.

position vector vector that defines the position of a point with respect to the origin.

positive integer any whole number from 0 upwards. 0 is included so that the properties of positive integers

include those related to the *identity for addition.

positron antiparticle of the electron; an *elementary particle having the same mass as an electron but exhibiting a positive charge. The positron was discovered in 1932 by US physicist Carl Anderson at the California Institute of Technology, USA, its existence having been predicted by the British physicist Paul Dirac in 1928.

positron emission tomography PET, imaging technique that enables doctors to observe the metabolic activity of the human body by following the progress of a radioactive chemical that has been inhaled or injected, detecting *gamma radiation given out when *positrons emitted by the chemical are annihilated. The technique has been used to study a wide range of conditions, including schizophrenia, Alzheimer's disease, and *Parkinson's disease.

POSIX acronym for Portable Operating System Interface for UniX, ANSI standard, developed to describe how the programming interfaces and other features of *Unix worked, in order to remove control from the developers, AT&T Bell Laboratories. Subsequently many other (proprietary) *operating systems were modified in order to become POSIX-compliant, that is, they provide an *open systems interface, so that they can communicate with other POSIX-compliant systems, even though the operating systems themselves are internally quite different.

post in computing, to send a message to a newsgroup or *bulletin board for others to read.

post-coital contraception in medicine, contraception used to prevent conception after intercourse has taken place.

POS terminal acronym for *point of sale terminal, a cash register linked to a computer.

posting in computing, another word for article.

postmaster in computing, systems administrator in charge of a mail server. The term is especially used for people who manage the e-mail system in a *local area network (LAN) or other local network.

post-traumatic stress disorder PTSD, psychiatric disorder caused by the experiencing of a trauma.

Diagnosis is based on the presence of a number of symptoms such as recurrent recollections of the trauma, emotional numbness, lack of motivation, and irrational outbursts. Symptoms are intense enough to prevent the sufferer from functioning normally. Treatment is by *psychotherapy. Sufferers include soldiers, rape victims, emergency services staff, and children living in a violent environment.

post-viral fatigue syndrome in medicine, another name for *chronic fatigue syndrome.

potash general name for any potassium-containing mineral, most often applied to potassium carbonate (K_2CO_3) or potassium hydroxide (KOH). Potassium carbonate, originally made by roasting plants to ashes in earthenware pots, is commercially produced from the mineral sylvite (potassium chloride, KCl) and is used mainly in making artificial fertilizers, glass, and soap.

The potassium content of soils and fertilizers is also commonly expressed as potash, although in this case it usually refers to potassium oxide (K_2O).

potassium chemical symbol K, (Dutch *potassa* 'potash') soft, waxlike, silver-white, metallic element, atomic number 19, relative atomic mass 39.0983. Its chemical symbol comes from the Latin *kalium*. It is one of the *alkali metals (in Group 1 of the *periodic table of the elements), and has a very low density – it floats on water, and is the second lightest metal (after lithium). It is one of the most reactive in the *reactivity series of metals, oxidizing rapidly when exposed to air and reacting violently with water. Of great abundance in the Earth's crust, it is widely distributed. It is found in salt and mineral deposits in the form of potassium aluminium silicates and potassium nitrate (saltpetre). Potassium has to be extracted from its compounds by *electrolysis because of its high reactivity.

Potassium is the main base ion of the fluid in the body's cells. Along with *sodium, it is important to the electrical potential of the nervous system and, therefore, for the efficient functioning of nerve and muscle. Shortage, which may occur with excessive fluid loss (prolonged diarrhoea, vomiting), may lead to muscular paralysis; potassium overload may result in *cardiac arrest. It is also required by plants for growth. The element was discovered and named in 1807 by English chemist Humphry Davy, who isolated it from potash in the first instance of a metal being isolated by electrolysis.

potassium–argon dating or K–Ar **dating**, isotopic dating method based on the radioactive decay of potassium-40 (^{40}K) to the stable isotope argon-40 (^{40}Ar). Ages are based on the known half-life of ^{40}K, and the ratio of ^{40}K to ^{40}Ar. The method is routinely applied to rock samples about 100,000 to 30 million years old. The method is used primarily to date volcanic layers in stratigraphic sequences with archaeological deposits, and the palaeomagnetic-reversal timescale. Complicating factors, such as sample contamination by argon absorbed from the atmosphere, and argon gas loss by diffusion out of the mineral, limit application of this technique.

potassium dichromate $K_2Cr_2O_7$, orange, crystalline solid, soluble in water, that is a strong *oxidizing agent in the presence of dilute sulphuric acid. As it oxidizes other compounds it is itself reduced to potassium chromate (K_2CrO_4), which is green. Industrially it is used in the manufacture of dyes and glass and in tanning, photography, and ceramics.

potassium manganate(VII) or **potassium permanganate**; $KMnO_4$, dark purple, crystalline solid, soluble in water, that is a strong *oxidizing agent in the presence of dilute sulphuric acid. In the process of oxidizing other compounds it is itself reduced to manganese(II) salts (containing the Mn^{2+} ion), which are colourless.

potato blight disease of the potato caused by a parasitic fungus

Phytophthora infestans. It was the cause of the 1845 potato famine in Ireland. New strains of *P. infestans* continue to arise. The most virulent version so far is *P. infestans US-8*, which arose in Mexico in 1992 , spreading to North America in 1994.

potential difference PD, difference in the electrical potential (see *potential, electric) of two points, being equal to the electrical energy converted by a unit electric charge moving from one point to the other. Electrons flow in a conducting material towards the part that is relatively more positive (fewer negative charges). The SI unit of potential difference is the volt (V). The potential difference between two points in a circuit is commonly referred to as voltage (and can be measured with a voltmeter). See also *Ohm's law.

One volt equals one joule of energy used for each coulomb of charge. In equation terms, potential difference V may be defined by: $V = E \div Q$, where E is the electrical energy converted in joules and Q is the charge in coulombs. Chemical energy from a battery is converted to electrical energy in a circuit, this energy being given by rearranging the above formula: $E = Q \times V$.

potential divider or **voltage divider**, two resistors connected in series in an electrical circuit in order to obtain a fraction of the potential difference, or voltage, across the battery or electrical source. The potential difference is divided across the two resistors in direct proportion to their resistances.

potential, electric energy required to bring a unit electric charge from infinity to the point at which potential is defined. The SI unit of potential is the volt (V). Positive electric charges will flow 'downhill' from a region of high potential to a region of low potential.

A charged *conductor, for example, has a higher potential than the Earth, the potential of which is taken by convention to be zero. An electric cell (battery) has a potential difference between its terminals, which can make current flow in an external circuit. The difference in potential – *potential

difference (pd) – is expressed in volts so, for example, a 12 V battery has a pd of 12 volts between its negative and positive terminals.

potential energy PE, *energy possessed by an object by virtue of its relative position or state (for example, as in a compressed spring or a muscle). It can be thought of as 'stored' energy. An object that has been raised up has energy stored due to its height. It is described as having *gravitational potential energy.

If a ball is raised to a certain height and released, the ball falls to the ground. The potential energy changes to kinetic energy. As the ball hits the ground some of the kinetic energy is lost as sound and elastic energy. A stretched spring has stored elastic energy; this is known as elastic potential energy. Springs and elastics are designed to store energy and release it either rapidly or slowly. For example, a mechanical toy works by an unwinding spring coil inside the toy. As the spring coil turns, elastic potential energy changes into kinetic and sound energy as the toy operates.

potentiometer electrical *resistor that can be divided so as to compare, measure, or control voltages. In radio circuits, any rotary variable resistance (such as volume control) is referred to as a potentiometer. A simple type of potentiometer consists of a length of uniform resistance wire (about 1 m/ 3 ft long) carrying a constant current provided by a battery connected across the ends of the wire. The source of potential difference (voltage) to be measured is connected (to oppose the cell) between one end of the wire, through a *galvanometer (instrument for measuring small currents), to a contact free to slide along the wire. The sliding contact is moved until the galvanometer shows no deflection. The ratio of the length of potentiometer wire in the galvanometer circuit to the total length of wire is then equal to the ratio of the unknown potential difference to that of the battery.

pound abbreviation lb, imperial unit of mass. The commonly used avoirdupois pound, also called the **imperial standard pound** (7,000 grains/0.45 kg), differs from the **pound troy** (5,760 grains/0.37 kg), which is used for weighing precious metals. It derives from the Roman *libra*, which weighed 0.327 kg.

powder metallurgy method of shaping heat-resistant metals such as tungsten. Metal is pressed into a mould in powdered form and then sintered (heated to very high temperatures).

power in mathematics, that which is represented by an *exponent or index, denoted by a superior numeral. A number or symbol raised to the power of 2 – that is, multiplied by itself – is said to be squared (for example, 3^2, x^2), and when raised to the power of 3, it is said to be cubed (for example, 2^3, y^3). Any number to the power zero always equals 1.

Powers can be negative. Negative powers produce fractions, with the numerator as one, as a number is divided by itself, rather than being multiplied by itself, so for example $2^{-1} = 1/2$ and $3^{-3} = 1/27$.

power in optics, a measure of the amount by which a lens will deviate light rays. A powerful converging lens will converge parallel rays strongly, bringing them to a focus at a short distance from the lens. The unit of power is the **dioptre**, which is equal to the reciprocal of focal length in metres. By convention, the power of a converging (or convex) lens is positive and that of a diverging (or concave) lens negative.

power in physics, the rate of doing work or transferring energy from one form to another. It is measured in watts (joules per second) or other units of work per unit time.

power station building where electrical energy is generated from a fuel or from another form of energy. Fuels used include *fossil fuels such as coal, gas, and oil, and the nuclear fuel uranium. Renewable sources of energy include *gravitational potential energy, used to produce *hydroelectric power, and *wind power.

PPP abbreviation for point-to-point protocol, newer of the two standard methods for connecting a computer to the Internet via a modem and the public switched telephone network (*PSTN). Unlike the earlier *SLIP, PPP can handle both *synchronous and asynchronous communication and provides error detection.

prairie central North American plain, formerly grass-covered, extending over most of the region between the Rocky Mountains to the west, and the Great Lakes and Ohio River to the east.

The term was first applied by French explorers to vast, largely level grasslands in central North America, centred on the Mississippi River valley, which extend from the Gulf of Mexico to central Alberta, Canada, and from west of the Appalachian system into the Great Plains. When first seen by explorers, the prairies were characterized by unbroken, waist-high, coarse grasses. Trees were common only along rivers and streams, or in occasional depressions in the land. This prairie is now almost gone, altered by farming into the 'Corn Belt', much of the 'Wheat Belt', and other ploughed lands. Its humus-rich black loess soils, adequate rainfall, and warm summers foster heavily productive agriculture. In the west – west Kansas, Nebraska, and the Dakotas – is the **short-grass prairie**, occupying large parts of the Great Plains. Higher, drier land here has been used primarily for wheat production (aided by deep-well irrigation) and stock raising. The prairies were formerly the primary habitat of the American bison; other prominent species include prairie dogs, deer and antelope, grasshoppers, and a variety of prairie birds.

praseodymium chemical symbol Pr, (Greek *prasios* 'leek-green' + *didymos* 'twin') silver-white, malleable, metallic element of the *lanthanide series, atomic number 59, relative atomic mass 140.907. It occurs in nature in the minerals monzanite and bastnaesite, and its green salts are used to colour glass and ceramics. It was named in 1885 by Austrian chemist Carl von Welsbach.

He fractionated it from dydymium (originally thought to be an element but actually a mixture of rare-earth metals consisting largely of neodymium, praseodymium, and cerium) and named it for its green salts and spectroscopic line.

Precambrian in geology, the time from the formation of the Earth (4.6 billion years ago) up to 570 million years ago. Its boundary with the succeeding Cambrian period marks the time when animals first developed hard outer parts (exoskeletons) and so left abundant fossil remains. It comprises about 85% of geological time and is divided into two eons: the Archaean and the Proterozoic.

precipitation in chemistry, the formation of an insoluble solid in a liquid as a result of a reaction within the liquid between two or more soluble substances. For example, if solutions of lead nitrate and potassium iodide are added together, bright yellow, insoluble lead iodide appears as a precipitate in the solution, making it cloudy. The **precipitation reaction** is: lead nitrate + potassium iodide → lead iodide + potassium nitrate. If the newly formed solid settles, it forms a **precipitate**; if the particles of solid are very small, they will remain in suspension, forming a *colloidal precipitate.

Precipitation reactions are used in the preparation of *salts; in *qualitative analysis reactions; and in the formation of some *sedimentary rocks.

pregnancy in humans, the process during which a developing embryo grows within the woman's womb. It begins at conception and ends at birth, and the normal length is 40 weeks, or around nine months.

Menstruation usually stops on conception. About one in five pregnancies fails, but most of these failures occur very early on, so the woman may notice only that her period is late. After the second month, the breasts become tender, and the areas round the nipples become darker. Enlargement of the uterus can be felt at about the end of the third month, and after this the abdomen enlarges progressively. Fetal movement can be felt at about 18 weeks; a heartbeat may be heard during the sixth month. Pregnancy in animals is called *gestation.

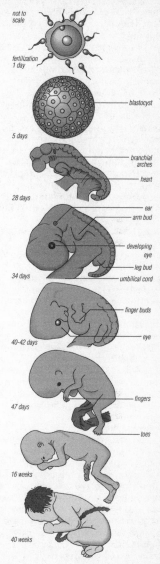

not to scale

fertilization
1 day

5 days — blastocyst

branchial arches

heart

28 days

ear

arm bud

developing eye

leg bud

34 days — umbilical cord

finger buds

eye

40–42 days

47 days — fingers

toes

16 weeks

40 weeks

pregnancy The development of a human embryo. Division of the fertilized egg, or ovum, begins within hours of conception. Within a week a ball of cells – a blastocyst – has developed. After the third week, the embryo has changed from a mass of cells into a recognizable shape. From the end of the second month, the embryo is almost fully formed and further development is mainly by growth.

premenstrual tension PMT; or **premenstrual syndrome**, medical condition caused by hormone changes and comprising a number of physical and emotional features that occur cyclically before menstruation and disappear with its onset. Symptoms include mood changes, breast tenderness, a feeling of bloatedness, and headache.

pressure in a fluid, the force exerted normally (at right angles) on the surface of a body immersed in the fluid. The SI unit of pressure is the pascal (Pa), equal to a pressure of one newton per square metre. In the atmosphere, the pressure declines with increasing height from about 100 kPa at sea level to zero where the atmosphere fades into space. Pressure is commonly measured with a *barometer, *manometer, or *Bourdon gauge. Other common units of pressure are the bar and the torr.

pressure law law stating that the pressure of a fixed mass of gas at constant volume is directly proportional to its absolute temperature. The law may be expressed as: pressure/temperature = constant or, more usefully, as: $P_1/T_1 = P_2/T_2$ where P_1 and T_1 are the initial pressure and absolute temperature of a gas, and P_2 and T_2 are its final pressure and temperature. See also *gas laws.

pressurized water reactor PWR, *nuclear reactor design used in nuclear power stations in many countries, and in nuclear-powered submarines. In the PWR, water under pressure is used in a closed system, as a coolant, passing through the reactor to the generator. Boric acid is added as a moderator. It circulates through a steam generator, where its heat boils water to provide steam to drive power *turbines.

primary sexual characteristic endocrine gland producing maleness and femaleness. In males, the primary sexual characteristic is the *testis; in females it is the *ovary. Both are endocrine glands that produce hormones responsible for secondary sexual characteristics, such as facial hair and a deep voice in males, and breasts in females.

prime factor any factor of a number that is a *prime number. The fundamental theorem of arithmetic states that every number is either prime or can be expressed as a unique product of primes.

prime number number that can be divided only by 1 and itself, that is, having no other factors. There is an infinite number of primes, the first ten of which are 2, 3, 5, 7, 11, 13, 17, 19, 23, and 29 (by definition, the number 1 is excluded from the set of prime numbers). The number 2 is the only even prime because all other even numbers have 2 as a factor. Numbers other than primes can be expressed as a product of their *prime factors.

printed circuit board PCB, electrical circuit created by laying (printing) 'tracks' of a conductor such as copper on one or both sides of an insulating board. The PCB was invented in 1936 by Austrian scientist Paul Eisler, and was first used on a large scale in 1948.

prion acronym, by rearrangement, for PROteinaceous INfectious particle, infectious agent, a hundred times smaller than a virus. Composed of protein, and without any detectable nucleic acid (genetic material), it is strongly linked to a number of fatal degenerative brain diseases in mammals, such as bovine spongiform encephalopathy (BSE) in cattle, scrapie in sheep, and *Creutzfeldt–Jakob disease (CJD) and kuru in humans.

The existence of prions was postulated by US neurologist Stanley Prusiner in 1982, when he and his colleagues isolated a single infectious agent for scrapie that consisted only of protein and had no associated nucleic acid (RNA or DNA). His theory remains unproven but has been upheld by subsequent research, which has identified the protein concerned as well as a mechanism for its action.

A US researcher proved that prions are capable of spreading disease, in July 2000, three years after Prusiner was awarded the Nobel Prize for Physiology or Medicine for his prion theory.

prism in mathematics, a solid figure whose cross-section is the same along its length. When a slice is cut through a prism, the size and shape of the cross-section is always the same. A cube, for example, is a rectangular prism with all faces (bases and sides) the same shape and size. A cylinder is a prism with a circular cross section. The name of a prism is often derived from the shape of its prism, such as a triangular prism and a cuboid.

prism in optics, a triangular block of transparent material (plastic, glass, or silica) commonly used to 'bend' a ray of light or split a light beam (for example, white light) into its component colours. Prisms are used as mirrors to define the optical path in binoculars, camera viewfinders, and periscopes. The dispersive property of prisms is used in the *spectroscope.

privacy on the Internet, generally used to mean the right to control who has access to the personal information generated by interaction with computers.

probability likelihood, or chance, that an event will occur, often expressed as odds, or in mathematics, numerically as a fraction or decimal.

In general, the probability that n particular events will happen out of a total of m possible events is n/m. A certainty has a probability of 1; an impossibility has a probability of 0.

$$\text{Probability} = \frac{\text{number of successful events}}{\text{total possible number of events}}$$

procedural programming programming in which programs are written as lists of instructions for the computer to obey in sequence. It closely matches the computer's own sequential operation.

procedure in computing, a small part of a computer program that performs a specific task, such as clearing the screen or sorting a file. A **procedural language**, such as BASIC, is one in which the programmer describes a task in terms of how it is to be done, as opposed to a **declarative language**, such as PROLOG, in which it is described in terms of the required result.

process control automatic computerized control of a manufacturing process, such as glassmaking. The computer receives *feedback information from sensors about the performance of the machines involved, and compares this with ideal performance data stored in its control program. It then outputs instructions to adjust automatically the machines' settings.

processor in computing, another name for the *central processing unit or microprocessor of a computer.

productivity, biological in an ecosystem, the amount of material in the food chain produced by the primary producers (plants) that is available for consumption by animals. Plants turn carbon dioxide and water into sugars and other complex carbon compounds by means of photosynthesis. Their net productivity is defined as the quantity of carbon compounds formed, less the quantity used up by the respiration of the plant itself.

progesterone *steroid hormone that occurs in vertebrates. In mammals, it regulates the menstrual cycle and pregnancy. Progesterone is secreted by the corpus luteum (the ruptured Graafian follicle of a discharged ovum).

progestogen in medicine, drug closely related to progesterone, a female sex hormone that stimulates growth and secretion of the endometrial glands of the uterus in the first half of the menstrual cycle. Oral contraceptives contain progestogens, either alone or in combination with oestrogens.

program in computing, a set of instructions that controls the operation of a computer. There are two main kinds: applications programs, which carry out tasks for the benefit of the user – for example, word processing; and *systems programs, which control the internal workings of the computer. A *utility program is a systems program that carries out specific tasks for the user. Programs can be written in any of a number of *programming languages but are always translated into machine code before they can be executed by the computer.

program counter in computing, an alternative name for *sequence-control register.

program documentation
documentation that provides a complete technical description of a program, built up as the software is written, and is intended to support any later maintenance or development of the program.

program files in computing, files which contain the code used by a computer program.

program flow chart type of flow chart used to describe the flow of data through a particular computer program.

program loop part of a computer program that is repeated several times. The loop may be repeated a fixed number of times (**counter-controlled loop**) or until a certain condition is satisfied (**condition-controlled loop**). For example, a counter-controlled loop might be used to repeat an input routine until exactly ten numbers have been input; a condition-controlled loop might be used to repeat an input routine until the *data terminator 'XXX' is entered.

Program Manager main screen, or 'front end', of the software product Microsoft Windows 3.x. All Windows 3.x operations can be accessed from Program Manager. Although versions of Windows from Windows 95 onwards made use of a different graphical user interface (GUI), Program Manager was supplied with Windows 95, 98, and ME to ease the transition to the new style of Windows desktop.

programmer job classification for computer personnel. Programmers write the software needed for any new computer system or application.

programming language in computing, a special notation in which instructions for controlling a computer are written. Programming languages are designed to be easy for people to write and read, but must be capable of being mechanically translated (by a *compiler or an *interpreter) into the *machine code that the computer can execute. Programming languages may be classified as *high-level languages or *low-level languages. See also *source language.

program trading in finance, buying and selling a group of shares using a computer program to generate orders automatically whenever there is an appreciable movement in prices.

progression sequence of numbers each occurring in a specific relationship to its predecessor. An **arithmetic progression** has numbers that increase or decrease by a common sum or difference (for example, 2, 4, 6, 8); a **geometric progression** has numbers each bearing a fixed ratio to its predecessor (for example, 3, 6, 12, 24); and a **harmonic progression** has numbers whose *reciprocals are in arithmetical progression, for example $1, \frac{1}{2}, \frac{1}{3}, \frac{1}{4}$.

Project Gutenberg electronic 'library' containing hundreds of 'etexts' – books made freely accessible via the *World Wide Web and downloadable via *FTP. The project started in 1971 at the University of Illinois. For copyright reasons, most of the books available are classics dating from before the 20th century, but there are also some current reference books. By 2002, over 3,500 etexts had been input to Project Gutenberg.

projectile particle that travels with both horizontal and vertical motion in the Earth's gravitational field. If the frictional forces of air resistance are ignored, the two components of its motion can be analysed separately: its vertical motion will be accelerated due to its weight in the gravitational field; its horizontal motion may be assumed to be at constant velocity. In a uniform gravitational field and in the absence of frictional forces the path of a projectile is a parabola.

prokaryote in biology, an organism whose cells lack organelles (specialized segregated structures such as nuclei, mitochondria, and chloroplasts). Prokaryote DNA is not arranged in chromosomes but forms a coiled structure called a **nucleoid**. The prokaryotes comprise only the **bacteria** and **cyanobacteria** (see *blue-green algae); all other organisms are eukaryotes.

PROLOG contraction of programming in logic, high-level computer programming language based on logic.

Invented in 1971 at the University of Marseille, France, it did not achieve widespread use until more than ten years later. It is used mainly for *artificial intelligence programming.

PROM acronym for Programmable Read-Only Memory, in computing, a memory device in the form of an integrated circuit (chip) that can be programmed after manufacture to hold information permanently. PROM chips are empty of information when manufactured, unlike ROM (read-only memory) chips, which have information built into them. Other memory devices are *EPROM (erasable programmable read-only memory) and *RAM (random-access memory).

promethium chemical symbol Pm, radioactive, metallic element of the *lanthanide series, atomic number 61, relative atomic mass 145. It occurs in nature only in minute amounts, produced as a fission product/by-product of uranium in *pitchblende and other uranium ores; for a long time it was considered not to occur in nature. The longest-lived isotope has a half-life of slightly more than 20 years.

Promethium is synthesized by neutron bombardment of neodymium and is a product of the fission of uranium, thorium, or plutonium; it can be isolated in large amounts from the fission-product debris of uranium fuel in nuclear reactors. It is used in phosphorescent paints and as an X-ray source.

prompt symbol displayed on a screen indicating that the computer is ready for input. The symbol used will vary from system to system and application to application. The current cursor position is normally next to the prompt. Generally prompts only appear in *command line interfaces.

proof in mathematics, a set of arguments used to deduce a mathematical theorem from a set of axioms.

propane C_3H_8, gaseous hydrocarbon of the *alkane series, found in petroleum and used as a fuel and as a refrigerant.

propanol or **propyl alcohol**, third member of the homologous series of *alcohols. Propanol is usually a mixture of two isomeric compounds (see *isomer): propan-1-ol ($CH_3CH_2CH_2OH$) and propan-2-ol ($CH_3CHOHCH_3$). Both are colourless liquids that can be mixed with water and are used in perfumery.

propanone common name **acetone**; CH_3COCH_3, colourless flammable liquid used extensively as a solvent, as in nail-varnish remover, and for making acrylic plastics. It boils at 56.5°C/133.7°F, mixes with water in all proportions, and has a characteristic odour.

propene common name **propylene**; $CH_3CH=CH_2$, second member of the alkene series of hydrocarbons. A colourless, flammable gas, it is widely used by industry to make organic chemicals, including *polypropylene plastics.

propenoic acid common name **acrylic acid**; $H_2C=CHCOOH$, acid obtained from the aldehyde propenal (acrolein) derived from glycerol or fats. Glasslike thermoplastic resins are made by polymerizing *esters of propenoic acid or methyl propenoic acid and used for transparent components, lenses, and dentures. Other acrylic compounds are used for adhesives, artificial fibres, and artists' acrylic paint.

proper fraction or **simple fraction** or **common fraction**, *fraction whose value is less than 1. The numerator has a lower value than the denominator in a proper fraction.

properties in chemistry, the characteristics a substance possesses by virtue of its composition.

Physical properties of a substance can be measured by physical means, for example boiling point, melting point, hardness, elasticity, colour, and physical state. **Chemical properties** are the way it reacts with other substances; whether it is acidic or basic, an oxidizing or a reducing agent, a salt, or stable to heat, for example.

prophylaxis any measure taken to prevent disease, including exercise and *vaccination. Prophylactic (preventive) medicine is an aspect of public-health provision that is receiving increasing attention.

proportion relation of a part to the whole (usually expressed as a *fraction or *percentage). In mathematics two variable quantities x and y are proportional if, for all values of x, $y = kx$, where k is a constant. This means that if x increases, y increases in a linear fashion.

proportional font font in which individual letters of the alphabet take up different amounts of space according to their shape.

proprioceptor one of the sensory nerve endings that are located in muscles, tendons, and joints. They relay information on the position of the body and the state of muscle contraction.

propyl alcohol common name for *propanol.

propylene common name for *propene.

prostaglandin any of a group of complex fatty acids present in the body that act as messenger substances between cells. Effects include stimulating the contraction of smooth muscle (for example, of the womb during birth), regulating the production of stomach acid, and modifying hormonal activity. In excess, prostaglandins may produce inflammatory disorders such as arthritis. Synthetic prostaglandins are used to induce labour in humans and domestic animals.

prostate cancer *cancer of the *prostate gland. It is a slow progressing cancer, and about 60% of cases are detected before *metastasis (spreading), so it can be successfully treated by surgical removal of the gland and radiotherapy. It is, however, the second commonest cancer-induced death in males (after lung cancer). It kills 32,000 men a year in the USA alone.

prostatectomy surgical removal of the *prostate gland. In many men over the age of 60 the prostate gland enlarges, causing obstruction to the urethra. This causes the bladder to swell with retained urine, leaving the sufferer more prone to infection of the urinary tract.

prostate gland gland surrounding and opening into the *urethra at the base of the bladder in male mammals.

protactinium chemical symbol Pa, (Latin *protos* 'before' + *aktis* 'first ray') silver-grey, radioactive, metallic element of the *actinide series, atomic number 91, relative atomic mass 231.036. It occurs in nature in very small quantities in *pitchblende and other uranium ores. It has 14 known isotopes; the longest-lived, Pa-231, has a half-life of 32,480 years.

protease general term for a digestive enzyme capable of splitting proteins. Examples include pepsin, found in the stomach, and trypsin, found in the small intestine.

protected mode operating mode of Intel microprocessors (80286 and above), which allows multitasking and provides other features such as extended memory and *virtual memory (above 1 Gbyte). Protected mode operation also improves *data security.

protein large, complex, biologically-important molecules composed of amino acids joined by *peptide bonds. The number of amino acids used can be many hundreds. There are 20 different amino acids and they can be joined in any order. Proteins are essential to all living organisms. As *enzymes they regulate all aspects of metabolism. Structural proteins such as keratin and collagen make up skin, claws, bones, tendons, and ligaments; muscle proteins produce movement; haemoglobin transports oxygen; and membrane proteins regulate the movement of substances into and out of cells. For humans, protein is an essential part of the diet, and is found in greatest quantity in soy beans and other grain legumes, meat, eggs, and cheese. During digestion protein molecules are broken down into amino acids which are then easily absorbed into the body.

Protein synthesis occurs in cells. The information describing the order in which the different amino acids are joined is found in *DNA in the form of a code. The part of the DNA that carries the code for making one protein is called a *gene. Each protein described in the code has an effect on the appearance and characteristics of the organism.

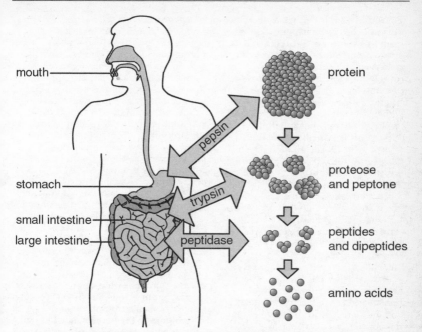

mouth

stomach

small intestine

large intestine

pepsin

trypsin

peptidase

protein

proteose
and peptone

peptides
and dipeptides

amino acids

protein The digestion of protein. Protein is broken down into amino acids by the enzymes pepsin (secreted by the stomach) and trypsin and peptidase (in the small intestine).

During digestion in the body, proteases (any enzymes that break down a protein) are produced by the *stomach, the *pancreas, and the small *intestine. These enzymes catalyse the breakdown of protein into amino acids. Pepsin is an enzyme which is produced by the walls of the stomach. It digests large protein molecules into smaller protein molecules (smaller polypeptides). The conditions in the stomach are very acid and pepsin works at its fastest rate in such conditions.

The amino acids produced by the digestion of proteins are absorbed into the blood in the small intestine. They are transported around the body by the *blood for use by cells to make the proteins they need. This protein synthesis needs energy and this energy is provided by *respiration.

protein engineering creation of synthetic proteins designed to carry out specific tasks. For example, an enzyme may be designed to remove grease from soiled clothes and remain stable at the high temperatures in a washing machine.

protein synthesis manufacture, within the cytoplasm of the cell, of the proteins an organism needs. The building blocks of proteins are *amino acids, of which there are 20 types. The pattern in which the amino acids are linked decides what kind of protein is produced. In turn it is the genetic code, contained within *DNA, that determines the precise order in which the amino acids are linked up during protein manufacture.

Interestingly, DNA is found only in the nucleus, yet protein synthesis occurs only in the cytoplasm. The information necessary for making the proteins is carried from the nucleus to the cytoplasm by another nucleic acid, *RNA.

proteinuria presence of protein in the urine. It indicates kidney disease or damage.

Proterozoic Eon eon of geological time, 3.5 billion to 570 million years ago, the second division of the

Precambrian. It is defined as the time of simple life, since many rocks dating from this eon show traces of biological activity, and some contain the fossils of bacteria and algae.

prothallus short-lived *gametophyte of many ferns and other *pteridophytes (such as horsetails or clubmosses). It bears either the male or female sex organs, or both. Typically it is a small, green, flattened structure that is anchored in the soil by several *rhizoids (slender, hairlike structures, acting as roots) and needs damp conditions to survive. The reproductive organs are borne on the lower surface close to the soil. See also *alternation of generations.

prothrombin substance in the plasma from which *thrombin is derived during blood clotting.

protocol in computing, an agreed set of **standards** for the transfer of data between different devices. They cover transmission speed, format of data, and the signals required to synchronize the transfer. See also *interface.

protogyny in a flower, the state where the female reproductive organs reach maturity before those of the male. Like protandry, in which the male organs reach maturity first, this is a method of avoiding self-fertilization, but it is much less common.

proton (Greek 'first') in physics, a positively charged subatomic particle, a constituent of the nucleus of all *atoms. It carries a unit positive charge equal to the negative charge of an *electron. Its mass is almost 1,836 times that of an electron, or 1.673×10^{-24} g. The number of protons in the atom of an *element is equal to the *atomic number of that element.

A proton belongs to the baryon group of *hadrons and is composed of two up quarks and one down quark. A proton is extremely long-lived, with a lifespan of at least 10^{32} years.

proton number alternative name for *atomic number.

protoplasm contents of a living cell. Strictly speaking it includes all the discrete structures (organelles) in a cell, but it is often used simply to mean the jellylike material in which these float.

The contents of a cell outside the nucleus are called *cytoplasm.

protozoa group of single-celled organisms without rigid cell walls. Some, such as amoeba, ingest other cells, but most are *saprotrophs or parasites. The group is polyphyletic (containing organisms which have different evolutionary origins). **Xenophyophores** are giant protozoa that measure 20 cm/8 in in diameter and live on the deep-ocean floor.

protractor instrument used to measure a flat *angle.

proxy server on the *World Wide Web, a server which 'stands in' for another server, storing and forwarding files on behalf of a computer that might be slower or too busy to deal with the request itself. Many *URLs (Web addresses) redirect the enquirer to a proxy server that then supplies the requested page.

Prozac or **fluoxetine**, antidepressant drug that functions mainly by boosting levels of the neurotransmitter *serotonin in the brain. Side effects include nausea and loss of libido. It is also used to treat some eating disorders, such as *bulimia. It is one of a class of drugs known as *selective serotonin-reuptake inhibitors (SSRIs).

pruritus itching, usually caused by irritation of the skin.

prussic acid former name for *hydrocyanic acid.

pseudocarp fruitlike structure that incorporates tissue that is not derived from the ovary wall. The additional tissues may be derived from floral parts such as the *receptacle and *calyx. For example, the coloured, fleshy part of a strawberry develops from the receptacle and the true fruits are small *achenes – the 'pips' embedded in its outer surface. Rose hips are a type of pseudocarp that consists of a hollow, fleshy receptacle containing a number of achenes within. Different types of pseudocarp include pineapples, figs, apples, and pears.

pseudocopulation attempted copulation by a male insect with a flower. It results in *pollination of the flower and is common in the orchid family, where the flowers of many species resemble a particular species of female bee. When a male bee attempts to mate with a flower, the pollinia (groups of pollen grains) stick to its body. They are transferred to the stigma of another flower when the insect attempts copulation again.

pseudomorph mineral that has replaced another in situ and has retained the external crystal shape of the original mineral.

psi in parapsychology, a hypothetical faculty common to humans and other animals, said to be responsible for *extrasensory perception, telekinesis, and other paranormal phenomena.

PSP abbreviation for *payment service provider.

PSTN abbreviation for public switched telephone network, telephone network used by the general public, and sometimes used as the medium to link *LANs. PSTNs are minor roads which lead to the *information superhighway.

psychiatry branch of medicine dealing with the diagnosis and treatment of mental disorder, normally divided into the areas of **neurotic conditions**, including anxiety, depression, and hysteria, and **psychotic disorders**, such as schizophrenia. Psychiatric treatment consists of drugs, analysis, or electroconvulsive therapy.

psychoanalysis theory and treatment method for neuroses, developed by Austrian physician Sigmund Freud (1856–1939) in the 1890s. Psychoanalysis asserts that the impact of early childhood sexuality and experiences, stored in the *unconscious, can lead to the development of adult emotional problems. The main treatment method involves the free association of ideas, and their interpretation by patient and analyst, in order to discover these long-buried events and to grasp their significance to the patient, linking aspects of the patient's historical past with the present relationship to the analyst. Psychoanalytic treatment aims to free the patient from specific symptoms and from irrational inhibitions and anxieties.

psychology systematic study of human and animal behaviour. The first

psychology laboratory was founded in 1879 by Wilhelm Wundt at Leipzig, Germany. The subject includes diverse areas of study and application, among them the roles of instinct, heredity, environment, and culture; the processes of sensation, perception, learning, and memory; the bases of motivation and emotion; and the functioning of thought, intelligence, and language. Significant psychologists have included Gustav Fechner, founder of psychophysics; Wolfgang Köhler, one of the *Gestalt or 'whole' psychologists; Sigmund Freud (1856–1939) and his associates Carl Jung (1875–1961) and Alfred Adler (1870–1937),; William James, Jean Piaget (1896–1980); Carl Rogers; Hans Eysenck (1916–1997); J B Watson; and B F Skinner (1904–1990).

psychopathy personality disorder characterized by chronic antisocial behaviour (violating the rights of others, often violently) and an absence of feelings of guilt about the behaviour.

psychopharmacology discipline concerned with the effects of drugs (particularly *psychotropic drugs) on mood and behaviour and their use to treat mental illness.

psychosis or **psychotic disorder**, general term for a serious mental disorder in which the individual commonly loses contact with reality and may experience hallucinations (seeing or hearing things that do not exist) or delusions (fixed false beliefs). For example, in a paranoid psychosis, an individual may believe that others are plotting against him or her. A major type of psychosis is *schizophrenia.

psychosomatic of a physical symptom or disease thought to arise from emotional or mental factors.

psychotherapy any treatment for psychological problems that involves talking rather than surgery or drugs. Examples include cognitive therapy and *psychoanalysis.

psychotic disorder another name for *psychosis.

psychotropic drug any drug that affects mood. The term covers a range of drugs, including stimulants, antidepressants, and tranquillizers.

pt symbol for *pint.

pteridophyte simple type of *vascular plant. The pteridophytes comprise four classes: the Psilosida, including the most primitive vascular plants, found mainly in the tropics; the Lycopsida, including the club mosses; the Sphenopsida, including the horsetails; and the Pteropsida, including the ferns. They do not produce seeds.

PTFE abbreviation for *polytetrafluoroethene.

ptomaine any of a group of toxic chemical substances (alkaloids) produced as a result of decomposition by bacterial action on proteins.

puberty stage in human development when the individual becomes sexually mature. It may occur from the age of ten upwards, but each person is individual. The sexual organs take on their adult form and pubic hair grows. In girls, menstruation begins, and the breasts develop; in boys, the voice breaks and becomes deeper, and facial hair develops. Both boys and girls will experience emotional as well as physical changes.

public-domain software any computer program that is not under copyright and can therefore be used freely without charge. Much of this software has been written in US universities, under government contract. Public-domain software should not be confused with *shareware, which is under copyright, and may be freely distributed for evaluation purposes, but requires purchasing to use in the longer term.

public key in *public-key cryptography, a string of *bits that is associated with a particular person and that may be used to decrypt messages from that person or to encrypt messages to him/her.

public-key cryptography system of *cryptography that allows remote users to exchange encrypted data without the need to transmit a secret digital 'key' in advance. The system was first proposed by Whitfield Diffie and Martin Hellman in a widely read and influential paper 'New Directions in Cryptography' (1976).

public key infrastructure PKI, core framework required to provide *public key cryptography and *digital signature services for *electronic commerce transactions via the Internet. A PKI provides confidentiality (to keep information private), integrity (to prove it has not been altered), authentication (to prove the identity of a person or a computer application), and non-repudiation (to prevent information being disowned). *SET is a form of PKI.

pull-down menu in computing, a list of options provided as part of a

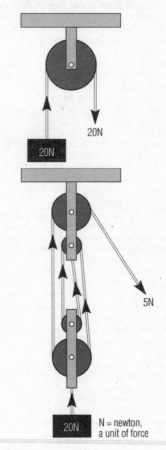

pulley The mechanical advantage of a pulley increases with the number of rope strands. If a pulley system has four ropes supporting the load, the mechanical advantage is four, and a 5 newton force will lift a 20 newton load.

*graphical user interface. The presence of pull-down menus is normally indicated by a row of single words at the top of the screen. When the user points at a word with a *mouse, a full menu appears (is pulled down) and the user can then select the required option. Compare with *pop-up menu.

pulley simple machine consisting of a fixed, grooved wheel, sometimes in a block, around which a rope or chain can be run. A simple pulley serves only to change the direction of the applied effort (as in a simple hoist for raising loads). The use of more than one pulley results in a mechanical advantage, so that a given effort can raise a heavier load.

pulmonary pertaining to the *lungs.

pulmonary embolism in medicine, condition in which a blood clot becomes lodged in the lungs. The clot usually migrates in the circulatory system from a vein in the lower abdomen or legs. The condition is characterized by a sudden pain in the chest. It can be fatal if the clot is large. Smaller clots can be removed surgically or dissolved by the use of fibrinolytic drugs.

pulse in biology, impulse transmitted by the heartbeat throughout the arterial systems of vertebrates. When the heart muscle contracts, it forces blood into the *aorta (the chief artery). Because the arteries are elastic, the sudden rise of pressure causes a throb or sudden swelling through them. The actual flow of the blood is about 60 cm/2 ft a second in humans. The average adult pulse rate is generally about 70 per minute. The pulse can be felt where an artery is near the surface, for example in the wrist or the neck.

pulse-code modulation PCM, in physics, a form of digital *modulation in which microwaves or light waves (the carrier waves) are switched on and off in pulses of varying length according to a binary code. It is a relatively simple matter to transmit data that are already in binary code, such as those used by computer, by these means. However, if an analogue audio signal is to be transmitted, it

must first be converted to a **pulse-amplitude modulated** signal (PAM) by regular sampling of its amplitude. The value of the amplitude is then converted into a binary code for transmission on the carrier wave.

pumice light volcanic rock produced by the frothing action of expanding gases during the solidification of lava. It has the texture of a hard sponge and is used as an abrasive.

punctuated equilibrium model evolutionary theory developed by Niles Eldredge and US palaeontologist and writer Stephen Jay Gould (1941–2002) in 1972 to explain discontinuities in the fossil record. It claims that evolution continues through periods of rapid change alternating with periods of relative stability (stasis), and that the appearance of new lineages is a separate process from the gradual

an analogue signal

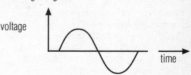

pulse-amplitude-modulated signal (PAM)

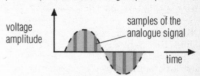

pulse-code modulation The amplitude, duration, and timing of a series of pulses are controlled in pulse-code modulation, which is relatively simple for digital data already in binary code. Analogue signals need to be converted into a recognizable binary code (a pulse-amplitude modulated signal) by regular sampling of its amplitude. Morse code is a very simple example of pulse-code modulation.

evolution of adaptive changes within a species.

pupa nonfeeding, largely immobile stage of some insect life cycles, in which larval tissues are broken down, and adult tissues and structures are formed.

pupil circular aperture in the *iris through which light falls onto the lens of the eye.

push-button in computing, in a *dialog box or *toolbar, a square, oval, or oblong button which presents the user with an option. By clicking on the button, the user opts to initiate an action such as centering text or saving a file. Most programs offer keyboard *shortcuts as an alternative to using a push-button, for example one can often hit the return key instead of clicking the 'OK' button in a dialogue box. Compare *radio button.

push technology automatic transmission of information to Internet users, without their having to find it and 'pull' it to themselves. Push technology was introduced in 1996 by PointCast (now EntryPoint) and Marimba, and in 1997, Microsoft and Netscape built push 'channels' into their Web *browsers. The technology did not prove popular with consumers, however, as the principle was too similar to 'junk e-mail' (see *spamming). Marimba now aims its technology at large companies and *application service providers, for corporate distribution and management of software.

putrefaction decomposition of organic matter by micro-organisms.

PVC abbreviation for *polyvinyl chloride.

P-wave abbreviation of primary wave, in seismology, a class of *seismic wave that passes through the Earth in the form of longitudinal pressure waves at speeds of 6–7 kps/3.7–4.4 mps in the crust and up to 13 kps/8 mps in deeper layers, the speed depending on the density of the rock. P-waves from an earthquake travel faster than S-waves and are the first to arrive at monitoring stations (hence primary waves). They can travel through both solid rock and the liquid outer core of the Earth.

PWR abbreviation for *pressurized water reactor, a type of nuclear reactor.

pyramid in geometry, a solid shape with triangular side-faces meeting at a common vertex (point) and with a *polygon as its base. The volume V of a pyramid is given by $V = 1/3Bh$, where B is the area of the base and h is the perpendicular height.

pyramid of numbers in ecology, a diagram that shows quantities of plants and animals at different levels (steps) of a *food chain. This may be measured in terms of numbers (how many animals) or biomass (total mass of living matter), though in terms of showing transfer of food, biomass is a more useful measure. Where biomass is measured, the diagram is often termed a pyramid of biomass. There is always far less biomass, or fewer organisms, at the top of the chain than at the bottom, because only about 10% of the food (energy) an animal eats is turned into flesh – the rest is lost through metabolism and excretion. The amount of food flowing through the chain therefore drops with each step up the chain, supporting fewer organisms, hence giving the characteristic 'pyramid' shape.

pyrexia another name for *fever.

pyridine C_5H_5N, heterocyclic compound (see *cyclic compounds). It is a liquid with a sickly smell and occurs in coal tar. It is soluble in water, acts as a strong *base, and is used as a solvent, mainly in the manufacture of plastics.

pyridoxine or **vitamin B$_6$**; $C_8H_{11}NO_3$, water-soluble *vitamin of the B complex. There is no clearly identifiable disease associated with deficiency but its absence from the diet can give rise to malfunction of the central nervous system and general skin disorders. Good sources are liver, meat, milk, and cereal grains. Related compounds may also show vitamin B_6 activity.

pyrite or **fool's gold**, iron sulphide FeS_2. It has a yellow metallic lustre and a hardness of 6–6.5 on the *Mohs scale. It is used in the production of sulphuric acid.

pyroclastic describing fragments of solidified volcanic magma, ranging in

pyramids of numbers

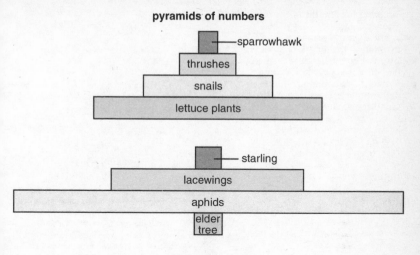

pyramid of biomass

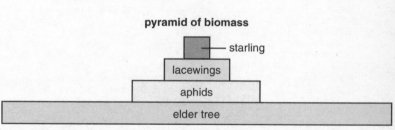

pyramid of numbers Pyramids of numbers and pyramid of biomass. The pyramid of numbers is a useful way of representing a food chain as it shows how the number of consumers at each level decreases, with plants being the most numerous at the base of the pyramid and top carnivores the smallest group. Where the plant being eaten is a tree, however, the pyramid no longer works as a useful model. This is rectified by the use of the pyramid of biomass, where it is the mass of the levels rather than numbers that are represented.

size from fine ash to large boulders, that are extruded during an explosive volcanic eruption; also the rocks that are formed by consolidation of such material. Pyroclastic rocks include tuff (ash deposit) and agglomerate (volcanic breccia).

pyroclastic deposit deposit made up of fragments of rock, ranging in size from fine ash to large boulders, ejected during an explosive volcanic eruption.

pyrolysis decomposition of a substance by heating it to a high temperature in the absence of air. The process is used to burn and dispose of old tyres, for example, without contaminating the atmosphere.

pyroxene any one of a group of minerals, silicates of calcium, iron, and magnesium with a general formula X,YSi_2O_6, found in igneous and metamorphic rocks. The internal structure is based on single chains of silicon and oxygen. Diopside ($X = Ca$, $Y = Mg$) and augite ($X = Ca$, $Y = Mg$,Fe,Al) are common pyroxenes.

Pythagoras' theorem in geometry, formulated by Greek mathematician and philosopher Pythagoras (c. 580–500 BC). a theorem stating that in a right-angled triangle, the square of the *hypotenuse (the longest side) is equal to the sum of the squares of the other two sides. If the hypotenuse is h units long and the lengths of the other sides are a and b, then $h^2 = a^2 + b^2$.

The theorem provides a way of calculating the length of any side of a *right-angled triangle if the lengths of the other two sides are known. Pythagoras' theorem is also used to determine certain *trigonometric identities such as $\sin^2 \theta + \cos^2 \theta = 1$.

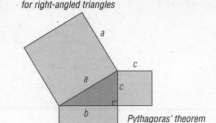

for right-angled triangles

Pythagoras' theorem
$$a^2 = b^2 + c^2$$

Pythagoras' theorem This states that the area of a square drawn on the longest side of a right-angled triangle (the hypotenuse), will be equal to the sum of the areas of the squares drawn on the other two sides. The theorem is likely to have been known long before the time of Pythagoras. It was probably used by the ancient Egyptians to lay out the pyramids.

quadrant one quarter of a circle. When plotting *coordinates on a *graph, the x- and y-axes may intersect to create four quadrants. The first quadrant lies in the region where both x and y are positive. The remaining three quadrants are numbered in an anticlockwise direction. Thus in the second quadrant, x is negative and y is positive, in the third quadrant, both are negative and in the fourth quadrant x is positive and y negative.

quadrat in environmental studies, a square structure used to study the distribution of plants in a particular place, for instance a field, rocky shore, or mountainside. The size varies, but is usually 0.5 or 1 metre square, small enough to be carried easily. The quadrat is placed on the ground and the abundance of species estimated. By making such measurements a reliable understanding of species distribution is obtained.

quadratic equation in mathematics, a polynomial *equation of second degree (that is, an equation containing as its highest power the square of a variable, such as x^2). The general *formula of such equations is

$$ax^2 + bx + c = 0$$

in which a, b, and c are real numbers, and only the *coefficient a cannot equal 0.

In *coordinate geometry, a quadratic function represents a *parabola. Some quadratic equations can be solved by factorization (see *factor (algebra)), or the values of x can be found by using the formula for the general solution

$$x = \frac{[-b + \sqrt{(b^2 - 4ac)}]}{2a} \quad \text{or}$$

$$x = \frac{[-b - \sqrt{(b^2 - 4ac)}]}{2a}$$

Depending on the value of the discriminant, $b^2 - 4ac$, a quadratic equation has two real, two equal, or two complex roots (solutions). When $b^2 - 4ac > 0$, there are two distinct real roots. When $b^2 - 4ac = 0$, there are two equal real roots. When $b^2 - 4ac < 0$, there are two distinct complex roots.

quadrilateral   plane (two-dimensional) figure with four straight sides. The sum of all *interior angles is 360°. The following are all quadrilaterals, each with distinguishing properties: ***square** with four equal angles and sides, and four axes of *symmetry; ***rectangle** with four equal angles, opposite sides equal, and two axes of symmetry; ***rhombus** with four equal sides, and two axes of symmetry; ***parallelogram** with two pairs of parallel sides, and rotational symmetry; ***kite** with two pairs of adjacent equal sides, and one axis of symmetry; and ***trapezium** with one pair of parallel sides.

quad speed   in computing, a *CD-ROM drive that transfers data at 600 *kilobytes per second – four times as fast as the first CD-ROM drives on the market. Because access time remains a key factor in the movement of data, the extra speed does not necessarily translate into a fourfold improvement in performance.

qualitative analysis   in chemistry, a procedure for determining the identity of the component(s) of a single substance or mixture. A series of simple reactions and tests can be carried out on a compound to determine the elements present.

quantitative analysis   in chemistry, a procedure for determining the precise amount of a known component present in a single substance or mixture. A known amount of the substance is subjected to particular procedures.

Gravimetric analysis determines the mass of each constituent present; ***volumetric analysis** determines the concentration of a solution by *titration against a solution of known concentration.

quantum chromodynamics   QCD, in physics, a theory describing the interactions of *quarks, the *elementary particles that make up all *hadrons (subatomic particles such as protons and neutrons). In quantum chromodynamics, quarks are considered to interact by exchanging particles called gluons, which carry the *strong nuclear force, and whose role is to 'glue' quarks together.

quantum computing   use of particles such as atoms, ions, and photons to perform computations, initally suggested by physicist Richard

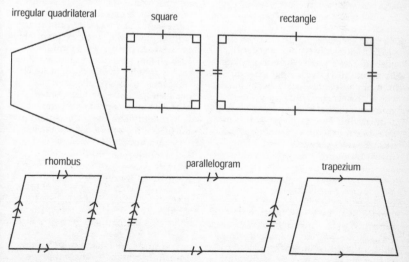

irregular quadrilateral   square   rectangle

rhombus   parallelogram   trapezium

quadrilateral Different types of quadrilateral.

Feynman in 1982. In 1985, David Deutsch of the University of Oxford described a 'universal quantum computer' that would be able to perform feats beyond the capabilities of conventional computers. No such computer has been built to date, but quantum computing is thought to show great promise in certain areas, such as cryptography.

quantum electrodynamics QED, in physics, a theory describing the interaction of charged subatomic particles within electric and magnetic fields. It combines *quantum theory and *relativity, and considers charged particles to interact by the exchange of photons. QED is remarkable for the accuracy of its predictions; for example, it has been used to calculate the value of some physical quantities to an accuracy of ten decimal places, a feat equivalent to calculating the distance between New York and Los Angeles to within the thickness of a hair. The theory was developed by US physicists Richard Feynman and Julian Schwinger and by Japanese physicist Sin-Itiro Tomonaga in 1948.

quantum mechanics branch of physics dealing with the interaction of *matter and *radiation, the structure of the *atom, the motion of atomic particles, and with related phenomena (see *elementary particle and *quantum theory).

quantum number in physics, one of a set of four numbers that uniquely characterize an *electron and its state in an *atom. The **principal quantum number** n defines the electron's main energy level. The **orbital quantum number** l relates to its angular momentum. The **magnetic quantum number** m describes the energies of electrons in a magnetic field. The **spin quantum number** m_s gives the spin direction of the electron.

quantum theory or **quantum mechanics**, in physics, the theory that *energy does not have a continuous range of values, but is, instead, absorbed or radiated discontinuously, in multiples of definite, indivisible units called quanta. Just as earlier theory showed how light, generally seen as a wave motion, could also in some ways be seen as composed of discrete particles (*photons), quantum theory shows how atomic particles such as electrons may also be seen as having wavelike properties. Quantum theory is the basis of *particle physics, modern theoretical chemistry, and the solid-state physics that describes the behaviour of the silicon chips used in computers.

The theory began with the work of German physicist Max Planck (1858–1947) in 1900 on radiated energy, and was extended by German-born US physicist Albert Einstein (1879–1955) to electromagnetic radiation generally, including light. Danish physicist Niels Bohr (1885–1962) used it to explain the *spectrum of light emitted by excited hydrogen atoms. Bohr postulated that an atom may exist in only a certain number of stable states, each with a certain amount of energy, in which electrons orbit the nucleus without emitting or absorbing energy. He proposed that emission or absorption of energy occurs only with a transition from one stable state to another. When a transition occurs, an electron moving to a higher orbit absorbs energy and an electron moving to a lower orbit emits energy. In so doing, a set number of quanta of energy are emitted or absorbed at a particular frequency. Later work by Erwin Schrödinger (1887–1961), Werner Heisenberg (1901–1976), Paul Dirac (1902–1984), and others elaborated the theory to what is called quantum mechanics (or wave mechanics).

quark in physics, the *elementary particle that is the fundamental constituent of all *hadrons (subatomic particles that experience the strong nuclear force and divided into baryons, such as neutrons and protons, and mesons). Quarks have electric charges that are fractions of the electronic charge ($+\frac{2}{3}$ or $-\frac{1}{3}$ of the electronic charge). There are six types, or 'flavours': up, down, top, bottom, strange, and charmed, each of which has three varieties, or 'colours': red, green, and blue (visual colour is not meant, although the analogy is useful in many ways). To each quark there is an antiparticle, called an antiquark. See *quantum chromodynamics (QCD).

quart imperial liquid or dry measure, equal to two pints or 1.136 litres. In the USA, a liquid quart is equal to 0.946 litre, while a dry quart is equal to 1.101 litres.

quartile in statistics, any one of the three values that divide data into four equal parts. They comprise the **lower quartile**, below which lies the lowest 25% of the data; the *median, which is the middle 50%, half way through the data; and the **upper quartile**, above which lies the top 25%. The difference of value between the upper and lower quartiles is known as the *interquartile range, which is a useful measure of the dispersion of a statistical distribution because it is not affected by freak extreme values (see *range). These values are usually found using a *cumulative frequency diagram.

quartz crystalline form of *silica SiO_2, one of the most abundant minerals of the Earth's crust (12% by volume). Quartz occurs in many different kinds of rock, including sandstone and granite. It ranks 7 on the *Mohs scale of hardness and is resistant to chemical or mechanical breakdown. Quartzes vary according to the size and purity of their crystals. Crystals of pure quartz are coarse, colourless, transparent, show no cleavage, and fracture unevenly; this form is usually called rock crystal. Impure coloured varieties, often used as gemstones, include *agate, citrine quartz, and *amethyst. Quartz is also used as a general name for the cryptocrystalline and noncrystalline varieties of silica, such as chalcedony, chert, and opal.

Quartz is used in ornamental work and industry, where its reaction to electricity makes it valuable in electronic instruments (see *piezoelectric effect). Quartz can also be made synthetically.

quartzite *metamorphic rock consisting of pure quartz sandstone that has recrystallized under increasing heat and pressure.

In sedimentology, quartzite is an unmetamorphosed sandstone

composed chiefly of quartz grains held together by silica that was precipitated after the original sand was deposited.

quasar contraction of **quasi-stellar object**; or **QSO**, one of the most distant extragalactic objects known, discovered in 1963. Quasars appear starlike, but each emits more energy than 100 giant galaxies. They are thought to be at the centre of galaxies, their brilliance emanating from the stars and gas falling towards an immense *black hole at their nucleus. The *Hubble Space Telescope revealed in 1994 that quasars exist in a remarkable variety of galaxies.

Quasar light shows a large *red shift, indicating that the quasars are very distant. The furthest are over 10 billion light years away. A few quasars emit radio waves (see *radio astronomy), which is how they were first identified.

quasi-atom particle assemblage resembling an atom, in which particles not normally found in atoms become bound together for a brief period. Quasi-atoms are generally unstable structures, either because they are subject to matter–antimatter annihilation (positronium), or because one or more of their constituents is unstable (muonium).

Quaternary Period period of geological time from 1.64 million years ago through to the present. It is divided into the *Pleistocene (1.64 million to 10,000 years ago) and *Holocene (last 10,000 years) epochs.

quenching *heat treatment used to harden metals. The metals are heated to a certain temperature and then quickly plunged into cold water or oil.

query in computing, a question asked of the database. Queries form the bulk of programmed processes and are run as an intermediate stage between data input and the output report. A query needs to be set up and executed in order to generate information that will then be passed to the report generator.

queue in computing, back-up of *packets of data awaiting processing, or of *e-mail waiting to be read.

quicksilver another name for the element *mercury.

quinine antimalarial drug extracted from the bark of the cinchona tree. Peruvian Indians taught French missionaries how to use the bark in 1630, but quinine was not isolated until 1820. It is a bitter alkaloid, with the formula $C_{20}H_{24}N_2O_2$.

quotient result of dividing one number or variable into another.

R

rabies or **hydrophobia**, (Greek 'fear of water') viral disease of the central nervous system that can afflict all warm-blooded creatures. It is caused by a *lyssavirus. It is almost invariably fatal once symptoms have developed. Its transmission to humans is generally by a bite from an infected animal. Rabies continues to kill hundreds of thousands of people every year; almost all these deaths occur in Asia, Africa, and South America.

raceme in botany, a type of *inflorescence.

rad unit of absorbed radiation dose, now replaced in the SI system by the gray (one rad equals 0.01 gray). It is defined as the dose when one kilogram of matter absorbs 0.01 joule of radiation energy.

radar acronym for **radio direction and ranging**, device for locating objects in space, direction finding, and navigation by means of transmitted and reflected high-frequency radio waves.

The direction of an object is ascertained by transmitting a beam of short-wavelength (1–100 cm/0.5–40 in), short-pulse radio waves, and picking up the reflected beam. Distance is determined by timing the journey of the radio waves (travelling at the speed of light) to the object and back again. Radar is also used to detect objects underground, for example service pipes, and in archaeology. Contours of remains of ancient buildings can be detected down to 20 m/66 ft below ground. Radar is essential to navigation in darkness, cloud, and fog, and is widely used in warfare to detect enemy aircraft and missiles. To avoid detection, various devices, such as modified shapes (to reduce their radar cross-section), radar-absorbent paints, and electronic jamming are used. To pinpoint small targets *laser 'radar', instead of microwaves, has been developed. Developed independently in Britain, France, Germany, and the USA in the 1930s, it was first put to practical use for aircraft detection by the British, who had a complete coastal chain of radar stations installed by September 1938. It proved invaluable in the Battle of Britain in 1940, when the ability to spot incoming German aircraft did away with the need to fly standing patrols. Chains of ground radar stations are used to warn of enemy attack – for example, North Warning System 1985, consisting of 52 stations across the Canadian Arctic and northern Alaska. Radar is also used in *meteorology and *astronomy.

radial circuit circuit used in household electric wiring in which all electrical appliances are connected to cables that radiate out from the main supply point or fuse box. In more modern systems, the appliances are connected in a ring, or *ring circuit, with each end of the ring connected to the fuse box.

radial velocity in astronomy, velocity of an object, such as a star or galaxy, along the line of sight, moving towards or away from an observer. The amount of Doppler (named after Austrian physicist Christian Doppler (1803–1853)) shift (apparent change in wavelength) of the light reveals the object's velocity. If the object is approaching, the *Doppler effect causes a blue shift in its light. That is, the wavelengths of light coming from the object appear to be shorter, tending toward the blue end of the *spectrum. If the object is receding, there is a *red shift, meaning the wavelengths appear to be longer, toward the red end of the spectrum.

radian symbol rad, SI unit of plane angles, an alternative unit to the *degree. It is the angle at the centre of a circle when the centre is joined to the two ends of an arc (part of the circumference) equal in length to the radius of the circle. There are 2π (approximately 6.284) radians in a full circle (360°).

radiant heat energy that is radiated by all warm or hot bodies. It belongs to the *infrared part of the *electromagnetic spectrum and causes heating when absorbed. Radiant heat is invisible and should not be confused with the red glow associated with very hot objects, which belongs to the visible part of the spectrum.

radiation emission of radiant *energy as particles or waves – for example, heat, light, alpha particles, and beta particles (see diagram for sources of radioactive radiation in the environment). See also *atomic radiation. All hot objects radiate heat. Radiated heat does not need a medium through which to travel (it can travel in a vacuum).

Most of the energy received on Earth arrives by radiation of *heat energy from the Sun. Of the radiation given off by the Sun, only a tiny fraction of it, called insolation, reaches the Earth's surface; much of it (for example, radio waves, ultraviolet rays, and X-rays) is absorbed and scattered as it passes through the *atmosphere. Visible light and infrared rays pass through the atmosphere, the infrared rays causing a rise in temperature. The radiation given off by the Earth itself is called **ground radiation**.

How much a surface radiates heat depends on its temperature and the type of surface. Dull black surfaces absorb more heat and therefore radiate more heat that polished shiny surfaces, which reflect heat and are therefore poor radiators of heat. For example,

engine-cooling mantles in cars are black so that they radiate heat from the engine. A vacuum flask has a polished, silvery surface so as to keep hot liquids hot and cold liquids cold as no heat is radiated or absorbed.

The human body radiates heat at a rate of 100 joules every second. This is the same energy as radiated by a 100-watt light bulb.

radiation biology study of how living things are affected by radioactive (ionizing) emissions (see *radioactivity) and by electromagnetic (nonionizing) radiation (*electromagnetic waves). Both are potentially harmful and can cause mutations as well as leukaemia and

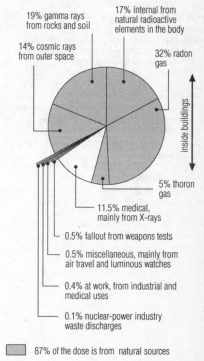

19% gamma rays from rocks and soil

17% internal from natural radioactive elements in the body

14% cosmic rays from outer space

32% radon gas

inside buildings

5% thoron gas

11.5% medical, mainly from X-rays

0.5% fallout from weapons tests

0.5% miscellaneous, mainly from air travel and luminous watches

0.4% at work, from industrial and medical uses

0.1% nuclear-power industry waste discharges

87% of the dose is from natural sources

13% of the dose is from artificial sources

radiation Pie chart showing the various sources of radioactve radiation in the environment. Most radiation is from natural sources, such as radioactive minerals, but 13% comes from the by-products of human activities.

other cancers; even low levels of radioactivity are very dangerous. Both, however, are used therapeutically, for example to treat cancer, when the radiation dose is very carefully controlled (*radiotherapy or X-ray therapy).

radiation monitoring system
network of monitors to detect any rise in background gamma radiation and to warn of a major nuclear accident within minutes of its occurrence. The accident at Chernobyl in Ukraine in 1986 prompted several Western European countries to begin installation of such systems locally, and in 1994 work began on a pilot system to provide a **gamma curtain**, a dense net of radiation monitors, throughout Eastern and Western Europe.

radiation sickness   sickness resulting from exposure to radiation, including X-rays, gamma rays, neutrons, and other nuclear radiation, as from weapons and fallout. Such radiation ionizes atoms in the body and causes nausea, vomiting, diarrhoea, and other symptoms. The body cells themselves may be damaged even by very small doses, causing leukaemia and other cancers.

radiation units   units of measurement for radioactivity and radiation doses. In SI units, the activity of a radioactive source is measured in becquerels (symbol Bq), where one becquerel is equal to one nuclear disintegration per second (an older unit is the curie). The exposure is measured in coulombs per kilogram (C kg^{-1}); the amount of ionizing radiation (X-rays or gamma rays) that produces one coulomb of charge in one kilogram of dry air (replacing the roentgen). The absorbed dose of ionizing radiation is measured in grays (symbol Gy) where one gray is equal to one joule of energy being imparted to one kilogram of matter (the rad is the previously used unit). The dose equivalent, which is a measure of the effects of radiation on living organisms, is the absorbed dose multiplied by a suitable factor that depends upon the type of radiation. It is measured in sieverts (symbol Sv), where one sievert is a dose equivalent

of one joule per kilogram (an older unit is the rem).

radical   in chemistry, a group of atoms forming part of a molecule, which acts as a unit and takes part in chemical reactions without disintegration, yet often cannot exist alone for any length of time; for example, the methyl radical $-CH_3$, or the carboxyl radical $-COOH$.

radicle   part of a plant embryo that develops into the primary root. Usually it emerges from the seed before the embryonic shoot, or *plumule, its tip protected by a root cap, or calyptra, as it pushes through the soil. The radicle may form the basis of the entire root system, or it may be replaced by adventitious roots (positioned on the stem).

radio   transmission and reception of radio waves. In radio transmission a microphone converts *sound waves (pressure variations in the air) into a varying electric current, which is amplified and used to modulate a carrier wave which is transmitted as *electromagnetic waves, which are then picked up by a receiving aerial, amplified, and fed to a loudspeaker, which converts them back into sound waves.

The theory of electromagnetic waves was first developed by Scottish physicist James Clerk Maxwell (1831–1879) in 1864, given practical confirmation in the laboratory in 1888 by German physicist Heinrich Hertz (1857–1894), and put to practical use by Italian inventor Guglielmo Marconi, who in 1901 achieved reception of a signal in Newfoundland, Canada, transmitted from Cornwall, England.

To carry the transmitted electrical signal, an *oscillator produces a carrier wave of high frequency; different stations are allocated different transmitting carrier frequencies. A modulator superimposes the audio-frequency signal on the carrier. There are two main ways of doing this: *amplitude modulation (AM), used for long- and medium-wave broadcasts, in which the strength of the carrier is made to fluctuate in time with the audio signal; and *frequency modulation (FM), as used for VHF broadcasts, in which the frequency of the carrier is made to fluctuate. The transmitting aerial emits the modulated electromagnetic waves, which travel outwards from it.

radioactive decay process of disintegration undergone by the nuclei of radioactive elements, such as radium and various isotopes of uranium and the transuranic elements, in order to produce a more stable nucleus. The three most common forms of radioactive decay are alpha, beta, and gamma decay.

In **alpha decay** (the loss of a helium nucleus – two protons and two neutrons) the atomic number decreases by two and a new nucleus is formed, for example, an atom of uranium isotope of mass 238, on emitting an alpha particle, becomes an atom of thorium, mass 234. In **beta decay** the loss of an electron from an atom is accomplished by the transformation of a neutron into a proton, thus resulting in an increase in the atomic number of one. For example, the decay of the carbon-14 isotope results in the formation of an atom of nitrogen (mass 14, atomic number 7) and the emission of a high-energy electron. **Gamma emission** usually occurs as part of alpha or beta emission. In gamma emission high-speed electromagnetic radiation is emitted from the nucleus, making it more stable during the loss of an alpha or beta particle. Certain lighter, artificially created isotopes also undergo radioactive decay. The associated radiation consists of alpha rays, beta rays, or gamma rays (or a combination of these), and it takes place at a constant rate expressed as a specific half-life, which is the time taken for half of any mass of that particular isotope to decay completely. Less commonly occurring decay forms include heavy-ion emission, electron capture, and spontaneous fission (in each of these the atomic number decreases). The original nuclide is known as the parent substance, and the product is a daughter nuclide (which may or may not be radioactive). The final product in all modes of decay is a stable element.

radioactive tracer any of various radioactive *isotopes added to fluids in order to monitor their flow and

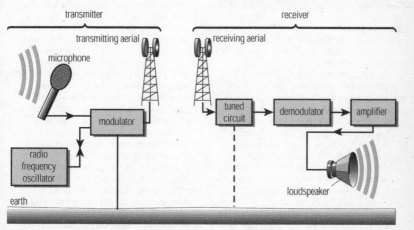

radio Radio transmission and reception. The radio frequency oscillator generates rapidly varying electrical signals, which are sent to the transmitting aerial. In the aerial, the signals produce radio waves (the carrier wave), which spread out at the speed of light. The sound signal is added to the carrier wave by the modulator. When the radio waves fall on the receiving aerial, they induce an electrical current in the aerial. The electrical current is sent to the tuning circuit, which picks out the signal from the particular transmitting station desired. The demodulator separates the sound signal from the carrier wave and sends it, after amplification, to the loudspeaker.

therefore identify leaks or blockages. See *tracer.

radioactive waste any waste that emits radiation in excess of the background level. See *nuclear waste.

radioactivity spontaneous change of the nuclei of atoms, accompanied by the emission of radiation. Such atoms are called radioactive. It is the property exhibited by the radioactive *isotopes of stable elements and all isotopes of radioactive elements, and can be either natural or induced. See *radioactive decay.

A radioactive material decays by releasing radiation, and transforms into a new substance. The energy is released in the form of *alpha particles and *beta particles or in the form of high-energy electromagnetic waves known as *gamma radiation. Natural radioactive elements are those with an atomic number of 83 and higher. Artificial radioactive elements can also be formed.

Devices such as the Geiger–Müller tube, a photographic plate, or an electroscope can detect radioactivity. An electronic counter on the Geiger–Müller instrument displays a digital reading of the amount of radiation detected.

radio astronomy study of radio waves emitted naturally by objects in space, by means of a *radio telescope. Radio emission comes from hot gases (**thermal radiation**); electrons spiralling in magnetic fields (**synchrotron radiation**); and specific wavelengths (**lines**) emitted by atoms and molecules in space, such as the 21-cm/8.3-in line emitted by hydrogen gas.

radio button in computing, in a *dialog box, a round button denoting an option. Users are offered a choice of radio buttons, and can choose only one.

radiocarbon cycle production and recycling of the radioisotope carbon-14 (^{14}C). The radioisotope occurs when a neutron flux, caused by cosmic radiation bombarding the upper atmosphere, reacts efficiently with nitrogen present. Carbon-14 intake by living organisms eventually returns to the atmosphere when dead vegetation

or animal flesh decomposes, except when it is locked in preserved organic artefacts and remains. Radioactive decay occurs, forming the basis of the radiocarbon dating method.

Initially the concentration of carbon-14 is nonuniform (levels being higher over polar regions where the Earth's magnetic field is least effective in deflecting cosmic radiation) but air currents at about 10 km/6 mi soon redistribute the newly formed carbon-14 as part of *carbon dioxide gas. Around 7.5 kg/16 lb of carbon-14 is added to the Earth's carbon reservoir each year and distributed throughout the oceans, the biosphere, and the atmosphere, although variations in the magnetic field and sunspot activity can alter the intensity of cosmic radiation, affecting carbon-14 production.

radiochemistry chemical study of radioactive isotopes and their compounds (whether produced from naturally radioactive or irradiated materials) and their use in the study of other chemical processes.

When such isotopes are used in labelled compounds, they enable the biochemical and physiological functioning of parts of the living body to be observed. They can help in the testing of new drugs, showing where the drug goes in the body and how long it stays there. They are also useful in diagnosis – for example cancer, fetal abnormalities, and heart disease.

radio frequencies and wavelengths see *electromagnetic waves.

radio galaxy galaxy that is a strong source of electromagnetic waves of radio wavelengths. All galaxies, including our own, emit some radio waves, but radio galaxies are up to a million times more powerful.

radiography branch of science concerned with the use of radiation (particularly *X-rays) to produce images on photographic film or fluorescent screens. X-rays penetrate matter according to its nature, density, and thickness. In doing so they can cast shadows on photographic film, producing a radiograph. Radiography is widely used in medicine for examining bones and tissues and in industry for examining solid materials; for example, to check welded seams in pipelines.

radioimmunoassay RIA, in medicine, technique for measuring small quantities of circulating hormones. The assay depends upon the ability of a hormone to inhibit the binding of the same hormone (which has been labelled with a radioactive isotope) to a specific antibody by competition for the binding sites.

radioisotope or **radioactive isotope**, naturally occurring or synthetic radioactive form of an element. Most radioisotopes are made by bombarding a stable element with neutrons in the core of a nuclear reactor (see *fission). The radiations given off by radioisotopes are easy to detect (hence their use as *tracers), can in some instances penetrate substantial thicknesses of materials, and have profound effects (such as genetic *mutation) on living matter.

Most natural isotopes of relative atomic mass below 208 are not radioactive. Those from 210 and up are all radioactive. Radioisotopes have many uses in medicine, for example in *radiotherapy and *radioisotope scanning. The use of radioactive isotopes in the diagnosis, investigation, and treatment of disease is called **nuclear medicine**.

radioisotope decay The nucleus of a radioisotope is unstable and undergoes changes by breaking down into a more stable form. A radioisotope decays over a period of time into a new element. As it decays it emits radiation energy in the form of alpha and beta particles and gamma radiation. The time taken for half of the original atoms to decay is known as the half-life. The product of the radioactive decay is called a 'daughter' atom.

radioisotope scanning use of radio-active materials (radioisotopes or radionucleides) to pinpoint disease. It reveals the size and shape of the target organ and whether any part of it is failing to take up radioactive material, usually an indication of disease.

radiology medical speciality concerned with the use of radiation, including X-rays, and radioactive materials in the diagnosis and treatment of injury and disease.

radiometric dating method of dating rock by assessing the amount of *radioactive decay of naturally occurring *isotopes. The dating of rocks may be based on the gradual decay of uranium into lead. The ratio of the amounts of 'parent' to 'daughter' isotopes in a sample gives a measure of the time it has been decaying, that is, of its age. Different elements and isotopes are used depending on the isotopes present and the age of the rocks to be dated. Once-living matter can often be dated by radiocarbon dating, employing the half-life of the isotope carbon-14, which is naturally present in organic tissue.

Radiometric methods have been applied to the decay of long-lived isotopes, such as potassium-40, rubidium-87, thorium-232, and uranium-238, which are found in rocks. These isotopes decay very slowly and this has enabled rocks as old as 3,800 million years to be dated accurately. Carbon dating can be used for material between 1,000 and 100,000 years old. **Potassium** dating is used for material more than 100,000 years old, **rubidium** for rocks more than 10 million years old, and **uranium** and **thorium** dating is suitable for rocks older than 20 million years.

radio telescope instrument for detecting radio waves from the universe in *radio astronomy. Radio telescopes usually consist of a metal bowl that collects and focuses radio waves the way a concave mirror collects and focuses light waves. Radio telescopes are much larger than optical telescopes, because the wavelengths they are detecting are much longer than the wavelength of light. The largest single dish is at Arecibo Observatory, Puerto Rico.

radiotherapy treatment of disease by *radiation from X-ray machines or radioactive sources. Radiation in the correct dosage can be used to kill cancerous cells and prevent their spreading.

radio wave electromagnetic wave possessing a long wavelength (ranging from about 10^{-3} to 10^4 m) and a low frequency (from about 10^5 to 10^{11} Hz) that travels at the speed of light. Included in the radio wave part of the spectrum are: *microwaves, used for both communications and for cooking; ultra high- and very high-frequency waves, used for television and FM (*frequency modulation) radio communications; and short, medium, and long waves, used for AM (*amplitude modulation) radio communications. Radio waves that are used for communications have all been modulated (see *modulation) to carry information. Certain astronomical objects emit radio waves, which may be detected and studied using *radio telescopes.

There is a layer in the atmosphere, called the ionosphere, where gas molecules are separated from their electrons by radiation from the Sun. When radio waves reach the ionosphere, they produce movements or oscillations of the electrons. As the electrons oscillate they produce electromagnetic waves that are identical to the radio waves with which the electrons were stimulated. The radio waves are reflected back and can be detected by a receiver. It is important to note that the reflected radio waves come from the oscillating electrons in the ionosphere.

radium chemical symbol Ra, (Latin *radius* 'ray') white, radioactive, metallic element, atomic number 88, relative atomic mass 226. It is one of the *alkaline-earth metals, found in nature in *pitchblende and other uranium ores. Of the 16 isotopes, the commonest, Ra-226, has a half-life of 1,620 years. The element was discovered and named in 1898 by French scientists Pierre and Marie Curie, who were investigating the residues of pitchblende.

Radium decays in successive steps to produce radon (a gas), polonium, and finally a stable isotope of lead. The isotope Ra-223 decays through the uncommon mode of heavy-ion emission, giving off carbon-14 and transmuting directly to lead. Because

radium luminesces, it was formerly used in paints that glowed in the dark; when the hazards of radioactivity became known its use was abandoned, but factory and dump sites remain contaminated and many former workers and neighbours contracted fatal cancers.

radius in biology, one of the two bones in the lower forelimb of tetrapod (four-limbed) vertebrates.

radius straight line from the centre of a circle to its circumference, or from the centre to the surface of a sphere.

radon chemical symbol Rn, colourless, odourless, gaseous, radioactive, non-metallic element, atomic number 86, relative atomic mass 222. It is grouped with the *noble gases (rare gases) and was formerly considered nonreactive, but is now known to form some compounds with fluorine. Of the 20 known isotopes, only three occur in nature; the longest half-life is 3.82 days (Rn-222).

Radon is the densest gas known and occurs in small amounts in spring water, streams, and the air, being formed from the natural radioactive decay of radium. New Zealand-born British physicist Ernest Rutherford (1871–1937) discovered the isotope Rn-220 in 1899, and Friedrich Dorn (1848–1916) in 1900; after several other chemists discovered additional isotopes, Scottish chemist William Ramsay (1852–1916) and R W Whytlaw-Gray isolated the element, which they named niton in 1908. The name radon was adopted in the 1920s.

RAID acronym for Redundant Array of Independent (or Inexpensive) Disks, in computing, arrays of disks are each connected to a controller that can be configured in different ways, depending on the application. RAID 1 is, for example, disk mirroring, while RAID 5 spreads every character between disks. RAID is intended to improve data security, and can also improve performance.

rain form of precipitation in which separate drops of water fall to the Earth's surface from clouds. The drops are formed by the accumulation of fine droplets that condense from water vapour in the air. The *condensation is usually brought about by rising and subsequent cooling of air.

Rain can form in three main ways: frontal (or cyclonic) rainfall, orographic (or relief) rainfall, and convectional rainfall. **Frontal rainfall** takes place at the boundary, or *front, between a mass of warm air from the tropics and a mass of cold air from the poles. The water vapour in the warm air is chilled and condenses to form clouds and rain. **Orographic rainfall** occurs when an airstream is forced to rise over a mountain range. The air becomes cooled and precipitation takes place. In the UK, the Pennine hills, which extend southwards from Northumbria to Derbyshire in northern England, interrupt the path of the prevailing southwesterly winds, causing orographic rainfall. Their presence is partly responsible for the west of the UK being wetter than the east. **Convectional rainfall**, associated with hot climates, is brought about by rising and abrupt cooling of air that has been warmed by the extreme heat of the ground surface. The water vapour carried by the air condenses, producing heavy rain. Convectional rainfall is usually accompanied by a thunderstorm, and it can be intensified over urban areas due to higher temperatures there.

rainbow arch in the sky displaying the colours of the *spectrum formed by the refraction and reflection of the Sun's rays through rain or mist. Its cause was discovered by Theodoric of Freiburg in the 14th century.

rainforest dense forest usually found on or near the *Equator where the climate is hot and wet. Moist air brought by the converging trade winds rises because of the heat and produces heavy rainfall. More than half the tropical rainforests are in Central and South America, primarily the lower Amazon and the coasts of Ecuador and Columbia. The rest are in Southeast Asia (Malaysia, Indonesia, and New Guinea) and in West Africa and the Congo.

Tropical rainforests once covered 14% of the Earth's land surface, but are now being destroyed at an increasing rate as their valuable timber is harvested and the land cleared for agriculture, causing problems of *deforestation. Although by 1991 over 50% of the world's rainforests had been removed, they still comprise about 50% of all growing wood on the planet, and harbour at least 40% of the Earth's species (plants and animals).

The vegetation in tropical rainforests typically includes an area of dense forest called **selva**; a **canopy** formed by high branches of tall trees providing shade for lower layers; an intermediate layer of shorter trees and tree roots; lianas; and a ground cover of mosses and ferns. The lack of **seasonal rhythm** causes adjacent plants to flower and shed leaves simultaneously. Chemical weathering and leaching take place in the iron-rich soil due to the high temperatures and humidity.

Rainforests comprise some of the most complex and diverse *ecosystems on the planet, deriving their energy from the Sun and photosynthesis. In a hectare (10,000 sq m/107,640 sq ft) of rainforest there are an estimated 200–300 tree species compared with 20–30 species in a hectare of temperate forest. The trees are the main **producers**. *Herbivores such as insects, caterpillars, and monkeys feed on the plants and trees and in turn are eaten by the *carnivores, such as ocelots and puma. Fungi and bacteria, the primary *decomposers, break down the dead material from the plants, herbivores, and carnivores with the help of heat and humidity. This decomposed material provides the **nutrients** for the plants and trees.

The rainforest ecosystem helps to regulate global weather patterns – especially by taking up CO_2 (carbon dioxide) from the atmosphere – and stabilizes the soil. Rainforests provide most of the oxygen needed for plant and animal respiration. When deforestation occurs, the microclimate of the mature forest disappears; soil erosion and flooding become major problems since rainforests protect the deep tropical soils. Once an area is cleared it is very difficult for shrubs and bushes to re-establish because soils are poor in nutrients. This causes problems for plans to convert rainforests into agricultural land – after two or three years the crops fail and the land is left bare. Clearing of the rainforests may lead to *global warming of the atmosphere, and contribute to the *greenhouse effect.

RAM acronym for **random-access memory**, in computing, a memory device in the form of a collection of integrated circuits (chips), frequently used in microcomputers. Unlike *ROM (read-only memory) chips, RAM chips can be both read from and written to by the computer, but their contents are lost when the power is switched off.

RAMdisk *RAM that has been configured to appear to the operating system as a disk. It is much faster to access than an actual hard disk and therefore can be used for applications that need frequent read-and-write operations. However, as the data is stored in RAM, it will be lost when the computer is turned off.

ramp another name for an inclined plane, a slope used as a simple machine.

random access in computing, an alternative term for *direct access.

range in physical geography, a line of mountains (such as the Alps or Himalayas). In human geography, the distance that people are prepared to travel (often to a central place) to obtain various goods or services. In mathematics, the range of a set of numbers is the difference between the largest and the smallest number; for example, 5, 8, 2, 9, 4 = 9 − 2 = 7; this sense is used in terms like 'tidal range' and 'temperature range'. Range is also a name for an open piece of land where cattle are ranched.

range in statistics, a measure of dispersion in a *frequency distribution, equalling the difference between the largest and smallest values of the variable. The range is sensitive to extreme values in the sense that it will give a distorted picture of the dispersion if one measurement is unusually large or small. The *interquartile range is often preferred.

range check in computing, a *validation check applied to a numerical data item to ensure that its value falls in a sensible range.

rare-earth element alternative name for *lanthanide.

rarefaction region of a *sound wave where the particles of the medium through which it is travelling are spread out, that is at a low density. Sound waves consist of alternate regions of *compressions and rarefactions travelling away from the source.

rare gas alternative name for *noble gas.

raster graphics computer graphics that are stored in the computer memory by using a map to record data (such as colour and intensity) for every *pixel that makes up the image. When transformed (enlarged, rotated, stretched, and so on), raster graphics become ragged and suffer loss of picture resolution, unlike *vector graphics. Raster graphics are typically used for painting applications, which allow the user to create artwork on a computer screen much as if they were painting on paper or canvas.

raster image processor full name for printer program *RIP.

rate of change change of a variable per unit of time. See *distance–time graph.

rate of reaction speed at which a chemical reaction proceeds. It is usually expressed in terms of the concentration (usually in *moles per litre) of a reactant consumed, or product formed, in unit time; so the units would be moles per litre per second ($mol\,l^{-1}\,s^{-1}$). The rate of a reaction may be affected by the concentration of the reactants, the temperature of the reactants (or the amount of light in the case of a photochemical reaction), and the presence of a *catalyst. If the reaction is entirely in the gas state, the rate is affected by pressure, and, where one of the reactants is a solid, it is affected by the particle size.

During a reaction at constant temperature the concentration of the reactants decreases and so the rate of reaction gradually slows down. These changes can be represented by drawing graphs.

For an *endothermic reaction (one that absorbs heat) increasing the

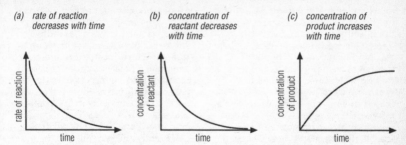

(a) rate of reaction decreases with time

(b) concentration of reactant decreases with time

(c) concentration of product increases with time

rate of reaction The rate of reaction decreases with time while the concentration of product increases.

temperature may produce large increases in the rate of reaction. A 10°C rise can double the rate while a 40°C rise can produce a 50- to 100-fold increase in the rate.

*Collision theory is used to explain these effects. Increasing the concentration or the pressure of a gas means there are more particles per unit volume, therefore there are more collisions and more successful collisions. Increasing the temperature makes the particles move much faster, resulting in more collisions per unit time and more successful collisions; consequently the rate increases.

ratio measure of the relative size of two quantities or of two measurements (in similar units), expressed as a *proportion. For example, the ratio of vowels to consonants in the alphabet is 5 : 21. As a *fraction $^5/_{26}$ of the letters are vowels. The ratio of 500 m to 2 km is 500 : 2,000, or in its simplest integer form 1 : 4 (dividing both sides of the ratio by 500).

Ratios are normally expressed as whole numbers, so 2 : 3.5 would become 4 : 7 (the ratio remains the same provided both parts of the ratio are multiplied or divided by the same number).

rational number in mathematics, any number that can be expressed as an exact fraction (with a denominator not equal to 0), that is, as $a \div b$ where a and b are integers; or an exact decimal. For example, $^2/_1$, $^1/_4$, $15/_4$, $-^3/_5$ are all rational numbers, whereas π (which represents the constant 3.141592 ...) is not. Numbers such as π are called *irrational numbers.

Raynaud's disease chronic condition in which the blood supply to the extremities is reduced by periodic spasm of the blood vessels on exposure to cold. It is most often seen in young women.

rays original name for radiation of all types, such as *X-rays and gamma rays.

ray-tracing in computer graphics, method of rendering sharp, detailed images. Designers specify the size, shape, colour, and texture of objects and the type and location of light sources, and use a program to devise a mathematical model tracing how light rays would bounce off the surfaces. The results, complete with shading, shadows, and reflections, depict 'virtual worlds' with near-photographic clarity.

RDP abbreviation for *Remote Desktop Protocol.

reaction in chemistry, the coming together, or interaction, of two or more atoms, ions, or molecules with the result that a chemical change takes place and a new substance is formed, with a different chemical composition. The nature of the reaction is described by a *chemical equation. For example, in the chemical reaction that occurs when magnesium burns in oxygen, a new substance, magnesium oxide is made:

$$2Mg_{(s)} + O_{2(g)} \rightarrow 2MgO_{(s)}$$

Chemical equations show the reactants and products of a chemical reaction by using chemical symbols and formulae. State symbols and the energy symbol (ΔH) can be used to show whether reactants and products are solids (s), liquids (l), gases (g), or are in solution (aq); and whether energy has been released or absorbed during the reaction.

reaction force in physics, the equal and opposite force described by Newton's third law of motion (see *Newton's laws of motion) that arises whenever one object applies a force (**action force**) to another. For example, if a magnet attracts a piece of iron, then that piece of iron will also attract the magnet with a force that is equal in magnitude but opposite in direction. When any object rests on the ground the downwards contact force applied to the ground always produces an equal, upwards reaction force.

reactivity series chemical series produced by arranging the metals in order of their ease of reaction with reagents such as oxygen, water, and acids. An example of such an arrangement, starting with the most reactive, is: potassium, sodium, calcium, magnesium, aluminium, zinc, iron, tin, lead, copper, silver, gold. This arrangement aids the understanding of the properties of metals, helps to explain differences between them, and enables predictions to be made about a metal's behaviour, based on a knowledge of its position or properties. It also allows prediction of the relative stability of the compounds formed by an element: the more reactive the metal, the more stable its compounds are likely to be.

The position of a metal in the series determines the reactions of the metal with various reagents, the displacement of one metal from its compound by another metal, and the method of extraction of a metal from its ore.

real number in mathematics, any of the *rational numbers (which include the integers) or *irrational numbers. Real numbers exclude *imaginary numbers, found in *complex numbers of the general form $a + bi$ where $i = \sqrt{-1}$, although these do include a real component a.

real-time system in computing, a program that responds to events in the world as they happen. For example, an automatic-pilot program in an aircraft must respond instantly in order to correct deviations from its course. Process control, robotics, games, and many military applications are examples of real-time systems.

receiver, radio component of a radio communication system that receives and processes radio waves. It detects and selects modulated radio waves (see *modulation) by means of an aerial and tuned circuit, and then separates the transmitted information from the carrier wave by a process that involves rectification.

receptacle the enlarged end of a flower stalk to which the floral parts are attached. Normally the receptacle is rounded, but in some plants it is flattened or cup-shaped. The term is also used for the region on that part of some seaweeds which becomes swollen at certain times of the year and bears the reproductive organs.

receptor in biology, receptors are discrete areas of cell membranes or areas within cells with which neuro-transmitters, hormones, and drugs interact. Such interactions control the activities of the body. For example, adrenaline transmits nervous impulses to receptors in the sympathetic nervous system, which initiates the characteristic response to excitement and fear in an individual.

recessive gene *allele (alternative form of a gene) that will show in the (*phenotype observed characteristics of an organism) only if its partner allele on the paired chromosome is similarly recessive. Such an allele will not show if its partner is dominant, that is if the organism is *heterozygous for a particular characteristic. Alleles for blue eyes in humans and for shortness in pea plants are recessive. Most mutant alleles are recessive and therefore are only rarely expressed (see *haemophilia).

For every characteristic of a plant or animal that is inherited, there are two genes present in the cells determining this characteristic in all but a few examples. By 'characteristic' is meant 'height' or 'eye colour' or 'ability to make a particular enzyme'. If the two genes are identical (homozygous state) the characteristic seen in the organism is determined by the two genes. However, one gene may be different from the other (heterozygous state). If so, the two genes are alleles – contrasting genes for a characteristic. In this case it is possible that one of them determines the characteristic seen and the other does not. The characteristic seen in this case is said to be dominant. The other allele not expressed in this case will only be expressed when present in the homozygous state. This characteristic is said to be recessive. Sometimes the allele that produces the dominant characteristic is described as being a dominant allele and the one that tends to produce the recessive characteristic as being the recessive allele. This is not really the correct use of the terms dominant and recessive. An allele is one of two or more alternative forms of a gene. This is best explained with examples. A gene which tends to produce blue eyes in a person will have an alternative allele that tends to produce brown eye colour. In a plant that may be found in tall and short forms may have an allele that tends to produce tall plants though its alternative allele produces short plants.

reciprocal result of dividing a given quantity into 1. Thus the reciprocal of 2 is $^1/_2$; the reciprocal of $^2/_3$ is $^3/_2$; and the reciprocal of x^2 is $1/x^2$ or x^{-2}. Reciprocals are used to replace *division by *multiplication, since multiplying by the reciprocal of a number is the same as dividing by that number. On a calculator, the reciprocals of all numbers except 0 are obtained by using the button marked $^1/_x$.

recombinant DNA in genetic engineering, *DNA formed by splicing together genes from different sources into new combinations.

recombination in genetics, any process that recombines, or 'shuffles', the genetic material, thus increasing genetic variation in the offspring. The two main processes of recombination both occur during meiosis (reduction division of cells). One is ***crossing over**, in which chromosome pairs

exchange segments; the other is the random reassortment of chromosomes that occurs when each gamete (sperm or egg) receives only one of each chromosome pair.

record in computing, a collection of related data items or **fields**. A record usually forms part of a file. Records may be of either **fixed** or **variable** length; variable records require a separator at the end of the field, in order that the end of the record can be detected by the computer.

rectangle quadrilateral (four-sided plane figure) with opposite sides equal and parallel and with each interior angle a right angle (90°). Its area A is the product of the length l and height h; that is, $A = l \times h$. A rectangle with all four sides equal is a *square.

rectifier device for obtaining one-directional current (DC) from an alternating source of supply (AC). (The process is necessary because almost all electrical power is generated, transmitted, and supplied as alternating current, but many devices, from television sets to electric motors, require direct current.) Types include plate rectifiers, thermionic *diodes, and *semiconductor diodes.

rectum lowest part of the large intestine of animals, which stores faeces prior to elimination (defecation).

recursion in computing and mathematics, a technique whereby a *function or *procedure calls itself into use in order to enable a complex problem to be broken down into simpler steps. For example, a function that finds the factorial of a number n (calculates the product of all the whole numbers between 1 and n) would obtain its result by multiplying n by the factorial of $n - 1$.

recycling processing of industrial and household waste (such as paper, glass, and some metals and plastics) so that the materials can be reused. This saves expenditure on scarce raw materials, slows down the depletion of *non-renewable resources, and helps to reduce pollution. Aluminium is frequently recycled because of its value and special properties that allow it to be melted down and re-pressed without loss of quality, unlike paper and glass, which deteriorate when recycled.

red blood cell or **erythrocyte**, most common type of blood cell, and responsible for transporting *oxygen around the body. They contain haemoglobin, a red protein, which combines with oxygen from the lungs to form oxyhaemoglobin. When transported to the tissues the oxyhaemoglobin splits into its original constituents, and the cells are able to release the oxygen. There are about 6 million red cells in every cubic centimetre of blood.

The red cell is a highly specialized cell with a distinctive shape. In mammals they are in the shape of a disc with a depression in the face of the disc and lose their *nucleus before they work as oxygen transporters in order to make maximum space for haemoglobin. In other vertebrates they are oval and nucleated. They are manufactured in the bone marrow. In humans, red cells last for only four months before being destroyed in the *liver. Haemoglobin contains iron, which is why this *mineral must be included as part of a balanced diet. However, the liver can store and re-use iron from old red cells.

red dwarf any star that is cool, faint, and small (about one-tenth the mass and diameter of the Sun). Red dwarfs burn slowly, and have estimated lifetimes of 100 billion years. They may be the most abundant type of star, but are difficult to see because they are so faint. Two of the closest stars to the Sun, Proxima Centauri and Barnard's Star, are red dwarfs.

red giant any large bright star with a cool surface. It is thought to represent a late stage in the evolution of a star like the Sun, as it runs out of hydrogen fuel at its centre and begins to burn heavier elements, such as helium, carbon, and silicon. Because of more complex nuclear reactions that then occur in the red giant's interior, it eventually becomes gravitationally unstable and begins to collapse and heat up. The result is either explosion of the star as a *supernova, leaving behind a *neutron star, or loss of mass by more gradual means to produce a *white dwarf.

redox reaction chemical change where one reactant is reduced and the other reactant oxidized. The reaction can only occur if both reactants are present and each changes simultaneously. For example, hydrogen reduces copper(II) oxide to copper while it is itself oxidized to water. The corrosion of iron and the reactions taking place in electric and electrolytic cells are just a few examples of redox reactions.

red shift in astronomy, lengthening of the wavelengths of light from an object as a result of the object's motion away from us. It is an example of the *Doppler effect. The red shift in light from galaxies is evidence that the universe is expanding.

reduce to lowest terms in mathematics, to cancel a fraction until numerator and denominator have no further factors in common.

reduction in chemistry, the gain of electrons, loss of oxygen, or gain of hydrogen by an atom, ion, or molecule during a chemical reaction.

Reduction may be brought about by reaction with another compound, which is simultaneously oxidized (reducing agent), or electrically at the cathode (negative electrode) of an

view from above   side view

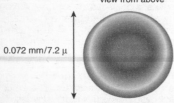

0.072 mm/7.2 μ

0.022 mm/2.2 μ

concave shape gives large surface area

red blood cell The structure of a human red blood cell. Its concave surfaces gives it a large surface area for the transportation of haemoglobin.

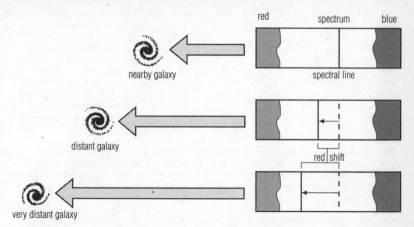

red   spectrum   blue

nearby galaxy

spectral line

distant galaxy

red shift

very distant galaxy

red shift The red shift causes lines in the spectra of galaxies to be shifted towards the red end of the spectrum. More distant galaxies have greater red shifts than closer galaxies. The red shift indicates that distant galaxies are moving apart rapidly, as the universe expands.

electric cell. Examples include the reduction of iron(III) oxide to iron by carbon monoxide:

$Fe_2O_3 + 3CO \rightarrow 2Fe + 3CO_2$

the hydrogenation of ethene to ethane:

$CH_2=CH_2 + H_2 \rightarrow CH_3-CH_3$

and the reduction of a sodium ion to sodium:

$Na^+ + e^- \rightarrow Na$

redundancy in computing, duplication of information. Redundancy is often used as a check, when an additional check digit or bit is included. See also *validation.

reflecting telescope *telescope in which light is collected and brought to a focus by a concave mirror. Cassegrain and Newtonian telescopes are examples.

reflection in geometry, a *transformation that reflects every point on a shape to a point that is the same distance on the opposite side of a fixed line – the mirror line, or line of symmetry. Reflections in two perpendicular axes produce a rotation of 180° (a half turn). Shapes can also be transformed by *translation, *rotation, and *enlargement.

reflection throwing back or deflection of waves, such as *light or *sound

waves, when they hit a surface. Reflection occurs whenever light falls on an object. The laws of reflection state

1. The incident ray, the reflected ray and the normal (a perpendicular line drawn to the surface at the point of incidence) all lie in the same plane.

2. The angle of incidence (the angle between the ray and the normal) is equal to the angle of reflection (the angle between the reflected ray and the normal).

An example of light rays reflecting towards the observer is seen when looking at an image on the surface of the water in a lake. Reflection of light takes place from all materials. Some materials absorb a small amount of light and reflect most of it back, for example, a shiny, silvery surface. Other materials absorb most of the light and reflect only a small amount back, for example, a dark, dull surface. Reflected light gives objects their visible texture and colour.

Light reflected from a surface can be either regular (plane), where the surface is flat and smooth and light rays are reflected without any scattering; or scattered, where the surface is irregular (in effect, many different surfaces). The colour of the sky is due to scattering of sunlight by

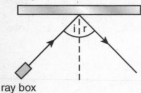

angle of incidence equals
angle of reflection

ray box

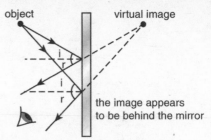

object

virtual image

the image appears
to be behind the mirror

reflection Light rays reflected from a regular (plane) mirror. The angle of incidence is the angle between the ray and a perpendicular line drawn to the surface and the angle of reflection is the angle between the reflected ray and a perpendicular to the surface. The image of an object in a plane mirror is described as virtual or imaginary because it appears to be the position from which the rays are formed.

particles in the atmosphere, such as dust and gas particles, water droplets, or ice crystals. On a clear day the sky appears blue due to the scattering of shorter wavelength light.

When light passes from a dense medium to a less dense medium, such as from water to air, both *refraction and reflection can occur. If the angle of incidence is small, the reflection will be relatively weak compared to the refraction. But as the angle of incidence increases the relative degree of reflection will increase. At some critical angle of incidence the angle of refraction is 90°. Since refraction cannot occur above 90°, the light is totally reflected at angles above this critical angle of incidence. This condition is known as *total internal reflection. Total internal reflection is used in *fibre optics to transmit data over long distances, without the need of amplification.

reflex in animals, a very rapid involuntary response to a particular stimulus. It is controlled by the *nervous system. A reflex involves only a few nerve cells, unlike the slower but more complex responses produced by the many processing nerve cells of the brain.

reflex angle angle greater than 180° but less than 360°.

refraction bending of a wave when it passes from one transparent material into another. It is the effect of the different speeds of wave propagation in two materials that have different

densities. When light hits the denser material, its *frequency remains constant, but its velocity decreases due to the influence of electrons in the denser medium. Constant frequency means that the same number of light waves must pass by in the same amount of time. If the waves are slowing down, wavelength must also decrease to maintain the constant frequency. The waves become more closely spaced, bending toward the **normal** (a line perpendicular to the surface of the material at the point of incidence) as if they are being dragged.

Refraction occurs with all types of progressive waves – *electromagnetic waves, sound waves, and water waves – and differs from *reflection, which involves no change in velocity.

refraction of light The degree of refraction depends on the angle at which the light hits the surface of a material and on the densities of the materials through which it passes. It is also affected by the amount of bending and change of velocity corresponding to the wave's frequency (dispersion).

The angle between an light ray striking a surface and the normal is called the **angle of incidence**. The angle between the refracted ray and the normal is called the **angle of refraction**. The angle of refraction cannot exceed 90°. The relationship between the angle of incidence and the angle of refraction is given by *Snell's law of refraction, one of the **laws of**

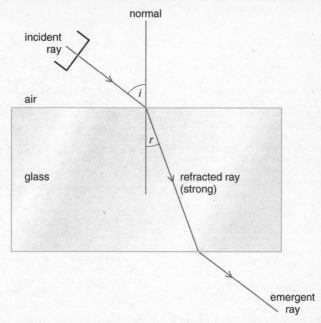

refraction The refraction of light through glass. When the light ray strikes the glass (a denser medium than the air) it is bent towards the normal. When it leaves the glass and re-enters the less dense medium it is bent away from the normal.

refraction, and is related to the *refractive index of the materials concerned. The **absolute refractive index** of a material is the velocity of light in that material relative to the velocity of light in a vacuum. The other law of refraction states that the incident ray, the refracted ray and the normal all lie in the same plane.

An example of refraction is light hitting a glass pane. When light in air enters glass (a denser medium than air), it slows down (from 300 million to 200 million metres per second) and is bent toward the normal. When light passes out of the glass into the air it is bent away from the normal. and is refracted. The incident light will be parallel to the emerging light if the two faces of the glass are parallel. If the two faces of a material through which light passes are not parallel, as with a prism, the emerging light will not be parallel to the incident light. The angle between the incident ray and the emerging ray is called the **angle of**

deviation. The amount of bending and change in velocity of the refracted wave is due to the amount of *dispersion corresponding to the wave's frequency, and the refractive index of the material.

refractive index   measure of the refraction of a ray of light as it passes from one transparent medium to another, symbol n. (See *Snell's law of refraction). If the angle of incidence is i and the angle of refraction is r for light travelling from medium 1 to medium 2, the ratio of the two refractive indices is given by $n_2/n_1 = \sin i /\sin r$. If medium 1 is air or a vacuum, the equation becomes $n_2 = \sin i /\sin r$. It is also equal to the speed of light in the medium 1 divided by the speed of light in medium 2, and it varies with the wavelength of the light.

refrigeration   use of technology to transfer heat from cold to warm, against the normal temperature gradient, so that an object can remain substantially colder than its surroundings. Refrigeration equipment is used for the chilling and deep-freezing of food in food technology, and in air conditioners and industrial processes.

Refrigeration is commonly achieved by a vapour-compression cycle, in which a suitable chemical (the refrigerant) travels through a long circuit of tubing, during which it changes from a vapour to a liquid and back again. A compression chamber makes it condense, and thus give out heat. In another part of the circuit, called the evaporator coils, the pressure is much lower, so the refrigerant evaporates, absorbing heat as it does so. The evaporation process takes place near the central part of the refrigerator, which therefore becomes colder, while the compression process takes place near a ventilation grille, transferring the heat to the air outside. The most commonly used refrigerants in modern systems were *chlorofluorocarbons (CFCs), but these are now being replaced by coolants that do not damage the ozone layer.

regeneration in biology, regrowth of a new organ or tissue after the loss or removal of the original. It is common in plants, where a new individual can often be produced from a 'cutting' of the original. In animals, regeneration of major structures is limited to lower organisms; certain lizards can regrow their tails if these are lost, and new flatworms can grow from a tiny fragment of an old one. In mammals, regeneration is limited to the repair of tissue in wound healing and the regrowth of peripheral nerves following damage.

regurgitation in medicine, the return of food that has already been swallowed from the oesophagus or the stomach to the mouth.

rejection immune response in the recipient of a tissue or organ transplant. If rejection does not respond to treatment, the donated tissue is destroyed. Immunosuppressive drugs are used to try and prevent rejection.

relational database database in which data are viewed as a collection of linked tables. It is the most popular of the three basic database models, the others being **network** and **hierarchical**.

relative in computing (of a value), variable and calculated from a base value. For example, a **relative address** is a memory location that is found by adding a variable to a base (fixed) address, and a **relative cell reference** locates a cell in a spreadsheet by its position relative to a base cell – perhaps directly to the left of the base cell or three columns to the right of the base cell. The opposite of relative is *absolute.

relative atomic mass mass of an atom relative to one-twelfth the mass of an atom of carbon-12. It depends primarily on the number of protons and neutrons in the atom, the electrons having negligible mass. If more than one *isotope of the element is present, the relative atomic mass is calculated by taking an average that takes account of the relative proportions of each isotope, resulting in values that are not whole numbers. The term **atomic weight**, although commonly used, is strictly speaking incorrect.

relative biological effectiveness RBE, relative damage caused to living tissue by different types of radiation. Some radiations do much more damage than others; alpha particles, for example, cause 20 times as much destruction as electrons (beta particles).

relative density density (at 20°C/68°F) of a solid or liquid relative to (divided by) the maximum density of water (at 4°C/39.2°F). The relative density of a gas is its density divided by the density of hydrogen (or sometimes dry air) at the same temperature and pressure.

relative humidity concentration of water vapour in the air. It is expressed as the ratio of the partial pressure of the water vapour to its saturated vapour pressure at the same temperature. The higher the temperature, the higher the saturated vapour pressure.

relative molecular mass mass of a molecule, calculated relative to one-twelfth the mass of an atom of carbon-12. It is found by adding the relative atomic masses of the atoms that make up the molecule.

The term **molecular weight**, although commonly used, is strictly speaking incorrect.

relativity in physics, theory of the relative rather than absolute character of mass, time, and space, and their interdependence, as developed by German-born US physicist Albert Einstein (1879–1955) in two phases:

special theory of relativity (1905) Starting with the premises that (1) the laws of nature are the same for all observers in unaccelerated motion and (2) the speed of light is independent of the motion of its source, Einstein arrived at some rather unexpected consequences. Intuitively familiar concepts, like mass, length, and time, had to be modified. For example, an object moving rapidly past the observer will appear to be both shorter and more massive than when it is at rest (that is, at rest relative to the observer), and a clock moving rapidly past the observer will appear to be running slower than when it is at rest. These predictions of relativity theory seem to be foreign to everyday experience merely because the changes are quite negligible at speeds less than about 1,500 kps/930 mps and only become appreciable at speeds approaching the speed of light.

Einstein showed that, while Newton's laws (see *Newton's laws of motion) still hold good at ordinary velocities, for consistency with the above premises (1) and (2), the principles of dynamics as established by Newton needed modification; the most celebrated new result was the equation $E = mc^2$, which expresses an equivalence between mass (m) and *energy (E), c being the speed of light in a vacuum. In 'relativistic mechanics', conservation of mass is replaced by the new concept of conservation of 'mass-energy'.

general theory of relativity (1915) This theory makes predictions concerning light and gravitation. The geometrical properties of space-time were to be conceived as modified locally by the presence of a body with mass; and light rays should bend when they pass by a massive object. General relativity theory was inspired by the simple idea that it is impossible in a small region to distinguish between acceleration and gravitation effects (as in a lift one feels heavier when it accelerates upwards).

A planet's orbit around the Sun (as observed in three-dimensional space) arises from its natural trajectory in modified space-time. Einstein's general theory accounts for a peculiarity in the behaviour of the motion of the perihelion of the orbit of the planet Mercury that cannot be explained in Newton's theory.

The general theory predicted that a red shift is produced if light passes through an intense gravitational field. This was subsequently detected in astronomical observations in 1925. The theory also predicted that the apparent positions of stars would shift when they are seen near the Sun because the Sun's intense gravity would bend the light rays from the stars as they pass the Sun. Einstein was triumphantly vindicated when observations of a solar eclipse in 1919 showed apparent shifts of exactly the amount he had predicted.

General relativity is central to modern *astrophysics and *cosmology; it predicts, for example, the possibility of *black holes. The mathematical development of the general theory is formidable, unlike the special theory, which a non-expert can follow up to $E = mc^2$ and beyond.

relaxation therapy development of regular and conscious control of physiological processes and their related emotional and mental states, and of muscular tensions in the body, as a way of relieving stress and its results. Meditation, hypnotherapy, *autogenics, and *biofeedback are techniques commonly employed.

relaxin hormone produced naturally by women during pregnancy that assists childbirth. It widens the pelvic opening by relaxing the ligaments, inhibits uterine contractility, so preventing premature labour, and causes dilation of the cervix. A synthetic form was pioneered by the Howard Florey Institute in Australia, and this drug has possible importance in helping the birth process and avoiding surgery or forceps delivery.

relay in electrical engineering, an electromagnetic switch. A small current passing through a coil of wire wound around an iron core attracts an *armature whose movement closes a pair of sprung contacts to complete a secondary circuit, which may carry a large current or activate other devices. The solid-state equivalent is a thyristor switching device.

relay neuron nerve cell in the spinal cord, connecting motor neurons to sensory neurons. Relay neurons allow information to pass straight through the spinal cord, bypassing the brain. In humans such reflex actions, which are extremely rapid, cause the sudden removal of a limb from a painful stimulus.

reload in computing, command which asks a *browser to reopen a currently-displayed *URL. Reloading may 'unstall' a partially-loaded page or bring a faster download from a busy server.

remainder part left over when one number cannot be exactly divided by another. For example, the remainder of 11 divided by 3 is 2; the remainder

may be represented as a fraction or decimal. Decimal remainders are either recurring (0.66666...), cyclic (0.37373737), or terminating (0.125).

remote access term used to describe the issue of, and the computer *hardware and *software designed to be used by, employees needing access to their company *network and data whilst away from the office, either at home or travelling. The main problems relate to security, because public networks like the *Internet are often used. An industry has emerged to provide special security devices and networking equipment to make remote access secure and reliable.

Remote Desktop Protocol RDP, communications protocol through which servers running Microsoft's Windows NT Terminal Server software communicate with thin clients which may be running Microsoft Windows CE operating system. While Microsoft provides a Microsoft-to-Microsoft system, firms like Citrix provide connectivity with non-Microsoft systems.

remote sensing gathering and recording information from a distance. Aircraft and satellites can observe a

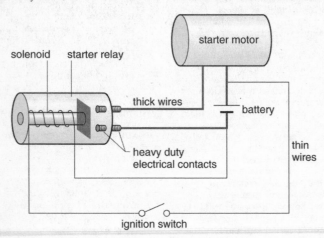

relay A relay is a device that allows a small amount of electrical current to control a large amount of current. A car starter motor uses a relay to solve the problem that a car has in needing a large amount of current to start the engine. A starter relay is installed in series between the battery and the starter. Some cars use a starter solenoid (as shown here) to accomplish the same purpose of allowing a small amount of current from the ignition switch to control a high current flow from the battery to the starter.

planetary surface or atmosphere, and space probes have sent back photographs and data about planets as distant as Neptune. Remote sensing usually refers to gathering data of the electromagnetic spectrum (such as visible light, ultraviolet light, and infrared light). In archaeology, surface survey techniques provide information without disturbing subsurface deposits.

remote terminal   in computing, a terminal that communicates with a computer via a modem (or acoustic coupler) and a telephone line.

REM sleep   acronym for **rapid-eye-movement sleep**, phase of sleep that recurs several times nightly in humans and is associated with dreaming. The eyes flicker quickly beneath closed lids.

renal   pertaining to the *kidneys.

rendering   using a computer to draw an image on a computer screen. In graphics, this often means using *ray-tracing, *Phong shading, or a similar program to turn an outline sketch into a detailed image of a solid object.

renewable energy   power from any source that can be replenished. Most renewable systems rely on *solar energy directly or through the weather cycle as *wave power, *hydroelectric power, wind power via *wind turbines, or solar energy collected by plants (alcohol fuels, for example). In addition, the gravitational force of the Moon can be harnessed through tidal power stations, and the heat trapped in the centre of the Earth is used via *geothermal energy systems. Other examples are energy from *biofuel and *fuel cells. Renewable energy resources have the advantage of being non-polluting. However, some (such as wind energy) can be unreliable and therefore lose their effectiveness in providing a constant supply of energy.

renewable resource   natural resource that is replaced by natural processes in a reasonable amount of time. Soil, water, forests, plants, and animals are all renewable resources as long as they are properly conserved. Solar, wind, wave, and geothermal energies are based on renewable resources.

renin   enzyme produced by the kidney in response to stress. It is involved in

the production of angiotensin, which causes blood vessels to constrict, thus contributing to a rise in blood pressure.

rennin   or **chymase**, enzyme found in the gastric juice of young mammals, used in the digestion of milk.

repetitive strain injury   RSI, inflammation of tendon sheaths, mainly in the hands and wrists, which may be disabling. It is found predominantly in factory workers involved in constant repetitive movements, and in those who work with computer keyboards. The symptoms include aching muscles, weak wrists, tingling fingers, and in severe cases, pain and paralysis. Some victims have successfully sued their employers for damages. In 1999 RSI affected more than a million people annually in Britain and the USA.

replication   in biology, production of copies of the genetic material DNA; it occurs during cell division (*mitosis and *meiosis). Most mutations are caused by mistakes during replication.

repression   in psychology, a mental process that ejects and excludes from consciousness ideas, impulses, or memories that would otherwise threaten emotional stability.

reproduction   in biology, the process by which a living organism produces other organisms more or less similar to itself. The ways in which species reproduce differ, but the two main methods are by *asexual reproduction and *sexual reproduction. Asexual reproduction involves only one parent without the formation of *gametes: the parent's cells divide by *mitosis to produce new cells with the same number and kind of *chromosomes as its own. Thus offspring produced asexually are clones of the parent and there is no variation. Sexual reproduction involves two parents, one male and one female. The parents' sex cells divide by *meiosis producing gametes, which contain only half the number of chromosomes of the parent cell. In this way, when two sets of chromosomes combine during *fertilization, a new combination of genes is produced. Hence the new organism will differ from both parents, and variation is introduced. The ability to reproduce is considered one of the fundamental attributes of living things.

reserpine psychoactive drug extracted from the root of Southeast Asian plants of the genus *Rauwolfia*, especially *R. serpentina* or serpent wood. It was once used as a tranquillizer for some mental illnesses and to treat hypertension. Digestive upsets and depression are possible side effects.

reserved word word that has a meaning special to a programming language. For example, 'if' and 'for' are reserved words in most high-level languages.

reservoir natural or artificial lake that collects and stores water for community water supplies and for irrigation.

resin substance exuded from pines, firs, and other trees in gummy drops that harden in air. Varnishes are common products of the hard resins, and ointments come from the soft resins.

Rosin is the solid residue of distilled turpentine, a soft resin. The name 'resin' is also given to many synthetic products manufactured by polymerization; they are used in adhesives, plastics, and varnishes.

resistance in physics, that property of a conductor that restricts the flow of electricity through it, associated with the conversion of electrical energy to heat; also the magnitude of this property.

Materials that are good conductors of electricity have electrons held loosely in the outer shells of their atoms. Current can flow easily and these materials have low resistance. In poor conductors of electricity, the electrons in the outer shells of their atoms are more strongly attracted by the positively-charged nucleus. Such materials restrict the flow of electrons and have a high resistance. Resistance (R) is related to current (I) and voltage (V) by the formula: $R = V/I$. The statement that current is proportional to voltage (resistance is constant) at constant temperature is known as *Ohm's law. It is approximately true for many materials that are accordingly described as 'ohmic'.

Resistance depends on many factors, such as the nature of the material, its temperature, its length and cross-sectional area, and its thermal properties.

resistivity in physics, a measure of the ability of a material to resist the flow of an electric current. It is numerically equal to the *resistance of a sample of unit length and unit cross-sectional area, and its unit is the ohm metre (symbol Ωm). A good conductor has a low resistivity (1.7×10^{-8} Ωm for copper); an insulator has a very high resistivity (10^{15} Ωm for polyethane).

resistor in physics, any component in an electrical circuit used to introduce *resistance to a current by restricting the flow of electrons. Resistors are often made from wire-wound coils (higher resistance) or pieces of carbon (lower resistance). *Rheostats and *potentiometers are variable resistors.

When resistors R_1, R_2, R_3,... are connected in a *series circuit, the total resistance of the circuit is $R_1 + R_2 + R_3 +...$. When resistors R_1, R_2, R_3,... are connected in a *parallel circuit, the total resistance of the circuit is R, given by $1/R = 1/R_1 + 1/R_2 + 1/R_3 +...$.

resolution in computing, the number of dots per unit length in which an image can be reproduced on a screen or printer. A typical screen resolution for colour monitors is 72–96 dpi (dots per inch). A laser printer will typically have a printing resolution of 600 dpi, ink-jet printers have a minimum of 300 dpi, and dot matrix printers typically have resolutions from 60 to 180 dpi. Photographs in books and magazines have a resolution of 1,200 dpi or 2,400 dpi.

resolution of forces in mechanics, the division of a single force into two parts that act at right angles to each other. The two parts of a resolved force, called its **components**, have exactly the same effect when acting together on an object as the single force which they replace.

For example, the weight W of an object on a slope, tilted at an angle θ, can be resolved into two components: one acting at a right angle to the slope, equal to $W \cos \theta$, and one acting parallel to and down the slope, equal to $W \sin \theta$. The component acting down the slope (minus any friction force that may be acting in the

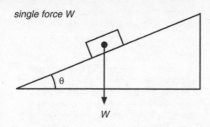

single force W

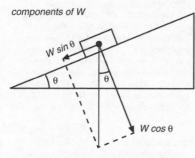

components of W

resolution of forces In mechanics, the resolution of forces is the division of a single force into two parts that act at right angles to each other. In the diagram, the weight W of an object on a slope, tilted at an angle q, can be resolved into two parts or components: one acting at a right angle to the slope, equal to W cos q, and one acting parallel to and down the slope, equal to W sin q.

opposite direction) is responsible for the acceleration of the object.

resolver in computing, see *domain name server.

resonance rapid amplification of a vibration when the vibrating object is subject to a force varying at its *natural frequency. In a trombone, for example, the length of the air column in the instrument is adjusted until it resonates with the note being sounded. Resonance effects are also produced by many electrical circuits. Tuning a radio, for example, is done by adjusting the natural frequency of the receiver circuit until it coincides with the frequency of the radio waves falling on the aerial.

Resonance has many physical applications. Children use it to increase the size of the movement on a swing, by giving a push at the same point during each swing. Soldiers marching across a bridge in step could cause the bridge to vibrate violently if the frequency of their steps coincided with its natural frequency. Resonance caused of the collapse of the Tacoma Narrows Bridge, USA, in 1940, when the frequency of the wind gusts coincided with the natural frequency of the bridge.

resources materials that can be used to satisfy human needs. Because human needs are varied and extend from basic physical requirements, such as food and shelter, to spiritual and emotional needs that are hard to define, resources cover a vast range of items. The intellectual resources of a society – its ideas and technologies – determine which aspects of the environment meet that society's needs, and therefore become resources. For example, in the 19th century uranium was used only in the manufacture of coloured glass. Today, with the development of nuclear technology, it is a military and energy resource. Resources are often divided into **human resources**, such as labour, supplies, and skills, and **natural resources**, such as climate, *fossil fuels, and water. Natural resources are divided into *non-renewable resources and *renewable resources.

respiration process that occurs inside cells in which carbohydrate, particularly glucose, is broken down to release *energy that the cell can use. The word 'respiration' should not be used to refer to the air movements in the air passageways. These air movements are called ventilation (*breathing).

Respiration at the cellular level is termed **internal respiration**, and in all higher organisms occurs in the *mitochondria. This takes place in two stages: the first stage, which does not require oxygen, is a form of *anaerobic respiration; the second stage is the main energy-producing stage and does require oxygen (*aerobic respiration). This is termed the *Krebs cycle. In some bacteria the oxidant is the nitrate or sulphate ion.

In **anaerobic respiration**, glucose is only partially broken down, and the end products are energy and either lactic acid or ethanol (alcohol) and carbon dioxide; this process is termed *fermentation.

glucose → lactic acid + energy

glucose → ethanol + carbon dioxide + energy

In humans, anaerobic respiration can only carry on for a short time. The *muscles producing lactic acid will stop working as it builds up. However, many micro-organisms can respire anaerobically for long periods of time or all the time. Yeast respires aerobically if oxygen is present, but, if there is no oxygen, it respires anaerobically. In anaerobic respiration it produces *alcohol.

Aerobic respiration is a complex process of chemical reactions in which oxygen is used to break down glucose into *carbon dioxide and *water.

glucose + oxygen → carbon dioxide + water + energy

Energy is released in the form of energy-carrying molecules (*ATP). Most of the released energy is used to drive various processes in the cell, such as growth or movement.

ATP production Many of the metabolic processes taking place inside cells are dependent upon the use of *enzymes. Respiration – the release of energy within cells – is a complex series of reactions which employs about 70 different enzymes that act as catalysts. The energy is released at several stages during this process, about three-quarters of it in the form of heat. Heat energy that

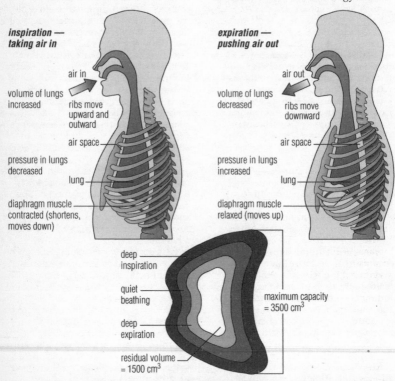

inspiration — taking air in

air in

volume of lungs increased

ribs move upward and outward

air space

pressure in lungs decreased

lung

diaphragm muscle contracted (shortens, moves down)

expiration — pushing air out

air out

volume of lungs decreased

ribs move downward

air space

pressure in lungs increased

lung

diaphragm muscle relaxed (moves up)

deep inspiration

quiet breathing

deep expiration

residual volume = 1500 cm^3

maximum capacity = 3500 cm^3

subdivisions of lung air

respiration The two phases of the process of respiration. *Gas exchange occurs in the alveoli, tiny air tubes in the lungs.

cannot be used by the cell is lost, but the non-heat energy released is stored by the cell as a readily available substance called adenosine triphosphate (ATP).

In aerobic respiration one molecule of glucose can produce 38 molecules of ATP, in anaerobic respiration only two molecules of ATP are produced for each molecule of glucose.

glucose supplies Glucose is essential for energy production. Plants are able to produce it for themselves through *photosynthesis, but animals are unable to do this and need to obtain their glucose second-hand from the carbohydrates produced by plants. Hence, in this respect, plants are **producers** and animals are **consumers**.

oxygen supplies Plants and animals obtain their oxygen directly from their surroundings. The part of the organism through which oxygen enters the body is called the 'respiratory surface'. Carbon dioxide, a by-product of respiration, is expelled from the organism also at the respiratory surface. This process is termed *gas exchange.

transport systems in unicellular organisms In single-cell organisms, such as amoebas that live in water, the respiratory surface is the cell membrane. Since oxygen dissolves in water, there is normally oxygen available in the surrounding water. The process of respiration inside the cell results in a lower concentration of oxygen in the cell than outside, and so oxygen diffuses into the cell across the cell membrane, while carbon dioxide diffuses out.

transport systems in multicellular organisms In single-cell organisms the oxygen is quickly able to reach the centre of the organism, but in larger and more complex organisms, such as humans, which contain vast numbers of cells, this is not practicable. In these organisms, the blood system acts as a transport system carrying oxygen to all the body's cells and removing carbon dioxide.

lungs An additional problem for larger organisms is the provision of sufficient oxygen to meet the energy demands of the body. Since the surface area to volume ratio decreases with increased body size, this problem is overcome in larger organisms by the provision of respiratory surfaces within the body – namely, the *lungs.

oxygen debt The oxygen produced in the lungs is carried by the bloodstream to all cells in every part of the body, which need it in order to produce energy. The ATP produced by the mitochondria in muscles provides the power to enable them to work. Where a set of muscles is required to work extra hard, such as leg muscles in running, energy production has to be increased to match the demand. If that demand exceeds the supply from aerobic respiration, extra energy is produced by anaerobic respiration. When this happens glucose in the blood is broken down without combining with oxygen, resulting in the production of lactic acid. Once the activity ceases and energy needs return to normal levels, the lactic acid that has been produced must be broken down by oxygen; consequently the increase in breathing rate and heartbeat has to be sustained until this process has been completed. This additional energy that has been acquired without oxygen is termed the oxygen debt – in other words, it has been 'borrowed' and has to be 'repaid'.

disorders The optimal operation of the respiratory system is dependent upon both an adequate supply of the basic materials, oxygen and glucose, and the efficient working of all the body organs, tissues, and metabolic processes involved. Change or disruption in any one aspect will affect the overall process, and consequently the well-being of the body. Change can come about as a result of an alteration in oxygen or sugar intake, or the degeneration of, or damage to, organs and tissues, either as a result of accident, ageing, or *disease.

respiratory tract in animals, the air passageways that lead from the nose and mouth to the lungs, including the air sacs of the lungs. A specialized epithelial tissue lines much of these air passages. This is cilated *epithelium and contains cells with beating hairs called cilia that sweep mucus and trapped particles away from the air sacs in the lungs. Smoking damages this epithelium.

rest mass   in physics, the mass of a body when its velocity is zero or considerably below that of light. According to the theory of *relativity, at very high velocities, there is a relativistic effect that increases the mass of the particle.

restriction enzyme   or **endonuclease**, bacterial *enzyme that breaks a chain of *DNA into two pieces at a specific point; used in *genetic engineering. The point along the DNA chain at which the enzyme can work is restricted to places where a specific sequence of base pairs occurs. Different restriction enzymes will break a DNA chain at different points. The overlap between the fragments is used in determining the sequence of base pairs in the DNA chain.

resultant force   in mechanics, a single force acting on a particle or body whose effect is equivalent to the combined effects of two or more separate forces. The resultant of two forces acting at one point on an object can be found using the *parallelogram of forces method.

retina   light-sensitive area at the back of the *eye connected to the brain by the optic nerve. It has several layers and in humans contains over a million rods and cones, sensory cells capable of converting light into nervous messages that pass down the optic nerve to the brain.

The **rod cells**, about 120 million in each eye, are distributed throughout the retina. They are sensitive to low levels of light, but do not provide detailed or sharp images, nor can they detect colour. The **cone cells**, about 6 million in number, are mostly concentrated in a central region of the retina called the **fovea**, and provide both detailed and colour vision. The cones of the human eye contain three visual pigments, each of which responds to a different primary colour (red, green, or blue). The brain can interpret the varying signal levels from the three types of cone as any of the different colours of the visible spectrum.

The image actually falling on the retina is highly distorted; research into the eye and the optic centres within the brain has shown that this poor quality image is processed to improve its quality. The retina can become separated from the wall of the eyeball as a result of a trauma, such as a boxing injury. It can be reattached by being 'welded' into place by a laser.

retinol   or **vitamin A**, fat-soluble chemical derived from b-carotene and found in milk, butter, cheese, egg yolk, and liver. Lack of retinol in the diet leads to the eye disease **xerophthalmia**.

retrograde   in astronomy, describing the orbit or rotation of a *planet or *satellite if the sense of rotation is opposite to the general sense of rotation of the Solar System. On the *celestial sphere, it refers to motion from east to west against the background of stars.

retrovirus   any of a family of *viruses (Retroviridae) containing the genetic material *RNA rather than the more usual *DNA. For the virus to express itself and multiply within an infected cell, its RNA must be converted to DNA. It does this by using a built-in enzyme known as reverse transcriptase (since the transfer of genetic information from DNA to RNA is known as *transcription, and retroviruses do the reverse of this). Retroviruses include those causing *AIDS and some forms of leukaemia. See *immunity. Retroviruses are used as vectors in *genetic engineering, but they cannot be used to target specific sites on the chromosome. Instead they incorporate their genes at random sites.

reverberation   in acoustics, the multiple reflections, or echoes, of sounds inside a building that merge and persist a short time (up to a few seconds) before fading away. At each reflection some of the sound energy is absorbed, causing the amplitude of the sound wave and the intensity of the sound to reduce a little.

reverse engineering   in computing, analysing an existing piece of computer hardware or software by finding out what it does and then working out how it does it. Companies perform this process on their own products in order to iron out faults, and on their competitors' products in order to find out how they work. For example, the microchips in the first IBM PCs were reverse engineered by other computer firms to make compatible machines without infringing IBM's copyright.

reverse osmosis movement of solvent (liquid) through a semipermeable membrane from a more concentrated solution to a more dilute solution. The solvent's direction of movement is opposite to that which it would experience during *osmosis, and is achieved by applying an external pressure to the solution on the more concentrated side of the membrane. The technique is used in desalination plants, when water (the solvent) passes from brine (a salt solution) into fresh water via a semipermeable filtering device.

reversible reaction chemical reaction that proceeds in both directions at the same time, as the product decomposes back into reactants as it is being produced. Such reactions do not run to completion, provided that no substance leaves the system. Examples include the manufacture of ammonia from hydrogen and nitrogen, and the oxidation of sulphur dioxide to sulphur trioxide.

Reynolds number number used in *fluid mechanics to determine whether fluid flow in a particular situation (through a pipe or around an aircraft body or a fixed structure in the sea) will be turbulent or smooth. The Reynolds number is calculated using the flow velocity, density, and viscosity of the fluid, and the dimensions of the flow channel. It is named after Irish engineer Osborne Reynolds (1842–1912).

rhe unit of fluidity equal to the reciprocal of the *poise.

rhenium chemical symbol Re, (Latin *Rhenus* 'Rhine') heavy, silver-white, metallic element, atomic number 75, relative atomic mass 186.21. It has chemical properties similar to those of manganese and a very high melting point (3,180°C/5,756°F), which makes it valuable as an ingredient in alloys. It was identified and named in 1925 by German chemists Walter Noddack (1893–1960), Ida Tacke, and Otto Berg from the Latin name for the River Rhine.

rheostat variable *resistor, usually consisting of a high-resistance wire-wound coil with a sliding contact. It is used to vary electrical resistance without interrupting the current (for example, when dimming lights). The circular type, which can be used, for example, as the volume control of an amplifier, is also known as a *potentiometer.

rhesus disease serious condition affecting some babies See *rhesus factor.

rhesus factor group of *antigens on the surface of red blood cells of humans which characterize the rhesus blood group system. Most individuals possess the main rhesus factor (Rh+), but those without this factor (Rh−) produce *antibodies if they come into contact with it. The name comes from rhesus monkeys, in whose blood rhesus factors were first found.

rheumatic fever or **acute rheumatism**, acute or chronic illness characterized by fever and painful swelling of joints. Some victims also experience involuntary movements of the limbs and head, a form of *chorea. It is now rare in the developed world.

rheumatism nontechnical term for a variety of ailments associated with inflammation and stiffness of the joints and muscles.

rheumatoid arthritis inflammation of the joints; a chronic progressive disease, it begins with pain and stiffness in the small joints of the hands and feet and spreads to involve other joints, often with severe disability and disfigurement. There may also be damage to the eyes, nervous system, and other organs. The disease is treated with a range of drugs and with surgery, possibly including replacement of major joints.

rhizoid hairlike outgrowth found on the *gametophyte generation of ferns, mosses, and liverworts. Rhizoids anchor the plant to the substrate and can absorb water and nutrients. They may be composed of many cells, as in mosses, where they are usually brownish, or may be unicellular, as in liverworts, where they are usually colourless. Rhizoids fulfil the same functions as the *roots of higher plants but are simpler in construction.

rhizome or **rootstock,** horizontal underground plant stem. It is a perennating organ in some species, where it is generally thick and fleshy, while in other species it is mainly a means of *vegetative reproduction, and is therefore long and slender, with buds all along it that send up new plants. The potato is a rhizome that has two distinct parts, the tuber being the swollen end of a long, cordlike rhizome.

rhodium chemical symbol Rh, (Greek *rhodon* 'rose') hard, silver-white, metallic element, atomic number 45, relative atomic mass 102.905. It is one of the so-called platinum group of metals and is resistant to tarnish, corrosion, and acid. It occurs as a free metal in the natural alloy osmiridium and is used in jewellery, electroplating, and thermocouples.

rhombic sulphur allotropic form of sulphur. At room temperature, it is the stable *allotrope, unlike monoclinic sulphur.

rhombus equilateral (all sides equal) *parallelogram. As with a parallelogram, the rhombus has diagonally opposed angles of equal size. Its diagonals bisect each other at right angles, and its area is half the product of the lengths of the two diagonals. The shape is sometimes called a diamond. A rhombus whose internal angles are 90° is called a *square.

rhyolite *igneous rock, the fine-grained volcanic (extrusive) equivalent of granite.

ria long narrow sea inlet, usually branching and surrounded by hills. A ria is deeper and wider towards its mouth, unlike a *fjord. It is formed by the flooding of a river valley due to either a rise in sea level or a lowering of a landmass.

ribbon lake long, narrow lake found on the floor of a *glacial trough. A ribbon lake will often form in an elongated hollow carved out by a glacier, perhaps where it came across a weaker band of rock. Ribbon lakes can also form when water ponds up behind a terminal moraine or a landslide. The English Lake District is named after its many ribbon lakes, such as Lake Windermere and Coniston Water.

riboflavin or **vitamin B$_2$,** *vitamin of the B complex important in cell respiration. It is obtained from eggs, liver, and milk. A deficiency in the diet causes stunted growth.

ribonucleic acid full name of *RNA.

ribosome protein-making machinery of the cell. Ribosomes are located on the endoplasmic reticulum (ER) of eukaryotic cells, and are made of proteins and a special type of *RNA, ribosomal RNA. They receive messenger RNA (copied from the *DNA) and *amino acids, and 'translate' the messenger RNA by using its chemically coded instructions to link amino acids in a specific order, to make a strand of a particular protein.

Richter scale quantitative scale of earthquake magnitude based on the measurement of seismic waves, used to indicate the magnitude of an *earthquake at its epicentre. The Richter scale is logarithmic, so an earthquake of 6.0 is ten times greater than one of 5.0. The magnitude of an earthquake differs from its intensity, measured by the *Mercalli scale, which is qualitative and varies from place to place for the same earthquake. The scale is named after US seismologist Charles Richter (1900–1985).

An earthquake's magnitude is a function of the total amount of energy released, and each point on the Richter scale represents a thirtyfold increase in energy over the previous point. One of the greatest earthquakes ever recorded, in 1920 in Gansu, China, measured 8.6 on the Richter scale.

rich text format in computing, file format usually abbreviated to *RTF.

ricin extremely poisonous extract from the seeds of the castor-oil plant. When incorporated into *monoclonal antibodies, ricin can attack cancer cells, particularly in the treatment of lymphoma and leukaemia.

rickets defective growth of bone in children due to an insufficiency of calcium deposits. The bones, which do not harden adequately, are bent out of shape. It is usually caused by a lack of vitamin D and insufficient exposure to sunlight. Renal rickets, also a condition of malformed bone, is associated with kidney disease.

ridge of high pressure elongated area of high atmospheric pressure extending from an anticyclone. On a synoptic weather chart it is shown as a pattern of lengthened isobars. The weather under a ridge of high pressure is the same as that under an anticyclone.

rift valley valley formed by the subsidence of a block of the Earth's *crust between two or more parallel *faults. Rift valleys are steep-sided and form where the crust is being pulled apart, as at *mid-ocean ridges, or in the Great Rift Valley of East Africa. In cross-section they can appear like a widened gorge with steep sides and a wide floor.

right-angled triangle triangle in which one of the angles is a right angle (90°). It is the basic form of triangle for defining trigonometrical ratios (for example, sine, cosine, and tangent) and for which *Pythagoras' theorem holds true. The longest side of a right-angled triangle is called the hypotenuse; its area is equal to half the product of the lengths of the two shorter sides. Any triangle constructed with its hypotenuse as the diameter of a circle, with its opposite vertex (corner) on the circumference, is a right-angled triangle. This is a fundamental theorem in geometry, first credited to the Greek mathematician Thales about 580 BC.

right-hand rule in physics, a memory aid used to recall the relative directions of motion, magnetic field, and current in an electric generator. It was devised by English physicist John Fleming. See *Fleming's rules.

ring circuit household electrical circuit in which appliances are connected in series to form a ring with each end of the ring connected to the power supply. It superseded the *radial circuit.

ringworm any of various contagious skin infections due to related kinds of fungus, usually resulting in circular, itchy, discoloured patches covered with scales or blisters. The scalp and feet (athlete's foot) are generally involved. Treatment is with antifungal preparations.

RIP acronym for Raster Image Processor, program in a laser printer (or other high-resolution printer) that converts the stream of printing instructions from a computer into the pattern of dots that make up the printed page. A separate program is required for each type of printer and for each page description language (such as PostScript or *PCL).

RiPEM acronym for Riordan's Internet Privacy Enhanced Mail, in computing, *public domain software for sending and receiving e-mail using Pretty Good Privacy (PGP) for improved security. The name refers to Mark Riordan, who wrote most of the program.

ripple tank in physics, shallow water-filled tray used to demonstrate various properties of waves, such as reflection, refraction, diffraction, and interference.

RISC acronym for Reduced Instruction-Set Computer, in computing, a microprocessor (processor on a single chip) that carries out fewer instructions than other (*CISC) microprocessors in common use in the 1990s. Because of the low number and the regularity of *machine code instructions, the processor carries out those instructions very quickly.

river large body of water that flows down a slope along a channel restricted by adjacent banks and levees. A river starts at a point called its **source**, and enters a sea or lake at its **mouth**. Along its length it may be joined by smaller rivers called **tributaries**; a river and its tributaries are contained within a drainage basin. The point at which two rivers join is called the *confluence.

Rivers are formed and moulded over time chiefly by the processes of *erosion, and by the **transport** and

deposition of *sediment. Rivers are able to work on the landscape through erosion, transport, and deposition. The amount of potential energy available to a river is proportional to its initial height above sea level. A river follows the path of least resistance downhill, and deepens, widens and lengthens its channel by erosion. Up to 95% of a river's potential energy is used to overcome friction.

One way of classifying rivers is by their stage of development. An upper course is typified by a narrow V-shaped valley with numerous *waterfalls, *lakes, and rapids. Because of the steep gradient of the topography and the river's height above sea level, the rate of erosion is greater than the rate of deposition, and downcutting occurs by **vertical corrasion** (erosion or abrasion of the bed or bank caused by the load carried by the river).

In the middle course of a river, the topography has been eroded over time and the river's course has a shallow gradient. Such a river is said to be graded. Erosion and deposition are delicately balanced as the river meanders (gently curves back and forth) across the extensive *flood plain. **Horizontal corrasion** is the main process of erosion. The flood plain is an area of periodic flooding along the course of a river valley where fine silty material called *alluvium is deposited by the flood water. Features of a mature river (or the **lower course** of a river) include extensive *meanders, oxbow lakes, and braiding.

Many important flood plains, such as the Nile flood plain in Egypt, occur in arid areas where their exceptional fertility is very important to the local economy. However, using flood plains as the site of towns and villages involves a certain risk, and it is safer to use flood plains for other uses, such as agriculture and parks. Water engineers can predict when flooding is likely and take action to prevent it by studying hydrographs, which show how the discharge of a river varies with time.

river blindness another name for *onchocerciasis, a disease prevalent in some countries of the developing world.

rms in physics, abbreviation for *root-mean-square.

RNA abbreviation for ribonucleic acid, nucleic acid involved in the process of translating the genetic material *DNA into proteins. It is usually single-stranded, unlike the double-stranded DNA, and consists of a large number of nucleotides strung together, each of which comprises the sugar ribose, a phosphate group, and one of four bases (uracil, cytosine, adenine, or guanine). RNA is copied from DNA by the formation of *base pairs, with uracil taking the place of thymine.

RNA occurs in three major forms, each with a different function in the synthesis of protein molecules. **Messenger RNA** (mRNA) acts as the template for protein synthesis. Each *codon (a set of three bases) on the RNA molecule is matched up with the corresponding amino acid, in accordance with the *genetic code. This process (translation) takes place in the ribosomes, which are made up of proteins and **ribosomal RNA** (rRNA). **Transfer RNA** (tRNA) is responsible for combining with specific amino acids, and then matching up a special 'anticodon' sequence of its own with a codon on the mRNA.

Although RNA is normally associated only with the process of protein synthesis, it makes up the hereditary material itself in some viruses, such as *retroviruses.

Roaring Forties nautical expression for regions south of latitude 40°S, in the southern oceans, where strong westerly winds prevail.

rock constituent of the Earth's crust composed of *minerals or materials of organic origin that have consolidated into hard masses. There are three basic types of rock: *igneous, *sedimentary, and *metamorphic. Rocks are composed of a combination (or aggregate) of minerals, and the property of a rock will depend on its components. Where deposits of economically valuable minerals occur they are termed *ores. As a result of *weathering, rock breaks down into very small particles that combine with organic materials from plants and animals to form *soil. In *geology the

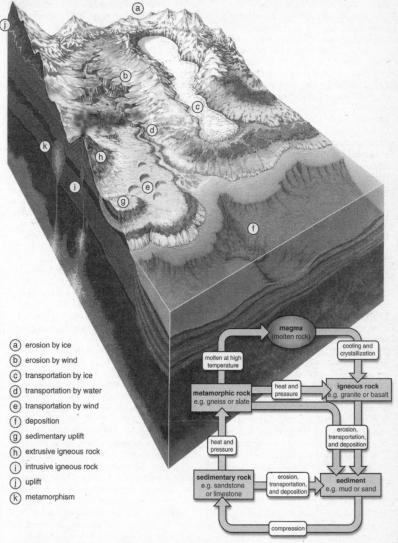

(a) erosion by ice
(b) erosion by wind
(c) transportation by ice
(d) transportation by water
(e) transportation by wind
(f) deposition
(g) sedimentary uplift
(h) extrusive igneous rock
(i) intrusive igneous rock
(j) uplift
(k) metamorphism

rock The rock cycle. Rocks are not as permanent as they seem but are being constantly destroyed and renewed. When a rock becomes exposed on the Earth's surface, it starts to break down through weathering and erosion. The resulting debris is washed or blown away and deposited, for example in sea or river beds, or in deserts, where it eventually becomes buried by yet more debris. Over time, this debris is compressed and compacted to form sedimentary rock, which may in time become exposed and eroded once more. Alternatively the sedimentary rock may be pushed further towards the Earth's centre where it melts and solidifies to form igneous rock or is heated and crushed to such a degree that its mineral content alters and it becomes metamorphic rock. Igneous and metamorphic rock may also become exposed and eroded by the same processes as sedimentary rock, and the cycle continues.

term 'rock' can also include unconsolidated materials such as *sand, mud, *clay, and *peat.

The change of one type of rock to another is called the rock cycle.

rocket projectile driven by the reaction of gases produced by a fast-burning fuel. Unlike jet engines, which are also reaction engines, rockets carry their own oxygen supply to burn their fuel and do not require any surrounding atmosphere. For warfare, rocket heads carry an explosive device.

Rockets have been valued as fireworks since the middle ages, but their intensive development as a means of propulsion to high altitudes, carrying payloads, started only in the interwar years with the state-supported work in Germany (primarily by German-born US rocket engineer Wernher von Braun) and the work of US rocket pioneer Robert Hutchings Goddard. Being the only form of propulsion available that can function in a vacuum, rockets are essential to exploration in outer space. Multistage rockets have to be used, consisting of a number of rockets joined together.

Two main kinds of rocket are used: one burns liquid propellants, the other solid propellants. The fireworks rocket uses gunpowder as a solid propellant. The *space shuttle's solid rocket boosters use a mixture of powdered aluminium in a synthetic rubber binder. Most rockets, however, have liquid propellants, which are more powerful and easier to control. Liquid hydrogen and kerosene are common fuels, while liquid oxygen is the most common oxygen provider, or oxidizer. One of the biggest rockets ever built, the Saturn V Moon rocket, was a three-stage design, standing 111 m/365 ft high. It weighed more than 2,700 tonnes on the launch pad, developed a take-off thrust of some 3.4 million kg/7.5 million lb, and could place almost 140 tonnes into low Earth orbit. In the early 1990s, the most powerful rocket system was the Soviet Energiya, capable of placing 190 metric tonnes into low Earth orbit. The US space shuttle can only carry up to 29 metric tonnes of equipment into orbit.

rod type of light-sensitive cell in the *retina of most vertebrates. Rods are highly sensitive and provide only black and white vision. They are used when lighting conditions are poor and are the only type of visual cell found in animals active at night.

roentgen or **röntgen**; symbol R, unit of radiation exposure, used for X-rays and gamma rays. It is defined in terms of the number of ions produced in one cubic centimetre of air by the radiation. Exposure to 1,000 roentgens gives rise to an absorbed dose of about 870 rads (8.7 grays), which is a dose equivalent of 870 rems (8.7 sieverts).

ROM acronym for **read-only memory**, in computing, a memory device in the form of a collection of integrated circuits (chips), frequently used in microcomputers. ROM chips are loaded with data and programs during manufacture and, unlike *RAM (random-access memory) chips, can subsequently only be read, not written to, by a computer. However, the contents of the chips are not lost when the power is switched off, as happens in RAM.

Roman numerals ancient European number system using symbols different from Arabic numerals (the ordinary numbers 1, 2, 3, 4, 5, and so on). The seven key symbols in Roman numerals, as represented today, are I (1), V (5), X (10), L (50), C (100), D (500), and M (1,000). There is no zero, and therefore no place-value as is fundamental to the Arabic system. The first ten Roman numerals are I, II, III, IV (or IIII), V, VI, VII, VIII, IX, and X. When a Roman symbol is preceded by a symbol of equal or greater value, the values of the symbols are added (XVI = 16). When a symbol is preceded by a symbol of less value, the values are subtracted (XL = 40). A horizontal bar over a symbol indicates a multiple of 1,000 Although addition and subtraction are fairly straightforward using Roman numerals, the absence of a zero makes other arithmetic calculations (such as multiplication) clumsy and difficult.

root in computing, the account used by system administrators and other superusers in *Unix systems. Users logged in as root (or in some systems, avatar) have permission to access and change all the files in the system.

root the part of a plant that is usually underground, and whose primary functions are anchorage and the absorption of water and dissolved mineral salts. Roots usually grow downwards and towards water (that is, they are positively geotropic and hydrotropic; see *tropism). Plants such as epiphytic orchids, which grow above ground, produce aerial roots that absorb moisture from the atmosphere. Others, such as ivy, have climbing roots arising from the stems, which serve to attach the plant to trees and walls.

root in mathematics, a number which when multiplied by itself will equal a given number (the inverse of an *exponent or power. The root of an equation is a value that satisfies the equality. For example, $x = 0$ and $x = 5$ are roots of the equation $x^2 - 5x = 0$.

root cap cap at the tip of a growing plant root. It gives protection to the zone of actively dividing cells as the root pushes through the soil.

root directory in computing, the top directory in a *tree-and-branch filing system. It contains all the other directories.

root hair tiny, hair-like outgrowth of some surface cells of plant roots that greatly increases the area available for the absorption of water and other materials such as *minerals. It is a delicate structure, which survives for a few days only and does not develop into a root. Root hair cells are found just behind the root tip where they are continually being formed. They grow out into the soil for a few millimetres to produce the root hair. Each root hair is made of a single cell. At the root tip *mitosis cell division is very active making new cells for growth to replace the root hair cells that die.

root mean square **rms,** value obtained by taking the *square root of the mean

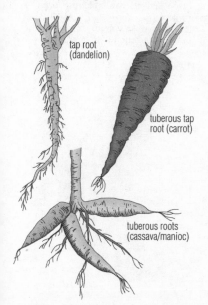

tap root (dandelion)

tuberous tap root (carrot)

tuberous roots (cassava/manioc)

fibrous roots (grass)

prop roots (maize)

aerial roots (epiphyte, e.g. orchid)

root Types of root. Many flowers (dandelion) and vegetables (carrot) have swollen tap roots with smaller lateral roots. The tuberous roots of the cassava are swollen parts of an underground stem modified to store food. The fibrous roots of the grasses are all of equal size. Prop roots grow out from the stem and then grow down into the ground to support a heavy plant. Aerial roots grow from stems but do not grow into the ground; many absorb moisture from the air.

of the squares of a set of values; for example the rms value of four quantities a, b, c, and d is

$$\sqrt{[(a^2 + b^2 + c^2 + d^2)/4]}.$$

It provides a measure of the effective magnitude of a cyclic or randomly varying quantity whose average value is zero. For a cyclic quantity, it is equal to the square root of the average of the squares of the quantity at each instant over a complete cycle; where the variation is a *sine wave, the rms value is equal to the peak value divided by $\sqrt{2}$. In electrical engineering, for example, the rms value of the current or voltage in an AC circuit is used in power calculations. The rms value is also used as a measure of the amplitude of randomly varying quantities, such as the noise in a telecommunications channel.

root nodule   clearly visible swelling that develops in the roots of members of the bean family, the Leguminosae. The cells inside this tumourous growth have been invaded by the bacteria Rhizobium, a soil microbe capable of converting gaseous nitrogen into nitrate. The nodule is therefore an association between a plant and a bacterium, with both partners benefiting. The plant obtains nitrogen compounds while the bacterium obtains nutrition and shelter.

Rorschach test   in psychology, a method of diagnosis involving the use of inkblot patterns that subjects are asked to interpret, to help indicate personality type, degree of intelligence, and emotional stability. It was invented by the Swiss psychiatrist Hermann Rorschach.

rotation   in geometry, a *transformation in which a figure is turned about a given point, known as the **centre of rotation**. A rotation of 180° is known as a half turn. Three things are needed to describe a rotation: the **angle of rotation**, the **direction of rotation** (clockwise or anticlockwise), and the **centre of rotation**. When a shape is rotated but appears not to have moved it is said to have **rotational symmetry**. *Translation, *reflection, and *enlargement can also transform shapes.

rotation   in astronomy, movement of a planet rotating about its own axis. For

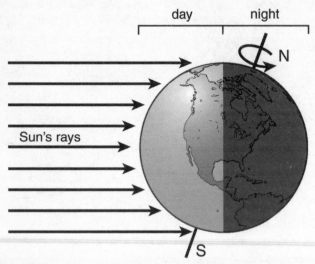

rotation The Earth rotates in an anticlockwise direction (looking along its axis from north to south) and takes approximately 24 hours to rotate on its own axis. Day and night depend on the Earth's rotation; the half of the Earth facing the Sun is in daylight and the half facing away from the Sun is in darkness.

the *Earth, one complete rotation takes 23 hours and 56 minutes. The Earth rotates in an anticlockwise direction (as viewed looking along its axis from north to south), leading to the appearance from Earth of the Sun moving from east to west in the daily cycle. The rotation of the Earth produces a surface speed at the Equator of about 1,600 kph/1,000 mph. This speed decreases further north and south of the Equator. Artificial *satellites use the Earth's natural rotation to orbit the Earth.

As the Earth's axis is tilted with respect to the Sun, the lengths of day and night vary across the globe. At the *Equator, day and night are both 12 hours long. During the northern-hemisphere winter, days become shorter with increasing latitude north, until a point where there is continuous night. At the same time, during the southern-hemisphere summer, days become longer with increasing latitude south, until a point where there is continuous daylight. The situation is reversed during the northern-hemisphere summer.

rotavirus in medicine, one of a group of viruses that infect the small intestine and are a common cause of gastroenteritis in young children.

roughage alternative term for dietary fibre, material of plant origin that cannot be digested by enzymes normally present in the human *gut.

rounding in mathematics, a process by which a number is approximated to the nearest above or below with fewer *decimal places. For example, 34.3583 might be rounded to 3 decimal places – 34.358 (the 3 is below 5), whereas 34.3587 would be rounded to 34.359 (the 7 is above 5). Similarly, 3,587 might be rounded to the nearest thousand (or to 1 *significant figure), giving 4,000. When unwanted decimals are simply left out, the process is known as **truncating**.

roundworm parasitic *nematode worm, genus *Toxocara*, found in dogs, cats, and other animals.

router in computing, a device that pushes traffic through a packet-switched network. On the Internet, traffic travels through a series of routers that relay each **packet** of data to its destination by the best possible route.

row in mathematics, a *matrix, a horizontal line of numbers. Matrices are made up of rows and columns. A matrix that consists of one row only is called a **row matrix**, while a matrix with one column only is called a **column matrix**.

RS-232 interface standard type of computer *interface used to connect computers to serial devices. It is used for modems, mice, screens, and serial printers.

RSA in computing, name of both an encryption *algorithm and the company (RSA Inc) set up to exploit that algorithm in commercial encryption products. First described 1977 by its inventors, Ronald Rivest, Adi Shamir, and Leonard Adelman, and published in *Scientific American*, the RSA algorithm is used in the free program Pretty Good Privacy (PGP) and in commercial products released by RSA Laboratories.

RSI abbreviation for *repetitive strain injury, a condition that can affect people who repeatedly perform certain movements with their hands and wrists for long periods of time, such as typists, musicians, or players of computer games.

RTF abbreviation for rich text format, in computing, file format designed to facilitate the exchange of documents between different word-processing programs. RTF text files make it possible to transfer formatting such as font styles or paragraph indents from one program to another. A further advantage is that Microsoft Word files sent as RTF cannot carry macro viruses.

rubber coagulated latex of a variety of plants, mainly from the New World. Most important is Para rubber, which comes from the tree *Hevea brasiliensis*, belonging to the spurge family. It was introduced from Brazil to Southeast Asia, where most of the world supply is now produced, the chief exporters being Peninsular Malaysia, Indonesia, Sri Lanka, Cambodia, Thailand, Sarawak, and Brunei. At about seven years the tree, which may grow to

20 m/60 ft, is ready for tapping. Small cuts are made in the trunk and the latex drips into collecting cups. In pure form, rubber is white and has the formula $(C_5H_8)_n$.

Early uses of rubber were limited by its tendency to soften on hot days and harden on colder ones, a tendency that was eliminated by Charles Goodyear's invention of *vulcanization in 1839. In the 20th century, world production of rubber increased a hundredfold, and World War II stimulated the production of synthetic rubber to replace the supplies from Malaysian sources overrun by the Japanese. There are an infinite variety of synthetic rubbers adapted to special purposes, but economically foremost is SBR (styrene-butadiene rubber). Cheaper than natural rubber, it is preferable for some purposes, for example in car tyres, where its higher abrasion resistance is useful, and it is either blended with natural rubber or used alone for industrial moulding and extrusions, shoe soles, hoses, and latex foam.

rubella   technical term for *German measles.

rubidium   chemical symbol Rb, (Latin *rubidus* 'red') soft, silver-white, metallic element, atomic number 37, relative atomic mass 85.47. It is one of the *alkali metals in Group 1 of the *periodic table. Reactivity of the alkali metals increases down the group and so rubidium is more reactive than lithium, sodium, and potassium. Rubidium ignites spontaneously in air and reacts violently with water. It is used in photocells and vacuum-tube filaments. Rubidium was discovered spectroscopically by German physicists Robert Bunsen and Gustav Kirchhoff (1824–1887) in 1861 and named after the red lines in its spectrum.

ruby   red transparent gem variety of the mineral corundum Al_2O_3, aluminium oxide. Small amounts of chromium oxide, Cr_2O_3, substituting for aluminium oxide, give ruby its colour. Natural rubies are found mainly in Myanmar (Burma), but rubies can also be produced artificially and such synthetic stones are used in *lasers.

ruminant   any even-toed hoofed mammal with a rumen, the 'first stomach' of its complex digestive system. Plant food is stored and fermented before being brought back to the mouth for chewing (chewing the cud) and then is swallowed to the next stomach. Ruminants include cattle, antelopes, goats, deer, and giraffes, all with a four-chambered stomach. Camels are also ruminants, but they have a three-chambered stomach.

run-time version   in computing, copy of a program that is provided with another application, so that the latter can be run, although it does not provide the full functionality of the program. An example is the provision of run-time versions of Microsoft Windows with Windows applications for those users who do not have the full version of Windows.

rust   reddish-brown oxide of iron formed by the action of moisture and oxygen on the metal. It consists mainly of hydrated iron(III) oxide ($Fe_2O_3.H_2O$) and iron(III) hydroxide ($Fe(OH)_3$). Rusting is the commonest form of *corrosion.

   rust prevention There are two main approaches to protect against rusting. **Barrier methods** introduce a barrier

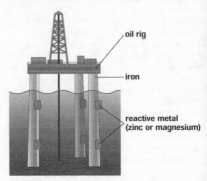

oil rig

iron

reactive metal (zinc or magnesium)

rust Sacrificial metal is used to protect pipes and other exposed metal on oil rigs from corrosion by rust. A more reactive metal than iron, such as zinc or magnesium, is attached in large lumps by conducting wires, at intervals along the pipes. Electrons released during the reaction travel along the conducting wires to the pipes, preventing rust from occurring. It is cheaper to replace the sacrificial metal than it is to replace the pipes.

between the metal and the air and moisture to minimize the reaction. This is the commonest method of rust prevention and the barrier may consist of a layer of grease, paint, plastic, or an unreactive metal, such as tin, copper, or chromium.

In **sacrificial protection**, the iron is actually covered by a more reactive metal, such a zinc (galvanization), or connected to a more reactive metal, such as magnesium, so that as long as the 'sacrificial' metal is present it will corrode first.

ruthenium chemical symbol Ru, hard, brittle, silver-white, metallic element, atomic number 44, relative atomic mass 101.07. It is one of the so-called platinum group of metals; it occurs in platinum ores as a free metal and in the natural alloy osmiridium. It is used as a hardener in alloys and as a catalyst; its compounds are used as colouring agents in glass and ceramics.

rutherfordium chemical symbol Rf, synthesized, radioactive, metallic element. It is the first of the *transactinide series, atomic number 104, relative atomic mass 262. It is produced by bombarding californium with carbon nuclei and has ten isotopes, the longest-lived of which, Rf-262, has a half-life of 70 seconds. Two institutions claim to be the first to have synthesized it: the Joint Institute for Nuclear Research in Dubna, Russia, in 1964; and the University of California at Berkeley, USA, in 1969. Named after New Zealand-born British physicist Ernest Rutherford (1871–1937).

Rydberg constant in physics, a constant that relates atomic spectra to the *spectrum of hydrogen. Its value is 1.0977×10^7 per metre.

S in physics, symbol for *siemens, the SI unit of electrical conductance, equal to a conductance of 1 ohm^{-1}.

S abbreviation for **south**.

SAA abbreviation for *systems application architecture.

saccharin or **ortho-sulpho benzimide**; $C_7H_5NO_3S$, sweet, white, crystalline solid derived from coal tar and substituted for sugar. Since 1977 it has been regarded as potentially carcinogenic. Its use is not universally permitted and it has been largely replaced by other sweetening agents.

sacrum compound triangular bone at the base of the human spine, formed of five fused vertebrae between the last lumbar vertebra above and the coccyx below. The sacrum forms the back of the *pelvis.

safety glass glass that does not splinter into sharp pieces when smashed. **Toughened glass** is made by heating a glass sheet and then rapidly cooling it with a blast of cold air; it shatters into rounded pieces when smashed. **Laminated glass** is pressed and heated; when struck, it simply cracks.

safety lamp portable lamp designed for use in places where flammable gases such as methane may be encountered; for example, in coal mines. The electric head lamp used as a miner's working light has the bulb and contacts in protected enclosures. The flame safety lamp, now used primarily for gas detection, has the wick enclosed within a strong glass cylinder surmounted by wire gauzes. English chemist Humphrey Davy (1815) and English engineer George Stephenson each invented flame safety lamps.

St Elmo's fire bluish, flamelike electrical discharge that sometimes occurs above ships' masts and other pointed objects or about aircraft in stormy weather. Although of high voltage, it is of low current and therefore harmless. St Elmo (or St Erasmus) is the patron saint of sailors.

sal ammoniac former name for *ammonium chloride.

salicylic acid HOC_6H_4COOH, active chemical constituent of aspirin, an analgesic drug. The acid and its salts (salicylates) occur naturally in many plants; concentrated sources include willow bark and oil of wintergreen. When purified, salicylic acid is a white solid that crystallizes into prismatic needles at 159°C/318°F. It is used as an antiseptic, in food preparation and dyestuffs, and in the preparation of aspirin.

salinization accumulation of salt in water or soil; it is a factor in *desertification.

saliva in vertebrates, an alkaline secretion from the salivary glands that aids the swallowing and digestion of food in the mouth. In mammals, it contains the enzyme amylase, which converts starch to sugar. The salivary glands of mosquitoes and other blood-sucking insects produce *anticoagulants.

salivary gland or **parotid gland**, in mammals, one of two glands situated near the mouth responsible for the manufacture of saliva and its secretion into the mouth. The salivary glands are stimulated to produce saliva during a meal. Saliva contains an enzyme, ptyalin, and mucous which are essential for the mastication and initial digestion of food.

salmonella any of a very varied group of bacteria, genus *Salmonella*, that colonize the intestines of humans and some animals. Some strains cause typhoid and paratyphoid fevers, while others cause salmonella *food poisoning, which is characterized by stomach pains, vomiting, diarrhoea, and headache. It can be fatal in elderly people, but others usually recover in a few days without antibiotics. Most cases are caused by contaminated animal products, especially poultry meat.

salt in chemistry, any compound formed from an acid and a base through the replacement of all or part of the hydrogen in the acid by a metal or electropositive radical. **Common**

*****salt** is sodium chloride. A salt may be produced by a chemical reaction between an acid and a base, or by the displacement of hydrogen from an acid by a metal (see *displacement reaction). As a solid, the ions normally adopt a regular arrangement to form crystals. Some salts form only stable crystals as hydrates (when combined with water). Most inorganic salts readily dissolve in water to give an electrolyte (a solution that conducts electricity).

saltation in earth science, bouncing of rock particles along a river bed. It is the means by which bedload (material that is too heavy to be carried in suspension) is transported downstream. The term is also used to describe the movement of sand particles bounced along by the wind.

salt, common or **sodium chloride**; NaCl, white crystalline solid, found dissolved in seawater and as rock salt (the mineral halite) in large deposits and salt domes. Common salt is used extensively in the food industry as a preservative and for flavouring, and in the chemical industry in the making of chlorine and sodium. While common salt is an essential part of our diet, some medical experts believe that excess salt can lead to high blood pressure and increased risk of heart attacks and strokes. A high level of salt in the diet presents the greatest risk to those who are overweight, according to a 1999 US study. Salt has historically been considered a sustaining substance, often taking on religious significance in ancient cultures. Roman soldiers were paid part of their wages as salt allowance (Latin *salerium argentinium*), hence the term 'salary'.

salt marsh wetland with halophytic vegetation (tolerant to seawater). Salt marshes develop around *estuaries and on the sheltered side of sand and shingle spits. They are formed by the deposition of mud around salt-tolerant vegetation. This vegetation must tolerate being covered by seawater as well as being exposed to the air. It also traps mud as the tide comes in and out. This helps build up the salt marsh. Salt marshes usually have a network of creeks and drainage channels by which tidal waters enter and leave the marsh.

saltpetre or **potassium nitrate**; KNO_3, compound used in making gunpowder (from about 1500). It occurs naturally, being deposited during dry periods in places with warm climates, such as India.

sal volatile another name for *ammonium carbonate.

samarium chemical symbol Sm, hard, brittle, grey-white, metallic element of the *lanthanide series, atomic number 62, relative atomic mass 150.36. It is widely distributed in nature and is obtained commercially from the minerals monzanite and bastnaesite. It is used only occasionally in industry, mainly as a catalyst in organic reactions. Samarium was discovered by spectroscopic analysis of the mineral samarskite and named in 1879 by French chemist Paul Lecoq de Boisbaudran (1838–1912) after its source.

sampling measurement of an *analogue signal (such as an audio or video signal) at regular intervals. The result of the measurement can be converted into a digital signal that can be electronically enhanced, edited, or processed.

sand loose grains of rock, 0.0625–2.00 mm /0.0025–0.08 in in diameter, consisting most commonly of *quartz, but owing their varying colour to mixtures of other minerals. Sand is used in cement-making, as an abrasive, in glass-making, and for other purposes. Sands are classified into marine, freshwater, glacial, and terrestrial. Some 'light' soils contain up to 50% sand. Sands may eventually consolidate into *sandstone.

sandbar ridge of sand built up by the currents across the mouth of a river or bay. A sandbar may be entirely underwater or it may form an elongated island that breaks the surface. A sandbar stretching out from a headland is a **sand spit**.

sandstone *sedimentary rocks formed from the consolidation of sand, with sand-sized grains (0.0625–2 mm/ 0.0025–0.08 in) in a matrix or cement. Their principal component is quartz. Sandstones are commonly permeable and porous, and may form freshwater aquifers. They are mainly used as building materials. Sandstones are classified according to the matrix or cement material (whether derived from clay or silt).

Santa Ana periodic warm Californian *wind.

sap fluids that circulate through *vascular plants, especially woody ones. Sap carries water and food to plant tissues. Sap contains alkaloids, protein, and starch; it can be milky (as in rubber trees), resinous (as in pines), or syrupy (as in maples).

saponification in chemistry, the *hydrolysis (splitting) of an *ester by treatment with a strong alkali, resulting in the liberation of the alcohol from which the ester had been derived and a salt of the constituent fatty acid. The process is used in the manufacture of soap.

sapphire deep-blue, transparent gem variety of the mineral corundum Al_2O_3, aluminium oxide. Small amounts of iron and titanium give it its colour. A corundum gem of any colour except red (which is a ruby) can be called a sapphire; for example, yellow sapphire.

saprophyte in botany, an obsolete term for a *saprotroph, an organism that lives in dead or decaying matter.

saprotroph formerly **saprophyte**, organism that feeds on the excrement or the dead bodies or tissues of others. They include most fungi (the rest being parasites); many bacteria and protozoa; animals such as dung beetles and vultures; and a few unusual plants, including several orchids. Saprotrophs cannot make food for themselves, so they are a type of *heterotroph. They are useful scavengers, and in sewage farms and refuse dumps break down organic matter into nutrients easily assimilable by green plants.

sarcoma malignant *tumour arising from the fat, muscles, bones, cartilage, or blood and lymph vessels and connective tissues. Sarcomas are much less common than *carcinomas.

sarin poison gas 20 times more lethal to humans than potassium cyanide. It impairs the central nervous system, blocking the action of an enzyme that removes acetylcholine, the chemical that transmits signals. Sarin was developed in Germany during World

War II. Sarin was used in 1995 in a terrorist attack on the Tokyo underground by a Japanese sect. It is estimated that the USA had a stockpile of 15,000 tonnes of sarin, and more than 1,000 US rockets with sarin warheads were found to be leaking in 1995. There are no known safe disposal methods.

satellite any small body that orbits a larger one. **Natural satellites** that orbit planets are called **moons**. The first **artificial satellite**, Sputnik 1, was launched into orbit around the Earth by the USSR in 1957. Artificial satellites can transmit data from one place on Earth to another, or from space to Earth. *Satellite applications include science, communications, weather forecasting, and military use.

satellite applications uses to which artificial satellites are put. These include:

scientific experiments and observation Many astronomical observations are best taken above the disturbing effect of the atmosphere. Satellite observations have been carried out by the Infrared Astronomy Satellite (1983) which made a complete infrared survey of the skies, and Solar Max (1980), which observed solar flares. The Hipparcos satellite, launched in 1989, measured the positions of many stars. The Röntgen Satellite, launched in 1990, examined ultraviolet and X-ray radiation. In 1992, the Cosmic Background Explorer satellite detected details of the Big Bang that mark the first stage in the formation of galaxies. Medical experiments have been carried out aboard crewed satellites, such as the Russian space station *Mir* and the US *Skylab*.

reconnaissance, land resource, and mapping applications Apart from military use and routine mapmaking, the US Landsat, the French Satellite Pour l'Observation de la Terre, and equivalent Russian satellites have provided much useful information about water sources and drainage, vegetation, land use, geological structures, oil and mineral locations, and snow and ice.

weather monitoring The US National Oceanic and Atmospheric Administration series of satellites, and others launched by the European Space Agency, Japan, and India, provide continuous worldwide observation of the atmosphere.

navigation The US Global Positioning System (GPS) uses 24 Navstar satellites that enable users (including walkers and motorists) to find their position to within 100 m/328 ft. The US military can make full use of the system, obtaining accuracy to within 1.5 m/4 ft 6 in. The Transit system, launched in the 1960s, with 12 satellites in orbit, locates users to within 100 m/328 ft.

communications A complete worldwide communications network is now provided by satellites such as the US-run Intelsat system.

satellite image image of the Earth or any other planet obtained from instruments on a satellite. Satellite images can provide a variety of information, including vegetation patterns, sea surface temperature, weather, and geology.

Landsat 4, launched in 1982, orbits at 705 km/438 mi above the Earth's surface. It completes nearly 15 orbits per day, and can survey the entire globe in 16 days. The instruments on Landsat scan the planet's surface and record the brightness of reflected light. The data is transmitted back to Earth and translated into a satellite image.

saturated compound organic compound, such as propane, that contains only single covalent bonds. *Alkanes are saturated *hydrocarbons that contain only carbon to carbon single bonds. Saturated organic compounds can only undergo further reaction by *substitution reactions, as in the production of chloropropane from propane.

saturated fatty acid *fatty acid in which there are no double bonds in the hydrocarbon chain.

saturated solution in physics and chemistry, a solution obtained when a solvent (liquid) can dissolve no more of a solute (usually a solid) at a particular temperature. Normally, a slight fall in temperature causes some of the solute to crystallize out of solution. If this does not happen the phenomenon is called supercooling, and the solution is said to be **supersaturated**.

savannah or **savanna**, extensive open tropical grasslands, with scattered trees and shrubs. Savannahs cover large areas of Africa, North and South America, and northern Australia. The soil is acidic and sandy and generally considered suitable only as pasture for low-density grazing. A new strain of rice suitable for savannah conditions was developed in 1992. It not only grew successfully under test conditions in Colombia but also improved pasture quality so that grazing numbers could be increased twentyfold.

scabies contagious infection of the skin caused by the parasitic itch mite *Sarcoptes scabiei*, which burrows under the skin to deposit eggs. Treatment is by antiparasitic creams and lotions.

scalability ability of a software or hardware system or a network to grow without breaking down or requiring an expensive redesign. Scalable systems may thus be tested and perfected at a modest size and then expanded to meet future needs.

scalar quantity in mathematics and science, a quantity that has magnitude but no direction, as distinct from a *vector quantity, which has a direction as well as a magnitude. Temperature, mass, volume, and *speed are scalar quantities.

scale or **limescale**, in chemistry, *calcium carbonate precipitates that form on the inside of a kettle or boiler as a result of boiling *hard water (water containing concentrations of soluble calcium and magnesium salts). The salts present in hard water also precipitate out by reacting with soap molecules. Scale may cause damage to water pipes and appliances such as washing machines and water boilers. The build-up of scale causes blockage of pipes, and electrical appliances have to use more electrical energy to operate efficiently.

scandium chemical symbol Sc, silver-white, metallic element of the *lanthanide series, atomic number 21, relative atomic mass 44.956. Its compounds are found widely distributed in nature, but only in minute amounts. The metal has little industrial importance.

Scandium oxide (scandia) is used as a catalyst, in making crucibles and other ceramic parts, and scandium sulphate (in very dilute aqueous solution) is used in agriculture to improve seed germination. Scandium is relatively more abundant in the Sun and other stars than on Earth.

scanning electron microscope SEM, electron microscope that produces three-dimensional images, magnified 10–200,000 times. A fine beam of electrons, focused by electromagnets, is moved, or scanned, across the specimen. Secondary radiation reflected from the specimen is collected by a detector, giving rise to an electrical signal, which is then used to generate a point of brightness on a television-like screen. As the point moves rapidly over the screen, in phase with the scanning electron beam, an image of the specimen is built up.

scanning tunnelling microscope STM, microscope that produces a magnified image by moving a tiny tungsten probe across the surface of the specimen. The tip of the probe is so fine that it may consist of a single atom, and it moves so close to the specimen surface that electrons jump (or tunnel) across the gap between the tip and the surface.

scarlet fever or **scarlatina**, acute infectious disease, especially of children, caused by the bacteria in the *Streptococcus pyogenes* group. It is marked by fever, vomiting, sore throat, and a bright red rash spreading from the upper to the lower part of the body. The rash is followed by the skin peeling in flakes. It is treated with antibiotics.

Scart socket acronym for *Syndicat des Constructeurs des Appareils Radiorécepteurs et Téléviseurs*, 'syndicate of radio receiver and television equipment manufacturers', 21-pin audio/video connector used in consumer electronics equipment such as television sets and video recorders. Scart was defined by a group of French television manufacturers, and Scart cables are often the preferred method for connecting video games consoles to TV sets.

scatter diagram or **scattergram**, diagram whose purpose is to establish whether or not a relationship or *correlation exists between two variables; for example, between life expectancy and gross national product. Each observation is marked with a dot in a position that shows the value of both variables. The pattern of dots is then examined to see whether they show any underlying trend by means of a *line of best fit (a straight line drawn so that its distance from the various points is as short as possible).

scattering in physics, the random deviation or reflection of a stream of particles or of a beam of radiation such as light, by the particles in the matter through which it passes.

schistosomiasis another name for *bilharzia.

schizophrenia mental disorder, a psychosis of unknown origin, which can lead to profound changes in personality, behaviour, and perception, including delusions and hallucinations. It is more common in males and the early-onset form is more severe than when the illness develops in later life. Modern treatment approaches include drugs, family therapy, stress reduction, and rehabilitation.

sciatica persistent pain in the back and down the outside of one leg, along the sciatic nerve and its branches. Causes of sciatica include inflammation of the nerve or pressure of a displaced disc on a nerve root leading out of the lower spine.

scientific law principles that are taken to be universally applicable.

scientific method belief that experimentation and observation, properly understood and applied, can avoid the influence of cultural and social values and so build up a picture of a reality independent of the observer. Techniques and mechanical devices that improve the reliability of measurements may seem to support this theory; but the realization that observations of subatomic particles influence their behaviour has undermined the view that objectivity is possible in science (see *uncertainty principle).

scientific notation alternative term for *standard form.

scintillation counter instrument for measuring very low levels of radiation. The radiation strikes a scintillator (a device that emits a unit of light when a charged elementary particle collides with it), whose light output is 'amplified' by a *photomultiplier; the current pulses of its output are in turn counted or added by a scaler to give a numerical reading.

sclerosis any abnormal hardening of body tissues, especially the nervous system or walls of the arteries. See *multiple sclerosis and *atherosclerosis.

screen in computing, another name for *monitor.

screen saver in computing, program designed to prevent a static image from 'burning' itself into the phosphor screen of an idle computer monitor. If the user leaves the computer alone for more than a few minutes, the screen saver automatically displays a moving or changing image – perhaps a sequence of random squiggles, or an animation – on the screen. When the user touches any key, the computer returns to its previous state. Screen savers can be password protected, so that unauthorized users cannot access the machine in the operator's absence. In Windows 98 and above, screen savers can be part of a desktop theme.

script in communications, a series of instructions for a computer. For example, when users log on to an ISP (Internet Service Provider) or other service, their computers follow a script containing passwords and other information to tell the ISP's server who they are. Scripts may be used in office productivity software (such as word processors and spreadsheets), and in Web pages to perform automatic activities.

scripting language simple programming language used to issue a set of commands (a script), often to control a particular software application. A popular scripting languages commonly used on the Internet is *JavaScript.

scrollbar in computing, a narrow box along two sides of a *window, enabling users to move its contents up, down, left or right. Each end of the scrollbar represents the same end of the document on display, and the *mouse is used to move a small 'scrollbox' up and down the bar to scroll the contents or to click on the directional arrows.

scrotum paired sac in male mammals enclosing the testes and the epididymes (coiled tubes in which the sperm mature).

SCSI acronym for Small Computer System Interface; pronounced 'scuzzy', in computing, one of the methods for connecting peripheral devices (such as printers, scanners, and CD-ROM drives) to a computer. A group of peripherals linked in series to a single SCSI *port is called a **daisy-chain**.

scurvy disease caused by deficiency of vitamin C (ascorbic acid), which is contained in fresh vegetables and fruit. The signs are weakness and aching joints and muscles, progressing to bleeding of the gums and other spontaneous haemorrhage, and drying-up of the skin and hair. It is reversed by giving the vitamin.

SDK abbreviation for software development kit, in computing, suite of programs supplied to software developers to help them develop applications for environments such as Microsoft Windows and Microsoft Office.

SDRAM abbreviation for **synchronous dynamic random-access memory**, in computing, the latest generation of *DRAM memory modules, rapidly replacing *EDO-RAM technology. SDRAM *memory chips are able to transfer data in a single clock cycle, because their operations are synchronized with the system clock.

seaborgium chemical symbol Sg, synthesized radioactive element of the *transactinide series, atomic number 106, relative atomic mass 266. It was first synthesized in 1974 in the USA and given the temporary name unnilhexium. The discovery was not confirmed until 1993. It was officially named in 1997 after US nuclear chemist Glenn Seaborg. The University of California, Berkeley, bombarded californium with oxygen nuclei to get isotope 263; the Joint Institute for Nuclear Research, Dubna, Russia, bombarded lead with chromium nuclei to obtain isotopes 259 and 260.

sea breeze gentle coastal wind blowing off the sea towards the land. It is most noticeable in summer when the warm land surface heats the air above it and causes it to rise. Cooler air from the sea is drawn in to replace the rising air, so causing an onshore breeze. At night and in winter, air may move in the opposite direction, forming a land breeze.

sea level average height of the surface of the oceans and seas measured throughout the tidal cycle at hourly intervals and computed over a 19-year period. It is used as a datum plane from which elevations and depths are measured. Factors affecting sea level include temperature of seawater (warm water is less dense and therefore takes up a greater volume than cool water) and the topography of the ocean floor.

search engine in computing, remotely accessible program to help users find information on the Internet. Commercial search engines such as Google and Lycos comprise databases of documents, *URLs, *Usenet articles, and more, which can be searched by keying in a key word or phrase. The databases are compiled by a mixture of automated agents (*spiders) and webmasters registering their sites.

search for extraterrestrial intelligence full name of *SETI.

season period of the year having a characteristic climate. The change in seasons is mainly due to the Earth's axis being tilted in relation to the Sun, and hence the position of the Sun in the sky at a particular place changes as the Earth orbits the Sun.

When the northern hemisphere is tilted away from the Sun (winter) the Sun's rays have further to travel through the atmosphere (they strike the Earth at a shallower angle) and so have less heating effect, resulting in colder weather. The days are shorter and the nights are longer. At the same time, the southern hemisphere is tilted towards the Sun (summer) and experiences warmer weather, with longer days and shorter nights. The opposite occurs when the northern hemisphere is tilted towards the Sun and the southern hemisphere away from the Sun.

seasonal affective disorder SAD, form of depression that occurs in winter and is relieved by the coming of spring. Its incidence decreases closer to the Equator. One type of SAD is associated with increased sleeping and appetite.

sebaceous gland gland in the skin producing waxlike sebum. It opens into a hair follicle. Distribution of sebaceous glands varies in different parts of the body.

sec or **s**, abbreviation for **second**, a unit of time.

secant in trigonometry, the function of a given angle in a right-angled triangle, obtained by dividing the length of the hypotenuse (the longest side) by the length of the side adjacent to the angle. It is the *reciprocal of the *cosine (sec = $1/\cos$).

second symbol sec or s, basic *SI unit of time, one-sixtieth of a minute. It is defined as the duration of 9,192,631,770 cycles of regulation (periods of the radiation corresponding to the transition between two hyperfine levels of the ground state) of the caesium-133 isotope. In mathematics, the second is a unit (symbol ") of angular measurement, equalling one-sixtieth of a minute, which in turn is one-sixtieth of a degree.

secondary emission emission of electrons from the surface of certain substances when they are struck by high-speed electrons or other particles from an external source. It can be detected with a *photomultiplier.

secondary growth or **secondary thickening**, increase in diameter of the roots and stems of certain plants (notably shrubs and trees) that results from the production of new cells by the *cambium. It provides the plant with additional mechanical support and new conducting cells, the secondary *xylem and *phloem. Secondary growth is generally confined to *gymnosperms and, among the *angiosperms, to the dicotyledons. With just a few exceptions, the monocotyledons (grasses, lilies) exhibit only primary growth, resulting from cell division at the apical *meristems.

secondary sexual characteristic external feature of an organism that is indicative of its gender (male or female), but not the reproductive organs themselves. They include facial hair in men and breasts in women, combs in cockerels, brightly coloured plumage in many male birds, and manes in male lions. In many cases, they are involved in displays and contests for mates and have evolved by sexual selection. Their development is stimulated by sex *hormones – in humans the change in concentrations of these hormones at puberty results not only in the development of the physical secondary sexual characteristics, but also emotional changes.

secretin *hormone produced by the small intestine of vertebrates that stimulates the production of digestive secretions by the pancreas and liver.

sector in computing, part of the magnetic structure created on a disk surface during disk formatting so that data can be stored on it. The disk is first divided into circular tracks and then each circular track is divided into a number of sectors.

sector in geometry, part of a circle enclosed by two radii and the arc that joins them. A **minor sector** has an angle at the centre of the circle of less than 180°. A **major sector** has an angle at the centre of the circle of more than 180°.

secure HTTP in computing, communications protocol that provides the basis for privacy-enhanced or encrypted communications between a Web *browser and a server. Secure HTTP enables users to send private information such as credit card numbers and addresses over the Internet.

secure socket layer SSL, in computing, standard protocol, designed by Netscape Communications, allowing secure communications over the Internet. The use of SSL involves the transmission of a secure message from Web *browser to Web server. Both sender and recipient have a digital security certificate that enables an encrypted message to be decrypted. This process is automatic. SSL technology is commonly used in electronic commerce transactions where users are purchasing products over the Internet, with the information,

including credit card numbers and
passwords, being encrypted.

security protection against loss or
misuse of data; see *data security.

sedative any drug that has a calming
effect, reducing anxiety and tension.
Sedatives will induce sleep in larger
doses. Examples are *barbiturates,
*narcotics, and benzodiazepines.

sediment any loose material that has
'settled' after deposition from
suspension in water, ice, or air,
generally as the water current or wind
speed decreases. Typical sediments are,
in order of increasing coarseness: clay,
mud, silt, sand, gravel, pebbles,
cobbles, and boulders. Sediments differ
from *sedimentary rocks, in which
deposits are fused together in a solid
mass of rock by a process called
*lithification (solidification). Pebbles
are cemented into *conglomerates;
sands become sandstones; muds
become mudstones or shales; peat is
transformed into coal.

sedimentary rock rock formed by the
accumulation and cementation of
deposits that have been laid down by
water, wind, ice, or gravity.
Sedimentary rocks cover more than
two-thirds of the Earth's surface and
comprise three major categories: clastic,
chemically precipitated, and organic (or
biogenic). Clastic sediments are the
largest group and are composed of
fragments of pre-existing rocks; they
include clays, sands, and gravels.
Chemical precipitates include some
limestones and evaporated deposits
such as gypsum and halite (rock salt).
Coal, oil shale, and limestone made of
fossil material are examples of organic
sedimentary rocks.

Most sedimentary rocks show
distinct layering (stratification),
because they are originally deposited
as more or less horizontal layers.
Sedimentary rocks are categorized by
the grain sizes of the particles:
*conglomerate rocks may contain rock
pieces over 2 mm in diameter;
*sandstone particles are up to 0.2 mm;
siltstone up to 0.02 mm; and claystone
less than 0.02 mm in diameter.

Seebeck effect generation of a voltage
in a circuit containing two different
metals, or semiconductors, by keeping

the junctions between them at different
temperatures. Discovered by the
German physicist Thomas Seebeck
(1770–1831), it is also called the
thermoelectric effect, and is the basis of
the *thermocouple. It is the opposite of
the *Peltier effect (in which current
flow causes a temperature difference
between the junctions of different
metals).

seed reproductive structure of higher
plants (*angiosperms and
*gymnosperms). It develops from a
fertilized ovule and consists of an
embryo and a food store, surrounded
and protected by an outer seed coat,
called the testa. The food store is
contained either in a specialized
nutritive tissue, the *endosperm, or in
the *cotyledons of the embryo itself. In
angiosperms the seed is enclosed within
a *fruit, whereas in gymnosperms it is
usually naked and unprotected, once
shed from the female cone.

Following *germination the seed
develops into a new plant.

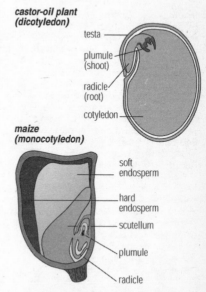

seed The structure of seeds. The castor-oil
plant is a dicotyledon, a plant in which the
developing plant has two leaves, developed from
the cotyledon. In maize, a monocotyledon, there
is a single leaf developed from the scutellum.

segment in geometry, part of a circle cut off by a straight line or *chord, running from one point on the circumference to another. All angles in the same segment are equal.

seismic gap theory theory that along faults that are known to be seismically active, or in regions of high seismic activity, the locations that are more likely to experience an *earthquake in the relatively near future are those that have not shown seismic activity for some time. When records of past earthquakes are studied and plotted onto a map, it becomes possible to identify **seismic gaps** along a fault or plate margin. According to the theory, an area that has not had an earthquake for some time will have a great deal of stress building up, which must eventually be released, causing an earthquake.

Although the seismic gap theory can suggest areas that are likely to experience an earthquake, it does not enable scientists to predict when that earthquake will occur.

seismic wave energy wave generated by an *earthquake or an artificial explosion. There are two types of seismic waves: **body waves** that travel through the Earth's interior; and **surface waves** that travel through the surface layers of the crust and can be felt as the shaking of the ground, as in an earthquake. Seismic waves show similar properties of reflection and refraction as light and sound waves. Seismic waves change direction and speed as they travel through different densities of the Earth's rocks.

body waves There are two types of body waves: P-waves and S-waves, so-named because they are the primary and secondary waves detected by a seismograph. **P-waves**, or compressional waves, are longitudinal waves (wave motion in the direction the wave is travelling), whose compressions and rarefactions resemble those of a sound wave. **S-waves** are transverse waves or shear waves, involving a back-and-forth shearing motion at right angles to the direction the wave is travelling (see *wave). Because liquids have no resistance to shear and cannot sustain a shear wave, S-waves cannot travel through liquid material. The Earth's outer core is believed to be liquid because S-waves disappear at the mantle-core boundary, while P-waves do not.

surface waves Surface waves travel in the surface and subsurface layers of the crust. **Rayleigh waves** travel along the free surface (the uppermost layer) of a solid material. The motion of particles is elliptical, like a water wave, creating the rolling motion often felt during an earthquake. **Love waves** are transverse waves trapped in a subsurface layer due to different densities in the rock layers above and below. They have a horizontal side-to-side shaking motion transverse (at right angles) to the direction the wave is travelling.

seismogram or **seismic record**, trace, or graph, of ground motion over time, recorded by a seismograph. It is used to determine the magnitude and duration of an earthquake.

seismograph instrument used to record ground motion. A heavy inert weight is suspended by a spring and attached to this is a pen that is in contact with paper on a rotating drum. During an earthquake the instrument frame and drum move, causing the pen to record a zigzag line on the paper; held steady by inertia, the pen does not move.

seismology study of *earthquakes, the seismic waves they produce, the processes that cause them, and the effects they have. By examining the global pattern of waves produced by an earthquake, seismologists can deduce the nature of the materials through which they have passed. This leads to an understanding of the Earth's internal structure.

On a smaller scale, artificial earthquake waves, generated by explosions or mechanical vibrators, can be used to search for subsurface features in, for example, oil or mineral exploration. Earthquake waves from underground nuclear explosions can be distinguished from natural waves by their shorter wavelength and higher frequency.

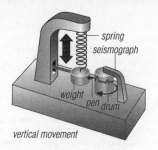

vertical movement

sideways movement

a seismogram recorded by a seismograph

first rumbles
of earthquake

most violent shaking
of earthquake

quiet and stable
before earthquake

quiet again

◄──────── time 5 seconds approximately ────────►

seismograph A seismogram, or recording made by a seismograph. Such recordings are used to study earthquakes and in prospecting.

selective serotonin-reuptake inhibitor
SSRI, in medicine, one of a group of drugs that prevent the uptake of *serotonin in neurones, resulting in an increase in its concentration in the brain. They are used in the treatment of depression. Fluoxetine (Prozac) is an example of a commonly prescribed selective serotonin-reuptake inhibitor.

selenium chemical symbol Se, (Greek *Selene* 'Moon') grey, non-metallic element, atomic number 34, relative atomic mass 78.96. It belongs to the sulphur group and occurs in several allotropic forms that differ in their physical and chemical properties. It is an essential trace element in human nutrition. Obtained from many sulphide ores and selenides, it is used as a red colouring for glass and enamel. Because its electrical conductivity varies with the intensity of light, selenium is used extensively in photoelectric devices. It was discovered in 1817 by Swedish chemist Jöns Berzelius and named after the Moon because its properties follow those of tellurium, whose name derives from Latin *Tellus* 'Earth'.

self-inductance or **self-induction**, in physics, the creation of an

electromotive force opposing the current. See *inductance.

semen fluid containing *sperm from the testes and secretions from various sex glands (such as the prostate gland) that is ejaculated by male animals during copulation. The secretions serve to nourish and activate the sperm cells, and prevent them clumping together.

semiconductor material with electrical conductivity intermediate between metals and insulators and used in a wide range of electronic devices. Certain crystalline materials, most notably silicon and germanium, have a small number of free electrons that have escaped from the bonds between the atoms. The atoms from which they have escaped possess vacancies, called holes, which are similarly able to move from atom to atom and can be regarded as positive charges. Current can be carried by both electrons (negative carriers) and holes (positive carriers). Such materials are known as **intrinsic semiconductors**.

Conductivity of a semiconductor can be enhanced by doping the material with small numbers of impurity atoms which either release free electrons (making an **n-type semiconductor** with more electrons than holes) or capture them (a **p-type semiconductor** with more holes than electrons). When p-type and n-type materials are brought together to form a p–n junction, an electrical barrier is formed that conducts current more readily in one direction than the other. This is the basis of the *semiconductor diode, used for rectification, and numerous other devices including *transistors, rectifiers, and *integrated circuits (silicon chips).

semiconductor diode   or **p–n junction *diode**, in electronics, a two-terminal *semiconductor device that allows electric current to flow in only one direction, the **forward-bias** direction. A very high resistance prevents current flow in the opposite, or **reverse-bias**, direction. It is used as a *rectifier, converting alternating current (AC) to direct current (DC).

Semiconductor diodes are made from silicon-based materials primarily used in electronic circuits to allow an electrical current to flow in one direction only. Silicon on its own is not useful and it is therefore doped (mixed) with other materials in order to achieve the desired physical and chemical properties. The impurity atoms occupy various positions in the silicon crystal. Two types of materials are produced by doping: p-type and n-type.

p-type Silicon has four valence (free) electrons in its outer shell. It is doped with boron atoms, with only three valence electrons in their outer shells. The silicon atoms will share their valence electrons with the valence electrons of the boron atoms. This results in a matrix of silicon atoms doped with boron atoms, which are short of one electron in order to form a fourth chemical bond with the silicon atoms. This is called a p-type junction (p refers to the boron atom being positive as it is short of one electron).

The p-type hole can be occupied by an electron moving into it from a neighbouring atom. The p-type hole then moves on to a new position in the matrix. As the p-type hole moves from one position to another, it is carrying a current with electrons also moving.

n-type Phosphorous has five valence electrons, so if phosphorous atoms are added to wafers of silicon they can donate extra electrons. This is called an n-type junction (n refers to phosphorous being negative due to an extra electron).

p–n junction In a semiconductor diode, wafers of p-type and n-type are fused together by heating the wafers and the doping substances in a furnace. An n-type silicon wafer is placed on top of a p-type silicon wafer to form a junction where the two types come into contact. Electrons flow through the silicon semiconductor in one direction only.

There are many different types of semiconductor diodes such as light-emitting diodes (LEDs), made from gallium arsenide phosphide and emitting electromagnetic radiation, and light-dependent diodes (photodiodes), allowing a flow of current in the presence of light.

sendmail   in computing, Unix program for sending *e-mail via *TCP/IP (Transport Control Protocol/Internet Protocol) using *SMTP (simple mail transfer protocol).

senescence   in biology, the deterioration in physical and (sometimes) mental capacities that occurs with ageing.

senile dementia   *dementia associated with old age, often caused by *Alzheimer's disease.

sense organ   any organ that an animal uses to gain information about its surroundings. All sense organs have specialized receptors (such as light receptors in the eye) and some means of translating their response into a nerve impulse that travels to the brain. The main human sense organs are the eye, which detects light and colour (different wavelengths of light); the ear, which detects sound (vibrations of the air) and gravity; the nose, which detects some of the chemical molecules in the air; and the tongue, which detects some of the chemicals in food, giving a sense of taste. There are also many small sense organs in the skin, including pain, temperature, and pressure sensors, contributing to our sense of touch.

sensor in computing, a device designed to detect a physical state or measure a physical quantity, and produce an input signal for a computer. For example, a sensor may detect the fact that a printer has run out of paper or may measure the temperature in a kiln.

sepal part of a flower, usually green, that surrounds and protects the flower in bud. The sepals are derived from modified leaves, and are collectively known as the *calyx.

sepsis general term for infectious change in the body caused by bacteria or their toxins.

septicaemia general term for any form of *blood poisoning.

sequence set of elements, arranged in order according to some rule. See *arithmetic progression, *geometric progression, and *series.

sequence-control register or **program counter**, in computing, a special memory location used to hold the address of the next instruction to be fetched from the immediate access memory for execution by the computer. It is located in the control unit of the *central processing unit.

sequencing in biochemistry, determining the sequence of chemical subunits within a large molecule. Techniques for sequencing amino acids in proteins were established in the 1950s, insulin being the first for which the sequence was completed. The *Human Genome Project is attempting to determine the sequence of the 3 billion base pairs within human *DNA. A rough draft of the whole genome had been sequenced by June 2000. In July 2001 the Global Musa Genomics Consortium, which is made up of publicly funded institutes, announced that the banana was to become the first fruit to have its genome sequenced. The process was expected to take five years and all sequences were to be made available on the Internet.

sequential file in computing, a file in which the records are arranged in order of a key field and the computer can use a searching technique, like a *binary search, to access a specific record.

serial device in computing, a device that communicates binary data by sending the bits that represent each character one by one along a single data line, unlike a *parallel device.

serial file in computing, a file in which the records are not stored in any particular order and therefore a specific record can be accessed only by reading through all the previous records.

serial interface in computing, an *interface through which data is transmitted one bit at a time.

Serial Line Internet Protocol in computing, method of connecting a computer to the Internet; usually abbreviated to *SLIP.

series sum of the terms of a sequence. Series may be convergent (the limit of the sum is a finite number) or divergent (the limit is infinite). For example, $1 + 1/2 + 1/4 + 1/8 + ...$ is a convergent series because the limit of the sum is 2; $1 + 2 + 3 + 4 + ...$ is a divergent series.

series circuit electrical circuit in which the components are connected end to end, so that the current flows through them all one after the other.

current in a series circuit The current flowing through each of the components is the same. This can be written as $I_T = I_1 = I_2$, where I_T is the total current flow, I_1 is the current flowing through component 1, and I_2 is the current flowing through component 2. To measure the current flow an ammeter is connected to the components in series. For example, if the current at component 1 is 2 amps

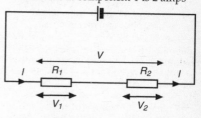

series circuit In a series circuit, the components, R_1 and R_2, of the circuit are connected end to end, so that the current passes through each component one after the other, without division or branching into parallel circuits.

and at component 2 is 2 amps, then the total current is 2 amps.

voltage in a series circuit As the current flowing through each of the components is the same, the energy required (voltage) in moving this equal amount of current through each of the components is also the same. If V_1 is the voltage across component 1 and V_2 is the voltage across component 2, then the total voltage, V_T, is given by $V_T = V_1 + V_2$. The voltage can be measured across each component by connecting a voltmeter in parallel with the component. For example, if the voltage across component 1 is 3 volts and across component 2 is 3 volts, then the total voltage is 6 volts.

serif font typeface, such as Times or Palatino, the strokes of which terminate in ornamental curves or cross-strokes. These are said to aid legibility.

serotonin or **5-hydroxytryptamine (5-HT)**, chemical widely distributed in the body tissues – in the blood platelets, the wall of the intestine, and the *central nervous system. It is believed to be implicated in the inflammatory process and, in the nervous system, it acts as a *neurotransmitter, controlling sleep. Serotonin is derived from the amino acid tryptophan.

serotype in medicine, serotyping is the classification of substances according to the antigens they contain or the antibodies they provoke. Bacterial serotyping is used to classify bacteria of the same species that produce different immunological responses.

serum clear fluid that separates out from clotted blood. It is blood plasma with the anticoagulant proteins removed, and contains *antibodies and other proteins, as well as the fats and sugars of the blood. It can be produced synthetically, and is used to protect against disease.

server in computing, computer or device that manages *network resources. For example, file servers are computers used to store files. Print servers manage printing devices and network servers deal with network traffic. Servers may be dedicated to one function or coexist on a multiprocessing system.

sessile in botany, describing a leaf, flower, or fruit that lacks a stalk and sits directly on the stem, as with the sessile acorns of certain oaks. In zoology, it is an animal that normally stays in the same place, such as a barnacle or mussel. The term is also applied to the eyes of crustaceans when these lack stalks and sit directly on the head.

SET acronym for Secure Electronic Transfer, form of *public key infrastructure developed by the credit card companies Visa and Mastercard. SET provides confidentiality and integrity of information, and authentication of customer and merchant, but is expensive and difficult to implement.

set or **class**, in mathematics, any collection of defined things (elements), provided the elements are distinct and that there is a rule to decide whether an element is a member of a set. It is usually denoted by a capital letter and indicated by curly brackets {}.

SETI acronym for **search for extraterrestrial intelligence**, in astronomy, programme originally launched by *NASA in 1992, using powerful *radio telescopes to search the skies for extraterrestrial signals. NASA cancelled the SETI project in 1993, but other privately funded SETI projects continue.

set-top box box containing decoding equipment for satellite or cable television broadcasts. Such boxes represent a means of linking television sets to a network such as the *Internet, enabling people to browse the *World Wide Web using their televisions as the monitor, or to view video-on-demand.

sewage disposal the disposal of human excreta and other waterborne waste products from houses, streets, and factories. Conveyed through sewers to sewage works, sewage has to undergo a series of treatments to be acceptable for discharge into rivers or the sea, according to various local laws. Raw sewage, or sewage that has not been treated adequately, is one serious source of water pollution and a cause of eutrophication.

sex chromosome *chromosome that differs between the sexes and serves to determine the sex of the individual. In humans, whether a person is male or female is determined by the combination of the two sex chromosomes in the body cells. In females both chromosomes are the same – two X chromosomes (XX). In males the two chromosomes are different – one X chromosome and one Y chromosome (XY). The Y chromosome is shorter than the X. *Genes on these chromosomes determine sex.

As a result of *meiosis gametes from a female each contain one X chromosome. However, gametes from a male are of two kinds. Half of the gametes contain an X chromosome and half contain a Y chromosome. If an X-

carrying gamete from a male fertilizes a female gamete the result will be a female. If a Y-carrying gamete from a male fertilizes a female gamete, the result will be a male.

sex determination   process by which the sex of an organism is determined. In many species, the sex of an individual is dictated by the two sex chromosomes (X and Y) it receives from its parents. In mammals, some plants, and a few insects, males are XY, and females XX; in birds, reptiles, some amphibians, and butterflies the reverse is the case. In bees and wasps, males are produced from unfertilized eggs, females from fertilized eggs. In 1991 it was shown that maleness is caused by a single gene, 14 base pairs

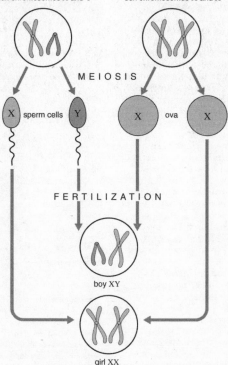

sex determination In humans, sex is determined by the male. Sperm cells contain an X or a Y chromosome but egg cells contain only X chromosomes. If a sperm cell carrying an X chromosome fertilizes the egg the resulting baby will be female; if the sperm cell is carrying a Y chromosome then the baby will be male.

long, on the Y chromosome.

Environmental factors can affect some fish and reptiles, such as turtles, where sex is influenced by the temperature at which the eggs develop.

sex hormone steroid hormone produced and secreted by the gonads (testes and ovaries). Sex hormones control development and reproductive functions and influence sexual and other behaviour.

sex linkage in genetics, the tendency for certain characteristics to occur exclusively, or predominantly, in one sex only. Human examples include red-green colour blindness and haemophilia, both found predominantly in males. In both cases, these characteristics are *recessive and are determined by genes on the *X chromosome.

Since females possess two X chromosomes, any such recessive *allele on one of them is likely to be masked by the corresponding allele on the other. In males (who have only one X chromosome paired with a largely inert *Y chromosome) any gene on the X chromosome will automatically be expressed. Colour blindness and haemophilia can appear in females, but only if they are *homozygous for these traits, due to inbreeding, for example.

sexuality attribute or characteristic of being male or female, usually taken to involve more than the ability or disposition to play the appropriate role in sexual reproduction. Today, as much emphasis is placed on an individual's awareness of and response to culturally and socially derived *gender differences as on biological factors in the development of sexuality.

sexually transmitted disease STD, any disease transmitted by sexual contact, involving transfer of body fluids. STDs include not only traditional *venereal disease, but also a growing list of conditions, such as *AIDS and scabies, which are known to be spread primarily by sexual contact. Other diseases that are transmitted sexually include viral *hepatitis. The World Health Organization (WHO) estimates that there are 356,000 new cases of STDs daily worldwide.

sexual reproduction reproductive process in organisms that requires the union, or *fertilization, of gametes (such as eggs and sperm). These are usually produced by two different individuals, although self-fertilization occurs in a few *hermaphrodites such as tapeworms. Most organisms other than bacteria and cyanobacteria (*blue-green algae) show some sort of sexual process. Except in some lower organisms, the gametes are of two distinct types called the egg (*ovum) and the *sperm. The organisms producing the eggs are called females,

female reproductive system

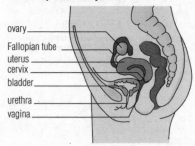

ovary
Fallopian tube
uterus
cervix
bladder
urethra
vagina

male reproductive system

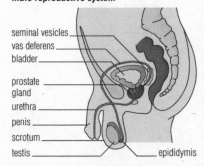

seminal vesicles
vas deferens
bladder
prostate gland
urethra
penis
scrotum
testis
epididymis

sexual reproduction The human reproductive organs. In the female, gametes called ova are released regularly in the ovaries after puberty. The Fallopian tubes carry the ova to the uterus or womb, in which the fetus will develop. In the male, sperm is produced inside the testes after puberty; about 10 million sperm cells are produced each day, enough to populate the world in six months. The sperm duct or vas deferens, a continuation of the epididymis, carries sperm to the urethra during ejaculation.

and those producing the sperm, males. The fusion of a male and female gamete produces a **zygote**, from which a new individual develops. See *reproduction.

Gametes are produced by *meiosis. If the male and female gametes are produced by two different individuals, sexual reproduction combines inherited information from the two parents. In mammals, the male gametes, sperm, are made in the testes, while the female gametes, eggs (ova), are made in the ovaries.

Most animals and plants reproduce sexually, though quite a few plants also reproduce asexually. The male and female sex organs of a plant are usually found in a flower. Many flowers contain both male and female organs. The male gametes of a plant are inside the pollen grains, and the female gametes are inside the ovules.

In micro-organisms, *asexual reproduction is very common. Yeasts and a few bacteria may reproduce sexually, although most reproduce asexually.

SGML abbreviation for standard generalized markup language, *International Standards Organization standard describing how the structure (features such as headers, columns, margins, and tables) of a text can be identified so that it can be used, probably via filters, in applications such as desktop publishing and *electronic publishing. *HTML and *VRML are both types of SGML.

shadow area of darkness behind an opaque object that cannot be reached by some or all of the light coming from a light source in front. Its presence may be explained in terms of light rays travelling in straight lines and being unable to bend around obstacles. The light in front of the object is blocked. A point source of light produces an *umbra, a completely black shadow with sharp edges. An extended source of light produces both a central umbra and a penumbra, a region of semidarkness with blurred edges where darkness gives way to light.

shale fine-grained and finely layered *sedimentary rock composed of silt and clay. It is a weak rock, splitting easily along bedding planes to form thin, even slabs (by contrast, mudstone splits into irregular flakes). Oil shale contains kerogen, a solid bituminous material that yields *petroleum when heated.

shared memory bus architecture SMBA, in computing, system of *buses that allows parallel computers to share *RAM for greater processing power.

shareware software distributed free via the Internet or on disks given away with magazines. Users have the opportunity to test its functionality and ability to meet their requirements before paying a small registration fee directly to the author. This may bring additional functionality, documentation, and occasional upgrades. Shareware is not copyright-free. In the 1980s and early 1990s, many shareware libraries existed, but shareware has since fallen in popularity. Compare with *public-domain software.

shelf sea relatively shallow sea, usually no deeper than 200 m/650 ft, overlying the continental shelf around the coastlines. Most fishing and marine mineral exploitations are carried out in shelf seas.

shell in computing, program that mediates access to a particular system or server. Microsoft Windows is a shell that interposes a *graphical user interface and other utilities between the operator and MS-DOS. In DOS itself, the file COMMAND.COM is a shell that makes the operating system display a *prompt and enables it to interpret user instructions. Shells can also be used to improve computer security.

shell in zoology, hard outer covering of a wide variety of invertebrates. The covering is usually mineralized, normally with large amounts of calcium. The shell of birds' eggs is also largely made of calcium.

shell account in computing, cheap method of accessing the *Internet via another computer, usually a *Unix machine. Users with a shell account are given text-only access (via the telephone) to another computer connected to the Net.

shell script in computing, the Unix equivalent of batch files created for a program *shell. Essentially, shell scripts allow users to create their own commands by creating a file that contains the sequence of commands they want to run and then designating the file as executable. Thereafter, typing the name of the file executes the sequence of commands.

shell shock or **combat neurosis** or **battle fatigue**, any of the various forms of mental disorder that affect soldiers exposed to heavy explosions or extreme *stress. Shell shock was first diagnosed during World War I.

SHF in physics, abbreviation for superhigh *frequency. SHF radio waves have frequencies in the range 3–30 GHz.

shield volcano broad, flat *volcano formed at a *constructive margin between tectonic plates or over a *hot spot. The *magma (molten rock) associated with shield volcanoes is usually basalt – thin and free-flowing. An example is Mauna Loa on the Pacific island of Hawaii. A *composite volcano, on the other hand, is formed at a destructive margin.

shingles common name for *herpes zoster, a disease characterized by infection of sensory nerves, with pain and eruption of blisters along the course of the affected nerves.

shock wave narrow band or region of high pressure and temperature in which the flow of a fluid changes from subsonic to supersonic.

Shock waves are produced when an object moves through a fluid at a supersonic speed. See *sonic boom.

shoot in botany, the parts of a *vascular plant growing above ground, comprising a stem bearing leaves, buds, and flowers. The shoot develops from the *plumule of the embryo.

short circuit unintended direct connection between two points in an electrical circuit. *Resistance is proportional to the length of wire through which current flows. By bypassing the rest of the circuit, the short circuit has low resistance and a large current flows through it. This may cause the circuit to overheat dangerously.

shortcut in computing, keyboard combination or icon that activates a procedure otherwise available only through pull-down menus and *dialog boxes. Most commercial software comes with built-in keyboard shortcuts and *pushbuttons, and many allow users to create their own custom shortcuts. In Microsoft Windows, a shortcut is an icon that launches a program direct from the desktop.

short-sightedness nontechnical term for *myopia.

shunt in electrical engineering, a conductor of very low resistance that is connected in parallel to an *ammeter in order to enable it to measure larger electric currents. Its low resistance enables it to act like a bypass, diverting most of the current through itself and away from the ammeter.

SI abbreviation for **Système International d'Unités** (French 'International System of Metric Units'); see *SI units.

siamese twin identical *twin born physically joined to his or her twin.

sick building syndrome malaise diagnosed in the early 1980s among office workers and thought to be caused by such pollutants as formaldehyde (from furniture and insulating materials), benzene (from paint), and the solvent trichloroethene, concentrated in air-conditioned buildings. Symptoms include headache, sore throat, tiredness, colds, and flu. Studies have found that it can cause a 40% drop in productivity and a 30% rise in absenteeism.

sickle-cell disease or **sickle-cell anaemia**, hereditary chronic blood disorder common among people of black African descent; also found in the eastern Mediterranean, parts of the Gulf, and in northeastern India. It is characterized by distortion and fragility of the red blood cells, which are lost too rapidly from the circulation. This often results in *anaemia.

sidereal period orbital period of a planet around the Sun, or a moon around a planet, with reference to a background star. The sidereal period of a planet is in effect a 'year' for that planet. A *synodic period is a full circle as seen from Earth.

sidereal time in astronomy, time measured by the rotation of the Earth with respect to the stars. A sidereal day is the time taken by the Earth to turn once with respect to the stars, namely 23 hours 56 minutes 4 seconds. It is divided into sidereal hours, minutes, and seconds, each of which is proportionally shorter than the corresponding SI unit.

siemens symbol S, SI unit of electrical conductance, the reciprocal of the *resistance of an electrical circuit. One siemens equals one ampere per volt. It was formerly called the mho or reciprocal ohm.

sievert symbol Sv, SI unit of radiation dose equivalent. It replaces the rem (1 Sv equals 100 rem). Some types of radiation do more damage than others for the same absorbed dose – for example, an absorbed dose of alpha radiation causes 20 times as much biological damage as the same dose of beta radiation. The equivalent dose in sieverts is equal to the absorbed dose of radiation in grays multiplied by the relative biological effectiveness. Humans can absorb up to 0.25 Sv without immediate ill effects; 1 Sv may produce radiation sickness; and more than 8 Sv causes death.

sigma bond type of chemical bond in which an electron pair (regarded as being shared by the two atoms involved in the bond) occupies a molecular orbital situated between the two atoms; the orbital is located along a hypothetical line linking the atoms' nuclei. See also *hybridization.

signal processing in computing, the digitizing of an analogue signal such as a voice stream.

significant figures figures in a number that, by virtue of their position, express the magnitude of that number to a specified degree of accuracy. In a number, the first significant figure is the first figure that is not a 0. The final significant figure is rounded up if the following digit is greater than 5. For example, 5,463,254 to three significant figures is 5,460,000; 3.462891 to four significant figures is 3.463; 0.00347 to two significant figures is 0.0035.

silica silicon dioxide, SiO_2, the composition of the most common mineral group, of which the most familiar form is quartz. Other silica forms are chalcedony, chert, opal, tridymite, and cristobalite. Common sand consists largely of silica in the form of quartz.

silicate one of a group of minerals containing silicon and oxygen in tetrahedral units of SiO_4, bound together in various ways to form specific structural types. Silicates are the chief rock-forming minerals. Most rocks are composed, wholly or in part, of silicates (the main exception being limestones). Glass is a manufactured complex polysilicate material in which other elements (boron in borosilicate glass) have been incorporated. Generally, additional cations are present in the structure, especially Al^{3+}, Fe^{2+}, Mg^{2+}, Ca^{2+}, Na^+, K^+, but quartz and other polymorphs of SiO_2 are also considered to be silicates; stishovite (a high-pressure form of SiO_2) is a rare exception to the usual tetrahedral coordination of silica and oxygen.

In **orthosilicates**, the oxygens are all ionically bonded to cations such as Mg^{2+} or Fe^{2+} (as olivines), and are not shared between tetrahedra. All other silicate structures involve some degree of oxygen sharing between adjacent tetrahedra. For example, beryl is a **ring silicate** based on tetrahedra linked by sharing oxygens to form a circle. Pyroxenes are single **chain silicates**, with chains of linked tetrahedra extending in one direction through the structure; amphiboles are similar but have double chains of tetrahedra. In micas, which are **sheet silicates**, the tetrahedra are joined to form continuous sheets that are stacked upon one another. **Framework silicates**, such as feldspars and quartz, are based on three-dimensional frameworks of tetrahedra in which all oxygens are shared.

silicon chemical symbol Si, (Latin *silex* 'flint') brittle, non-metallic element, atomic number 14, relative atomic mass 28.086. It is the second-most abundant element (after oxygen) in the Earth's crust and occurs in amorphous and crystalline forms. In nature it is found only in combination with other elements, chiefly with oxygen in silica (silicon dioxide, SiO_2) and the silicates.

These form the mineral *quartz, which makes up most sands, gravels, and beaches.

Pottery glazes and glassmaking are based on the use of silica sands and date from prehistory. Today the crystalline form of silicon is used as a deoxidizing and hardening agent in steel, and has become the basis of the electronics industry because of its *semiconductor properties, being used to make 'silicon chips' for microprocessors.

The element was isolated by Swedish chemist Jöns Berzelius in 1823, having been named in 1817 by Scottish chemist Thomas Thomson by analogy with boron and carbon because of its chemical resemblance to these elements.

silicon chip *integrated circuit with microscopically small electrical components on a piece of silicon crystal only a few millimetres square.

silicone member of a class of synthetic polymers whose molecular structure is based on long chains of alternating silicon and oxygen atoms, usually with connecting organic groups attached at silicon atoms at various points on the chain. Silicones form synthetic rubbers and lubricants.

silicosis chronic disease of miners and stone cutters who inhale *silica dust, which makes the lung tissues fibrous and less capable of aerating the blood. It is a form of pneumoconiosis.

sill sheet of igneous rock created by the intrusion of magma (molten rock) between layers of pre-existing rock. (A dyke, by contrast, is formed when magma cuts *across* layers of rock.) An example of a sill in the UK is the Great Whin Sill, which forms the ridge along which Hadrian's Wall was built.

silt sediment intermediate in coarseness between clay and sand; its grains have a diameter of 0.002–0.02 mm/ 0.00008–0.0008 in. Silt is usually deposited in rivers, and so the term is often used generically to mean a river deposit, as in the silting-up of a channel.

Silurian Period period of geological time 439–409 million years ago, the third period of the Palaeozoic era. Silurian sediments are mostly marine and consist of shales and limestone. Luxuriant reefs were built by coral-like organisms. The first land plants began to evolve during this period, and there were many ostracoderms (armoured jawless fishes). The first jawed fishes (called acanthodians) also appeared.

silver chemical symbol Ag, white, lustrous, extremely malleable and ductile, metallic element, atomic number 47, relative atomic mass 107.868. Its chemical symbol comes from the Latin *argentum*. It occurs in nature in ores and as a free metal; the chief ores are sulphides, from which the metal is extracted by smelting with lead. It is the best metallic conductor of both heat and electricity; its most useful compounds are the chloride and bromide, which darken on exposure to light and are the basis of photographic emulsions.

Silver is used ornamentally, for jewellery and tableware, for coinage, in electroplating, electrical contacts, and dentistry, and as a solder. It has been mined since prehistory; its name is an ancient non-Indo-European one, *silubr*, borrowed by the Germanic branch as *silber*.

simple harmonic motion SHM, oscillatory or vibrational motion in which an object (or point) moves so that its acceleration towards a central point is proportional to its distance from it. A simple example is a pendulum, which also demonstrates another feature of SHM, that the maximum deflection is the same on each side of the central point. A graph of the varying distance with respect to time is a sine curve, a characteristic of the oscillating current or voltage of an alternating current (AC), which is another example of SHM.

Simple Mail Transfer Protocol SMTP, in computing, protocol for transferring e-mail between computers, commonly abbreviated to *SMTP.

Simple Network Management Protocol SNMP, in computing, agreed method of managing a computer network, commonly abbreviated to *SNMP.

simplify of a fraction, to reduce to lowest terms by dividing both numerator and denominator by any number that is a factor of both, until there are no common factors between the numerator and denominator. Also, in algebra, to condense an algebraic expression by grouping similar terms and reducing constants to their lowest terms. For example, the expression $a + 2b + b + 2a - 2(a + b)$ can be simplified to $a + b$.

simulation short for computer simulation.

simultaneous equations two or more algebraic equations that contain two or more unknown quantities that may have a unique solution. For example, in the case of two linear equations with two unknown variables, such as:

(i) $x + 3y = 6$

and

(ii) $3y - 2x = 6$

the solution will be those unique values of x and y that are valid for both equations. Linear simultaneous equations can be solved by using algebraic manipulation to eliminate one of the variables, *coordinate geometry, or matrices (see *matrix).

For example, by using algebra, both sides of equation (i) could be multiplied by 2, which gives

(i) $2x + 6y = 12$.

Adding this to equation (ii) gives

$9y = 18$

which is easily solved: $y = 2$.

The variable x can now be found by inserting the known y value into either original equation and solving for x.

Simultaneous equations can be solved using coordinate geometry because the two equations represent straight lines in coordinate geometry and the coordinates of their point of intersection are the values of x and y that are true for both of them. If the equations represent either two parallel lines or the same line, then there will be no solutions or an infinity of solutions, respectively.

A third method of solving linear simultaneous equations involves manipulating matrices.

sine sin, in trigonometry, a function of an angle in a right-angled *triangle that is defined as the ratio of the length of the side opposite the angle to the length of the hypotenuse (the longest side). This function can be used to find either angles or sides in a right-angled triangle.

sine rule in trigonometry, a rule that relates the sides and angles of a triangle, stating that the ratio of the length of each side and the sine of the angle opposite is constant (twice the radius of the circumscribing circle). If the sides of a triangle are a, b, and c, and the angles opposite are A, B, and C, respectively, then the sine rule may be expressed as

$$\frac{a}{\sin A} = \frac{b}{\sin B} = \frac{c}{\sin C}$$

sine wave or **sine curve**, in physics, graph demonstrating properties that vary sinusoidally. It is obtained by plotting values of angles against the values of their *sines. Examples include *simple harmonic motion, such as the way alternating current (AC) electricity varies with time.

Single In-line Memory Module SIMM, in computing, small printed circuit board carrying multiple *memory chips. Although there are electrical contacts on both sides of the board they are electrically connected and do not function independently.

sink hole funnel-shaped hollow in an area of limestone. A sink hole is usually formed by the enlargement of a joint, or crack, by *carbonation (the dissolving effect of water). It should not be confused with a swallow hole, or swallet, which is the opening through which a stream disappears underground when it passes onto limestone.

sirocco hot, normally dry and dust-laden wind that blows from the deserts of North Africa across the Mediterranean into southern Europe. It occurs mainly in the spring. The name 'sirocco' is also applied to any hot oppressive wind.

site location at which computers are used. If a company uses only *IBM computers, for example, it is known as an IBM site. The term is also used for a computer that acts as a *server for files that can be accessed via the *World Wide Web, also called a *Web site.

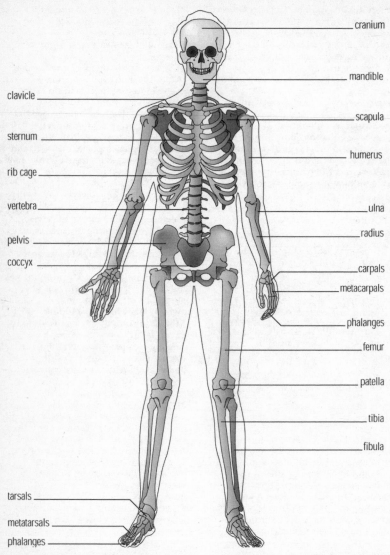

cranium

mandible

clavicle

scapula

sternum

humerus

rib cage

vertebra

ulna

pelvis

radius

coccyx

carpals

metacarpals

phalanges

femur

patella

tibia

fibula

tarsals

metatarsals

phalanges

skeleton The human skeleton is made up of 206 bones and provides a strong but flexible supportive framework for the body.

site licence in computing, licence issued with commercial software entitling the purchaser to install the program onto machines within a particular site or campus.

site of special scientific interest SSSI, in the UK, land that has been identified as having animals, plants, or geological features that need to be protected and conserved. From 1991 these sites were designated and administered by English Nature, Scottish Natural Heritage, and the Countryside Council for Wales.

Numbers fluctuate, but Britain had almost 5,000 SSSIs in 1998, two-thirds of which were privately owned. Although SSSIs enjoy some legal protection, this does not in practice always prevent damage or destruction; during 1992, for example, 40% of SSSIs were damaged by development, farming, public access, and neglect. A report by English Nature estimated that a quarter of the total area of SSSIs, over 1 million acres, had been damaged by acid rain. Around 1% of SSSIs are irreparably damaged each year. In 1995–96 7% of Welsh SSSIs and 4.2% of English SSSIs experienced damage.

SI units French **Système International d'Unités**, standard system of scientific units used by scientists worldwide. Originally proposed in 1960, it replaces the *m.k.s., * c.g.s., and f.p.s. systems. It is based on seven basic units: the metre (m) for length, kilogram (kg) for mass, second (s) for time, ampere (A) for electrical current, kelvin (K) for temperature, mole (mol) for amount of substance, and candela (cd) for luminosity. Details are given at the end of this book.

skeleton framework of bones that supports and gives form to the body, protects its internal organs, and provides anchorage points for its *muscles. It is composed of about 200 bones. Each bone is made of a *mineral, calcium phosphate, and *protein. Bones of the skeleton are joined to each other by ligaments. In the human body, walking, running, arm and leg movements, hand actions, and even just standing, are all achieved by the operation of muscles attached to bones of the skeleton. Movement of the body is brought about by the moveable joints of the body. The elbow joint is a good example. Muscles are attached to bones by tendons, and contractions of the muscle bring about movement.

skew distribution in statistics, a distribution in which frequencies are not balanced about the mean. For example, low wages are earned by a great number of people, while high wages are earned by very few. However, because the high wages can be very

high they pull the average up the scale, making the average wage look unrepresentatively high.

skin covering of the body of a vertebrate. In mammals, the outer layer (epidermis) is dead and its cells are constantly being rubbed away and replaced from below; it helps to protect the body from infection and to prevent dehydration. The lower layer (dermis) contains blood vessels, *nerves, hair roots, and sweat and sebaceous glands (producing oil), and is supported by a network of fibrous and elastic cells. (See diagram, page 544.)

The skin helps to protect the body from drying out. It is waterproof and covered with dry, dead cells, so little water is lost from skin cells. However, water is lost from a human body when the body sweats. The skin helps to regulate body temperature. Body temperature is monitored and controlled by the thermoregulatory centre in the *brain. This centre has special cells sensitive to the temperature of blood flowing through the brain (receptors). Temperature receptors in the skin also send nerve impulses to this centre giving information about skin temperature.

If the body temperature is too high blood vessels supplying the skin capillaries expand (dilate) so that more blood flows through the capillaries and more heat is lost. To further assist heat loss sweat glands release sweat, which cools the body as it evaporates. If the body temperature is too low blood vessels supplying the skin capillaries constrict to reduce the flow of blood through the capillaries.

Other than measuring the temperature, the skin senses the environment in several other ways. Some receptors in the skin are sensitive to touch and pressure. If they are stimulated, nerve impulses are sent to the brain carrying information. It also protects the body from disease-causing organisms, which find it hard to penetrate the skin.

slag in chemistry, the molten mass of impurities that is produced in the smelting or refining of metals. The slag produced in the manufacture of iron in

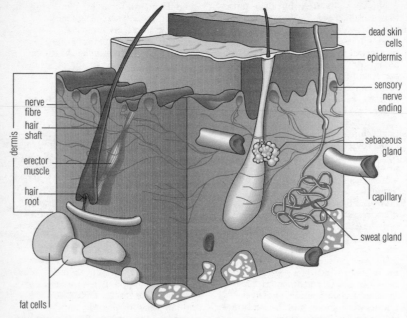

dead skin cells
epidermis
sensory nerve ending
sebaceous gland
capillary
sweat gland

nerve fibre
hair shaft
erector muscle
hair root
dermis
fat cells

skin The skin of an adult man covers about 1.9 sq m/20 sq ft; a woman's skin covers about 1.6 sq m/ 17 sq ft. During our lifetime, we shed about 18 kg/40 lb of skin.

a *blast furnace floats on the surface above the molten iron. It contains mostly silicates, phosphates, and sulphates of calcium. When cooled, the solid is broken up and used as a core material in the foundations of roads and buildings.

slaked lime common name for *calcium hydroxide.

slate fine-grained, usually grey metamorphic rock that splits readily into thin slabs along its *cleavage planes. It is the metamorphic equivalent of *shale.

sleep state of natural unconsciousness and activity that occurs at regular intervals in most mammals and birds, though there is considerable variation in the amount of time spent sleeping. Sleep differs from hibernation in that it occurs daily rather than seasonally, and involves less drastic reductions in metabolism. The function of sleep is unclear. People deprived of sleep become irritable, uncoordinated, forgetful, hallucinatory, and even psychotic.

sleep apnoea in medicine, temporary cessation of breathing during sleep. It is usually due to a transient obstruction of the airway between the soft palate and the larynx when the airway dilator muscles relax too much. Muscle relaxation is accentuated by the use of sedative drugs and alcohol.

sleep-walking or **somnambulism**, walking around and performing other activities during sleep with no recollection of it on waking. Common in childhood, it may continue into the adult years; or it may arise spontaneously, particularly if the person concerned is experiencing stress.

sleet precipitation consisting of a mixture of water and ice.

slime mould or **myxomycete**, extraordinary organism that shows some features of *fungus and some of *protozoa. Slime moulds are not closely related to any other group, although they are often classed, for convenience, with the fungi. There are two kinds, cellular slime moulds and plasmodial slime moulds, differing in their complex life cycles.

SLIP acronym for Serial Line Internet Protocol, in computing, the older of two standard methods for connecting a computer to the Internet via a modem and telephone line. Unlike PPP (Point-to-Point Protocol), a SLIP connection needs to have its *IP address reset every time it is used, and offers no error detection.

slip-off slope gentle slope forming the inner bank of a *meander (bend in a river). It is formed by the deposition of fine silt, or alluvium, by slow-flowing water.

slipped disc or **prolapsed intervertebral disc**, protrusion of the soft inner substance of an intervertebral disc through the outer covering, causing pressure on nerve roots. It is caused by heavy or awkward lifting or sudden twisting of the spine or bending. Symptoms are severe pain, with *sciatica, and immobility. There is a range of treatments available, including bed-rest, traction or osteopathy, with *analgesics for the pain.

slow virus in medicine, one of a group of viruses that cause disorders of the central nervous system and take many years to develop. Gradual and widespread damage of nerve tissue is followed eventually by a loss of brain function and death. Slow viruses are implicated in *Creutzfeldt-Jakob disease and a type of *meningitis.

slug obsolete unit of mass, equal to 14.6 kg/32.17 lb. It is the mass that will have an acceleration of one foot per second when under a force of one pound weight.

smallpox acute, highly contagious viral disease, marked by aches, fever, vomiting, and skin eruptions leaving pitted scars. Widespread vaccination programmes have wiped out this often fatal disease.

Smalltalk first high-level programming language used in *object-oriented applications.

smart in computing, term for any piece of equipment that works with the help of a microprocessor: a 'smart' carburettor, for example, maintains the correct proportion of air-to-petrol

vapour in a car engine by electronically monitoring engine temperature, acceleration, and other variables. Designers are incorporating smart technology into an increasing range of products, such as smart toasters, which can prevent toast from burning. Smart furniture, such as chairs with cushions that adjust themselves according the size and weight of the person sitting in them, is a typical area of current research.

smart card plastic card with an embedded microprocessor and memory. It can store, for example, personal data, identification, and bank-account details, to enable it to be used as a credit or debit card. The card can be loaded with credits, which are then spent electronically, and reloaded as needed. Possible other uses range from hotel door 'keys' to passports.

smart drug any drug or combination of nutrients (vitamins, amino acids, minerals, and sometimes herbs) said to enhance the functioning of the brain, increase mental energy, lengthen the span of attention, and improve the memory. As yet there is no scientific evidence to suggest that these drugs have any significant effect on healthy people.

smart fluid or **electrorheological fluid**, liquid suspension that solidifies to form a jellylike solid when a high-voltage electric field is applied across it and that returns to the liquid state when the field is removed. Most smart fluids are *zeolites or metals coated with polymers or oxides.

smelling salts mixture of ammonium carbonate, bicarbonate, and carbamate together with other strong-smelling substances, formerly used as a restorative for dizziness or fainting.

smelting processing a metallic ore in a furnace to produce the metal. Oxide ores such as iron ore are smelted with coke (carbon), which reduces the ore into metal and also provides fuel for the process. A substance such as limestone is often added during smelting to facilitate the melting process and to form a slag, which dissolves many of the impurities present.

smog natural fog containing impurities, mainly nitrogen oxides (NO_x) and volatile organic compounds (VOCs) from domestic fires, industrial furnaces, certain power stations, and internal-combustion engines (petrol or diesel). It can cause substantial illness and loss of life, particularly among chronic bronchitics, and damage to wildlife.

smokeless fuel fuel that does not give off any smoke when burned, because all the carbon is fully oxidized to carbon dioxide (CO_2). Natural gas, oil, and coke are smokeless fuels.

smoker hot fissure in the ocean floor, known as a *hydrothermal vent.

smoothing capacitor large electronic *capacitor connected across the output of a rectifier circuit that has the effect of smoothing out the voltage variations to give a nearly steady DC voltage supply. The voltage and current output from a rectifier circuit fitted with a smoothing capacitor is similar to that provided by a battery.

smooth muscle involuntary muscle capable of slow contraction over a period of time. It is present in hollow organs, such as the intestines, stomach, bladder, and blood vessels. Its presence in the wall of the alimentary canal allows slow rhythmic movements known as *peristalsis, which cause food to be mixed and forced along the gut. Smooth muscle has a microscopic structure distinct from other forms.

SMPTE abbreviation for Society of Motion Picture and Television Engineers, US organization founded in 1916 to advance the theory and application of motion-imaging technology including film, television, video, computer imaging, and telecommunications.

SMS abbreviation for short message service, service for sending and receiving text messages to and from mobile phones. SMS was created when it was incorporated into the Global System for Mobiles (GSM) digital mobile phone standard. The length of a single message can be up to 160 characters, and this limit has forced users to adapt the English language to create an abbreviated language peculiar to SMS. It is also possible to deliver an SMS message through an Internet Web site, called an SMS gateway.

SMTP abbreviation for simple mail transfer protocol, in computing, the basic protocol for transferring e-mail between computers. SMTP is an agreed procedure for identifying the *host, sending and receiving data, and checking e-mail addresses. It is the e-mail delivery mechanism for almost all Internet based e-mail.

SNA abbreviation for IBM's *Systems Network Architecture.

snail mail in the computing community, nickname for the conventional postal service. E-mail can deliver messages within minutes while conventional postal services take at least a day. One's postal address is therefore a 'snail mail address'.

Snell's law of refraction in optics, the rule that when a ray of light passes from one medium to another, the sine of the angle of incidence divided by the sine of the angle of refraction is equal to the ratio of the indices of refraction in the two media. For a ray passing from medium 1 to medium 2: $n_2/n_1 = \sin i/\sin r$ where n_1 and n_2 are the refractive indices of the two media. The law was devised by the Dutch physicist, Willebrord Snell.

sniffer in computing, software tool that analyses the transport data attached to *packets sent across a network, used to monitor the network's efficiency and level of usage. Hackers (see *hacking) also use sniffers to collect people's passwords for *Telnet connections.

SNMP abbreviation for simple network management protocol, in computing, a *protocol which gathers information from *network *hardware to monitor its performance. An updated version of SNMP, called Remote Network Monitoring Specification (RNMS), enables network hardware devices to send alerts to network management software systems when different types of errors exceed prescribed limits.

snow precipitation in the form of soft, white, crystalline flakes caused by the condensation in air of excess water vapour below freezing point. Light reflecting in the crystals, which have a basic hexagonal (six-sided) geometry, gives snow its white appearance.

snow blindness pain in the cornea of the eye due to exposure to ultraviolet light reflected off snow. It resolves within a day or so of avoiding the glare. It is prevented by sunglasses or protective goggles.

soap mixture of the sodium salts of various *fatty acids: palmitic, stearic, and oleic acid. It is made by the action of sodium hydroxide (caustic soda) or potassium hydroxide (caustic potash) on fats of animal or vegetable origin. Soap makes grease and dirt disperse in water in a similar manner to a *detergent.

Soap was mentioned by Greek physician and anatomist Galen (*c.* 129–*c.* 200) in the 2nd century for washing the body, although the Romasn seem to have washed with a mixture of sand and oil. Soap was manufactured in Britain from the 14th century, but better-quality soap was imported from Castile or Venice. The Soapmakers' Company, London, was incorporated in 1638. Soap was taxed in England from the time of Cromwell in the 17th century to 1853.

SOAP acronym for Simple Object Access Protocol, method of bi-directional communication developed by Microsoft, DevlopMentor, and Userland Software. SOAP allows programs running under different operating systems to communicate with each other across *firewalls using *HTTP and *XML as the exchange mechanisms.

soapstone compact, massive form of impure *talc.

social psychology branch of *psychology concerned with the behaviour of individuals in groups and the ways in which they relate to one another and to the societies of which they are a part.

sociobiology study of the biological basis of all social behaviour, including the application of *population genetics to the evolution of behaviour. It builds on the concept of *inclusive fitness, contained in the notion of the 'selfish gene'. Contrary to some popular interpretations, it does not assume that all behaviour is genetically determined.

socket in computing, mechanism for creating a connection to an application on another computer. A socket combines an *IP address (denoting the host computer on a network) with a port number describing the application (perhaps *FTP or *SMTP) the user requires.

soda ash former name for *sodium carbonate.

soda lime powdery mixture of calcium hydroxide and sodium hydroxide or potassium hydroxide, used in medicine and as a drying agent.

sodium chemical symbol Na, soft, waxlike, silver-white, metallic element, atomic number 11, relative atomic mass 22.989. Its chemical symbol comes from the Latin *natrium*. It is one of the *alkali metals (in Group 1 of the *periodic table) and has a very low density, being light enough to float on water. It is the sixth-most abundant element (the fourth-most abundant metal) in the Earth's crust. Sodium is highly reactive, oxidizing rapidly when exposed to air and reacting violently with water. It is one of the most reactive metals in the *reactivity series of metals. Its most familiar compound is sodium chloride (common salt), which occurs naturally in the oceans and in salt deposits left by dried-up ancient seas.

sodium carbonate or **soda ash**; Na_2CO_3, anhydrous white solid. The hydrated, crystalline form $(Na_2CO_3.10H_2O)$ is also known as washing soda.

It is made by the *Solvay process and used as a mild alkali, as it is hydrolysed in water:

$$CO_3{}^{2-}{}_{(aq)} + H_2O_{(l)} \rightarrow HCO_3{}^{-}{}_{(aq)} + OH^-{}_{(aq)}$$

It is used to neutralize acids, in glass manufacture, and in water softening.

sodium chloride or **common salt** or **table salt**; NaCl, white, crystalline compound found widely in nature. The crystals are cubic in shape. It is a typical ionic solid with a high melting point (801°C/1,474°F); it is soluble in water, insoluble in organic solvents, and is a strong electrolyte when molten or in aqueous solution. Found in concentrated deposits as the mineral halite, it is widely used in the food industry as a flavouring and preservative, and in the chemical industry in the manufacture of sodium, chlorine, and sodium carbonate.

Chlorine and sodium hydroxide are produced by the *electrolysis of brine, a solution of sodium chloride in water. Salt for use in the food industry includes added iodine to reduce the incidence of the iodine deficiency diseases such as goitre and cretinism.

sodium hydrogencarbonate $NaHCO_3$, chemical name for *bicarbonate of soda.

sodium hydroxide or **caustic soda**; $NaOH$, commonest alkali. The solid and the solution are corrosive. It is used to neutralize acids, in the manufacture of soap, and in drain and oven cleaners. It is prepared industrially from sodium chloride by the *electrolysis of concentrated brine.

software in computing, a collection of programs and procedures for making a computer perform a specific task, as opposed to *hardware, the physical components of a computer system. Software is created by programmers and is either distributed on a suitable medium, such as the *floppy disk, or built into the computer in the form of firmware. Examples of software include *operating systems, *compilers, and applications programs such as payrolls or word processors. No computer can function without some form of software.

software piracy in computing, unauthorized duplication of computer software. Although some software piracy is done by companies for financial gain, most piracy is done by private individuals who lend disks to friends or copy programs from the workplace to their computers at home.

software project life cycle various stages of development in the writing of a major program (software), from the identification of a requirement to the installation, maintenance, and support of the finished program. The process includes *systems analysis and *systems design.

software suite in computing, set of complementary programs which can be bought separately, or (at a considerable saving) as a bundled

package. Office suites are an especially common form of software suite.

soft water water that contains very few dissolved metal ions such as calcium (Ca^{2+}) or magnesium (Mg^{2+}). It lathers easily with soap, and no *scale is formed inside kettles or boilers. It has been found that the incidence of heart disease is higher in soft-water areas.

soil loose covering of broken rocky material and decaying organic matter overlying the bedrock of the Earth's surface. It is composed of minerals (formed from *physical weathering and *chemical weathering of rocks), organic matter (called *humus) derived from decomposed plants and organisms, living organisms, air, and water. Soils differ according to climate, parent material, rainfall, relief of the bedrock, and the proportion of organic material. The study of soils is **pedology**.

Soils influence the type of agriculture employed in a particular region – light well-drained soils favour arable farming, whereas heavy clay soils give rise to lush pasture land. Plant roots take in nutrients (in the form of *ions) dissolved in the water in soil. The main elements that plants need to absorb through their roots are *nitrogen, *phosphorus, and *potassium.

soil erosion wearing away and redistribution of the Earth's soil layer. It is caused by the action of water, wind, and ice, and also by improper methods of agriculture. If unchecked, soil erosion results in the formation of deserts (*desertification). It has been estimated that 20% of the world's cultivated topsoil was lost between 1950 and 1990.

If the rate of erosion exceeds the rate of soil formation (from rock and decomposing organic matter), then the land will become infertile. The removal of forests (*deforestation) or other vegetation often leads to serious soil erosion, because plant roots bind soil, and without them the soil is free to wash or blow away, as in the American dust bowl. The effect is worse on hillsides, and there has been devastating loss of soil where forests have been cleared from mountainsides, as in Madagascar.

Improved agricultural practices such as contour ploughing are needed to combat soil erosion. Windbreaks, such as hedges or strips planted with coarse grass, are valuable, and organic farming can reduce soil erosion by as much as 75%.

Soil degradation and erosion are becoming as serious as the loss of the rainforest. It is estimated that more than 10% of the world's soil lost a large amount of its natural fertility during the latter half of the 20th century. Some of the worst losses are in Europe, where 17% of the soil is damaged by human activity such as mechanized farming and fallout from acid rain. In Mexico and Central America, 24% of soil is highly degraded, mostly as a result of deforestation.

soil mechanics branch of engineering that studies the responses of soils with different water and air contents, to loads. Soil is investigated during construction work to ensure that it has the mechanical properties necessary to support the foundations of dams, bridges, and roads.

sol *colloid of very small solid particles dispersed in a liquid that retains the physical properties of a liquid.

solar cycle variation of activity on the *Sun over an 11-year period indicated primarily by the number of *sunspots visible on its surface. The next period of maximum activity is expected round 2011.

solar energy energy derived from the light and heat from the Sun. The amount of energy falling on just 1 sq km/0.4 sq mi is about 4,000 megawatts, enough to heat and light a small town. In one second the Sun gives off 13 million times more energy than all the electricity used in the USA in one year. **Solar heaters** usually consist of concave mirrors or reflective parabolic surfaces that concentrate the Sun's rays onto a black (heat-absorbing) panel containing pipes through which air or water is circulated, either by thermal *convection or by a pump. The heat energy of the air or water is converted into *electrical energy via a *turbine and a *generator. Hot water for industrial and domestic use can be produced by circulating water through panels, the water absorbing heat from the Sun as it passes through the panels.

Solar energy may also be harnessed indirectly using **solar cells** (photovoltaic cells) made of panels of *semiconductor material (usually silicon), which generate electricity when illuminated by sunlight. Although it is difficult to generate a high output from solar energy compared to sources such as nuclear or fossil fuels, it is a major nonpolluting and renewable energy source used as far north as Scandinavia as well as in the southwestern USA and in Mediterranean countries.

solar flare brilliant eruption on the Sun above a *sunspot, thought to be caused by release of magnetic energy. Flares reach maximum brightness within a few minutes, then fade away over about an hour. They eject a burst of atomic particles into space at up to 1,000 kps/600 mps. When these particles reach Earth they can cause radio blackouts, disruptions of the Earth's magnetic field, and *auroras.

solar pond natural or artificial 'pond', such as the Dead Sea, in which salt becomes more soluble in the Sun's heat. Water at the bottom becomes saltier and hotter, and is insulated by the less salty water layer at the top. Temperatures at the bottom reach about 100°C/212°F and can be used to generate electricity.

solar radiation radiation given off by the Sun, consisting mainly of visible light, *ultraviolet radiation, and *infrared radiation, although the whole spectrum of *electromagnetic waves is present, from radio waves to X-rays. High-energy charged particles, such as electrons, are also emitted, especially from solar *flares. When these reach the Earth, they cause magnetic storms (disruptions of the Earth's magnetic field), which interfere with radio communications.

Solar System *Sun (a star) and all the bodies orbiting it: the nine *planets (Mercury, Venus, Earth, Mars, Jupiter, Saturn, Uranus, Neptune, and Pluto), their moons, and smaller objects such as *asteroids and *comets. The Sun contains 99.86% of the mass of the Solar System. The planets orbit the Sun in elliptical paths, and in the same direction as the Sun itself rotates. The planets nearer the Sun have shorter orbital times than those further away since the distance they travel in each orbit is less.

The inner planets (Mercury, Venus, Earth, and Mars) have solid, rocky surfaces; relatively slow periods of rotation (Mercury takes 59 days to complete one rotation, Venus 243 days, Earth nearly 24 hours, and Mars 24.5 hours); very few natural *satellites; and diameters less than 13,000 km/8,000 mi. Venus can be seen with the unaided eye, appearing in the evening as the brightest 'star' in the sky. In contrast, the outer planets (Jupiter, Saturn, Uranus, and Neptune) have denser, gaseous atmospheres composed mainly of hydrogen and helium; fast periods of rotation (Jupiter takes 10 hours for one rotation, Saturn nearly 10.5 hours, and Uranus 11 hours); and many natural satellites (Jupiter and Saturn have more than 30 between them and Uranus has 15). Uranus, Neptune, and Pluto were discovered after the development of the telescope.

solar time time of day as determined by the position of the *Sun in the sky.

solar wind stream of atomic particles, mostly protons and electrons, from the Sun's corona, flowing outwards at speeds of between 300 kps/200 mps and 1,000 kps/600 mps.

solenoid coil of wire, usually cylindrical, in which a magnetic field is created by passing an electric current through it (see *electromagnet). This field can be used to temporarily magnetize, and so move, an iron rod placed on its axis. Mechanical valves attached to the rod can be operated by switching the current on or off, so converting electrical energy into mechanical energy. Solenoids are used to relay energy from the battery of a car to the starter motor by means of the ignition switch.

The solenoid has a magnetic field, one end being the north pole and the other end the south pole. A solenoid

behaves as any other magnet; by changing the direction of the current the position of the north and south poles can be switched. If a larger current is passed through the coil it becomes a stronger magnet.

solid in physics, a state of matter that holds its own shape (as opposed to a liquid, which takes up the shape of its container, or a gas, which totally fills its container). According to the *kinetic theory of matter, the atoms or molecules in a solid are packed closely together in a regular arrangement, and are not free to move but merely vibrate about fixed positions, such as those in crystal lattices.

solidification change of state from liquid (or vapour) to solid that occurs at the *freezing point of a substance.

solid-state circuit electronic circuit in which all the components (resistors, capacitors, transistors, and diodes) and interconnections are made at the same time, and by the same processes, in or on one piece of single-crystal silicon. The small size of this construction accounts for its use in electronics for space vehicles and aircraft.

solstice either of the days on which the Sun is farthest north or south of the celestial equator each year. In the northern hemisphere, the **summer solstice**, when the Sun is farthest north, occurs around 21 June and the **winter solstice** around 22 December.

solubility measure of the amount of solute (usually a solid or gas) that will dissolve in a given amount of solvent (usually a liquid) at a particular temperature. Solubility may be expressed as grams of solute per 100 grams of solvent or, for a gas, in parts per million (ppm) of solvent.

solubility curve graph that indicates how the solubility of a substance varies with temperature. Most salts increase their solubility with an increase in temperature, as they dissolve endothermically. These curves can be used to predict which salt, and how much of it, will crystallize out of a mixture of salts.

solute substance that is dissolved in another substance (see *solution).

solution two or more substances mixed to form a single, homogenous

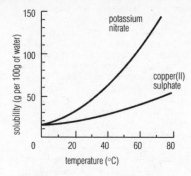

solubility curve Comparative aqueous solubility curves for copper(II) sulphate and potassium nitrate.

phase. One of the substances is the **solvent** and the others (**solutes**) are said to be dissolved in it.

The constituents of a solution may be solid, liquid, or gaseous. The solvent is normally the substance that is present in greatest quantity; however, if one of the constituents is a liquid this is considered to be the solvent even if it is not the major substance.

solution or **dissolution**, in earth science, process by which the minerals in a rock are dissolved in water. Solution is one of the processes of *erosion as well as *weathering (in which the breakdown of rock occurs without transport of the dissolved material). An example of this is when weakly acidic rainfall dissolves calcite.

Solvay process industrial process for the manufacture of sodium carbonate. It is a multistage process in which carbon dioxide is generated from limestone and passed through *brine saturated with ammonia. Sodium hydrogencarbonate is isolated and heated to yield sodium carbonate. All intermediate by-products are recycled so that the only ultimate by-product is calcium chloride.

solvent substance, usually a liquid, that will dissolve another substance (see *solution). Although the commonest solvent is water, in popular use the term refers to low-boiling-point organic liquids, which are harmful if vapour from them is breathed in as a result of their use in a confined space,

or from skin contact with the liquid. They can give rise to respiratory problems, liver damage, and neurological complaints.

Typical organic solvents are petroleum distillates (in glues), xylol (in paints), alcohols (for synthetic and natural resins such as shellac), esters (in lacquers, including nail varnish), ketones (in cellulose lacquers and resins), and chlorinated hydrocarbons (in paint stripper and dry-cleaning fluids). The fumes of some solvents, when inhaled (glue-sniffing), affect mood and perception. In addition to damaging the brain and lungs, repeated inhalation of solvent from a plastic bag can cause death by asphyxiation.

somnambulism another name for *sleepwalking.

sonar acronym for SOund Navigation And Ranging, method of locating underwater objects by the reflection of ultrasonic waves. The time taken for an acoustic beam to travel to the object and back to the source enables the distance to be found since the velocity of sound in water is known. Sonar devices, or **echo sounders**, were developed in 1920, and are the commonest means of underwater navigation.

sonic boom noise like a thunderclap that occurs when an aircraft passes through the *sound barrier, or begins to travel faster than the speed of sound. It happens when the cone-shaped shock wave caused by the plane touches the ground.

sonoluminescence emission of light by a liquid that is subjected to high-frequency sound waves. The rapid changes of pressure induced by the sound cause minute bubbles to form in the liquid, which then collapse. Light is emitted at the final stage of the collapse, probably because it squeezes and heats gas inside the bubbles.

sort in computing, a specific function used in a *spreadsheet program to arrange selected items into ascending or descending order. Information contained in tables (for example in Microsoft Word) can also be sorted.

sorting in computing, arranging data in sequence. When sorting a collection, or file, of data made up of several different *fields, one must be chosen as the **key field** used to establish the correct sequence. For example, the data in a company's mailing list might include fields for each customer's first names, surname, address, and telephone number. For most purposes the company would wish the records to be sorted alphabetically by surname; therefore, the surname field would be chosen as the key field.

sound physiological sensation received by the ear, originating in a vibration causing sound waves. The sound waves are pressure variations in the air and travel in every direction, spreading out as an expanding sphere. Sound energy cannot travel in a vacuum.

All sound waves in air travel with a speed dependent on the temperature; under ordinary conditions, this is about 330 m/1,080 ft per second. The pitch of the sound depends on the number of vibrations imposed on the air per second (*frequency), but the speed is unaffected. The loudness of a sound is dependent primarily on the amplitude of the vibration of the air.

sound absorption in acoustics, the conversion of sound energy to heat energy when sound waves strike a surface. The process reduces the amplitude of each reflected sound wave (echo) and thus the degree to which *reverberation takes place. Materials with good sound-absorbing properties are often fitted on walls and ceilings in buildings such as offices, factories, and concert halls in order to reduce or control sound levels.

sound barrier concept that the speed of sound, or sonic speed (about 1,220 kph/760 mph at sea level), constitutes a speed limit to flight through the atmosphere, since a badly designed aircraft suffers severe buffeting at near sonic speed owing to the formation of shock waves. US test pilot Chuck Yeager first flew through the 'barrier' in 1947 in a Bell X-1 rocket plane. Now, by careful design, such aircraft as Concorde can fly at supersonic speed with ease, though they create in their wake a *sonic boom.

sound card printed circuit board that, coupled with a set of speakers, enables a computer to reproduce music and sound effects.

sound wave longitudinal wave motion with which sound energy travels through a medium. It carries energy away from the source of the sound

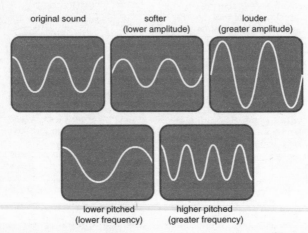

original sound   softer (lower amplitude)   louder (greater amplitude)

lower pitched (lower frequency)   higher pitched (greater frequency)

sound wave Some patterns created by sound waves of varying loudness and pitch. Loudness is related to the amplitude of a sound wave; a softer sound has a smaller amplitude and a louder sound a greater amplitude. Pitch is related to the frequency of a sound wave; a higher pitched sound will have a higher frequency and a lower pitched sound a lower frequency.

without carrying the material itself with it. Sound waves are mechanical; unlike electromagnetic waves, they require vibration of their medium's molecules or particles (manifested in air as compressions and rarefactions of the air), and this is why sound cannot travel through a vacuum.

In air, the pressure variations as an object vibrates travel at a speed of 330 m/1,080 ft per second, are detected by the ear, and interpreted by the brain as sound. A person with normal hearing can detect sounds with frequencies in the range of 20 to 20,000 hertz.

The energy of the air vibrating travels along the wave without transferring matter. A loosely-coiled spring can be used to demonstrate how sound waves travel through air. The disturbance produced by the vibrating object causes compressions and rarefactions of air particles to move in the same direction as the waves; they are called longitudinal waves. The pattern of longitudinal waves is produced when the spring is given a forward 'push'. Sound can be produced by the vibrations of objects such as the stretched strings of a violin or air particles in wind instruments; the sounds of the voice are produced by air causing the 'strings' of the vocal cords to vibrate.

Sound can also travel through solids and liquids. For example, voices can be heard through a wall between one room and another. This effect is used when listening with a stethoscope to hear the sounds of heart and lungs through the walls of the chest. Porpoises use ultrasonic sound in water as an echo-guiding system.

Sound waves can be reflected to produce echoes, and diffracted to produce interference patterns of louder and softer sound.

source code in computing, the original instructions written by computer programmers. Before these instructions can be understood, they must be processed by a *compiler and turned into *machine code.

source language in computing, the language in which a program is written, as opposed to *machine code, which is the form in which the program's instructions are carried out by the computer. Source languages are classified as either *high-level languages or *low-level languages, according to whether each notation in the source language stands for many or only one instruction in machine code.

source program in computing, a program written in a *source language.

source resistance alternative term for *internal resistance, the resistance inside an electric power supply.

southern lights common name for the *aurora australis, coloured light in southern skies.

space or **outer space**, void that exists beyond Earth's atmosphere. Above 120 km/75 mi, very little atmosphere remains, so objects can continue to move quickly without extra energy. The space between the planets is not entirely empty, but filled with the tenuous gas of the *solar wind as well as dust.

space adaptation syndrome alternative term for *space sickness.

space probe any instrumented object sent beyond Earth to collect data from other parts of the Solar System and from deep space. The first probe was the Soviet Lunik 1, which flew past the Moon in 1959. The first successful planetary probe was the US Mariner 2, which flew past Venus in 1962, using a transfer orbit. The first space probe to leave the Solar System was Pioneer 10 in 1983. Space probes include Galileo, Giotto, Magellan, Mars Observer, Ulysses, the *Moon probes, and the Mariner, Pioneer, Viking, and Voyager series.

space shuttle in full **space shuttle orbiter**, reusable crewed spacecraft developed by NASA to reduce the cost of using space for commercial, scientific, and military purposes. The orbiter, the part that goes into space, is 37.2 m/122 ft long and weighs 68 tonnes. The first, *Columbia*, was launched on 12 April 1981. After leaving its payload in space, the space shuttle can be flown back to Earth to land on a special runway 4.5 km/2.8 mi long and 91 m/100 yd wide, and is then available for reuse.

space sickness or **space adaptation syndrome**, feeling of nausea, sometimes accompanied by vomiting, experienced by about 40% of all astronauts during their first few days in space. It is akin to travel sickness, and is thought to be caused by confusion of the body's balancing mechanism, located in the inner ear, by weightlessness. The sensation passes after a few days as the body adapts.

space station any large structure designed for human occupation in space for extended periods of time. Space stations are used for carrying out astronomical observations and surveys of Earth, as well as for biological studies and the processing of materials in weightlessness. The first space station was the Soviet *Salyut 1* (1971). In 1973, NASA launched *Skylab*. The core of the Soviet space station *Mir* was launched in 1986. In 1998 the first component of the *International Space Station*, being constructed by the USA, Russia, and 14 other nations, was launched.

space-time in physics, combination of space and time used in the theory of *relativity. When developing relativity, German-born US phyicist Albert Einstein (1879–1955) showed that time was in many respects like an extra dimension (or direction) to space. Space and time can thus be considered as entwined into a single entity, rather than two separate things.

spamming advertising on the *Internet by broadcasting to many or all newsgroups regardless of relevance; **spam** is the junk e-mail received. Spamming is contrary to netiquette, the Net's conduct code, and is likely to result in the advertiser being bombarded by flames (angry messages), and 'dumping' (the downloading of large, useless files).

Sparc brand name formed from **Scalable Processor Architecture** to describe the design of *RISC processors designed by Sun Microsystems. Sparc chips are used in computers running *Solaris, Sun's version of *Unix.

spark chamber electronic device for recording tracks of charged subatomic particles, decay products, and rays. In combination with a stack of condenser plates, a spark chamber enables the point where an interaction has taken place to be located, to within a cubic centimetre. At its simplest, it consists of two smooth threadlike *electrodes that are positioned 1–2 cm/0.5–1 in apart, the space between being filled with a mixture of neon and helium gas. Sparks jump through the gas along the ionized path created by the radiation. See *particle detector.

Spearman's rank in statistics, an index of the strength of the relationship between two sets of information; for example, altitude and temperature. The value of the index varies between +1 (a perfect positive correlation) and –1 (a perfect negative correlation); 0 = no correlation.

speciation emergence of a new species during evolutionary history. One cause of speciation is the geographical separation of populations of the parent species, followed by reproductive isolation and selection for different environments so that they no longer produce viable offspring when they interbreed. Other causes are assortative mating and the establishment of a *polyploid population.

species in biology, a distinguishable group of organisms that resemble each other or consist of a few distinctive types (as in *polymorphism), and that can all interbreed to produce fertile offspring. Species are the lowest level in the system of biological classification.

specific gravity alternative term for *relative density.

specific heat capacity quantity of heat required to raise unit mass (1 kg) of a substance by one *kelvin (1 K). The unit of specific heat capacity in the SI system is the *joule per kilogram per kelvin ($J kg^{-1} K^{-1}$).

specific latent heat heat that changes the physical state of a unit mass (one kilogram) of a substance without causing any temperature change.

The **specific latent heat of fusion** of a solid substance is the heat required to change one kilogram of it from solid to liquid without any temperature change. The **specific latent heat of vaporization** of a liquid substance is the heat required to change one kilogram of it from liquid to vapour without any temperature change.

spectator ion in a chemical reaction that takes place in solution, an ion that remains in solution without taking part in the chemical change. For example, in the precipitation of barium sulphate from barium chloride and sodium sulphate, the sodium and chloride ions are spectator ions.

$$BaCl_2\ _{(aq)} + Na_2SO_4\ _{(aq)} \rightarrow BaSO_4\ _{(s)} + 2NaCl_{(aq)}$$

spectral classification in astronomy, classification of stars according to their surface temperature and luminosity, as determined from their spectra. Stars are assigned a spectral type (or class) denoted by the letters O, B, A, F, G, K, and M, where O stars (about 40,000 K/39,700°C/71,500°F) are the hottest and M stars (about 3,000 K/2,700°C/5,000°F) are the coolest.

spectrometer in physics and astronomy, instrument used to study the composition of light emitted by a source. The range, or *spectrum, of wavelengths emitted by a source depends upon its constituent elements, and may be used to determine its chemical composition.

The simpler forms of spectrometer analyse only visible light. A **collimator** receives the incoming rays and produces a parallel beam, which is then split into a spectrum by either a *diffraction grating or a prism mounted on a turntable. As the turntable is rotated each of the constituent colours of the beam may be seen through a **telescope**, and the angle at which each has been deviated may be measured on a circular scale. From this information the wavelengths of the colours of light can be calculated. Spectrometers are used in astronomy to study the electromagnetic radiation emitted by stars and other celestial bodies. The spectral information gained may be used to determine their chemical composition, or to measure the *red shift of light associated with the expansion of the universe and thereby calculate the speed with which distant stars are moving away from the Earth.

spectrometry in analytical chemistry, a technique involving the measurement of the spectrum of energies (not necessarily electromagnetic radiation) emitted or absorbed by a substance.

spectroscopic binary binary star in which two stars are so close together that they cannot be seen separately, but their separate light spectra can be distinguished by a spectroscope.

spectroscopy study of spectra (see *spectrum) associated with atoms or molecules in the solid, liquid, or gaseous phase. Spectroscopy can be used to identify unknown compounds and is an invaluable tool in science, medicine, and industry (for example, in checking the purity of drugs).

Emission spectroscopy is the study of the characteristic series of sharp lines in the spectrum produced when an *element is heated. Thus an unknown mixture can be analysed for its component elements. Related is **absorption spectroscopy**, dealing with atoms and molecules as they absorb energy in a characteristic way. Again, dark lines can be used for analysis. More detailed structural information can be obtained using **infrared spectroscopy** (concerned with molecular vibrations) or **nuclear magnetic resonance (NMR) spectroscopy** (concerned with interactions between adjacent atomic nuclei). **Supersonic jet laser beam spectroscopy** enables the isolation and study of clusters in the gas phase. A laser vaporizes a small sample, which is cooled in helium, and ejected into an evacuated chamber. The jet of clusters expands supersonically, cooling the clusters to near absolute zero, and stabilizing them for study in a *mass spectrometer.

spectrum plural **spectra**, in physics, the pattern of frequencies or wavelengths obtained when electromagnetic radiations are separated into their constituent parts. Visible light is part of the *electromagnetic spectrum and most sources emit waves over a range of wavelengths that can be broken up or 'dispersed'; white light can be separated (for example, using a triangular prism) into red, orange,

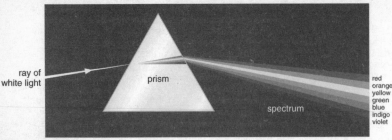

spectrum A prism (a triangular block of transparent material such as plastic, glass, or silica) is used to split a ray of white light into its spectral colours.

yellow, green, blue, indigo, and violet. The visible spectrum was first studied by English physicist Isaac Newton (1642–1727), who showed in 1666 how white light could be broken up into different colours.

speech recognition software computing system that enables data to be input by voice. It includes a microphone and *sound card that plugs into the computer and converts the analogue signals of the voice to digital signals. Examples include Dragon Naturally Speaking and IBM's Via Voice.

speed rate at which an object moves, or how fast an object moves. The average speed v of an object may be calculated by dividing the distance s it has travelled by the time t taken to do so, and may be expressed as:

$$v = s/t$$

The usual units of speed are metres per second or kilometres per hour.

Speed is a *scalar quantity in which direction of motion is unimportant (unlike the vector quantity *velocity, in which both magnitude and direction must be taken into consideration). Movement can be described by using motion graphs. Plotting distance against time in a *distance–time graph allows the total distance covered to be worked out. See also *speed–time graph.

speed of light speed at which light and other *electromagnetic waves travel in a vacuum. Its value is 299,792,458 m per second/186,282 mi per second but for most calculations 3 × 10^8 m s^{-1} (300 million metres per second) suffices. In glass the speed of light is two-thirds of its speed in air, about 200 million metres per second. The speed of light is the highest speed possible, according to the theory of *relativity, and its value is independent of the motion of its source and of the observer. It is impossible to accelerate any material body to this speed because it would require an infinite amount of energy.

speed of reaction alternative term for *rate of reaction.

speed of sound speed at which sound travels through a medium, such as air or water. In air at a temperature of 0°C/32°F, the speed of sound is 331 m/1,087 ft per second. At higher temperatures, the speed of sound is greater; at 18°C/64°F it is 342 m/1,123 ft per second. It is also affected by the humidity of the air. It is greater in liquids and solids; for example, in water it is about 1,440 m/4,724 ft per second, depending on the temperature.

speed–time graph graph used to describe the motion of a body by illustrating how its speed or velocity changes with time. The gradient of the graph gives the object's *acceleration. If the gradient is zero (the graph is horizontal) then the body is moving with constant speed or uniform velocity (the acceleration is zero); if the gradient is constant, the body is moving with uniform acceleration. If the gradient is positive then the body is accelerating; if the gradient is negative then the body is decelerating. The area under the graph gives the total distance travelled by the body.

sperm or **spermatozoon**, in biology, the male *gamete of animals before fertilization in *sexual reproduction.

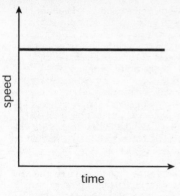

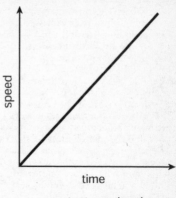

constant speed   constant acceleration

speed–time graph A speed–time graph shows how an object's speed (or velocity) changes over time. The gradient of the line indicates the acceleration of the object; if the line is horizontal, the acceleration is zero and the object is travelling at constant speed. If the gradient is positive, the object is accelerating and if negative, decelerating. The area under the graph measures the distance travelled.

Each sperm cell has a head capsule containing a nucleus, a middle portion containing *mitochondria (which provide energy), and a long tail (flagellum). In mammals sperm cells are produced in the testes of a male. They are produced by a special kind of cell division called *meiosis, which

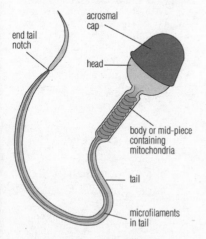

sperm Only a single sperm is needed to fertilize an egg, or ovum. Yet up to 500 million may start the journey towards the egg. Once a sperm has fertilized an egg, the egg's wall cannot be penetrated by other sperm. The unsuccessful sperm die after about three days.

halves the number of *chromosomes present.

spermatophyte another name for a seed plant.

spermatozoon in medicine, another name for a *sperm.

sphalerite mineral composed of zinc sulphide with a small proportion of iron, formula (Zn,Fe)S. It is the chief ore of zinc. Sphalerite is brown with a non-metallic lustre unless an appreciable amount of iron is present (up to 26% by weight). Sphalerite usually occurs in ore veins in limestones, where it is often associated with galena. It crystallizes in the cubic system but does not normally form perfect cubes.

sphere perfectly round object with all points on its surface the same distance from the centre. This distance is the radius of the sphere. For a sphere of radius r, the volume $V = \frac{4}{3}\pi r^3$ and the surface area $A = 4\pi r^2$.

sphincter ring of muscle, such as is found at various points in the *alimentary canal, that contracts and relaxes to open and close the canal and control the movement of food. The **pyloric sphincter**, at the base of the stomach, controls the release of the gastric contents into the *duodenum. After release the sphincter contracts, closing off the stomach. The **external anal sphincter** closes the anus; the

internal anal sphincter constricts the rectum; the **sphincter vesicae** controls the urethral orifice of the bladder. In the eye the **sphincter pupillae** contracts the pupil in response to bright light.

spider in computing, program that combs the *Internet for new documents such as Web pages and *FTP files. Spiders start their work by retrieving a document such as a Web page and then following all the links and references contained in it. They repeat the process with the followed links, supplying all the references they find to a database that can be searched via a *search engine.

spin in physics, the intrinsic *angular momentum of a subatomic particle, nucleus, atom, or molecule, which continues to exist even when the particle comes to rest. A particle in a specific energy state has a particular spin, just as it has a particular electric charge and mass. According to *quantum theory, this is restricted to discrete and indivisible values, specified by a spin *quantum number. Because of its spin, a charged particle acts as a small magnet and is affected by magnetic fields.

spinel any of a group of 'mixed oxide' minerals consisting mainly of the oxides of magnesium and aluminium, $MgAl_2O_4$ and $FeAl_2O_4$. Spinels crystallize in the cubic system, forming octahedral crystals. They are found in high-temperature igneous and metamorphic rocks. The aluminium oxide spinel contains gem varieties, such as the ruby spinels of Sri Lanka and Myanmar (Burma).

spiracle in insects, the opening of a *trachea, through which oxygen enters the body and carbon dioxide is expelled. In cartilaginous fishes (sharks and rays), the same name is given to a circular opening that marks the remains of the first gill slit.

spiral galaxy one of the main classes of *galaxy in the *Hubble classification comprising up to 30% of known galaxies. Spiral galaxies are characterized by a central bulge surrounded by a flattened disc containing (normally) two spiral arms composed of hot young stars and clouds of dust and gas. In about half of spiral galaxies (barred spirals) the arms originate at the ends of a bar across the central bulge. The bar is not a rigid object but consists of stars in motion about the centre of the galaxy.

spleen organ in vertebrates, part of the reticuloendothelial system, which helps to process *lymphocytes. It also regulates the number of red blood cells in circulation by destroying old cells, and stores iron. It is situated on the left side of the body, behind the stomach.

spondylosis degenerative changes in the spine due to ageing or secondary to injury or other disease.

spongiform encephalopathy in medicine, another term for *transmissible spongiform encephalopathy.

spontaneous combustion burning that is not initiated by the direct application of an external source of heat. A number of materials and chemicals, such as hay and sodium chlorate, can react with their surroundings, usually by oxidation, to produce so much internal heat that combustion results. Special precautions must be taken for the storage and handling of substances that react violently with moisture or air. For example, phosphorus ignites spontaneously in the presence of air and must therefore be stored under water; sodium and potassium are stored under kerosene in order to prevent their being exposed to moisture.

spontaneous generation or **abiogenesis**, erroneous belief that living organisms can arise spontaneously from nonliving matter. This survived until the mid-19th century, when the French chemist Louis Pasteur (1822–1895) demonstrated that a nutrient broth would not generate micro-organisms if it was adequately sterilized. The theory of *biogenesis holds that spontaneous generation cannot now occur; it is thought, however, to have played an essential role in the origin of life on this planet 4 billion years ago.

sporangium structure in which *spores are produced.

spore small reproductive or resting body, usually consisting of just one cell. Unlike a *gamete, it does not need to fuse with another cell in order to develop into a new organism. Spores are produced by the lower plants, most fungi, some bacteria, and certain protozoa. They are generally light and easily dispersed by wind movements.

Plant spores are haploid and are produced by the sporophyte, following *meiosis; see *alternation of generations.

sporophyte diploid spore-producing generation in the life cycle of a plant that undergoes *alternation of generations.

spreadsheet in computing, a program that mimics a sheet of ruled paper, divided into columns down the page, and rows across. The user enters values into *cells within the worksheet, then instructs the program to perform some operation on them, such as totalling a column or finding the average of a series of numbers. Calculations are made by using a *formula. Highly complex numerical analyses may be built up from these simple steps.

spring in geology, a natural flow of water from the ground, formed at the point of intersection of the water table and the ground's surface. The source of water is rain that has percolated through the overlying rocks. During its underground passage, the water may have dissolved mineral substances that may then be precipitated at the spring (hence, a mineral spring). A spring may be continuous or intermittent, and depends on the position of the water table and the topography (surface features).

spring balance instrument for measuring weight that relates the weight of an object to the extent to which it stretches or compresses a vertical spring. According to *Hooke's law, the extension or compression will be directly proportional to the weight, providing that the spring is not overstretched. A pointer attached to the spring indicates the weight on a scale, which may be calibrated in newtons (the SI unit of force) for physics experiments, or in grams, kilograms, or pounds (units of mass) for everyday use.

sprite in computing, a graphics object made up of a pattern of *pixels (picture elements) defined by a computer programmer. Some *high-level languages and applications programs contain routines that allow a user to define the shape, colours, and other characteristics of individual graphics objects. These objects can then be manipulated and combined to produce animated games or graphic screen displays.

sprite in earth science, rare thunderstorm-related luminous flash. Sprites occur in the mesosphere, at altitudes of 50–90 km/30–55 mi. They are electrical, like lightning, and arise when the electrical field that occurs between the thunder cloud top and the ionosphere (ionized layer of the Earth's atmosphere) draws electrons upwards from the cloud. If the air is thin and this field is strong the electrons accelarate rapidly, transferring kinetic energy to molecules they collide with. The excited molecules then discharge this energy as a light flash.

sq abbreviation for **square** (measure).

SQL abbreviation for structured query language, high-level computer language designed for use with *relational databases. Although it can be used by programmers in the same way as other languages, it is often used as a means for programs to communicate with each other. Typically, one program (called the 'client') uses SQL to request data from a database 'server'.

square in geometry, a quadrilateral (four-sided) plane figure with all sides equal and each angle a right angle. Its diagonals bisect each other at right angles. The area A of a square is the length l of one side multiplied by itself ($A = l \times l$). Also, any quantity multiplied by itself is termed a square, represented by an *exponent of power 2; for example, $4 \times 4 = 4^2 = 16$ and $6.8 \times 6.8 = 6.8^2 = 46.24$.

square root number that when it is *squared (multiplied by itself) equals a given number. For example, the square root of 25 (written as $\sqrt{25}$) is +5 or –5. This is because $+5 \times +5 = 25$, and $-5 \times -5 = 25$.

sr in physics, symbol for *steradian, the SI unit of solid angle. For example, a sphere subtends a solid angle of 4π steradians at its centre.

SRAM acronym for **static random-access memory**, computer memory device in the form of a silicon chip used to provide *immediate access memory. SRAM is faster but more expensive than *DRAM (dynamic random-access memory), and does not require such frequent refreshing.

SSSI abbreviation for *Site of Special Scientific Interest.

stability measure of how difficult it is to move an object from a position of balance or *equilibrium with respect to gravity.

stable equilibrium state of equilibrium possessed by a body that will return to its original rest position if displaced slightly. See *stability.

stack in computing, a method of storing data in which the most recent item stored will be the first to be retrieved. The technique is commonly called 'last in, first out'.

stain coloured compound that will bind to other substances. Stains are used extensively in microbiology to colour micro-organisms and in histochemistry to detect the presence and whereabouts in plant and animal tissue of substances such as fats, cellulose, and proteins.

stainless steel widely used *alloy, in which chromium is dominant with traces of nickel, that resists rusting. Its chromium content also gives it a high tensile strength. It is used for cutlery and kitchen fittings, and in surgical instruments. Stainless steel was first produced in the UK in 1913 and in Germany in 1914.

stalactite and stalagmite cave structures formed by the deposition of calcite dissolved in ground water. **Stalactites** grow downwards from the roofs or walls and can be icicle-shaped, straw-shaped, curtain-shaped, or formed as terraces. **Stalagmites** grow upwards from the cave floor and can be conical, fir-cone-shaped, or resemble a stack of saucers. Growing stalactites and stalagmites may meet to form a continuous column from floor to ceiling.

Stalactites are formed when ground water, hanging as a drip, loses a proportion of its carbon dioxide into the air of the cave. This reduces the amount of calcite that can be held in solution, and a small trace of calcite is deposited. Successive drips build up the stalactite over many years. In stalagmite formation the calcite comes out of the solution because of agitation – the shock of a drop of water hitting the floor is sufficient to remove some calcite from the drop. The different shapes result from the splashing of the falling water.

stamen male reproductive organ of a flower. The stamens are collectively referred to as the androecium. A typical stamen consists of a stalk, or filament, with an anther, the pollen-bearing organ, at its apex, but in some

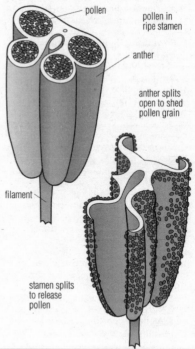

stamen The stamen is the male reproductive organ of a flower. It has a thin stalk called a filament with an anther at the tip. The anther contains pollen sacs, which split to release tiny grains of pollen.

primitive plants, such as *Magnolia*, the stamen may not be markedly differentiated.

The number and position of the stamens are significant in the classification of flowering plants. Generally the more advanced plant families have fewer stamens, but they are often positioned more effectively so that the likelihood of successful pollination is not reduced.

stand-alone computer self-contained computer, usually a microcomputer, that is not connected to a network of computers and can be used in isolation from any other device.

standard measure against which others are compared. For example, until 1960 the standard for the metre was the distance between two lines engraved on a platinum–iridium bar kept at the International Bureau of Weights and Measures in Sèvres, France.

standard atmosphere alternative term for *atmosphere, a unit of pressure.

standard deviation in statistics, a measure (symbol σ or s) of the spread of data. The deviation (difference) of each of the data items from the mean is found, and their values squared. The mean value of these squares is then calculated. The standard deviation is the square root of this mean.

standard form or **scientific notation**, method of writing numbers often used by scientists, particularly for very large or very small numbers. The numbers are written with one digit before the decimal point and multiplied by a power of 10. The number of digits given after the decimal point depends on the accuracy required. For example, the *speed of light is 2.9979×10^8 m/ 1.8628×10^5 mi per second.

standard gravity acceleration due to gravity, generally taken as 9.81274 m/ 32.38204 ft per second per second. See also *g scale.

standard illuminant any of three standard light intensities, A, B, and C, used for illumination when phenomena involving colour are measured. A is the light from a filament at 2,848 K (2,575°C/4,667°F), B is noon sunlight, and C is normal daylight. B and C are defined with respect to A. Standardization is necessary because colours appear different when viewed in different lights.

standard model in physics, modern theory of *elementary particles and their interactions. According to the standard model, elementary particles are classified as leptons (light particles, such as electrons), *hadrons (particles, such as neutrons and protons, that are formed from quarks), and gauge bosons. Leptons and hadrons interact by exchanging *gauge bosons, each of which is responsible for a different fundamental force: photons mediate the electromagnetic force, which affects all charged particles; gluons mediate the strong nuclear force, which affects quarks; gravitons mediate the force of gravity; and the intermediate vector bosons mediate the weak nuclear force. See also *forces, fundamental, *quantum electrodynamics, and *quantum chromodynamics.

standards in computing, any agreed system or protocol that helps different pieces of software or different computers to work together.

standard temperature and pressure STP, in chemistry, a standard set of conditions for experimental measurements, to enable comparisons to be made between sets of results. Standard temperature is 0°C/32°F (273 K) and standard pressure 1 atmosphere (101,325 Pa).

standard time time in any of the 24 time zones, each an hour apart, into which the Earth is divided. The respective times depend on their distances, east or west of Greenwich, England. In North America the eight zones (Atlantic, Eastern, Central, Mountain, Pacific, Alaska, Hawaii-Aleutian, and Samoa) use the mean solar times of meridians 15° apart, starting with 60° longitude. (See also *time.)

standard volume in physics, the volume occupied by one kilogram molecule (the molecular mass in kilograms) of any gas at standard temperature and pressure. Its value is approximately 22.414 cubic metres.

standing crop in ecology, the total number of individuals of a given species alive in a particular area at any moment. It is sometimes measured as the weight (or *biomass) of a given species in a sample section.

standing wave in physics, a wave in which the positions of *nodes (positions of zero vibration) and antinodes (positions of maximum vibration) do not move. Standing waves result when two similar waves travel in opposite directions through the same space.

For example, when a sound wave is reflected back along its own path, as when a stretched string is plucked, a standing wave is formed. In this case the antinode remains fixed at the centre and the nodes are at the two ends. Water and *electromagnetic waves can form standing waves in the same way.

staphylococcus spherical bacterium that occurs in clusters. It is found on the skin and mucous membranes of humans and other animals. It can cause abscesses and systemic infections that may prove fatal.

star luminous globe of gas, mainly hydrogen and helium, which produces its own heat and light by nuclear reactions. Although stars shine for a very long time – many billions of years – they change in appearance at different stages in their lives (they are said to have a 'life cycle'). Stars seen at night belong to our *galaxy, the Milky Way. The Sun is the nearest star to Earth; other stars in the Milky Way are large distances away (to get to the nearest would take about 4 years travelling at the speed of light).

starch widely distributed, high-molecular-mass *carbohydrate, produced by plants as a food store; main dietary sources are cereals, legumes, and tubers, including potatoes. It consists of varying proportions of two *glucose polymers (*polysaccharides): straight-chain (amylose) and branched (amylopectin) molecules.

Purified starch is a white powder used to stiffen textiles and paper and as a raw material for making various chemicals. It is used in the food industry as a thickening agent. Chemical treatment of starch gives rise to a range of 'modified starches' with varying properties. Hydrolysis (splitting) of starch by acid or enzymes generates a variety of 'glucose syrups' or 'liquid glucose' for use in the food industry. Complete hydrolysis of starch with acid generates the *monosaccharide glucose only. Incomplete hydrolysis or enzymic hydrolysis yields a mixture of glucose, maltose, and nonhydrolysed fractions called dextrins.

The chemical test for starch consists of adding a drop of iodine solution to the substance. It will turn bright blue if starch is present.

star cluster group of related stars, usually held together by gravity. Members of a star cluster are thought to form together from one large cloud of gas in space. **Open clusters** such as the Pleiades contain from a dozen to many hundreds of young stars, loosely scattered over several light years. Globular clusters are larger and much more densely packed, containing perhaps 10,000–1,000,000 stars.

start bit *bit used in asynchronous communications to indicate the beginning of a piece of data.

startup screen in computing, screen displayed by a PC while it loads its *operating system and other resident software. Well-known startup screens include the Windows 95 'flying window' motif and the Macintosh 'smiling face'. It is also possible to create a custom start-up screen, perhaps incorporating a favourite image or corporate logo.

stasis in biology, the cessation of blood flow in the blood vessels or the passage of food in the gastrointestinal tract.

states of matter forms (solid, liquid, or gas) in which material can exist. Whether a material is solid, liquid, or gaseous depends on its temperature and pressure. The transition between states takes place at definite temperatures, called the melting point and boiling point.

*Kinetic theory describes how the state of a material depends on the movement and arrangement of its atoms or molecules. The atoms or molecules of **gases** move randomly in

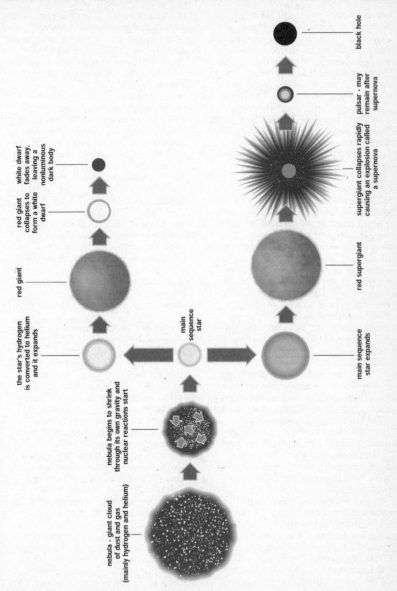

star The life cycle of a star. New stars are being formed all the time when nebulae (giant clouds of dust and gas) contract due to the action of gravity. As the star contracts and heats up eventually nuclear reactions begin and the star becomes a main sequence star. If the star is less than 1.2 times the mass of the Sun, it eventually forms a white dwarf that finally fades to a dark body. If it is a massive star, then the main sequence star expands to become a red supergiant that eventually explodes as a supernova. It leaves part of the core as a neutron star (pulsar), or as a black hole if the mass of the collapsing supernova core is three times greater than the Sun.

otherwise empty space, filling any size or shape of container. Gases can be liquefied by cooling as this lowers the speed of the molecules and enables attractive forces between them to bind them together. A **liquid** forms a level surface and assumes the shape of its container; its atoms or molecules do not occupy fixed positions, nor do they have total freedom of movement. **Solids** hold their own shape as their atoms or molecules are not free to move about but merely vibrate about fixed positions, such as those in *crystal lattices.

state symbol symbol used in chemical equations to indicate the physical state of the substances present. The symbols are: (s) for solid, (l) for liquid, (g) for gas, and (aq) for aqueous.

static electricity *electric charge that is stationary, usually acquired by a body by means of electrostatic induction or friction. Rubbing different materials can produce static electricity, as seen in the sparks produced on combing one's hair or removing a nylon shirt. The frictional force causes electrons to move out of their orbits. The electrons are then transferred to another material. The material that gains electrons becomes negatively charged and the material that loses electrons becomes positively charged. In some processes static electricity is useful, as in paint spraying where the parts to be sprayed are charged with electricity of opposite polarity to that on the paint droplets, and in xerography.

statics branch of mechanics concerned with the behaviour of bodies at rest and forces in equilibrium, and distinguished from *dynamics.

statistical mechanics branch of physics established by Austrian physicist Ludwig Boltzmann (1844–1906) in which the properties of large collections of particles are predicted by considering the motions of the constituent particles. It is closely related to *thermodynamics.

statistics branch of mathematics concerned with the collection and interpretation of data. For example, to determine the *mean age of the children in a school, a statistically acceptable answer might be obtained by calculating an average based on the ages of a representative sample, consisting, for example, of a random tenth of the pupils from each class. *Probability is the branch of statistics dealing with predictions of events.

STD abbreviation for *sexually transmitted disease.

steady-state theory in astronomy, rival theory to that of the *Big Bang, which claims that the universe has no origin but is expanding because new matter is being created continuously throughout the universe. The theory was proposed in 1948 by Austrian-born British cosmologist Hermann Bondi, Austrian-born US astronomer Thomas Gold, and English astronomer, cosmologist, and writer Fred Hoyle, but it was dealt a severe blow in 1965 by the discovery of *cosmic background radiation (radiation left over from the Big Bang and the formation of the universe) and is now largely rejected.

steam dry, invisible gas formed by vaporizing water. The visible cloud that normally forms in the air when water is vaporized is due to minute suspended water particles. Steam is widely used in chemical and other industrial processes and for the generation of power.

steam engine engine that uses the power of steam to produce useful work. The first successful steam engine was built in 1712 by English inventor Thomas Newcomen at Dudley, West Midlands; it was developed further by Scottish instrument maker James Watt from 1769 and by English mining engineer Richard Trevithick, whose high-pressure steam engine of 1802 led to the development of the steam locomotive.

In Newcomen's engine, steam was admitted to a cylinder as a piston moved up, and was then condensed by a spray of water, allowing air pressure to force the piston downwards. James Watt improved Newcomen's engine in 1769 by condensing the steam outside the cylinder (thus saving energy formerly used to reheat the cylinder) and by using steam to move the piston. Watt also introduced the **double-acting engine**, in which steam is alternately sent to each side of the piston forcing it up and down. The **compound engine** (1781) uses the exhaust from one cylinder to drive the piston of another. A later development was the steam *turbine, still used today to power ships and generators in power stations. In other contexts, the steam engine was superseded by the *internal-combustion engine or the *electric motor.

stearic acid $CH_3(CH_2)_{16}COOH$, saturated long-chain *fatty acid, soluble in alcohol and ether but not in water. It is found in many fats and oils, and is used to make soap and candles and as a lubricant. The salts of stearic acid are called stearates.

stearin mixture of stearic and palmitic acids, used to make soap.

steel alloy or mixture of iron and up to 1.7% carbon, sometimes with other elements, such as manganese, phosphorus, sulphur, and silicon. The USA, Russia, Ukraine, and Japan are the main steel producers. Steel has innumerable uses, including ship and car manufacture, skyscraper frames, and machinery of all kinds.

Steels with only small amounts of other metals are called **carbon steels**. These steels are far stronger than pure iron, with properties varying with the composition. **Alloy steels** contain greater amounts of other metals. Low-alloy steels have less than 5% of the alloying material; high-alloy steels have more. Low-alloy steels containing up to 5% silicon with relatively little carbon have a high electrical resistance and are used in power transformers and motor or generator cores, for example. **Stainless steel** is a high-alloy steel containing at least 11% chromium. Steels with up to 20% tungsten are very hard and are used in high-speed cutting tools. About 50% of the world's steel is now made from scrap.

Steel is produced by increasing the carbon content of wrought iron, or decreasing that of cast iron, produced by a *blast furnace. The main industrial process is the *basic–oxygen process, in which molten pig iron and scrap steel is placed in a container lined with heat-resistant, alkaline

(basic) bricks. A pipe or lance is lowered near to the surface of the molten metal and pure oxygen blown through it at high pressure. The surface of the metal is disturbed by the blast and the impurities are oxidized (burned out). The **open-hearth process** is an older steelmaking method in which molten iron and limestone are placed in a shallow bowl or hearth (such as an *open-hearth furnace). Burning oil or gas is blown over the surface of the metal, and the impurities are oxidized. High-quality steel is made in an **electric furnace**. A large electric current flows through electrodes in the furnace, melting a charge of scrap steel and iron. The quality of the steel produced can be controlled precisely because the temperature of the furnace can be maintained exactly and there are no combustion by-products to contaminate the steel. Electric furnaces are also used to refine steel, producing the extra-pure steels used, for example, in the petrochemical industry.

The steel produced is cast into ingots, which can be worked when hot by hammering (forging) or pressing between rollers to produce sheet steel. Alternatively, the **continuous-cast process**, in which the molten metal is fed into an open-ended mould cooled by water, produces an unbroken slab of steel.

Stefan–Boltzmann constant symbol σ, in physics, a constant relating the energy emitted by a black body (a hypothetical body that absorbs or emits all the energy falling on it) to its temperature. Its value is 5.6697×10^{-8} W m^{-2} K^{-4}.

Stefan–Boltzmann law in physics, a law that relates the energy, E, radiated away from a perfect emitter (a *black body), to the temperature, T, of that body. It has the form $E = \sigma T^4$, where E is the energy radiated per unit area per second, T is the temperature, and σ is the **Stefan–Boltzmann constant**. Its value is 5.6697×10^{-8} W m^{-2} K^{-4}. The law was derived by the Austrian physicists Josef Stefan (1835–1893) and Ludwig Boltzmann (1844–1906).

steganography in computing, camouflaging messages in large computer files, especially those carrying audio, video, or graphics, by appropriating a small percentage of their constituent data. For example, a graphics file measuring 500×500 *pixels, using 32 *bits to represent each pixel, contains 8 million bits. A single bit of each pixel (perhaps the 1st, the 15th, or the 32nd) could be used to insert some 5,000 words of text, chopped into individual bits, without making any perceptible difference to the image. The text message itself can be encrypted using Pretty Good Privacy (PGP) for added security.

stellar population in astronomy, classification of stars according to their chemical composition as determined by *spectroscopy.

stem cell embryonic cell that can develop into different tissues. Stem cells are used in medical research but their use is ethically controversial as the donor *embryo is destroyed. There are three basic kinds of stem cell, which are known as totipotent, pluripotent, and multipotent. **Totipotent** stem cells form when a fertilized egg first divides, and can form a complete organism by forming all the necessary tissues, such as bone, muscle, or nerve tissue, and also form the placenta. **Pluripotent** stem cells form as part of the *blastocyst and can form most kinds of tissue, but are not able to generate the whole organism. **Multipotent** cells are generated by an embryo and give rise to only specific kinds of cells.

steradian symbol sr, SI unit of measure of solid (three-dimensional) angles, the three-dimensional equivalent of the *radian. One steradian is the angle at the centre of a sphere when an area on the surface of the sphere equal to the square of the sphere's radius is joined to the centre.

stereoscopic display in computing, a display which achieves a 3-dimensional (3D) image effect using only one *monitor, and a pair of special glasses. The effect is achieved by running the monitor at double the normal frame rate and arranging to show only every other frame to each eye by means of special polarising glasses which black out each eye alternately. This enables

each eye to see a normal frame rate image, and the two images differ sufficiently to cause a 3D image to be seen. These kind of displays are used for industrial 3D visualisation, and in 3D cinemas.

sterility inability to reproduce. It may be due to infertility or it may be induced, *sterilization.

sterilization killing or removal of living organisms such as bacteria and fungi. A sterile environment is necessary in medicine, food processing, and some scientific experiments. Methods include heat treatment (such as boiling), the use of chemicals (such as disinfectants), irradiation with gamma rays, and filtration. See also *asepsis.

sterilization in medicine, any surgical operation to terminate the possibility of reproduction. In women, this is normally achieved by sealing or tying off the *Fallopian tubes (tubal ligation) so that fertilization can no longer take place. In men, the transmission of sperm is blocked by *vasectomy.

sternum or **breastbone**, large flat bone, 15–20 cm/5.9–7.8 in long in the adult, at the front of the chest, joined to the ribs. It gives protection to the heart and lungs. During open-heart surgery the sternum must be split to give access to the thorax.

steroid any of a group of cyclic, unsaturated alcohols (lipids without fatty acid components), which, like sterols, have a complex molecular structure consisting of four carbon rings. Steroids include the sex hormones, such as *testosterone, the corticosteroid hormones produced by the *adrenal gland, bile acids, and *cholesterol. The term is commonly used to refer to *anabolic steroid. In medicine, synthetic steroids are used to treat a wide range of conditions. Steroids are also found in plants. The most widespread are the **brassinosteroids**, necessary for normal plant growth.

sterol any of a group of solid, cyclic, unsaturated alcohols, with a complex structure that includes four carbon rings; cholesterol is an example. Steroids are derived from sterols.

stigma in a flower, the surface at the tip of a *carpel that receives the *pollen. It often has short outgrowths, flaps, or hairs to trap pollen and may produce a sticky secretion to which the grains adhere.

Stirling engine A hot-air external combustion engine invented by Scottish priest Robert Stirling 1816. Modern versions use helium. The engine operates by adapting to the fact that the air in its two cylinders, the working cylinder and the displacement, heats up when it is compressed and cools when it expands. Expanding gas provides the power for the working cylinder. The engine will operate on any fuel, is nonpolluting and relatively quiet. It was used fairly widely in the 19th century before the appearance of small, powerful, and reliable electric motors. Attempts have also been made in recent times to use Stirling's engine to power a variety of machines.

stokes symbol St, c.g.s. unit of kinematic viscosity (a liquid's resistance to flow). Named after Irish physicist George Gabriel Stokes (1819–1903).

Stokes' law A law discovered by Irish physicist George Gabriel Stokes (1819–1903), giving the force (F) acting on a sphere falling through a liquid $F = 6\pi\eta rv$ where η is the liquid's viscosity and r and v are the radius and velocity of the sphere.

stoma plural **stomata**, in botany, a pore (tiny hole) in the epidermis (outer layer of tissue) of a plant. There are lots of these holes, usually in the lower surface of the *leaf. A leaf contains several layers of *tissue. The outer layer is the epidermis and is only one cell thick. Stomata occur in the lower epidermis. Each stoma is surrounded by a pair of guard cells that are crescent-shaped when the stoma is open but can collapse to an oval shape, thus closing off the opening between them. Stomata allow the exchange of carbon dioxide and oxygen (needed for *photosynthesis and *respiration) between the internal tissues of the plant and the outside atmosphere. They are also the main route by which water is lost from the plant, and they can be closed to conserve water, the

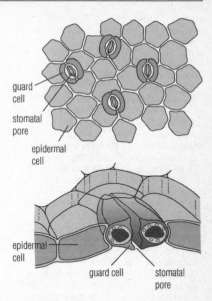

stoma The stomata, tiny openings in the epidermis of a plant, are surrounded by pairs of crescent-shaped cells, called guard cells. The guard cells open and close the stoma by changing shape.

movements being controlled by changes in turgidity of the guard cells. At night the stomata may allow oxygen to diffuse in and carbon dioxide out because only respiration is taking place. However in many plants the stomata close at night, preventing this gas exchange.

stomach organ that forms the first cavity in the digestive system of animals. In mammals it is a bag of *muscle situated just below the diaphragm. Food enters it from the *oesophagus, is digested by the acid and *enzymes secreted by the stomach lining. The wall of the stomach contracts to mix the food with the digestive juice to help digestion of *protein. After a while, partly digested food is then passed into the small *intestine (duodenum). Some plant-eating mammals have multi-chambered stomachs that harbour bacteria in one of the chambers to assist in the digestion of *cellulose.

The gizzard is part of the stomach in birds.

stone plural **stone**; abbreviation st, imperial unit of mass. One stone is 14 pounds (6.35 kg).

stool in medicine, alternative term for faeces.

stop bit *bit used in asynchronous communications to indicate the end of a piece of data.

storm extreme weather condition characterized by strong winds, rain, hail, thunder, and lightning.

storm surge abnormally high tide brought about by a combination of a deep atmospheric *depression (very low pressure) over a shallow sea area, high spring tides, and winds blowing from the appropriate direction. It can be intensified by snowmelt and/or river flooding. A storm surge can cause severe flooding of lowland coastal regions and river estuaries.

story board in film and television, technique for reviewing a particular story or scene before it is expensively filmed, animated, or scripted. The scene is broken down into key frames or moments, and sketched in varying detail onto boards with accompanying text outlining the plot's progress.

straight line line that does not bend or curve. The graph of a linear relationship is a straight line and is often presented in the form $y = mx + c$, where m is the slope, or gradient, of the line and c is the y-intercept (the point at which the line cuts the y-axis).

strain in the science of materials, the extent to which a body is distorted when a deforming force (stress) is applied to it. It is a ratio of the extension or compression of that body (its length, area, or volume) to its original dimensions (see *Hooke's law). For example, linear strain is the ratio of the change in length of a body to its original length.

strata singular **stratum**, layers or beds of sedimentary rock.

stratosphere that part of the atmosphere 10–40 km/6–25 mi from the Earth's surface, where the temperature slowly rises from a low of −55°C/−67°F to around 0°C/32°F. The air is rarefied and at around 25 km/15 mi much *ozone is concentrated.

stratovolcano another term for *composite volcano, a type of volcano made up of alternate ash and lava layers.

streaming in computing, sending data, for example video frames or radio broadcasts, in a steady flow over the Internet. Streaming requires data to pass through a special channel or dedicated connection; conventional *packets, which travel by a multiplicity of routes, may arrive in the wrong order or be duplicated on the way. Streaming is used in multimedia presentations, such as RealAudio, RealVideo, RealText, and Flash animations; it enables presentations to be commenced before files are finished downloading. The data is placed in a temporary memory area until it is downloaded.

strength of acids and bases in chemistry, the ability of *acids and *bases to dissociate in solution with water, and hence to produce a low or high *pH, respectively. A strong acid is fully dissociated in aqueous solution, whereas a weak acid is only partly dissociated. Since the *dissociation of acids generates hydrogen ions, a solution of a strong acid will have a high concentration of hydrogen ions and therefore a low pH. A strong base will have a high pH, whereas a weaker base will not dissociate completely and will have a pH of nearer 7.

streptococcus any one of a genus of round or oval Gram-positive bacteria that have a tendency to form pairs or chains. They are widely distributed in nature, living mainly as parasites in the bodies of animals and humans. Some are harmless, but others are implicated in a number of infections, including *scarlet fever. Streptococcal bacteria found in the human intestinal tract are called **enterococci**. The number of enterococcal infections showing resistance to vancomycin (a toxic 'last resort' antibiotic) are increasing. In New Jersey, USA, where hospitals must report cases of drug-resistant micro-organisms, reports of vancomycin-resistant enterococci rose from 99 in 1992 to 278 in 1994.

streptokinase enzyme produced by *Streptococcus* bacteria that is capable of digesting fibrin, the protein making up blood clots. It is used to treat pulmonary *embolism and heart attacks, reducing mortality.

streptomycin antibiotic drug discovered in 1944, active against a wide range of bacterial infections. Streptomycin is derived from a soil bacterium *Streptomyces griseus* or synthesized.

stress in psychology, any event or situation that makes heightened demands on a person's mental or emotional resources. Stress can be caused by overwork, anxiety about exams, money, job security, unemployment, bereavement, poor relationships, marriage breakdown, sexual difficulties, poor living or working conditions, and constant exposure to loud noise.

stress and strain in the science of materials, measures of the deforming force applied to a body (stress) and of the resulting change in its shape (*strain). For a perfectly elastic material, stress is proportional to strain (*Hooke's law).

striation scratch formed by the movement of a glacier over a rock surface. Striations are caused by the scraping of rocky debris embedded in the base of the glacier (corrasion), and provide an useful indicator of the direction of ice flow in past *ice ages. They are common features of roche moutonnées.

stridulatory organs in insects, organs that produce sound when rubbed together. Crickets rub their wings together, but grasshoppers rub a hind leg against a wing. Stridulation is thought to be used for attracting mates, but may also serve to mark territory.

strike compass direction of a horizontal line on a planar structural surface, such as a fault plane, bedding plane, or the trend of a structural feature, such as the axis of a fold. Strike is 90° from dip.

strike-slip fault common name for a lateral fault in which the motion is sideways in the direction of the *strike of the fault.

string in computing, a group of characters manipulated as a single object by the computer. In its simplest form a string may consist of a single

letter or word – for example, the single word SMITH might be established as a string for processing by a computer. A string can also consist of a combination of words, spaces, and numbers – for example, 33 HIGH STREET ANYTOWN ALLSHIRE could be established as a single string.

string theory mathematical theory developed in the 1980s to explain the behaviour of *elementary particles; see *superstring theory.

string, vibrations of standing waves set up in a stretched string or wire when it is plucked, or stroked with a bow. They are formed by the reflection of progressive waves at the fixed ends of the string. Waves of many different *frequencies can be established on a string at the same time. Those that match the natural frequencies of the string will, by a process called *resonance, produce large-amplitude vibrations. The vibration of lowest frequency is called the *fundamental vibration; vibrations of frequencies that are multiples of the fundamental frequency are called harmonics.

stroboscope instrument for studying continuous periodic motion by using light flashing at the same frequency as that of the motion; for example, rotating machinery can be optically 'stopped' by illuminating it with a stroboscope flashing at the exact rate of rotation.

stroke or **cerebrovascular accident** or **apoplexy**, interruption of the blood supply to part of the brain due to a sudden bleed in the brain (cerebral haemorrhage) or *embolism or *thrombosis. Strokes vary in severity from producing almost no symptoms to proving rapidly fatal. In between are those (often recurring) that leave a wide range of impaired function, depending on the size and location of the event.

strong nuclear force one of the four fundamental *forces of nature, the other three being the gravitational force or gravity, the electromagnetic force, and the weak nuclear force. The strong nuclear force was first described by the Japanese physicist Hideki Yukawa in 1935. It is the strongest of all the forces, acts only over very small distances within the nucleus of the atom (10^{-13} cm), and is responsible for binding together *quarks to form *hadrons, and for binding together protons and neutrons in the atomic nucleus. The particle that is the carrier of the strong nuclear force is the *gluon, of which there are eight kinds, each with zero mass and zero charge.

strontium chemical symbol Sr, soft, ductile, pale-yellow, metallic element, atomic number 38, relative atomic mass 87.62. It is one of the *alkaline-earth metals, widely distributed in small quantities only as a sulphate or carbonate. Strontium salts burn with a red flame and are used in fireworks and signal flares.

The radioactive isotopes Sr-89 and Sr-90 (half-life 25 years) are some of the most dangerous products of the nuclear industry; they are fission products in nuclear explosions and in the reactors of nuclear power plants. Strontium is chemically similar to calcium and deposits in bones and other tissues, where the radioactivity is damaging. The element was named in 1808 by English chemist Humphry Davy, who isolated it by electrolysis, after Strontian, a mining location in Scotland where it was first found.

structured programming in computing, the process of writing a program in small, independent parts. This makes it easier to control a program's development and to design and test its individual component parts. Structured programs are built up from units called **modules**, which normally correspond to single *procedures or *functions. Some programming languages, such as *Pascal and Modula-2, are better suited to structured programming than others.

strychnine $C_{21}H_{22}O_2N_2$, bitter-tasting, poisonous alkaloid. It is a poison that causes violent muscular spasms, and is usually obtained by powdering the seeds of plants of the genus *Strychnos* (for example *S. nux vomica*). Curare is a related drug.

style in flowers, the part of the *carpel bearing the *stigma at its tip. In some flowers it is very short or completely lacking, while in others it may be long and slender, positioning the stigma in the most effective place to receive the pollen. Usually the style withers after fertilization but in certain species, such as traveller's joy *Clematis vitalba*, it develops into a long feathery plume that aids dispersal of the fruit.

subatomic particle in physics, a particle that is smaller than an atom. Such particles may be indivisible *elementary particles, such as the *electron and *quark, or they may be composites, such as the *proton, *neutron, and *alpha particle. See also *particle physics.

subcutaneous tissue in medicine, loose cellular tissue beneath the skin. Injection into the subcutaneous tissue is a route of administration for some drugs, such as insulin.

subduction zone in *plate tectonics, a region where two plates of the Earth's rigid *lithosphere collide, and one plate descends below the other into the weaker asthenosphere. Subduction results in the formation of ocean trenches, most of which encircle the Pacific Ocean.

subglacial beneath a glacier. Subglacial rivers are those that flow under a glacier; subglacial material is debris that has been deposited beneath glacier ice. Features formed subglacially include drumlins and eskers.

sublimation conversion of a solid to vapour without passing through the liquid phase. It is one of the *changes of state of matter. Sublimation depends on the fact that the boiling point of the solid substance is lower than its melting point at atmospheric pressure. Thus by increasing pressure, a substance that sublimes can be made to go through a liquid stage before passing into the vapour state. Some substances that do not sublime at atmospheric pressure can be made to do so at low pressures. This is the principle of freeze-drying, during which ice sublimes at low pressure.

subliminal message any message delivered beneath the human conscious threshold of perception. It may be visual (words or images flashed between the frames of a cinema or TV film), or aural (a radio message broadcast constantly at very low volume).

subset in mathematics, set that is part of a larger set. For example, the girls in a class make up a subset if the class contains both boys and girls.

subsidence downward movement of a block of rock. Subsidence is usually due to the removal of material from below the surface, and can be caused by faults, *erosion, or by human activities such as mining.

substitution reaction in chemistry, the replacement of one atom or *functional group in an organic molecule by another.

substrate in biochemistry, a compound or mixture of compounds acted on by an enzyme. The term also refers to a substance such as *agar that provides the nutrients for the metabolism of micro-organisms. Since the enzyme systems of micro-organisms regulate their metabolism, the essential meaning is the same.

subtraction taking one number or quantity away from another, or finding the difference between two quantities; it is one of the four basic operations of arithmetic. Subtraction is neither commutative:

$$a - b \neq b - a$$

nor associative:

$$a - (b - c) \neq (a - b) - c.$$

For example,

$$8 - 5 \neq 5 - 8$$

$$7 - (4 - 3) \neq (7 - 4) - 3.$$

succession in ecology, a series of changes that occur in the structure and composition of the vegetation in a given area from the time it is first colonized by plants (**primary succession**), or after it has been disturbed by fire, flood, or clearing (**secondary succession**).

succulent plant thick, fleshy plant that stores water in its tissues; for example, cacti and stonecrops *Sedum*. Succulents live either in areas where water is very scarce, such as deserts, or in places where it is not easily obtainable because of the high concentrations of salts in the soil, as in salt marshes. Many desert plants are *xerophytes.

sucrose or **cane sugar** or **beet sugar**; $C_{12}H_{22}O_{11}$, sugar found in the pith of sugar cane and in sugar beet. It is popularly known as sugar. Sucrose is a disaccharide sugar, each of its molecules being made up of two simple sugar (monosaccharide) units: glucose and fructose.

sudden infant death syndrome SIDS, in medicine, the technical term for cot death.

sulphate SO_4^{2-}, salt or ester derived from sulphuric acid. Most sulphates are water soluble (the chief exceptions are lead, calcium, strontium, and barium sulphates) and require a very high temperature to decompose them. The commonest sulphates seen in the laboratory are copper(II) sulphate ($CuSO_4$), iron(II) sulphate ($FeSO_4$), and aluminium sulphate ($Al_2(SO_4)_3$). The ion is detected in solution by using barium chloride or barium nitrate to precipitate the insoluble barium sulphate.

sulphide compound of sulphur and another element in which sulphur is the more electronegative element (see *electronegativity). Sulphides occur in many minerals. Some of the more volatile sulphides have extremely unpleasant odours (hydrogen sulphide smells of bad eggs).

sulphite SO_3^{2-}, salt or ester derived from sulphurous acid.

sulphonamide any of a group of compounds containing the chemical group sulphonamide (SO_2NH_2) or its derivatives, which were, and still are in some cases, used to treat bacterial diseases. Sulphadiazine ($C_{10}H_{10}N_4O_2S$) is an example. Sulphonamide was the first commercially available antibacterial drug, the forerunner of a range of similar drugs. Toxicity and increasing resistance have limited their use chiefly to the treatment of urinary-tract infection.

sulphur chemical symbol S, brittle, pale-yellow, non-metallic element, atomic number 16, relative atomic mass 32.065. It occurs in three allotropic forms: two crystalline (called rhombic and monoclinic, following the arrangements of the atoms within the crystals) and one amorphous. It burns in air with a blue flame and a stifling odour. Insoluble in water but soluble in carbon disulphide, it is a good electrical insulator. Sulphur is widely

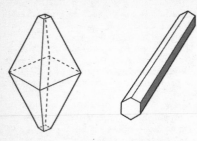

sulphur Two common allotropes of sulphur: rhombic and monoclinic crystals. A reactive element, sulphur combines with most other elements and has a wide range of industrial uses. It often occurs around hot springs and in volcanic regions, and there are large deposits in the USA (Texas and Louisiana), Japan, Sicily, and Mexico.

used in the manufacture of sulphuric acid (used to treat phosphate rock to make fertilizers) and in making paper, matches, gunpowder and fireworks, in vulcanizing rubber, and in medicines and insecticides.

sulphur dioxide SO_2, pungent gas produced by burning sulphur or sulphide ores in air or oxygen. It is widely used for making sulphuric acid and for disinfecting food vessels and equipment for bleaching paper, and as a preservative in some food products. It occurs in industrial flue gases and is a major cause of *acid rain.

sulphuric acid or **oil of vitriol**; H_2SO_4, dense, viscous, colourless liquid that is extremely corrosive. It gives out heat when added to water and can cause severe burns. Sulphuric acid is used extensively in the chemical industry, in the refining of petrol, and in the manufacture of fertilizers, detergents, explosives, and dyes. It forms the acid component of car batteries.

sulphurous acid H_2SO_3, solution of sulphur dioxide (SO_2) in water. It is a weak acid.

sulphur trioxide SO_3, colourless solid prepared by reacting sulphur dioxide and oxygen in the presence of a vanadium(V) oxide catalyst in the *contact process. It reacts violently with water to give sulphuric acid.

The violence of its reaction with water makes it extremely dangerous. In the contact process, it is dissolved in concentrated sulphuric acid to give oleum ($H_2S_2O_7$).

Sun *star at the centre of our Solar System. It is about 5 billion years old, with a predicted lifetime of 10 billion years; its diameter is 1.4 million km/865,000 mi; its temperature at the surface (the *photosphere) is about 5,800 K/5,530°C/9,986°F, and at the centre 15 million K/about 15 million°C/about 27 million°F. It is composed of about 70% hydrogen and 30% helium, with other elements making up less than 1%. The Sun's energy is generated by nuclear fusion reactions that turn hydrogen into helium, producing large amounts of light and heat that sustain life on Earth.

Space probes to the Sun have included NASA's series of Orbiting Solar Observatory satellites, launched between 1963 and 1975, the Ulysses space probe, launched in 1990, and Genesis, launched in 2001.

sunspot dark patch on the surface of the Sun, actually an area of cooler gas, thought to be caused by strong magnetic fields that block the outward flow of heat to the Sun's surface. Sunspots consist of a dark central **umbra**, about 4,000 K (3,700°C/6,700°F), and a lighter surrounding **penumbra**, about 5,500 K (5,200°C/9,400°F). They last from several days to over a month, ranging in size from 2,000 km/1,250 mi to groups stretching for over 100,000 km/62,000 mi.

superactinide any of a theoretical series of superheavy, radioactive elements, starting with atomic number 113, that extend beyond the *transactinide series in the periodic table. They do not occur in nature and none has yet been synthesized. It is postulated that this series has a group of elements that have half-lives longer than those of the transactinide series. This group, centred on element 114, is referred to as the 'island of stability', based on the nucleon arrangement. The longer half-lives will, it is hoped, allow enough time for their chemical and physical properties to be studied when they have been synthesized.

superbug popular name given to an infectious bacterium that has

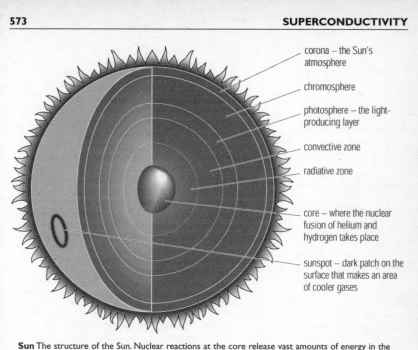

corona – the Sun's atmosphere

chromosphere

photosphere – the light-producing layer

convective zone

radiative zone

core – where the nuclear fusion of helium and hydrogen takes place

sunspot – dark patch on the surface that makes an area of cooler gases

Sun The structure of the Sun. Nuclear reactions at the core release vast amounts of energy in the form of light and heat that radiate out to the photosphere and corona. Surges of glowing gas rise as prominences from the surface of the Sun and cooler areas, known as sunspots, appear as dark patches on the star's surface.

developed resistance to most or all known *antibiotics.

supercluster in astronomy, grouping of several clusters of galaxies to form a structure about 100–300 million light years across. Our own Galaxy and its neighbours lie on the edge of the local supercluster of which the Virgo cluster is the dominant member.

supercomputer fastest, most powerful type of computer, capable of performing its basic operations in picoseconds (trillionths of a second), rather than nanoseconds (billionths of a second), like most other computers. Cray Research, formed by US computer scientist and pioneer in the field of supercomputing Seymour Roger Cray (1925–1996), released the first popular supercomputer, the Cray-1, in 1976.

superconductivity increase in electrical conductivity at low temperatures. The resistance of some metals and metallic compounds decreases uniformly with decreasing temperature until at a critical temperature (the superconducting point), within a few degrees of absolute zero (0 K/–273.15°C/ –459.67°F), the resistance suddenly falls to zero. The phenomenon was discovered by Dutch scientist Heike Kamerlingh Onnes in 1911.

Some metals, such as platinum and copper, do not become superconductive; as the temperature decreases, their resistance decreases to a certain point but then rises again. Superconductivity can be nullified by the application of a large magnetic field. In the superconducting state, an electric current will continue indefinitely once started, provided that the material remains below the superconducting point. In 1986 IBM researchers achieved superconductivity with some ceramics at –243°C/–405°F, opening up the possibility of **'high-temperature' superconductivity**; a year later Paul Chu at the University of Houston, Texas, achieved superconductivity at –179°C/–290°F, a temperature that can be sustained using liquid nitrogen.

Researchers are trying to find a material that will be superconducting at room temperature; in 2001 the material magnesium diboride (MgB_2) was found to become a superconductor when cooled to only 39 K/−234.15°C/−389.47°F.

supercooling cooling of a liquid below its freezing point without freezing taking place; or the cooling of a *saturated solution without crystallization taking place, to form a supersaturated solution. In both cases supercooling is possible because of the lack of solid particles around which crystals can form. Crystallization rapidly follows the introduction of a small crystal (seed) or agitation of the supercooled solution.

supercritical fluid fluid that combines the properties of a gas and a liquid, see *fluid, supercritical.

superego in Freudian psychology, the element of the human mind concerned with the ideal, responsible for ethics and self-imposed standards of behaviour. It is characterized as a form of conscience, restraining the *ego, and responsible for feelings of guilt when the moral code is broken.

superfluid fluid that flows without viscosity or friction and has a very high thermal conductivity. Liquid helium at temperatures below 2 K (−271°C/−456°F) is a superfluid: it shows unexpected behaviour. For instance, it flows uphill in apparent defiance of gravity and, if placed in a container, will flow up the sides and escape.

supergiant largest and most luminous type of star known, with a diameter of up to 1,000 times that of the Sun and an apparent magnitude of between 0.4 and 1.3. Supergiants are likely to become *supernovae.

superior planet planet that is farther away from the Sun than the Earth is: that is, Mars, Jupiter, Saturn, Uranus, Neptune, and Pluto.

SuperJANET acronym for Super Joint Academic NETtwork, in computing, high speed telecommunications network linking academic sites in the UK. The government-funded project, a successor to *JANET, was started in 1989. Many institutions are now linked to SuperJANET, which has ample *bandwidth to support *multimedia educational projects, such as the televisation of operations for medical students. SuperJANET has gateways to the Internet via mainland Europe and the USA. This enables all of the institutions connected to SuperJANET to have Internet access.

supernova explosive death of a star, which temporarily attains a brightness of 100 million Suns or more, so that it can shine as brilliantly as a small galaxy for a few days or weeks. Very approximately, it is thought that a supernova explodes in a large galaxy about once every 100 years. Many supernovae – astronomers estimate some 50% – remain undetected because of obscuring by interstellar dust.

supernova remnant SNR, in astronomy, glowing remains of a star that has been destroyed in a *supernova explosion. The brightest and most famous example is the Crab nebula.

superphosphate phosphate fertilizer made by treating apatite (calcium phosphate mineral) with sulphuric or phosphoric acid. The commercial mixture contains largely monocalcium phosphate. Single-superphosphate obtained from apatite and sulphuric acid contains 16–20% available phosphorus, as P_2O_5; triple-superphosphate, which contains 45–50% phosphorus, is made by treating apatite with phosphoric acid.

supersaturation state of a solution that has a higher concentration of *solute than would normally be obtained in a *saturated solution. Many solutes have a higher *solubility at high temperatures. If a hot saturated solution is cooled slowly, sometimes the excess solute does not come out of solution. This is an unstable situation and the introduction of a small solid particle will encourage the release of excess solute.

supersonic speed speed greater than that at which sound travels, measured in *Mach numbers. In dry air at 0°C/32°F, sound travels at about 1,170 kph/727 mph, but decreases its speed with altitude until, at 12,000 m/39,000 ft, it is only 1,060 kph/658 mph.

superstring theory in physics and cosmology, a mathematical theory developed in the 1980s to explain the properties of *elementary particles and the forces between them (in particular, gravity and the nuclear forces) in a way that combines *relativity and *quantum theory. In string theory, the fundamental objects in the universe are not pointlike particles but extremely small stringlike objects. These objects exist in a universe of ten dimensions, but since the earliest moments of the Big Bang six of these have been compacted or 'rolled up', so that now, only three space dimensions and one dimension of time are discernible. There are many unresolved difficulties with superstring theory, but some physicists think it may be the ultimate 'theory of everything' that explains all aspects of the universe within one framework.

supraglacial on top of a glacier. A supraglacial stream flows over the surface of the glacier; supraglacial material collected on top of a glacier may be deposited to form *lateral and *medial moraines.

suprarenal gland alternative name for the *adrenal gland.

surd in mathematics, expression containing the root of an *irrational number that can never be exactly expressed – for example, $\sqrt{3}$ = 1.732050808... . $\sqrt{3}$ can be expressed in *index notation as $3^{1/2}$ or 3 to the *power half.

Expressions involving surds can be factorized numerically to simplify them. For example, to simplify $\sqrt{20} + \sqrt{5}$:

$$\sqrt{20} = \sqrt{(4 \times 5)} = \sqrt{4} \times \sqrt{5} = 2\sqrt{5},\text{ so}$$

$$\sqrt{20} + \sqrt{5} = 2\sqrt{5} + \sqrt{5} = 3\sqrt{5}$$

surface area area of the outside surface of a solid. (See diagram, page 576.)

surface-area-to-volume ratio in biology, the ratio of an animal's surface area (the area covered by its skin) to its total volume. This is high for small animals, but low for large animals such as elephants.

surface tension property that causes the surface of a liquid to behave as if it were covered with a weak elastic skin; this is why a needle can float on water.

It is caused by the exposed surface's tendency to contract to the smallest possible area because of cohesive forces between *molecules at the surface. Allied phenomena include the formation of droplets, the concave profile of a meniscus, and the *capillary action by which water soaks into a sponge.

surfactant contraction of surface-active agent, substance added to a liquid in order to increase its wetting or spreading properties. Detergents are surfactants.

surgical spirit *ethanol to which has been added a small amount of methanol to render it unfit to drink. It is used to sterilize surfaces and to cleanse skin abrasions and sores.

suspension mixture consisting of small solid particles dispersed in a liquid or gas, which will settle on standing. An example is milk of magnesia, which is a suspension of magnesium hydroxide in water.

sustainable capable of being continued indefinitely. For example, the sustainable yield of a forest is equivalent to the amount that grows back. Environmentalists made the term a catchword, in advocating the sustainable use of resources.

SUSY in physics, an abbreviation for supersymmetry.

Sv in physics, symbol for *sievert, the SI unit of dose equivalent of ionizing radiation, equal to the dose in grays multiplied by a factor that depends mainly on the type of radiation and its effects.

SVGA abbreviation for super video graphics array, in computing, a graphic display standard providing higher resolution than *VGA. SVGA screens have resolutions of either 800×600 or $1,024 \times 768$ pixels.

swamp region of low-lying land that is permanently saturated with water and usually overgrown with vegetation; for example, the everglades of Florida, USA. A swamp often occurs where a lake has filled up with sediment and plant material. The flat surface so formed means that run-off is slow, and the water table is always close to the surface. The high humus content of swamp soil means that good

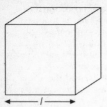

surface area of a **cube**
(faces are identical)
= 6 × area of each surface
= $6l^2$

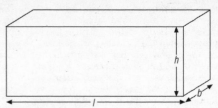

surface area of a **cuboid**
(opposite faces are identical)
= area of two end faces + area of two sides
  + area of top and base
= $2lh + 2hb + 2lb$
= $2(lh + hb + lb)$

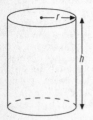

surface area of a **cylinder**
= area of a curved surface + area of top
  and base
= 2π × (radius of cross-section × height)
  + 2π (radius of cross-section)²
= $2\pi rh + 2\pi r^2$
= $2\pi r(h + r)$

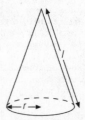

surface area of a **cone**
= area of a curved surface + area base
= π × (radius of cross-section × slant height)
  + π (radius of cross-section)²
= $\pi rl + \pi r^2$
= $\pi r(l + r)$

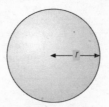

surface area of a **sphere**
= 4π × radius²
= $4\pi r^2$

surface area Surface area of common three-dimensional shapes.

agricultural soil can be obtained by draining.

swap in computing, to move segments of data in and out of memory. For fast operation as much data as possible is required in main memory, but it is generally not possible to include all data at the same time. Swapping is the operation of writing and reading from the back-up store, often a special space on the disk.

S-wave abbreviation of secondary wave, in seismology, a class of *seismic wave that passes through the Earth in the form of transverse shear waves. S-waves from an earthquake travel at roughly half the speed of P-waves (about 3 kps/1.9 mps), the speed depending on the density of the rock, and arrive later at monitoring stations (hence secondary waves) though with greater amplitude. They can travel through solid rock but not through the liquid outer core of the Earth. Rocks move perpendicular (at right angles) to the direction of travel of the wave, and as such an S-wave is an example of a *transverse wave.

sweat gland *gland within the skin of mammals that produces surface perspiration. In primates, sweat glands are distributed over the whole body, but in most other mammals they are more localized; for example, in cats and dogs they are restricted to the feet and around the face.

swine flu virulent, highly contagious form of influenza, infecting pigs and communicable to people.

switch device used to turn electrical systems on and off, usually by closing or opening a circuit, or to change the route of a particular current. For example, switches are used to turn a light on or off, to select a channel on a television set, or to set the cooking time on an electric oven.

switch in computing, a network device which allows the interconnection of a number of computers or *local area network (LAN) segments. Switches differ from more traditional *bus devices in that when a computer on one socket of the switch needs to talk to another a dedicated path is made through the switch between the two sockets. This allows the computers to communicate at the full speed of the switch, rather than having to compete for a share of the *bandwidth on the bus.

SXGA abbreviation for super extended graphics array, computer colour display system, providing true colour (24-bit colour depth, giving over 16 million shades) at a *resolution of 1280×1024. A widescreen format, SXGA+, provides 1400×1050 pixels.

symbiosis any close relationship between two organisms of different species, and one where both partners benefit from the association. A well-known example is the pollination relationship between insects and flowers, where the insects feed on nectar and carry pollen from one flower to another. This is sometimes known as *mutualism.

symbol, chemical letter or letters used to represent a chemical element, usually derived from the beginning of its English or Latin name. Symbols derived from English include B, boron; C, carbon; Ba, barium; and Ca, calcium. Those derived from Latin include Na, sodium (Latin *natrium*); Pb, lead (Latin *plumbum*); and Au, gold (Latin *aurum*). See Appendix.

symbolic address in computing, a symbol used in *assembly language programming to represent the binary address of a memory location.

symbolic processor computer purpose-built to run so-called symbol-manipulation programs rather than programs involving a great deal of numerical computation. They exist principally for the *artificial intelligence language *LISP, although some have also been built to run *PROLOG.

symmetric-key cryptography system of *cryptography whereby both sender and recipient have identical digital 'keys' used to encrypt and decrypt messages. Although simpler and faster than *public-key cryptography, this method depends on the secure transmission of a key to the recipient before an encrypted message can be read. The most popular example of symmetric-key cryptography is the *Data Encryption Standard (DES), used by the US government.

symmetry exact likeness in shape about a given line (axis), point, or plane. A figure has symmetry if one half can be rotated and/or reflected onto the other. (Symmetry preserves length, angle, but not necessarily orientation.) In a wider sense, symmetry exists if a change in the system leaves the essential features of the system unchanged; for example, reversing the sign of electric charges does not change the electrical behaviour of an arrangement of charges. (See diagram, page 578.)

sympathetic nervous system in medicine, a division of the *autonomic nervous system.

synapse junction between two *nerve cells, or between a nerve cell and a muscle (a neuromuscular junction), across which a nerve impulse is

plane symmetry

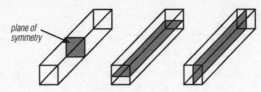

cuboid has three planes of symmetry

square-based pyramid has four planes of symmetry

sphere has infinite number of planes of symmetry

axes of symmetry

square

axis of symmetry

rectangle

rhombus

four axes of symmetry

two axes of symmetry

two axes of symmetry

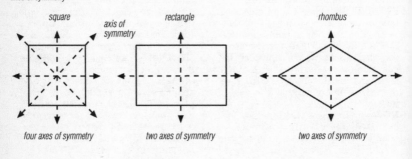

equilatral triangle

circle

three axes of symmetry

infinite number of axes of symmetry

symmetry

transmitted. The two cells are separated by a narrow gap called the **synaptic cleft**. The gap is bridged by a chemical *neurotransmitter, released by the nerve impulse.

The threadlike extension, or *axon, of the transmitting nerve cell has a slightly swollen terminal point, the **synaptic knob**. This forms one half of the synaptic junction and houses membrane-bound vesicles, which contain a chemical neurotransmitter. When nerve impulses reach the knob, the vesicles release the transmitter and this flows across the gap and binds itself to special receptors on the receiving cell's membrane. If the receiving cell is a nerve cell, the other half of the synaptic junction will be one or more extensions called *dendrites; these will be stimulated by the neurotransmitter to set up an impulse, which will then be conducted along the length of the nerve cell and on to its own axons. If the receiving cell is a muscle cell, it will be stimulated by the neurotransmitter to contract.

synchronous regular. Most communication within a computer system is synchronous, controlled by the computer's own internal clock, while communication between computers is usually asynchronous. Synchronous telecommunications are, however, becoming more widely used.

synchronous orbit another term for geostationary orbit.

synchronous rotation in astronomy, another name for *captured rotation.

synchrotron particle *accelerator in which particles move, at increasing speed, around a hollow ring. The particles are guided around the ring by electromagnets, and accelerated by electric fields at points around the ring. Synchrotrons come in a wide range of sizes, the smallest being about 1 m/ 3.3 ft across while the largest is 27 km/ 17 mi across. The Tevatron synchrotron at *Fermilab is some 6 km/4 mi in circumference and accelerates protons and antiprotons to 1 TeV.

synergy in medicine, the 'cooperative' action of two or more drugs, muscles, or organs; applied especially to drugs whose combined action is more powerful than their simple effects added together.

synodic period time taken for a planet or moon to return to the same position in its orbit as seen from the Earth; that is, from one opposition to the next. It

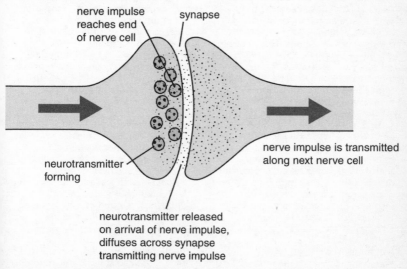

synapse How the nerve impulse is transmitted across the synapse. When the impulse reaches the end of the nerve cell, neurotransmitter is released. This diffuses across the synapse to the next nerve cell, which then continues to transmit the impulse.

differs from the *sidereal period because the Earth is moving in orbit around the Sun.

synovial fluid viscous colourless fluid that bathes movable joints between the bones of vertebrates. It nourishes and lubricates the *cartilage at the end of each bone.

synthesis in chemistry, the formation of a substance or compound from more elementary compounds. The synthesis of a drug can involve several stages from the initial material to the final product; the complexity of these stages is a major factor in the cost of production.

synthetic any material made from chemicals. Since the 1900s, more and more of the materials used in everyday life are synthetics, including plastics (polythene, polystyrene), *synthetic fibres (nylon, acrylics, polyesters), synthetic resins, and synthetic rubber. Most naturally occurring organic substances can now be made synthetically, especially pharmaceuticals.

synthetic fibre fibre made by chemical processes, unknown in nature. There are two kinds. One, a regenerated synthetic fibre, is made from natural materials that have been chemically processed in some way; rayon, for example, is made by processing the cellulose in wood pulp. The other type is the true synthetic fibre, made entirely from chemicals. *Nylon was the original true synthetic fibre, made from chemicals obtained from petroleum (crude oil).

syphilis sexually transmitted disease caused by the spiral-shaped bacterium (spirochete) *Treponema pallidum*. Untreated, it runs its course in three stages over many years, often starting with a painless hard sore, or chancre, developing within a month on the area of infection (usually the genitals). The second stage, months later, is a rash with arthritis, hepatitis, and/or meningitis. The third stage, years later, leads eventually to paralysis, blindness, insanity, and death. The Wassermann test is a diagnostic blood test for syphilis.

Syquest manufacturer of removable *hard disk drives and high-capacity

floppy disk drives, which may be used to transport large files from one location to another.

Système International d'Unités official French name for *SI units.

system flow chart type of flow chart used to describe the flow of data through a particular computer system.

systemic in medicine, relating to or affecting the body as a whole. A systemic disease is one where the effects are present throughout the body, as opposed to local disease, such as *conjunctivitis, which is confined to one part.

system implementation in computing, the process of installing a new computer system.

SYSTEM.INI abbreviation for system initialization, in computing, file used by Microsoft Windows to store information about which parts of Windows to load and how to set itself up on the PC on which it is running, for example SYSTEM.INI specifies drivers for the keyboard, graphics card, and sound card, if any. Most of the settings managed by the SYSTEM.INI file have been developed so that they can be found in the Windows Registry.

system requirements in computing, minimum specification necessary in order to use a particular piece of hardware or software.

systems analysis in computing, the investigation of a business activity or clerical procedure, with a view to deciding if and how it can be computerized. The analyst discusses the existing procedures with the people involved, observes the flow of data through the business, and draws up an outline specification of the required computer system. The next step is *systems design. A recent system is Unified Modeling Language (UML), which is specifically designed for the analysis and design of *object-oriented programming systems.

Systems Application Architecture SAA, in computing, IBM model for client–server computing, introduced in 1987. SAA was a grandiose attempt to reduce the incompatibilities between IBM's many ranges of hardware, including mainframes, minicomputers, and PCs. It uses *CUA (common user access) standards to ensure that commands and keystrokes are used consistently in different applications.

systems design in computing, the detailed design of an *applications package. The designer breaks the system down into component programs, and designs the required input forms, screen layouts, and printouts. Systems design forms a link between systems analysis and programming.

Systems Network Architecture SNA, set of communication protocols developed by IBM and incorporated in hardware and software implementations. See also *TCP/IP.

systems program in computing, a program that performs a task related to the operation and performance of the computer system itself. For example, a systems program might control the operation of the display screen, or control and organize backing storage. In contrast, an applications program is designed to carry out tasks for the benefit of the computer user.

System X in communications, a modular, computer-controlled, digital switching system used in telephone exchanges.

systole in biology, the contraction of the heart. It alternates with diastole, the resting phase of the heart beat.

systolic pressure in medicine, the measurement due to the pressure of blood against the arterial wall during systole (the contraction of the heart). It is the highest *blood pressure during the cardiac cycle.

t symbol for *tonne, *ton.

T in physics, symbol for *tesla, the SI unit of magnetic flux density, equal to a flux density of one weber of flux per square metre.

T1 link US term for a digital telephone line which can transfer data at 1.544 megabits per second. T1 lines are a type of *Integrated Services Digital Network (ISDN) communication.

tag formatting command that tells a browser how it should display an element in a Web page or other document. In *HTML, tags are separated from text by angle brackets and travel in pairs: one for *on* and one for *off*. <whisper> The same technique is used for effect in geeky e-mail </whisper>.

taiga or **boreal forest**, Russian name for the forest zone south of the *tundra, found across the northern hemisphere. Here, dense forests of conifers (spruces and hemlocks), birches, and poplars occupy glaciated regions punctuated with cold lakes, streams, bogs, and marshes. Winters are prolonged and very cold, but the summer is warm enough to promote dense growth.

talc $Mg_3Si_4O_{10}(OH)_2$, mineral, hydrous magnesium silicate. It occurs in tabular crystals, but the massive impure form, known as **steatite** or **soapstone**, is more common. It is formed by the alteration of magnesium compounds and is usually found in metamorphic rocks. Talc is very soft, ranked 1 on the *Mohs scale of hardness. It is used in powdered form in cosmetics, lubricants, and as an additive in paper manufacture.

tamoxifen *oestrogen-blocking drug used to treat breast cancer and also infertility. It kills cancer cells by preventing formation of the tiny blood vessels that supply the tumour. Without a blood supply the tumour shrinks as its cells die. Side effects include hot flushes, skin rashes, vaginal discharge, and more rarely, endometrial cancer. Trials are under way to investigate its potential for preventing breast cancer in healthy women with a family history of the disease.

tangent line that touches a *circle at only one point. A tangent is at right angles to the radius at the point of contact. From any point outside a circle, the lines of two tangents drawn to the circle will be of equal length.

tangent in graphs, a straight line that touches a curve and gives the *gradient of the curve at the point of contact. At a maximum, minimum, or point of inflection, the tangent to a curve has zero gradient. To find the gradient of the tangent of the curve at point P, extend the tangent AC and construct the right-angled triangle ABC. The gradient of the tangent at the point P is $^{AB}/_{BC}$.

tangent tan, in trigonometry, a function of an acute angle in a right-angled *triangle, defined as the ratio of the length of the side opposite the angle to the length of the side adjacent to it; a way of expressing the gradient of a line. This function can be used to find either sides or angles in a right-angled triangle.

See also *trigonometry.

tantalum symbol Ta, hard, ductile, lustrous, grey-white, metallic element, atomic number 73, relative atomic mass 180.948. It occurs with niobium in tantalite and other minerals. It can be drawn into wire with a very high melting point and great tenacity, useful for lamp filaments subject to vibration. It is also used in alloys, for corrosion-resistant laboratory apparatus and chemical equipment, as a catalyst in manufacturing synthetic rubber, in tools and instruments, and in rectifiers and capacitors.

Tantalum was discovered and named in 1802 by Swedish chemist Anders Ekeberg (1767–1813) after the mythological Greek character Tantalus. It is mined as tantalite ore and most of the world's supply comes from Australia, with significant deposits in Brazil, Canada, and Nigeria. Tantalum is used in mobile phone capacitors, and as a consequence demand for the element has risen dramatically.

TAPI acronym for Telephony Application Programming Interface, in computing, program included in Windows 95 to enable applications to use the telephone. The TAPI standard was developed by Microsoft and Intel in 1993.

tar dark brown or black viscous liquid obtained by the destructive distillation of coal, shale, and wood. Tars consist of a mixture of hydrocarbons, acids, and bases. Creosote and *paraffin oil are produced from wood tar.

tar in computing, a compression routine in common use on the Internet. Originally developed for *Unix operating systems, the tar utility archives files and directories by grouping them together into one large file, which can then be compressed and stored off-line. It is often used to distribute software for Unix systems, and tar files bear the extension .tar. A file with the extension tar.Z is a tar archive that has also been compressed with the Unix compression utility.

taskbar in computing, strip at the bottom of a Windows 95, 98, 2000, or NT screen containing icons ('task buttons') of all programs launched in the current session. The taskbar makes it possible to switch between applications simply by clicking the mouse on a task button.

tau *elementary particle with the same electric charge as the electron but a mass nearly double that of a proton. It has a lifetime of around 3×10^{-13} seconds and belongs to the *lepton family of particles – those which interact via the electromagnetic, weak nuclear, and gravitational forces, but not the strong nuclear force.

tautomerism form of isomerism in which two interconvertible *isomers are in equilibrium. It is often specifically applied to an equilibrium between the keto ($-CH_2-C-O$) and enol ($-CH=C(OH)-$) forms of carbonyl compounds.

taxis plural **taxes**; or **tactic movement**, in botany, the movement of a single cell, such as a bacterium, protozoan, single-celled alga, or gamete, in response to an external stimulus. A movement directed towards the stimulus is described as positive taxis, and away from it as negative taxis. The alga *Chlamydomonas*, for example, demonstrates positive **phototaxis** by swimming towards a light source to increase the rate of photosynthesis. **Chemotaxis** is a response to a chemical stimulus, as seen in many bacteria that move towards higher concentrations of nutrients.

taxonomy another name for the *classification of living organisms.

Tay–Sachs disease inherited disorder, due to a defective gene, causing an enzyme deficiency that leads to blindness, retardation, and death in infancy. It is most common in people of Eastern European Jewish descent.

TB abbreviation for the infectious disease *tuberculosis.

T-carrier high-speed communications service supplied by one of the USA's telecommunications companies. Lines with different capacities are identified by number, such as T-1, T-2, T-3, T-4, and so on.

T cell or **T lymphocyte**, immune cell (see *immunity and *lymphocyte) that plays several roles in the body's defences. T cells are so called because they mature in the *thymus. There are three main types of T cells: T helper cells (Th cells), which allow other immune cells to go into action; T suppressor cells (Ts cells), which stop specific immune reactions from occurring; and T cytotoxic cells (Tc cells), which kill cells that are cancerous or infected with viruses. Like *B cells, to which they are related, T cells have surface receptors that make them specific for particular antigens.

TCO abbreviation for Tjänstemännens Centralorganisation; **Swedish Confederation of Professional Employees**, an association of 18 trade unions, whose activities include the specification and management of an international quality and environmental labelling scheme for the computer industry. Their scheme covers workstation efficiency, monitors, printers, and most recently, keyboards. The scheme has been most widely used by monitor manufacturers, who display TCO92 or TCO95 labels to show their conformity with electromagnetic emissions standards.

TCP/IP abbreviation for transmission or transfer control protocol/Internet protocol, set of network protocols, developed principally by the US Department of Defense. TCP/IP is the protocol used by the Internet, and is the technology that underpins Internet services like the *World Wide Web, Internet Relay Chat, and *e-mail. TCP/IP has always been the principal networking protocol used by *Unix, and is now supported by almost all types of operating system.

technetium chemical symbol Tc, (Greek *technetos* 'artificial') silver-grey, radioactive, metallic element, atomic number 43, relative atomic mass 98.906. It occurs in nature only in extremely minute amounts, produced as a fission product from uranium in *pitchblende and other uranium ores. Its longest-lived isotope, Tc-99, has a half-life of 216,000 years. It is a superconductor and is used as a hardener in steel alloys and as a medical tracer.

It was synthesized in 1937 (named in 1947) by Italian physicists Carlo Perrier and Emilio Segrè, who bombarded molybdenum with deuterons, looking to fill a gap in the *periodic table of the elements (at that time it was considered not to occur in nature). It was later isolated in larger amounts from the fission product debris of uranium fuel in nuclear reactors.

tectonics in geology, the study of the movements of rocks on the Earth's surface. On a small scale tectonics involves the formation of *folds and *faults, but on a large scale *plate tectonics deals with the movement of the Earth's surface as a whole.

Teflon trade name for polytetrafluoroethene (PTFE), a tough, waxlike, heat-resistant plastic used for coating nonstick cookware and in gaskets and bearings. In 2001 Canadian scientists discovered that Teflon degraded when heated to form trifluoroacetic acid (TFA), a pollutant that persists in the environment for many years.

telecommunications communications over a distance, generally by electronic means. Long-distance voice communication was pioneered in 1876 by Scottish scientist Alexander Graham Bell when he invented the telephone. The telegraph, radio, and television followed. Today it is possible to communicate internationally by telephone cable or by satellite or microwave link, with over 100,000 simultaneous conversations and several television channels being carried by the latest satellites.

history The first mechanical telecommunications systems were semaphore and the heliograph (using flashes of sunlight), invented in the mid-19th century, but the forerunner of the present telecommunications age was the electric telegraph. The earliest practicable telegraph instrument was invented by William Cooke and Charles Wheatstone in Britain in 1837 and used by railway companies. In the USA, Samuel Morse invented a signalling code, Morse code, which is still used, and a recording telegraph, first used commercially between England and France in 1851.

Following the discovery of electromagnetic waves by German physicist Heinrich Hertz (1857–1894), Italian inventor Guglielmo Marconi pioneered a 'wireless' telegraph, ancestor of the radio. He established wireless communication between England and France in 1899 and across the Atlantic in 1901.

The modern telegraph uses teleprinters to send coded messages along telecommunications lines. Telegraphs are keyboard-operated machines that transmit a five-unit Baudot code. The receiving teleprinter automatically prints the received message. The modern version of the telegraph is *e-mail in which text messages are sent electronically from computer to computer via network connections such as the *Internet.

communications satellites The chief method of relaying long-distance calls on land is microwave radio transmission. The drawback to long-distance voice communication via microwave radio transmission is that the transmissions follow a straight line from tower to tower, so that over the sea the system becomes impracticable. A solution was put forward in 1945 by the science fiction writer Arthur C Clarke, when he proposed a system of communications satellites in an orbit 35,900 km/22,300 mi above the Equator, where they would circle the Earth in exactly 24 hours, and thus appear fixed in the sky. Such a system is now in operation internationally, by Intelsat. The satellites are called geostationary satellites (syncoms). The first to be successfully launched, by Delta rocket from Cape Canaveral, was *Syncom 2* in July 1963. Many such satellites are now in use, concentrated over heavy traffic areas such as the Atlantic, Indian, and Pacific oceans. Telegraphy, telephony, and television transmissions are carried simultaneously by high-frequency radio waves. They are beamed to the satellites from large dish antennae or Earth stations, which connect with international networks.

*Integrated Services Digital Network (ISDN) makes videophones and high-quality fax possible; the world's first large-scale centre of ISDN began operating in Japan in 1988. ISDN is a system that transmits voice and image data on a single transmission line by changing them into digital signals.

Fibre-optic cables consisting of fine glass fibres present an alternative to the usual copper cables for telephone lines. The telecommunications signals are transmitted along the fibres in digital form as pulses of laser light.

telegraphy transmission of messages along wires by means of electrical signals. The first modern form of telecommunication, it now uses printers for the transmission and receipt of messages. Telex is an international telegraphy network.

Overland cables were developed in the 1830s, but early attempts at underwater telegraphy were largely unsuccessful until the discovery of the insulating gum gutta-percha in 1845 enabled a cable to be successfully laid across the English Channel in 1851. **Duplex telegraph** was invented in the 1870s, enabling messages to be sent in both directions simultaneously. Early telegraphs were mainly owned by the UK: 72% of all submarine cables were British-owned in 1900.

telemetry measurement at a distance, in particular the systems by which information is obtained and sent back by instruments on board a spacecraft. See also *remote sensing.

telephone instrument for communicating by voice along wires, developed by Scottish-US inventor Alexander Graham Bell in 1876, consisting of an earpiece that receives electrical signals and a mouthpiece that sends electrical signals. The transmitter (mouthpiece) consists of a carbon microphone, with a diaphragm that is vibrated by sound waves when a person speaks into it. The diaphragm vibrations compress grains of carbon to a greater or lesser extent, altering their resistance to an electric current passing through them. This sets up variable electrical signals, which travel along the telephone lines to the receiver of the person being called. The earpiece contains an electromagnet attached to a diaphragm. As the incoming electrical signal varies, the strength of the electromagnet also varies, resulting in a variable movement of an armature of the electromagnet. These movements cause the diaphragm to vibrate, producing the pattern of sound waves that originally entered the mouthpiece.

The standard instrument has a handset, which houses the transmitter (mouthpiece), and receiver (earpiece), resting on a base, which has a dial or push-button mechanism for dialling a telephone number. Cordless telephones combine a push-button mechanism, a mouthpiece, and an earpiece in one unit, and are connected to the base unit not by wires but by radio. They can be used at distances of up to 100 m/330 ft from the base unit. In the 1990s, digital networks provided wireless mobility with the mobile phone, and enabled the adaptation of personal computers for telephone use with the Internet phone.

Telephony Application Programming Interface in computing, Windows program, commonly abbreviated to *TAPI.

telescope optical instrument that magnifies images of faint and distant objects; any device for collecting and

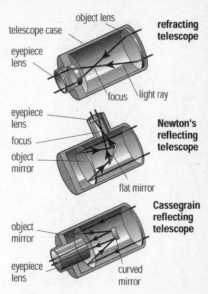

telescope Three kinds of telescope.

focusing light and other forms of electromagnetic radiation. A telescope with a large aperture, or opening, can distinguish finer detail and fainter objects than one with a small aperture. The **refracting telescope** uses lenses, and the **reflecting telescope** uses mirrors. A third type, the **catadioptric telescope**, is a combination of lenses and mirrors. See also *radio telescope.

teletype TTY, in computing, originally a 'hard copy terminal' that printed text slowly in capital letters on rolls of paper. Teletypes were made by Teletype Corporation. When DEC introduced visual display terminals these could operate in Teletype mode, like a 'glass typewriter'. Many communications programs still include a TTY mode to provide the simplest level of communication with a remote computer.

television TV, reproduction of visual images at a distance using radio waves. For transmission, a television camera converts the pattern of light it takes in into a pattern of electrical charges. This is scanned line by line by a beam of electrons from an electron gun, resulting in variable electrical

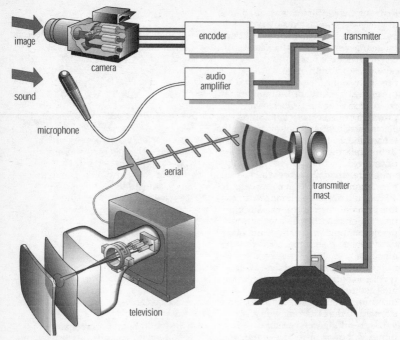

television Simplified block diagram of a complete colour television system – transmitting and receiving. The camera separates the picture into three colours – red, blue, green – by using filters and different camera tubes for each colour. The audio signal is produced separately from the video signal. Both signals are transmitted from the same aerial using a special coupling device called a diplexer. There are four sections in the receiver: the aerial, the tuners, the decoders, and the display. As in the transmitter, the audio and video signals are processed separately. The signals are amplified at various points.

signals that represent the picture. These signals are combined with a radio carrier wave and broadcast as electromagnetic waves. The TV aerial picks up the wave and feeds it to the receiver (TV set). This separates out the vision signals, which pass to a cathode-ray tube where a beam of electrons is made to scan across the screen line by line, mirroring the action of the electron gun in the TV camera. The result is a recreation of the pattern of light that entered the camera.

tellurium chemical symbol Te, (Latin *Tellus* 'Earth') silver-white, semi-metallic (*metalloid) element, atomic number 52, relative atomic mass 127.60. Chemically it is similar to sulphur and selenium, and it is considered one of the sulphur group. It occurs naturally in telluride minerals, and is used in colouring glass blue-brown, in the electrolytic refining of zinc, in electronics, and as a catalyst in refining petroleum. It was discovered in 1782 by the Austrian mineralogist Franz Müller (1740–1825), and named in 1798 by the German chemist Martin Klaproth.

Telnet in computing, Internet utility that enables a user to work on a remote computer as if directly connected. Telnet connections to a remote computer system are typically much cheaper than long-distance telephone calls; the user makes a local call to an Internet access provider and the rest of the connection is handled via the Internet at no additional cost. *Bulletin board systems usually work via Telnet.

telomere (Greek *telos* 'end'; *meros* 'part') chromosome tip. Telomeres prevent chromosomes sticking together. Like DNA they are made up of nucleotides, usually rich in thymine and guanine. Every time a cell divides, the telomeres shorten. They trigger the cell's senescence (inability to reproduce) when they reach a threshold length. This process is prevented from happening during the replication of cancer cells by the presence of the enzyme **telomerase**. Telomerase replaces the segments of the telomeres, maintaining their length so that cell replication is no longer controlled.

temperate deciduous forest type of vegetation mass which covers over 4.5% of the Earth's land surface and is found mainly in Europe, the USA, and China. Common tree species are oak, maple, and beech and the annual leaf fall means that soils are rich in nutrients.

temperate grasslands a vegetation type, dominated by deep-rooted native grasses, found in areas which have a dry continental climate with a summer drought. Temperate grasslands cover 6% of the Earth's land surface and are found in the USA (*prairies), Central Asia (steppe), and South America (pampas)

temperature measure of how hot an object is. It is temperature difference that determines whether heat transfer will take place between two objects and in which direction it will flow, that is from warmer object to cooler object. The temperature of an object is a measure of the average kinetic energy possessed by the atoms or molecules of which it is composed. The SI unit of temperature is the kelvin (symbol K) used with the Kelvin scale. Other measures of temperature in common use are the Celsius scale and the Fahrenheit scale. The Kelvin scale starts at absolute zero (0 K = –273°C). The Celsius scale starts at the freezing point of water (0°C = 273 K). 1 K is the same temperature interval as 1°C.

temperature regulation ability of an organism to control its internal body temperature. Animals that rely on their environment for their body temperature (and therefore have a variable temperature) are known as **ectotherms** or 'cold-blooded' animals

(for example, lizards). Animals with a constant body temperature irrespective of their environment are known as **endotherms** or 'warm-blooded' animals (for example, birds and mammals). Their temperature is regulated by the hypothalamus in the brain.

tempering heat treatment for improving the properties of metals, often used for steel alloys. The metal is heated to a certain temperature and then quenched (cooled suddenly) in a water or oil bath to fix its state.

template in computing, file that lays down a document's format. Templates are used in word processing, spreadsheet, and other programs to specify all the styles used in a document, such as fonts, margins, macros, formulae and so on. They are widely used to automate the production of documents such as memos, mailings, and reports, making sure that they have a uniform appearance.

temporary hardness hardness of water that is removed by boiling (see *hard water).

tendon or **sinew**, in vertebrates, a cord of very strong, fibrous connective tissue that joins muscle to bone. Tendons are largely composed of bundles of fibres made of the protein collagen, and because of their inelasticity are very efficient at transforming muscle power into movement.

tension in physics, the stress (force) set up in a stretched material. In a stretched string or wire it exerts a pull that is equal in magnitude but opposite in direction to the stress being applied at the string ends. Tension is measured in newtons.

terabyte in computing, 1,024 *gigabytes, or 1,099,511,627,776 *bytes.

teraflops in computing, jargon for 1 trillion floating point operations (computations) per second. Despite earlier predictions, there was still no single computer in 2000 that could boast teraflop performance. Just over one teraflop has been achieved by utilizing thousands of smaller computers, all working in parallel over the Internet.

teratogen any substance or agent that can induce deformities in the fetus if absorbed by the mother during pregnancy. Teratogens include some drugs (notably alcohol and *thalidomide), other chemicals, certain disease organisms, and radioactivity.

teratoma tumour composed of cells not normally found in that part of the body. It mostly occurs in the ovary or testis. Malignant teratoma is the commonest of the testicular cancers.

terbium chemical symbol Tb, soft, silver-grey, metallic element of the *lanthanide series, atomic number 65, relative atomic mass 158.925. It occurs in gadolinite and other ores, with yttrium and ytterbium, and is used in lasers, semiconductors, and television tubes. It was named in 1843 by Swedish chemist Carl Mosander (1797–1858) after the town of Ytterby, Sweden, where it was first found.

terminal in computing, a device consisting of a keyboard and display screen (*VDU) to enable the operator to communicate with the computer. The terminal may be physically attached to the computer or linked to it by a telephone line (remote terminal). A *dumb terminal has no processor of its own, whereas an *intelligent terminal has its own processor and takes some of the processing load away from the main computer.

terminal emulation in computing, communications program such as *Telnet that allows a computer to emulate a terminal or workstation of a remote host. The host accepts instructions from the remote computer as if it were one of its own workstations.

terminal moraine linear, slightly curved ridge of rocky debris deposited at the front end, or snout, of a glacier. It represents the furthest point of advance of a glacier, being formed when deposited material (till), which was pushed ahead of the snout as it advanced, became left behind as the glacier retreated.

terminal velocity or **terminal speed**, the maximum velocity that can be reached by a given object moving through a fluid (gas or liquid) under the action of an applied force. As the speed of the object increases so does the total magnitude of the forces resisting its motion. Terminal velocity is reached when the resistive forces exactly balance the applied force that has caused the object to accelerate; because there is now no resultant force, there can be no further acceleration. For example, an object falling through air experiences air resistance. It will reach a terminal velocity and cease to accelerate under the influence of gravity when the air resistance equals the force of gravity (the object's weight). Parachutes are designed to increase air resistance so that the acceleration of a falling person or package ceases more rapidly, thereby limiting terminal velocity to a safe level.

terminal voltage potential difference (pd) or voltage across the terminals of a power supply, such as a battery of cells. When the supply is not connected in circuit its terminal voltage is the same as its *electromotive force (emf); however, as soon as it begins to supply current to a circuit its terminal voltage falls because some electric potential energy is lost in driving current against the supply's own *internal resistance. As the current flowing in the circuit is increased the terminal voltage of the supply falls.

terminate and stay resident TSR, term given to an MS-DOS program that remains in the memory – for example, a clock, calculator, or thesaurus. The program is recalled by the use of a *hot key.

terrestrial planet any of the four small, rocky inner *planets of the Solar System: *Mercury, Venus, *Earth, and *Mars. The *Moon is sometimes also included, although it is a satellite of the Earth and not strictly a planet.

Tertiary period period of geological time 65 to 1.64 million years ago, divided into five epochs: Palaeocene, Eocene, Oligocene, Miocene, and Pliocene. During the Tertiary period, mammals took over all the ecological niches left vacant by the extinction of the dinosaurs, and became the prevalent land animals. The continents took on their present positions, and climatic and vegetation zones as we know them became established. Within the geological time column the Tertiary follows the Cretaceous period and is succeeded by the Quaternary period.

tesla symbol T, SI unit of *magnetic flux density. One tesla represents a flux density of one *weber per square metre, or 10^4 *gauss. It is named after the Croatian-born US physicist Nikola Tesla.

testa outer coat of a seed, formed after fertilization of the ovule. It has a protective function and is usually hard and dry. In some cases the coat is adapted to aid dispersal, for example by being hairy. Humans have found uses for many types of testa, including the fibre of the cotton seed.

test cross in genetics, a breeding experiment used to discover the genotype of an individual organism. By crossing with a double recessive of the same species, the offspring will indicate whether the test individual is homozygous or heterozygous for the characteristic in question. In peas, a tall plant under investigation would be crossed with a double recessive short plant with known genotype tt. The results of the cross will be all tall plants if the test plant is TT. If the test plant is in fact Tt then there will be some short plants (genotype tt) among the offspring.

testis plural **testes**, organ that produces *sperm in male (and hermaphrodite) animals. In vertebrates it is one of a pair of oval structures that are usually internal, but in mammals (other than elephants and marine mammals), the paired testes (or testicles) descend from the body cavity during development, to hang outside the abdomen in a scrotal sac. The testes also secrete the male sex hormone *androgen.

testosterone *hormone secreted chiefly by the testes, but also by the ovaries and the cortex of the adrenal glands. It promotes the development of *secondary sexual characteristics in males at puberty. It is also needed for the development of the male sex organs and for male fertility. The hormone is partly responsible for the difference in behaviour that may be seen between males and females. In animals with a breeding season, the onset of breeding behaviour is accompanied by a rise in the level of testosterone in the blood.

tetanus or **lockjaw**, acute disease caused by the toxin of the bacillus *Clostridium tetani*, which usually enters the body through a wound. The bacterium is chiefly found in richly manured soil. Untreated, in seven to ten days tetanus produces muscular spasm and rigidity of the jaw spreading to other parts of the body, convulsions, and death. There is a vaccine, and the disease may be treatable with tetanus antitoxin and antibiotics.

tetrachloromethane or **carbon tetrachloride**; CCl_4, chlorinated organic compound that is a very efficient solvent for fats and greases, and was at one time the main constituent of household dry-cleaning fluids and of fire extinguishers used with electrical and petrol fires. Its use became restricted after it was discovered to be carcinogenic and it has now been largely removed from educational and industrial laboratories.

tetracycline one of a group of antibiotic compounds having in common the four-ring structure of chlortetracycline, the first member of the group to be isolated. They are prepared synthetically or obtained from certain bacteria of the genus *Streptomyces*. They are broad-spectrum antibiotics, effective against a wide range of disease-causing bacteria.

tetraethyl lead $Pb(C_2H_5)_4$, compound added to leaded petrol as a component of *antiknock to increase the efficiency of combustion in car engines. It is a colourless liquid that is insoluble in water but soluble in organic solvents such as benzene, ethanol, and petrol.

tetrahedron plural **tetrahedra**, in geometry, a solid figure (*polyhedron) with four triangular faces; that is, a *pyramid on a triangular base. A regular tetrahedron has equilateral triangles as its faces.

tetrapod (Greek 'four-legged') type of *vertebrate. The group includes mammals, birds, reptiles, and amphibians. Birds are included because they evolved from four-legged ancestors, the forelimbs having become modified to form wings. Even snakes are tetrapods, because they are descended from four-legged reptiles.

T_EX (pronounced 'tek') *public domain text formatting and typesetting system, developed by Donald Knuth and widely used for producing mathematical and technical documents. Unlike desktop publishing applications, T_EX is not WYSIWYG, although in some implementations a screen preview of pages is possible.

text editor in computing, a program that allows the user to edit text on the screen and to store it in a file. Text editors are similar to word processors, except that they lack the ability to format text into paragraphs and pages and to apply different typefaces and styles.

TFT LCD another name for *active matrix LCD, a type of colour liquid crystal display commonly used in laptop computers. TFT screens have one or more transistors to control each pixel on the screen, as opposed to the less expensive passive matrix displays where each pixel is not matched with a transistor. TFT LCDs have a faster refresh rate than other types of LCD, reducing the lag between computer and display.

thalamus one of two oval masses of grey matter lying in the cerebral hemispheres. It is an integrating centre for sensory information on its way to the *cerebral cortex and is also important in motor control.

thalidomide *hypnotic drug developed in the 1950s for use as a sedative. When taken in early pregnancy, it caused malformation of the fetus (such as abnormalities in the limbs) in over 5,000 recognized cases, and the drug was withdrawn.

thallium chemical symbol Tl, (Greek *thallos* 'young green shoot') soft, bluish-white, malleable, metallic element, atomic number 81, relative atomic mass 204.38. It is a poor conductor of electricity. Its compounds are poisonous and are used as insecticides and rodent poisons; some are used in the optical-glass and infrared-glass industries and in photocells.

Discovered spectroscopically by its green line, thallium was isolated and named by William Crookes in 1861.

thallus any plant body that is not divided into true leaves, stems, and roots. It is often thin and flattened, as in the body of a seaweed, lichen, or liverwort, and the gametophyte generation (*prothallus) of a fern.

theorem mathematical proposition that can be deduced by logic from a set of axioms (basic facts that are taken to be true without proof). Advanced mathematics consists almost entirely of theorems and proofs, but even at a simple level theorems are important.

theory of everything ToE, another name for *grand unified theory.

therm unit of energy defined as 10^5 British thermal units; equivalent to 1.055×10^8 J. It is no longer in scientific use.

thermal capacity another name for *heat capacity

thermal conductivity ability of a substance to conduct heat. Good thermal conductors, like good electrical conductors, are generally materials with many free electrons, such as *metals. A poor conductor, called an *insulator, has low conductivity. Thermal conductivity is expressed in units of joules per second per metre per kelvin ($J\ s^{-1}\ m^{-1}\ K^{-1}$). For a block of material of cross-sectional area a and length l, with temperatures T_1 and T_2 at its end faces, the thermal conductivity λ equals $Hl/at(T_2 - T_1)$, where H is the amount of heat transferred in time t.

thermal dissociation reversible breakdown of a chemical compound into simpler substances by heating it (see *dissociation). The splitting of ammonium chloride into ammonia and hydrogen chloride is an example. On cooling, they recombine to form the salt.

thermal expansion expansion that is due to a rise in temperature. It can be expressed in terms of linear, area, or volume expansion. The coefficient of linear expansion α is the fractional increase in length per degree temperature rise; area, or superficial, expansion β is the fractional increase in area per degree; and volume, or cubic, expansion γ is the fractional increase in volume per degree. To a close approximation, β = 2α and γ = 3α.

Thermal Oxide Reprocessing Plant THORP, nuclear plant built at Sellafield, UK, for reprocessing spent fuel from countries around the world, with plutonium as the end product. The plant began operating 1994, despite a court action by Greenpeace to prevent this.

thermal reactor nuclear reactor in which the neutrons released by fission of uranium-235 nuclei are slowed down in order to increase their chances of being captured by other uranium-235 nuclei, and so induce further fission. The material (commonly graphite or heavy water) responsible for doing so is called a **moderator**. When the fast newly emitted neutrons collide with the nuclei of the moderator's atoms, some of their kinetic energy is lost and their speed is reduced. Those that have been slowed down to a speed that matches the thermal (heat) energy of the surrounding material are called **thermal neutrons**, and it is these that are most likely to induce fission and ensure the continuation of the chain reaction. See *nuclear reactor and *nuclear energy.

thermionics branch of electronics dealing with the emission of electrons from matter under the influence of heat.

thermionic tube electronic tube, used in telegraphy and telephony and in radio and radar, using space conduction by thermionically emitted electrons from an electrically heated cathode. Classification is into diode, triode, and multi-electrode tubes, but in most applications they have been replaced by *transistors.

thermistor resistor whose *resistance changes significantly when its temperature changes. The resistance of a *semiconductor thermistor decreases with increase of temperature. As temperature rises, the resistance of the material decreases so that the current passing through a thermistor increases rapidly. They are used to electronically detect changes in temperature (electrical thermometers), as in fire alarms and thermostats. They also are

used in lamp filaments and electric motors to stop large currents flowing through them when they are initially turned on.

When a lamp is switched on, the thermistor is cold initially and conducts poorly and has a high resistance. As the bulb heats up, the **intrinsic conduction** of the thermistor improves because the increasing thermal energy of vibrating atoms liberates electrons. These electrons flow more freely as current as the resistance of the lamp's thermistor decreases. The resistance of metals increases with temperature. When a metal is heated, the atoms vibrate more, thus impeding the flow of electrons.

thermocouple electric temperature measuring device consisting of a circuit having two wires made of different metals welded together at their ends. A current flows in the circuit when the two junctions are maintained at different temperatures (*Seebeck effect). The electromotive force generated – measured by a millivoltmeter – is proportional to the temperature difference.

thermodynamics branch of physics dealing with the transformation of heat into and from other forms of energy. It is the basis of the study of the efficient working of engines, such as the steam and internal combustion engines. The three laws of thermodynamics are:

1. Energy can be neither created nor destroyed, heat and mechanical work being mutually convertible;

2. It is impossible for an unaided self-acting machine to convey heat from one body to another at a higher temperature;

3. It is impossible by any procedure, no matter how idealized, to reduce any system to the *absolute zero of temperature (0 K/–273.15°C/ –459.67°F) in a finite number of operations.

Fundamental to the three laws of thermodynamics is the zeroth law of thermodynamics. This states that, if two systems are separately in thermal equilibrium with a third system, all three systems are in thermal equilibrium with each other.

thermoluminescence TL, release, in the form of a light pulse, of stored nuclear energy in a mineral substance when heated to perhaps 500°C/930° F. The energy originates from the radioactive decay of uranium and thorium, and is absorbed by crystalline inclusions within the mineral matrix, such as quartz and feldspar. The release of TL from these crystalline substances is used in archaeology to date pottery, and by geologists in studying terrestrial rocks and meteorites.

thermometer instrument for measuring temperature. There are many types, designed to measure different temperature ranges to varying degrees of accuracy. Each makes use of a different physical effect of temperature. Expansion of a liquid is employed in common **liquid-in-glass thermometers**, such as those containing mercury or alcohol. The more accurate **gas thermometer** uses the effect of temperature on the pressure of a gas held at constant volume. A **resistance thermometer** takes advantage of the change in resistance of a conductor (such as a platinum wire) with variation in temperature. Another electrical thermometer is the *thermocouple. Mechanically, temperature change can be indicated by the change in curvature of a **bimetallic strip** (as commonly used in a *thermostat).

thermopile instrument for measuring radiant heat, consisting of a number of *thermocouples connected in series with alternate junctions exposed to the radiation. The current generated (measured by an *ammeter) is proportional to the radiation falling on the device.

thermoplastic or **thermosoftening plastic**, type of *plastic that always

Monomer	Polymer	Polymer name	Uses
$CH_2=CH_2$ ethene	$[CH_2-CH_2]_n$	poly(ethene), polythene	bottles, packaging, insulation, pipes
$CH_2=CH-CH_3$ propene	$[CH_2-CH]_n$ $\quad\quad CH_3$	poly(propene), polypropylene	mouldings, film, fibres
$CH_2=CH-CL$ chloroethene (vinyl chloride)	$[CH_2-CH]_n$ $\quad\quad Cl$	polyvinylchloride (PVC), poly(chloroethene)	insulation, flooring, household fabric
$CH_2=CH-C_6H_5$ phenylethene (styrene)	$[CH_2-CH]_n$ $\quad\quad C_6H_5$	polystyrene, poly(phenylethene)	insulation, packaging
$CF_2=CF_2$ tetrafluoroethene	$[CF_2-CF_2]_n$ $(n = 1000+)$	poly(tetrafluoroethene) (PTFE)	high resistance to chemical and electrical reaction, low-friction applications

thermoplastic Examples of thermosoftening plastics, their basic monomer origins, polymer names, and everyday uses.

Monomer I	Monomer II	Polymer name	Uses
formaldehyde (methanal)	phenol	PF resins (Bakelites)	electrical fittings, radio cabinets
formaldehyde	urea	UF resins	electrical fittings, insulation, adhesives
formaldehyde	melamine	melamines	laminates for furniture

thermoset Unlike thermoplastics, thermosets remain rigid when set and do not soften when heated.

softens on repeated heating. Thermoplastics include polyethene (polyethylene), polystyrene, nylon, and polyester.

thermoset type of *plastic that remains rigid when set, and does not soften with heating. Thermosets have this property because the long-chain polymer molecules cross-link with each other to give a rigid structure. Examples include Bakelite, resins, melamine, and urea–formaldehyde resins.

thermosphere layer in the Earth's *atmosphere above the mesosphere and below the exosphere. Its lower level is about 80 km/50 mi above the ground, but its upper level is undefined. The ionosphere is located in the thermosphere. In the thermosphere the temperature rises with increasing height to several thousand degrees Celsius. However, because of the thinness of the air, very little heat is actually present.

thermostat temperature-controlling device that makes use of feedback. It employs a temperature sensor (often a bimetallic strip) to operate a switch or valve to control electricity or fuel supply. Thermostats are used in central heating, ovens, and engine cooling systems.

thiamine or **vitamin B$_1$**, a water-soluble vitamin of the B complex. It is found in seeds and grain. Its absence from the diet causes the disease *beriberi.

thigh the lower limb between the pelvis and the knee. The thigh-bone (femur), is the longest bone in the human body. It articulates with the hip-bone above, and the tibia (shin-bone) below.

thorax part of the body in four-limbed vertebrates containing the *heart and *lungs, and protected by the ribcage. It is separated from the abdomen by the diaphragm. During *breathing (ventilation) the volume inside the thorax is changed. This then causes air to move in or out of the air passages that lead to the lungs. The volume of the thorax is altered by the contraction of *muscles in the diaphragm and the contraction of muscles between the ribs – the intercostal muscles.

thorium chemical symbol Th, dark-grey, radioactive, metallic element of the *actinide series, atomic number 90, relative atomic mass 232.038. It occurs throughout the world in small quantities in minerals such as thorite and is widely distributed in monazite beach sands. It is one of three fissile elements (the others are uranium and plutonium) and its longest-lived isotope has a half-life of 1.39×10^{10} years. Thorium is used to strengthen alloys. It was discovered by Jöns Berzelius in 1828 and was named by him after the Norse god Thor.

thread in computing, subject line of electronic messages within an online topic or conference. Most online conferencing systems use some kind of threading; one advantage is that it makes it easy for readers of a particular conference or forum to skip over sections that do not interest them. Threading is an important feature of off-line readers, as otherwise it is difficult to tell how individual messages relate to one another.

throat in human anatomy, the passage that leads from the back of the nose and mouth to the *trachea and *oesophagus. It includes the *pharynx and the *larynx, the latter being at the top of the trachea. The word 'throat' is also used to mean the front part of the neck, both in humans and other vertebrates; for example, in describing the plumage of birds. In engineering, it is any narrowing entry, such as the throat of a carburettor.

thrombin substance that converts the soluble protein fibrinogen to the insoluble fibrin in the final stage of blood clotting. Its precursor in the bloodstream is *prothrombin.

thrombocyte in medicine, another name for a *platelet.

thrombosis condition in which a blood clot forms in a vein or artery, causing loss of circulation to the area served by the vessel. If it breaks away, it often travels to the lungs, causing pulmonary embolism.

thrush infection usually of the mouth (particularly in infants), but also sometimes of the vagina, caused by a yeastlike fungus (*Candida*). It is seen as white patches on the mucous membranes.

thulium chemical symbol Tm, soft, silver-white, malleable and ductile, metallic element of the *lanthanide series, atomic number 69, relative atomic mass 168.93. It is the least abundant of the rare earth metals, and was first found in gadolinite and various other minerals. It is used in arc lighting. The X-ray-emitting isotope Tm-170 is used in portable X-ray units. Thulium was named by Swedish chemist Per Cleve in 1879 after the northland, Thule.

thumbnail in computing, a small version of a larger image used for reference. A *PhotoCD or clip art collection might initially present images as thumbnails, while publishing programs include the facility for designers to produce thumbnail page layouts.

thunderstorm severe storm of very heavy rain, thunder, and *lightning. Thunderstorms are usually caused by the intense heating of the ground surface during summer. The warm air rises rapidly to form tall cumulonimbus clouds with a characteristic anvil-shaped top. Electrical charges accumulate in the clouds and are discharged to the ground as flashes of lightning. Air in the path of lightning becomes heated and expands rapidly, creating shock waves that are heard as a crash or rumble of thunder.

thymine in biochemistry, a colourless crystalline solid, also known as 5-methyl-2,6-dioxytetrahydropyrimidine. Thymine is one of four molecules that form the *base pairs in the DNA molecule, where it is always paired with the purine *adenine. Its base pairing with adenine occurs through hydrogen bonding.

thymus organ in vertebrates, situated in the upper chest cavity in humans. The thymus processes *lymphocyte cells to produce T-lymphocytes (T denotes 'thymus-derived'), which are responsible for binding to specific invading organisms and killing them or rendering them harmless.

The thymus reaches full size at puberty, and shrinks thereafter; the stock of T-lymphocytes is built up early in life, so this function diminishes in adults, but the thymus continues to function as an *endocrine gland, producing the hormone thymosin, which stimulates the activity of the T-lymphocytes.

thyristor type of *rectifier, an electronic device that conducts electricity in one direction only. The thyristor is composed of layers of *semiconductor material sandwiched between two electrodes called the anode and cathode. The current can be switched on by using a third electrode called the gate.

thyroid *endocrine gland of vertebrates, situated in the neck in front of the trachea. It secretes several hormones, principally thyroxine, an iodine-containing hormone that stimulates growth, metabolism, and other functions of the body. The thyroid gland may be thought of as the regulator gland of the body's metabolic rate. If it is overactive, as in hyperthyroidism, the sufferer feels hot and sweaty, has an increased heart rate, diarrhoea, and weight loss. Conversely, an underactive thyroid leads to myxoedema, a condition characterized by sensitivity to the cold, constipation, and weight gain. In infants, an underactive thyroid leads to cretinism, a form of mental retardation.

thyroxine hormone containing iodine that is produced by the thyroid gland. It is used to treat conditions that are due to deficiencies in thyroid function, such as myxoedema.

TIA abbreviation for **transient ischaemic attack**.

tibia anterior of the pair of bones in the leg between the ankle and the knee. In humans, the tibia is the shinbone. It articulates with the *femur above to form the knee joint, the *fibula externally at its upper and lower ends, and with the talus below, forming the ankle joint.

tidal wave common name for a *tsunami.

tide rhythmic rise and fall of the *sea level in the Earth's oceans and their inlets and estuaries due to the gravitational attraction of the Moon and, to a lesser extent, the Sun, affecting regions of the Earth unequally as it rotates. Water on the side of the Earth nearest to the Moon feels the Moon's pull and accumulates directly below the Moon, producing a high tide.

TIFF acronym for Tagged Image File Format, a *graphics file format.

tiling in computing, arrangement of *windows in a *graphical user interface system so that they do not overlap.

time continuous passage of existence, recorded by division into hours, minutes, and seconds. Formerly the measurement of time was based on the Earth's rotation on its axis, but this was found to be irregular. Therefore the second, the standard *SI unit of time, was redefined in 1956 in terms of the Earth's annual orbit of the Sun, and in 1967 in terms of a radiation pattern of the element caesium.

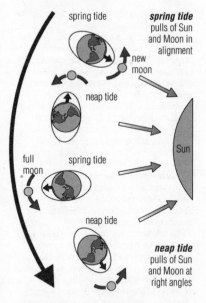

tide The gravitational pull of the Moon is the main cause of the tides. Water on the side of the Earth nearest the Moon feels the Moon's pull and accumulates directly under the Moon. When the Sun and the Moon are in line, at new and full Moon, the gravitational pull of Sun and Moon are in line and produce a high spring tide. When the Sun and Moon are at right angles, lower neap tides occur.

tin chemical symbol Sn, soft, silver-white, malleable and somewhat ductile, metallic element, symbol Sn, atomic number 50, relative atomic mass 118.71. Its chemical symbol comes from the Latin *stannum*. Tin exhibits *allotropy, having three forms: the familiar lustrous metallic form above 13.2°C/55.8°F, a brittle form above 161°C/321.8°F, and a grey powder form below 13.2°C/55.8°F (commonly called tin pest or tin disease). The metal is quite soft (slightly harder than lead) and can be rolled, pressed, or hammered into extremely thin sheets; it has a low melting point. In nature it occurs rarely as a free metal. It resists corrosion and is therefore used for coating and plating other metals.

Tin and copper smelted together form the oldest desired alloy, bronze; since the Bronze Age (3500 BC) that alloy has been the basis of both useful and decorative materials. Tin is also alloyed with metals other than copper to make solder and pewter. It was recognized as an element by French chemist Antoine Lavoisier (1743–1794), but the name is very old and comes from the Germanic form *zinn*. The mines of Cornwall were the principal Western source of tin until the 19th century, when rich deposits were found in South America, Africa, South-East Asia, and Australia. Tin production is concentrated in Malaysia, Indonesia, Brazil, and Bolivia.

tinnitus in medicine, constant buzzing or ringing in the ears. The phenomenon may originate from prolonged exposure to noisy conditions (drilling, machinery, or loud music) or from damage to or disease of the middle or inner ear. The victim may become overwhelmed by the relentless noise in the head.

tin ore mineral from which tin is extracted, principally cassiterite, SnO_2. The world's chief producers are Malaysia, Thailand, and Bolivia.

tissue in biology, any kind of cellular fabric that occurs in an organism's body. It is a group of similar *cells that are carrying out a function in a plant or animal. Several kinds of tissue can usually be distinguished, each consisting of cells of a particular kind bound together by cell walls (in plants) or extracellular matrix (in animals). Thus, nerve and muscle are different kinds of tissue in animals, as are *parenchyma and sclerenchyma in plants. Tissues of different kinds may be found in a distinct structure, which is then called an *organ. The leaf of a plant or the heart of a mammal is an organ.

tissue culture process by which cells from a plant or animal are removed from the organism and grown under controlled conditions in a sterile medium containing all the necessary nutrients. Tissue culture can provide information on cell growth and differentiation, and is also used in plant propagation and drug production.

tissue plasminogen activator tPA, naturally occurring substance in the body tissues that activates the enzyme plasmin that is able to dissolve blood clots. Human tPA, produced in bacteria by genetic engineering, has, like *streptokinase, been used to dissolve blood clots in the coronary arteries of heart-attack victims. It has been shown to be more effective than streptokinase when used in conjunction with heparin, but it is much more expensive.

titanium chemical symbol Ti, strong, lightweight, silver-grey, metallic element, atomic number 22, relative atomic mass 47.87. The ninth most abundant element in the Earth's crust, its compounds occur in practically all igneous rocks and their sedimentary deposits. It is very strong and resistant to corrosion, so it is used in building high-speed aircraft and spacecraft; it is also widely used in making alloys, as it unites with almost every metal except copper and aluminium. Titanium oxide is used in high-grade white pigments. Titanium bonds with bone in a process called **osseointegration**. As the body does not react to the titanium it is valuable for permanent implants such as prostheses.

The element was discovered in 1791 by English mineralogist William Gregor (1761–1817) and was named by German chemist Martin Klaproth in 1796 after the Titans, the giants of Greek mythology. It was not obtained in pure form until 1925.

titanium ore any mineral from which titanium is extracted, principally ilmenite ($FeTiO_3$) and rutile (TiO_2). Brazil, India, and Canada are major producers. Both these ore minerals are found either in rock formations or concentrated in heavy mineral sands.

titration in analytical chemistry, a technique to find the concentration of one compound in a solution by determining how much of it will react with a known amount of another compound in solution. One of the solutions is measured by *pipette into the reaction vessel. The other is added a little at a time from a *burette. The end-point of the reaction is determined with an *indicator or an electrochemical device.

TNT abbreviation for **trinitrotoluene**, $CH_3C_6H_2(NO_2)_3$, a powerful high explosive. It is a yellow solid, prepared in several isomeric forms from *toluene by using sulphuric and nitric acids.

tocolytic drug drug that prevents premature labour.

tocopherol or **vitamin E**, fat-soluble chemical found in vegetable oils. Deficiency of tocopherol leads to multiple adverse effects on health. In rats, vitamin E deficiency has been shown to cause sterility.

ToE abbreviation for theory of everything, another name for *grand unified theory.

tog unit of measure of thermal insulation used in the textile trade; a light summer suit provides 1.0 tog.

toggle in computing, to switch between two settings. In software a toggle is usually triggered by the same code, so it is important that this code only has two meanings. An example is the use of the same character in a text file to indicate both opening and closing quotation marks; if the same character is also used to mean an apostrophe, then conversion, via a toggle switch, for a desktop publishing system that uses different opening and closing quotation marks, will not be carried out correctly.

tokamak Russian acronym for toroidal magnetic chamber, an experimental machine conceived by Soviet physicist Andrei Sakharov and developed in the Soviet Union to investigate controlled nuclear fusion. It consists of a doughnut-shaped chamber surrounded by electromagnets capable of exerting very powerful magnetic fields. The fields are generated to confine a very hot (millions of degrees) *plasma of ions and electrons, keeping it away from the chamber walls. See also *JET.

tolerance in medicine, gradual increase of resistance to a drug effects on the patient, when is taken over a long period. This means that more of the drug must be taken to achieve the desired effect.

toluene or **methyl benzene**; $C_6H_5CH_3$, colourless, inflammable liquid, insoluble in water, derived from petroleum. It is used as a solvent, in aircraft fuels, in preparing phenol (carbolic acid, used in making resins for adhesives, pharmaceuticals, and as a disinfectant), and the powerful high explosive *TNT.

ton symbol t, imperial unit of mass. The **long ton**, used in the UK, is 1,016 kg/2,240 lb; the **short ton**, used in the USA, is 907 kg/2,000 lb. The **metric ton** or **tonne** is 1,000 kg/2,205 lb.

ton in shipping, unit of volume equal to 2.83 cubic metres/100 cubic feet.

Gross tonnage is the total internal volume of a ship in tons; **net register tonnage** is the volume used for carrying cargo or passengers.

Displacement tonnage is the weight of the vessel, in terms of the number of imperial tons of seawater displaced when the ship is loaded to its load line; it is used to describe warships.

tonne symbol t, metric ton of 1,000 kg/2,204.6 lb; equivalent to 0.9842 of an imperial *ton.

tonsillitis inflammation of the *tonsils.

tonsils in higher vertebrates, masses of lymphoid tissue situated at the back of the mouth and throat (palatine tonsils), and on the rear surface of the tongue (lingual tonsils). The tonsils contain many *lymphocytes and are part of the body's defence system against infection.

toolbar in computing, area usually at the top or side of a screen with *push-buttons and other features to perform frequently used tasks. For example, the toolbar of a paint program might offer quick access to different brushes, spray cans, erasers, and other useful tools. 'Floating' and 'smart' toolbars have become more common, where a toolbar only appears when a relevant object is selected and can be repositioned at will.

tooth in vertebrates, one of a set of hard, bonelike structures in the

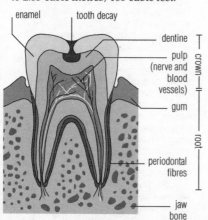

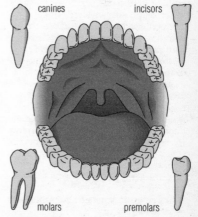

tooth Adults have 32 teeth: two incisors, one canine, two premolars, and three molars on each side of each jaw. Each tooth has three parts: crown, neck, and root. The crown consists of a dense layer of mineral, the enamel, surrounding hard dentine with a soft centre, the pulp.

mouth, used for biting and chewing food, and in defence and aggression. In humans, the first set (20 milk teeth) appear from age six months to two and a half years. The permanent *dentition replaces these from the sixth year onwards, the wisdom teeth (third molars) sometimes not appearing until the age of 25 or 30. Adults have 32 teeth: two incisors, one canine (eye tooth), two premolars, and three molars on each side of each jaw. Each tooth consists of an enamel coat (hardened calcium deposits), dentine (a thick, bonelike layer), and an inner pulp cavity, housing nerves and blood vessels. Mammalian teeth have roots surrounded by cementum, which fuses them into their sockets in the jawbones. The neck of the tooth is covered by the *gum, while the enamel-covered crown protrudes above the gum line.

toothache in medicine, dental pain commonly due to inflammation of the pulp or the tooth or the fibrous tissue supporting the tooth in the bone socket. It is most frequently due to *caries when the cavity is close to the pulp.

topaz mineral, aluminium fluorosilicate, $Al_2(F_2SiO_4)$. It is usually yellow, but pink if it has been heated, and is used as a gemstone when transparent. It ranks 8 on the *Mohs scale of hardness.

topography surface shape and composition of the landscape, comprising both natural and artificial features, and its study. Topographical features include the relief and contours of the land; the distribution of mountains, valleys, and human settlements; and the patterns of rivers, roads, and railways.

topology branch of geometry that deals with those properties of a figure that remain unchanged even when the figure is transformed (bent, stretched) – for example, when a square painted on a rubber sheet is deformed by distorting the sheet.

Topology has scientific applications, as in the study of turbulence in flowing fluids.

tor isolated mass of rock, often granite, left upstanding on a hilltop after the

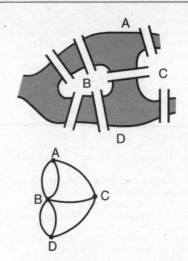

topology The Könisberg bridge problem. The challenge is to cross each of the seven bridges (top) over the River Pregol'a in Königsberg (now Kaliningrad, Russia) once only and return to one's starting point. Representing the puzzle as a topological network (bottom) shows that this is impossible.

surrounding rock has been broken down. Weathering takes place along the joints in the rock, reducing the outcrop into a mass of rounded blocks.

tornado extremely violent revolving storm with swirling, funnel-shaped clouds, caused by a rising column of warm air propelled by strong wind. A tornado can rise to a great height, but with a diameter of only a few hundred metres or less. Tornadoes move with wind speeds of 160–480 kph/100–300 mph, destroying everything in their path. They are common in the central USA and Australia.

torque turning effect of force on an object. A turbine produces a torque that turns an electricity generator in a power station. Torque is measured by multiplying the force by its perpendicular distance from the turning point.

torr unit of pressure equal to 1/760 of an *atmosphere, used mainly in high-vacuum technology.

torsion in physics, the state of strain set up in a twisted material; for example, when a thread, wire, or rod is twisted, the torsion set up in the material tends to return the material to its original state. The **torsion balance**, a sensitive device for measuring small gravitational or magnetic forces, or electric charges, balances these against the restoring force set up by them in a torsion suspension.

torus in nuclear physics, ring-shaped chamber used to contain *plasma in nuclear fusion reactors such as the Joint European Torus (*JET) reactor.

total internal reflection complete reflection of a beam of light that occurs from the surface of an optically 'less dense' material. For example, a beam from an underwater light source can be reflected from the surface of the water, rather than escaping through the surface. Total internal reflection can only happen if a light beam hits a surface at an angle greater than the *critical angle for that particular pair of materials.

Tourette's syndrome or **Gilles de la Tourette syndrome**, rare neurological condition characterized by multiple tics and vocal phenomena such as grunting, snarling, and obscene speech, named after French physician Georges Gilles de la Tourette (1859–1904). It affects one to five people per 10,000, with males outnumbering females by four to one, and the onset is usually around the age of six. There are no convincing explanations of its cause, and it is usually resistant to treatment.

toxaemia another term for *blood poisoning; **toxaemia of pregnancy** is another term for pre-eclampsia.

toxicology branch of medicine dealing with the study of poisons. This includes the chemical nature of poisons, their origin and preparation, their physiological action, tests to recognize them, pathological changes due to their presence, their antidotes, and the recognition of them by post-mortem evidence.

toxic shock syndrome rare condition marked by rapid onset of fever, vomiting, and low blood pressure, sometimes leading to death. It is caused by a toxin of the bacterium *Staphylococcus aureus*, normally harmlessly present in the body. It is seen most often in young women using tampons during menstruation.

toxic waste *hazardous waste, especially when it has been dumped.

toxin any poison produced by another living organism (usually a bacterium) that can damage the living body. In vertebrates, toxins are broken down by *enzyme action, mainly in the liver.

trace element chemical element necessary in minute quantities for the health of a plant or animal. For example, magnesium, which occurs in chlorophyll, is essential to photosynthesis, and iodine is needed by the thyroid gland of mammals for making hormones that control growth and body chemistry.

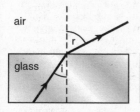 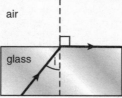 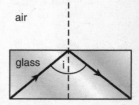

if the angle of incidence is less than the critical angle, the light refracts away from the normal

if the angle of incidence is equal to the critical angle, the light refracts at 90° to the normal

if the angle of incidence is greater than the critical angle, total internal reflection occurs

total internal reflection Total internal reflection occurs at the boundary of two transparent substances when a) the incident ray is in the substance with the higher refractive index and b) the angle of incidence exceeds a particular angle known as the critical angle. The critical angle for glass is 42° and for water, about 48°.

tracer in science, a small quantity of a radioactive *isotope (form of an element) used to follow the path of a substance. Certain chemical reactions or physical or biological processes can then be monitored. The location (and possibly concentration) of the tracer is usually detected by using a Geiger–Müller counter. For example, the activity of the thyroid gland can be monitored by giving the patient an injection containing a small dose of a radioactive isotope of iodine, which is selectively absorbed from the bloodstream by the gland.

traceroute in computing, network *utility program which allows the user to find out the *bang path taken by *packets of data sent across the Internet. Traceroute can help to debug a network, or check how it works.

trachea tube that forms an airway in air-breathing animals. In land-living *vertebrates, including humans, it is also known as the **windpipe** and runs

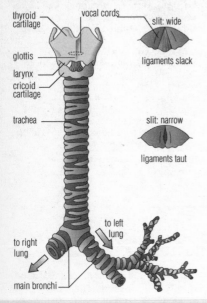

thyroid cartilage
vocal cords
slit: wide
glottis
ligaments slack
larynx
cricoid cartilage
trachea
slit: narrow
ligaments taut
to left lung
to right lung
main bronchi

trachea The human trachea, or windpipe. The larynx, or voice box, lies at the entrance to the trachea. The two vocal cords are membranes that normally remain open and still. When they are drawn together, the passage of air makes them vibrate and produce sounds.

from the larynx to the upper part of the chest. Its diameter is about 1.5 cm/0.6 in and its length 10 cm/4 in. It is strong and flexible, and reinforced by rings of *cartilage. In the upper chest, the trachea branches into two tubes: the left and right bronchi, which enter the lungs. Insects have a branching network of tubes called tracheae, which conduct air from holes (*spiracles) in the body surface to all the body tissues. The finest branches of the tracheae are called tracheoles.

track in computing, part of the magnetic structure created on a disk surface during disk formatting so that data can be stored on it. The disk is first divided into circular tracks and then each circular track is divided into a number of sectors.

trackball *input device that carries out the same function as a *mouse, but remains stationary. In a trackball the ball controlling the cursor position is operated directly with the fingers.

tracking in computing, amount of space between text characters. Many word processing and page layout programs allow users to adjust tracking for a wide-spaced or slightly condensed appearance.

tracking agent chemical compound (most usually NPPD, nitrophenyl pentadiene) used in espionage. It is applied to people, clothes, and cars to trace their movements and contacts.

trade wind prevailing wind that blows towards the Equator from the northeast and southeast. Trade winds are caused by hot air rising at the Equator and the consequent movement of air from north and south to take its place. The winds are deflected towards the west because of the Earth's west-to-east rotation.

The unpredictable calms known as the doldrums lie at their convergence.

traffic in computing, messages sent over a network such as the Internet.

trance mental state in which the subject loses the ordinary perceptions of time and space, and even of his or her own body.

tranquillizer common name for any drug for reducing anxiety or tension (*anxiolytic), such as benzodiazepines, barbiturates, antidepressants, and beta-blockers. The use of drugs to control anxiety is becoming much less popular, because most of the drugs available are capable of inducing dependence.

transactinide element any of a series of eight radioactive, metallic elements with atomic numbers that extend beyond the *actinide series, those from 104 (rutherfordium) upwards. They are grouped because of their expected chemical similarities (they are all bivalent), the properties differing only slightly with atomic number. All have *half-lives of less than two minutes.

transaction file in computing, a file that contains all the additions, deletions, and amendments required during file updating to produce a new version of a master file.

transcription in living cells, the process by which the information for the synthesis of a protein is transferred from the *DNA strand on which it is carried to the messenger *RNA strand involved in the actual synthesis.

It occurs by the formation of *base pairs when a single strand of unwound DNA serves as a template for assembling the complementary nucleotides that make up the new RNA strand.

transducer device that converts one form of energy into another. For example, a thermistor is a transducer that converts heat into an electrical voltage, and an electric motor is a transducer that converts an electrical voltage into mechanical energy. Transducers are important components in many types of *sensor, converting the physical quantity to be measured into a proportional voltage signal.

transduction transfer of genetic material between cells by an infectious mobile genetic element such as a virus. Transduction is used in *genetic engineering to produce new varieties of bacteria.

transference in psychoanalysis, the patient's transfer of feelings and wishes experienced in earlier relationships into the relationship with the analyst.

transfermium element or **superheavy element**, chemical element with an atomic number greater than 100 (the atomic number of fermium). Transfermium elements are all highly unstable and have existed only briefly, and because of this there is a continuing controversy over who should be credited with their discoveries. See also *transuranic element.

transfer RNA tRNA, relatively small molecule of ribonucleic acid, the function of which is to carry *amino acids to *ribosomes where *protein synthesis occurs. Each amino acid is borne by a different tRNA molecule. tRNA is complementary to *messenger RNA (mRNA).

transformation in mathematics, a mapping or *function, especially one which causes a change of shape or position in a geometric figure. *Reflection, *rotation, *enlargement, and *translation are the main geometrical transformations.

Two or more transformations performed on one shape are called **combined transformations**. The transformation of a shape from A to B can be reversed by an **inverse transformation** from B back onto A.

transformer device in which, by electromagnetic induction, an alternating current (AC) of one voltage is transformed to another voltage, without change of *frequency. Transformers are widely used in electrical apparatus of all kinds, and in particular in power transmission where high voltages and low currents are utilized. (See diagram, page 602.)

transfusion intravenous delivery of blood or blood products (plasma, red cells) into a patient's circulation to make up for deficiencies due to disease, injury, or surgical intervention.

Cross-matching is carried out to ensure the patient receives the right blood group. Because of worries about blood-borne disease, there is a growing interest in autologous transfusion with units of the patient's own blood 'donated' over the weeks before an operation.

transgenic organism plant, animal, bacterium, or other living organism that has had a foreign gene added to it by means of *genetic engineering. By January 2002, around 5.5 million

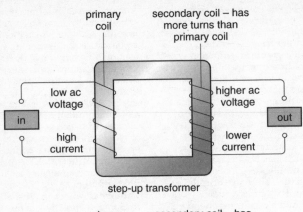

primary coil

secondary coil – has more turns than primary coil

low ac voltage

higher ac voltage

in

out

high current

lower current

step-up transformer

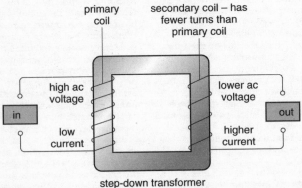

primary coil

secondary coil – has fewer turns than primary coil

high ac voltage

lower ac voltage

in

out

low current

higher current

step-down transformer

transformer A step-up transformer increases voltage and has more turns on the secondary coil than on the primary. A step-down transformer decreases voltage and has more turns on the primary coil than on the secondary. If the numbers of turns in the primary and secondary coils are n_1 and n_2, the primary and secondary voltages are V_1 and V_2, and the primary and secondary currents are T_1 and T_2, then the relationships between these quantities may be expressed as: $V_1/V_2 = I_2/I_1 = n_1/n_2$.

farmers worldwide were growing transgenic crops. These crops were grown mainly in the USA, Argentina, Canada, and China, and covered an estimated 50 million ha/124 million acres. The first genetically-modified primate, a rhesus monkey, was born in January 2001. The monkey had a gene from a jellyfish, that was inserted into an unfertilized monkey egg at the Regional Primate Research Centre, Portland, Oregon.

transistor solid-state electronic component, made of *semiconductor material, with three or more electrical contacts that can regulate a current passing through it. A transistor can act as an amplifier, *oscillator, *photocell, or switch, and (unlike earlier thermionic valves) usually operates on a very small amount of power. Transistors commonly consist of a tiny sandwich of *germanium or *silicon, alternate layers having different electrical properties because they are impregnated with minute amounts of different impurities.

transistor–transistor logic TTL, in computing, the type of integrated circuit most commonly used in building electronic products. In TTL chips the bipolar transistors are directly connected (usually collector to base). In mass-produced items, large numbers of TTL chips are commonly replaced by a small number of *uncommitted logic arrays (ULAs), or logic gate arrays.

transition in computing, way in which one image changes to another in a slide show, animation, or multimedia presentation.

transition metal any of a group of metallic *elements that have incomplete inner electron shells and exhibit variable valency – for example, cobalt, copper, iron, and molybdenum. They form a long block in the middle of the *periodic table of the elements, between groups 2 and 3. They are excellent conductors of electricity, and generally form highly coloured compounds.

They include most of the hard, dense, and less reactive metals used as building materials, such as iron, lead, and copper. The precious metals, gold, platinum, and silver are also transition metals. They are rare, beautiful and costly, and so unreactive that they can be used as jewellery. Transition metals and their compounds are often used in industry as catalysts. For example, iron is used as a catalyst in the *Haber process to produce ammonia. They are also used to make *alloys.

translation in living cells, the process by which proteins are synthesized. During translation, the information coded as a sequence of nucleotides in messenger *RNA is transformed into a sequence of amino acids in a peptide chain. The process involves the 'translation' of the *genetic code. See also *transcription.

translation in mathematics, the movement of a point, line, or shape along a straight line. Translations may be combined with *rotations and *reflections. The amount and direction of a translation can be described using vector notation.

translation program in computing, a program that translates another program written in a high-level language or assembly language into the machine-code instructions that a computer can obey. See *compiler and *interpreter.

translocation in genetics, the exchange of genetic material between chromosomes. It is responsible for

congenital abnormalities, such as Down's syndrome.

transmissible spongiform encephalopathy term used for a group of human and animal diseases characterized by the development of spongy changes in the brain. All fatal, they can be passed to other individuals in contaminated tissue. They are caused by *prions, and about 10% of the human diseases seem to be inherited. The group includes: *bovine spongiform encephalopathy, scrapie in sheep, and *Creutzfeldt–Jakob disease and kuru in humans.

transmission electron microscope TEM, most powerful type of *electron microscope, with a resolving power ten times better than that of a *scanning electron microscope and a thousand times better than that of an optical microscope. A fine electron beam passes through the specimen, which must therefore be sliced extremely thinly – typically to about one-thousandth of the thickness of a sheet of paper (100 nanometres). The TEM can resolve objects 0.001 micrometres (0.04 millionth of an inch) apart, a gap that is 100,000 times smaller than the unaided eye can see.

transpiration loss of water from a plant by evaporation. Most water is lost by *diffusion of water vapour from the leaves through pores known as *stomata to the outside air. The primary function of stomata is to allow *gas exchange between the plant's internal tissues and the atmosphere. Transpiration from the leaf surfaces causes a continuous upward flow of water from the roots via the *xylem, which is known as the transpiration stream. This replaces the water that is lost, and allows minerals absorbed from the soil to be transported through the xylem to the leaves. This is important because many plant cells need the minerals as nutrients. (See diagram, page 604.)

A single maize plant has been estimated to transpire 245 l/54 gal of water in one growing season.

transport in plants and animals, method by which substances such as nutrients and water move into and out of organisms and into and out of cells

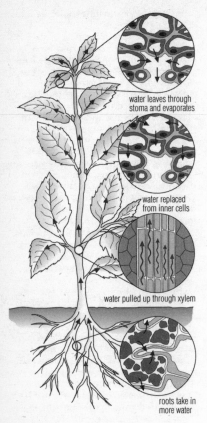

water leaves through stoma and evaporates

water replaced from inner cells

water pulled up through xylem

roots take in more water

transpiration A plant loses most of its water through the surface openings, or stomata, on the leaves. The evaporation produces what is known as the transpiration stream, a tension that draws water up from the roots through the xylem, water-carrying vessels in the stem.

in the body. Water entering and leaving cells usually does so by *osmosis. In both animals and plants there is a *cell membrane around every body cell and this is partially permeable so osmosis can occur. When water is taken up into the body of a plant or animal, the water has to enter a cell and must do so by osmosis. However, with plants, the presence of a strong cell wall in addition to the cell membrane means that plant cells do not take up too much water. The cells just become fully filled or turgid (see

*turgor). Loss of water from plants is by *transpiration, which is evaporation of water from leaves. The movement of nutrients across cell membranes may involve a transport process called *active transport. This can be thought of as a 'pump' because it requires an input of *energy. Plant cells use active transport to absorb *minerals from the soil. Humans will use the same process for the same reasons to absorb minerals from food in the *alimentary canal.

Within plants and animals there are great differences between the ways in which fluids are transported. In mammals there is a *circulatory system using *blood to transport *mineral salts and gases for *respiration. In plants there are two systems. One uses *xylem tissue and carries water and minerals from roots to leaves. The other uses *phloem tissue and carries sugars in solution from leaves to all parts of the plant.

transposon or **jumping gene**, segment of DNA able to move within or between chromosomes. Transposons trigger changes in gene expression by shutting off genes or causing insertion *mutations. The origins of transposons are obscure, but geneticists believe some may be the remnants of viruses that have permanently integrated their genes with those of their hosts. They were first identified by US geneticist Barbara McClintock in 1947.

transputer in computing, a member of a family of microprocessors designed for parallel processing, developed in the UK by Inmos. In the circuits of a standard computer the processing of data takes place in sequence; in a transputer's circuits processing takes place in parallel, greatly reducing computing time for those programs that have been specifically written for it.

transsexual person who identifies himself or herself completely with the opposite sex, believing that the wrong sex was assigned at birth. Unlike **transvestites**, who desire to dress in clothes traditionally worn by the opposite sex, transsexuals think and feel emotionally in a way typically considered appropriate to members of the opposite sex, and may undergo surgery to modify external sexual characteristics.

transuranic element or **transuranium element**, chemical element with an atomic number of 93 or more – that is, with a greater number of protons in the nucleus than has uranium. All transuranic elements are radioactive. Neptunium and plutonium are found in nature; the others are synthesized in nuclear reactions.

transverse wave *wave in which the displacement of the medium's particles, or in electromagnetic waves the direction of the electric and magnetic fields, is at right angles to the direction of travel of the wave motion. Various methods are used to reproduce waves, such as a ripple tank or a rope, in order to understand their properties. If one end of a rope is moved in an up and down motion (the other end being fixed), a wave travels along the rope. The particles in the rope oscillate at right angles to the direction of the wave, moving up and down as the wave travels along the rope. This is known as a transverse wave. The rope remains as it was after the wave has travelled along the rope; waves carry energy from one place to another but they do not transfer matter. Examples of transverse waves are water waves and electromagnetic waves.

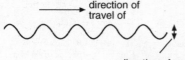

direction of travel of

direction of displacement of particles

transverse wave The diagram illustrates the motion of a transverse wave. Light waves are examples of transverse waves: they undulate at right angles to the direction of travel and are characterized by alternating crests and troughs. Simple water waves, such as the ripples produced when a stone is dropped into a pond, are also examples of transverse waves.

trapezium US **trapezoid**, in geometry, a quadrilateral (a shape with four sides) with two of its sides parallel. If the parallel sides have lengths a and b and

the perpendicular distance between them is h (the height of the trapezium), its area $A = \frac{1}{2}h(a + b)$. An isosceles trapezium has its sloping sides equal, is symmetrical about a line drawn through the midpoints of its parallel sides, and has equal base angles:

tree-and-branch filing system in computing, a filing system where all files are stored within directories, like folders in a filing cabinet. These directories may in turn be stored within further directories. The root directory contains all the other directories and may be thought of as equivalent to the filing cabinet. Another way of picturing the system is as a tree with branches from which grow smaller branches, ending in leaves (individual files).

treeline highest point on a mountain where trees will grow, and above which it is too cold for them to survive.

trematode parasitic flatworm with an oval non-segmented body, of the class Trematoda, including the fluke.

tremor minor *earthquake.

triangle in geometry, a three-sided plane figure, the sum of whose *interior angles is 180°. Triangles can be classified by the relative lengths of their sides. A **scalene triangle** has three sides of unequal length. An *** isosceles triangle** has at least two equal sides; it has one line of *symmetry. An *** equilateral triangle** has three equal sides (and three equal angles of 60°); it has three lines of symmetry.

Triassic Period period of geological time 245–208 million years ago, the first period of the Mesozoic era. The present continents were fused together in the form of the world continent *Pangaea. Triassic sediments contain remains of early dinosaurs and other animals now extinct. By late Triassic times, the first mammals had evolved. There was a mass extinction of 95% of plants at the end of the Triassic possibly caused by rising temperatures. The climate was generally dry; desert sandstones are typical Triassic rocks.

tribology science of friction, lubrication, and lubricants. It studies the origin of frictional forces, the wearing of interacting surfaces, and the problems of efficient lubrication.

trichina worm parasitic nematode worm *Trichinella spiralis*, found in cats, dogs, pigs, and rats. The adults live in the gut and the larvae migrate to the muscles. If consumed in undercooked meat they cause *trichinosis in humans.

trichinosis disease caused by larvae of the nematode worm, *Trichinella spiralis*, which are found in cats and dogs, pigs, hares and rats. It is transmitted to humans in contaminated meat. First symptoms, which occur within 72 hours, include headache, vomiting, diarrhoea and abdominal pain. These are followed, within ten days to three weeks, by eye problems, fever, muscle pain and cramps; pleurisy may also occur and occasionally the heart muscle and *central nervous system are affected. Treatment is with bed-rest and drugs, although symptoms may linger for many months. Trichinosis is mostly seen in Europe and North America; it is rare in tropical countries.

trichloromethane technical name for *chloroform.

tricuspid valve flap of tissue situated on the right side of the *heart between the atrium and the ventricle. It prevents blood flowing backwards when the ventricle contracts.

triglyceride chemical name for *fat comprising three fatty acids reacted with a glycerol.

trigonometry branch of mathematics that concerns finding lengths and angles in *triangles. In a right-angled triangle the sides and angles are related by three trigonometric ratios: *sine, *cosine, and *tangent. Trigonometry is used frequently in navigation, surveying, and simple harmonic motion in physics.

Using trigonometry, it is possible to calculate the lengths of the sides and the sizes of the angles of a right-angled triangle as long as one angle and the length of one side are known, or the lengths of two sides. The longest side, which is always opposite to the right angle, is called the **hypotenuse**. The other sides are named depending on their position relating to the angle that is to be found or used: the side opposite this angle is always termed

opposite and that adjacent is the **adjacent**. So the following trigonometric ratios are used:

$$\text{sine} = \frac{\text{opposite}}{\text{hypotenuse}}$$

$$\text{cosine} = \frac{\text{adjacent}}{\text{hypotenuse}}$$

$$\text{tangent} = \frac{\text{opposite}}{\text{adjacent}}$$

for any right-angled triangle with angle θ as shown the trigonometrical ratios are

$$\sin(\text{e})\,\theta = \frac{BC}{AB} = \frac{\text{opposite}}{\text{hypotenuse}}$$

$$\cos\theta = \frac{AC}{AB} = \frac{\text{adjacent}}{\text{hypotenuse}}$$

$$\tan\theta = \frac{BC}{AC} = \frac{\text{opposite}}{\text{adjacent}}$$

trigonometry At its simplest level, trigonometry deals with the relationships between the sides and angles of triangles. Unknown angles or lengths are calculated by using trigonometrical ratios such as sine, cosine, and tangent. The earliest applications of trigonometry were in the fields of navigation, surveying, and astronomy, and usually involved working out an inaccessible distance such as the distance of the Earth from the Moon.

triiodomethane technical name for *iodoform.

triode three-electrode thermionic *valve containing an anode and a cathode (as does a *diode) with an additional negatively-biased control grid. Small variations in voltage on the grid bias result in large variations in the output current. The triode was commonly used in amplifiers but has now been almost entirely superseded by the *transistor.

triple bond three covalent bonds between adjacent atoms in a chemical compound, as in the *alkynes (–C≡C–).

triploblastic in biology, having a body wall composed of three layers. The outer layer is the **ectoderm**, the middle layer the **mesoderm**, and the inner layer the **endoderm**. This pattern of development is shown by most multicellular animals (including humans).

tritium radioactive isotope of hydrogen, three times as heavy as ordinary hydrogen, consisting of one proton and two neutrons. It has a half-life of 12.5 years.

troilite FeS, iron sulphide, mineral abundant in meteorites and probably present in the Earth's core.

trophic level in ecology, the position occupied by a species (or group of species) in a *food chain. The main levels are **primary producers** (photosynthetic plants), **primary consumers** (herbivores), **secondary consumers** (carnivores), and *decomposers** (bacteria and fungi).

tropical cyclone another term for *hurricane.

tropical (hot) climate There are three distinctive tropical climates: equatorial, tropical continental, and hot deserts. The mean monthly temperature in these climate areas never falls below 21°C/70°F.

Areas with an **equatorial** climate are located within 5° north and south of the Equator. These areas include the Amazon and Congo basins and the coastal lands of Ecuador and West Africa. Temperatures are high and constant throughout the year because the Sun is always high in the sky. Each day has approximately 12 hours of daylight and 12 hours of darkness. Annual rainfall totals usually exceed 2,000 mm/6.5 ft; most afternoons there are heavy showers. There is high daytime humidity and winds are light and variable.

Areas with a **tropical continental** climate are mainly located between latitudes 5° and 15° north and south of the Equator, and within central parts of continents. These areas include the Campos (Brazilian highlands), most of Central Africa surrounding the Congo basin, and parts of northern Australia. Temperatures are high throughout the year but there is a short season, slightly cooler than the equatorial climate, when the sun is overhead at either the Tropic of Cancer or the Tropic of Capricorn. The annual temperature range is slightly greater than that of the equatorial climate due to the sun being at a slightly lower angle in the sky for part of the year. The higher temperature is also due to tropical continental areas being at a greater distance from the sea and having less cover from cloud and vegetation. The main feature of this climate is the alternate wet and dry seasons.

Hot deserts are usually found on the west coast of continents between 15° and 30° north or south of the Equator and in the *trade wind belt. The exception is the extensive Sahara-Arabian-Thar desert which owes its existence to the size of the Afro-Asian continent. Desert temperatures are characterized by their extremes. The annual range is often 20–30°C/36–54°F and the diurnal (daytime) range over 50°C/90°F. Due to the lack of cloud cover and the bare rocks or sand surface, daytimes receive intense insolation (exposure) from the overhead sun. In contrast, nights may be extremely cold with temperatures likely to fall below 0°C/32°F. No deserts are truly dry even though they suffer from extreme water shortages. The amount of precipitation is extremely unreliable; some desert areas may receive rain only once every two or three years. Rain, when it does fall, is heavy.

tropical storm intense low pressure system which forms over the warm oceans of the world's tropical areas, occurring in late summer and early autumn. Warm, moist air rises in a spiral to form a storm that can measure several kilometres across. The weather conditions vary considerably throughout the duration of the storm. Coastal areas are usually the worst affected by tropical storms, which can bring about both human and financial damage. Areas affected by a tropical storm often lose their electricity, water, and sewerage facilities. Tropical storms in different parts of the world are known by various names: *hurricane, *tropical cyclone, or *typhoon.

tropics area between the tropics of Cancer and Capricorn, defined by the parallels of latitude approximately 23°30¢ north and south of the Equator. They are the limits of the area of Earth's surface in which the Sun can be directly overhead. The mean monthly temperature is over 20°C/68°F.

Climates within the tropics lie in parallel bands. Along the Equator is the *intertropical convergence zone, characterized by high temperatures and year-round heavy rainfall. Tropical rainforests are found here. Along the tropics themselves lie the tropical high-pressure zones, characterized by descending dry air and desert conditions. Between these, the conditions vary seasonally between wet and dry, producing the tropical grasslands.

tropism or **tropic movement**, directional growth of a plant, or part of a plant, in response to an external stimulus such as gravity or light. If the movement is directed towards the stimulus it is described as positive; if away from it, it is negative.
Geotropism for example, the response of plants to gravity, causes the root (positively geotropic) to grow downwards, and the stem (negatively geotropic) to grow upwards.

troy system system of units used for precious metals and gems. The pound troy (0.37 kg) consists of 12 ounces (each of 120 carats) or 5,760 grains (each equal to 65 mg).

truncation error in computing, an error that occurs when a decimal result is cut off (truncated) after the maximum number of places allowed by the computer's level of accuracy.

trypanosome any parasitic flagellate protozoan of the genus *Trypanosoma* that lives in the blood of vertebrates, including humans. They often cause serious diseases, called *trypanosomiases.

trypanosomiasis any of several debilitating long-term diseases caused by a trypanosome (protozoan of the genus *Trypanosoma*). They include sleeping sickness in Africa, transmitted by the bites of tsetse flies, and Chagas's disease in Central and South America, spread by assassin bugs.

trypsin enzyme in the vertebrate gut responsible for the digestion of protein molecules. It is secreted by the pancreas but in an inactive form known as trypsinogen. Activation into working trypsin occurs only in the small intestine, owing to the action of another enzyme, enterokinase, secreted by the wall of the duodenum. Unlike the digestive enzyme pepsin, found in the stomach, trypsin does not require an acid environment.

TSR abbreviation for *terminate and stay resident.

tsunami (Japanese 'harbour wave') ocean wave generated by vertical movements of the sea floor resulting from *earthquakes or volcanic activity or large submarine landslides. Unlike waves generated by surface winds, the entire depth of water is involved in the wave motion of a tsunami. In the open ocean the tsunami takes the form of several successive waves, rarely in excess of 1 m/3 ft in height but travelling at speeds of 650–800 kph/400–500 mph. In the coastal shallows tsunamis slow down and build up producing huge swells over 15 m/45 ft high in some cases and over 30 m/90 ft in rare instances. The waves sweep inland causing great loss of life and damage to property.

TTL abbreviation for *transistor–transistor logic, a family of integrated circuits.

TTP abbreviation for trusted third party.

TTY in computing, contraction of *teletype.

tuber swollen region of an underground stem or root, usually modified for storing food. The potato is a **stem tuber**, as shown by the presence of terminal and lateral buds, the 'eyes' of the potato. **Root tubers**, for example dahlias, developed from adventitious roots (growing from the stem, not from other roots) lack these. Both types of tuber can give rise to new individuals and so provide a means of *vegetative reproduction.

tuberculosis TB; formerly known as **consumption** or **phthisis**, infectious disease caused by the bacillus *Mycobacterium tuberculosis*. It takes several forms, of which pulmonary

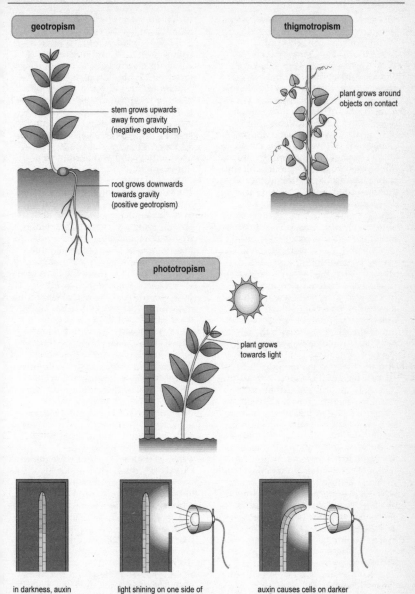

geotropism

stem grows upwards
away from gravity
(negative geotropism)

root grows downwards
towards gravity
(positive geotropism)

thigmotropism

plant grows around
objects on contact

phototropism

plant grows
towards light

in darkness, auxin
is evenly concentrated
around shoot tip

light shining on one side of
the shoot causes auxin to be
destroyed on that side but
remains present on darker side

auxin causes cells on darker
side to elongate and shoot bends,
thus growing towards the light

tropism Tropisms are plant movements in response to external stimuli. Geotropism is the movement
in response to gravity, either towards it in most root systems (positive geotropism), or away as in
most stems (negative geotropism). Most plants exhibit positive phototropism and grow towards light.
Thigmotropism is found in most climbing plants.

tuberculosis is by far the most common. A vaccine, *BCG, was developed around 1920 and the first anti-tuberculosis drug, streptomycin, in 1944. The bacterium is mostly kept in check by the body's immune system; about 5% of those infected develop the disease. Treatment of patients with a combination of anti-TB medicines for 6–8 months produces a cure rate of 80%. In 1999 there were 8 million new cases of TB and 2 million deaths. Only 5% of cases are in developed countries. Worldwide there are 16 million people with TB and 2 billion (a third of the global population) are infected with *Mycobacterium tuberculosis*.

Tullgren funnel in biology, device used to extract mites, springtails, fly larvae, and other small invertebrates from a sample of soil. Soil resting on a net inside a funnel is illuminated from above, so that the top layers of the soil warm and begin to dry out. The invertebrates, trying to escape the desiccating effect of the heat burrow downwards and in the process drop out of the soil and through the net, to be collected in a beaker resting below the funnel.

tumour overproduction of cells in a specific area of the body, often leading to a swelling or lump. Tumours are classified as **benign** or **malignant** (see *cancer). Benign tumours grow more slowly, do not invade surrounding tissues, do not spread to other parts of the body, and do not usually recur after removal. However, benign tumours can be dangerous in areas such as the brain. The most familiar types of benign tumour are warts on the skin. In some cases, there is no sharp dividing line between benign and malignant tumours.

tundra region in high latitudes with almost no trees – they cannot grow because the ground is permanently frozen (*permafrost). The vegetation consists mostly of grasses, sedges, heather, mosses, and lichens. Tundra stretches in a continuous belt across northern North America and Eurasia. Tundra is also used to describe similar conditions at high altitudes.

tungsten chemical symbol W, (Swedish *tung sten* 'heavy stone') hard, heavy, grey-white, metallic element, atomic number 74, relative atomic mass 183.85. Its chemical symbol comes from the German *Wolfram*. It occurs in the minerals wolframite, scheelite, and hubertite. It has the highest melting point of any metal ($3,410°C/6,170°F$) and is added to steel to make it harder, stronger, and more elastic; its other uses include high-speed cutting tools, electrical elements, and thermionic couplings. Its salts are used in the paint and tanning industries.

Tungsten was first recognized in 1781 by Swedish chemist Karl Scheele in the ore scheelite. It was isolated in 1783 by Spanish chemists Fausto D'Elhuyar (1755–1833) and his brother Juan José (1754–1796).

tungsten ore either of the two main minerals, wolframite ($FeMn)WO_4$ and scheelite, $CaWO_4$, from which tungsten is extracted. Most of the world's tungsten reserves are in China, but the main suppliers are Bolivia, Australia, Canada, and the USA.

turbidity current gravity-driven current in air, water, or other fluid resulting from accumulation of suspended material, such as silt, mud, or volcanic ash, and imparting a density greater than the surrounding fluid. Marine turbidity currents originate from tectonic movement, storm waves, tsunamis (tidal waves) or earthquakes and move rapidly downward, like underwater avalanches, leaving distinctive deposits called **turbidites**. They are thought to be one of the mechanisms by which submarine canyons are formed.

turbine engine in which steam, water, gas, or air is made to spin a rotating shaft by pushing on angled blades, like a fan. There are two sets of blades, the stator (does not rotate) and the rotor (does rotate). The rotating turbine shaft can be connected to an electricity generator. Turbines are among the most powerful machines. Steam turbines are used to drive generators in power stations and ships' propellers; water turbines spin the generators in hydroelectric power plants; and gas turbines (as jet engines; see *jet propulsion) power most aircraft and drive machines in industry.

turgor rigid condition of a plant caused by the fluid contents of a plant *cell exerting a mechanical pressure against the cell wall. Turgor supports plants that do not have woody stems. Plants lacking in turgor visibly wilt. The process of *osmosis plays an important part in maintaining the turgidity of plant cells.

Turing machine abstract model of an automatic problem-solving machine, formulated by English mathematician and logician Alan Turing (1912–1954) in 1936. It provides the theoretical basis of modern digital computing.

turnaround document output document produced by a computer that is later, after additional data has been added, used as an input document.

Turner's syndrome genetic condition in females in which one of the X chromosomes is missing. Victims are of short stature and infertile.

turnkey system in computing, a system that the user has only to switch on to have direct access to application software that is usually specific to a particular application area. Turnkey systems often use menus. The user is expected to follow instructions on the screen and to have no knowledge of how the system operates.

turquoise mineral, hydrous basic copper aluminium phosphate, $CuAl_6(PO_4)_4(OH)_8.5H_2O$. Blue-green, blue, or green, it is a gemstone. Turquoise is found in Australia, Egypt, Ethiopia, France, Germany, Iran, Turkestan, Mexico, and southwestern USA. It was originally introduced into Europe through Turkey, from which its name is derived.

twin one of two young produced from a single pregnancy. Human twins may be genetically identical (monozygotic), having been formed from a single fertilized egg that splits into two cells, both of which became implanted. Nonidentical (fraternal or dizygotic) twins are formed when two eggs are fertilized at the same time.

two-stroke cycle operating cycle for internal combustion piston engines. The engine cycle is completed after just two strokes (up or down) of the piston, which distinguishes it from the more common *four-stroke cycle. Some power mowers and lightweight motorcycles use two-stroke petrol engines, which are cheaper and simpler than four-strokes.

Most marine diesel engines are also two-stroke. In a typical two-stroke motorcycle engine, fuel mixture is drawn into the crankcase as the piston moves up on its first stroke to compress the mixture above it. Then the compressed mixture is ignited, and hot gases are produced, which drive the piston down on its second stroke. As it moves down, it uncovers an opening (port) that allows the fresh fuel mixture in the crankcase to flow into the combustion space above the piston. At the same time, the exhaust gases leave through another port.

typhoid fever acute infectious disease of the digestive tract, caused by the bacterium *Salmonella typhi*, and usually contracted through a contaminated water supply. It is characterized by bowel haemorrhage and damage to the spleen. Treatment is with antibiotics.

typhoon violent revolving storm, a *hurricane in the western Pacific Ocean.

typhus any one of a group of infectious diseases caused by bacteria transmitted by lice, fleas, mites, and ticks. Symptoms include fever, headache, and rash. The most serious form is epidemic typhus, which also affects the brain, heart, lungs, and kidneys and is associated with insanitary overcrowded conditions. Treatment is by antibiotics.

tyrosine in medicine, an *amino acid. It is important in the production of *catecholamines, such as adrenaline, noradrenaline, and dopamine, which are involved in the transmission of nerve impulses. It is also important in the production of thyroid hormones, such as thyroxine.

U

UART abbreviation for universal asynchronous receiver–transmitter, in computing, integrated circuit that converts computer data into asynchronous signals suitable for transmission via a telephone line, and vice versa. UARTs combine a transmitter (parallel-to-serial converter) and a receiver (serial-to-parallel converter) to provide a 'bridge' between the parallel signals used by the computer and the serial signals used by communications networks.

UDP abbreviation for *User Datagram Protocol.

UDSL abbreviation for *Universal Digital Subscriber Line.

UID abbreviation for unique identifier, general computing term used to refer to a number, such as a product code, registration code, or a text entry. In databases, the UID can also be referred to as the **key field**, a unique identifier to a particular record. If the user does not specify a key field, the software must generate a UID so that each entry is individually tagged.

ULA abbreviation for *uncommitted logic array, a type of integrated circuit.

ulcer any persistent breach in a body surface (skin or mucous membrane). It may be caused by infection, irritation, or tumour and is often inflamed. Common ulcers include aphthous (mouth), gastric (stomach), duodenal, decubitus ulcers (pressure sores), and those complicating varicose veins.

ulna one of the two bones found in the lower limb of the tetrapod (four-limbed) vertebrate. It articulates with the shorter *radius and *humerus (upper arm bone) at one end and with the radius and wrist bones at the other.

ultrafiltration process by which substances in solution are separated on the basis of their molecular size. A solution is forced through a membrane with pores large enough to permit the passage of small solute molecules but not large ones. Ultrafiltration is a vital mechanism in the vertebrate kidney: the cell membranes lining the Bowman's capsule act as semipermeable membranes, allowing water and substances of low molecular weight such as urea and salts to pass through into the urinary tubules but preventing the larger proteins from being lost from the blood.

ultrasonics branch of physics dealing with the theory and application of ultrasound: sound waves occurring at frequencies too high to be heard by the human ear (that is, above about 20 kHz).

ultrasound pressure waves, known as ultrasonic waves, similar in nature to sound waves but occurring at frequencies above 20,000 Hz (cycles per second), the approximate upper limit of human hearing (15–16 Hz is the lower limit). Ultrasonics is concerned with the study and practical application of these phenomena.

ultrasound scanning or **ultrasonography**, in medicine, the use of ultrasonic pressure waves to create a diagnostic image. It is a safe, noninvasive technique that often eliminates the need for exploratory surgery.

ultraviolet astronomy study of cosmic ultraviolet emissions using artificial satellites. The USA launched a series of satellites for this purpose, receiving the first useful data in 1968. Only a tiny percentage of solar ultraviolet radiation penetrates the atmosphere, this being the less dangerous longer-wavelength ultraviolet radiation. The dangerous shorter-wavelength radiation is absorbed by gases in the ozone layer high in the Earth's upper atmosphere.

ultraviolet radiation electromagnetic radiation of wavelengths from about 400 to 10 nanometres (where the *X-ray range begins). Physiologically, ultraviolet radiation is extremely powerful, producing sunburn and causing the formation of vitamin D in the skin. Ultraviolet radiation is invisible to the human eye, but its effects can be demonstrated.

umbilical cord connection between the *embryo (or fetus) and the *placenta inside the *uterus of placental mammals. It has one vein and two arteries, transporting *oxygen and nutrients to the developing young, and removing waste products. *Blood is carried from the fetus along the umbilical cord and into the placenta. Here it is brought close to the mother's blood. Oxygen, nutrients, and antibodies from the mother diffuse (see *diffusion) into the fetal blood. Waste materials from the fetus pass into the mother's blood. The fetal blood, which has been enriched with nutrients, oxygenated, and cleaned of waste, is carried back to the fetus by another blood vessel in the umbilical cord. At birth, the connection between the young and the placenta is no longer necessary. The umbilical cord drops off or is severed, leaving a scar called the navel.

umbra central region of a *shadow that is totally dark because no light reaches it, and from which no part of the light source can be seen. In astronomy, it is a region of the Earth from which a complete *eclipse of the Sun can be seen.

UMTS abbreviation for universal mobile telecommunications system, third generation (3G) of mobile phone networks. UMTS networks are expected to arrive in Europe in late 2002 or 2003, with data transmission speeds of up to 2 megabits per second (mbps) per channel. Videoconferencing and video-on-demand are expected to be available over UMTS, as well as services that utilize its ability to calculate the location of a mobile phone to within a few metres.

uncertainty principle or **indeterminacy principle**, in quantum mechanics, the principle that it is impossible to know with unlimited accuracy the position and momentum of a particle. The principle arises because in order to locate a particle exactly, an observer must bounce light

(in the form of a *photon) off the particle, which must alter its position in an unpredictable way.

It was established by German physicist Werner Heisenberg (1901–1976), and gave a theoretical limit to the precision with which a particle's momentum and position can be measured simultaneously: the more accurately the one is determined, the more uncertainty there is in the other.

uncommitted logic array ULA; or **gate array**, in computing, a type of semicustomized integrated circuit in which the logic gates are laid down to a general-purpose design but are not connected to each other. The interconnections can then be set in place according to the requirements of individual manufacturers. Producing ULAs may be cheaper than using a large number of TTL (*transistor–transistor logic) chips or commissioning a fully customized chip.

unconscious in psychoanalysis, a part of the personality of which the individual is unaware, and which contains impulses or urges that are held back, or repressed, from conscious awareness.

UnCover in computing, service comprised of a free searchable database of periodicals known as CARL and a fee-based fax-back facility. Researchers either dial into or access the database via Telnet or the World Wide Web and search for articles from approximately 15,000 periodicals, many not indexed elsewhere. They then select the ones they want, and pay approximately $15 each to have the articles faxed to the location of their choice, usually within 24 hours.

unicellular organism animal or plant consisting of a single cell. Most are invisible without a microscope but a few, such as the giant amoeba, may be visible to the naked eye. The main groups of unicellular organisms are bacteria, protozoa, unicellular algae, and unicellular fungi or yeasts. Some become disease-causing agents (*pathogens).

Unicode 16-bit character encoding system, intended to cover all characters in all languages (including Chinese and similar languages) and to be backwards compatible with *ASCII.

unified field theory in physics, the theory that attempts to explain the four fundamental forces (strong nuclear, weak nuclear, electromagnetic, and gravity) in terms of a single unified force (see *particle physics). Research was begun by German-born US physicist Albert Einstein (1879–1955), and by 1971 a theory developed by US physicists Steven Weinberg and Sheldon Glashow, Pakistani physicist Abdus Salam, and others, had demonstrated the link between the weak and electromagnetic forces. The next stage is to develop a theory (called the *grand unified theory) that combines the strong nuclear force with the electroweak force. The final stage will be to incorporate gravity into the scheme. Work on the *superstring theory indicates that this may be the ultimate 'theory of everything'.

uninterruptible power supply (UPS) power supply that includes a battery, so that in the event of a power failure it is possible to continue operations. UPSs are normally used to provide time either for a system to be shut down in the usual way (so that files are not corrupted) or for an alternative power supply to be connected. For large systems these operations are usually carried out automatically.

union in mathematics, a set which is formed by joining up two or more other sets.

unit standard quantity in relation to which other quantities are measured. There have been many systems of units. Some ancient units, such as the day, the foot, and the pound, are still in use. *SI units, the latest version of the metric system, are widely used in science.

Universal Digital Subscriber Line UDSL, low-cost version of xDSL having been standardized in 1998. UDSL is backed by a consortium of companies led by Compaq, Intel, and Microsoft.

universal indicator in chemistry, a mixture of *pH indicators, used to gauge the acidity or alkalinity of a solution. Each component changes colour at a different pH value, and so the indicator is capable of displaying a range of colours, according to the pH of the test solution, from red (at pH 1, strong acid) through green (neutral) to purple (at pH 13, strong alkali).

Universal Serial Bus USB, in computing, a royalty-free connector intended to replace the out-of-date COM and parallel printer ports that have been used in PCs since 1981. The USB allows up to 127 peripherals – including joysticks, scanners, printers, and keyboards – to be daisy-chained from a single socket, offering higher speeds and improved plug-and-play facilities.

universal set in mathematics, the set from which subsets are taken, the set of all the objects under consideration in a problem. The odd numbers 1, 3, 5, 7 ... and the square numbers 1, 4, 9, 16 ... are taken from the universal set of natural numbers.

Universal Time UT, another name for *Greenwich Mean Time. It is based on the rotation of the Earth, which is not quite constant. Since 1972, UT has been replaced by Universal Time Coordinated (UTC), which is based on uniform atomic time. See *time.

universe all of space and its contents, the study of which is called *cosmology. The universe is thought to be between 10 billion and 20 billion years old, and is mostly empty space, dotted with *galaxies for as far as telescopes can see. These galaxies are moving further apart as the universe expands. Several theories attempt to explain how the universe came into being and evolved; for example, the *Big Bang theory of an expanding universe originating in a single explosive event (creating hydrogen and helium gases), and the contradictory *steady-state theory.

Unix multiuser *operating system designed for minicomputers but becoming increasingly popular on microcomputers, workstations, mainframes, and supercomputers.

unleaded petrol petrol manufactured without the addition of *antiknock. It has a slightly lower octane rating than leaded petrol, but has the advantage of not polluting the atmosphere with lead compounds. Many cars can be converted to run on unleaded petrol by altering the timing of the engine, and most new cars are designed to do so.

Cars fitted with a *catalytic converter must use unleaded fuel.

Aromatic hydrocarbons and alkenes are added to unleaded petrol instead of lead compounds to increase the octane rating. After combustion the hydrocarbons produce volatile organic compounds. These have been linked to cancer, and are involved in the formation of phytochemical smog. A low-lead fuel is less toxic than unleaded petrol for use in cars that are not fitted with a catalytic converter.

unnilennium temporary name that was assigned to the element *meitnerium, atomic number 109.

unnilhexium temporary name that was assigned to the element *seaborgium, atomic number 106.

unniloctium temporary name that was assigned to the element *hassium, atomic number 108.

unnilpentium temporary name that was assigned to the element *dubnium, atomic number 105.

unnilquadium temporary name that was assigned to the element *rutherfordium, atomic number 104.

unnilseptium temporary name that was assigned to the element *bohrium, atomic number 107.

unsaturated compound chemical compound in which two adjacent atoms are linked by a double or triple covalent bond. *Alkenes, such as *ethene, are unsaturated *hydrocarbons that contain a carbon to carbon double bond.

Other examples are *ketones, where the unsaturation exists between atoms of different elements (carbon and oxygen). The laboratory test for unsaturated compounds is the addition of bromine water; if the test substance is unsaturated, the bromine water will be decolorized.

unsaturated solution solution that is capable of dissolving more solute than it already contains at the same temperature.

unshielded twisted pair UTP, in computing, a form of cabling used for *local area networks (LAN), now commonly used as an alternative to *coaxial cable. The benefits of UTP over coaxial cables are that it is cheaper to install and can be used by a variety of LAN standards, for example *Ethernet and Token Ring, as well as being used for telephone circuits.

unstable equilibrium state of equilibrium possessed by a body that will not remain at rest if displaced slightly, but will topple over into a new position; it will not return to its original rest position. See *stability.

ununbium symbol Uub, temporary name for a synthesized radioactive element of the *transactinide series, atomic number 112, relative atomic mass of most stable isotope 277. It was discovered in 1996 at the GSI heavy-ion cyclotron, Darmstadt, Germany, when lead was bombarded with zinc.

ununhexium symbol Uuh, temporary name for a synthesized radioactive element of the *transactinide series, atomic number 116, relative atomic of most stable isotope 289. It was discovered in 1999 by a research team at the Lawrence Berkeley National Laboratory in California, when plutonium was bombarded with calcium. It lasted for 30 seconds, much longer than most synthesized elements.

ununnilium symbol Uun, temporary name for a synthesized radioactive element of the *transactinide series, atomic number 110, relative atomic mass 269. It was discovered in October 1994, detected for a millisecond, at the GSI heavy-ion cyclotron, Darmstadt, Germany, while lead atoms were bombarded with nickel atoms.

ununoctium symbol Uuo, temporary name for a synthesized radioactive element of the *transactinide series, atomic number 118. It was claimed in 1999 by a research team at the Lawrence Berkeley National Laboratory in California, when lead was bombarded with krypton. The claim was retracted in 2001, and no other laboratory has replicated the result.

ununquadium symbol Uuq, temporary name for a synthesized radioactive element of the *transactinide series, atomic number 114, relative atomic mass of most stable isotope 289. It was discovered in 1998 by a Russian research team at the Joint Institute for Nuclear Research at Dubna, when lead was bombarded with krypton.

unununium symbol Uuu, temporary name for a synthesized radioactive element of the *transactinide series, atomic number 111, relative atomic mass 272. It was detected at the GSI heavy-ion cyclotron, Darmstadt, Germany, in December 1994, when bismuth-209 was bombarded with nickel.

upper bound in mathematics, a value that is greater than or equal to all the values of a given set. The upper bound of a measurement is taken as the top extreme of the possible values. For example, for a length given as 3.4 cm correct to one decimal place, the upper bound is 3.45 correct to two decimal places. The limiting value of an infinite set is the **least upper bound**. See also *approximation.

UPS abbreviation for *uninterruptible power supply.

uracil in biochemistry, pyrimidine base that is part of the genetic code of ribonucleic acid (RNA). It pairs with *adenine during the process of *transcription, in which a strand of messenger RNA (mRNA) is produced against a sense strand of deoxyribonucleic acid (DNA) acting as a template. Uracil is replaced by thymine in DNA.

uranium chemical symbol U, hard, lustrous, silver-white, malleable and ductile, radioactive, metallic element of the *actinide series, atomic number 92, relative atomic mass 238.029. It is the most abundant radioactive element in the Earth's crust, its decay giving rise to essentially all radioactive elements in nature; its final decay product is the stable element lead. Uranium combines readily with most elements to form compounds that are extremely poisonous. The chief ore is *pitchblende, in which the element was discovered by German chemist Martin Klaproth in 1789; he named it after the planet Uranus, which had been discovered in 1781.

Small amounts of certain compounds containing uranium have been used in the ceramics industry to make orange-yellow glazes and as mordants in dyeing; however, this practice was discontinued when the dangerous effects of radiation became known.

Uranium is one of three fissile elements (the others are thorium and plutonium). It was long considered to be the element with the highest atomic number to occur naturally on Earth. The isotopes U-238 and U-235 have been used to help determine the age of the Earth.

Uranium-238, which comprises about 99% of all naturally occurring uranium, has a half-life of 4.51×10^9 years. Because of its abundance, it is the isotope from which fissile plutonium is produced in breeder *nuclear reactors. The fissile isotope U-235 has a half-life of 7.13×10^8 years and comprises about 0.7% of naturally occurring uranium; it is used directly as a fuel for nuclear reactors and in the manufacture of nuclear weapons.

uranium ore material from which uranium is extracted, often a complex mixture of minerals. The main ore is uraninite (or pitchblende), UO_2, which is commonly found with sulphide minerals. The USA, Canada, and South Africa are the main producers in the West.

Uranus seventh planet from the Sun, discovered by German-born British astronomer William Herschel in 1781. It is twice as far out as the sixth planet, Saturn. Uranus has a mass 14.5 times that of Earth. The spin axis of Uranus is tilted at 98°, so that one pole points towards the Sun, giving extreme seasons.

mean distance from the Sun: 2.9 billion km/1.8 billion mi

equatorial diameter: 50,800 km/31,600 mi

rotation period: 17 hours 12 minutes

year: 84 Earth years

atmosphere: deep atmosphere composed mainly of hydrogen and helium

surface: composed primarily of rock and various ices with only about 15% hydrogen and helium, but may also contain heavier elements, which might account for Uranus's mean density being higher than that of Saturn

satellites: 18 moons (two discovered in 1997, one in 1999)

rings: 11 rings, composed of rock and dust, around the planet's equator, were detected by the US space probe

Voyager 2 in 1977. The rings are charcoal black and may be debris of former 'moonlets' that have broken up. The ring furthest from the planet centre (51,000 km/31,800 mi), Epsilon, is 100 km/62 mi at its widest point. In 1995, US astronomers determined that the ring particles contained long-chain hydrocarbons. Looking at the brightest region of Epsilon, they were also able to calculate the precession of Uranus as 264 days, the fastest known precession in the Solar System

urban ecology study of the *ecosystems, animal and plant communities, soils and microclimates found within an urban landscape.

Parks are important for many organisms, such as song birds, while birds of prey (the kestrel being a notable example) find ample food in the wasteland around estates and offices. Mammals, including foxes and badgers, are regular visitors, especially if there is an undisturbed corridor penetrating into the town, such as a disused railway.

urea $CO(NH_2)_2$, waste product formed in the mammalian liver when nitrogen compounds are broken down. It is filtered from the blood by the kidneys, and stored in the bladder as urine prior to release. When purified, it is a white, crystalline solid. Most of the urea, including foxes and make urea-formaldehyde plastics (or resins), pharmaceuticals, and fertilizers.

urea cycle biochemical process discovered by German-born British biochemist Hans Krebs (1900–1981) and German physician Kurt Henseleit in 1932, by which nitrogenous waste is converted into urea, which is easily excreted. When *proteins and *amino acids break down, *ammonia, which is highly toxic, is formed. Most of the ammonia is converted into glutamate, and becomes usable for the synthesis of more amino acids and proteins. Any excess is converted into the water-soluble compound *urea, which can be excreted as *urine. Free ammonia, carbon dioxide, and *ATP react to form carbamyl phosphate. This compound then forms *citrulline by reacting with the amino acid ornithine, which can then accept another amino group

giving arginine. Arginine can then break down, giving urea, for excretion, and ornithine, which can take part in the cycle again.

ureter tube connecting the kidney to the bladder. Its wall contains fibres of smooth muscle whose contractions aid the movement of urine out of the kidney.

urethra in mammals, a tube connecting the bladder to the exterior. It carries urine and, in males, semen.

uric acid $C_5H_4N_4O_3$, nitrogen-containing waste substance, formed from the breakdown of food and body protein.

It is only slightly soluble in water. Uric acid is the normal means by which most land animals that develop in a shell (birds, reptiles, insects, and land gastropods) deposit their waste products. The young are unable to get rid of their excretory products while in the shell and therefore store them in this insoluble form.

urinary system system of organs that removes nitrogenous waste products and excess water from the bodies of animals. In vertebrates, it consists of a pair of kidneys, which produce urine; ureters, which drain the kidneys; and (in bony fishes, amphibians, some reptiles, and mammals) a bladder that stores the urine before its discharge. In mammals, the urine is expelled through the urethra; in other vertebrates, the urine drains into a common excretory chamber called a cloaca, and the urine is not discharged separately.

urine amber-coloured fluid filtered out by the kidneys from the blood. It contains excess water, salts, proteins, waste products in the form of urea, a pigment, and some acid.

URL abbreviation for uniform resource locator, series of letters and/or numbers specifying the location of a document on the *World Wide Web. Every URL consists of a domain name, a description of the document's location within the host computer, and the name of the document itself, separated by full stops and backslashes. Thus *The Times* Web site can be found at http://www. timesonline.co.uk, and a tribute to Elvis Presley is at http://ibiblio.org/elvis/elvishom.html.

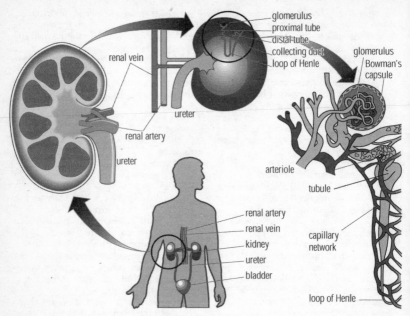

urinary system The human urinary system. At the bottom, the complete system in outline; on the left, the arrangement of blood vessels connected to the kidney; at the top right, a detail of the network of vessels within a kidney.

The complexity of URLs explains why bookmarks and links, which save the user from the chore of typing them in, are so popular.

urology medical speciality concerned with diseases of the urinary tract.

Ursa Major (Latin 'Great Bear') third-largest constellation in the sky, in the north polar region. Its seven brightest stars make up the familiar shape or asterism of the **Big Dipper** or **Plough**. The second star of the handle of the dipper, called Mizar, has a companion star, Alcor.

Two stars forming the far side of the dipper bowl act as pointers to the north pole star, Polaris. Dubhe, one of them, is the constellation's brightest star.

Ursa Minor (Latin 'Little Bear') small constellation of the northern hemisphere, popularly known as the Little Dipper. It is shaped like a dipper, with the bright north pole star Polaris at the end of the handle.

USB in computing, abbreviation for *Universal Serial Bus.

Usenet contraction of users' network, world's largest *bulletin board system, which brings together people with common interests to exchange views and information. It consists of *e-mail messages and articles organized into newsgroups. Usenet is uncensored and governed by the rules of netiquette.

User Datagram Protocol UDP, communications protocol that, like the Transmission Control Protocol (TCP), runs on top of the Internet Protocol (IP). UDP tolerates packets of information being lost or corrupted en route to their destination. This tolerance is an advantage if the message is one where drop-outs are less damaging than delays caused by attempts at error recovery, for example, when speech is broadcast across the Internet.

user-friendly term used to describe the ease of use of a computer system, particularly for those with little understanding or familiarity with computers. Even for experienced users, user-friendly programs are quicker to learn.

user ID abbreviation for user identification, name or nickname that identifies the user of a computer system or network.

user interface in computing, the procedures and methods through which the user operates a program. These might include menus, input forms, error messages, and keyboard procedures. A *graphical user interface (GUI or WIMP) is one that makes use of icons (small pictures) and allows the user to make menu selections with a mouse.

UT abbreviation for *Universal Time.

UTC abbreviation for coordinated universal time, an alternative name for universal time coodinated, the standard measurement of *time.

uterus hollow muscular organ of female mammals, located between the bladder and rectum, and connected to the Fallopian tubes above and the vagina below. The embryo develops within the uterus, and in placental mammals is attached to it after implantation via the *placenta and umbilical cord. The lining of the uterus changes during the *menstrual cycle to prepare it for pregnancy. In humans and other higher primates, it is a single structure, but in other mammals it is paired.

utility program in computing, a systems program designed to perform a specific task related to the operation of the computer when requested to do so by the computer user. For example, a utility program might be used to complete a screen dump, format a disk, or convert the format of a data file so that it can be accessed by a different applications program.

UTP abbreviation for *unshielded twisted pair.

UUCP abbreviation for Unix to Unix copy program, in computing, protocol which allows *Unix users to share files, read *Usenet articles, and exchange *e-mail. The system is based on computers regularly 'polling' (connecting to) each other to swap data. Polling can take place via an ordinary telephone connection or over the Internet.

UUencode in computing, *utility program that converts a *binary file (typically, a program or graphics file) into *ASCII text suitable for inclusion in *e-mail or *Usenet messages. The recipient then **UUdecodes** the text file, reconverting it from ASCII to the original binary file.

UUNET Technologies, Inc US-based provider of Internet access. UUNET was the first commercial Internet Service Provider, founded in 1987; the company now has an international network of *PoPs and is a major *backbone provider. It acquired *PIPEX from Unipalm in 1996 (and spun it off again in 1998) and is now itself part of WorldCom Inc.

UV in physics, abbreviation for **ultraviolet**.

v in physics, symbol for *velocity.

V in physics, symbol for *volt, the SI unit of emf or potential difference, equal to the potential difference between two points in an electrical circuit when a current of 1 ampere flowing between them dissipates a power of 1 watt.

vaccine any preparation of modified pathogens (viruses or bacteria) that is introduced into the body, usually either orally or by a hypodermic syringe, to induce the specific *antibody reaction that produces *immunity against a particular disease.

In 1796 Edward Jenner was the first

to inoculate a child successfully with cowpox virus to produce immunity to smallpox. His method, the application of an infective agent to an abraded skin surface, is still used in smallpox inoculation.

Vactor contraction of virtual actor, in computing, animated character moved and voiced by an actor behind the scenes using a Waldo and dataglove to control the character.

vacuole in biology, a fluid-filled, membrane-bound cavity inside a cell. It may be a reservoir for fluids that the cell will secrete to the outside, or may be filled with excretory products or essential nutrients that the cell needs to store.

Plant cells usually have a large central vacuole containing call sap (sugar and salts in solution) which serves both as a store of food and as a key factor in storing water and in maintaining *turgor. Absorbing more water to make a bigger vacuole adds bulk to the plant. This expansion of cells is very important in plant *growth. In amoebae (single-celled animals), vacuoles are the sites of digestion of engulfed food particles. Animal cells may only have small vacuoles, which are usually called vesicles.

vacuum in general, a region completely empty of matter; in physics, any enclosure in which the gas pressure is considerably less than atmospheric pressure (101,325 pascals).

vagina lower part of the reproductive tract in female mammals, linking the uterus to the exterior. It admits the penis during sexual intercourse, and is the birth canal down which the baby passes during delivery.

valence in chemistry, the measure of an element's ability to combine with other elements, expressed as the number of atoms of hydrogen (or any other standard univalent element) capable of uniting with (or replacing) its atoms. The number of electrons in the outermost shell of the atom dictates the combining ability of an element.

The elements are described as uni-, di-, tri-, and tetravalent when they unite with one, two, three, and four univalent atoms respectively. Some elements have **variable valence**: for

example, nitrogen and phosphorus have a valence of both three and five. The valency of oxygen is two: hence the formula for water, H_2O (hydrogen being univalent).

valence electron electron in the outermost shell of an *atom. It is the valence electrons that are involved in the formation of ionic and covalent bonds (see *molecule). The number of electrons in this outermost shell represents the maximum possible valence for many elements and matches the number of the group that the element occupies in the *periodic table of the elements.

valence shell outermost shell of electrons in an *atom. It contains the *valence electrons. Elements with four or more electrons in their outermost shell can show variable valence. Chlorine can show valences of 1, 3, 5, and 7 in different compounds.

valerian any of a group of perennial plants native to the northern hemisphere, with clustered heads of fragrant tubular flowers in red, white, or pink. The root of the common valerian or garden heliotrope *V. officinalis* is used in medicine to relieve wind and to soothe or calm patients. (Genera *Valeriana* and *Centranthus*, family Valerianaceae.)

validation in computing, the process of checking input data to ensure that it is complete, accurate, and reasonable. Although it would be impossible to guarantee that only valid data are entered into a computer, a suitable combination of validation checks should ensure that most errors are detected.

valley a long, linear depression sloping downwards towards the sea or an inland drainage basin. Types of valleys include the V-shaped valley, U-shaped valley, hanging valley, dry valley, misfit valley, asymmetric valley, and *rift valley.

value in mathematics, a number or other fixed quantity applied to a *variable. The value of an expression will depend on the numbers which are substituted for the variables. For example, $x^2 + y^2$ has the value 25 when $x = 3$ and $y = 4$.

valve in animals, a structure for controlling the direction of the blood flow. In humans and other vertebrates, the contractions of the beating heart cause the correct blood flow into the arteries because a series of valves prevents back flow. Diseased valves, detected as 'heart murmurs', have decreased efficiency. The tendency for low-pressure venous blood to collect at the base of limbs under the influence of gravity is counteracted by a series of small valves within the veins. It was the existence of these valves that prompted the English physician William Harvey (1578–1657) to suggest that the blood circulated around the body.

valve or **electron tube**, in electronics, a glass or metal tube containing gas at low pressure, which is used to control the flow of electricity in a circuit. The electron tube valve was developed by US radio engineer Lee de Forest (1873–1961) and is used to modify electrical signals. Three or more metal electrodes are inset into the tube. By varying the voltage on one of them, called the **grid electrode**, the current through the valve can be controlled, and the valve can act as an amplifier. It is defined as a valve because it allows a undirectional flow of electrons. Valves have been replaced for most applications by *transistors. However, they are still used in high-power transmitters and amplifiers, and in some hi-fi systems.

vanadium chemical symbol V, silver-white, malleable and ductile, metallic element, atomic number 23, relative atomic mass 50.942. It occurs in certain iron, lead, and uranium ores and is widely distributed in small quantities in igneous and sedimentary rocks. It is used to make steel alloys, to which it adds tensile strength. Spanish mineralogist Andrés del Rio (1764–1849) and Swedish chemist Nils Sefström (1787–1845) discovered vanadium independently, the former in 1801 and the latter in 1831. Del Rio named it 'erythronium', but was persuaded by other chemists that he had not in fact discovered a new element; Sefström gave it its present name, after the Norse goddess of love and beauty, Vanadis (or Freya).

vanadium(V) oxide or **vanadium pentoxide**; V_2O_5, crystalline compound used as a catalyst in the *contact process for the manufacture of sulphuric acid.

Van Allen radiation belts two zones of charged particles around the Earth's magnetosphere, discovered in 1958 by US physicist James Van Allen (1914–). The atomic particles come from the Earth's upper atmosphere and the *solar wind, and are trapped by the Earth's magnetic field. The inner belt lies 1,000–5,000 km/620–3,100 mi above the Equator, and contains *protons and *electrons. The outer belt lies 15,000–25,000 km/9,300–15,500 mi above the Equator, but is lower around the magnetic poles. It contains mostly electrons from the solar wind.

van de Graaff generator electrostatic generator capable of producing a voltage of over a million volts. It consists of a continuous vertical conveyor belt that carries electrostatic charges (resulting from friction) up to a large hollow sphere supported on an insulated stand. The lower end of the belt is earthed, so that charge accumulates on the sphere. The size of the voltage built up in air depends on the radius of the sphere, but can be increased by enclosing the generator in an inert atmosphere, such as nitrogen.

van der Waals' forces in chemistry, short-range, weak attractive forces between molecules. They are thought to arise from the formation of temporary *dipoles caused by variations in the electronic configuration of the molecules.

van der Waals' law modified form of the *gas laws that includes corrections for the non-ideal behaviour of real gases (the molecules of ideal gases occupy no space and exert no forces on each other). It is named after Dutch physicist J D van der Waals (1837–1923). The equation derived from the law states that:

$$(P + \frac{a}{V^2})(V - b) = RT$$

where P, V, and T are the pressure, volume, and temperature (in kelvin) of the gas, respectively; R is the *gas constant; and a and b are constants for that particular gas.

vapour one of the three states of matter (see also *solid and *liquid). The molecules in a vapour move randomly and are far apart, the distance between them, and therefore the volume of the vapour, being limited only by the walls of any vessel in which they might be contained. A vapour differs from a *gas only in that a vapour can be liquefied by increased pressure, whereas a gas cannot unless its temperature is lowered below its *critical temperature; it then becomes a vapour and may be liquefied.

vapour density density of a gas, expressed as the *mass of a given volume of the gas divided by the mass of an equal volume of a reference gas (such as hydrogen or air) at the same temperature and pressure. If the reference gas is hydrogen, it is equal to half the relative molecular weight (mass) of the gas.

vapour pressure pressure of a vapour given off by (evaporated from) a liquid or solid, caused by atoms or molecules continuously escaping from its surface. In an enclosed space, a maximum value is reached when the number of particles leaving the surface is in equilibrium with those returning to it; this is known as the **saturated vapour pressure** or **equilibrium vapour pressure**.

variable in computing, a quantity that can take different values. Variables can be used to represent different items of data in the course of a program.

variable in mathematics, a changing quantity (one that can take various values), as opposed to a *constant. For example, in the algebraic expression $y = 4x^3 + 2$, the variables are x and y, whereas 4 and 2 are constants. A variable may be dependent or independent. Thus if y is a *function of x, written $y = f(x)$, such that $y = 4x^3 + 2$, the domain of the function includes all values of the **independent variable** x while the range (or co-domain) of the function is defined by the values of the **dependent variable** y.

variable star star whose brightness changes, either regularly or irregularly, over a period ranging from a few hours to months or years. The Cepheid variables regularly expand and contract in size every few days or weeks.

variance in statistics, the square of the *standard deviation, the measure of spread of data. Population variance and sample variance are denoted by σ^2 and s^2, respectively. Variance provides a measure of the dispersion of a set of statistical results about the mean or average value.

variation in mathematics, a practical relationship between two *variables. **Direct variation** (when the values of the variables maintain a constant ratio) corresponds to $y = kx$ on a straight line graph, for example distance travelled at a steady speed. **Inverse variation** (when an increase in the value of one variable results in a decrease in that of the other) corresponds to $y = k/x$ on a rectangular hyperbolic graph, for example the price of an article and the quantity that can be bought for a fixed sum of money. In problems of direct and inverse variation, the first step is to find the value of k.

variation in biology, a difference between individuals of the same species, found in any sexually reproducing population. Variations may be almost unnoticeable in some cases, obvious in others, and can concern many aspects of the organism. Typically, variations in size, behaviour, biochemistry, or colouring may be found. The cause of the variation is genetic (that is, inherited), environmental, or more usually a combination of the two. Some variation is the result of the environment modifying inherited characteristics. The origins of variation can be traced to the recombination of the genetic material during the formation of the gametes, and, more rarely, to mutation.

variety in biology, a stable group of organisms within a single species, clearly different from the rest of the species. Such a group would generally be called a variety for plants and a breed for animals. The differences lie in their genetic make-up and could have arisen naturally – through *natural selection or as a result of selective breeding by humans. Most varieties have been produced by selective breeding, for example 'Cox's', 'Golden Delicious', and 'Bramley' apple varieties.

vascular bundle in botany, strand of primary conducting tissue (a 'vein') in vascular plants, consisting mainly of water-conducting tissues, metaxylem and protoxylem, which together make up the primary *xylem, and nutrient-conducting tissue, *phloem. It extends from the roots to the stems and leaves. Typically the phloem is situated nearest to the epidermis and the xylem towards the centre of the bundle. In plants exhibiting *secondary growth, the xylem and phloem are separated by a thin layer of vascular *cambium, which gives rise to new conducting tissues.

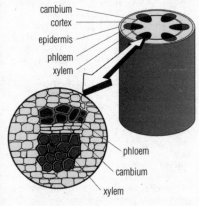

vascular bundle The fluid-carrying tissue of most plants is normally arranged in units called vascular bundles. The vascular tissue is of two types: xylem and phloem. The xylem carries water up through the plant; the phloem distributes food made in the leaves to all parts of the plant.

vascular plant plant containing vascular bundles. *Pteridophytes (ferns, horsetails, and club mosses), *gymnosperms (conifers and cycads), and *angiosperms (flowering plants) are all vascular plants.

vas deferens in male vertebrates, a tube conducting sperm from the testis to the urethra. The sperm is carried in a fluid secreted by various glands, and can be transported very rapidly when the smooth muscle in the wall of the vas deferens undergoes rhythmic contraction, as in sexual intercourse.

vasectomy male sterilization; an operation to cut and tie the ducts (see *vas deferens) that carry sperm from the testes to the penis. Vasectomy does not affect sexual performance, but the semen produced at ejaculation no longer contains sperm.

VBI abbreviation for *vertical blanking interval.

vCard abbreviation for virtual business card, emerging standard format through which users can exchange name and address information even if they use different and incompatible contact management programs from different companies (or sometimes programs from the same company, such as Microsoft). Users with handheld computers can *beam* cards to one another using *IrDA links. A vCard can also replace a sig at the end of an e-mail message. If the vCard data is saved as a .vcf file, it can be loaded into a compatible program such as the address book supplied with Microsoft's Internet Explorer 4 Web browser.

VD abbreviation for *venereal disease.

VDU abbreviation for *visual display unit.

vector in medicine, disease-carrying agent. The term is usually applied to insects, ticks, and the like, that transmit parasites from one person to another or from animals to human beings. The *Anopheles* mosquito is a vector, transmitting the *malaria parasite in its salivary glands.

vector graphics computer graphics that are stored in the computer memory by using geometric formulae. Vector graphics can be transformed (enlarged, rotated, stretched, and so on) without loss of picture resolution. It is also possible to select and transform any of the components of a vector-graphics display because each is separately defined in the computer memory. In these respects vector graphics are superior to *raster graphics. Vector graphics are typically used for drawing applications, allowing the user to create and modify technical diagrams such as designs for houses or cars.

vector quantity any physical quantity that has both magnitude (size) and direction, such as velocity, acceleration, or force, as distinct from a *scalar quantity such as speed, density, or mass, which has magnitude but no direction. A vector is represented either geometrically by an arrow whose length corresponds to its *magnitude and points in an appropriate direction, or by two or three numbers representing the magnitude of its components. Vectors can be added graphically by constructing a parallelogram of vectors (such as the *parallelogram of forces commonly employed in physics and engineering). This will give a **resultant vector**.

The *position vector of a point A(x,y) represents the move from the origin to A, that is a *translation. A free vector has magnitude and direction but no fixed position in space. If two forces p and q are acting on a body at A, then the parallelogram of forces is drawn to determine the resultant force and direction r. p, q, and r are vectors. In technical writing, a vector is denoted by **bold** type, underlined, or overlined.

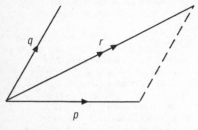

vector quantity A parallelogram of vectors. Vectors can be added graphically using the parallelogram rule. According to the rule, the sum of vectors p and q is the vector r which is the diagonal of the parallelogram with sides p and q.

vegetative reproduction type of *asexual reproduction in plants that relies not on spores, but on multicellular structures formed by the parent plant. Some of the main types are stolons and runners, gemmae, bulbils, sucker shoots produced from roots (such as in the creeping thistle *Cirsium arvense*), *tubers, *bulbs, corms, and *rhizomes. Vegetative reproduction

has long been exploited in horticulture and agriculture, with various methods employed to multiply stocks of plants. See also *plant propagation.

vegetative state in medicine, another name for *persistent vegetative state.

vein vessel that carries *blood from the body to the *heart in animals with a circulatory system. Veins contain valves that prevent the blood from running back when moving against gravity. They carry blood at low pressure, so their walls are thinner than those of arteries. They always carry deoxygenated blood, with the exception of the pulmonary vein, leading from the lungs to the heart in birds and mammals, which carries newly oxygenated blood.

The term is also used more loosely for any system of channels that strengthens living tissues and supplies them with nutrients – for example, leaf veins (see *vascular bundle), and the veins in insects' wings. In leaves they make up the network that can normally be seen, especially from the underside of the leaf. These veins are made up of the two transport tissues of a plant, *xylem and *phloem.

veldt subtropical grassland in South Africa, equivalent to the Pampas of South America.

velocity symbol v, speed of an object in a given direction, or how fast an object changes its position in a given direction. Velocity is a *vector quantity, since its direction is important as well as its magnitude (or speed). For example, a car could have a speed of 48 kph/30 mph and a velocity of 48 kph/30 mph northwards. Velocity = change in position/time taken.

The velocity at any instant of a particle travelling in a curved path is in the direction of the tangent to the path at the instant considered. The velocity v of an object travelling in a fixed direction may be calculated by dividing the distance s it has travelled by the time t taken to do so, and may be expressed as: $v = s/t$

vena cava either of the two great veins of the trunk, returning deoxygenated blood to the right atrium of the *heart. The **superior vena cava**, beginning where the arches of the two innominate veins join high in the chest, receives blood from the head, neck, chest, and arms; the **inferior vena cava**, arising from the junction of the right and left common iliac veins, receives blood from all parts of the body below the diaphragm.

venereal disease   VD, any disease mainly transmitted by sexual contact, although commonly the term is used specifically for gonorrhoea and syphilis, both occurring worldwide, and chancroid ('soft sore') and lymphogranuloma venerum, seen mostly in the tropics. The term *sexually transmitted disease (STD) is more often used to encompass a growing list of conditions passed on primarily, but not exclusively, by sexual contact.

Venn diagram   in mathematics, a diagram representing a *set or sets and the logical relationships between them.

The sets are drawn as circles. An area of overlap between two circles (sets) contains elements that are common to both sets, and thus represents a third set. Circles that do not overlap represent sets with no elements in common (disjoint sets). The method is named after the English logician John Venn (1834–1923).

ventral surface   front of an animal. In vertebrates, the side furthest from the backbone; in invertebrates, the side closest to the ground. The positioning of the main nerve pathways on the ventral side is a characteristic of invertebrates.

ventricle   in zoology, either of the two lower chambers of the heart that force blood to circulate by contraction of their muscular walls. The term also refers to any of four cavities within the brain in which cerebrospinal fluid is produced.

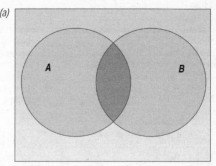

= set of whole numbers from 1 to 20
O = set of odd numbers
P = set of prime numbers

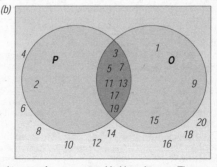

Venn diagram Sets and their relationships are often represented by Venn diagrams. The sets are drawn as circles – the area of overlap between the circles shows elements that are common to each set, and thus represent a third set. (Top) A Venn diagram of two intersecting sets and (bottom) a Venn diagram showing the set of whole numbers from 1 to 20 and the subsets P and O of prime and odd numbers, respectively. The intersection of P and O contains all the prime numbers that are also odd.

vertebral column backbone, giving support to an animal and protecting its spinal cord. It is made up of a series of bones or vertebrae running from the skull to the tail, with a central canal containing the nerve fibres of the spinal cord.

vertebrate any animal with a backbone. The 41,000 species of vertebrates include mammals, birds, reptiles, amphibians, and fishes. They include most of the larger animals, but in terms of numbers of species are only a tiny proportion of the world's animals. The zoological taxonomic group Vertebrata is a subgroup of the *phylum Chordata.

vertex plural **vertices**, in geometry, a point shared by three or more sides of a solid figure; the point farthest from a figure's base; or the point of intersection of two sides of a plane figure or the two rays of an angle.

vertical at right angles to the horizontal plane.

vertical blanking interval VBI, brief space between the drawing of television frames when no picture information is transmitted. The VBI may appear as black bars at the top and bottom of the screen on badly-adjusted TV sets. Several systems have exploited the VBI to broadcast data signals for other purposes. Examples include teletext pages (Ceefax, Teletext), subtitles for deaf viewers, and Intel's Intercast.

vertically opposite angles pair of angles that lie vertically opposite each other on the same side of a transveral (a line that intersects two or more lines).

vertical spam in computing, on *Usenet, spam which consists of many, often repetitive, messages per day posted to the same newsgroup or small set of newsgroups. The effect is to drown out other, more useful, conversation in the newsgroup.

Very Large Array VLA, largest and most complex single-site *radio telescope in the world. It is located on the Plains of San Augustine, 80 km/ 50 mi west of Socorro, New Mexico. It consists of 27 dish antennae, each 25 m/82 ft in diameter, arranged along three equally spaced arms forming a Y-shaped array. Two of the arms are

21 km/13 mi long, and the third, to the north, is 19 km/11.8 mi long. The dishes are mounted on railway tracks enabling the configuration and size of the array to be altered as required.

Pairs of dishes can also be used as separate interferometers, each dish having its own individual receivers that are remotely controlled, enabling many different frequencies to be studied. There are four standard configurations of antennae ranging from A (the most extended) through B and C to D. In the A configuration the antennae are spread out along the full extent of the arms and the VLA can map small, intense radio sources with high resolution. The smallest configuration, D, uses arms that are just 0.6 km/0.4 mi long for mapping larger sources. Here the resolution is lower, although there is greater sensitivity to fainter, extended fields of radio emission.

VESA local bus in computing, hardware configuration laid down by the Video Electronics Standards Association for computers based on the Intel 486 chip. It was rendered obsolete by the arrival of Pentium chips and the PCI bus.

vesicle small sac containing liquid. It is the medical term for a blister or elevation of the outer layer of the *skin (epidermis) containing serous fluid.

vestigial organ in biology, an organ that remains in diminished form after it has ceased to have any significant function in the adult organism. In humans, the appendix is vestigial, having once had a digestive function in our ancestors.

veterinary science the study, prevention, and cure of disease in animals. More generally, it covers animal anatomy, breeding, and relations to humans.

VGA abbreviation for video graphics array, in computing, a colour display system that provides either 16 colours on screen and a resolution of 640 × 480, or 256 colours with a resolution of 320 × 200. *SVGA (Super VGA) provides even higher resolution and more colours.

Viagra drug used to treat impotence, approved by the US Food and Drug Administration (FDA) in March 1998.

Viagra works by blocking the enzyme phosphodiesterase, which breaks down cGMP, a relaxant of smooth muscle. The cGMP allows for increased dilation of blood vessels (dilated by nitric oxide) and it is the increase in blood flow that enables two thirds of men suffering from erectile dysfunction to achieve an erection. It must be taken about an hour before intercourse. Side effects include headaches and fainting (due to dilation of blood vessels), and blue-tinted vision. The inhibition of phophodiesterase in the eye may result in damage to the retina as a side effect of taking Viagra.

vibration in physics, a periodic or oscillatory motion about a position. Sound is produced by vibrations of objects such as the stretched strings of a violin or the air particles in a wind instrument. The sound of the human voice is produced by the motion of air causing the 'strings' of the vocal cords to vibrate.

villus plural **villi**, small fingerlike projection extending into the interior of the small intestine and increasing the absorptive area of the intestinal wall. Digested nutrients, including sugars and amino acids, pass into the villi and are carried away by the circulating blood.

virion the smallest unit of a mature *virus.

virtual in computing, without physical existence. Most computers have *virtual memory, making their immediate-access memory seem larger than it is. *Virtual reality is a computer simulation of a whole physical environment.

virtual community in computing, group of people joined by using the same electronic conferencing system. The sense of virtual community can be extremely strong, going beyond simply exchanging mutually useful information to helping with real-life events such as family illnesses and financial crises.

virtual corporation company with no real-life headquarters but whose employees and/or individual contract-ors are linked via telecommunications.

virtual memory in computing, a technique whereby a portion of the computer backing storage, or external, *memory is used as an extension of its immediate-access, or internal, memory. The contents of an area of the immediate-access memory are stored on, say, a hard disk while they are not needed, and brought back into main memory when required.

virtual private network VPN, corporate computer network where data is routed via the Internet rather than, or as well as, via more expensive dedicated lines. There are several ways of implementing a VPN but the data will usually be encrypted to prevent them from being read as they pass across the public parts of the Internet.

virtual reality advanced form of computer simulation, in which a participant has the illusion of being part of an artificial environment. The participant views the environment through two tiny television screens (one for each eye) built into a visor. Sensors detect movements of the participant's head or body, causing the apparent viewing position to change. Gloves (datagloves) fitted with sensors may be worn, which allow the participant seemingly to pick up and move objects in the environment.

Virtual Reality Modelling Language in computing, method of displaying three-dimensional images on a Web page, usually abbreviated to *VRML.

virus in computing, a piece of *software that can replicate and transfer itself from one computer to another, without the user being aware of it. Some viruses are relatively harmless, but others can damage or destroy data.

virus infectious particle consisting of a core of nucleic acid (*DNA or RNA) enclosed in a protein shell. They are extremely small and cause disease. They differ from all other forms of life in that they are not cells – they are acellular. They are able to function and reproduce only if they can invade a living cell to use the cell's system to replicate themselves. In the process they may disrupt or alter the host cell's own DNA. They use the cell they invade to make more virus particles that are then released. This usually kills the cell. The healthy human body reacts by producing an antiviral protein, *interferon, which prevents the infection spreading to adjacent cells. There are around 5,000 species of virus known to science (1998), although there may be as many as half a million actually in existence.

Examples of diseases in humans caused by viruses are the common cold, chickenpox, influenza, AIDS, herpes, mumps, measles, and rubella. Recent evidence implicates viruses in the development of some forms of cancer (see *oncogenes). Antibiotics do not work against viruses. The best protection against diseases caused by viruses is *immunization.

Viruses can change by mutation. When they do so, a human body is sometimes unable to fight the new virus very well. This happens regularly with the influenza virus. A small change results in a small influenza epidemic, but a big change results in a pandemic that can kill millions of people worldwide. Many viruses mutate continuously so that the host's body has little chance of developing permanent resistance; others transfer between species, with the new host similarly unable to develop resistance. The viruses that cause *AIDS and Lassa fever are both thought to have 'jumped' to humans from other mammalian hosts.

types of virus Bacteriophages are viruses that infect bacterial cells. Retroviruses are of special interest because they have an RNA genome and can produce DNA from this RNA by a process called reverse transcription. Viroids, discovered in 1971, are even smaller than viruses; they consist of a single strand of nucleic acid with no protein coat. They may cause stunting in plants and some rare diseases in animals, including humans.

It is debatable whether viruses and viroids are truly living organisms, since they are incapable of an independent existence. Outside the cell of another organism they remain completely inert. The origin of viruses is also unclear, but it is believed that they are degenerate forms of life, derived from cellular organisms, or pieces of nucleic acid that have broken away from the genome of some higher organism and taken up a parasitic existence.

antiviral drugs Antiviral drugs are difficult to develop because viruses

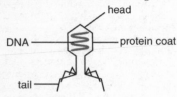

head

DNA — protein coat

tail

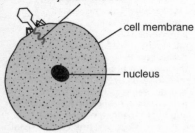

viral DNA injected into cell

cell membrane

nucleus

virus attacking cell

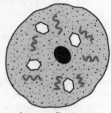

virus replicates

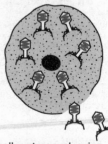

cell ruptures releasing new viruses into bloodstream

virus How a virus replicates itself to spread infection through the body.

replicate by using the genetic machinery of host cells, so that drugs tend to affect the host cell as well as the virus. Acyclovir (used against the herpes group of diseases) is one of the few drugs so far developed that is successfully selective in its action. It is converted to its active form by an enzyme that is specific to the virus, and it then specifically inhibits viral replication. Some viruses have shown developing resistance to the few antiviral drugs available.

occurrence of viruses in water
Viruses have recently been found to be very abundant in seas and lakes, with between 5 and 10 million per millilitre of water at most sites tested, but up to 250 million per millilitre in one polluted lake. These viruses infect bacteria and, possibly, single-celled algae. They may play a crucial role in controlling the survival of bacteria and algae in the plankton.

viscose yellowish, syrupy solution made by treating cellulose with sodium hydroxide and carbon disulphide or with cuprammonium liquor and sodium hydroxide. The solution is then regenerated as continuous filament (rayon) by forcing it through very small holes and solidifying it (in a process known as spinning). Forcing viscose through a slit instead of small holes produces cellophane.

viscosity the resistance of a fluid to flow, caused by its internal friction, which makes it resist flowing past a solid surface or other layers of the fluid. Treacle and other thick, sticky liquids are highly viscous liquids. Water and petrol are runny liquids and have low viscosity.

vision system computer-based device for interpreting visual signals from a video camera. Computer vision is important in robotics where sensory abilities would considerably increase the flexibility and usefulness of a robot.

visual acuity sharpness of vision. It is assessed using the Snellen test chart.

Visual Basic Microsoft computer language based on *BASIC. Visual Basic provides a rich set of visual programming tools that allow programmers to copy and paste common Windows components on to forms, in order to build Windows applications. BASIC is the language that is used to combine all the graphical components, and to code subroutines.

visual display unit VDU, computer terminal consisting of a keyboard for input data and a screen for displaying output. The screen and its housing are now more usually termed a *monitor.

visual field area that can be seen without moving the head or eyes. When both eyes are open and looking straight ahead it is known as the binocular visual field; if only one eye is used it is uni-ocular. Away from the centre of the visual field white objects show up much better than do coloured ones.

visualization in computing, turning numerical data into graphics. A simple example is to create a bar chart from a set of sales figures; more complex types of visualization include *fractals and other forms of computer-generated art.

visual programming in computing, programming method that uses a system of graphics, instead of text, to build software.

vitamin any of various chemically-unrelated organic (carbon-containing) compounds that are necessary in small quantities for the normal functioning of the mammalian body. Many act as coenzymes, small molecules that enable *enzymes to function effectively. Vitamins must be supplied by the diet because the body generally cannot make them. Deficiency of a vitamin may lead to a metabolic disorder ('deficiency disease'), which can be remedied by sufficient intake of the vitamin. Vitamins are generally classified as **water-soluble** (B and C) or **fat-soluble** (A, D, E, and K).

Vitamin A (retinol) is found in milk, cheese, butter, eggs, liver, kidney, oily fish (such as herring), and cod liver oil. However, the chemical carotene (found in most plants) can be converted to vitamin A in the human body. This vitamin is needed to make a chemical used by rod cells in the retina of the eye that respond in dim light. Shortage of this vitamin results in poor vision, especially at night.

Vitamin D is found in the greatest amounts in fish, but is present in tiny amounts in milk, cheese, butter, eggs, and liver. However, it can be made in the human skin when it is exposed to sunlight and this is sufficient for most people. It is needed to maintain bone in the body. Without the vitamin, bones are weak and can bend (rickets). Young children, pregnant women, and women breastfeeding a baby need much more than most other people. People who are housebound and do not get into the sun may become short of this vitamin.

Most people have balanced diets and so a real need to obtain extra vitamins from tablets is not common. Some vitamins are poisonous in high doses and there have been cases of people dying as a result of taking excessive amounts of vitamin pills.

vitamin A another name for *retinol.

vitamin B₁ another name for *thiamine.

vitamin B₆ another name for *pyridoxine.

vitamin B₁₂ another name for *cyanocobalamin.

vitamin B₂ another name for *riboflavin.

vitamin C another name for *ascorbic acid.

vitamin D another name for *cholecalciferol.

vitamin E another name for *tocopherol.

vitamin H another name for *biotin.

vitamin K another name for *phytomenadione.

vitreous humour transparent jellylike substance behind the lens of the vertebrate *eye. It gives rigidity to the spherical form of the eye and allows light to pass through to the retina.

vitriol former name for any of a number of sulphate salts. Blue, green, and white vitriols are copper, ferrous, and zinc sulphate, respectively. **Oil of vitriol** is a former name for sulphuric acid.

viviparous in animals, a method of reproduction in which the embryo develops inside the body of the female from which it gains nourishment (in contrast to *oviparous and *ovoviviparous). Vivipary is best developed in placental mammals, but also occurs in some arthropods, fishes, amphibians, and reptiles that have

placentalike structures. In plants, it is the formation of young plantlets or bulbils instead of flowers. The term also describes seeds that germinate prematurely, before falling from the parent plant.

vivisection literally, cutting into a living animal. Used originally to mean experimental surgery or dissection practised on a live subject, the term is often used by antivivisection campaigners to include any experiment on animals, surgical or otherwise.

VLF in physics, abbreviation for **very low *frequency**. VLF radio waves have frequencies in the range 3–30 kHz.

VLSI abbreviation for very large-scale integration, in electronics, the early-1990s level of advanced technology in the microminiaturization of *integrated circuits, and an order of magnitude smaller than *LSI (large-scale integration).

VMS in computing, operating system created in 1978 by *DEC for its VAX minicomputers. VMS has since been rewritten for DEC Alpha based machines, and renamed OpenVMS. OpenVMS provides a high level of availability, scalability, and data integrity, and is still a very good operating system for operations that use large volumes of data. Applications written for Windows NT can also run without difficulty in OpenVMS.

V numbers in computing, series of *protocols issued by the CCITT defining the rate at which modems transfer data. The numbers have come to designate a modem's speed: the current generation of modems are rated V.90. They transmit at up to 56,600 *bits per second (bps).

vocal cords paired folds, ridges, or cords of tissue within a mammal's larynx, and a bird's syrinx. Air expelled from the lungs passes between these folds or membranes and makes them vibrate, producing sounds. Muscles in the larynx change the pitch of the sounds produced, by adjusting the tension of the vocal cords.

voice-to-MIDI converter microphone that sends human vocal input to a synthesizer. This system of singing to run a synthesizer does not work well unless the singer has perfect pitch, so it is not commonly used.

vol. abbreviation for **volume**.

volatile describing a substance that readily passes from the liquid to the vapour phase. Volatile substances have a high *vapour pressure.

volatile memory in computing, *memory that loses its contents when the power supply to the computer is disconnected.

volcanic rock another name for *extrusive rock, igneous rock formed on the Earth's surface.

volcano crack in the Earth's crust through which hot *magma (molten rock) and gases well up. The magma is termed **lava** when it reaches the surface. A volcanic mountain, usually cone-shaped with a crater on top, is formed around the opening, or vent, by the build-up of solidified lava and ash (rock fragments). Most volcanoes occur on plate margins (see *plate tectonics), where the movements of plates generate magma or allow it to rise from the mantle beneath. However, a number are found far from plate-margin activity, on *hot spots where the Earth's crust is thin, for example in Hawaii. There are two main types of volcano: *composite volcanoes and *shield volcanoes.

volt symbol V, SI unit of electromotive force or electric potential (see *potential, electric), named after Italian physicist Alessandro Volta (1745–1827). A small battery has a potential of 1.5 volts, while a high-tension transmission line may carry up to 765,000 volts. The domestic electricity supply in the UK is 230 volts (lowered from 240 volts in 1995); it is 110 volts in the USA.

voltage commonly used term for *potential difference (PD).

voltage amplifier electronic device that increases an input signal in the form of a voltage or *potential difference, delivering an output signal that is larger than the input by a specified ratio.

voltaic cell another name for an electric cell.

voltmeter instrument for measuring the *potential difference (voltage) between two points in a circuit. It should not be confused with an

*ammeter, which measures current. A voltmeter has a high internal resistance (so that it passes only a small current), and is connected in parallel with the component across which potential difference is to be measured – that is, the current divides and passes through both the voltmeter and the component at the same time.

volume in geometry, the space occupied by a three-dimensional (3D) solid object. A *prism such as a *cube, *cuboid, or a *cylinder has a volume equal to the *area of the base multiplied by the height. For a *pyramid or *cone, the volume is equal to one-third of the area of the base multiplied by the perpendicular height. The volume of a *sphere is equal to $4/3 \times \pi r^3$, where r is the radius. Volumes of irregular solids may be calculated by the technique of *integration.

volumetric analysis chemical procedure used for determining the concentration of a solution. A known volume of a solution of unknown concentration is reacted with a known volume of a solution of known concentration (standard). The standard solution is delivered from a *burette so the volume added is known. This technique is known as *titration. Often an indicator is used to show when the correct proportions have reacted. This procedure is used for acid–base, *redox, and certain other reactions involving solutions.

voluntary muscles or **skeletal muscles**, in biology, *muscles that an individual controls consciously, for example the muscles that control walking and swallowing.

VPN abbreviation for *virtual private network.

vPoP acronym for virtual Point of Presence, telephone link which enables users to connect to a distant point of presence (*PoP) for the price of a local call.

VR abbreviation for **velocity ratio**.

VRAM acronym for Video Random-Access Memory, form of *RAM that allows simultaneous access by two different devices, so that graphics can be handled at the same time as data are updated. VRAM improves graphic display performance.

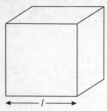

volume of a **cube**
= length^3
= l^3

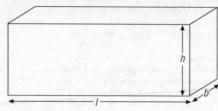

volume of a **cuboid**
= length × breadth × height
= $l \times b \times h$

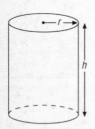

volume of a **cylinder**
= π × (radius of cross section)^2 × height
= $\pi r^2 h$

volume of a **cone**
= $\frac{1}{3}$ π × (radius of cross section)^2 × height
= $\frac{1}{3} \pi r^2 h$

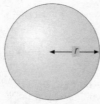

volume of a **sphere**
= $\frac{4}{3}$ π radius^3
= $\frac{4}{3} \pi r^3$

volume Volume of common three-dimensional shapes.

VR browser in computing, application that enables PC users to 'walk through' a *virtual reality scene on their monitors.

VRML abbreviation for virtual reality modelling language, in computing, method of displaying three-dimensional images on a *Web page. VRML, which functions as a counterpart to *HTML, is a platform-independent language that creates a *virtual reality scene which users can 'walk' through and follow links much like a conventional Web page.

vulcanization technique for hardening rubber by heating and chemically combining it with sulphur. The process also makes the rubber stronger and more elastic. If the sulphur content is increased to as much as 30%, the product is the inelastic solid known as ebonite. More expensive alternatives to sulphur, such as selenium and tellurium, are used to vulcanize rubber for specialized products such as vehicle tyres. The process was discovered by US inventor Charles Goodyear in 1839 and patented in 1844.

vulgar fraction or **common fraction**; or **simple fraction**, fraction comprising natural or whole numbers and written as a ratio, for example $^a/_b$, such that a is less than b.

vulva female external genitalia. Two folds of fleshy tissue, the labia majora and labia minora, extend downwards from the clitoris to surround the entrances to the *urethra and the vagina.

W in physics, symbol for *watt, the SI unit of power, equal to a power output of 1 joule per second. Multiple units include the kilowatt (kW, 1,000 watts) and megawatt (MW, 1,000,000 watts).

W abbreviation for **west**; in physics, symbol for **watt**.

WAIS abbreviation for wide area information server, software tool for searching for and retrieving information from a range of archives on the *Internet.

wall pressure in plants, the mechanical pressure exerted by the cell contents against the cell wall. The rigidity (turgor) of a plant often depends on the level of wall pressure found in the cells of the stem. Wall pressure falls if the plant cell loses water.

WAN abbreviation for wide area *network.

WAP acronym for Wireless Application Protocol, initiative started in the 1990s by Unwired Planet and mobile phone manufacturers Motorola, Nokia, and Ericsson to develop a standard for delivering Web-like applications on mobile phones and other wireless devices. In theory WAP phones can be used for e-mail and messaging, reading Web pages, shopping, booking tickets, and making other financial transactions, as well as for phone calls. WAP as been superseded to some extent by 3G (third generation) mobile phone technology.

wAreZ in computing, slang for pirated games or other applications that can be downloaded using *FTP.

warfarin poison that induces fatal internal bleeding in rats; neutralized with sodium hydroxide, it is used in medicine as an anticoagulant in the treatment of *thrombosis: it prevents blood clotting by inhibiting the action of vitamin K. It can be taken orally and begins to act several days after the initial dose.

warm temperate (subtropical) climate Warm temperate climates can be classified into two areas: *Mediterranean climate and the Eastern margin climate.

Eastern margin climates, which are found in southeast and eastern Asia, are dominated by the *monsoon. Temperature figures and rainfall distributions are similar to those of tropical continental areas although annual amounts of rain are much higher.

washing soda chemical name **sodium carbonate decahydrate**; $Na_2CO_3.10H_2O$, substance added to washing water to 'soften' *hard water. The calcium and magnesium ions in hard water react with the carbonate ions from the sodium carbonate and a precipitate of calcium carbonate results.

$$Na_2CO_{3(aq)} + CaSO_{4(aq)} \rightarrow CaCO_{3(s)} + Na_2SO_{4(aq)}$$

Washington Convention alternative name for *CITES, the international agreement that regulates trade in endangered species.

water chemical compound of hydrogen and oxygen elements – H_2O. It can exist as a solid (ice), liquid (water), or gas (water vapour). Water is the most common compound on Earth and vital to all living organisms. It covers 70% of the Earth's surface, and provides a habitat for large numbers of aquatic organisms. It is the largest constituent of all living organisms – the human body consists of about 65% water. It is found in all cells and many chemicals involved in processes such as respiration and photosynthesis need to be in solution in water in order to react. Pure water is a colourless, odourless, tasteless liquid which freezes at 0°C/32°F, and boils at 100°C/212°F. Natural water in the environment is never pure and always contains a variety of dissolved substances. Some 97% of the Earth's water is in the oceans; a further 2% is in the form of snow or ice, leaving only 1% available as freshwater for plants and animals. The recycling and circulation of water through the biosphere is termed the **water cycle**, or 'hydrological cycle'; regulation of the water balance in organisms is termed *osmoregulation. Water becomes more dense when it cools but reaches maximum density at 4°C/39°F. When cooled below this temperature the cooler water floats on the surface, as does ice formed from it. Animals and plants can survive under the ice.

The recycling and circulation of water on Earth is called the water cycle. Water occurs on the Earth's surface as standing water in oceans and lakes, as running water in rivers and streams, as rain, and as water vapour in the atmosphere. Together these sources comprise the hydrosphere which is in a constant state of flux – water vapour condenses as it cools to form clouds, droplets of water in the clouds merge to form raindrops that fall to earth (precipitation), and after flowing through rivers and streams into lakes and oceans water is returned to the atmosphere by evaporation; and so the cycle continues. Since the hydrological cycle is a closed system, the amount of water in the Earth's hydrosphere is constant. The cycle is powered by solar radiation which provides the energy to maintain the flow through the processes of evaporation, transpiration, precipitation, and run-off.

water-borne disease disease associated with poor water supply. In developing world countries four-fifths of all illness is caused by water-borne diseases, with diarrhoea being the leading cause of childhood death. Malaria, carried by mosquitoes that are dependent on stagnant water for breeding, affects 400 million people every year and kills 5 million. Polluted water is also a problem in industrialized nations, where industrial dumping of chemical, hazardous, and radioactive wastes causes a range of medical conditions from headache to cancer.

water cycle or **hydrological cycle**, natural circulation of water through the upper part of the Earth. It is a complex system involving a number of physical and chemical processes (such as *evaporation, precipitation, and infiltration) and stores (such as rivers, oceans, and soil).

Water is lost from the Earth's surface to the atmosphere by evaporation caused by the Sun's heat on the surface of lakes, rivers, and oceans, and through the *transpiration of plants. This atmospheric water is carried by the air moving across the Earth, and **condenses** as the air cools to form clouds, which in turn deposit moisture on the land and sea as precipitation. The water that collects on land flows to the ocean overland – as streams, rivers, and glaciers – or through the soil (infiltration) and rock (*groundwater). The boundary that marks the upper limit of groundwater is called the *water table.

The oceans, which cover around 70% of the Earth's surface, are the source of most of the moisture in the atmosphere.

waterfall cascade of water in a river or stream. It occurs when a river flows over a bed of rock that resists erosion; weaker rocks downstream are worn away, creating a steep, vertical drop and a plunge pool into which the water falls. Over time, continuing erosion causes the waterfall to retreat upstream forming a deep valley, or

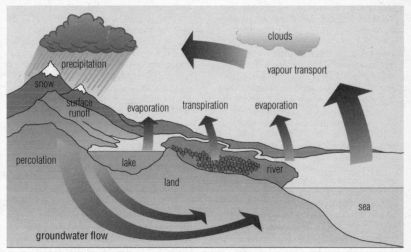

water cycle About one-third of the solar energy reaching the Earth is used in evaporating water. About 380,000 cubic km/95,000 cubic mi is evaporated each year. The entire contents of the oceans would take about one million years to pass through the water cycle.

*gorge. Good examples of waterfalls include Victoria Falls (Zimbabwe/Zambia), Niagara Falls (USA/Canada), and Angel Falls (Venezuela).

water gas fuel gas consisting of a mixture of carbon monoxide and hydrogen, made by passing steam over red-hot coke. The gas was once the chief source of hydrogen for chemical syntheses such as the *Haber process for making ammonia, but has been largely superseded in this and other reactions by hydrogen obtained from natural gas.

water of crystallization water chemically bonded to a salt in its crystalline state. For example, in copper(II) sulphate, there are five moles of water per mole of copper sulphate: hence its formula is $CuSO_4.5H_2O$. This water is responsible for the colour and shape of the crystalline form. When the crystals are heated gently, the water is driven off as steam and a white powder of the anhydrous salt is formed.

$$CuSO_4.5H_2O_{(s)} \rightarrow CuSO_{4\ (s)} + 5H_2O_{(g)}$$

water pollution any addition to fresh or seawater that interferes with biological processes or causes a health hazard. Common pollutants include nitrates, pesticides, and sewage (resulting from poor *sewage disposal methods), although a huge range of industrial contaminants, such as chemical by-products and residues created in the manufacture of various goods, also enter water – legally, accidentally, and through illegal dumping.

In 1980 the UN launched the 'Drinking Water Decade', aiming for cleaner water for all by 1990. However, in 1994 it was estimated that roughly half of all people in the developing world did not have safe drinking water. A 1995 World Bank report estimated that some 10 million deaths in developing countries were caused annually by contaminated water.

watershed boundary formed by height differences in the land, situated between two rivers or drainage basins.

water softener any substance or unit that removes the hardness from water. Hardness is caused by the presence of calcium and magnesium ions, which combine with soap to form an insoluble scum, prevent lathering, and cause deposits to build up in pipes and cookware (kettle fur). A water softener replaces these ions with sodium ions, which are fully soluble and cause no scum.

waterspout funnel-shaped column of water and cloud that is drawn from the surface of the sea or a lake by a *tornado.

water table upper level of *groundwater (water collected underground in porous rocks). Water that is above the water table will drain downwards; a spring forms where the water table meets the surface of the ground. The water table rises and falls in response to rainfall and the rate at which water is extracted, for example for irrigation and industry.

In many irrigated areas the water table is falling due to the extraction of water. Below northern China, for example, the water table is sinking at a rate of 1 m/3 ft a year. Regions with a high water table and dense industrialization have problems with pollution of the water table. In the USA, New Jersey, Florida, and Louisiana have water tables that are contaminated by both industrial wastes and saline seepage from the ocean.

watt symbol W, SI unit of power (the rate of expenditure or transformation of energy from one form to another) defined as one joule per second. A light bulb, for example, may use 40, 60, 100, or 150 watts of power; an electric heater will use several kilowatts (thousands of watts). The watt is named after the Scottish engineer James Watt.

WAV abbreviation for Windows waveform, in computing, audio file format for *IBM-compatible PCs, widely used to distribute sounds over the Internet. WAV files, which contain a digitized recording of a sound, bear the suffix .wav.

wave in the oceans, a ridge or swell formed by wind or other causes. The power of a wave is determined by the strength of the wind and the distance of open water over which the wind blows (the fetch). Waves are the main agents of *coastal erosion and deposition: sweeping away or building up beaches, creating spits and berms, and wearing down cliffs by their *hydraulic action and by the *corrosion of the sand and shingle that they carry. A *tsunami (misleadingly called a 'tidal wave') is formed after a submarine earthquake.

wave in physics, oscillation that is propagated from a source. Mechanical waves require a medium through which to travel. Electromagnetic waves do not; they can travel through a vacuum. Waves carry energy but they do not transfer matter. The medium (for example the Earth, for seismic waves) is not permanently displaced by the passage of a wave. The model of waves as a pattern is used to help understand the properties of light and sound. Experiments conducted in a ripple tank with water waves can explain how waves slow down as water becomes shallower, how waves change direction when travelling through another medium, and how waves are reflected from different surfaces. See also *standing wave.

types of wave There are various ways of classifying wave types. One of these is based on the way the wave travels. In a *transverse wave, the displacement of the medium is perpendicular to the direction in which the wave travels. An example of this type of wave is a mechanical wave projected along a tight string. The string moves at right angles to the wave motion. *Electromagnetic waves are another example of transverse waves. The directions of the electric and magnetic fields are perpendicular to the wave motion. In a *longitudinal wave the disturbance takes place parallel to the wave motion. A longitudinal wave consists of a series of compressions and rarefactions (states of maximum and minimum density and pressure, respectively). Such waves are always mechanical in nature and thus require a medium through which to travel. *Sound waves are an example of longitudinal waves. Waves that result from a stone being dropped into water appear as a series of circles. These are called circular waves and can be generated in a *ripple tank for study. Waves on water that appear as a series of parallel lines are called plane waves.

characteristics of waves All waves have a *wavelength. This is measured as the distance between successive crests (or successive troughs) of the wave. It is given the Greek symbol λ. The *frequency of a wave is the number of vibrations per second. The reciprocal of this is the wave period. This is the time taken for one complete cycle of the wave oscillation. The speed of the wave is measured by multiplying wave frequency by the wavelength.

properties of waves When a wave moves from one medium to another (for example a light wave moving from air to glass) it moves with a different speed in the second medium. This change in speed causes it to change direction. This property is called *refraction. The *angle of refraction depends on whether the wave is speeding up or slowing down as it changes medium. *Reflection occurs whenever a wave hits a barrier. The wave is sent back, or reflected, into the medium. The *angle of incidence (the angle between the ray and a perpendicular line drawn to the surface) is equal to the *angle of reflection (the angle between the reflected ray and a perpendicular to the surface). See also *total internal reflection. An echo is the repetition of a sound wave by reflection from a surface. All waves spread slightly as they travel. This is called *diffraction and it occurs chiefly when a wave interacts with a solid object. The degree of diffraction depends on the relationship between the wavelength and the size of the object (or gap through which the wave travels). If the two are similar in size, diffraction occurs and the wave can be seen to spread out. Large objects cast *shadows because the difference between their size and the wavelength is so large that light waves are not diffracted around the object. A dark shadow results. When two or more waves meet at a point, they interact and combine to produce a resultant wave of larger or smaller amplitude (depending on whether the combining waves are in or out of phase with each other). This is called *interference. Transverse waves can exhibit polarization. If the oscillations of the wave take place in many different directions (all at right angles to the directions of the wave) the wave is unpolarized. If the oscillations occur in one plane only, the wave is polarized. Light, which consists of transverse waves, can be polarized.

wavelength distance between successive crests or troughs of a *wave. The wavelength of a light wave determines its colour; red light has a wavelength of about 700 nanometres, for example. The complete range of wavelengths of electromagnetic waves is called the electromagnetic *spectrum.

wavelength division multiplexing WDM, technique for sending multiple streams of light down a single fibre-optic cable. The different signals have slightly different wavelengths, or colours. WDM can greatly increase the capacity of existing fibre-optic links and thus dramatically lower the cost of long-distance communications.

wave power power obtained by harnessing the energy of water waves. Various schemes have been advanced since 1973 when oil prices rose dramatically and an energy shortage threatened. In 1974 the British engineer Stephen Salter developed the 'duck' – a floating boom, the segments of which nod up and down with the waves. The nodding motion can be used to drive pumps and spin generators. Another device, developed in Japan, uses an oscillating water column to harness wave power. A major technological breakthrough will be required if wave power is ever to contribute significantly to the world's energy needs, although several ideas have reached prototype stage.

wave refraction distortion of waves as they reach the coast, due to variations in the depth of the water and shape of the coastline. It is particularly evident where there are headlands and bays.

The bending of a wave crest as it approaches a headland concentrates the energy of the wave in the direction of that headland, and increases its power of erosion. By contrast, the bending that a wave crest experiences when it moves into a bay causes its energy to be dissipated away from the direction of the shore. As a result the wave loses its erosive power and becomes more likely to deposit sediment on the shore.

wavetable synthesizer *MIDI synthesizer that uses sampling – recordings of actual musical instruments – to create sounds. The authenticity of the sound source means that wavetable synthesizers can achieve highly realistic results.

wax solid fatty substance of animal, vegetable, or mineral origin. Waxes are composed variously of *esters, *fatty acids, free *alcohols, and solid hydrocarbons. **Mineral waxes** are obtained from petroleum and vary in hardness from the soft petroleum jelly (or petrolatum) used in ointments to the hard paraffin wax employed for making candles and waxed paper for drinks cartons. **Animal waxes** include beeswax, the wool wax lanolin, and spermaceti from sperm-whale oil; they are used mainly in cosmetics, ointments, and polishes. Another animal wax is tallow, a form of suet obtained from cattle and sheep's fat, once widely used to make candles and soap. Sealing wax is made from lac or shellac, a resinous substance obtained from secretions of scale insects.

Vegetable waxes, which usually occur as a waterproof coating on plants that grow in hot, arid regions, include carnauba wax (from the leaves of the carnauba palm) and candelilla wax, both of which are components of hard polishes such as car waxes.

Wb in physics, symbol for *weber, the SI unit of magnetic flux, equal to the flux that (linking a circuit of one turn) produces an emf (*electromotive force) of 1 volt when the flux is reduced at a constant rate to zero in 1 second.

WBT abbreviation for a *Windows-Based Terminal. WBT is sometimes pronounced 'wabbit'.

WDM abbreviation for *wave division multiplexing.

weak acid acid that only partially ionizes in aqueous solution (see *dissociation). Weak acids include ethanoic acid and carbonic acid. The pH of such acids lies between pH 3 and pH 6.

weak base base that only partially ionizes in aqueous solution (see *dissociation); for example, ammonia.

The pH of such bases lies between pH 8 and pH 10.

weak interacting massive particle WIMP, hypothetical subatomic particle found in the Galaxy's *dark matter. These particles could constitute the 80% of dark matter unaccounted for by massive astrophysical compact halo objects.

weak nuclear force or **weak interaction**, one of the four fundamental *forces of nature, the other three being the gravitational force or gravity, the electromagnetic force, and the strong nuclear force. It causes radioactive beta decay and other subatomic reactions. The particles that carry the weak force are called *weakons (or intermediate vector bosons) and comprise the positively and negatively charged W particles and the neutral Z particle.

weakon or **intermediate vector boson**, member of a group of elementary particles, see * intermediate vector boson.

weather variation of atmospheric conditions at any one place over a short period of time. Such conditions include humidity, precipitation, temperature, cloud cover, visibility, and wind. Weather differs from *climate in that the latter is a composite of the average weather conditions of a locality or region over a long period of time (at least 30 years). *Meteorology is the study of short-term weather patterns and data within a particular area; climatology is the study of weather over longer timescales on a zonal or global basis. *Weather forecasts are based on current meterological data, and predict likely weather for a particular area.

weather forecast prediction for changes in the *weather. The forecast is based on several different types of information from several sources. Weather stations are situated around the world, on land, in the air (on aircraft and air balloons), and in the sea (on ships, buoys, and oilrigs). Forecasters also use satellite photographs which show cloud patterns, and information from the Meteorological Office, such as a synoptic chart which shows *atmospheric pressure and changes in *air masses (known as a weather *front). The information is used to predict the weather over a 24-hour period, and to suggest likely changes over a period of three or four days.

weathering process by which exposed rocks are broken down on the spot (in situ) by the action of rain, frost, wind, and other elements of the weather. It differs from *erosion in that no movement or transportion of the broken-down material takes place. Two types of weathering are recognized: *physical (or mechanical) weathering and *chemical weathering. They usually occur together, and are an important part of the development of landforms.

Web authoring tool in computing, software for creating *Web pages. The basic Web authoring tool is *HTML, the source code that determines how a Web page is constructed and how it looks. Other programs, such as *Java and *VRML, can also be incorporated to enhance Web pages with animations and interactive features. Commercial authoring tools include HoTMetaL PRO, NetObjects' Fusion, and Microsoft's Front Page.

Web browser in computing, client software that allows access to the World Wide Web. See *browser.

Webcam any camera connected to the Internet, usually for the purpose of displaying an image on a Web page. Webcams are trained on a wide range of famous sights such as London's Tower Bridge. They have also been placed inside birds' nesting boxes, refrigerators, strip clubs, and bedrooms. What may have been the first Webcam was aimed at a flask of coffee in the Cambridge University Computer Laboratory.

weber symbol Wb, SI unit of *magnetic flux (the magnetic field strength multiplied by the area through which the field passes). It is named after German chemist Wilhelm Weber. One weber equals 10^8 *maxwells.

Web page in computing, a *hypertext document on the *World Wide Web.

Web server computer system that stores the *HTML pages of a *Web site and transmits those pages over the *Internet to other computers when requested by Web browser software.

Web site collection of *Web pages belonging to the same company, organization, or individual. The first Web site page to be displayed on a *browser is known as the *home page. Sites designed to be a gateway to the rest of the *World Wide Web are called *portals. A Web site may be physically located on a local Web *server or hosted by an *Internet Service Provider.

weight force exerted on an object by *gravity. The weight of an object depends on its mass – the amount of material in it – and the strength of the Earth's gravitational pull (the acceleration due to gravity), which decreases with height. Consequently, an object weighs less at the top of a mountain than at sea level. On the surface of the Moon, an object has only one-sixth of its weight on Earth (although its mass is unchanged), because the Moon's surface gravity is one-sixth that of the Earth's.

weightlessness apparent loss in weight of a body in *free fall. Astronauts in an orbiting spacecraft do not feel any weight because they are falling freely in the Earth's gravitational field (not because they are beyond the influence of Earth's gravity). The same phenomenon can be experienced in a falling lift or in an aircraft imitating the path of a freely falling object.

weights and measures see under * c.g.s. system, f.p.s. system, *m.k.s. system, *SI units.

Weil's disease or **leptospirosis**, infectious disease of animals that is occasionally transmitted to human beings, usually by contact with water contaminated with rat urine. It is characterized by acute fever, and infection may spread to the brain, liver, kidneys, and heart. It has a 10% mortality rate.

Werner's syndrome in medicine, a rare genetic disorder characterized by premature ageing. In their late twenties sufferers show signs of advanced ageing, with greying hair, wrinkles, and cataracts. They develop cancer, heart disease, and other medical problems usually associated with the elderly, and do not usually survive beyond 50. Werner's gene, located 1996, codes for the enzyme helicase that is involved in the unwinding of DNA before division. The syndrome results from two defective copies of this gene.

wetland permanently wet land area or habitat. Wetlands include areas of *marsh, fen, *bog, flood plain, and shallow coastal areas. Wetlands are extremely fertile. They provide warm, sheltered waters for fisheries, lush vegetation for grazing livestock, and an abundance of wildlife. Estuaries and seaweed beds are more than 16 times as productive as the open ocean.

whirlwind rapidly rotating column of air, often synonymous with a *tornado. On a smaller scale it produces the dust-devils seen in deserts.

white blood cell or **leucocyte**, one of a number of different cells that play a part in the body's defences and give immunity against disease. Some (neutrophils and *macrophages) engulf invading micro-organisms, others kill infected cells, while *lymphocytes produce more specific immune responses. White blood cells are colourless, with clear or granulated cytoplasm, and are capable of independent amoeboid movement. They occur in the blood, *lymph, and elsewhere in the body's tissues. Unlike mature red blood cells they contain a nucleus.

Human blood contains about 11,000 leucocytes to the cubic millimetre – about one to every 500 red blood cells. White blood cell numbers may be reduced (leucopenia) by starvation, pernicious anaemia, and certain infections, such as typhoid and malaria. An increase in their numbers (leucocytosis) is a reaction to normal events such as digestion, exertion, and pregnancy, and to abnormal ones such as loss of blood, cancer, and most infections.

Some white cells can produce antibodies. Antibodies are specific chemicals that can attach to chemicals that do not belong to the body, such as toxins or chemicals on disease-causing bacteria. Chemicals that do not belong to the body are said to be 'foreign'. If these chemicals cause the body to produce antibodies, they are called antigens. Once these foreign chemicals have contacted the blood, some white blood cells can retain the memory of the particular antibody that is needed for defence. The next time the antibody is needed it is produced quickly and in large amounts to prevent the body being harmed – the body shows immunity to the disease. This is also the basis of *immunization by a vaccine.

white dwarf small, hot *star, the last stage in the life of a star such as the Sun. White dwarfs make up 10% of the stars in the Galaxy; most have a mass 60% of that of the Sun, but only 1% of the Sun's diameter, similar in size to the Earth. Most have surface temperatures of 8,000°C/14,400°F or more, hotter than the Sun. However, being so small, their overall luminosities may be less than 1% of that of the Sun. The Milky Way contains an estimated 50 billion white dwarfs.

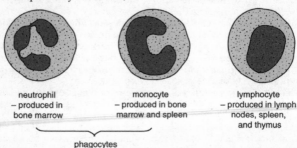

neutrophil
– produced in bone marrow

monocyte
– produced in bone marrow and spleen

lymphocyte
– produced in lymph nodes, spleen, and thymus

phagocytes

white blood cell Three kinds of white blood cell. The phagocytes (neutrophils and monocytes) engulf invading bacteria and other alien particles. There are two types of lymphocyte, T cells and B cells. T cells destroy infected cells and B cells produce antibodies that persist in the blood stream.

whiteout 'fog' of grains of dry snow caused by strong winds in temperatures of between −18°C/0°F and −1°C/30°F. The uniform whiteness of the ground and air causes disorientation in humans.

white spirit colourless liquid derived from petrol; it is used as a solvent and in paints and varnishes.

whois in computing, searchable database of every registered *domain and the names of their users. A special application, also called whois, is needed to search the database.

whooping cough or **pertussis**, acute infectious disease, seen mainly in children, caused by colonization of the air passages by the bacterium *Bordetella pertussis*. There may be catarrh, mild fever, and loss of appetite, but the main symptom is violent coughing, associated with the sharp intake of breath that is the characteristic 'whoop', and often followed by vomiting and severe nose bleeds. The cough may persist for weeks.

wide area network WAN, in computing, a *network that connects computers distributed over a wide geographical area. 'Dumb' terminals and microcomputers act as workstations, which connect to remote systems via a local host computer.

WIDs acronym for *Wireless Information Devices.

Wien's displacement law in physics, a law of radiation stating that the wavelength carrying the maximum energy is inversely proportional to the absolute temperature of a *black body: the hotter a body is, the shorter the wavelength. It has the form $\lambda_{max}T =$ constant, where λ_{max} is the wavelength of maximum intensity and T is the temperature. The law is named after German physicist Wilhelm Wien (1864–1928).

wild card in computing, character which represents 'any character' in a search or command. When comparing *strings, the computer does not seek a precise match for a wild card character. The most useful wildcards are ?, which matches any single character, and *, which matches any number of characters, including zero. Hence the *DOS commands DEL *.* – delete all files – and COPY *.DOC – copy all files with the filename extension 'DOC'.

wildlife corridor passage between habitats. See *corridor, wildlife.

wildlife trade international trade in live plants and animals, and in wildlife products such as skins, horns, shells, and feathers. The trade has made some species virtually extinct, and whole ecosystems (for example, coral reefs) are threatened. Wildlife trade is to some extent regulated by *CITES (Convention on International Trade in Endangered Species).

Species almost eradicated by trade in their products include many of the largest whales, crocodiles, marine turtles, and some wild cats. Until recently, some 2 million snake skins were exported from India every year. Populations of black rhino and African elephant have collapsed because of hunting for their horns and tusks (ivory), and poaching remains a problem in cases where trade is prohibited.

wild type in genetics, the naturally occurring gene for a particular character that is typical of most individuals of a given species, as distinct from new genes that arise by mutation.

will-o'-the-wisp light sometimes seen over marshy ground, believed to be burning gas containing methane from decaying organic matter.

WIMP acronym for *weak interacting massive particle.

WIM acronym for **Windows, icons, menus, pointing device**, in computing, another name for *graphical user interface (GUI).

Winchester drive in computing, an old-fashioned term for *hard disk.

wind lateral movement of the Earth's atmosphere from high-pressure areas (anticyclones) to low-pressure areas (depressions). Its speed is measured using an anemometer or by studying its effects on, for example, trees by using the *Beaufort scale. Although modified by features such as land and water, there is a basic worldwide system of *trade winds, westerlies, and polar easterlies.

wind-chill factor or **wind-chill index**, estimate of how much colder it feels when a wind is blowing. It is arrived at by combining the actual temperature and wind speed and is given as a different temperature. Wind chill can be calculated in Celsius using the formula: wind chill = $0.045(5. \times \sqrt{V} + 10.45 - 0.28V) (T - 33) + 33$ where V is in the wind speed in kilometres per hour and T is the temperature in degrees Celsius; and in Fahrenheit: wind chill = $0.0817(3.71 \times \sqrt{V} + 5.81 - 0.25V)(T - 91.4) + 91.4$ where V is in the wind speed in miles per hour and T is the temperature in degrees Fahrenheit.

window in computing, a rectangular area on the screen of a *graphical user interface. A window is used to display data and can be manipulated in various ways by the computer user.

Windows-Based Terminal WBT, network computer designed to display Microsoft Windows programs, such as Word and Excel, which are being run not on the local machine but on a server computer. WBTs have their own processor and memory but do not necessarily have a hard drive and usually will not have a floppy disk drive. WBTs typically use Microsoft's *RDP protocol and/or ICA, from Citrix.

Windows Open Services Architecture WOSA, set of interfaces defined by Microsoft to enable Windows programs from different vendors to communicate with one another.

Windows WAVeform in computing, audio file format commonly abbreviated to *WAV.

wind power power produced from the harnessing of wind energy. The wind has long been used as a source of energy: sailing ships and windmills are ancient inventions. After the energy crisis of the 1970s *wind turbines began to be used to produce electricity on a large scale.

wind turbine windmill of advanced aerodynamic design connected to a generator producing *electrical energy and used in wind-power installations. Wind is a form of *renewable energy that is used to turn the *turbine blades of the windmill. Wind turbines can be either large propeller-type rotors mounted on a tall tower, or flexible metal strips fixed to a vertical axle at

top and bottom. The world's largest wind turbine is on Hawaii, in the Pacific Ocean. It has two blades 50 m/160 ft long on top of a tower 20 storeys high. An example of a propeller turbine is found at Tvind in Denmark; it has an output of some 2 MW. Other machines use novel rotors, such as the 'egg-beater' design developed at Sandia Laboratories in New Mexico, USA.

WIN.INI acronym for Windows initialization, in computing, file used by Microsoft Windows to store a range of settings that in general govern the appearance of Windows and some Windows 3 applications. Changes made via Windows 3's Control Panel program are often stored in WIN.INI. In Windows 95/98, control has been moved to the Registry.

Winsock contraction of Windows socket, in computing, program that supplies an interface between Windows software and a *TCP/IP application.

wired gloves in computing, interface worn on the hands for *virtual reality applications. The gloves detect the movement of the hands, enabling the user to 'touch' and 'move' objects in a virtual environment.

wire frame in computing, method of creating three-dimensional computerized animations by drawing a series of frames showing the moving image in outline, like a moving skeleton. When the designer is satisfied with the action of the wire-frame figure, he or she adds the 'skin', superimposing textures to give the final effect.

Wireless Information Devices WIDS, generic label for a growing class of portable systems such as mobile phones, handheld computers, digital cameras, and electronic organizers.

wizard in computing, an interactive tool developed by Microsoft that 'talks' program users through a complex operation, such as creating a *template or a presentation. The wizard presents the user with a series of *dialog boxes asking simple questions in ordinary language, which the user answers by choosing *radio buttons, checking boxes, and entering information by keyboard.

wold (Old English *wald* 'forest') open, hilly country. The term refers specifically to certain areas in England, notably the Yorkshire and Lincolnshire Wolds and the Cotswold Hills.

wolfram alternative name for *tungsten.

womb common name for the *uterus.

work in physics, a measure of the result of transferring energy from one system to another to cause an object to move. Work should not be confused with *energy (the capacity to do work, which is also measured in joules) or with *power (the rate of doing work, measured in joules per second). Work is equal to the product of the force used and the distance moved by the object in the direction of that force. If the force is F newtons and the distance moved is d metres, then the work W is given by: $W = Fd$. For example, the work done when a force of 10 newtons moves an object 5 m against resistance is 50 joules (50 newton-metres).

workstation high-performance desktop computer with strong graphics capabilities, traditionally used for engineering (*CAD and *CAM), scientific research, and desktop publishing. From 1985 to 1995 workstations were frequently based on fast RISC (reduced instruction-set computer) chips running the Unix operating system. However, the market is under attack from 'Wintel' PCs with Intel Pentium processors running Microsoft Windows NT, which are cheaper and run PC software as well as workstation programs. By 1997, four of the five leading workstation manufacturers – DEC, Hewlett-Packard, IBM, and Silicon Graphics Inc., but not Sun Microsystems – had committed themselves to supporting NT.

World Wide Web WWW, *hypertext system for publishing information on the *Internet. World Wide Web documents ('Web pages') are text files coded using *HTML to include text and graphics, and are stored on a Web server connected to the Internet. Web pages may also contain dynamic objects and *Java applets for enhanced animation, video, sound, and interactivity.

World Wide Web Consortium W3C, computing industry group which seeks to promote standards and coordinate developments in the World Wide Web. Founded in 1994 and based at the Massachusetts Institute of Technology (MIT), the Consortium is directed by Tim Berners-Lee, inventor of the Web.

WORM acronym for **write once read many times**, in computing, a storage device, similar to a *CD-ROM. The computer can write to the disk directly, but cannot later erase or overwrite the same area. WORMs are mainly used for archiving and back-up copies.

worm in computing, *virus designed to spread from computer to computer across a network. Worms replicate themselves while 'hiding' in a computer's memory, causing systems to 'crash' or slow down, but do not infect other programs or destroy data directly.

WOSA acronym for *Windows Open Services Architecture.

W particle *elementary particle, one of the * intermediate vector bosons responsible for transmitting the *weak nuclear force. The W particle exists as both W^+ and W^-

write-once technology in computing, technology that allows a user to write data onto an optical disk once. After that the data are permanent and can be read any number of times.

write protection device on computer disks and tapes that provides *data security by allowing data to be read but not deleted, altered, or overwritten.

wrought iron fairly pure iron containing some beads of slag, widely used for construction work before the days of cheap steel. It is strong, tough, and easy to machine. It is made in a puddling furnace, invented by Henry Colt in England in 1784. Pig iron is remelted and heated strongly in air with iron ore, burning out the carbon in the metal, leaving relatively pure iron and a slag containing impurities. The resulting pasty metal is then hammered to remove as much of the remaining slag as possible. It is still used in fences and gratings.

wt abbreviation for **weight**.

x in computing, *wild card character often used to describe versions of hardware or software. One might, therefore, refer to Windows 3.x (any version of Windows, from 3.0 to 3.31), or an x86 chip (any of the chips manufactured by Intel with serial numbers ending in 86).

X.25 in computing, communications protocol for sending *packets of data over a network.

X.400 in computing, standard maintained by the ITU (International Telecommunications Union, formerly the *CCITT) which forms the basis for a message handling system. X.400 is used as a shorthand term for a number of recommendations and standards involved in running some e-mail systems over telecommunications lines.

X.500 directory standards for network addresses, issued by the Comité Consultatif International Téléphonique et Télégraphique (CCITT).

xanthophyll yellow pigment in plants that, like *chlorophyll, is responsible for the production of carbohydrates by photosynthesis.

X chromosome larger of the two *sex chromosomes, the smaller being the *Y chromosome. These two chromosomes are involved in sex determination. In humans, whether a person is male or female is determined by the particular combination of the two sex chromosomes in the body cells. In females both the sex chromosomes are the same – two X chromosomes (XX). In males the two are different – one X chromosome and one Y chromosome (XY). The Y chromosome is shorter than the X. *Genes on these chromosomes determine a person's sex. Genes carried on the X chromosome produce the phenomenon of *sex linkage.

As a result of *meiosis gametes from a female each contain one X chromosome. However, gametes from a male are of two kinds. Half of the gametes contain an X chromosome and half contain a Y chromosome. If an X carrying gamete from a male fertilizes a female gamete the result will be a female. If a Y carrying gamete from a male fertilizes a female gamete, the result will be a male.

Early in the development of a female embryo, one of the X chromosomes becomes condensed so that most of its genes are inactivated. If this inactivation is incomplete, skeletal defects and mental retardation result.

xenon chemical symbol Xe, (Greek *xenos* 'stranger') colourless, odourless, gaseous, non-metallic element, atomic number 54, relative atomic mass 131.30. It is grouped with the *noble gases (rare gases) and was long believed not to enter into reactions, but is now known to form some compounds, mostly with fluorine. It is a heavy gas present in very small quantities in the air (about one part in 20 million). Xenon is used in bubble chambers, light bulbs, vacuum tubes, and lasers. It was discovered in 1898 in a residue from liquid air by Scottish chemists William Ramsay (1852–1916) and Morris Travers (1872–1961).

xerophyte plant adapted to live in dry conditions. Common adaptations to reduce the rate of *transpiration include a reduction of leaf size, sometimes to spines or scales; a dense covering of hairs over the leaf to trap a layer of moist air (as in edelweiss); water storage cells; sunken stomata; and permanently rolled leaves or leaves that roll up in dry weather (as in marram grass). Many desert cacti are xerophytes.

XGA abbreviation for extended graphics array, in computing, colour display system that provides either 256 colours on screen and a resolution of 1,024 × 768 *pixels or 25,536 colours with a resolution of 640 × 480. This gives a much sharper image than, for example, *VGA, which can display only 16 colours at 480 lines of 640 pixels.

XHTML abbreviation for **extensible hypertext markup language**, latest version of HTML to be released by the *W3 Consortium (World Wide Web Consortium). Most of the markup tags that were available in HTML 4.0 are still available in XHTML. The major development is that the entire language has been rewritten in *XML (eXtensible Markup Language). The design is modular, with different modules for different classes of markup tags; for example, there are separate modules for form tags, and for frame tags. This modularity means that devices that do not need all the functions of HTML can be designed to handle only the core modules. This makes their programming easier and more compact, and suits the new breed of mobile Web browsers, and devices like TV set-top boxes.

XML abbreviation for **eXtensible markup language**, in computing, a simplified subset of *SGML for defining languages for specific purposes or specific industries for use on the World Wide Web. XML is more powerful than *HTML, because the formatting *tags are user-defined, but less cumbersome than SGML. XML has been developed through the *W3 Consortium, who published XML 1.0 in December 1997.

Xmodem in computing, an *FTP *protocol designed to make transmitting files via telephone speedy and error-free.

XON/XOFF in computing, control commands used when two devices handshake using a modem connection. XON starts or resumes transmission of data and XOFF pauses it. XON and XOFF can be manually activated by control-Q and control-S respectively.

XOR abbreviation for exclusive or, in computing, search filter meaning 'A or B, but not both'. Thus a search for 'chocolate xor biscuit' might yield 'chocolate', 'biscuit', 'chocolate cake' and 'shortbread biscuit' but never 'chocolate biscuit'. See also *Boolean algebra.

X-ray band of electromagnetic radiation in the wavelength range 10^{-12} to 10^{-8} m (between gamma rays and ultraviolet radiation; see *electromagnetic waves). Applications of X-rays make use of their short wavelength (as in *X-ray diffraction) or their penetrating power (as in medical X-rays of internal body tissues). X-rays are dangerous and can cause cancer.

X-ray astronomy detection of X-rays from intensely hot gas in the universe. Such X-rays are prevented from reaching the Earth's surface by the atmosphere, so detectors must be placed in rockets and satellites. The first celestial X-ray source, Scorpius X-1, was discovered by a rocket flight in 1962.

X-ray diffraction method of studying the atomic and molecular structure of crystalline substances by using *X-rays. X-rays directed at such substances spread out as they pass through the crystals owing to *diffraction (the slight spreading of waves around the edge of an opaque object) of the rays around the atoms. By using measurements of the position and intensity of the diffracted waves, it is possible to calculate the shape and size of the atoms in the crystal. The method has been used to study substances such as *DNA that are found in living material.

X-ray telescope in astronomy, telescope designed to receive *electromagnetic waves in the X-ray part of the spectrum. X-rays cannot be focused by lenses or mirrors in the same way as visible light, and a variety of alternative techniques is used to form images. Because X-rays cannot penetrate the Earth's atmosphere, X-ray telescopes are mounted on *satellites, *rockets, or high-flying balloons.

X Windows in computing, a networked window management system developed as part of Project Athena at the Massachusetts Institute of Technology (MIT) in 1984. It has been adopted as a standard by the Unix community but is platform-independent and versions exist for many different operating systems. X Windows enables a user to open windows into a number of different computers at the same time, either using an X Terminal or via X software running on a PC, a Unix workstation, or another computer. Although X Windows is not a graphical user interface, it provides software tools to support such interfaces: Motif is a popular Unix GUI built on X.

xylem transport tissue found in *vascular plants, whose main function is to conduct water and dissolved mineral nutrients from the roots to other parts of the plant. The water is ultimately lost by *transpiration from the leaves (see *leaf). Xylem is composed of a number of different types of cell, and may include long, thin, usually dead cells known as tracheids; fibres (schlerenchyma); thin-walled *parenchyma cells; and conducting vessels.

xylene or **dimethylbenzene**; $C_6H_4(CH_3)_2$, volatile organic liquid that exists as three isomers, 1,2-xylene, 1,3-xylene, and 1,4-xylene. It is used in the production of polyester synthetic fibres, for example Terylene.

yard symbol yd, unit of length, equivalent to 3 feet (0.9144 m).

Y chromosome smaller of the two *sex chromosomes. In male mammals it occurs paired with the other type of sex chromosome (X), which carries far more genes. The Y chromosome is the smallest of all the mammalian chromosomes and is considered to be largely inert (that is, without direct effect on the physical body), apart from containing the genes that control the development of the testes. There are only 20 genes discovered so far on the human Y chromosome, far fewer than on all other human chromosomes. In humans, whether a person is male or female is determined by the particular combination of the two sex chromosomes in the body cells. In females both the sex chromosomes are the same – two X chromosomes (XX). In males the two are different – one X chromosome and one Y chromosome (XY). The Y chromosome is shorter than the X. *Genes on these chromosomes determine a person's sex (*sex determination).

As a result of *meiosis gametes from a female each contain one X chromosome. However, gametes from a male are of two kinds. Half of the gametes contain an X chromosome and half contain a Y chromosome. If an X carrying gamete from a male fertilizes a female gamete the result will be a female. If a Y carrying gamete from a male fertilizes a female gamete, the result will be a male.

In humans, about one in 300 males inherits two Y chromosomes at conception, making him an XYY triploid. Few if any differences from normal XY males exist in these individuals, although at one time they were thought to be emotionally unstable and abnormally aggressive. In 1989 the gene determining that a human is male was found to occur on the X as well as on the Y chromosome; however, it is not activated in the female.

yd abbreviation for *yard.

year unit of time measurement, based on the orbital period of the Earth around the Sun. The **tropical year** (also called equinoctial and solar year), from one spring *equinox to the next, lasts 365.2422 days (nearly $365\frac{1}{4}$ days). It governs the occurrence of the seasons, and is the period on which the calendar year is based. Every four years is a leap year, when the four quarters of a day are added as one extra day. A year on Mercury is only 88 days; a year on Mars is 23 months.

yeast one of various single-celled fungi (see *fungus) that form masses of tiny round or oval cells by budding. When placed in a sugar solution the cells multiply and convert the sugar into alcohol and carbon dioxide. Yeasts are used as fermenting agents in baking, brewing, and the making of wine and spirits. Brewer's yeast (*S. cerevisiae*) is a rich source of vitamin B. (Especially genus *Saccharomyces*; also other related genera.)

yeast artificial chromosome YAC, fragment of *DNA from the human genome inserted into a yeast cell. The yeast replicates the fragment along with its own DNA. In this way the fragments are copied to be preserved in a gene library. YACs are characteristically between 250,000 and 1 million base pairs in length. A cosmid works in the same way.

yield point or **elastic limit**, stress beyond which a material deforms by a relatively large amount for a small increase in stretching force. Beyond this stress, the material no longer obeys *Hooke's law.

ytterbium chemical symbol Yb, soft, lustrous, silvery, malleable, and ductile metallic element of the *lanthanide series, symbol Yb, atomic number 70, relative atomic mass 173.04. It occurs with (and resembles) yttrium in gadolinite and other minerals, and is used in making steel and other alloys. In 1878 Swiss chemist Jean-Charles de Marignac gave the name ytterbium (after the Swedish town of Ytterby, near where it was found) to what he believed to be a new element. French chemist Georges Urbain (1872–1938) discovered in 1907 that this was in fact a mixture of two elements: ytterbium and lutetium.

yttrium chemical symbol Y, silver-grey, metallic element, atomic number 39, relative atomic mass 88.905. It is associated with and resembles the rare-earth elements (*lanthanides), occurring in gadolinite, xenotime, and other minerals. It is used in colour-television tubes and to reduce steel corrosion. The name derives from the Swedish town of Ytterby, near where it was first discovered in 1788. Swedish chemist Carl Mosander (1797–1858) isolated the element in 1843.

Z in physics, the symbol for **impedance** (electricity and magnetism).

z-buffer in computing, *buffer for storing depth information for displaying three-dimensional graphics. (Two-dimensional images may be displayed using x, y coordinates but the third dimension implies x, y, and z.) In a graphics card, z-buffer memory keeps track of which onscreen elements are visible and which are hidden behind other objects.

zenith uppermost point of the celestial horizon, immediately above the observer; the *nadir is below, diametrically opposite. See *celestial sphere.

zeolite any of the hydrous aluminium silicates, also containing sodium, calcium, barium, strontium, or potassium, chiefly found in igneous rocks and characterized by a ready loss or gain of water. Zeolites are used as 'molecular sieves' to separate mixtures because they are capable of selective absorption. They have a high ion-exchange capacity and can be used to separate petrol, benzene, and toluene from low-grade raw materials, such as coal and methanol.

zero number (written 0) that when added to any number leaves that number unchanged. It results when any number is subtracted from itself, or when any number is added to its negative. The product of any number with zero is itself zero.

zeroth law of thermodynamics see *thermodynamics

zidovudine formerly **AZT**, antiviral drug used in the treatment of *AIDS. It is not a cure for AIDS but is effective in prolonging life; it does not, however, delay the onset of AIDS in people carrying the virus.

ZIF socket acronym for Zero Insertion Force socket, socket on a computer's *motherboard that enables a chip to be easily removed or inserted by use of a lever. ZIF sockets are usually only used for expensive microprocessors that are designed to be upgraded.

zinc chemical symbol Zn, (Germanic *Zinke* 'point') hard, brittle, bluish-white, metallic element, atomic number 30, relative atomic mass 65.37. The principal ore is sphalerite or zinc blende (zinc sulphide, ZnS). Zinc is hardly affected by air or moisture at ordinary temperatures; its chief uses are in alloys such as brass, in coating metals (for example, galvanized iron), and in making batteries. Its compounds include zinc oxide, used in ointments (as an astringent) and cosmetics, paints, glass, and printing ink.

Zinc is an essential trace element in most animals; adult humans have 2–3 g/0.07–0.1 oz of zinc in their bodies. There are more than 300 known enzymes that contain zinc. Zinc has been used as a component of brass since the Bronze Age, but it was not recognized in Europe as a separate metal until the 16th century. It was isolated in 1746 by German chemist Andreas Sigismund Marggraf (1709–1782). The name derives from the shape of the crystals on smelting.

zinc ore mineral from which zinc is extracted, principally sphalerite (Zn,Fe)S, but also zincite, ZnO_2, and smithsonite, $ZnCO_3$, all of which occur in mineralized veins. Ores of lead and zinc often occur together, and are common worldwide; Canada, the USA, and Australia are major producers.

zinc oxide ZnO, white powder, yellow when hot, that occurs in nature as the mineral zincite. It is used in paints and as an antiseptic in zinc ointment; it is the main ingredient of calamine lotion.

zinc sulphide ZnS, yellow-white solid that occurs in nature as the mineral sphalerite (also called zinc blende). It is the principal ore of zinc, and is used in the manufacture of fluorescent paints.

Zip drive in computing, portable disk drive manufactured by or under license from Iomega. Zip drives can store 100 *megabytes (MB) or 250 MB on each disk.

zipped file in computing, a file that has been compressed using the PKZIP program.

zirconium chemical symbol Zr, (Germanic *zircon*, from Persian *zargun* 'golden') lustrous, greyish-white, strong, ductile, metallic element, atomic number 40, relative atomic mass 91.22. It occurs in nature as the mineral zircon (zirconium silicate), from which it is obtained commercially. It is used in some ceramics, alloys for wire and filaments, steel manufacture, and nuclear reactors, where its low neutron absorption is advantageous. It was isolated in 1824 by Swedish chemist Jöns Berzelius. The name was proposed by English chemist Humphry Davy in 1808.

Zmodem in computing, *FTP *protocol for transferring files across the *Internet or other communications link. It offers the facility to use *wild cards to search files and to resume interrupted transfers where they left off.

zodiac zone of the heavens containing the paths of the Sun, Moon, and planets. When this was devised by the ancient Greeks, only five planets were known, making the zodiac about 16° wide. In astrology, the zodiac is divided into 12 signs, each 30° in extent: Aries, Taurus, Gemini, Cancer, Leo, Virgo, Libra, Scorpio, Sagittarius, Capricorn, Aquarius, and Pisces. These do not cover the same areas of sky as the astronomical constellations.

zoological colony in zoology, colony formed when an organism gives rise to several buds which adhere to the parent and continue to reproduce by budding. Such colonies are common among sponges and corals.

zoology branch of biology concerned with the study of animals. It includes any aspect of the study of animal form and function – description of present-day animals, the study of evolution of animal forms, anatomy, *physiology, *embryology, behaviour, and geographical distribution.

Z particle in physics, an *elementary particle, one of the *intermediate vector bosons responsible for carrying the *weak nuclear force. The Z particle is neutral.

zwitterion ion that has both a positive and a negative charge, such as an *amino acid in neutral solution. For example, glycine contains both a basic amino group (NH_2) and an acidic carboxyl group (COOH); when these are both ionized in aqueous solution, the acid group loses a proton to the amino group, and the molecule is positively charged at one end and negatively charged at the other.

zygote *ovum (egg) after *fertilization but before it undergoes cleavage to begin embryonic development.

APPENDIX 1

The SI system

(a) SI base units

quantity	name	symbol	definition
length	metre	m	length of the path travelled by light in vacuum during a time interval of 1/299 792 458 second.
mass	kilogram	kg	mass of platinum-iridium international standard kept at Sèvres, France.
time	second	s	duration of 9,192,631,770 periods of the radiation corresponding to the transition between the two hyperfine levels of the ground state of the caesium-133 atom.
electric current	ampere	A	the constant current which, if maintained in two straight parallel conductors of infinite length, of negligible circular cross section, and placed 1 metre apart in vacuum would produce between these conductors a force equal to 2×10^{-7} newton per metre of length.
thermodynamic temperature	kelvin	K	the fraction 1/273.16 of the thermodynamic temperature of the triple point of water.
amount of substance	mole	mol	the amount of substance of a system which contains as many elementary entities as there are atoms in 0.012 kilogram of carbon-12.
luminous intensity	candela	cd	the luminous intensity, in a given direction, of a source that emits monochromatic radiation of frequency 540×10^{12} hertz and that has a radiant intensity in that direction of 1/683 watt per steradian.
plane angle	radian	rad	the plane angle between two radii of a circle that cut off on the circumference an arc equal in length to the radius.
solid angle	steradian	sr	solid angle that, having its vertex in the centre of a sphere, cuts off an area of the surface of the sphere equal to that of a square with sides of length equal to the radius of the sphere.

(b) SI derived units

quantity	name	symbol	definition
frequency	hertz	Hz	$1/s$
force	newton	N	$m\,kg/s^2$
pressure	pascal	Pa	N/m^2
energy	joule	J	N m
power	watt	W	J/s
electric charge	coulomb	C	A s
electric potential	volt	V	W A
capacitance	farad	F	C/V
resistance	ohm	Ω	V/A
conductance	siemens	S	A/V
magnetic flux	weber	Wb	V s
magnetic flux density	tesla	T	Wb/m^2
inductance	henry	H	Wb/A
luminous flux	lumen	lm	cd sr
illuminance	lux	lx	lm/m^2

(c) Prefixes used with SI units

10^{24}	yotta	Y
10^{21}	zetta	Z
10^{18}	exa	E
10^{15}	peta	P
10^{12}	tera	T
10^{9}	giga	G
10^{6}	mega	M
10^{3}	kilo	k
10^{2}	hecto	h
10^{1}	deca	da
10^{-1}	deci	d
10^{-2}	centi	c
10^{-3}	milli	m
10^{-6}	micro	μ
10^{-9}	nano	n
10^{-12}	pico	p
10^{-15}	femto	f
10^{-18}	atto	a
10^{-21}	zepto	z
10^{-24}	yocto	y

APPENDIX 2

Selected physical constants

(a) Fundamental constants

speed of light	c	3.0×10^8 m/s
charge on an electron	e	1.6×10^{-18} C
gravitational constant	G	6.7×10^{-11} N m^2/kg^2
Faraday constant	F	9.6×10^4 C/mol
electron rest mass	m_e	9.1×10^{-31} kg
proton rest mass	m_p	1.7×10^{-27} kg
Planck constant	h	6.6×10^{-34} J s
gas constant	R	8.3 J/K mol
Avogadro constant	L	6.02×10^{23}/mol

(b) Other useful constants

standard atmospheric pressure	P	1.0×10^5 Pa (1 bar)
astronomical unit	AU	1.5×10^{11} m
light year	ly	9.5×10^{15} m
parsec	pc	3.1×10^{16} m (3.3 ly)
Earth, acceleration due to gravity	g	9.8 N/kg or m/s^2

APPENDIX 3

The chemical elements

Note that the relative atomic mass is given to two places of decimals. Where an element has no stable nuclides, the value in brackets indicates the r.a.m. of the most stable isotope of the element.

Name	Symbol	At No	Relative atomic mass	Melting point (°C)	Boiling point (°C)	Date discovered
Actinium	Ac	89	[227]	1100	3200	1899
Aluminium	Al	13	26.98	658	1800	1824
Americium	Am	95	[243]	995	not known	1944
Antimony	Sb	51	121.76	630.5	1750	prehistoric
Argon	Ar	18	39.95	-189.4	-185.9	1894
Arsenic	As	33	74.92	613	613	c. 1250
Astatine	At	85	[210]	302	377	1940
Barium	Ba	56	137.32	725	1640	1808
Berkelium	Bk	97	[247]	986	not known	1949
Beryllium	Be	4	9.01	1285	2469	1797
Bismuth	Bi	83	208.98	271.3	1560	1753
Bohrium	Bh	107	[264]	not known	not known	1981
Boron	B	5	10.81	2079	2550	1808
Bromine	Br	35	79.90	-7.2	58.8	1826
Cadmium	Cd	48	112.41	321	765	1817
Caesium	Cs	55	132.91	28.4	678	1860
Calcium	Ca	20	40.08	839	1484	1808
Californium	Cf	98	[251]	not known	not known	1950
Carbon	C	6	12.01	3550	4200	prehistoric
Cerium	Ce	58	140.12	798	3257	1803
Chlorine	Cl	17	35.45	-101	-34.6	1774
Chromium	Cr	24	52.00	1890	2672	1797
Cobalt	Co	27	58.93	1495	2870	1737
Copper	Cu	29	63.55	1083	2567	prehistoric
Curium	Cm	96	[247]	1340	not known	1944
Dubnium	Db	105	[262]	not known	not known	1967
Dysprosium	Dy	66	162.50	1409	2235	1886
Einsteinium	Es	99	[252]	not known	not known	1952
Erbium	Er	68	167.26	1522	2863	1843
Europium	Eu	63	151.97	822	1597	1896
Fermium	Fm	100	[257]	not known	not known	1952
Fluorine	F	9	19.00	-220	-188	1771
Francium	Fr	87	[223]	not known	not known	1939

Name	Symbol	At No	Relative atomic mass	Melting point (°C)	Boiling point (°C)	Date discovered
Gadolinium	Gd	64	157.25	1311	3233	1880
Gallium	Ga	31	69.72	30	2403	1875
Germanium	Ge	32	72.64	937	2830	1886
Gold	Au	79	196.97	1062	2800	prehistoric
Hafnium	Hf	72	178.49	2227	4602	1923
Hassium	Hs	108	[277]	not known	not known	1984
Helium	He	2	4.00	-272	-269	1868
Holmium	Ho	67	164.93	1470	2720	1878
Hydrogen	H	1	1.01	-259	-253	1766
Indium	In	49	114.82	157	2080	1863
Iodine	I	53	126.90	113.5	184.5	1811
Iridium	Ir	77	192.22	2410	4130	1804
Iron	Fe	26	55.85	1535	2750	prehistoric
Krypton	Kr	36	83.80	-156.6	-152.3	1898
Lanthanum	La	57	138.91	920	3454	1839
Lawrencium	Lr	103	[262]	not known	not known	1839
Lead	Pb	82	207.19	237.5	1740	prehistoric
Lithium	Li	3	6.94	180.5	1347	1817
Lutetium	Lu	71	174.97	1656	3315	1906
Magnesium	Mg	12	24.31	649	1090	1808
Manganese	Mn	25	54.94	1244	1962	1774
Meitnerium	Mt	109	[268]	not known	not known	1982
Mendelevium	Md	101	[258]	not known	not known	1955
Mercury	Hg	80	200.59	-38.9	356.6	prehistoric
Molybdenum	Mo	42	95.94	2610	5560	1778
Neodymium	Nd	60	144.24	1010	3068	1885
Neon	Ne	10	20.18	-248.7	-246.1	1898
Neptunium	Np	93	[237]	640	3902	1940
Nickel	Ni	28	58.7	1453	2732	1751
Niobium	Nb	41	92.91	2468	4742	1801
Nitrogen	N	7	14.01	-209.9	-195.8	1772
Nobelium	No	102	[259]	not known	not known	1958
Osmium	Os	76	190.23	3045	5027	1804
Oxygen	O	8	16	-218.4	-183	1774
Palladium	Pd	46	106.42	1552	3140	1803
Phosphorus	P	15	30.97	44.1	280	1669
Platinum	Pt	78	195.08	1772	3800	1741
Plutonium	Pu	94	[244]	641	3232	1940
Polonium	Po	84	[209]	254	962	1898
Potassium	K	19	39.1	63.7	774	1807
Praseodymium	Pr	59	140.91	931	3512	1885

Name	Symbol	At No	Relative atomic mass	Melting point (°C)	Boiling point (°C)	Date discovered
Promethium	Pm	61	[145]	1080	2460	1945
Protactinium	Pa	91	231.04	1200	4000	1913
Radium	Ra	88	[226]	700	1140	1899
Radon	Rn	86	[222]	-71	-61.8	1899
Rhenium	Re	75	186.21	3180	5627	1925
Rhodium	Rh	45	102.91	1966	3727	1803
Rubidium	Rb	37	85.47	39	688	1861
Ruthenium	Ru	44	101.07	2310	3900	1827
Rutherfordium	Rf	104	[261]	not known	not known	1964
Samarium	Sm	62	150.36	1072	1791	1879
Scandium	Sc	21	44.96	1539	2832	1879
Seaborgium	Sg	106	[266]	not known	not known	1974
Selenium	Se	34	78.96	217	685	1817
Silicon	Si	14	28.09	1410	2355	1824
Silver	Ag	47	107.87	962	2212	prehistoric
Sodium	Na	11	22.99	98	882	1807
Strontium	Sr	38	87.62	769	1384	1808
Sulphur	S	16	32.07	113	445	prehistoric
Tantalum	Ta	73	180.95	2996	5425	1802
Technetium	Tc	43	[98]	2172	4877	1937
Tellurium	Te	52	127.60	450	990	1782
Terbium	Tb	65	158.93	1360	3041	1843
Thallium	Tl	81	204.38	304	1457	1861
Thorium	Th	90	232.04	1750	4790	1828
Thulium	Tm	69	168.93	1545	1947	1879
Tin	Sn	50	118.71	232	2270	prehistoric
Titanium	Ti	22	47.87	1660	3287	1791
Tungsten	W	74	183.84	3410	5660	1781
Ununbium	Uub	112	[285]	not known	not known	1996
Ununhexium	Uuh	116	[289]	not known	not known	1999
Ununnilium	Uun	110	[281]	not known	not known	1994
Ununoctium	Uuo	118		not known	not known	1999
Ununquadium	Uuq	114	[289]	not known	not known	1998
Unununium	Uuu	111	[272]	not known	not known	1994
Uranium	U	92	238.03	1132	3818	1789
Vanadium	V	23	50.94	1890	3380	1801
Xenon	Xe	54	131.29	-112	-107	1898
Ytterbium	Yb	70	173.04	824	1193	1878
Yttrium	Y	39	88.91	1523	3337	1828
Zinc	Zn	30	65.41	420	907	1746
Zirconium	Zr	40	91.22	1852	4377	1824

APPENDIX 4

Abbreviations of units and common symbols

'	foot	cl	centilitre
'	minute (angular measure)	cm	centimetre
"	inch	cm/s	centimetres per second
"	second (angular measure)	cm^2	square centimetre
°	degree	cm^3	cubic centimetre
a	are	cwt	hundredweight
a	acceleration	d	day
A	ampere	d	density; depth
A	area; mass number	dam	dekametre
A_r	relative atomic mass	dB	decibel
Å	ångström	DC	direct current
AC	alternating current	dl	decilitre
am	attometre	dm	decimetre
amu	atomic mass unit	dr	dram
aq	aqueous solution	dry pt	dry pint
as	attosecond	dry qt	dry quart
atm	atmosphere	dwt	pennyweight
AU	astronomical unit	e	electron
b	breadth	E	electric field strength; energy; tensile strength
b.p.	boiling point	emf	electromotive force
Btu	British thermal unit	eV	electronvolt
bu	bushel	f	frequency
c	centirad (angular measure)	F	farad
c	specific heat capacity	F	force
C	coulomb	°F	Fahrenheit
C	capacitance	fl dr	fluid dram
°C	Celsius	fl oz	fluid ounce
cal	calorie	fm	fathom
Cal	kilocalorie	fm	femtometre
cc	cubic centimetre	fs	femtosecond
cd	candela	ft	foot
cg	centigrades (angular measure)	ft/min	feet per minute
ch	chain	ft/s	feet per second
ch^2	square chain	ft^2	square foot

ft^3	cubic foot	km/s	kilometres per second
fur	furlong	km^2	square kilometre
g	gram; gas (state symbol)	km^3	cubic kilometre
g	acceleration of free fall; gravitational field strength	kn	international knot
		ks	kilosecond
g/cm^2	grams per square centimetre	kt	kiloton
g/cm^3	grams per cubic centimetre	kW	kilowatt
gal	gallon	kWh	kilowatt-hour
GHz	gigahertz	l	litre; liquid (state symbol)
gi	gill	L	Avogadro constant
Gm	gigametre	L	specific latent heat
gr	grain	λ	thermal conductivity; wavelength
Gs	gigasecond		
h	hour	lb	pound
h	height	lb/ft^2	pounds per square foot
H	enthalpy	lb/ft^3	pounds per cubic foot
ha	hectare	lt	long (UK) ton
hm	hectometre	ly	light year
hp	horsepower	m	metre; minute
hr	hour	m	mass
Hz	hertz	m/min	metres per minute
I	current	m/s	metres per second
in	inch	m^2	square metre
in/s	inches per second	m^3	cubic metre
in^2	square inch	M	molar unit of concentration (molarity)
in^3	cubic inch		
IR	infrared radiation	M$_r$	relative molecular mass
J	joule	mg	milligram
K	kelvin	mg	milligrades (angular measure)
kcal	kilocalorie	MHz	megahertz
kg	kilogram	mi	mile
kg/cm^2	kilograms per square centimetre	mi^2	square mile
kg/cm^3	kilograms per cubic centimetre	mi^3	cubic mile
kg/m^2	kilograms per square metre	mil	milli-inch
kg/m^3	kilograms per cubic metre	min	minim
kHz	kilohertz	min	minute
kJ	kilojoule	ml	millilitre
km	kilometre	mm	millimetre
km/h	kilometres per hour	Mm	Megametre

mm^2	square millimetre	${}^{\circ}R$	degrees Rankine
mm^3	cubic millimetre	rad	radian (angular measure)
mmHg	millimetres of mercury (unit of pressure)	r.a.m.	relative atomic mass
mol	mole	r.m.m.	relative molecular mass
μm	micrometre	ρ	density; resistivity
$μm^2$	square micrometre	s	second; solid (state symbol)
μs	microsecond	sec	second
m.p.	melting point	sh cwt	short hundredweight
mph	miles per hour	sh t	short (US) ton
Ms	megasecond	SI	International System of Units (Système International d'Unités)
ms	millisecond		
Mt	megaton	sr	steradian
μ	coefficient of friction	st	stone
n	refractive index	st	stère
N	newton	STP	standard temperature and pressure
n	neutron		
n mi	nautical mile	t	tonne
nm	nanometre	t	temperature; time
ns	nanosecond	T	period; thermodynamic temperature
NTP	normal temperature and pressure		
		Tm	terametre
ν	frequency	tn	ton
oz	ounce	Ts	terasecond
oz tr	ounce troy	u	speed
P	momentum; power; pressure	UK kn	British knot
Pa	pascal	UV	ultraviolet radiation
pc	parsec	v	velocity
p.d.	potential difference	V	volt
pm	Picometre	V	potential difference
ps	picosecond	W	watt
PSI	pounds per square inch	W	weight; work
pt	pint	w/m^2	watts per square metre
q	quintal	yd	yard
Q	charge; heat	yd/min	yards per minute
qt	quart	yd^2	square yard
r	radius	yd^3	cubic yard
R	resistance	ypm	yards per minute
${}^{\circ}r$	degrees Réaumur	Z	atomic number
		Ω	ohm

APPENDIX 5

Some common formulae and relationships

Kinematics – the study of motion

acceleration $\qquad$ $a = \dfrac{v - u}{t}$

speed $\qquad$ $v = \dfrac{s}{t}$

average speed $\qquad$ $v_{av} = \dfrac{v - u}{2}$

distance $\qquad$ s = average speed $\times$ time $= (\dfrac{u + v}{2})t$

distance travelled by a uniformly
   accelerated object $\qquad\qquad\qquad\qquad\qquad$ $s = ut + \frac{1}{2}at^2$
   if time is unknown use: $\qquad\qquad\qquad\qquad\qquad$ $v^2 - u^2 = 2as$

Dynamics – the study of forces and motion

force, mass and acceleration $\qquad$ $F = ma$

momentum $\qquad$ $P = mv$

impulse $\qquad$ $Ft = \Delta P = m\Delta v$

Gravitation

acceleration of free fall $\qquad$ g [m/s^2]

gravitational field strength $\qquad$ g [N/kg]

gravitational potential energy $\qquad$ mgh

gravitational force $\qquad$ $F = G\dfrac{Mm}{r^2}$

period of a simple pendulum $\qquad$ $T = 2\pi\sqrt{(l/g)}$

Matter

density $\qquad$ $d = \dfrac{m}{V}$

weight $\qquad$ $W = mg$

pressure $\qquad$ $P = \dfrac{F}{A}$

pressure in a fluid $\qquad$ $P = hdg$

gas law $\qquad$ $\dfrac{PV}{T} = constant$

Boyle's law $\qquad$ $PV = constant$

elasticity		$F = kx$
strain		$\dfrac{\text{change in length}}{\text{original length}}$
stress		$\dfrac{F}{A}$
moment (torque)		$T = Fs$
area	square	$A = l^2$
	rectangle	$A = bl$
	circle	$A = \pi r^2$
	triangle	$A = \frac{1}{2}bh$
volume	cube	$V = l^3$
	cuboid	$V = lbh$
	sphere	$V = \frac{4}{3}\pi r^3$
	cylinder	$V = \pi r^2 h$

Energy and power

work	$W = Fs$ or Fd
kinetic energy	$E_k = \frac{1}{2}mv^2$
gravitational potential energy	$E_p = mgh$
energy transferred electrically	VIt or QV or $\dfrac{V^2 t}{A}$
power	$P = \dfrac{\text{work done}}{\text{time}}$ or $\dfrac{\text{energy transferred}}{\text{time}}$
electrical power	$P = VI$ or I^2R or $\dfrac{V^2}{R}$
radiation energy, per photon	$H = hf$
efficiency	$\dfrac{\text{energy output}}{\text{energy input}} \times 100\%$
force ratio of machine (mechanical advantage)	$\dfrac{\text{load}}{\text{effort}}$
distance ratio (velocity ratio)	$\dfrac{\text{distance moved by effort}}{\text{distance moved by load}}$ or $\dfrac{\text{speed of effort}}{\text{speed of load}}$

Electric circuits

current	$I = Q/t$
resistance	$R = \dfrac{V}{I}$
voltage	$V = \dfrac{\text{energy}}{\text{change}}$
circuit with internal resistance	$E = Ir + IR$
resistivity	$\rho = \dfrac{RA}{l}$

resistors in series	$R = R_1 + R_2 + R_3 + \ldots$
resistors in parallel	$\dfrac{1}{R} = \dfrac{1}{R_1} + \dfrac{1}{R_2} + \dfrac{1}{R_3} + \ldots$
capacitance	$C = \dfrac{Q}{V}$
capacitance of a plane capacitor	$C = \dfrac{kA}{d}$
capacitors in series	$\dfrac{1}{C} = \dfrac{1}{C_1} + \dfrac{1}{C_2} + \dfrac{1}{C_3} + \ldots$
capacitors in parallel	$C = C_1 + C_2 + C_3 + \ldots$
electric field strength	$E = \dfrac{F}{Q}$
electric field strength in uniform field	$E = \dfrac{V}{d}$

Waves

wave speed	$v = f\lambda$
frequency	f = cycles or waves per second
period	$T = \dfrac{1}{f}$

Optics

lens and mirror formula	$\dfrac{1}{f} = \dfrac{1}{u} + \dfrac{1}{v}$
curved mirror	$f = \tfrac{1}{2}r$
'power' of a lens or mirror	$P = \dfrac{1}{f}$
refractive index, air to a medium	$n = \dfrac{\text{speed in air}}{\text{speed in medium}} = \dfrac{\sin i}{\sin r}$
critical angle	$\sin C = \dfrac{1}{n}$